T0319275

An Invitation to Applied Mathematics

An Invitation to Applied Mathematics: Differential Equations, Modeling, and Computation

Carmen Chicone

AMSTERDAM • BOSTON • HEIDELBERG • LONDON
NEW YORK • OXFORD • PARIS • SAN DIEGO
SAN FRANCISCO • SINGAPORE • SYDNEY • TOKYO
Academic Press is an imprint of Elsevier

Academic Press is an imprint of Elsevier
32 Jamestown Road, London NW1 7BY, UK
525 B Street, Suite 1800, San Diego, CA 92101-4495, USA
225 Wyman Street, Waltham, MA 02451, USA
The Boulevard, Langford Lane, Kidlington, Oxford OX5 1GB, UK

Copyright © 2017 Elsevier Inc. All rights reserved.

No part of this publication may be reproduced or transmitted in any form or by any means, electronic or mechanical, including photocopying, recording, or any information storage and retrieval system, without permission in writing from the publisher. Details on how to seek permission, further information about the Publisher's permissions policies and our arrangements with organizations such as the Copyright Clearance Center and the Copyright Licensing Agency, can be found at our website: www.elsevier.com/permissions.

This book and the individual contributions contained in it are protected under copyright by the Publisher (other than as may be noted herein).

Notices
Knowledge and best practice in this field are constantly changing. As new research and experience broaden our understanding, changes in research methods, professional practices, or medical treatment may become necessary.

Practitioners and researchers must always rely on their own experience and knowledge in evaluating and using any information, methods, compounds, or experiments described herein. In using such information or methods they should be mindful of their own safety and the safety of others, including parties for whom they have a professional responsibility.

To the fullest extent of the law, neither the Publisher nor the authors, contributors, or editors, assume any liability for any injury and/or damage to persons or property as a matter of products liability, negligence or otherwise, or from any use or operation of any methods, products, instructions, or ideas contained in the material herein.

British Library Cataloguing in Publication Data
A catalogue record for this book is available from the British Library

Library of Congress Cataloging-in-Publication Data
A catalog record for this book is available from the Library of Congress

ISBN: 978-0-12-804153-6

For information on all Academic Press publications visit our website at https://www.elsevier.com/

Working together to grow libraries in developing countries

www.elsevier.com • www.bookaid.org

Publisher: Nikki Levy
Acquisition Editor: Graham Nisbet
Editorial Project Manager: Susan Ikeda
Production Project Manager: Poulouse Joseph
Designer: Matthew Limbert

Typeset by SPi Books and Journals

TABLE OF CONTENTS

Preface xi

Acknowledgments xiii

To the Professor xv

To the Student xix

Chapter 1 Applied Mathematics and Mathematical Modeling...... 1
1.1 What is Applied Mathematics? 1
1.2 Fundamental and Constitutive Models 3
1.3 Descriptive Models 6
1.4 Applied Mathematics in Practice 8

Chapter 2 Differential Equations 11
2.1 The Harmonic Oscillator 12
2.2 Exponential and Logistic Growth 15
2.3 Linear Systems 16
2.4 Linear Partial Differential Equations 18
2.5 Nonlinear Ordinary Differential Equations 19
2.6 Numerics 21

Part I Conservation of Mass: Biology, Chemistry, Physics, and Engineering 31

Chapter 3 An Environmental Pollutant 31

Chapter 4 Acid Dissociation, Buffering, Titration, and Oscillation 39
4.1 A Model for Dissociation 39
4.2 Titration with a Base 54
4.3 An Improved Titration Model 58
4.4 The Oregonator: An Oscillatory Reaction 66

Chapter 5 Reaction, Diffusion, and Convection 85
5.1 Fundamental and Constitutive Model Equations 85
5.2 Reaction-Diffusion in One Spatial Dimension: Heat, Genetic Mutations, and Traveling Waves 88

5.3 Reaction-Diffusion Systems: The Gray–Scott Model and Pattern Formation 113
5.4 Analysis of Reaction-Diffusion Models: Qualitative and Numerical Methods 117
5.5 Beyond Euler's Method For Reaction-Diffusion PDE: Diffusion of Gas in a Tunnel, Gas in Porous Media, Second-Order in Time Methods, and Unconditional Stability 151

Chapter 6 Excitable Media: Transport of Electrical Signals on Neurons 195
6.1 The FitzHugh–Nagumo Model 196
6.2 Numerical Traveling Wave Profiles 211

Chapter 7 Splitting Methods 221
7.1 A Product Formula 221
7.2 Products for Nonlinear Systems 224

Chapter 8 Feedback Control 229
8.1 A Mathematical Model for Heat Control of a Chamber 229
8.2 A One-dimensional Heated Chamber With PID Control 232

Chapter 9 Random Walks and Diffusion 241
9.1 Basic Probability Theory 241
9.2 Random Walk 245
9.3 Continuum Limit of the Random Walk 249
9.4 Random Walk Generalizations and Applications 267

Chapter 10 Problems and Projects: Concentration Gradients, Convection, Chemotaxis, Cruise Control, Constrained Control, Pearson's Random Walk, Molecular Dynamics, Pattern Formation 277

Part II Newton's Second Law: Fluids and Elastic Solids 293

Chapter 11 Equations of Fluid Motion 293
11.1 Scaling: The Reynolds Number and Froude Number 313
11.2 The Zero Viscosity Limit 315
11.3 The Low Reynolds Number Limit 316

Chapter 12 Flow in a Pipe **321**

Chapter 13 Eulerian Flow **327**
13.1 Bernoulli's Form of Euler's Equations 327
13.2 Potential Flow 329
13.3 Potential Flow in Two Dimensions 332
13.4 Circulation, Lift, and Drag 335

Chapter 14 Equations of Motion in Moving Coordinate Systems **357**
14.1 Moving Coordinate Systems 357
14.2 Pure Rotation 363
14.3 Fluid Motion in Rotating Coordinates 367
14.4 Water Draining in Sinks Versus Hurricanes 375
14.5 A Counterintuitive Result: The Proudman–Taylor Theorem 379

Chapter 15 Water Waves **383**
15.1 The Ideal Water Wave Equations 383
15.2 The Boussinesq Equations 386
15.3 KdV 387
15.4 Boussinesq Steady State Water Waves 390
15.5 A Free-Surface Flow 397

Chapter 16 Numerical Methods for Computational Fluid Dynamics **403**
16.1 Approximations of Incompressible Navier–Stokes Flows 403
16.2 A Numerical Method for Water Waves 446
16.3 The Boundary Element Method (BEM) 450
16.4 Boundary Integral Representation 450
16.5 Boundary Integral Equation 455
16.6 Discretization for BEM 459
16.7 Smoothed Particle Hydrodynamics 475
16.8 Simulation of a Free-Surface Flow 506

Chapter 17 Channel Flow **511**
17.1 Conservation of Mass 513
17.2 Momentum Balance 515
17.3 Boundary Layer Theory 527
17.4 Flow in Prismatic Channels with Rectangular Cross Sections of Constant Width 536
17.5 Hydraulic Jump 539
17.6 Saint-Venant Model and Systems of Conservation Laws 542
17.7 Surface Waves 565

Chapter 18 Elasticity: Basic Theory and Equations of Motion 577
18.1 The Taut Wire: Separation of Variables and Fourier Series for the Wave Equation 589
18.2 Longitudinal Waves in a Rod With Varying Cross Section 611
18.3 Ultrasonics 616
18.4 A Three-Dimensional Elastostatics Problem: A Copper Block Bolted to a Steel Plate 620
18.5 A One-Dimensional Elasticity Model 624
18.6 Weak Formulation of One-Dimensional Boundary Value Problems 627
18.7 One-Dimensional Finite Element Method Discretization 635
18.8 Coding for the One-Dimensional Finite Element Method........ 641
18.9 Weak Formulation and Finite Element Method for Linear Elasticity 647
18.10 A Three-Dimensional Finite Element Application 653

Chapter 19 Problems and Projects: Rods, Plates, Panel Flutter, Beams, Convection-Diffusion in Tunnels, Gravitational Potential of a Galaxy, Taylor Dispersion, Cavity Flow, Drag, Low and High Reynolds Number Flows, Free-Surface Flow, Channel Flow 671
19.1 Problems: Fountains, Tapered Rods, Elasticity,Thermoelasticity, Convection-Diffusion, and Numerical Stability 671
19.2 Gravitational Potential of a Galaxy 675
19.3 Taylor Dispersion 677
19.4 Lid-Driven Cavity Flow 681
19.5 Aerodynamic Drag 682
19.6 Low Reynolds Number Flow 684
19.7 Fluid Motion in a Cylinder 684
19.8 Free-Surface Flow 685
19.9 Channel Flow Traveling Waves 700

Part III Electromagnetism: Maxwell's Laws and Transmission Lines 703

Chapter 20 Classical Electromagnetism........ 703
20.1 Maxwell's Laws and The Lorentz Force Law 703
20.2 Boundary Conditions 709
20.3 An Electromagnetic Boundary Value Problem 711
20.4 Comments on Maxwell's Theory 716
20.5 Time-Harmonic Fields 717

Chapter 21 Transverse Electromagnetic (TEM) Mode 719

Chapter 22 Transmission Lines 731
22.1 Time-Domain Reflectometry Model 731
22.2 TDR Matrix System 735
22.3 Initial Value Problem for the Ideal Transmission Line 736
22.4 The Initially Dead Ideal Transmission Line With Constant Dielectrics 738
22.5 The Riemann Problem 742
22.6 Reflected and Transmitted Waves 746
22.7 A Numerical Method for the Lossless Transmission Line Equation 749
22.8 The Lossy Transmission Line 757
22.9 TDR Applications 767
22.10 An Inverse Problem 770

Chapter 23 Problems and Projects: Waveguides, Lord Kelvin's Model 775
23.1 TE Modes in Waveguides with Circular Cross Sections 776
23.2 Rectangular Waveguides and Cavity Resonators 780

Mathematical and Computational Notes 793
A.1 Arzela–Ascoli Theorem 793
A.2 C^1 Convergence 793
A.3 Existence, Uniqueness, and Continuous Dependence 793
A.4 Green's Theorem and Integration by Parts 794
A.5 Gerschgorin's Theorem 795
A.6 Gram–Schmidt Procedure 795
A.7 Grobman–Hartman Theorem 795
A.8 Order Notation 796
A.9 Taylor's Formula 796
A.10 Liouville's Theorem 797
A.11 Transport Theorem 797
A.12 Least Squares and Singular Value Decomposition 798
A.13 The Morse Lemma 802
A.14 Newton's Method 802
A.15 Variation of Parameters Formula 811
A.16 The Variational Equation 812
A.17 Linearization and Stability 813
A.18 Poincaré–Bendixson Theorem 816
A.19 Eigenvalues of Tridiagonal Toeplitz Matrices 816
A.20 Conjugate Gradient Method 818
A.21 Numerical Computation and Programming Gems of Wisdom 823

Answers to Selected Exercises 827

References 841
Index 847

PREFACE

What is applied mathematics? Every answer to this question is likely to initiate a debate. My definition is the use of mathematics to solve problems or gain insight into phenomena that arise outside of mathematics. The prototypical example is the use of mathematics to solve problems in physics. Of course, the world of applied mathematics is much broader: important applications of mathematics occur in all areas of science, engineering, and technology.

The concept of this book is to introduce the reader to one aspect of applied mathematics: the use of differential equations to solve physical problems. To cover the full (ever expanding) range of applications of mathematics would require a series of books, which would include invitations to applied mathematics using the other branches of mathematics: calculus, linear algebra, differential geometry, graph theory, combinatorics, number theory, the calculus of variations, probability theory, and others. The application of statistics (especially in experimental science) is a branch of applied mathematics of great importance, but of a different character than the applied mathematics considered here.

Although there are already many books and articles devoted to applications of mathematical subjects, I believe that there is room for more introductory material accessible to advanced undergraduates and beginning graduate students. If my invitation is accepted, perhaps the reader will pursue further study, find a problem in applied mathematics, and make a contribution to technology or the understanding of the physical universe.

My invitation includes a tour through a few of the historically important uses of differential equations in science and technology. The relevant mathematics is presented in context where there is no question of its importance.

A typical scenario in many research papers by mathematicians is an introduction that includes such phrases as "our subject is important in the study of ...," "this problem arises in ..., " or "our subject has many applications to ...." The authors go on to state a precise mathematical problem, they prove a theorem—perhaps a very good theorem, and perhaps they give a mathematical example to illustrate their result, but all too often, their theorem does not solve a problem of interest in the scientific area that they used to advertise their work. This is not applied mathematics. The correct approach is joint work with an expert in some area of science: a physical problem is stated, a mathematical model is proposed, a prediction is made from the mathematical model—a step that might require some new mathematics including mathematical theorems—and the prediction is tested against a physical experiment. This point of view motivates the style of the presentation in all that follows.

Although the basics of mathematical modeling is discussed, the models to be considered arise from problems where the underlying science is easily accessible. The simple truth is that the construction of many important mathematical models requires a serious treatment of the corresponding science. This is one good reason for joint work between mathematicians and scientists or engineers on applied projects. Carefully chosen models, along with the essential science needed for their construction, are explored in this book.

Applied mathematics requires an understanding of mathematics, some familiarity with the subject area of application, creativity, hard work, and experience. The study of (pure) mathematics is essential. As an aspiring applied mathematician approaching this book, you should know at least what constitutes a mathematical proof and have a working knowledge of basic analysis and linear algebra. To proceed further toward competence in applied mathematics, you will need to know and understand more and deeper mathematics. Along the way, part of your mathematics education should include some study in an applied context. This book is intended to provide a wealth of this valuable experience.

Columbia, Missouri
March 4, 2016

Carmen Chicone

I thank all the people who have offered valuable suggestions for corrections of and additions to this book, especially Oksana Bihun, Michael Heitzman, Sean Sweany, and Samuel Walsh.

TO THE PROFESSOR

This book is suitable for courses in applied mathematics with numerics, basic fluid mechanics, basic mathematics of electromagnetism, or mathematical modeling. The prerequisites for students are vector calculus, basic differential equations, the rudiments of matrix algebra, knowledge of some programming language, and of course some mathematical maturity. No knowledge of partial differential equations or numerical analysis is assumed.

The author has used parts of this book while teaching courses in mathematical modeling at the University of Missouri where students (undergraduate and graduate) of engineering, the sciences, and mathematics enrolled. This heterogeneous mix of students should be expected in a course at the advanced undergraduate beginning graduate level with a title such as Mathematical Modeling I. Thus, the instructor must assess the abilities and background knowledge of the students who show up on the first day of class. Professors should be prepared and willing to modify their syllabus after a week or two of instruction to accommodate their students. In fact, the most likely modification is to cover less material at a slower pace. Perhaps learning a few concepts and techniques well is always more valuable than exposure to a survey of new ideas.

A typical 15-week semester course might consist of one lecture on Chapter 1, two weeks on Chapter 2 (mostly ODE), two weeks on Chapter 5 (fundamental physical modeling, reaction-diffusion systems, and basic numerics for simple parabolic PDE), one week on Chapter 6 (electrical signals on neurons and traveling wave solutions), and one week on Chapter 8 (basic PID control) to complete approximately half of the semester. Of course only parts of the material in these chapters (in particular Chapter 5) can be covered in detail in class. By this time in the semester at least three substantial homework assignments should be completed using exercises, problems, and projects suggested in the text. Of course, there is good reason to also include exercises designed by the instructor. At least, students should have written, tested, and reported applications to applied problems of a few basic codes for approximating solutions of ODEs and PDEs. Their work should be presented in (carefully) written reports (in English prose [or some other language]) where analysis and discussion of results are supplemented with references to output from numerical experiments in tabular or graphical formats. In-class exams are possible but perhaps not as appropriate to the material as homework assignments. The book does not contain many routine problems; in fact, many problems and all of the projects are open ended. How else will students experience challenges that anticipate realistic applied problems? Some of the projects introduce new concepts and are fleshed out accordingly. *A list of suggested projects is given in the index (see the entry Projects).* The second half of the semester might be devoted to continuum mechanics or electromagnetism. But, the usual choice is fluid mechanics. There will be sufficient time to derive the conservation of momentum equation and discuss the Euler and Navier–Stokes stress tensors as in Chapter 11. Standard applications include flow in a pipe (Chapter 12) followed by a discussion of potential flow with applications to circulation, lift, and drag in Chapter 13. Perhaps the end of the semester is reached with a discussion of the Coriolis effect on drains and hurricanes. The final exam can be replaced by a set of problems and projects taken from Chapters 10 and 19, with respect given to sufficient background material discussed in class. In addition, each student might be required to present a project—in the spirit of the course—taken directly from this book, related to their work in some other class, or related to their research.

A more advanced course might be devoted entirely to continuum mechanics with the intention of covering more sophisticated mathematics and numerics. In particular, basic water wave phenomena and free-surface flow can be addressed along with appropriate numerical methods. In Chapter 16, a complete treatment of Chorin's projection method is given in sufficient detail for students (and perhaps their professor) to write a basic CFD code that can be applied to a diverse set of applied problems. This is followed by the most mathematically sophisticated part of the book on the boundary element method, where classical potential theory is covered and all the ingredients of this numerical method are discussed in detail. This is followed by a treatment of smoothed particle hydrodynamics, again with sufficient detail to write a viable code. Channel flow provides a modeling experience along with a discussion and application of Prandtl's boundary layer theory, and a solid treatment of the theory and numerics of hyperbolic conservation laws. All of this material is written in context with applied problems. The chapter ends with a basic discussion of elastic solids, continuum mechanics, the weak formulation of PDEs, and sufficient detail to write a basic finite-element code that can be used to approximate the solutions of problems that arise in modeling elastic solids.

Likewise, an advanced course might be devoted to applied problems in electromagnetism. The material in Chapter 20 provides a basic (mathematically oriented) introduction to Maxwell's equations and the electromagnetic boundary value problem. An enlightening application of the theory is made to transverse electromagnetic waves and waveguides. This is specialized to the theory of transmission lines where the Riemann problem for hyperbolic conservation laws arises in context and its solution is used to construct a viable numerical method to approximate the electromagnetic waves. This theory is applied to the practical problem of time-domain reflectometry, which serves as an introduction to a basic inverse problem of wide interest: shine radiation on some object with the intent of identifying the object by analyzing the reflected electromagnetic waves.

The material in the book can be used to design undergraduate research projects and master's projects. Of course, it can also be used to help PhD students gain valuable experience before approaching an applied research problem.

TO THE STUDENT

This book was written for you. Perhaps you intend to read on your own, which is a good idea, or you are enrolled in a class at a college or university where a professor will help guide you through parts of the book. By this time in your education, you should understand a fundamental fact: you cannot learn from a mathematics text without confronting every sentence as a challenge to your understanding. I have tried my best to provide enough detail so that following discussions of new ideas, writing code, or checking calculations that appear in the text should be within your ability to understand without too much difficulty. But serious thinking, rereading, pencil and paper computations, and computer programming are required to understand the material. *Reading without checking details is a way to see what topics are discussed but definitely not sufficient to understand or use the material.* There is no royal road; reading a mathematics textbook demands a slow pace and a lot of effort. Don't be surprised by being lost in a sea of formulas and new concepts. Start over. Reread the text, think about the meaning of new concepts, check each formula, and ask questions. With enough effort you will experience wondrous breakthroughs to clarity, understanding, and knowledge.

Problems and projects, exercises, and questions are an integral part of the book. You should challenge yourself to solve some difficult problems. As you gain experience and knowledge, your personal toolkit will grow and eventually you will be prepared to work successfully on applied problems arising in science, engineering, and business. The motivation for writing this book is to give you some of the required experience and knowledge. Do your homework!

An essential ingredient of scientific and mathematical research is asking questions (and perhaps answering them). You should ask your own questions about the topics covered in the book as you progress. Some of your questions will be answered in the text once you fully understand what is written. You may also have a knowledgeable professor who is willing to help. Take advantage of the opportunities that are presented. A bit of advice is to prepare yourself with basic knowledge before asking questions so that you can understand and appreciate the answers. In science and mathematics, preparation includes understanding the language of the subject of inquiry (for instance, the meaning of mathematical concepts such as continuity, differentiability, convergence, iteration, ordinary differential equation, matrix multiplication, singularity, eigenvalue, and so on). If you don't understand the language, you are certainly not going to understand the answer. Of course, answers become more complex and require more understanding as questions are asked about more advanced material.

To begin reading this book, you should have a working knowledge of calculus (including vector calculus). You should also be familiar with differential equations and matrix algebra. A basic undergraduate course in differential equations is a requirement. Taking and learning the material in an undergraduate course in matrix theory is more than enough preparation in this important subject. Perhaps you have acquired some knowledge in matrix theory without having taken a formal course. You should at least know what is matrix multiplication, what is an eigenvalue, what is the determinate and trace of a matrix, what is a matrix inverse, what is an inner product, and what constitutes a basis of a vector space. In addition, you will need to be able to use a programming language to write simple codes and postprocess data to make graphs and tables. Ideally, you will already be proficient in at least one programming language. If not, a crash course on the rudiments of a language using widely available resources or some reading supplemented with the guidance of a professor or knowledgeable friend should be enough preparation to approach the introductory exercises and projects in this book. Your coding skills will improve as you work through the more advanced material in this book.

Writing out assignments in English prose (or some other language in case you are using this text in a non–English-speaking country) should be normal practice by this time in your education. Don't be one of the (poor) students who simply writes a few formulas with no explanation to answer a homework assignment. Pick up any book or article on mathematics (this book is an excellent choice) and notice how concepts and results are written out with complete sentences, how formulas are punctuated as parts of sentences, and how figures and tables are referenced. Don't include too many figures or tables in your reports and always explain to the intended readers what they are supposed to notice in a table and what they are supposed to see in your graphs. Emulate this style. You will soon see that expressing your thoughts and presenting your results as prose leads to better understanding (and better grades). Writing good reports and making good presentations are two of the most important skills you can acquire that will help you secure and keep a good job and be successful in public life. Now is the time to develop these skills. Write your homework assignments in complete sentences.

The utility of mathematics is amazing and powerful. By reading and understanding this book, you will certainly learn how to harness the power to solve some important problems. As new areas of the applied mathematical world open, perhaps you will be amazed. Enjoy.

REQUEST

Please send your corrections or comments.

E-mail: chiconeC@missouri.edu
Mail: C. Chicone
Department of Mathematics
University of Missouri
Columbia, MO 65211

CHAPTER 1

Applied Mathematics and Mathematical Modeling

1.1 WHAT IS APPLIED MATHEMATICS?

Applied mathematics is the use of mathematics to solve problems or gain insight into phenomena that arise *outside* of mathematics.

A prime number is an integer larger than 1 whose only divisors are itself and one. For example, 2, 3, 5, and 7 are the first four prime numbers. How many integers are prime numbers? This question arose *inside* mathematics.

Recall Newton's second law of motion: The rate of change of the linear momentum of a particle is the sum of the forces acting on it. Newton's law of universal gravitation may be described by the following two statements: (1) The magnitude of the gravitational force that one mass exerts on a second mass is directly proportional (with a universal constant of proportionality) to the product of the masses and inversely proportional to the square of the distance between their centers of mass; and (2) The direction of the force is along the line connecting the centers of mass toward the second mass. These laws prescribe the relative motion of two masses, each influenced only by the gravitational force of the other. The problem of determining the motion of two masses—the Newtonian two-body problem—is the prototype for applied mathematics. It arises *outside* of mathematics.

The two-body problem is a basic question in celestial mechanics. Using Newton's theory, we may build a mathematical model: Let m_1 and m_2 be point masses in three-dimensional Euclidean space, moving according to Newton's second law of motion and his law of universal gravitation. Denote their positions in space by the position vectors R_1 and R_2, define the vector $R = R_2 - R_1$, the distance between their centers $r = |R|$, and let G_0 denote the universal gravitational constant. The equations of motion for the two bodies are

$$m_1\ddot{R}_1 = \frac{G_0 m_1 m_2}{r^3}R, \qquad m_2\ddot{R}_2 = -\frac{G_0 m_1 m_2}{r^3}R.$$

An Invitation to Applied Mathematics: Differential Equations, Modeling, and Computation.
http://dx.doi.org/10.1016/B978-0-12-804153-6.50001-4, Copyright © 2017 Elsevier Inc. All rights reserved.

This mathematical model consists of a pair of second-order ordinary differential equations, which is typical in classical mechanics. We may now *make predictions* about two-body motion *with no further reference to physics* or observations of nature *by making mathematical deductions from these equations of motion*. When this model was first proposed—in not so compact language—Newton showed by mathematical deduction that these equations of motion predicted Kepler's three laws of planetary motion, which were derived directly from observations of the motion of the planets in the solar system. For example, Kepler stated that each planet moves in a plane on an elliptical orbit with the sun at one focus of the ellipse. Kepler *described* the motions of the planets; Newton *explained* their motion by making Kepler's laws special cases of a more general theory. Kepler's laws apply only to the motions of the observable planets; Newton's law of gravitation applies equally well to the motion of the moon or falling bodies near Earth, and his second law applies to all forces, not just the gravitational force. These astounding successes and many others verified that Newton's laws are (close approximations of) fundamental laws of nature. Of equal importance, these applications reinforced the notion that mathematical deductions from fundamental laws are predictive. Although these events were proceeded by the development and important applications of algebra, geometry, and probability, the development of calculus and Newton's laws (especially his second law) are the foundation of modern mathematical modeling and applied mathematics.

Exactly why mathematical deductions from physical laws are predictive of natural phenomena is a deep philosophical question, but this fact is bedrock. The rationality and determinism of nature lie at the heart of the scientific method, the power of mathematical modeling, and applied mathematics.

Although there are many compelling arguments for the value of pure mathematics as a subject worthy of study in and of itself, the effectiveness of mathematics applied to understand nature and make viable predictions legitimizes the entire mathematical enterprise.

The predictive power of mathematical deductions from Newton's laws cannot be overestimated. Halley's Comet appears in the sky. Using initial data supplied by observation, you solve the two-body problem for the sun and this celestial object and predict the comet will return in approximately 75 years. You wait for 75 years and the comet appears in the sky. What other methodology exists that can predict an event with certainty 75 years in advance? The combination of physical law, mathematical modeling, mathematical analysis, and computation can be used to make predictions of many other natural phenomena.

As you might know, Newton's laws are excellent approximations of reality but they are not correct. The true nature of gravity is much more complicated and Newton's second law is not valid for masses whose relative velocities approach the speed of light. An easy thought experiment should convince you that the law of universal gravitation is not a perfect model of gravity. Simply note that the gravitational force is felt instantaneously with a change in distance between two masses. If the sun started to oscillate, the motion of the Earth would be affected immediately. By the same reasoning, a message could be sent instantaneously anywhere in the universe: imagine shaking the sun for a second to represent a one and pausing the shaking for a second to represent a zero. Newton's law of universal gravitation predicts instantaneous action at a distance. Of course, Newton was well aware of this fact. Although the theory of gravity was modified by Albert Einstein and will likely be modified in the future to conform more closely with observations, Newton's model of motion due to gravitational interaction is predictive up to the precision of most practical measurements as long as the relative velocity of the masses is much less than the speed of light. It is the prototypical example of an excellent predictive model that is routinely used in many important applications; for instance, the planning of space missions. The main point here is that Newton's model is not exact, but it is useful. Utility is a measure of quality in the realm of applied mathematics. There are no perfect mathematical models of reality. Fortunately, utility does not require perfection. The prime objectives of applied mathematics are to develop, analyze, and use mathematical models to make useful predictions, test hypotheses, and explain natural phenomena.

1.2 FUNDAMENTAL AND CONSTITUTIVE MODELS

Although there is not a bright division line, mathematical models of physical phenomena are of two general types: fundamental and constitutive. Fundamental models are derived with fidelity to physical laws; for example, conservation of mass, conservation of momentum, the laws of electromagnetism, or the laws of gravity. Constitutive models mimic physical laws with simplifying assumptions that agree with experiment or observation over some limited range of applications.

As mentioned previously, Newton's laws are not truly laws of nature, but they are so widely applicable that for almost all practical science they can be considered fundamental. Thus, Newton's model for the motion of two massive bodies is considered a fundamental model; it is derived from two laws of nature: Newton's second law and his law of universal gravitation.

The reader might wonder about truly fundamental models of the two-body problem. There are at least three important cases: the motion of two massive bodies, the motion of two charged particles, and the motion of two massive charged particles. Fundamental models would use Einstein's theory of gravity (general realativity) or Maxwell's laws of electrodynamics and the Lorentz force law. This is not to mention the quantum nature of reality. No one knows how to write down such models in a manner that would be open to mathematical analysis. Thus, these problems—how do two massive or charged particles actually move according to fundamental physics—have not been solved. The complexity of applying fundamental physics to realistic situations is one reason why truly fundamental models are rarely used in practice.

Most useful models use constitutive laws. A familiar example is the usual model for the motion of a mass attached to the free end of a spring. Let m denote the mass and x the displacement of the spring from its equilibrium position. Newton's second law states that $md^2x/dt^2 = F$, where F is the sum of the forces on the mass. Although the total force may contain a gravitational summand, the most important summand is the restoring force of the spring. At a fundamental level this force is electromagnetic and it involves the atomic structure of the material in the spring. The restoring force is never modeled using the Lorentz force law and Maxwell's equations of electromagnetism; instead, models are constructed from the constitutive (also called a phenomenological) Hooke's law: The magnitude of the restoring force of the spring is proportional to its displacement from equilibrium and acts in the direction opposite the displacement. Hooke's law is not a fundamental law of nature. It leads to the mathematical model $md^2x/dt^2 = -kx$, where k is the constant of proportionality in Hooke's law. This model, often called the spring equation or the harmonic oscillator, is used extensively in physics and engineering. It is arguably the most important differential equation in these disciplines. Although it is not fundamental, predictions from the Hookean spring model closely approximate experimental measurements for small displacements.

Imagine the nature of a fundamental model for spring motion. It would involve, at least, a coupled system of partial differential equations to account for the electromagnetic force and perhaps coupled equations of motion for all the atoms in the spring. A correctly constructed model of this type would in theory yield more accurate predictions of spring motion. But, the added complexity of a fundamental model would certainly require sophisticated (perhaps yet unknown) mathematics or extensive numerical computations (perhaps beyond the limits of existing computers) to make predictions. Also, a fundamental model would likely depend on many parameters, some of which might not be easily measured. At present, no one knows how to construct a fundamental model for the motion of a spring. Modern elasticity theory, which includes the Hookean spring model, is based on constitutive laws. The theory is imperfect, but properly applied, predictions made from it agree with experimental measurements.

Except for theoretical physics, where the purpose of the discipline is to determine the fundamental laws of nature, constitutive models are ubiquitous in science because the fundamental laws are often too difficult to apply. For many situations of practical interest, no one knows how to construct a fundamental model. In other cases, where a fundamental model might be constructed, constitutive models are usually preferred because they are simpler, provide insight, and often are sufficiently close representations of reality to provide predictions that agree with experiments up to current experimental accuracy. The simplest model that provides insight and consistency with experiments is usually the best.

Many scientists say they *understand* a natural phenomenon that can be measured when there is a model based on fundamental or constitutive laws whose predictions always agree with experimental measurements. In other words, understanding in this sense means that measurements of the phenomenon can be predicted using a theory that applies more generally. Models derived from Newton's law of motion, the law of universal gravitation, or Hooke's law are prime examples.

When a *constitutive* model predicts behavior that does not agree with physical experiments, something should be changed. Usually, a more accurate model is required. For instance, the motion of a spring might not agree with the Hookean model to high accuracy using careful measurements.

Perhaps a nonlinear model of the form $m\ddot{x} = -kx - \ell x^3$, which generalizes Hooke's law to include nonlinear effects or one of the form $m\ddot{x} + \epsilon\dot{x} = -kx - \ell x^3$, which takes into account dissipation of energy also called damping, would be more accurate. Maybe the mass on a spring is moving too fast in air for viscous damping to be sufficiently accurate. Instead, dissipation of energy might be better modeled by an expression of the form $\epsilon\dot{x} + \delta|\dot{x}|\dot{x}$. Incorporating such modifications to improve accuracy does not signal a crisis in physics; rather, the process is one of refinement of the constitutive laws. The situation is different in case a prediction made from a fundamental model does not agree with an experiment. When this happens there *is* a crisis in physics; a new understanding of basic physics is required to construct models that agree with nature. A classic example already mentioned is Newton's second law $m\ddot{x} = F(x)$. This law is not fundamental: it simply does not agree with experiments when the velocity of the particle with position x is near the speed of light c. The new, more fundamental law (first given by Lorentz and Einstein in their development of the special theory of relativity), is

$$\frac{d}{dt}\Big(\frac{m\dot{x}}{\sqrt{1 - \dot{x}^2/c^2}}\Big) = F(x).$$

When models of the motion of electrons in atoms based on Newtonian physics did not agree with experiments, quantum mechanics was discovered, and so on.

Mathematical models are never exact representations of nature. They do not have to be faithful to fundamental physics to be useful. Indeed, making, analyzing, and drawing predictions from constitutive models is the core of applied mathematics and the main theme of this book.

1.3 DESCRIPTIVE MODELS

Kepler's laws are prototypical descriptive models: the planets move on ellipses, the radial vector from the sun to a planet sweeps out equal areas of the ellipse in equal time, and the square of the period of a planet is proportional to the cube of the semimajor axis of its elliptical orbit. His statements were not derived from fundamental or constitutive laws; they describe observational data.

time (sec)	distance (In)
4.33	5.0
11.6	5.25
22.0	6.5
33.0	7.25
45.0	9.0
56.0	10.038
60.33	11.75

Table 1.1 The data in this table was produced by observation of the distance from its origin of the diffusion front of a quantity of red ink deposited in a trough of water.

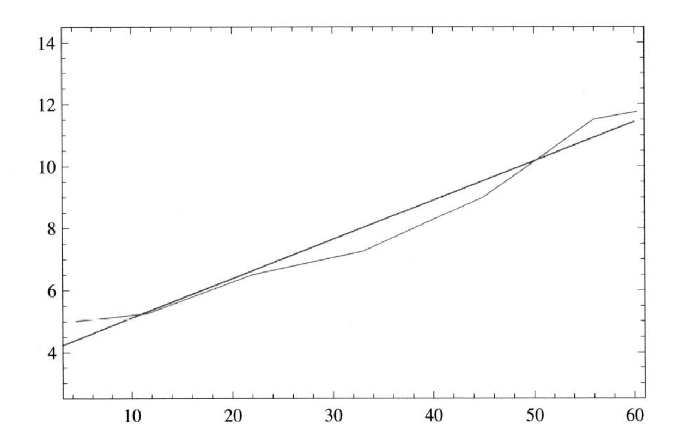

Fig. 1.1 A joined plot of distance versus time for the data in Table 1.1 together with the best fitting line is depicted.

A descriptive model is usually an equation chosen to fit experimental or observational data. For example, Kepler's law concerning the period of a planet's motion was obtained by fitting to observational data recorded by the astronomer Tycho Brahe.

Table 1.1 lists experimental data for a crudely constructed experiment on the diffusion of ink in pure water. A 14.5 inch trough (3.0 inches wide) was filled to a depth of 0.75 inches with water and left undisturbed for a period of time to diminish the strength of convection currents. Red ink was deposited in the water near one end of the trough and allowed to diffuse. Measurements of the changing position of the diffusion front were recorded as a function of elapsed time. The data is plotted in Fig. 1.1. The line in the figure is the graph of the linear function f given by $f(x) = 3.84 + 0.126x$; it is a descriptive model of the measured phenomenon.

This model can be used to make predictions. For instance, it implies that the ink front will be 12.7 inches from the origin after 70 seconds of elapsed time. Although this might be an accurate prediction, the model tells us nothing about *why* the front moves in the observed fashion. A constitutive model for this experiment is suggested in Exercise 5.11.

Descriptive models are ubiquitous and useful in many areas of science (especially the social sciences) and in engineering. From the point of view adopted in this book, descriptive models are precursors to fundamental or constitutive models. Organizing data via fitting to a function may provide some insight into the underlying phenomena being measured, but making such a model does not offer the predictive power of fundamental or constitutive models.

The reader should be aware of the differences among fundamental, constitutive, and descriptive models.

1.4 APPLIED MATHEMATICS IN PRACTICE

In an ideal world, scientists and engineers would create models and applied mathematicians would derive predictions from them. But, in practice, the boundaries between scientists, engineers, and mathematicians are blurred. Perhaps one person assumes all three roles.

The quality of applied mathematics is measured by the relevance of its predictions in the subject area of application. Often the best way to achieve quality results is interdisciplinary collaboration.

A major difficulty to overcome for aspiring applied mathematicians (and textbook authors) is the necessity of learning enough of some scientific discipline outside of mathematics to aid in the development of useful models and the formulation of research questions that address important science. Poor quality applied mathematics is often a result of insufficient knowledge about the scientific area of application; mathematics is developed and theorems are proved that do not answer questions posed by scientists working in the area of application. Applied mathematicians should at least be aware of the important questions in the science they seek to advance.

Fortunately, the apprentice can learn the tools of applied mathematics in context by studying the mathematics required to understand and appreciate known applications of mathematics to science. A journeyman does useful applied mathematics. To achieve this status requires a deep understanding of mathematics, understanding of the area of application, skill in computation, and a strong motivation to advance scientific knowledge. Masters of applied mathematics make important discoveries of lasting value. Newton was a master: he produced a fundamental model and developed the mathematics (calculus) that made it useful.

The usual goal for a working applied mathematician is to address a scientific problem by constructing a model of the underlying phenomenon and using it to make a useful prediction. Ideally, the model should be well-posed; that is, a unique solution should exist that depends continuously on the initial data, the boundary conditions, and the system parameters. Well-posedness is the hunting license required to seek the particular solution that would solve the original scientific problem. Unfortunately, realistic models are often too complicated for mathematical analysis. Sometimes numerical methods would require too much computer time to produce useful results. Thus, simplified models (which are designed to capture the main features of some phenomenon) are considered so that mathematical analysis can produce exact solutions or where theorems about the nature of their solutions can be proved. These results provide footholds for the climb to understanding the full model. Numerical algorithms, which should be designed to approximate solutions of the full model when possible, may be debugged and assessed by measuring their performance against known solutions for special cases or by comparing their qualitative features to proved properties of the solutions of a simplified model. Mathematical analysis of special cases, which has historically provided some of the best applied mathematics, is of great value for the stated reasons. With this essential step in place, insights and approximations from numerical simulations can be employed with high confidence to understand the original problem and make useful predictions. Many special cases and simplifications of fundamental models are considered in this book.

Difficult scientific problems often yield to the awesome power of clear thinking, mathematical modeling, mathematical analysis, and computation. This is the domain of applied mathematics.

CHAPTER 2

Differential Equations

A working knowledge of elementary differential equations (at the level of [11]) is a useful prerequisite for understanding many of the topics in this book. Some of the essential ideas and methods of the subject are mentioned in this chapter. Most of the material here is in the form of exercises designed to help the reader assess basic knowledge of differential equations and to provide a few challenges arising from perhaps unfamiliar elementary applications.

A first-order system of ordinary differential equations (ODEs) is an expression of the form $\dot{x} = f(x, t)$, where the overdot denotes differentiation with respect to time and f is a smooth[1] (vector valued) function of a vector variable x and scalar variable t. By a basic theorem of the subject, a first-order system of ODEs has a unique solution for each initial condition $x(t_0) = x_0$ as long as the point (x_0, t_0) is in the domain of f and the dimensions of the x and $f(x, t)$ are the same. Such a solution $t \mapsto x(t)$ exists at least until $x(t)$ reaches the boundary of the domain of f or $|x(t)|$ blows up to infinity. Moreover, solutions of the initial value problem depend smoothly on the initial data. In case the function f depends smoothly on a (vector) parameter λ, so that the system has the form $\dot{x} = f(x, t, \lambda)$, solutions also depend smoothly on this parameter. The existence and continuity of higher-order derivatives is assumed when needed. Thus, for ODEs, there is a general theory that ensures unique solutions exist for the initial value problem: the future state of a dynamical system is determined by its initial condition—the principle of determinism.

There is no general existence theory for ODE boundary value problems where some part of the initial data is left unspecified, and the problem is to determine the remainder of the initial data so that some other condition is satisfied. For example, consider a second-order differential equation of the form $\ddot{x} = f(x, \dot{x}, t)$, where x is a scalar. Given three numbers a, b, and T, a typical boundary value problem is to determine $\dot{x}(0)$ such that $x(0) = a$ and $x(T) = b$. There is no general method to prove that a solution exists and

[1] In this book, a function is called smooth if it is at least continuously differentiable.

An Invitation to Applied Mathematics: Differential Equations, Modeling, and Computation.
http://dx.doi.org/10.1016/B978-0-12-804153-6.50002-6, Copyright © 2017 Elsevier Inc. All rights reserved.

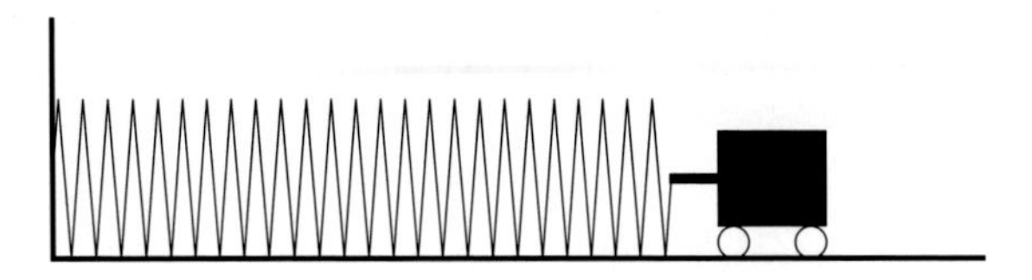

Fig. 2.1 The figure depicts a rectangular mass attached to a spring stretched between the mass and a fixed wall.

no general way to prove that the solution is unique. Indeed, there are such boundary value problems with more than one solution or no solution.

The situation for partial differential equations (PDEs) is much more complicated. Often both initial data and boundary conditions are imposed. Some of the most important mathematical theories and theorems are devoted to the existence and uniqueness problem for PDEs. Although great progress has been made, there is no general theory for the existence and uniqueness of solutions of PDEs.

A major part of applied mathematics is devoted to determining and approximating the solutions of differential equations that arise as mathematical models of physical processes. Thus, a working knowledge of the basic theory of differential equations is essential.

One of the primary goals of this book is to demonstrate the central role played by differential equations in applied mathematics.

2.1 THE HARMONIC OSCILLATOR

The fundamental ODE for physics, mechanical engineering, and electrical engineering is the periodically forced, damped, harmonic oscillator

$$m\ddot{x} + \epsilon\dot{x} + kx = A\sin(\Omega t + \rho), \tag{2.1}$$

where (in its mechanical applications) x is a coordinate measuring an unknown displacement, m is a given mass, ϵ a (viscous) damping coefficient, k the spring constant from Hooke's force law, and $A \sin \Omega t$ a sinusoidal external force with amplitude A, circular frequency Ω, and phase shift ρ. The model is derived from Newton's second law of motion: The time rate-of-change of linear momentum on a particle is equal to the sum of the forces acting on this particle. This statement is taken as a fundamental law of nature. In symbols

$$\frac{d}{dt}(mv) = F.$$

Imagine a spring attached to a wall and to a mass in a horizontal configuration so that the force of gravity does not affect the motion (see Fig. 2.1). Also, choose a horizontal coordinate system with its origin at the equilibrium position of the mass oriented so that the coordinate, say x, measures displacement with positive values corresponding to the stretched spring. The force F acting on the mass is due to the elasticity of the material used to make the spring. By the nature of a spring, the force acts in the direction that would restore the spring to its equilibrium position. What is this force? Recall that there are four fundamental forces: the weak and strong nuclear forces, gravity, and electromagnetism. The restoring force of the spring is clearly electromagnetic. To obtain a fundamental model, the laws of electromagnetism (Maxwell's laws) and the Lorentz force law would have to be used to determine the restoring force of the spring. No one knows how to make such a model. Instead, the force may be modeled by a constitutive law, which is meant to be a good approximation of reality. The usual force law is Hooke's law: the restoring force is proportional to the displacement and in the direction toward the equilibrium position of the mass. There is a constant k, depending on the material properties of the spring that are ultimately due to electromagnetism, such that $F = -kx$. Under the assumption of constant mass, a spring model is given by

$$m\ddot{x} + kx = 0,$$

where a dot over a variable denotes differentiation with respect to time, two dots denote the second derivative with respect to time, and so on. This model is an approximation to reality. What does it predict?

From Exercise 2.2 part (a), the general solution of this ODE involves periodic functions, which in this case are sines and cosines. Thus the displacement x is a periodic function of time. By observation of springs, this is the correct qualitative behavior, at least for a short amount of time. But, a real spring will eventually stop oscillating and return to equilibrium. Thus, our model does not take into account at least one force acting on the mass.

Why does the spring stop oscillating? The mass moves through air, the spring warms up, energy is radiated away due to heat, and perhaps other internal mechanisms are active. At a fundamental level, electromagnetic forces are acting. But the dynamical behavior of the system, which depends on forces at the molecular level, is so complicated that a fundamental model of damping forces is beyond current understanding. Also, imagine the complexity of a model that took into account molecular forces. Could predictions be made from such a model? The usual procedure for modeling macroscopic mechanical systems is to mimic fundamental forces with constitutive laws.

The simplest model for the damping force assumes that it is proportional to the velocity of the mass and acts in the direction opposite to the motion: $-\epsilon\dot{x}$, where ϵ is some positive constant. A more realistic model might be $-\epsilon\dot{x}^2$ or $\epsilon_1\dot{x} + \epsilon_2\dot{x}^2$, but these latter choices lead to nonlinear ODEs that are more difficult to analyze. With the linear damping force, we recover the basic model [Eq. (2.1)] with $A = 0$. This parameter is not zero when there is an external periodic force with circular frequency Ω and phase shift ρ, a case that is prevalent in many applications (for example in electrical engineering). In fact, the spring model [Eq. (2.1)] is accurate enough to make useful predictions for many physical phenomena.

To predict the outcome of an experiment by using this model, the parameters in the model must be identified and the initial position and velocity ($x(0)$ and $\dot{x}(0)$) of the mass must be specified. According to ODE theory, once the parameters and initial data are specified, the ODE has a unique solution. It is a prediction of the motion of the mass-spring system with the specified data. The accuracy of such predictions has been verified by many physical experiments. The harmonic oscillator model is widely used to simulate reality with no need to perform physical experiments.

Models of physical processes are never exact representations of reality. But, a good model produces predictions that agree with reality at some acceptable level of approximation. The existence of a model that produces physically correct predictions also provides evidence that the underlying physical intuition used to create the model is correctly understood.

Exercise 2.1. Use Newton's second law of motion to derive the harmonic oscillator model for a damped, vertically mounted, mass-spring system under the influence of gravity and an external sinusoidal force.

Exercise 2.2. (a) Solve the harmonic oscillator equation for the case $A = 0$ and $\epsilon = 0$ and describe the qualitative behavior of the corresponding solutions. (b) Solve the harmonic oscillator equation for the case $A = 0$ and $\epsilon \neq 0$ and describe the qualitative behavior of the corresponding solutions. (c) Solve the initial value problem

$$\ddot{x} + \frac{1}{2}\dot{x} + 9x = 2\sin t, \quad x(0) = 1, \quad \dot{x}(0) = 0$$

and determine the long-term behavior of the displacement x. In particular, what is its approximate amplitude in the long-term?

Exercise 2.3. (a) Suppose that incoming waves oscillating at γ cycles per second are to be detected by the motion of a mass-spring system with fixed mass m and damping coefficient ϵ. Imagine, for example, a mechanical earthquake detector. How should the spring constant k be chosen so that the response of the mass-spring system to the incoming wave has the greatest amplitude? (b) Is the result the same for a scenario where the spring constant is fixed and the frequency of the incoming wave is adjusted? Explain.

2.2 EXPONENTIAL AND LOGISTIC GROWTH

The fundamental ODE for chemistry, biology, and finance is the exponential growth equation

$$\dot{Q} = rQ,$$

where Q is a measure of the amount of some substance and r is a growth (or decay) rate. Additions (or subtractions) f to the quantity per time are included in the model via the equation

$$\dot{Q} = rQ + f(t). \tag{2.2}$$

The logistic growth model, which models limited growth, is

$$\dot{Q} = rQ(1 - \frac{Q}{k}),$$

where r and k are parameters. Here the growth rate $\dot{Q}/Q$ is dominated by r when Q is small and by $-rQ/k$ when Q is large.

Exercise 2.4. (a) Solve the initial value problem

$$\dot{Q} = -2Q, \quad Q(0) = 3.$$

(b) Solve the initial value problem

$$\dot{Q} = -2Q + 8, \quad Q(0) = 3.$$

(c) Solve the initial value problem

$$\dot{Q} = \frac{2}{t}Q + 8, \quad Q(1) = 3.$$

(d) Solve the initial value problem

$$\dot{Q} = 2Q(1 - \frac{Q}{10}), \quad Q(0) = 5$$

and determine the fate of the solution (that is, determine the behavior of the solution as t grows without bound).

Exercise 2.5. What is the monthly payment on a T year loan of P dollars with interest compounded continuously at the annual rate r?

2.3 LINEAR SYSTEMS

Linear systems occur in all areas of applied mathematics. A first-order linear system of ODEs has the form

$$\dot{x} = A(t)x + f(t),$$

where x in this equation is an n-dimensional vector variable, A is an $n \times n$-matrix, and f is an n-dimensional vector function of the independent scalar variable t (which is usually a coordinate measuring time). The exponential growth equation [Eq. (2.2)] is a linear system where $n = 1$.

Exercise 2.6. (a) Solve the initial value problem

$$\dot{x} = 2x - y, \qquad \dot{y} = x + 2y, \quad x(0) = 1, \quad y(0) = -1.$$

(b) Solve the initial value problem

$$\dot{x} = x + 4y, \qquad \dot{y} = 2x + y, \quad x(0) = 1, \quad y(0) = 1.$$

(c) Find the general solution of the system

$$\dot{x} = 4x - 2y + 2e^{-t}, \qquad \dot{y} = 3x - 3y + e^{-t}.$$

Hint: The key concept here is the variation of parameters formula (see Appendix A.15).

Exercise 2.7. Imagine two identical objects (perhaps wooden blocks on wheels) each with mass m riding on a horizontal track of length L. The object on the left is connected by a spring to the left end of the track; the object on the right is attached by an identical spring to the right end of the track. Also, the two objects are attached to each other by a spring, hereafter called the connecting spring. Let x denote the distance of the object on the left from the left end of the track and y denote the distance of the other object from the same left end of the track. (a) Show that the following system of differential equations is a reasonable differential equation model for the motions of the masses:

$$\begin{aligned} m\ddot{x} &= -\epsilon\dot{x} - K(x-\xi) - k(\ell - (y-x)), \\ m\ddot{y} &= -\epsilon\dot{y} + K((L-\xi) - y) + k(\ell - (y-x)), \end{aligned} \tag{2.3}$$

when K and k are the constants of proportionality using Hooke's law for the restoring force of the springs, ξ is the position of the left-hand object disconnected from the connecting spring, the equilibrium length of the free connecting spring is $\ell \leq L - 2\xi$, and ϵ is the viscous damping constant. What additional assumptions are made in the derivation of the model? (b) Determine the equilibrium positions of the two objects and verify that in equilibrium the mass with coordinate x is to the left of the mass with coordinate y, $x > 0$, and $y \leq L - \xi$. (c) The equations of motion are written for the positions of the objects along the track. They take a simpler and more symmetric form when the equations of motion are written for the displacements of the objects from their equilibrium positions before the connecting spring is attached. Show that the resulting system may be expressed in the form

$$\begin{aligned} m\ddot{u} &= -\epsilon\dot{u} - Ku + k(v-u), \\ m\ddot{v} &= -\epsilon\dot{v} + Kv - k(v-u), \end{aligned} \tag{2.4}$$

and write explicitly the transformation from (x, y) coordinates to (u, v) coordinates. Hint: Write the original system in matrix form and change coordinates by a translation. (d) Suppose there is no damping ($\epsilon = 0$). Find the general solution of system (2.4). (e) Suppose there is no damping ($\epsilon = 0$) for system (2.4) and the initial conditions are

$$u(0) = 0, \quad \dot{u}(0) = 0, \quad v(0) = \alpha, \quad \dot{v}(0) = 0.$$

Write the explicit solution. Show that this solution can be manipulated into the form

$$\begin{aligned} u(t) &= \alpha \sin\frac{\Omega - \omega}{2}t \sin\frac{\Omega + \omega}{2}t, \\ v(t) &= \alpha \cos\frac{\Omega - \omega}{2}t \cos\frac{\Omega + \omega}{2}t, \end{aligned}$$

Discuss the predicted motion. Hint: The circular frequency $\Omega - \omega$ is small compared with $\Omega + \omega$. Thus, both solutions can be viewed as amplitude modulated sinusoids. Also, it is possible to have the amplitude of u (respectively v) very near zero so that at such times most of the energy of the system corresponds to the motion with coordinate v (respectively u). The resulting motion is a beat phenomenon. (f) Determine initial data so that the two objects move with equal displacements from equilibrium.

2.4 LINEAR PARTIAL DIFFERENTIAL EQUATIONS

Applied mathematics is sometimes equated with the study of PDEs. *Skip this section if you are unfamiliar with PDEs; the subject will be discussed later in this book.*

Exercise 2.8. (a) Solve the initial boundary value problem

$$u_t = u_{xx}, \quad u(t,0) = 1, \quad u(t,1) = 1, \quad u(0,x) = 1 - \sin \pi x.$$

The PDE is called the heat equation. (b) Determine T such that $u(T, 1/2) = 3/4$.

Exercise 2.9. (a) Suppose that $f : \mathbb{R} \to \mathbb{R}$ is a twice continuously differentiable function. Solve the initial value problem

$$u_{tt} = u_{xx}, \quad u(0,x) = f(x), \quad u_t(0,x) = 0.$$

The PDE is called the wave equation. (b) Show that the function f given by $f(x) = -(10x-1)^3(10x+1)^3$ for x in the interval $(-1/10, 1/10)$ and zero otherwise is twice continuously differentiable. (c) Using f defined in (b), find the smallest time $t > 0$ when $u(t, 20) = 1/4$. (A numerical approximation correct within 1% is an acceptable answer.)

Exercise 2.10. Let Ω denote the open set bounded by the unit square in the plane and let $\bar{\Omega}$ denote its closure. The unit square has vertices with coordinates $(0,0)$, $(1,0)$, $(1,1)$, and $(0,1)$. Find the value of $u : \bar{\Omega} \to \mathbb{R}$ at the point $(1/2, 1/2)$ in case u is harmonic (that is, $u_{xx} + u_{yy} = 0$ in Ω) and $u(x,y) = 1$ everywhere on the boundary of Ω. The PDE is called Laplace's equation.

Exercise 2.11. (a) Find a nonconstant solution of the PDE

$$\sin^2 \phi \frac{\partial^2 f}{\partial \phi^2} + \frac{\partial^2 f}{\partial \theta^2} + \sin \phi \cos \phi \frac{\partial f}{\partial \phi} = 0$$

that is periodic in each variable separately and with the additional property that it is continuous at $\phi = 0$. Hint: Use separation of variables. (b) For the reader who has studied differential geometry, the PDE has a geometric interpretation (see [50]). The Laplace–Beltrami operator is a generalization of the Laplacian to Riemannian manifolds. The usual Riemannian metric on a sphere of radius R is given in spherical coordinates by $R^2(d\phi^2 + \sin^2 \phi\, d\theta^2)$. It is a covariant 2-tensor, which—in the present case—can be written more precisely as

$$E d\phi \otimes d\phi + F d\phi \otimes d\theta + F d\theta \otimes d\phi + G d\phi \otimes d\phi.$$

The Laplace–Beltrami operator Δ is given by

$$\Delta f = \frac{\partial}{\partial \phi}\Big(\frac{1}{\sqrt{EG - F^2}}\Big(G\frac{\partial f}{\partial \phi} - F\frac{\partial f}{\partial \theta}\Big)\Big) + \frac{\partial}{\partial \theta}\Big(\frac{1}{\sqrt{EG - F^2}}\Big(E\frac{\partial f}{\partial \theta} - F\frac{\partial f}{\partial \phi}\Big)\Big).$$

The PDE of part (a) is equivalent to the Laplace equation $\Delta f = 0$ on the round sphere. (c) Prove the last statement.

2.5 NONLINEAR ORDINARY DIFFERENTIAL EQUATIONS

Newton's second law states that a particle with constant mass influenced by forces moves so that its mass times its acceleration equals the sum of these forces. As an example, suppose the force law is Newton's law of universal gravitation. It states that the gravitational force on a body due to a second body is directly proportional to the product of their masses and inversely proportional to the square of the distance between them. The direction of the force is toward the center of mass of the second body. Clearly, the force is a nonlinear function of the particle's position. Thus, models that involve gravitational forces are fundamentally nonlinear.

As a review, recall that a force is called conservative if it is given as the negative gradient of a potential U. In this case, Newton's second law leads to the classical differential equation

$$m\ddot{x} = -\nabla U(x), \tag{2.5}$$

where x is the position of the particle and $U(x)$ is its potential energy at position x. In fact, this ODE is often called Newton's equation. The kinetic energy of the particle is, by definition, $\frac{1}{2}m\dot{x}^2$. Because potential energy is often a nonlinear function, this differential equation is a fundamental source of nonlinear ODEs.

Exercise 2.12. (a) Show that the (total) energy (the sum of the potential and kinetic energies) is constant on solutions of Newton's equation [Eq. (2.5)].
(b) Determine the total energy for Duffing's equation

$$\ddot{x} - x + x^3 = 0$$

and draw its phase portrait. Describe the qualitative behavior of the system.
(c) Determine the total energy for the mathematical pendulum

$$\ddot{\theta} + \sin\theta = 0$$

and draw its phase portrait. Describe the qualitative behavior of the system.
(d) Draw the phase portrait for (the alternate form of) Duffing's equation

$$\ddot{x} + x - x^3 = 0.$$

Describe the qualitative behavior of the solution with initial condition $x(0) = 0$ and $\dot{x}(0) = 7/10$.

(e) Determine the qualitative behavior of the solutions of the differential equation

$$\ddot{x} + \frac{1}{10}\dot{x} - x + x^3 = 0.$$

(f) Is the gravitational force conservative? If so, what is the gravitational potential?
(g) Suppose that two point masses fall from rest in the gravitational field due to a third mass, which might be Earth. And, suppose they move on the same line through the center of the third mass. Does the distance between them remain constant as they fall? If not, describe the relative distance as a function of time. Hint: A numerical approximation of the solution might be necessary. See the next section for a brief review of numerical methods for ODEs.
(h) Determine the fate (as time increases without bound) of the solution of the initial value problem

$$\ddot{x} + \frac{1}{10}\dot{x} - x + x^3 = 0, \quad x(0) = \frac{1}{2}, \quad \dot{x}(0) = 0.$$

(i) Draw the phase portrait (for $x \geq 0$) of the differential equation

$$\dot{x} = x - x\sqrt{x}$$

and find the general solution. Does your phase portrait agree with your solution? Does your solution include all orbits depicted in your phase portrait?

Exercise 2.13. A circular hoop with radius $a > 0$ made of thin wire is spinning with angular velocity Ω about a vertical axis that passes through the center of the hoop. A bead with mass m threaded on the hoop slides with viscous friction (that is, the bead's motion is opposed by a force proportional to its velocity *along the hoop*). Assume that the bead is free to move except through the highest point on the hoop, where a rod is fastened to the hoop and used (by connection to a motor) to maintain the constant angular velocity. Imagine the bead is held (by some external unknown force) at a position near but not at the bottom of the rotating hoop and released at time $t = 0$. Describe the qualitative motion of the bead.

Exercise 2.14. Imagine an object with mass m sliding on a horizontal plane connected by identical springs (each with spring constant K) to fixed positions on the plane at a distance $L = 2(\ell + \alpha)$ apart, where ℓ is the natural length of each spring and $\alpha > 0$ is the extra distance each spring is stretched to make its attachment to the mass. When the mass is pulled in the direction of the perpendicular bisector of the line connecting the attachments to the plane and let go from rest, it moves along the perpendicular bisector due to the symmetry of the apparatus. Suppose the mass is pulled out a distance d units from its rest position on the intersection of the line and the perpendicular bisector and released from rest. The idealization just described might be a crude model of a crossbow. At least two phenomena are of interest: The frequency of oscillation and (for the crossbow application) the velocity of the mass at the moment it passes through the equilibrium position of the mass-spring system. Both of these quantities are functions of all the parameters: L, ℓ, α, M, and d. (a) Show that the initial value problem for the displacement of the object from equilibrium (ignoring damping) can be expressed in the form

$$M\ddot{x} = -2Kx\Big(1 - \frac{\ell}{\sqrt{x^2 + (\ell + \alpha)^2}}\Big), \qquad x(0) = d, \quad \dot{x}(0) = 0.$$

(b) What can you say about the key questions using pencil and paper? At least consider the case where d is very small. What exactly is meant by saying d is small? Small compared to what? In a real application, the parameters would be assigned units. The spring constant is expressed in different units than the mass or the natural length of the spring. How can the sizes of the parameters be compared? Does asking such a question make sense? Hint: Make the model dimensionless by a change of variables. Linearize the differential equation near its equilibrium. (c) Incorporate viscous damping. Physical intuition suggests that the displacement of the damped system (perhaps caused by air resistance or friction on the horizontal plane) will approach the equilibrium position as time increases. Although physical intuition is very important, the purpose of mathematical models is to gain physical insight. Using physical intuition to describe a prediction of a mathematical model tacitly assumes the model agrees with reality. Perhaps it does not. If so, we learn that there is something wrong with the model. Either the model must be modified to more closely approximate the physics, or the physical assumptions used to construct the model must be incorrect. To make predictions, we must use mathematical deductions from the model itself. Can the model just constructed (with viscous damping included) be used to predict that every motion will decay toward equilibrium as time increases? State and prove a precise mathematical result.

Exercise 2.15. Imagine a spring attached to a peg on a vertical wall so that the spring can turn freely around the peg. A ball of mass m is attached to the other end of the spring. Assume the motion is confined to a plane parallel to the wall. Write the equations of motion for the ball taking into account the gravitational force but ignoring damping forces. (a) Determine the steady states of the system. Do these agree with your physical intuition? (b) Show that an initial position on the vertical line through the peg and no velocity component in the horizontal direction initiates a motion that is confined to the vertical line. (d) Show that the total energy is a constant of the motion. (e) Is the angular momentum a constant of the motion?

2.6 NUMERICS

The reader may be familiar with some methods for approximating solutions of ODEs. Key words here are Euler method, improved Euler method, trapezoidal method, Runge–Kutta method, Adams method, and others. Some of these methods are discussed in detail in this book. The purpose of this section is to provide some simple exercises designed to refresh (or initiate) a practical working knowledge of this important part of applied mathematics.

For the ODE $\dot{x} = f(x,t)$ with initial condition $x(t_0) = x_0$ (which may be a first-order system of ODEs), the most basic method for approximating the solution of the initial value problem is to choose a time discretization increment Δt and then iterate a procedure to approximate the solution at the discrete time steps $x_1 \approx x(t_0 + \Delta t)$, $x_2 \approx x(t_0 + 2\Delta t)$, ..., $x_n \approx x(t_0 + n\Delta t)$ according to the following and similar recipes.

1. Euler's method:

$$\begin{aligned} t_{n+1} &= t_n + \Delta t, \\ x_{n+1} &= x_n + \Delta t f(x_n, t_n); \end{aligned}$$

2. The improved Euler method:

$$\begin{aligned} t_{n+1} &= t_n + \Delta t, \\ y_{n+1} &= x_n + \Delta t f(x_n, t_n), \\ x_{n+1} &= x_n + \frac{\Delta t}{2}(f(x_n, t_n) + f(y_{n+1}, t_{n+1})); \end{aligned}$$

3. The trapezoidal method:

$$\begin{aligned} t_{n+1} &= t_n + \Delta t, \\ x_{n+1} &= x_n + \frac{\Delta t}{2}(f(x_n, t_n) + f(x_{n+1}, t_{n+1})). \end{aligned}$$

Although there is a wealth of knowledge on approximations of solutions of ODEs and much more sophisticated approximation methods exist, these simple algorithms are adequate for many applications. There is much to be gained by students of the subject who build personal libraries of numerical methods that are used to make predictions from mathematical models. Improvements to basic methods may be incorporated as the need arises. Using black box software instead will usually produce more accurate approximations, but reliance on black boxes does not enhance understanding.

Exercise 2.16. (1) Write a code to implement the Euler method, the improved Euler method, and the trapezoidal method. For the trapezoidal method you may wish to review Newton's method for finding roots of systems of equations (see Appendix A.14). Apply your codes to approximate $(x(3\pi/2), y(3\pi/2))$ for the system

$$\dot{x} = -y + x(1 - x^2 - y^2), \qquad \dot{y} = x + y(1 - x^2 - y^2), \tag{2.6}$$

with initial condition $x(0) = 1$ and $y(0) = 0$. The exact answer is $(0, -1)$. Also, apply your codes to approximate $(x(3\pi/2), y(3\pi/2))$ for the system

$$\dot{x} = -y + x(x^2 + y^2 - 1), \qquad \dot{y} = x + y(x^2 + y^2 - 1),$$

with initial condition $x(0) = 1$ and $y(0) = 0$. The exact answer is $(0, -1)$. Which system of differential equations is more amenable to the numerical methods? Explain the difference. Why does numerics work so well for one of the systems and not so well for the other? Hint: Change to polar coordinates. (2) Discuss, for each method, the largest step size Δt that will produce less than a 1% error in approximating the final value for system (2.6). (3) For system (2.6), make tables to show that the Euler method is first order and the other two methods are second order. The idea here is to start with some Δt and record in a table the absolute error (the absolute value of the difference between the exact solution and the approximation) produced by your code in computing the approximate final value for Δt, $\Delta t/2$, $\Delta t/4$, and so on. The absolute error when using a second order method should be (approximately) 1/4 of the previous absolute error each time the step size is divided by 2. Give evidence that your code has this property.

Exercise 2.17. (a) Consider the system

$$\dot{x} = 1, \qquad \dot{y} = axy,$$

where a is a parameter. Solve this system with initial data $x(0) = y(0) = -1$, and show that the exact value of the solution at $t = 2$ is $(x, y) = (1, -1)$ independent of a. (b) Generalize the result of part (a); that is, given $x(0) < 0$, show that there is a time $T > 0$ (which is independent of a) such that the solution starting at $(x(0), y(0))$ reaches the point with coordinates $(-x(0), y(0))$ at $t = T$.

Exercise 2.18. Use various numerical methods to approximate solutions of the system of ODEs in Exercise 2.17 at least for the parameter values $a = 1, 10^2$, and $a = 10^4$ with $x(0) \leq -1$. Do your computer codes produce correct results? Discuss your experiments (compare to Exercise 5.52).

Exercise 2.19. (a) Show that $t \mapsto (t, t^2)$ is a solution of the system of differential equations

$$\dot{x} = 1, \qquad \dot{y} = 2x + ax(y - x^2)$$

independent of the parameter a. (b) For $a = 10$ and initial data $x(0) = 0$ and $y(0) = 0$, the value of the solution at $t = 10$ is $x(10) = 10$, $y(10) = 100$. Use a numerical method to approximate the solution of the initial value problem and compare the output with the exact value of the solution. (c) The ODE in this exercise is the same as the ODE in Exercise 2.17 via a change of coordinates. What is this change of coordinates?

Exercise 2.20. Determine the fate of the solution of the initial value problem

$$\ddot{x} + \frac{1}{10}\dot{x} - x + x^3 = 0, \qquad x(0) = 0, \qquad \dot{x}(0) = 1$$

as time grows without bound.

Exercise 2.21. Consider the system of differential equations

$$\begin{aligned}\dot{x}_1 &= -k_1 x_1,\\ \dot{x}_i &= k_{i-1}x_{i-1} - k_i x_i, \quad i = 2, 3, 4, \ldots, n-1,\\ \dot{x}_n &= k_{n-1}x_{n-1}.\end{aligned}$$

It arises in situations that may schematically be described by a process $X_1 \to X_2 \to X_3 \to \cdots \to X_n$ where the amount or concentration x_i of some substance in a region X_i (perhaps a tank) is determined by the amount of the substance coming into X_i from X_{i-1} minus the amount going out. The parameter k_i is the rate constant for the amount of substance leaving X_i. Suppose initial data $x_i(0) = \xi_i$ is also given. (a) Show that the system can be solved explicitly. (b) Let $n = 10$, $k_i = i/1000$, and $\xi_i = 1 - 10i/101$. Determine x_{10} at time $t = 2000$. Compare the exact solution with approximations using numerical methods for ODEs.

Exercise 2.22. (a) In the context of Exercise 2.15, discuss the typical motion via a series of well-conceived numerical experiments. (b) The system certainly has periodic motions when the initial data is confined to the vertical line through the peg. Are there other periodic motions?

Exercise 2.23. In the context of Exercise 2.7, suppose that the two springs are different; that is, the springs may have different natural lengths or different spring constants. (a) Write a model for the displacement of the mass in the horizontal plane. (b) Predict typical motions that remain near the equilibrium position of the system. (c) Predict typical motions in the horizontal plane of the object connected to the springs. (d) Are there periodic motions? If so, how do the periods depend on the parameters. (e) It is possible to adjust the apparatus so that the motion is along a line as in Exercise 2.7. Is the symmetric configuration the only one that allows such motions? (f) Is it possible to adjust the apparatus and the initial data so that the motion stays on a circle? (g) What happens when viscous damping is taken into account?

Exercise 2.24. A network of 20 sensor stations is to be constructed on the surface of the moon. The sensors are required to be distributed uniformly to cover the entire surface. How should they be placed? Hint: Consider four sensors first, then eight sensors, and so on. Hint: One possible approach to this problem is to imagine the sensors are electrons confined to the surface of a sphere. The electrons all repel each other according to Coulomb's law. If the elections are placed on the sphere and allowed to move according to this law, they should come to rest in a desired configuration at least when their motions are damped by some force that acts in the direction opposite to the directions of their velocities on the surface of the sphere. (b) For a fixed number (say eight sensors) if a solution exists, there are infinitely many other configurations obtained by rigid rotation of the given solution. Can there be two solutions such that one cannot be rotated to the other?

Exercise 2.25. [Buckling] Suppose a compression force is applied to the ends of a thin flat strip of elastic material of length ℓ. Depending on the strength of the force and its material properties, the strip might deform in a direction normal to one of its faces.

Assuming no twist occurs in the strip, which might be the case when the ends of the strip are appropriately clamped, the position of the deformed strip is specified by the position of its deformed central axis. Once the deformed strip is in equilibrium, the forces acting on it must balance. A crude but informative model may be constructed by balancing the applied force with the restoring force that would tend to move the bent strip back to its flat state (see Chapter 18 for a more complete introduction to elasticity and Exercise 19.6 for a related problem). This second force may be assumed to be proportional to the (signed) curvature of the deformed strip. In the case considered here, where the strip is thin, the curvature of the strip is approximated by the curvature of its central axis. Note that in differential geometry, curvature is taken to be an intrinsic property of a curve; in particular, its value for a curve does not depend on the parameterization. For this force balance problem, directions are important. Suppose the undeformed strip resides in the (x, z) plane of a Cartesian coordinate system and its central axis is along the x-axis of these coordinates. Bending deflections are measured by deviations of the y-coordinate from zero. For convenience, define the unit tangent vector $T = (\cos\theta, \sin\theta)$ where θ is the angle between the positive direction of the x-axis and the velocity vector with respect to the parameterization. For definiteness, consider the parameterization $R(s) = (x(s), y(s))$, where s measures arc length from the left end of the clamped strip. In this case the normal vector N is defined to be $(-\sin\theta, \cos\theta)$, which is the positive 90° rotation of T with respect to the usual orientation of the plane. The unit tangent vector T is also given by $T = dR/ds$. Because $T \cdot T = 1$, the vector dT/ds is perpendicular to T. Thus, this vector is some scalar multiple of N. This scalar is defined to be the signed curvature κ; it can take on positive, negative, and zero real values. In symbols,

$$\frac{dT}{ds} = \kappa N.$$

To make the curvature intrinsic, the definitions are slightly different: The normal N is defined to be the unit vector in the direction of dT/ds and κ is the length of dT/ds.
(1) Show that $\kappa = d\theta/ds$.
A model for the deformed elastic strip is constructed using the following theory: The signed curvature at a point on the deformed strip is proportional to the bending moment, which is the magnitude of the applied force times the displacement of the point from its equilibrium position. For the compressed strip, the latter force with magnitude F acts in the direction of the usual basis vector e_1. To construct the model, the sign of the curvature and the bending moment must be the same. The constant of proportionality is taken to be the flexural rigidity defined to be the product of Young's modulus E (the ratio of stress to strain) for the deformed material and the second moment $I = \int_\Sigma y^2 \, dydz$ of the transverse cross section Σ of the physical strip. Using this theory, the model equation is

$$EI\frac{d\theta(s)}{ds} = -Fy(s).$$

As the position R is given by

$$R(s) = \int_0^s \frac{dR}{ds} \, ds = \int_0^s T(s) \, ds,$$

the second component of position y is recovered from the integral

$$y(s) = \int_0^s \sin \sigma \, d\sigma.$$

(2) Show that

$$EI\frac{d^2\theta}{ds^2} + F\sin\theta = 0.$$

There are many other ways to obtain the same result using different ideas from classical mechanics. Can you derive the same model using a different methodology? The physical reasoning used here is not based on fundamental principles, but it has the virtue of simplicity. Assume the strip is clamped at each end to the (x, z) plane. The corresponding boundary conditions are

$$\theta(0) = 0, \qquad \theta(\ell) = 0.$$

Under the assumption that F is a positive constant, the change of variables $s = \ell\tau$ renders the two-point boundary value problem dimensionless.
(3) Show that the dimensionless system is

$$\frac{d^2\Theta}{d\tau^2} + \lambda \sin\Theta = 0, \qquad \Theta(0) = 0, \quad \Theta(1) = 0,$$

with $\lambda = \ell^2 F/(EI)$ and $\Theta(\tau) = \theta(\ell\tau)$.
(4) The important question for the applied problem is easily stated: How large must λ be to allow a nontrivial solution of the boundary value problem? Note that $\theta = 0$ is always a solution. Are there other solutions? Although there is a pencil and paper answer to this problem, it requires a few new ideas that might not be familiar. Start with numerical experiments. A simple idea (which is not the best, but is adequate for this problem) is to employ the shooting method. The idea is very simple, use a numerical method, perhaps the trapezoidal rule, to solve the initial value problem for the ODE with initial data $\Theta(0) = 0$ and $\Theta'(0) = \sigma$, where σ is a real parameter. This initial value problem has a unique solution. The parameter σ is to be adjusted until some choice of σ determines a solution for which $\Theta(1) = 0$. Try it. Can you devise an adjustment algorithm that goes beyond trial and error? Hint: Recall Newton's method for solving nonlinear equations (see Appendix A.14). It can be used to solve implicitly for the value of the state at each step of the trapezoidal method. In the context of this problem, some algebraic manipulations can be used to reduce the system of equations to be solved at each step to one scalar equation in one unknown. Note that the solution of the initial value problem depends on σ. Thus, we may consider the solution in the form $\tau \mapsto (\Theta(\tau, \sigma), \dot{\Theta}(\tau, \sigma))$. The objective of shooting is to solve the equation $\Theta(1, \sigma) = 0$. Newton's method can be used a second time to approximate the solution of this equation:

$$\sigma_{i+1} = \sigma_i - \frac{\Theta(1, \sigma_i)}{\Theta_\sigma(1, \sigma_i)}.$$

The only difficulty is approximating the partial derivative of Θ with respect to σ at $(1, \sigma_i)$. This quantity may be found using an appropriate variational equation (see Appendix A.16). In fact, the desired quantities are obtained by approximating (via the trapezoidal method) the solution of the initial value problem

$$\begin{aligned}
\Theta' &= V, \\
V' &= -\lambda \sin \Theta, \\
U' &= W, \\
W' &= -\lambda U \cos \Theta, \\
\Theta(0) = 0, \quad V(0) &= \sigma_i, \quad U(0) = 0, \quad W(0) = 1,
\end{aligned}$$

where $U = \Theta_\sigma$ and $W = V_\sigma$. Note that derivatives of state variables in a differential equation model can be obtained by solving a variational equation simultaneously with the given differential equation. Here the model together with the variational equation is four-dimensional. Can you construct a better numerical method based on some idea other than shooting?

(5) With some experimentation with numerical approximations you should find evidence for the following scenario. If $\lambda > 0$ is sufficiently small, the only solution of the boundary value problem is the zero solution. There is a critical value of $\lambda > 0$ such that for every λ less than this value the only solution of the boundary value problem is the zero solution; and, for every λ exceeding the critical value, there is a nonzero solution. What is this critical value? The physical interpretation is clear: the critical value corresponds to the strength of the applied force necessary for the strip to buckle. Describe your findings in detail.

(6) What is the shape of the strip for an applied axial force whose strength is 5% larger than the critical value? Draw a graph.

(7) Consider the same strip configuration and an applied force G in the direction normal to the strip instead of the axial direction. What is the shape of the deformed surface when a constant normal force is applied? Does the strip deform for every nonzero applied force or must the strength of the force exceed some critical value?

(8) Suppose one or both ends of the strip are attached so they can pivot. What are the corresponding boundary conditions? How do the deformed configurations for the two possible forces change from the clamped case?

(9) Suppose there is no axial force and the applied normal force is not constant. For definiteness, suppose the force is constant on the interval $1/2 \leq \tau \leq 3/4$ and zero otherwise. Show the difference between the shape of this deformation and the deformation due to a force with the same total magnitude in case the force is distributed over the entire strip.

Exercise 2.26. Return to Exercise 2.14 and use numerical methods to gain insight (*which is the purpose of numerical approximations of dynamical systems*) into the key questions stated in the problem. At least (partially) answer the following questions and write a report on your findings. Be careful to state precisely the parameter values (or ranges of parameter values) used in your numerical experiments. Organize your work to make predictions about the behavior of the system. (a) How do the frequency and the velocity at the crossing of the equilibrium position depend on the parameters? (b) How would you write a data sheet for a manufactured version of the apparatus in a case where the tension on the springs (via changes in α) was adjustable? (c) How would you manufacture a version of the apparatus to achieve a desired velocity or a desired frequency for a specified initial displacement? As a specific case, consider the following data for the mass-spring system:

$$K = 200\,\mathrm{kg}\,/\,\mathrm{sec}^2, \quad M = 0.25\,\mathrm{kg}, \quad \ell = 0.4\,\mathrm{m}, \quad \alpha = 0.05\,\mathrm{m}.$$

The mass is to be pulled exactly d meters from its equilibrium position and released from rest. The problem is to determine d so that the mass crosses the equilibrium position at time $T = 0.1\,\mathrm{sec}$. What is d? Answer this question, but also discuss the problem and design your solution so that other parameters could be considered. Hint: Use a modification of the shooting method discussed in Exercise 2.25. Write code that automatically returns an approximation of the value of d so that your code could be used to approximate d for other choices of T. Is there a critical value (for some parameter or combination of parameters) that must be considered to ensure that a solution exists? Hint: Perhaps rescaling the problem to a dimensionless form would be useful in your analysis of this question.

Exercise 2.27. Second-order linear ODEs with nonconstant coefficients are important. For example, they play a fundamental role in the study of the solutions of some linear PDEs, especially when there is circular or spherical symmetry. An example is in Section 23.1 on waveguides. You should be familiar with solution methods for such equations. (a) Consider the ODE

$$ry'' + (2 - r)y' + 4y = 0.$$

Here the name r of the independent variable is supposed to suggest a radius (that is, the distance to the origin). Also, because r multiplies the highest-order derivative, setting r to zero changes the order of the ODE. The equation is singular at this value of r. For this reason, an important issue is the behavior of solutions at or near $r = 0$. Recall the power series method for representing solutions. The idea is simple and powerful: Look for a solution $r \mapsto y(r)$ of the differential equation in the form of a power series $y(r) = \sum_{n=0}^{\infty} a_n r^n$ where the coefficients a_n are to be determined, substitute the power series into the ODE, use the theorem that a power series represents the zero function if its coefficients are all zero, collect like powers of r on the left-hand side of the equation, and try to solve for the unknown coefficients by equating the coefficients of each power of r (starting with the lowest power) to zero. There might be free variables in this process. This is to be expected because a solution multiplied by a number is again a solution of the *linear* ODE. Also, to specify a solution uniquely requires two conditions; for example, the value of the independent variable y and its derivative y' at some value of the independent variable. Use the power series method to determine a nonzero solution of the ODE valid near $r = 0$. (b) The solution found in part (a) will be defined as a continuous function on an interval of real numbers containing the origin. A second-order linear ODE has a pair of solutions defined near each regular point—points that are not singular points—such that every solution is a linear combination of these two fundamental solutions. The solution of part (a) can be paired with another solution to form such a fundamental set. A second solution of the given ODE with this property is not defined at $r = 0$. In fact, the second solution is always unbounded as r approaches zero. Demonstrate this fact. Hint: It is possible (but not easy) to find a series representation of a second solution using the method of Frobenius. Another way to produce an appropriate second solution is reduction of order. You are not asked to represent the second solution as a convergent series. The problem is to show that a second solution is unbounded at the origin. (c) What happens to solutions when r is large (that is, as $r \to \infty$)? One might look at the equation, divide by r, and say that for large r the equation is nearly the same as the constant coefficient ODE $y'' - y' = 0$. Solving this later equation suggests that most solutions of the original ODE grow exponentially fast as r increases. Is this true? Gather evidence from analysis or numerics for an answer to this question. Remark: The ODE discussed in this problem belongs to a class of ODEs that is important in the quantum mechanical computation of the energy states of the hydrogen atom. (d) Approximate the value $y(1)$ of the solution of the ODE with initial data $y(2) = 1$ and $y'(2) = 0$. Use power series and compare with a numerical approximation.

Exercise 2.28. Suppose you are the captain of a ship and intend to steer a predetermined course. Let θ denote the deviation angle of the ship from the direction along the desired course. A crude model for the motion of the ship is

$$I\ddot{\theta} + \epsilon\dot{\theta} = f(t),$$

where I denotes the magnitude of the moment of inertia of the ship, $\epsilon\dot{\theta}$ is a sum of the forces (water pressure for example) that oppose the turning of the ship, and f is the sum of the external forces transverse to the ship heading. The ship is steered by a rudder. The turning force should be proportional to the rudder angle ψ (relative to the longitudinal axis of thc ship). Thus, a simple model for steering the ship is

$$I\ddot{\theta} + \epsilon\dot{\theta} = -k\psi + g(t)$$

where k is the constant of proportionality for the rudder force and $g(t)$ represents unknown and unexpected forces that might change the course of the ship. (a) Why is there a minus sign for the rudder force term? (b) What happens (according to the model) in case the ship is not steered and there is no external force? (c) Does this model predict the possibility to overcome every deviation from a straight heading by steering?

Exercise 2.29. [Automatic Control of Steering a Ship] As a continuation of Exercise 2.28, imagine the possibility of an automatic control system for the ship. The problem is to maintain the desired course. That is, the control system is designed to maintain the deviation at or near $\theta = 0$ by automatically moving the rudder in response to changes in the ship's current heading. To make a control system, there must be some sensor mechanism that detects the heading and an actuator to move the rudder. To keep the model simple, let us suppose that the sensor records the heading deviation and its rate of change. The controller is designed by feedback of this information to the actuator. A simple model is to assign two control gains a and b, which can be adjusted, and change the rudder angle according to the rule

$$\psi = a\theta + b\dot{\theta}.$$

The closed loop model is then

$$\ddot{\theta} + \epsilon\dot{\theta} = -k(a\theta + b\dot{\theta}) + g(t).$$

(a) Suppose there is no external force influencing the motion and the ship has some initial deviation (a specified $(\theta(0), \dot{\theta}(0))$). What happens under these circumstances to the future course deviation of the ship? How does the outcome depend on a and b? More precisely, for which choices of the gains does the model predict that the deviation will become small as time increases. Roughly speaking, the desired outcome is stabilization of the ship's motion. (b) Suppose $g(t) = A\sin(\omega t)$, perhaps for some relatively small A and ω so as to have a small amplitude slowly changing external force. How should the gains a and b be chosen to most efficiently keep the ship on course? (c) Try at least one other physically motivated choice for g. Explain why your g is a good choice and discuss the best choice of gains for the automatic control system. (d) In a real control system, some time elapses while the sensor determines the rate of change of heading and the actuator moves the rudder in response. Thus the control actuation is not instantaneous. Instead, there is some $\tau > 0$ (which for simplicity we may assume is fixed with respect to time) that enters the control system as a time delay. The corresponding model is

$$\ddot{\theta} + \epsilon\dot{\theta} = -k(a\theta(t-\tau) + b\dot{\theta}(t-\tau)) + g(t).$$

Is the inclusion of the delay significant in the stabilization problem? Hint: This is not a trivial question. The subject here is differential delay equations or more generally retarded functional differential equations. There is a useful theory for this type of dynamical equation. The reader familiar with this theory might use it to answer the stabilization question. Alternatively, perform some well-conceived numerical experiments to gather evidence for your conclusions about the significance of the time delay.

CHAPTER 3

An Environmental Pollutant

Consider a region in the natural environment where a waterborne pollutant enters and leaves by stream flow, rainfall, and evaporation. A plant species absorbs this pollutant and returns a portion of it to the environment after death. A herbivore species absorbs the pollutant by eating the plants and drinking the water. It returns a portion of the pollutant to the environment in excrement and after death. We are given (by ecologists) the initial concentrations of the pollutant in the ambient environment, in the plants, and in the animals, together with the rates at which the pollutant is transported by the water flow, the plants, and the animals. A basic problem is to determine the concentrations of the pollutant in the water, plants, and animals as a function of time and the parameters in the system.

Fig. 3.1 depicts the transport pathways for our pollutant, where the state variable x_1 denotes the pollutant concentration in the environment, x_2 its concentration in the plants, and x_3 its concentration in the herbivores. These three state variables all have dimensions of mass per volume (in some consistent units of measurement). The pollutant enters the environment at the rate A (with dimensions of mass per volume per time) and leaves at the rate ex_1 (where e has dimensions of inverse time). Likewise, the remaining rate constants a, b, c, d, and f all have dimensions of inverse time.

The scenario just described is typical. We will derive a model to track the environmental pollutant from *the fundamental law of conservation of mass: the rate of change of the amount of a substance (which is not created or destroyed) in some volume is its rate in (through the boundary) minus its rate out*, or in other words,

$$\text{time rate of change of amount of substance} = \text{rate in} - \text{rate out}.$$

The conservation of mass law is simply applied by taking the units of measurement into account. For example, we might be given the volume V of a container, compartment, or region measured in some units of volume; the concentration x of some substance in the compartment measured in amount (mass) per volume; and rates k of inflow and ℓ of outflow for some

An Invitation to Applied Mathematics: Differential Equations, Modeling, and Computation.
http://dx.doi.org/10.1016/B978-0-12-804153-6.50003-8, Copyright © 2017 Elsevier Inc. All rights reserved.

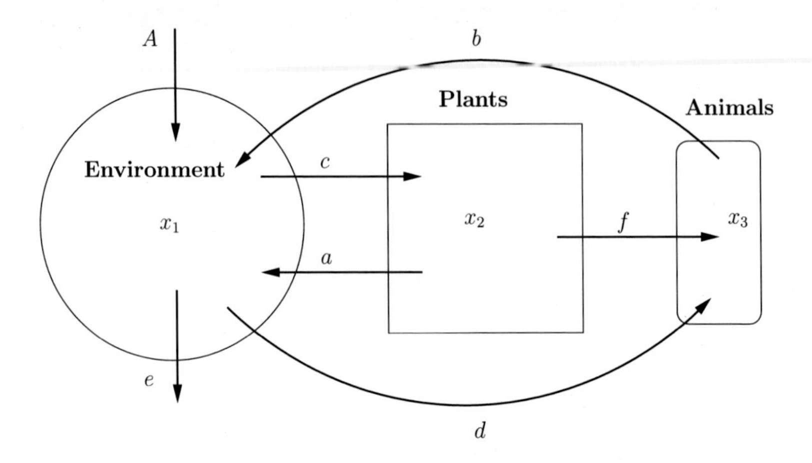

Fig. 3.1 Schematic representation of the transport of an environmental pollutant.

medium carrying the substance measured in units of volume per time. The rate of change dx/dt in this case, where t denotes time measured in some consistent unit, has units of amount per volume per time. The product of the rate of outflow ℓ and the concentration x has units of amount per time. In this situation, the time derivative must be multiplied by the total volume to achieve consistent units in the differential equation

$$\frac{d(Vx)}{dt} = \text{rate in} - \ell x,$$

where the inflow rate must have units of amount per time. When V is constant, both sides of the equation may be divided by V. The quantity ℓ/V has units of inverse time. It is the rate constant for the outflow of the substance.

For the environmental pollutant, the rate constants are given and the state variables x_1, x_2 and x_3 represent concentrations as in Fig. 3.1. Using conservation of mass, the transport equations are

$$\begin{aligned} \dot{x}_1 &= R + ax_2 + bx_3 - (c + d + e)x_1, \\ \dot{x}_2 &= cx_1 - (a + f)x_2, \\ \dot{x}_3 &= dx_1 + fx_2 - bx_3. \end{aligned} \tag{3.1}$$

This type of model, which is ubiquitous in mathematical biology, is called a compartment model.

The parameters in model (3.1) would be measured by the field work of ecologists. When these parameters are determined, the model can be used to make quantitative predictions of the future concentrations of the pollutant by solving the system of ordinary differential equations (ODEs) [Eqs. (3.1)].

Some of the rates might be difficult to measure directly (for example, the rate constant f that determines the transport of the pollutant from plants to the herbivore). These rates might be estimated using a two-step procedure: The pollutant concentrations are measured in the field over some suitable period of time and the parameters are chosen in the model to match these measurements. Exactly how to choose the parameters based on available data is the *parameter estimation problem*; it is one of the most important and difficult issues in mathematical modeling.

Our model may also be used to make qualitative predictions. For example, we might make a mathematical deduction from the form of the equations that determines the long-term substance concentrations for some range of parameter values. We might also use our model to predict the outcome of some intervention in the environment. For example, suppose that some portion of the plants are harvested by humans and removed from the region under study. The effect of this change can be predicted by solving our model equations with appropriate assumptions. The ability to make predictions without conducting new field studies is one of the most important motivations for developing a mathematical model. Another reason to develop a model is to test hypotheses about the underlying physical process. For example, a model that leads to predictions that do not agree with data obtained from field observations must be based on at least one false assumption.

Linear system [Eq. (3.1)] can be solved explicitly. But, since the eigenvalues of the system matrix are roots of a cubic polynomial, explicit formulas for the time evolution of the state variables x_1, x_2, and x_3 are complicated. This is to be expected. Differential equations with simple explicit solutions are rare. Most differential equations do not have explicit solutions and most explicit solutions are too complicated to be useful. For this reason, model validation and predictions are usually obtained by a combination of qualitative methods, analytic approximations, and numerical approximations. The most valuable information is usually obtained by qualitative analysis.

Physically meaningful concentrations are all nonnegative. Hence, our model should have the property that nonnegative initial concentrations remain nonnegative as time evolves. Geometrically, the positive octant of the three-dimensional state space for the variables u, v, and w should be positively invariant (that is, a solution starting in this set should stay in the set for all positive time or at least as long as the solution exists). To check the invariance, it suffices to show that the vector field given by the right-hand side of the system of differential equations is tangent to the boundary of the positive octant or points into the positive octant along its boundary, which consists of the nonnegative parts of the coordinate planes. In other words, positive invariance follows because $\dot{x}_3 \geq 0$ whenever $x_3 = 0$, $\dot{x}_2 \geq 0$ whenever $x_2 = 0$, and $\dot{x}_1 \geq 0$ whenever $x_1 = 0$.

What happens to the concentrations of the pollutant after a long time? Our expectation is that the system will evolve to a steady state (that is, a zero of the vector field that defines the ODE). To determine the steady state, we simply solve the system of algebraic equations obtained by setting the time derivatives of the system equal to zero. By a computation under the assumption that $be(a + f) \neq 0$, the solution of the linear system

$$\begin{aligned} R + ax_2 + bx_3 - (c + d + e)x_1 &= 0, \\ cx_1 - (a + f)x_2 &= 0, \\ dx_1 + fx_2 - bx_3 &= 0 \end{aligned} \tag{3.2}$$

is

$$x_1 = \frac{R}{e}, \quad x_2 = \frac{Rc}{e(a + f)}, \quad x_3 = \frac{R(ad + (c + d)f)}{be(a + f)}.$$

If $be(a + f) > 0$, then there is a steady state in the positive first octant. If $b = 0$, $e = 0$, or $(a + f) = 0$, then there is no steady state. This is reasonable on physical grounds. If the pollutant is not returned to the environment by the herbivores, then we would expect the pollutant concentrations in the herbivores to increase. If the pollutant does not leave the environment, then the concentration of the pollutant in our closed system will increase. If the plants do not transfer the pollutant to the environment or the herbivores, we would expect their pollutant concentration to increase.

Suppose that there is a steady state. Do the concentrations of the pollutant approach their steady state values as time passes? This basic question is answered for our linear system by determining the stability of the corresponding rest point in our dynamical system. The stability of a steady state can usually be determined from the signs of the real parts of the eigenvalues of the system matrix of the linearization of the system at the steady state (see Appendix A.17). In particular, if all real parts of these eigenvalues are negative, then the steady state is asymptotically stable.

The system matrix at our steady state is

$$\begin{pmatrix} -c-d-e & a & b \\ c & -(a+f) & 0 \\ d & f & -b \end{pmatrix}. \tag{3.3}$$

To show the asymptotic stability of the rest point, it suffices to prove that all the real parts of the eigenvalues are negative. This follows by a direct computation using the Routh–Hurwitz criterion (see Appendix A.17 and Exercise 3.1). Of course, this also agrees with physical intuition. Because the system is linear, we can make a strong statement concerning stability: If $e \neq 0$, then all initial concentrations evolve to the same steady state concentrations.

Our qualitative analysis makes numerical computation unnecessary for answering some questions about our system, at least in those cases where we can reasonably assume the system is in steady state. For example, if the rate at which the pollutant leaves the environment is decreased by 50%, then we can conclude that the pollutant concentration in the plants will be doubled. On the other hand, we may resort to numerical computation if we wish to predict some transient behavior of our system (see Exercise 3.2).

Exercise 3.1. (a) Prove that if a, b, c, d, e, and f are all positive, then all the roots of the characteristic polynomial of matrix (3.3) have negative real parts. (b) Show that the same result is true if a, b, e, and f are positive and the constants c and d are nonnegative.

Exercise 3.2. Suppose that the parameters in model (3.1) are

$$R = 10, \quad a = 1, \quad b = 1/100, \quad c = 3, \quad d = 1, \quad e = 1, \quad f = 4.$$

See Section 5.4 if you are not familiar with numerical methods for ODEs. (a) For the initial concentrations (at time $t = 0$) $x_1 = 0$, $x_2 = 0$, and $x_3 = 0$, determine the time $t = T$ such that $x_3(T) = 1/2$. (b) Suppose that the pollutant enters the environment periodically, $R = 10{+}5\sin(2\pi t)$ instead of the constant rate $R = 10$. Argue that the state concentrations fluctuate periodically and determine the amplitude of this fluctuation in the plant species.

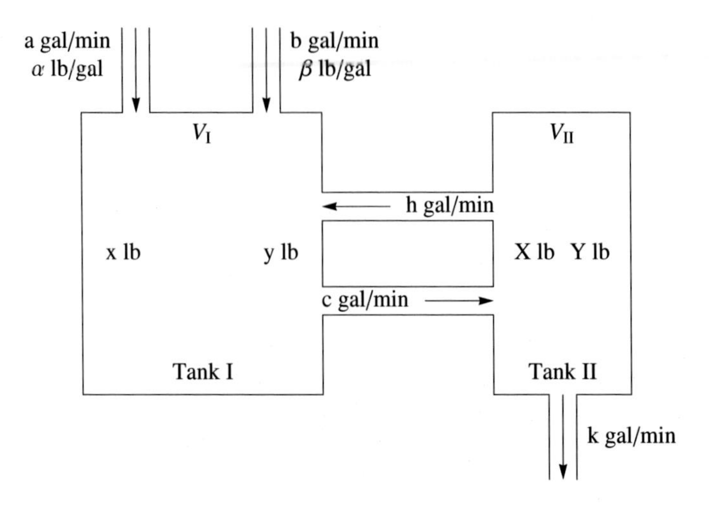

Fig. 3.2 Schematic representation of tanks with inflow, outflow, and connecting pipes.

Exercise 3.3. Consider the arrangement of tanks and pipes depicted in Fig. 3.2. Liquid is pumped into, out of, and between the tanks at the indicated rates. The first (respectively, the second) tank has initial liquid volume V_I (respectively, V_{II}). Two solutions enter the first tank via pipes that feed the system. The concentration of the solute x in one pipe is α lbs/gal, and the concentration of the second solute y in the other pipe is β lbs/gal. These enter at the flow rates of a gal/min and b gal/min (respectively). The solutions are exchanged between the two tanks in the indicated directions at the rates h and c gal/min. The concentrations of the two solutes in the second tank are denoted X and Y. The solution leaves the second tank at the rate k gal/min. Assume that both tanks are stirred so that the solutes have uniform concentrations at each instant of time in each tank, the liquid volume of each tank is constant during the process, and the initial amounts of the solutes in both tanks are given. (a) What constraints are imposed on the flow rates? (b) Determine the amounts of the solutes in each of the tanks as functions of time. (c) What happens to the concentrations of the solutes in each tank in the long run?

Exercise 3.4. Imagine three tanks each containing three different substances in solution. The tanks are connected by equal sized pipes to form a closed loop; that is, tank 1 feeds tank 2 and is fed by tank 3, tank 2 feeds tank 3 and is fed by tank 1, and tank 3 feeds tank 1 and is fed by tank 2. There are pumps in the pipes that maintain the circulation in the indicated direction at a fixed flow rate and the contents of the tanks are stirred so that the mixtures are homogeneous. (a) Make a model for the concentrations of the first substance in the three tanks and use it to predict concentrations as functions of time. What does the model predict? (Hint: Use pencil and paper to determine the steady states.) (b) Consider three bins each containing a mixture of red, blue, and green balls. The same number of balls (less than the minimum number of balls in a bin) is chosen at random from each bin and redistributed. Those chosen from the first bin are moved to the second, those from the second bin are moved to the third, and those from the third bin are moved to the first. This process may be repeated. Write a computer simulation of the process and track the concentration of red balls in each bin. (c) A simple model of the process in part (b) is given by the differential equations model of part (a). How well does the model predict the concentrations of red balls? Discuss the assumptions of the model and their validity for the redistribution process. (Hint: Make sure the initial states are the same for the process and the ODE model.) (d) Suppose a fixed amount of the first substance is created continuously in tank 2 and the same amount is destroyed in tank 1. What happens in the long run? Compare with a bin simulation. Note: The bin simulation is an example of a Markov process, for which there is a well-developed theory that is beyond the scope of this book.

Exercise 3.5. [PID Controller] (a) Imagine a tank partially filled with water. A pipe feeds water to the tank at a variable flow rate, and there is also a drain pipe with a computer-controlled variable flow valve connected to a sensor in the tank that measures the tank's volume. The valve opens exactly enough to let water drain from the tank at a rate proportional to the volume of the tank. The program allows the user to set one number: the constant of proportionality. Write a model for this physical problem. Be sure to define all the variables in your model. (b) Suppose the inflow rate is constant. How should the proportionality constant in the control mechanism be set to keep the tank near a constant desired volume? (c) Suppose the inflow rate is periodic. To be definite, take the flow rate to be sinusoidal and known exactly. How should the constant of proportionality be set for the controller to best keep the tank at a constant desired volume? Part of the problem is to define "best." Explain your choice. The abbreviation PID stands for proportional–integral–derivative. This exercise is about proportional control, which is implemented by a feedback controller, where the feedback is *proportional* to the volume as a function of time. Adding into the feedback a constant times the integral of the volume change would create a PI control. Adding into the feedback a constant times the derivative of the volume function would create a PD control. A feedback control given by a linear combination of the volume as a function of time, the integral of the volume, and the derivative of the volume creates a PID control. (d) Find the constants of proportionality for the PID control of the tank volume that tunes the control to best keep the tank volume constant for the case of constant inflow and for periodic inflow.

CHAPTER 4

Acid Dissociation, Buffering, Titration, and Oscillation

Dissociation is a basic chemical process that is amenable to a simple mathematical description. We will discuss the dissociation of acetic acid in water, titration with sodium hydroxide, and buffering. Two of the main purposes of the chapter are to introduce dynamic modeling of chemical reactions using the law of mass action and to further illustrate the important role played by autonomous systems of differential equations in applied mathematics.

4.1 A MODEL FOR DISSOCIATION

In a beaker of water containing a small amount of acetic acid, the acid dissociates into acetate ions and hydronium ions, which are properly denoted H_30^+, but often called protons and denoted H^+. Subscripted + or – signs denote the sign of the charge on the ion. In chemical symbols

$$CH_3COOH + H_2O \rightleftharpoons CH_3COO^- + H^+,$$

or in shorthand,

$$AcH \rightleftharpoons Ac^- + H^+ \tag{4.1}$$

with Ac^- used here to denote the acetate ion CH_3COO^-. The arrows indicate that the reaction occurs in both directions: the acid molecules dissociate and the ions recombine to form the acid.

Water molecules also ionize into protons and negatively charged ions consisting of an oxygen atom with one hydrogen attached. For simplicity, the reaction is usually written

$$H_2O \rightleftharpoons H^+ + OH^-;$$

but, more realistically and viewed as a dissociation of an acid, the chemical reaction is

$$H_2O + H_2O \rightleftharpoons H_3O^+ + OH^-, \tag{4.2}$$

An Invitation to Applied Mathematics: Differential Equations, Modeling, and Computation.
http://dx.doi.org/10.1016/B978-0-12-804153-6.50004-X, Copyright © 2017 Elsevier Inc. All rights reserved.

which is abbreviated by writing

$$H_2O + H_2O \rightleftharpoons H^+ + OH^-. \tag{4.3}$$

A simple chemistry experiment is to put some acetic acid in water. What happens?

The changing concentrations of four chemical species are of interest: acetic acid, acetate, protons, and OH ions:

$$AcH, \qquad Ac^-, \qquad H^+, \qquad OH^-.$$

To make a mathematical model, we will use conservation of mass. But this is not enough; the reactions among the species must be taken into account. A basic model from chemical kinetic theory is the principle of mass action: *the rate of reaction of two chemical species is proportional to powers of the products of their molar concentrations.* One mole (mol) is approximately 6×10^{23} molecules and molar concentrations are usually reported in moles per liter. For the principle of mass action in symbols, consider species A and B that combine in solution to produce C, written $A + B \to C$. Also, using the conventional notation $[A]$ for the (molar) concentration of A and the symbol $'$ to denote differentiation with respect to time, the rate of change of the concentration of C is

$$[C]' = k[A]^p[B]^q,$$

where k is the constant of proportionality and the powers p and q must be determined by experiment. For elemental reactions, $p = 1$ and $q = 1$. Such a reaction is called a first-order reaction. Acid dissociation reactions are all believed be of this type. The constant k depends on the temperature of the solution. But, for the simple models constructed here, this fact can safely be ignored.

For our reactions [Eqs. (4.1) and (4.3)], consider the rate of change of the concentration of the OH^+ ion. It is created and destroyed during the reaction. Using the conservation of mass (that is, taking into account the creation and destruction of the ion in the reaction [Eq. (4.3)]) and the principle of mass action, we have that

$$[OH^-]' = W_f[H_2O]^2 - W_b[H^+][OH^-],$$

where W_f and W_b are the forward and backward constants of proportionality for the dissociation of water, the products $[H_2O][H_2O]$ and $[H^+][OH^-]$ are due to the principle of mass action, and the difference on the right-hand side is rate in minus rate out for the OH^+ ions. Doing the same for the other species produces the system of differential equations.

$$\begin{aligned}
[OH^-]' &= W_f[H_2O]^2 - W_b[H^+][OH^-], \\
[AcH]' &= A_b[H^+][Ac^-] - A_f[AcH][H_2O], \\
[Ac^-]' &= A_f[AcH][H_2O] - A_b[H^+][Ac^-], \\
[H^+]' &= A_f[AcH][H_2O] - A_b[H^+][Ac^-] + W_f[H_2O]^2 - W_b[H^+][OH^-].
\end{aligned} \tag{4.4}$$

A complete model would include an equation for the rate of change of water concentration. But, under the assumption that there is much more water than acid, we may safely ignore the change in water concentration and treat this quantity as if it were constant. This is an example of a simplifying assumption in mathematical modeling. Under the assumption that water concentration is constant, the model is less realistic but perhaps more useful. To implement this idea, we define new parameters (quantities that *for our model* do not change with time)

$$w_f := W_f[H_2O]^2, \quad a_f := A_f[H_2O], \quad w_b = W_b, \quad a_b = A_b$$

and rewrite the model in the form

$$\begin{aligned}
[OH^-]' &= w_f - w_b[H^+][OH^-], \\
[AcH]' &= a_b[H^+][Ac^-] - a_f[AcH], \\
[Ac^-]' &= a_f[AcH] - a_b[H^+][Ac^-], \\
[H^+]' &= a_f[AcH] - a_b[H^+][Ac^-] + w_f - w_b[H^+][OH^-].
\end{aligned}$$

With this simplification, the model consists of four first-order ordinary differential equations in four unknowns. By the existence theory for ordinary differential equations, there is a unique solution of the system starting at each choice of initial concentrations (see Appendix A.3).

Mathematicians generally do not like extra decorations on variable names. Let us follow this convention by defining new state variables

$$W = [OH^-], \qquad X = [AcH], \qquad Y = [Ac^-], \qquad Z = [H^+]$$

and parameters

$$a = a_f, \qquad b = a_b \qquad c = w_f, \qquad d = w_b$$

so that our model has the more aesthetically pleasing form

$$\begin{aligned} W' &= c - dWZ, \\ X' &= bYZ - aX, \\ Y' &= aX - bYZ, \\ Z' &= aX - bYZ + c - dWZ. \end{aligned} \tag{4.5}$$

Unfortunately, the parameters in our model are difficult to measure directly. Instead, experimental measurements are made at equilibrium and involve ratios of concentrations.

At equilibrium, the concentrations of our species are not changing. For example, the rate of change of $[OH^-]$ is zero; that is,

$$W_f[H_2O]^2 - W_b[H^+][OH^-] = 0.$$

Define the dimensionless dissociation constant for water to be

$$\mathbb{K}_w = \frac{W_f}{W_b}; \tag{4.6}$$

and, by rearranging the steady state equality, note that

$$\mathbb{K}_w = \frac{[OH^-][H^+]}{[H_2O]^2}.$$

Warning: In the chemistry literature this number is usually replaced by the number

$$K_w := \mathbb{K}_w[H_2O]^2.$$

In addition to the possible confusion about exactly which number is reported, note that $\mathbb{K}_w$ is dimensionless but K_w is not. This later quantity can be measured; it has the value

$$K_w \approx 10^{-14}\Big(\frac{\text{mol}}{\text{liter}}\Big)^2, \tag{4.7}$$

which is usually written without units. To extract the exponent, the standard practice is to define the number.

$$pK_w := -\log_{10} K_w,$$

which is also treated as a dimensionless quantity. The underlying reason is that usually reactions take place by adding small amounts of some chemicals to water, and the water molar concentration remains nearly constant during the reaction. In fact, the molar concentration of water is approximately

$$\rho := [H_2 0] = 55.5\,\frac{\text{mol}}{\text{liter}}.$$

This constant value is used for $[H_2 0]$ in the dynamical equations. The approximate dimensionless value of $\mathbb{K}_w$ is

$$\mathbb{K}_w = \frac{10^{-14}}{\rho^2} \approx 3 \times 10^{-18}.$$

Similar conventions are used in defining K and pK constants in general. These numbers have dimensions, but usually they are reported without dimensions. Here, the corresponding dimensionless numbers are denoted using $\mathbb{K}$ with an appropriate subscript.

By observation, water does not dissociate easily. Thus, a reasonable choice of the model parameters might be

$$c = w_f = W_f[H_2O]^2 = 10^{-14}\frac{\text{mol}}{\text{m}^3\,\text{sec}}, \qquad d = w_b = W_b = 1\frac{\text{m}^3}{\text{mol sec}}.$$

For acetic acid, the dissociation constant is

$$\mathbb{K}_a = \frac{[Ac^-][H^+]}{[AcH][H_2O]}.$$

Again, in the chemistry literature this equilibrium constant is usually reported to be

$$K_a = \mathbb{K}_a[H_2O] \approx 10^{-4.75}$$

and

$$pK_a := -\log_{10} K_a \approx 4.75.$$

In both numbers the dimensions are usually ignored.

There is no obvious way to choose the forward and backward rate constants. Thus, some chemical intuition is required to make a reasonable choice; for example,

$$a = a_f = A_f[H_2O] = 10^5 \sec^{-1}, \qquad b = A_b = \frac{A_f[H_2O]}{\mathbb{K}_a[H_2O]}$$

so that again $\mathbb{K}_a = A_f/A_b$.

In summary, determination of the transient behavior of the system requires knowledge of the parameters a, b, c, and d, but the only quantities we can measure are the ratios $a/(b\rho)$ and $c/(d\rho^2)$. These ratios are sufficient to determine the steady state behavior of the system.

We are using parameters in our dynamic model derived from experiments at equilibrium. Is this justified? Perhaps this is a good time to emphasize that *no mathematical model is a perfect representation of reality.* If the model confirms other experimental data, we can be reasonably certain that the underlying chemistry (including rate constants, dissociation equations, elementary reactions [$p = 1$ and $q = 1$ in the law of mass action], and the constant water concentration assumption) is correct. We could then make predictions of the outcomes of new experiments with some confidence that the predictions will agree with nature. Of course, the final word in science is determined by experiment. On the other hand, we have very strong reasons to believe the universe is rational: if we start with fundamental laws and make logical deductions (for example, by applying mathematics), the conclusions will agree with new experiments. This is one of the main reasons why mathematics is important in science.

Rest points—solutions of our model that do not change with time—correspond to states in physical equilibrium. The rest points are exactly the solutions of the equations

$$c - dWZ = 0, \quad bYZ - aX = 0, \quad aX - bYZ = 0, \quad aX - bYZ + c - dWZ = 0.$$

Simple algebraic manipulation shows that this nonlinear system has infinitely many solutions that are given by the relations

$$X = \frac{bYZ}{a}, \qquad W = \frac{c}{dZ};$$

that is, there is a two-dimensional surface of rest points, which may be parametrized by the two variables Y and Z. These rest points cannot be asymptotically stable (see Appendix A.17) for a simple reason. Each rest point lies on an entire surface of rest points. Nearby rest points are not attracted to each other; they stay fixed. What about the evolution in time of initial states that are not rest points? Are they attracted to rest points as time approaches infinity? To obtain some insight, let us linearize as usual and see what happens.

The Jacobian matrix of the vector field is given by

$$\begin{pmatrix} -dZ & 0 & 0 & -dW \\ 0 & -a & bZ & bY \\ 0 & a & -bZ & -bY \\ -dZ & a & -bZ & -bY - dW \end{pmatrix}. \tag{4.8}$$

At a rest point we have that $W = c/(dZ)$; thus, after this substitution, we would like to find the eigenvalues of the matrix

$$J := \begin{pmatrix} -dZ & 0 & 0 & -\frac{c}{Z} \\ 0 & -a & bZ & bY \\ 0 & a & -bZ & -bY \\ -dZ & a & -bZ & -bY - \frac{c}{Z} \end{pmatrix}. \tag{4.9}$$

There are several ways to proceed. Perhaps the most instructive method uses basic geometry and linear algebra. The rest points lie on a two-dimensional surface of rest points in four-dimensional space. The function defined by

$$(Y, Z) \mapsto (\frac{c}{dZ}, \frac{bYZ}{a}, Y, Z)$$

parameterizes the surface. When Z is held fixed, the function traces out a curve, parameterized by Y on the surface of rest points, whose tangent vector must be tangent to the surface. Likewise, with Y fixed the tangent vectors of the curve parameterized by Z are all tangent to the surface. These vectors (obtained by differentiation with respect to Y and Z, respectively) are

$$\begin{pmatrix} 0 \\ \frac{bZ}{a} \\ 1 \\ 0 \end{pmatrix}, \quad \begin{pmatrix} -\frac{c}{dZ^2} \\ \frac{bY}{a} \\ 0 \\ 1 \end{pmatrix}.$$

The linearization of our system of differential equations approximates the dynamics of the nonlinear system. On the set of rest points, the nonlinear system does not move points. Thus, we might expect that the linear system does not move points in the directions of tangent vectors to the surface of rest points. This means these vectors should be eigenvectors corresponding to zero eigenvalues of the system matrix J for the linearization at each of the rest points. This fact is easily verified by simply multiplying each vector by J. Once we know the matrix J has two zero eigenvalues, it follows that there are exactly two more eigenvalues. These have to be roots of the characteristic polynomial p of the matrix J; in symbols, $p(\lambda) = \det(J - \lambda I)$. This characteristic polynomial has degree four. Because it has two zero roots, this polynomial is divisible by λ^2; therefore, the desired roots are the roots of a quadratic polynomial. By an easy computation, this polynomial p is found to be

$$p(\lambda) = ac+bcZ+adZ^2+bdYZ^2+bdZ^3+(c+aZ+bYZ+(b+d)Z^2)\lambda+Z\lambda^2. \tag{4.10}$$

Under our hypotheses, all the coefficients of this quadratic polynomial are positive. By Exercise 4.6, the real parts of its roots are negative.

We have not proved that every solution converges to a rest point, but this conclusion is supported by the linearization. In fact, every physically relevant solution of our model system is asymptotic to exactly one steady state as time grows without bound. This result requires some phase plane analysis and hypotheses derived from the underlying chemistry.

To make further progress, it is useful to compare the sizes of the system parameters. Although size comparisons can be made in the simple case studied here, such a comparison requires some thought simply because the parameters do not all have the same units of measurement. The remedy, which should be employed in the analysis of all physical models, is to change variables so that the model system is dimensionless; that is, the variables, the independent variable corresponding to time, and the system parameters are all dimensionless.

The state variables in the original model [Eq. (4.4)] are all concentrations, which we may assume are measured in moles per volume. The system parameters W_f, W_b, A_f, and A_b all have the same units, which may be abbreviated to be concentration per time. It is traditional to specify the units of a quantity α by $[\alpha]$. Square brackets also are used to denote concentrations in chemistry, but the appropriate meaning should always be clear from the context. By the definitions of a, b, c, and d, their units (using the abbreviation con for concentration) are easily determined to be

$$[a] = \frac{1}{\text{time}}, \quad [b] = \frac{1}{\text{con time}}, \quad [c] = \frac{\text{con}}{\text{time}}, \quad [d] = \frac{1}{\text{con time}}. \tag{4.11}$$

A natural rescaling to dimensionless form with the new dimensionless variables w, x, y, z, and s is given by defining

$$W = \rho w, \qquad X = \rho x, \qquad Y = \rho y, \qquad Z = \rho z, \qquad t = \frac{s}{a},$$

where ρ is the previously defined water concentration. By using the chain rule, for example, writing

$$\frac{dW}{dt} = \frac{dW}{ds}\frac{ds}{dt} = a\rho\frac{dw}{ds},$$

substituting for the state variables, and dividing by the coefficient $a\rho$ of each derivative, system (4.5) is recast in the dimensionless form

$$\begin{aligned}
w' &= \alpha(\mathbb{K}_w - wz),\\
x' &= \frac{1}{\mathbb{K}_a}yz - x,\\
y' &= x - \frac{1}{\mathbb{K}_a}yz,\\
z' &= x - \frac{1}{\mathbb{K}_a}yz + \alpha(\mathbb{K}_w - wz)
\end{aligned} \tag{4.12}$$

with the dimensionless parameters

$$\mathbb{K}_w = \frac{c}{d\rho^2}, \qquad \mathbb{K}_a = \frac{a}{b\rho}, \qquad \alpha := \frac{d\rho}{a}.$$

The physical meaning of the new dimensionless parameter α is simply W_b/A_f. Inspection of the dimensionless system (4.12) reveals that the transient behavior of this dissociation model depends on the dimensionless parameter α; the steady state behavior does not. It is not clear that α can be measured by experiment.

The set of nonnegative states $\mathcal{N}$ (corresponding to the concentrations) is positively invariant for system (4.12). For instance, on the coordinate hyperplane $\mathcal{H}$ corresponding to the coordinates x, y, and z (that is, the set of all states whose first coordinate w is zero), the tangent vectors to solutions of the differential equation at a point on $\mathcal{H}$ are transposed vectors of the form

$$(\alpha \mathbb{K}_w, \frac{1}{\mathbb{K}_a} yz - x, x - \frac{1}{\mathbb{K}_a} yz, x - \frac{1}{\mathbb{K}_a} yz + \alpha \mathbb{K}_w).$$

The first coordinate of such a vector is positive. Thus, this vector is not in the tangent space of the hyperplane $\mathcal{H}$; the vector points into the region $\mathcal{N}$. In other words, it is impossible for a solution that starts in $\mathcal{N}$ to exit this set through the hyperplane $\mathcal{H}$. A similar argument (which might require noting that on the boundary of $\mathcal{N}$ the coordinates are always nonnegative) applied to each of the coordinate hyperplanes bounding $\mathcal{N}$ can be used to complete the proof of a simple result: the vector field corresponding to the system of differential equations [Eqs. (4.12)] points into the region $\mathcal{N}$ on its boundary. Thus, this region is positively invariant, as it should be to reflect the correct chemistry: concentrations do not become negative as they evolve in time.

The second and third components of the vector field corresponding to the ODEs (4.12) are the same up to a sign. Addition of these two differential equations yields the identity $x' + y' = 0$. Hence, there must be a constant c_1 such that $x+y = c_1$. Similarly, there is a constant c_2 such that $z-y-w = c_2$. In other words, the functions

$$(w, x, y, z) \mapsto x + y, \qquad (w, x, y, z) \mapsto z - y - w$$

stay constant along all solutions. Such a function is called a first integral of the system.

Our model is designed to determine the concentration of the chemical species after a small amount of acid is added to water. At time $t = 0$, when the water has partially dissociated but the acid is all associated, the number of OH^- ions should be equal to the number of H^+ ions. In view of the scaling that makes all the dimensionless variables proportional with the *same* constant of proportionality to the unscaled variables, a correct choice for the initial data is

$$w(0) > 0, \quad w(0) = z(0), \quad x(0) = x_0 > 0, \quad y(0) = 0.$$

Note also that the acetate ion concentration cannot be larger than the original concentration of acetic acid, and this relation holds for the scaled variables as well. In symbols, $y \leq x(0)$ during the dissociation process.

Using the initial data, the constant values of the first integrals are determined: $c_1 = x(0)$ and $c_2 = 0$; therefore, the evolving states satisfy the relations

$$x(t) = x(0) - y(t), \qquad w(t) = z(t) - y(t). \tag{4.13}$$

The model system of four differential equations can therefore be reduced to the two equations

$$\begin{aligned} y' &= (x(0) - y) - \frac{1}{\mathbb{K}_a} yz, \\ z' &= (x(0) - y) - \frac{1}{\mathbb{K}_a} yz + \alpha(\mathbb{K}_w - (z - y)z); \end{aligned} \tag{4.14}$$

and, if desired, the states x and w may be recovered using the relations.

We have proved that the evolving states, with physical initial values, must remain positive. Repeating a similar argument for the system of ODEs (4.14) and using our assumption that $y \le x(0)$, it follows that the region bounded by the horizontal y-axis, the vertical axis, and the line with equation $y = x(0)$ is positively invariant. For z sufficiently large, $z' < 0$ in this strip because the dominant term is $-\alpha z^2$ (see Exercise 4.2). Thus, we have proved that system (4.14) has a positively invariant rectangle $\mathcal{R}$. Moreover, we can arrange the choice of the upper boundary of $\mathcal{R}$ so that all solutions starting above it eventually enter $\mathcal{R}$. This rectangle is the only possible location for rest points or periodic solutions.

Are all solutions of system (4.14) asymptotic to a rest point in $\mathcal{R}$?

To answer the question, let us first determine the existence of rest points in $\mathcal{R}$ by simultaneously solving the equations

$$(x(0) - y) - \frac{1}{\mathbb{K}_a} yz = 0, \qquad (x(0) - y) - \frac{1}{\mathbb{K}_a} yz + \alpha(\mathbb{K}_w - (z - y)z) = 0$$

or, equivalently, the system

$$y = \frac{\mathbb{K}_a x(0)}{\mathbb{K}_a + z}, \tag{4.15}$$

$$y = \frac{z^2 - \mathbb{K}_w}{z}. \tag{4.16}$$

By eliminating y, it suffices to find the roots of the cubic polynomial

$$\mathbb{K}_a \mathbb{K}_w + (\mathbb{K}_w + \mathbb{K}_a x_0) z - \mathbb{K}_a z^2 - z^3. \tag{4.17}$$

It has exactly one positive root (see Exercise 4.3). The corresponding value of y given by Eq. (4.15) is positive. Thus, this rest point must be in the rectangle $\mathcal{R}$.

Linearization at the rest point produces the system matrix

$$A = \begin{pmatrix} -1 - \frac{z}{\mathbb{K}_a} & -\frac{y}{\mathbb{K}_a} \\ -1 - \frac{z}{\mathbb{K}_a} + \alpha z & -\frac{y}{\mathbb{K}_a} + \alpha(y - 2z) \end{pmatrix}$$

with characteristic equation

$$\lambda^2 - \mathrm{tr}(A)\lambda + \det(A) = 0$$

where

$$\mathrm{tr}(A) = -(1 + \frac{y}{\mathbb{K}_a} + \frac{z}{\mathbb{K}_a} + \alpha z) + \alpha(y - z), \qquad (4.18)$$

$$\det(A) = -\alpha y + 2\alpha z + \frac{2\alpha z^2}{\mathbb{K}_a}.$$

Using Eq. (4.16) to substitute for y, the expression for the determinant becomes

$$\det(A) = \frac{\alpha}{\mathbb{K}_a z}(\mathbb{K}_a \mathbb{K}_w + \mathbb{K}_a z^2 + 2z^3) > 0.$$

Because $w = z - y$, the last term in the expression for the trace is negative. The same result is obtained by again using Eq. (4.16) to substitute for y in the expression $y - z$ to see that

$$y - z = -\frac{\mathbb{K}_w}{z}.$$

Thus, $\det(A) > 0$ and $\mathrm{tr}(A) < 0$ at the rest point. The roots of the characteristic polynomial are

$$\lambda = \frac{\mathrm{tr}(A) \pm \sqrt{\mathrm{tr}(A)^2 - 4\det(A)}}{2}.$$

If $\mathrm{tr}(A)^2 - 4\det(A) < 0$, the roots are complex conjugates with negative real parts because $\mathrm{tr}(A) < 0$. If $\mathrm{tr}(A)^2 - 4\det(A) > 0$, the square root of this quantity is less $|\mathrm{tr}(A)|$ because $\det(A) > 0$ and both roots are negative.

Thus, we have proved that there is exactly one rest point in the first quadrant (the physically realistic region) and this rest point is asymptotically stable.

A strategy that can be used to prove the global asymptotic stability of the rest point (that is, this rest point is stable and it is the limit as s goes to infinity of every solution) is to show that the system has no periodic orbits and then apply the Poincaré–Bendison theorem (see Appendix A.18, theorem A.12).

Suppose an autonomous system of differential equations

$$\dot{y} = P(y, z), \qquad \dot{z} = Q(y, z)$$

in the plane has a periodic orbit Γ. Let Ω be the bounded open region enclosed by the simple closed curve Γ. This statement contains a hidden assumption: A simple closed curve in the plane is the boundary of exactly two open regions, one bounded and the other unbounded. On first sight, this result may seem obvious; it is actually a deep theorem called the Jordan curve theorem. The difficulty arises because simple closed curves in the plane can have complicated shapes. Let us assume this important fact.

Write $\Gamma = \partial\Omega$ to denote that the boundary of Ω is Γ, let $X := (P, Q)$ denote the vector field corresponding to the differential equation, and let η denote the outer unit normal on $\partial\Omega$. Green's theorem states that

$$\int_{\Omega} \operatorname{div} X \, d\mathcal{A} = \int_{\partial\Omega} X \cdot \eta \, d\ell$$

The right-hand side of the last equality is the line integral around Γ of the inner product of X, which is tangent to Γ, and the normal vector field η on this curve. Clearly, $X \cdot \eta = 0$ everywhere on Γ. Thus, the area integral of the divergence of X must vanish. In other words, if the divergence of a vector field has a fixed sign in some (simply connected) region of the plane, then this region contains no periodic orbits of the corresponding system of differential equations. This result is called Bendixson's theorem (Ivar Otto Bendixson, 1900).

The divergence of X for the reduced model equation is the trace of the matrix A [Eq. (4.18)] not evaluated at the rest point; that is,

$$\begin{aligned} \operatorname{div} X &= -(1 + \frac{y}{\mathbb{K}_a} + \frac{z}{\mathbb{K}_a} + \alpha z) + \alpha(y - z) \\ &= -1 - \Big(\frac{1 - \alpha\mathbb{K}_a}{\mathbb{K}_a}\Big) y - \frac{z}{\mathbb{K}_a} - 2\alpha z. \end{aligned}$$

Although the sign of the divergence does not seem to be fixed in general, the sign is negative whenever $\alpha\mathbb{K}_a \leq 1$.

What happens for $\alpha\mathbb{K}_a > 1$?

There is a useful generalization of Bendixon's theorem called Dulac's Theorem (Henri Dulac, 1923; see Exercise 4.8). It states that the divergence of every positive function multiple of X must change sign in a region containing a periodic orbit. We can prove there are no periodic orbits by finding a positive function multiple—called a Dulac function—of X that has positive divergence. A useful function to try is an exponential of one of the state variables. The reason is that exponentials survive differentiation and the constant in the exponent can be used as an extra parameter to make it more likely the divergence has a fixed sign. Multiply X by the function e^{ry} and compute the divergence

$$\operatorname{div}(e^{ry}X) = -e^{ry}\big(1 - rx(0) + (r-\alpha)y + \frac{z+y}{\mathbb{K}_a} + \frac{r}{\mathbb{K}_a}yz + 2\alpha z\big).$$

Take $r = \alpha$ to eliminate the term $(r-\alpha)y$ so that the divergence is negative whenever $x(0)\alpha \leq 1$. This latter inequality is not always true, but recall the meaning of $x(0)$ and the physical process being modeled. The quantity $x(0)$ is the initial concentration of acetic acid divided by the water concentration. The model is constructed under the assumption that a *small amount* of acetic acid is put into a beaker of water. The desired result is obtained when the initial amount of acetic acid is so small that $x(0)\alpha < 1$.

We have outlined a proof of a useful fact: For physically meaningful initial data and system parameters, every solution of the system of ODEs (4.5) is asymptotic to a rest point where all state variables are positive. The steady state depends on the initial data; that is, the choice of $x(0)$ determines the steady state (compare to Exercise 4.7). Or, from a chemical perspective, a small amount of acetic acid in water dissociates and the corresponding ion concentrations reach positive steady states.

Exercise 4.1. Chemists will agree with mass action modeling of acid dissociation, but they would not see the necessity for writing the differential equations model. Instead, they would work with the equilibrium equations and not mention the rates of change of concentrations. Is their simplified modeling process justified? Discuss.

Exercise 4.2. Prove that the set $\mathcal{R}$ described in the text is a positively invariant rectangle for system (4.14) and that every solution with positive initial conditions enters this rectangle after some finite amount of time.

Exercise 4.3. Prove that cubic polynomial (4.17) has exactly one positive root. Hint: Use calculus to sketch the graph of this cubic.

Exercise 4.4. Consider the prototype reaction $A + B \to C$ with forward rate constant k. (a) Determine the steady state behavior of the concentration of C. (b) Find an explicit formula for the concentration of C as a function of time. (c) What is the steady state behavior in case the reverse reaction $C \to A + B$ with rate constant ℓ is included?

Exercise 4.5. [Michaelis–Menton Enzyme Kinetics] An enzyme, by definition, is a catalyst for a biochemical reaction. In a typical situation there is a substrate S, an enzyme E, and a product P. In the presence of the enzyme (perhaps a protein), biochemical elements that compose the substrate combine to form the product; in symbols, $S + E \to E + P$. This is an example of a situation where the principle of mass action leads to a prediction that is inconsistent with experimental evidence.
(a) By applying the principle of mass action, show that the rate of change of concentration of the product (called the velocity of the reaction) grows (without bound) as the concentration of substrate is increased without bound and the enzyme concentration is held constant.
A better model was proposed in 1913 by Lenor Michaelis and Maud Menton. They hypothesized the existence of an intermediate substance I produced by the combination of substrate and enzyme; it is included in the chain of reactions

$$S + E \rightleftharpoons I \to E + P.$$

(b) Suppose the forward and backward rate constants for the first reaction are k_f and k_b and the forward rate constant for the second reaction is ℓ. Write the rate equations for the complete reaction. Hint: There are four species and hence four rate equations.
We wish to know the rate of change of the product concentration as a function of the substrate concentration and the initial enzyme concentration.
(c) Show that the concentration of the intermediate substance plus the enzyme concentration is constant and this constant must be the initial enzyme concentration E_0. In 1925, G. E. Briggs and J. B. S. Haldane proposed a modification of the Michaelis–Menton substrate concentration assumption: The intermediate product concentration is (nearly) constant.
(d) Determine the rate of change of the product concentration as a function of the substrate concentration and the initial enzyme concentration under the Briggs–Haldane assumption.
(e) Conclude that the speed of the reaction is asymptotic to a maximum constant speed. In particular, the speed of the reaction does not increase without bound as the substrate concentration is increased.

Exercise 4.6. Prove that the roots of the polynomial for the dissociation model [Eq. (4.10)] have negative real parts. Are the roots real?

Exercise 4.7. (a) Specify a method to obtain the steady state of a solution of acetic acid and water (where all the pure acetic acid is assumed to be added all at once to water) that is already in steady state as predicted by model (4.5) without solving the system of differential equations. Your result should determine a function of the initial concentration of acetic acid at the instant it is added to the water. (b) Using a numerical method, approximate the length of time (in seconds) for the solution to reach steady state after the addition of one mole of acetic acid. According to the model, the solution never reaches steady state. Why? For this exercise, assume that (for practical purposes) steady state has been reached once the final concentrations are within 0.1% of their theoretical steady state values. (c) Discuss how the time to reach steady state is related to the amount of added acetic acid. Does the time to steady state increase or decrease with an increase in the amount of acetic acid? Discuss your answer. Does your result agree with physical intuition?

Exercise 4.8. Prove Dulac's generalization of Bendixon's theorem: *If a vector field has a periodic integral curve in the plane and the vector field is defined on the entire planar region bounded by this curve, then every positive function multiple of the vector field must have its divergence change sign in the region bounded by the curve.* This result is useful when the divergence of a given vector field X does not have a fixed sign in a region to be tested for the existence of periodic orbits. If there is a positive function f defined on the region, and the divergence of the new vector field fX has a fixed sign, then there are no periodic orbits in the region.

Exercise 4.9. Suppose that a certain environment contains a population of rabbits and foxes. The rabbits grow (in the absence of foxes) according to the law $\dot{R} = aR - bR^2$ and the foxes (absent rabbits) die off according to the law $\dot{F} = -dF$. The interaction between the rabbits and foxes may be modeled by the system of differential equations

$$\dot{R} = aR - bR^2 - cRF, \qquad \dot{F} = -dF + eRF,$$

where a, b, c, d, and e are positive constants. In case $b = 0$ this is called the Volterra-Lotka model. (a) Write a justification for the model. (b) Determine a rescaling of the system into the dimensionless form

$$\dot{x} = x - x^2 - xy, \qquad \dot{y} = -\alpha y + \beta xy,$$

and specify the definition of the dimensionless parameters α and β. (c) Show that a solution starting in the first quadrant (corresponding to the physically realistic values of the state variables) stays in the first quadrant for all positive time. In fact, the closed positive quadrant is an invariant set: solutions starting there, stay there for all positive and negative time. (d) Find the steady states of the dimensionless system and determine their local stability types. (e) Conjecture the fate of the rabbits and foxes according to the values of α and β. Will they coexist, or will the foxes go extinct? (f) Perhaps the population of rabbits or foxes goes to infinity. Prove that the model predicts this does not occur. (g) Perhaps the populations become periodic. In case $b > 0$, prove that the model predicts this does not occur. Hint: If $b = 0$ in the original model, then most orbits *are* periodic. Show this and try to use this fact to determine what happens when $b > 0$. Or, apply Dulac's criterion from Exercise 4.8.

4.2 TITRATION WITH A BASE

Imagine the following titration experiment: A beaker contains a solution of acetic acid and water made by combining a known amount of acetic acid with a known volume of water. Water with a known concentration of sodium hydroxide, stored in a graduated pipe (buret), is released slowly into the beaker; the solution is stirred continuously, and the pH of the solution is measured at specified time intervals. A rate equation model for the changing pH is discussed in this section.

The pH of a solution is a measure of the number of hydronium ions in the solution. Recall from the last section that a hydronium ion is also called a proton and denoted H^+. In modern chemistry there are at least two ways to define pH: via activity or concentrations. For simplicity of the chemistry, the classical definition is used here: pH is defined to be the negative logarithm, base ten, of the molar concentration of protons H^+. This number should be—as explained in the last section—the dimensionless quantity

$$p\mathbb{H} := -\log_{10} \frac{[H^+]}{[H_2O]},$$

but in practice it is taken to be

$$\text{pH} := -\log_{10}[H^+]. \tag{4.19}$$

As an example, consider the dissociation of water. A pair of ions H^+ and OH^- combine to form exactly two water molecules as in Eq. (4.3). Thus, their molar concentrations are equal. Taking the measured value $pK_w = 14$ from Eq. (4.7) and using definition (4.19), the pH of water must be half this number; that is, pH $= 7$. Of course, this is just an approximation in which the temperature of the solution is ignored, but this number gives the standard pH that divides acidic and basic solutions: $pH < 7$ is called acidic; $pH > 7$ is called basic.

For sodium hydroxide dissolved in water, the chemical equation is

$$NaOH \rightleftharpoons Na^+ + OH^-.$$

Sodium ions react with acetic acid to form sodium acetate. For pH calculation, only the OH^- ions are important during the titration as they may be protonated to form water. In case its concentration is not too high, almost all of the sodium hydroxide dissociates to sodium and OH^- ions. Thus, a solution of A mol / liter of $NaOH$ is assumed to have A mol / liter of OH^-.

The sodium hydroxide solution enters the beaker at a rate of B ml / sec. To keep consistent units, this concentration is measured as $B/1000$ liter / sec, and the OH^- ions enter the acetic acid solution at the rate of

$$AB\,\frac{\text{mol}}{\text{sec}}.$$

The concentration of this ion changes during the addition of $NaOH$ because the volume is changing from 100 ml via the function

$$\text{vol} = \frac{1}{10} + \frac{B}{1000}t.$$

A model of the ion concentration entering the solution (measured in mol / liter) as a function of time is

$$t \mapsto \frac{AB}{\left(\frac{1}{10} + \frac{B}{1000}t\right)}. \tag{4.20}$$

Using function (4.20) to model the addition of OH^- ions (during the addition of the $NaOH$ solution), the rate equation model [Eq. (4.5)] is modified to

$$\begin{aligned}
W' &= c - dWZ + \frac{AB}{\left(\frac{1}{10} + \frac{B}{1000}t\right)},\\
X' &= bYZ - aX,\\
Y' &= aX - bYZ,\\
Z' &= aX - bYZ + c - dWZ.
\end{aligned} \tag{4.21}$$

The flow rate B does not remain constant. In fact, the flow is turned on after the acetic acid solution is in steady state and turned off when the measurement device is stopped or the reservoir of $NaOH$ is empty. The model can be used to make predictions via numerical integration.

An experiment was performed where a solution of one mol / liter of $NaOH$ is added to a solution of one mol / liter of acetic acid at the rate of 0.5 ml / sec. The beaker containing the solution was stirred continuously and a pH sensor recorded the pH of the solution every 2 sec for an experimental duration of 74 sec (see Table. 4.1). The flow of $NaOH$ was started at 5 sec. The results of this experiment and a simulation using the model [Eq. (4.21)] are depicted in Fig. 4.1. The agreement is excellent (compare to Exercise 4.10).

sec	pH	sec	pH	sec	pH	sec	pH	sec	pH
0	2.567	16	3.590	32	4.196	48	4.490	64	4.677
2	2.563	18	3.738	34	4.240	50	4.514	68	4.717
4	2.563	20	3.852	36	4.283	52	4.540	70	4.737
6	2.563	22	3.923	38	4.327	54	4.564	72	4.755
8	2.689	24	3.979	40	4.363	56	4.590	74	4.771
10	3.038	26	4.042	42	4.396	58	4.607		
12	3.279	28	4.094	44	4.436	60	4.632		
14	3.418	30	4.148	46	4.462	62	4.656		

Table 4.1 The data in this table is produced by a pH meter while a one molar sodium hydroxide solution is added to a one molar solution of acetic acid and water.

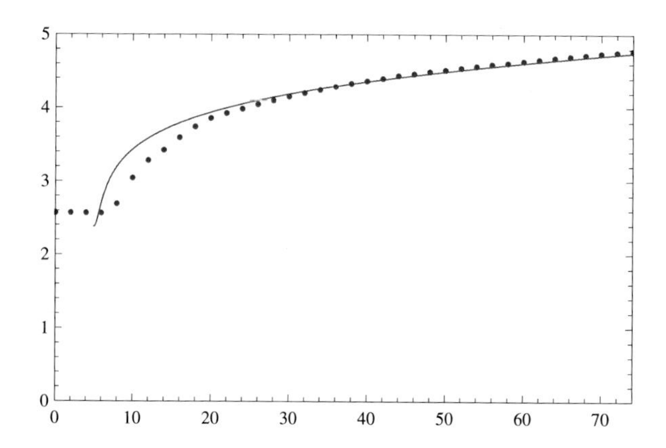

Fig. 4.1 Plots of pH versus time in seconds are depicted for the addition of $NaOH$ in acetic acid. The continuous graph is from the mathematical model [Eq. (4.21)]; the discrete graph is from the data in Table. 4.1.

We note that the solution pH does not rise significantly during the titration if not too much base is added. The presence of acid buffers the solution to the addition of the strong base. Buffering is an essential process in living organisms. The pH of essential fluids, for example blood, must remain within some narrow range for biochemical processes to work properly. Although our model is viable for small concentrations of the base, sodium hydroxide, it does not seem to predict the correct titration curve (pH versus the amount of added base) over a larger range of additional OH^-. An improved model is the subject of the next section.

Exercise 4.10. (a) The results depicted in Fig. 4.1 of physical experiments and numerical simulation using the mathematical model (4.21) show excellent agreement. On the other hand, the transient parts of the solutions of the model are very short. Good results can also be obtained by a static model based on the assumption that the mixtures are instantaneously in steady state. Discuss this alternative. Hint: This approach is used in textbooks on basic chemistry. (b) The accuracy of predictions from the model should depend on the accuracy of its parameters. How sensitive is the result to the given parameters? (c) Which choice of parameters best fits the experimental data?

4.3 AN IMPROVED TITRATION MODEL

The titration of a weak acid by a strong base (for example, titration of acetic acid by sodium hydroxide) stays in the buffer region, where the change in pH is not too great for small additions of the base until some critical amount of base is added. At this point a sharp rise in pH occurs. The pH levels off as more base is added. This general behavior is typical and is easily observed in experiments. Appropriate mathematical models, which are discussed in this section, predict the observed behavior.

Recall the chemical reactions for the dissociation of water, acetic acid, and sodium hydroxide. In shorthand, these are

$$H_2O \rightleftharpoons H^+ + OH^-,$$

$$AcH + H_2O \rightleftharpoons Ac^- + H^+,$$

and

$$NaOH \rightleftharpoons Na^+ + OH^-.$$

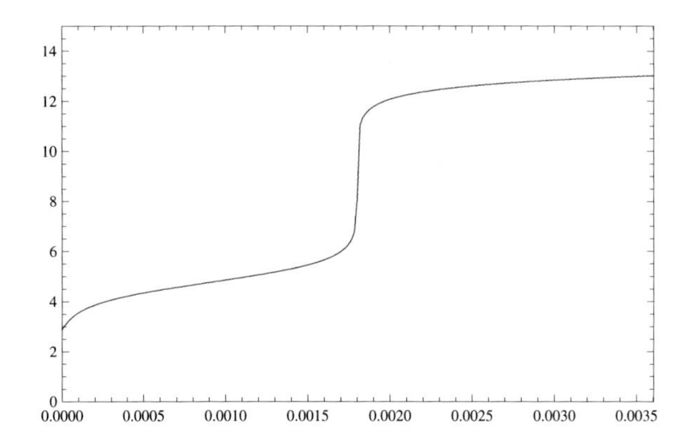

Fig. 4.2 The simulated titration curve (that is, pH versus initial molarity) of sodium hydroxide added to an 0.1 molar acetic acid solution is depicted.

The water concentration is considered to be very large compared with the concentrations of the other species. To account for the additional OH^- ions from the dissociated sodium hydroxide, one additional reaction should be included:

$$AcH + OH^- \rightleftharpoons Ac^- + H_2O. \tag{4.22}$$

In the previous section, we ignored the rate constants in this equation and simply changed the OH^- concentration in the mixture. This simplification leads to a model that does not agree with the experiment. It must be modified.

Using the notation of the system of ODEs (4.5), incorporating water concentration into the forward rate constant for the dissociation of water, and defining f to be the forward rate constant for reaction (4.22) and e to be the backward rate constant times the water concentration, we have the rate equations

$$\begin{aligned} W' &= c - dWZ + eY - fXW, \\ X' &= bYZ - aX + eY - fXW, \\ Y' &= aX - bYZ - eY + fXW, \\ Z' &= aX - bYZ + c - dWZ. \end{aligned} \tag{4.23}$$

Recall the units of parameters from display (4.11) and note the units of e and f to obtain the complete set

$$[a] = \frac{1}{\text{time}}, \quad [b] = \frac{1}{\text{con time}}, \quad [c] = \frac{\text{con}}{\text{time}}, \quad [d] = \frac{1}{\text{con time}},$$
$$[e] = \frac{1}{\text{time}}, \quad [f] = \frac{1}{\text{con time}}. \tag{4.24}$$

Also, using the concentration of water ($\rho = 55.5$ mol / liter), we have the values

$$\frac{10^{-4.75}}{\rho} = \mathbb{K}_a = \frac{a}{b\rho}, \quad \frac{10^{-14}}{\rho^2} = \mathbb{K}_w = \frac{c}{d\rho^2}.$$

The ratio of the rate constants e and f is determined at equilibrium using the first equation of system (4.23), which when written in full, is

$$W_f[H_2O] - W_b[H^+][OH^-] = f[OH^-][AcH] - e[Ac^-].$$

The left-hand side vanishes at equilibrium; therefore, using ρ as before to be the molar concentration of water, we have that

$$\frac{f\rho}{e} = \frac{[Ac^-]\rho}{[AcH][OH^-]} = \frac{\mathbb{K}_a}{\mathbb{K}_w} \approx 10^{11}.$$

Approximate values of the parameters (with the units just listed) are taken in this section to be

$$a = 1, \quad b = 5 \times 10^4 \quad c = 10^{-8}, \quad d = 10^6, \quad e = 10^{-5}, \quad f = 2 \times 10^4. \tag{4.25}$$

The titration curve depicted in Fig. 4.2 is computed using the following procedure. The parameters e and f are set to zero in system (4.23) to simulate the acid dissociation before the base is added. Starting at time $t = 0$ at the initial data

$$W(0) = 0, \qquad X(0) = 0.1, \qquad Y(0) = 0, \qquad Z(0) = 0$$

(which represents the moment at which a small amount of acetic acid is added), the system is integrated forward until a steady state is reached. The equilibrium pH is approximately 2.88. The equilibrium values of X, Y, and Z are used as the initial conditions during the titration simulation (see Exercise 4.12). A new initial OH^- concentration (that is, an initial value of W) is set and the system with e and f restored to their given values is evolved forward in time (by numerical integration) until (an approximation of) equilibrium is reached. The corresponding pH is computed at equilibrium. In other words, the pH is approximated as a function of the initial OH^- concentration. This function is plotted in Fig. 4.2. The figure shows a buffered region, where the pH remains relatively constant for a range of low concentrations of the added base, a region of rapid pH increase, and a region of high pH for the addition of high concentrations of the base. As mentioned, buffering is important in many applications, in particular in biochemistry. Solutions may also be buffered relative to the addition of additional acid.

Although the steep change in pH when the initial OH^- concentration is increased past a certain value is easily observed using numerical experiments (as in Fig. 4.2), this behavior is not an obvious prediction from the model.

To gain some insight into the shapes of titration curves, start by making the model dimensionless and reducing the dimension of the system of differential equations.

Using the rescaling

$$W = \rho w, \qquad X = \rho x, \qquad Y = \rho y, \qquad Z = \rho z, \qquad t = \frac{s}{a}$$

and the dimensionless groups

$$\mathbb{K}_w = \frac{c}{d\rho^2}, \quad \mathbb{K}_a = \frac{a}{b\rho}, \quad \alpha = \frac{d\rho}{a}, \quad \beta := \frac{e}{a}, \quad \gamma := \frac{f\rho}{a},$$

system (4.23) is equivalent to the dimensionless model

$$\begin{aligned}
w' &= \alpha\mathbb{K}_w + \beta y - \gamma x w - \alpha w z, \\
x' &= -x + \beta y - \gamma x w + \frac{1}{\mathbb{K}_a} y z, \\
y' &= x - \beta y + \gamma x w - \frac{1}{\mathbb{K}_a} y z, \\
z' &= \alpha\mathbb{K}_w + x - \alpha w z - \frac{1}{\mathbb{K}_a} y z.
\end{aligned} \tag{4.26}$$

In view of the dimensioned parameter values [Eq. (4.25)], the dimensionless parameters are approximately

$$\begin{gathered}
\mathbb{K}_a = 32 \times 10^{-8}, \quad \mathbb{K}_w = 3.25 \times 10^{-18}, \quad \alpha = 55.5 \times 10^6, \\
\beta = 10 \times 10^{-6}, \quad \gamma = 1.11 \times 10^6.
\end{gathered} \tag{4.27}$$

The ordering of the dimensionless parameters is approximately maintained by defining

$$\mu = 10^{-2}, \quad \mathbb{K}_a = 32\mu^4, \quad \mathbb{K}_w = 3\mu^9, \quad \alpha = 50\mu^{-3},$$
$$\beta = 10\mu^3, \quad \gamma = \mu^{-3}. \tag{4.28}$$

The process begins by preparing the acid bath. In dimensionless variables, the initial data is $w(0) = y(0) = z(0) = 0$ and $x(0)$ equal to the initial amount of acetic acid divided by ρ. This value for the experiment described here is approximately

$$x(0) = 20 \times 10^{-4} = 20\mu^2.$$

After the process reaches its steady state the dimensionless concentrations are w_{ss}, x_{ss}, y_{ss}, and z_{ss}. A variable dimensionless amount of the base ω is added and the steady state (dimensionless) proton concentration z is measured to determine the titration curve.

Note that the first integrals (see the discussion on page 48) for the system with the base added are

$$x + y = x_{\text{ss}} + y_{\text{ss}} = x(0), \qquad z - y - w = z_{\text{ss}} - y_{\text{ss}} - w_{\text{ss}} - \omega = -\omega.$$

Using them, the titration model reduces to

$$y' = x(0) - y - \frac{1}{\mathbb{K}_a} yz - \beta y + \gamma(x(0) - y)(\omega + z - y),$$
$$z' = x(0) - y - \frac{1}{\mathbb{K}_a} yz + \alpha \mathbb{K}_w - \alpha(\omega + z - y)z. \tag{4.29}$$

The steady state dependence of z on ω is determined by eliminating y from the equations

$$0 = x(0) - y - \frac{1}{\mathbb{K}_a} yz - \beta y + \gamma(x(0) - y)(\omega + z - y),$$
$$0 = x(0) - y - \frac{1}{\mathbb{K}_a} yz + \alpha \mathbb{K}_w - \alpha(\omega + z - y)z. \tag{4.30}$$

Fortunately, y appears linearly in the second equation. By solving for this variable and substituting the result into the first equation, a quartic equation for z is obtained. This equation can be solved explicitly (for example, by the Cardano–Tartaglia formula), but the result is complicated. There are four roots. Thus, the correct root must be determined. Also, as suggested by Fig. 4.2, the roots of this quartic are very small. The root size varies from approximately 10^{-3} down to 10^{-13}. There are at least two ways to make the sizes of the real roots more manageable: scale the variable z by $\lambda > 0$ and look for λz, or seek $\zeta := \lambda/z$ for some λ instead of z. Given the range of the root size, $\lambda = 10^{-10} = \mu^5$ is chosen here for the scaling factor in the change of variables $\zeta := \lambda/z$. Other choices are also viable. For $\zeta := \mu^5/z$, the problem reduces to finding the appropriate root of the quartic polynomial

$$\mathcal{F}(\zeta, \omega) := a(\omega)\zeta^4 + b(\omega)\zeta^3 + c(\omega)\zeta^2 + d(\omega)\zeta + e(\omega)$$

whose coefficients are functions of ω.

The leading coefficient of the quartic is

$$a(\omega) = \mathbb{K}_a^2(-\alpha\mathbb{K}_w - \beta x(0) - \alpha\beta\mathbb{K}_w + \alpha\gamma x(0)\mathbb{K}_w + \alpha^2\gamma\mathbb{K}_w^2 - \alpha\gamma\mathbb{K}_w\,\omega). \tag{4.31}$$

Using the approximate values of the parameters given by the Eqs. (4.3), the coefficient a is (up to a positive multiple)

$$56\mu^2 - 3\mu^3 + 420\mu^6 - 3\omega.$$

Thus, the coefficient of ζ^4 vanishes at

$$\omega_* := \frac{1}{3}(56\mu^2 - 3\mu^3 + 420\mu^6).$$

Using $\mu = 1/100$, this number is approximately

$$\omega_* \approx 0.0019.$$

The roots of a monic polynomial depend continuously on its coefficients (see Exercise 4.16). The dependence of the roots of a polynomial on its leading coefficient is problematic. Behavior near the parameter value corresponding to a zero of the leading coefficient may include abrupt changes or discontinuities. Inspection of Fig. 4.2 suggests that the steep rise in the titration curve is near the number ω_*.

The constant coefficient e is given by

$$e(\omega) = \alpha(1 - \mathbb{K}_a(\alpha - \gamma))\mu^{20} = 50\mu^{10}(1 - 1568\mu),$$

which is very small. Our goal is to explain the steep rise in the titration curve. It seems reasonable to simply ignore this small term. In doing so, the resulting explanation is no longer rigorous. Part of the purpose of this

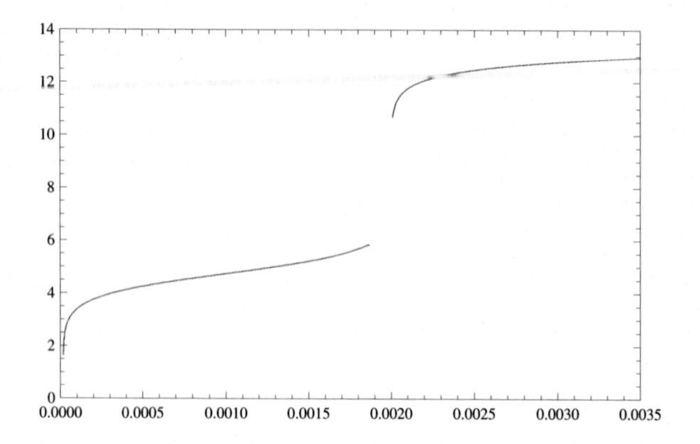

Fig. 4.3 The approximate titration curve (pH versus initial molarity) of sodium hydroxide added to an 0.1 molar acetic acid solution obtained as positive root of the quadratic $a(\omega)\zeta^2 + b(\omega)\zeta + c(\omega)$ *is depicted.*

section is to introduce the reader to this possibility in applied mathematics. In applied mathematics, useful insights are often gained from arguments that are not mathematical proofs. Order relations and series expansions often appear in reasoning of this type.

Because $\zeta = 0$ is not a physically relevant value, the function $\mathcal{F}$ may be replaced by the approximation

$$F(\zeta, \omega) := a(\omega)\zeta^3 + b(\omega)\zeta^2 + c(\omega)\zeta + d(\omega).$$

Because, $a(\omega_*) = 0$, the desired approximation of the physical root at this parameter value should be the solution of the quadratic equation

$$b(\omega_*)\zeta^2 + c(\omega_*)\zeta + d(\omega_*) = 0.$$

In fact, this quadratic has a positive root ζ_*, which can be computed using the quadratic formula. Thus, we have

$$F(\zeta_*, \omega_*) = 0.$$

Nearby roots are determined by an application of the implicit function theorem—which is an essential tool in mathematics (see, for example, [20]). A computation shows that $F_\zeta(\zeta_*, \omega_*) \neq 0$. By the implicit function theorem, the root ζ is a unique (smooth) function of ω; that is,

$$F(\zeta(\omega), \omega) \equiv 0.$$

Note that this formula can be used to obtain the derivative of ζ with respect to ω; in fact,

$$F_\zeta(\zeta(\omega), \omega)\zeta_\omega(\omega) + F_\omega(\zeta(\omega), \omega) = 0.$$

Recall that the titration curve is the graph of the function

$$\omega \to -\frac{\ln z(\omega)}{\ln 10}.$$

The derivative of this function at ω_* is equal to

$$\frac{1}{\zeta_* \ln 10}\zeta_\omega(\omega_*) = \frac{1}{\zeta_* \ln 10} \times \frac{-F_\omega(\zeta_*, \omega_*)}{F_\zeta(\zeta_*, \omega_*)}, \tag{4.32}$$

which (using the parameter values set in this section) has the approximate value 3465.6; certainly a large slope of the approximate titration curve at ω_*.

What counts in the chemistry? The dominant term in the expression

$$\omega_* = x(0) + \alpha \mathbb{K}_w - \frac{1+\beta}{\gamma} - \frac{\beta x(0)}{\alpha \gamma \mathbb{K}_w} \tag{4.33}$$

(obtained by solving for ω when $a(\omega)$ from Eq. (4.31) is equated to zero) is $x(0)$ (see Exercise 4.15). Thus, the pH rises rapidly in the titration as the concentration of OH^- ions begins to exceed the concentration of H^+ ions.

The titration curve away from the graph over a small interval around ω_* (where the slope is steep) is well approximated by roots of the quadratic equation

$$a(\omega)\zeta^2 + b(\omega)\zeta + c(\omega) = 0$$

obtained by ignoring the linear and constant terms of the quartic polynomial (see Exercise 4.17).

The kinetic theory expressed in the models discussed here predicts phenomena that are in complete agreement with physical experiments. Thus, we may have a high level of confidence that the underlying theory is correct. To make such a statement about a class of mathematical models is satisfying; the validation (and invalidation) of models is, of course, a strong motivation for doing applied mathematics.

Exercise 4.11. Redraw Fig. 4.1 using the improved titration model. Is there a difference?

Exercise 4.12. Redraw Fig. 4.2 using an alternative method: do not integrate the differential equations; solve for the steady states. Which method is more efficient?

Exercise 4.13. A titration of 0.1 molar acetic acid with sodium hydroxide is discussed in the section. The point of mutual neutralization occurs for the addition of approximately 0.1 mole of the base. Repeat the analysis, using the same methods, for titration of 0.3 molar acetic acid with sodium hydroxide.

Exercise 4.14. Consider the function

$$F(\zeta, \omega) := a_n(\omega)\zeta^n + a_{n-1}(\omega)\zeta^{n-1} + \cdots + a_0(\omega).$$

Suppose that $F(\zeta_*, \omega_*) = 0$ and $a(\omega_*) = 0$. State a condition on the polynomial

$$a_{n-1}(\omega)\zeta^{n-1} + \cdots + a_0(\omega)$$

that ensures the existence of a smooth function $\phi(\omega)$ defined on some interval containing ω_* such that $\phi(\omega_*) = \zeta_*$ and $F(\phi(\omega), \omega) \equiv 0$. Hint: Use the implicit function theorem.

Exercise 4.15. Show that the dominant term on the right-hand side of Eq. (4.33), for the choice of parameter values given in this section, is the term $x(0)$.

Exercise 4.16. (a) Prove that the roots of a monic polynomial depend continuously on its coefficients. Hint: Use the implicit function theorem. (b) What happens if the polynomial is not monic?

Exercise 4.17. (a) Show in detail how to recover the value 3465.6 for the steep part of the slope via Eq. (4.32). Why is it reasonable to ignore constant and linear terms in the quartic polynomial whose positive root determines the titration curve? (b) Write and discuss a computer code to recover Fig. 4.3.

Exercise 4.18. The titration curve depicted in Fig. 4.2 is obtained by a static model; that is, we set the new OH concentration, integrate to steady state, determine the pH, and iterate this process. A dynamic model would predict results for a titration where the sodium hydroxide solution is poured continuously into the acetic acid solution. Eq. (4.21) is a first approximation of such a model. Create a model that can be used to predict titration curves consistent with Fig. 4.2.

4.4 THE OREGONATOR: AN OSCILLATORY REACTION

The physically interesting chemistry previously discussed takes place in steady state. In fact, until the 1960s most chemists believed that oscillations in chemical reactions were impossible. A class of reactions discovered by B. P. Belousov and popularized by A. M. Zhabotinsky proved that oscillations are possible. Although the recipes for such reactions are widely available and easy to reproduce, the exact underlying chemistry is not completely understood. Mathematical models, which are based on the principle of mass action for the reactions that are believed to occur for the chemicals that appear as the ingredients of the recipes, predict good approximations of the observations. Thus, chemists know they are on the right path to understanding the reactions. These models also have rich mathematical structures. A simple model for oscillations of concentration of the chemical species in the Belousov–Zhabotinsky (BZ) reaction are discussed in this section.

One of many proposed models for the BZ reaction is called the Oregonator (because it was developed in Oregon). In one of its popular forms, it is the mass action model derived from the following reactions for chemical species A, B, P, X, Y, and Z and rates k_i:

$$\begin{aligned}
A + Y &\xrightarrow{k_1} X + P,\\
X + Y &\xrightarrow{k_2} 2P,\\
A + X &\xrightarrow{k_3} 2X + 2Z,\\
X + X &\xrightarrow{k_4} A + P,\\
B + Z &\xrightarrow{k_5} \rho Y.
\end{aligned} \tag{4.34}$$

The constant ρ is a dimensionless (stochiometric) parameter that determines the number of molecules of Y produced by the last reaction.

The concentrations of all six species change during the reaction. But, the concentrations of species A, B, and P remain nearly constant in comparison to the concentrations of species X, Y, and Z. Thus, after a chemist tells us that treating the first three species as constants is a close approximation to the dynamical part of the reaction, it suffices to write the three rate equations

$$\begin{aligned}
[X]' &= k_1[A][Y] - k_2[X][Y] + k_3[A][X] - k_4[X]^2,\\
[Y]' &= -k_1[A][Y] - k_2[X][Y] + \rho k_5[B][Z],\\
[Z]' &= 2k_3[A][X] - k_5[B][Z].
\end{aligned} \tag{4.35}$$

To make the system dimensionless, set

$$[X] = ax, \qquad [Y] = by, \qquad [Z] = cz, \qquad t = \tau s.$$

The equivalent system of differential equations is

$$\begin{aligned}\frac{dx}{ds} &= \frac{\tau k_1 b[A]}{a}y - \tau k_2 bxy + \tau k_3[A]x - \tau k_4 a x^2,\\ \frac{dy}{ds} &= -\tau k_1[A]y - \tau k_2 axy + \frac{\tau k_5 c[B]\rho}{2b}z,\\ \frac{dz}{ds} &= \frac{2\tau k_3[A]a}{c}x - \tau k_5[B]z.\end{aligned} \tag{4.36}$$

There are many ways to choose the scales in system (4.36). The most popular choice is derived after consideration of measured quantities for the BZ reaction; this scaling is

$$a := \frac{k_3[A]}{k_4}, \qquad b := \frac{k_3[A]}{k_2}, \qquad c := \frac{2k_3^2[A]^2}{k_4 k_5[B]}, \qquad \tau := \frac{1}{k_5[B]}.$$

With the additional dimensionless constants

$$\epsilon := \frac{k_5[B]}{k_3[A]}, \qquad \delta = \frac{k_4 k_5[B]}{k_2 k_3[A]}, \qquad \alpha := \frac{k_1 k_4}{k_2 k_3},$$

the dimensionless model is

$$\begin{aligned}\epsilon\frac{dx}{ds} &= \alpha y - xy + x - x^2,\\ \delta\frac{dy}{ds} &= -\alpha y - xy + \rho z,\\ \frac{dz}{ds} &= x - z.\end{aligned} \tag{4.37}$$

By measurement and chemical intuition, the parameters have approximate values

$$\epsilon = 10^{-2}, \qquad \delta = 2 \times 10^{-5}, \qquad \alpha = 10^{-4}, \qquad \rho \in [0, 3]. \tag{4.38}$$

The small parameter δ is much smaller than ϵ. This suggests setting $\delta = 0$, solving for $y = \rho z/(\alpha + x)$ in the second equation, and substituting to obtain the planar system

$$\begin{aligned}\epsilon\frac{dx}{ds} &= \rho\frac{\alpha - x}{\alpha + x}z + x - x^2,\\ \frac{dz}{ds} &= x - z.\end{aligned} \tag{4.39}$$

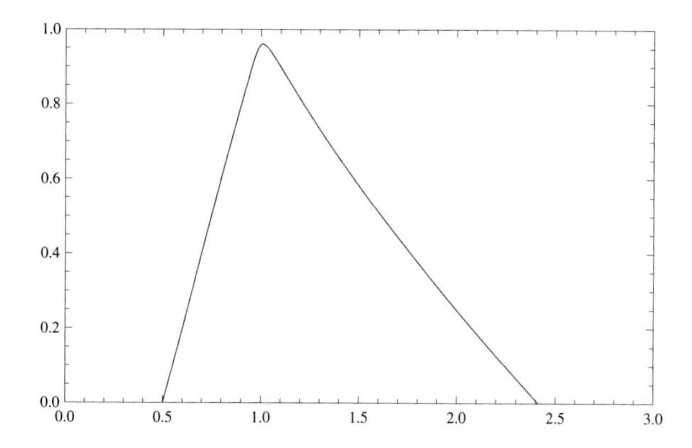

Fig. 4.4 The bifurcation diagram in the ρ-ϵ parameter space—$0 < \rho < 3$ and $0 < \epsilon < 1$—at $\alpha = 10^{-4}$ for system (4.39) is depicted. A Hopf bifurcation occurs as a curve in parameter space crosses the boundary of the wedge-shaped region. For parameter values inside the region there is an asymptotically stable limit cycle; outside the region the system has an asymptotically stable rest point.

Perhaps the product of the derivative of y and δ is not small because dy/ds is large. This is certainly a problem. But, the reason for initially ignoring this issue is clear: a two-dimensional system should be easier to analyze than a three-dimensional system.

The first quadrant—corresponding to nonnegative concentrations—is positively invariant. The rest points in this physically meaningful region are the origin $(x, z) = (0, 0)$ and a point with both coordinates equal to the positive solution (for x) of the equation

$$\rho\frac{x^2 - \alpha^2}{(x + \alpha)^2} = 1 - x, \tag{4.40}$$

which is obtained after substitution of $z = x$ and some algebraic manipulation of the right-hand side of the first equation of system (4.39) set to zero.

The rest point at the origin is a saddle point. More precisely it is a hyperbolic saddle point defined as the rest point of an autonomous system of first-order ordinary differential equations (ODEs) such that its linearization at this rest point has all real eigenvalues and at least two of opposite signs. In the present two-dimensional case, the stable manifold of the saddle corresponds to the negative eigenvalue and is contained in quadrants two and four, while its unstable manifold (corresponding to the positive eigenvalue) lies in quadrants one and three. Here, the stable and unstable manifolds are one-dimensional invariant curves passing through the rest point with their tangent lines at the rest point given by one of the eigenspaces of the system matrix of the linearized system. Solutions of the ODE approach the rest point along the stable manifold and recede from the rest point along the unstable manifold. In particular, solutions starting on the portion of the unstable manifold in the first quadrant stay in this quadrant for all $s > 0$.

The linearization at the interior rest point has system matrix

$$A = \begin{pmatrix} \frac{1-2x}{\epsilon} - \frac{2\alpha\rho z}{\epsilon(\alpha+x)^2} & \frac{\rho(\alpha-x)}{\epsilon(\alpha+x)} \\ 1 & -1 \end{pmatrix}.$$

with

$$\det(A) = \frac{1}{\epsilon}\left(2x - 1 + \frac{2\alpha\rho z}{(x+\alpha)^2} + \rho\frac{x^2-\alpha^2}{(x+\alpha)^2}\right).$$

Using Eq. (4.40) to substitute for the last term in this expression, the determinant of A is easily proved to be positive.

The trace of A is given by

$$\operatorname{tr}(A) = -1 + \frac{1-2x}{\epsilon} - \frac{2\alpha\rho z}{\epsilon(\alpha+x)^2}.$$

Recall that the characteristic polynomial for the eigenvalues of A is

$$\lambda^2 - \operatorname{tr}(A)\lambda + \det(A)$$

with roots

$$\frac{\operatorname{tr}(A) \pm \sqrt{\operatorname{tr}(A)^2 - 4\det(A)}}{2}.$$

The eigenvalues of A are pure imaginary exactly in case $\operatorname{tr}(A) = 0$; otherwise, the rest point is a source for $\operatorname{tr}(A) > 0$ and a sink for $\operatorname{tr}(A) < 0$. There are three parameters: α, ρ, and ϵ. The trace of A depends on all these parameters and the position of the rest point. As the position coordinates are the same ($z = x$) at rest points, Eq. (4.40) and $\operatorname{tr}(A) = 0$ must hold simultaneously at a rest point whose linearization has pure imaginary eigenvalues. Eq. (4.40) is essentially a quadratic in x; its positive root is

$$x = \frac{1}{2}\left(1 - \alpha - \rho + \sqrt{(1-\alpha-\rho)^2 + 4\alpha(1+\rho)}\right).$$

The result of substituting this expression into the equation $\mathrm{tr}(A) = 0$, is an equation for a hypersurface in the three-dimensional parameter space corresponding to parameter values such that the system of differential equations has a rest point with pure imaginary eigenvalues. For simplicity, this hypersurface is reduced to a curve in the ρ-ϵ parameter plane depicted in Fig. 4.4 by fixing $\alpha = 10^{-4}$. The equation for the hypersurface can be solved explicitly for ϵ as a function of α and ρ (see Exercise 4.19).

For parameter values inside the region bounded by the curve depicted in Fig. 4.4, the rest point with positive coordinates of system (4.39) is a source; outside the region this rest point is a sink.

The vector field given by the right-hand side of system (4.39) points to the left along the line $x = 1$ and points down along the line $z = 2$ as long as $0 < x < 1$. In other words, the rectangle bounded by sides along the coordinate axes, the line $x = 1$ and the line $z = 2$ is positively invariant. Solutions starting in this rectangle stay in the rectangle for all positive time. In case the rest point in the interior of this rectangle is a source, the Poincaré–Bendison theorem (see Appendix A.18, theorem A.12) implies that there is a stable limit cycle in the rectangle; thus, the (two-dimensional) Oregonator model predicts oscillations in the concentrations of the chemical reactants whenever the parameters are such that the interior rest point of the model is a source.

Clearly something interesting happens as parameter values cross the boundary into the bounded region in Fig. 4.4. Imagine that this change happens along a curve in the parameter space. The phase portrait changes from having a stable rest point to having an unstable rest point and a stable limit cycle. This is an excellent example of the Hopf bifurcation; it occurs along a curve in the parameter space for a system of ODEs when a pair of complex conjugate eigenvalues of the system matrix of the linearized system at a rest point crosses the imaginary axis in the complex plane with nonzero speed. Of course, the position of the rest point, the system matrix and its eigenvalues are all allowed to change with the parameter values as they move along the curve in parameter space. In this case, under an additional hypothesis (see, for example, [20]) on the nonlinear part of the system that is almost always satisfied, a limit cycle is born (in a supercritical Hopf bifurcation) at the rest point and grows in amplitude as the eigenvalues cross the imaginary axis or a limit cycle decreases in amplitude and dies (in a subcritical Hopf bifurcation) as the eigenvalues cross this axis.

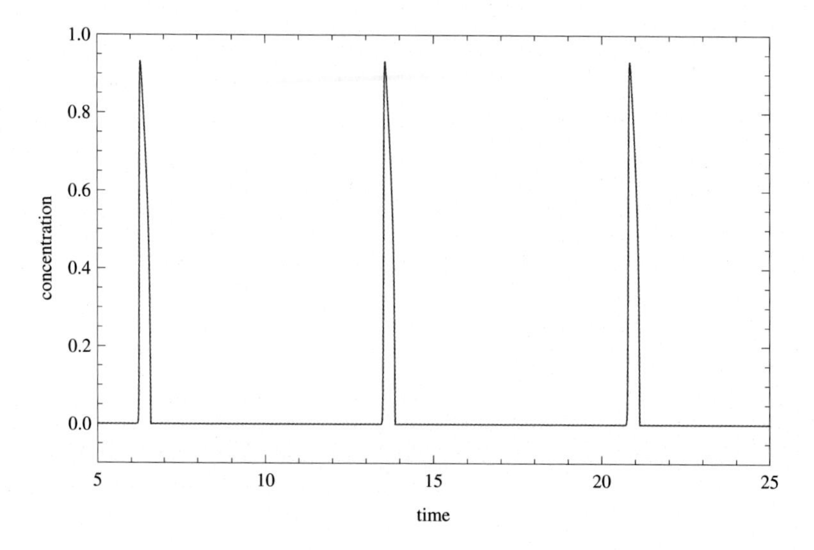

Fig. 4.5 The approximate oscillation of the x-concentration for system (4.39) with $\rho = 1.5$, and $\epsilon = 0.01$, and $\alpha = 10^{-4}$ is depicted.

The Hopf bifurcation is detected in autonomous systems of ODEs, in exactly the manner discussed for the Oregonator, by finding the curve(s) in the parameter space where the characteristic equation of the linearization at a rest point has pure imaginary eigenvalues. Generically, there is a Hopf bifurcation as a curve in the parameter spaces crosses this "Hopf curve."

Fig. 4.5 shows a computer-generated graph of the oscillation of concentration of the chemical species corresponding to x for a physically realistic choice of parameters inside the bounded region in Fig. 4.4. Note the nearly constant state punctuated with spikes of higher concentration.

The Hopf bifurcation analysis can be extended to n-dimensional systems with $n > 2$. The basic approach is the same as for two-dimensional systems: Locate and track a rest point of the system as it changes position with respect to a single parameter, and look for an interval of parameters over which the stability type of the rest point changes from a source to a sink or from a sink to a source. The signature of the Hopf bifurcation is a pair of complex conjugate eigenvalues of the system matrix of the linearized system at the rest point whose corresponding paths in the complex plane cross the imaginary axis with nonzero speed as the parameter changes on the interval. Under an additional requirement on the nonlinear part of the system of differential equations that is almost always satisfied, a limit cycle is born or dies at the rest point as the parameter changes through the critical value when the corresponding eigenvalues are pure imaginary. This approach can be used to prove the existence of a stable limit cycle for the three-dimensional Oregonator [Eq. (4.37)] for some choices of its parameters. Thus, the full Oregonator predicts the existence of oscillations in the concentrations of the chemical species for the reaction sequence it models.

An alternative approach to the dynamics of the Oregonator takes further advantage of the assumption that the parameter δ is small relative to the other parameters of the system. The case where δ is simply set to zero has been discussed. What is the behavior of the Oregonator for small nonzero values of this parameter? An answer is obtained using singular perturbation theory.

Perturbation is a fundamental idea in applied mathematics. Suppose the behavior of a system of differential equations is known when some parameter of the system is set to a special value (which, without loss of generality, is usually assumed to be zero). A natural question is to describe the behavior of the system when the parameter value is near this special value. We say the system with nearby parameter values is perturbed from the system with the special parameter value.

To give a simple example, consider a system $\dot{x} = f(x, \lambda)$, where $f : \mathbb{R}^n \times \mathbb{R}^k \to \mathbb{R}$ is a continuously differentiable function. The state variable is x, the (vector) parameter is λ. Perhaps $f(0, 0) = 0$; that is, $x = 0$ is a rest point for the system $\dot{x} = f(x, 0)$ with parameter value $\lambda = 0$. By changing λ to nonzero values, the corresponding systems may be viewed as perturbations of the special system $\dot{x} = f(x, 0)$. It might happen that the derivative $Df(0, 0)$ of the transformation $x \to f(x, 0)$ at $x = 0$ is an invertible $n \times n$ matrix. In this case, the implicit function theorem implies there is a function β mapping some neighborhood V of the origin in $\mathbb{R}^k$ to a neighborhood U of the origin in $\mathbb{R}^n$ such that $\beta(0) = 0$ and $f(\beta(\lambda), \lambda) \equiv 0$ for every $\lambda \in V$. Moreover, if $(\xi, \ell) \in U \times V$ and $f(\xi, \ell) = 0$, then $\beta(\ell) = \xi$. In other words, under the assumption that $Df(0, 0)$ is invertible (equivalently, the system matrix of the linearization at the rest point $x = 0$ for the system with parameter value $\lambda = 0$ has no zero eigenvalue), then every perturbation of the unperturbed system $\dot{x} = f(x, 0)$ has an isolated rest point near the origin.

The last example illustrates regular perturbation, which in the present context refers to the parameterized family of differential equations $\dot{x} = f(x, \lambda)$, where this family remains a differential equation as the parameter λ changes. In contrast, consider the system

$$\dot{x} = y^2 - x, \qquad \epsilon\dot{y} = -y, \tag{4.41}$$

where ϵ is a small parameter. At $\epsilon = 0$, this system reduces to the scalar ODE $\dot{x} = -x$. This one-dimensional dynamical system has a globally attracting rest point at the origin. With $\epsilon \neq 0$, there are two cases: If $\epsilon > 0$, then the original system has a globally asymptotic rest point at the origin, but if $\epsilon < 0$, then the original system has a rest point of saddle type at the origin. The perturbed systems ($\epsilon \neq 0$) do not all behave as might be predicted from the dynamics of the unperturbed system at $\epsilon = 0$. This result is typical of a singular perturbation problem, which in this case refers to two families of differential equations (one for $\epsilon > 0$, one for $\epsilon < 0$) connected at $\epsilon = 0$ by a differential algebraic equation (DAE). In other words, the system changes character at $\epsilon = 0$, usually because at least one of the highest-order derivatives is multiplied by ϵ.

Singular perturbation theory is a venerable subject with a long history in applied mathematics. One modern development in this theory is called geometric singular perturbation theory. The adjective "geometric" refers to consideration of perturbation problems in view of their geometry in phase space.

The phase space for system (4.41) is the (x, y) plane. One of the tools of geometric singular perturbation theory relies on the recognition that singular perturbation problems for differential equations usually involve at least two timescales. By rearranging the system of ODEs to the form

$$\dot{x} = y^2 - x, \qquad \dot{y} = -\frac{1}{\epsilon}y, \tag{4.42}$$

it is apparent that for small $\epsilon \neq 0$ the y component of the solution moves much faster than the x component because $\dot{y}$ is large compared with $\dot{x}$. Also, by the change of temporal variable $t = \epsilon s$ (which is defined for $\epsilon \neq 0$), the system is transformed to

$$\dot{x} = \epsilon(y^2 - x), \qquad \dot{y} = -y. \tag{4.43}$$

Thus, it appears that the singular perturbation problem (4.42) is transformed to the regular perturbation problem (4.43). This is true except at the critical

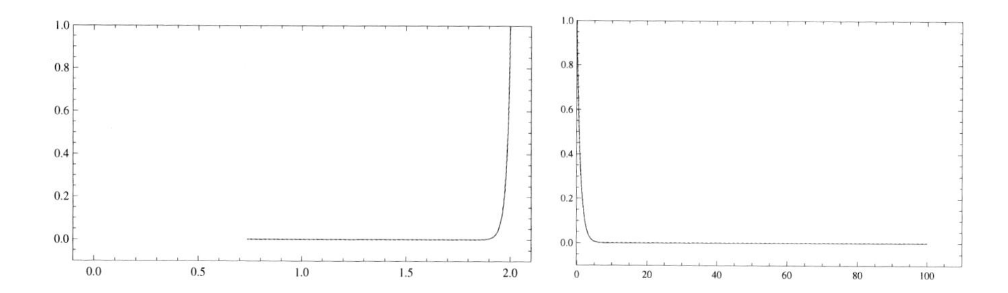

Fig. 4.6 Computer-generated approximations of solutions of system $\dot{x} = \epsilon(y^2 - x)$, $\dot{y} = -y$ for $\epsilon = 0.01$ with initial data $x(0) = 2$ and $y(0) = 1$ are depicted for $0 \le s \le 100$. The left-hand panel shows solution in the phase plane; the right-hand panel shows y versus s.

parameter value $\epsilon = 0$. The two systems are not equivalent as families: system (4.43) makes perfect sense at $\epsilon = 0$; system (4.42) does not. But, they are equivalent for $\epsilon \neq 0$. This is the essential point. System (4.43) may be analyzed as a regular perturbation problem near $\epsilon = 0$ to help determine the singular behavior of system (4.42) for nonzero values of the parameter. In other words, it is often useful to make singular changes of coordinates at singular points. Generally, there is no interest in a system *at* a singular parameter value; the important behavior is *near* the singular parameter value. Such singular coordinate transformations produce systems that *are equivalent* away from the singular value of the parameter.

Note that one unit of time measured in the original t variable is $1/\epsilon$ units of time in the s variable. Thus, s-time is moving fast relative to t-time. For this reason, system (4.42) is called the slow (or slow-time) system and (4.43) is called the fast (or fast-time) system.

The x-axis is invariant for the fast-time system at $\epsilon = 0$ and this invariant set consists entirely of rest points. Linearization at each of these rest points produces the system matrix

$$\begin{pmatrix} 0 & 0 \\ 0 & -1 \end{pmatrix}.$$

It has two eigenvalues: zero, whose eigenspace is parallel to the x-axis, and the eigenvalue -1, whose eigenspace is parallel to the y-axis. The linearized system has all fixed points along the x-axis, and every vertical line is invariant with a globally attracting sink at its intersection with the x-axis. For $\epsilon \neq 0$, the behavior of the regularly perturbed family (4.43) behaves as might be expected. The x-axis remains invariant but it now supports a slow flow toward the origin in case $\epsilon > 0$ and away from the origin in case $\epsilon < 0$. As shown in Fig. 4.6, the global flow exhibits two timescales: rapid motion toward the invariant x-axis followed by slow motion once the trajectory is near this invariant set. The graphs of the corresponding solutions for system (4.42) look exactly the same, but they are traversed on a slower timescale.

The analysis of system (4.41) is typical for a class of singular perturbation problems of the form

$$\dot{x} = f(x, y, \epsilon), \qquad \epsilon\dot{y} = g(x, y, \epsilon), \tag{4.44}$$

where the behavior of the system is desired near $\epsilon = 0$. The state variables x and y may be vector variables and the corresponding fast-time system (with the new temporal variable $s := t/\epsilon$) is

$$\dot{x} = \epsilon f(x, y, \epsilon), \qquad \dot{y} = g(x, y, \epsilon). \tag{4.45}$$

At $\epsilon = 0$, system (4.45) has the set of rest points $\{(x, y) : g(x, y, 0) = 0\}$. In a case where $g_y(x, y, 0)$ is an invertible linear transformation of the y space at each rest point (or on some connected subset of rest points), the implicit function theorem implies that there is a function β defined on an open subset of the x space such that $g(x, \beta(x), 0) = 0$. In other words, a subset $\mathcal{M}$ of the rest points is parameterized by the function $x \mapsto (x, \beta(x))$. Linearization at a rest point on $\mathcal{M}$ produces a system matrix whose dimension is the same as the dimension of the (x, y) space. Because g_y is invertible at points in $\mathcal{M}$, each such system matrix has the same number of zero eigenvalues as the dimension of the x space. The set $\mathcal{M}$ is called a normally hyperbolic invariant manifold if in addition the eigenvalues of the transformation g_y (which account for the remaining eigenvalues of the system matrix) all have nonzero real parts. If all these eigenvalues have negative real parts, $\mathcal{M}$ attracts nearby solutions; if all eigenvalues have positive real parts it repels; and, if it has some eigenvalues with positive and some with negative real parts, it behaves like a saddle rest point. Some solutions are attracted; others are repelled.

This is exactly the case in the last example where the x-axis, which plays the role of $\mathcal{M}$, is parameterized by $x \mapsto (x, \beta(x))$, where $\beta(x) \equiv 0$. The function g is given by $g(x, y, \epsilon) = -y$ and its partial derivative with respect to y is the 1×1 matrix (-1), which accounts for the nonzero eigenvalues of the system matrix of the linearization of each rest point on the x axis. For a real 1×1 matrix there are no complex eigenvalues to consider. The only eigenvalue is -1, which has nonzero real part. Thus the x axis is a normally hyperbolic invariant manifold.

An important theorem due to Neil Fenichel and independently by Morris Hirsch, Charles Pugh, and Stephen Smale states that normally hyperbolic invariant manifolds persist. In our simple case, persistence means that there is a continuous family of invariant sets $\mathcal{M}_\epsilon$ such that each member of this family is given as the graph of a function over the x space, $\mathcal{M}_0 = \mathcal{M}$, and the dynamical behavior in the normal direction (attracting, repelling, or saddle-like) also persists. There is a smooth function $(x, \epsilon) \mapsto B(x, \epsilon)$ such that $B(x, 0) = \beta$ and $\mathcal{M}_\epsilon$ is parameterized by the function $x \mapsto (x, B(x, \epsilon))$.

The perturbed manifold $\mathcal{M}_\epsilon$ is called the slow manifold. For small $|\epsilon|$, this manifold is close to the manifold $\mathcal{M}_0$ that consists entirely of rest points. Thus, the flow on the nearby manifold moves slowly with respect to the flow at $\epsilon = 0$ that does not move at all. The slow flow for a general singular perturbation problem can be arbitrarily complicated; for example, it can contain isolated rest points and periodic orbits. As in the example, the two timescales remain important in the global flow for $\epsilon \neq 0$. Orbits move toward or away from the slow manifold $\mathcal{M}_\epsilon$ rapidly; orbits on the slow manifold move slowly. Of course, the signs of the eigenvalues of the linear maps g_y determine the direction of the flow near, but not on, the slow manifold. In most cases of interest in applications, the slow manifold attracts orbits. Thus, after a fast transient, the dominant behavior of the system is determined by the dynamics on the slow manifold. The transient and dominant behaviors are the same for the original singularly perturbed system (for $\epsilon \neq 0$) except that the motion occurs on a slower timescale. Thus, this analysis gives one way to approach the dynamical behavior of some singular perturbation problems for systems of ODEs.

The general definition of normal hyperbolicity and the formulation and proof of the persistence theorem require a sophisticated mathematical analysis that is beyond the scope of this book. The results of this theory are used here to illustrate its utility in applied mathematics.

Returning to the Oregonator [Eq. (4.37)],

$$\epsilon \frac{dx}{ds} = \alpha y - xy + x - x^2,$$

$$\delta\frac{dy}{ds} = -\alpha y - xy + \rho z,$$
$$\frac{dz}{ds} = x - z, \tag{4.46}$$

rewritten for convenience, it should now be clear that this system, for fixed ϵ and α, yields a singular perturbation problem for δ near $\delta = 0$. Perhaps it should be considered as a singular perturbation problem in both ϵ and δ for fixed α, but only the former case is considered here.

The fast-time system (obtained using the change of temporal variable $s = \delta\tau$) is

$$\frac{dx}{d\tau} = \frac{\delta}{\epsilon}(\alpha y - xy + x - x^2),$$
$$\frac{dy}{d\tau} = -\alpha y - xy + \rho z,$$
$$\frac{dz}{d\tau} = \delta(x - z). \tag{4.47}$$

At $\delta = 0$, there is a two-dimensional invariant surface $\mathcal{M}$ (in the first octant) consisting entirely of rest points that is parameterized by the function $(x, z) \mapsto (x, \rho z/(\alpha + x), z)$ whose domain is the open first quadrant of the (x, z) plane. This invariant surface is a normally hyperbolic invariant manifold. In fact, the system matrix obtained by linearization at each point on $\mathcal{M}$ has two zero eigenvalues corresponding to the tangent directions along the two-dimensional surface of rest points and the negative eigenvalue $-(\alpha+x)$. Thus, $\mathcal{M}$ attracts all solutions starting near it in the positive octant of the (x, y, z) space. Actually, it attracts all solutions starting in this space. According to the geometric singular perturbation theory, this invariant two-dimensional slow manifold persists as a two-dimensional surface $\mathcal{M}_\delta$ (given by the graph of a function over the (x, y) space) that rapidly attracts nearby orbits. Solutions move slowly on the perturbed slow manifold $\mathcal{M}_\delta$ and it contains the important dynamical behavior, perhaps rest points or periodic orbits.

The perturbed slow manifold $\mathcal{M}_\delta$ and the flow on this manifold are determined by exploiting the persistence theory. In particular, $\mathcal{M}_\delta$ is given as the graph of a function over the (x, z) space. In fact, $\mathcal{M}_\delta$ is a regular perturbation of $\mathcal{M} = \mathcal{M}_0$. It is the graph of a function that must be expressible as a power series in δ at $\delta = 0$; that is, $\mathcal{M}_\delta$ is the graph of a function given by

$$y = \frac{\rho z}{\alpha + x} + B_1(x, z)\delta + B_2(x, z)\delta^2 + O(\delta^3), \tag{4.48}$$

where the functions $B_i(x, z)$, for $i = 1, 2, 3 \ldots, \infty$, are to be determined. The defining property of the graph is its invariance under the flow of the fast system (4.47). In particular, if $\tau \mapsto (x(\tau), y(\tau), z(\tau))$ is a solution of this system (at parameter value δ) that is on the manifold $\mathcal{M}_\delta$, then

$$\begin{aligned} y(\tau) &= B(x(\tau), z(\tau), \delta) \\ &= \frac{\rho z(\tau)}{\alpha + x(\tau)} + B_1(x(\tau), z(\tau))\delta + B_2(x(\tau), z(\tau))\delta + O(\delta^2). \end{aligned}$$

Using the smoothness of all functions in sight, differentiate with respect to τ at $\tau = 0$ to obtain the invariance relation

$$\frac{dy}{d\tau} = B_x(x, z, \delta)\frac{dx}{d\tau} + B_z(x, z, \delta)\frac{dz}{d\tau}. \tag{4.49}$$

By substitution for $dx/d\tau$, $dy/d\tau$, $dz/d\tau$, and y from Eq. (4.48) and by equating coefficients of like powers of δ in the resulting expression, it is possible to solve for the B_i in the order $i = 1, 2, 3 \ldots, \infty$; in fact, for $i = 1$,

$$\begin{aligned} B_1(x, z) = &-\frac{\rho}{(x+\alpha)^4\epsilon}(\alpha^2\epsilon(x - z) + 2\alpha\epsilon x^2 - \alpha(1 + 2\epsilon)xz - \alpha\rho z^2 \\ &+ \epsilon x^3 + (\alpha - 1 - \epsilon)x^2 z + \rho x z^2 + x^3 z). \end{aligned}$$

The flow on the slow manifold is simply the restriction of the fast system [Eq. (4.47)] to this invariant set; in effect, the vector field defining the system is evaluated at points in the (x, y, z) space where $y = B(x, z, \delta)$. The flow is determined by the x and z components of the resulting system. More precisely, consider the form of the fast system

$$x' = \delta p(x, y, z), \qquad y' = q(x, y, z), \qquad z' = \delta r(x, y, z).$$

Substitution as described yields

$$\begin{aligned} x' &= \delta p(x, B(x, z, \delta), z), \\ y' &= q(x, B(x, z, \delta), z), \\ z' &= \delta r(x, B(x, z, \delta), z). \end{aligned}$$

Note that the y component decouples from the system of ODEs. Thus, we may view the differential equation on the two-dimensional slow manifold to

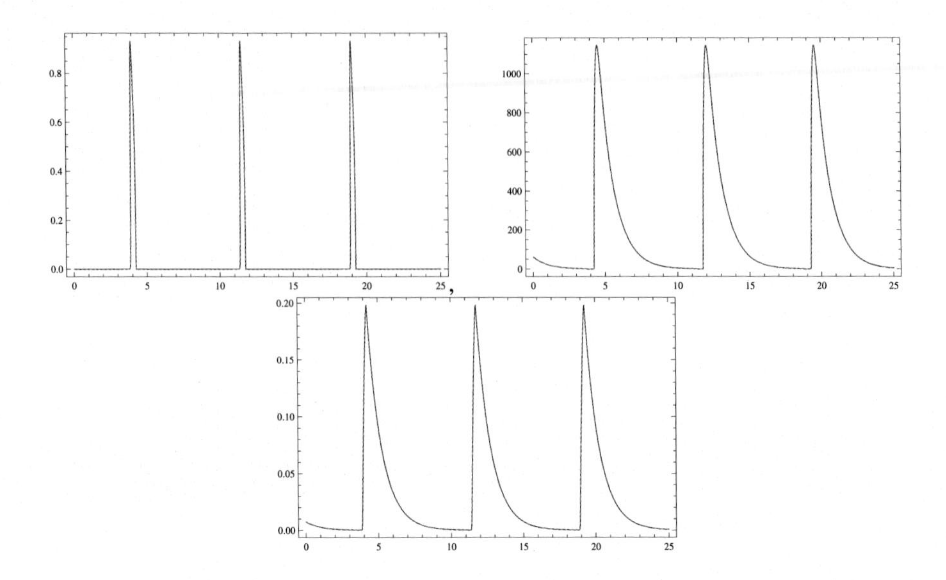

Fig. 4.7 Approximate graphs of the x, y, and z concentrations for the Oregonator (4.46) *with* $\rho = 1.5$, *and* $\epsilon = 0.01$, $\delta = 2 \times 10^{-5}$, *and* $\alpha = 10^{-4}$ *are depicted .*

be the two-dimensional ODE consisting of the x and z components of this system. The flow on the slow manifold is approximated by expanding the system of ODEs in δ at $\delta = 0$ and truncating.

At $\delta = 0$, the flow is already known on the unperturbed slow manifold $x' = 0$ and $z' = 0$, and every solution is a rest point. To obtain the first-order approximation in δ, expand the ODEs to first-order in δ and truncate at this order to obtain (with no surprise) the fast-time transformation of system (4.39):

$$\begin{aligned}\epsilon\frac{dx}{d\tau} &= \delta p(x, B(x,z,0), z) = \delta\rho\frac{\alpha - x}{\alpha + x}z + x - x^2,\\ \frac{dz}{d\tau} &= \delta r(x, B(x,z,0), z) = \delta(x - z).\end{aligned} \tag{4.50}$$

Thus, the system already investigated approximates (as it should) the full Oregonator on the slow manifold. The second-order approximation is

$$\epsilon\frac{dx}{d\tau} = \delta\rho\frac{\alpha - x}{\alpha + x}z + x - x^2 + \delta^2(\alpha - x)B_1(x,z),$$

$$\frac{dz}{d\tau} = \delta(x - z). \tag{4.51}$$

In principle, there is a function whose graph is exactly the slow manifold $\mathcal{M}_\delta$ and there is a system of differential equations for the exact flow on this manifold. The differential equations have the form

$$\begin{aligned} \epsilon \frac{dx}{d\tau} &= \delta\rho \frac{\alpha - x}{\alpha + x} z + x - x^2 + \delta^2 P(x, z, \delta) \\ \frac{dz}{d\tau} &= \delta(x - z), \end{aligned} \tag{4.52}$$

where P is a smooth function. Returning to the slow time, we have exactly the same system but with the right-hand sides of the ODEs divided by δ:

$$\begin{aligned} \epsilon \frac{dx}{ds} &= \rho \frac{\alpha - x}{\alpha + x} z + x - x^2 + \delta P(x, z, \delta) \\ \frac{dz}{ds} &= x - z. \end{aligned} \tag{4.53}$$

Using the implicit function theorem and these slow-time equations, it is easy to see that rest points for these equations with parameter value $\delta = 0$ persist in this family as rest points for members of the family with parameter values $|\delta| \neq 0$ and sufficiently small. Likewise but with more mathematics required, the full Hopf bifurcation scenario persists. In particular, limit cycles for the unperturbed system at $\delta = 0$ persist. Thus, it is possible to prove that the Orcgonator has periodic solutions; therefore, this model predicts the existence of oscillations in the concentrations of the chemical reactants in the BZ reaction. Fig. 4.7 depicts (numerical approximations of) concentration oscillations for physically realistic parameter values. The oscillations for small $|\delta|$ are very close to the oscillations obtained by perturbation theory at first-order in this parameter (compare Fig. 4.5).

Exercise 4.19. (a) Recreate Fig. 4.4 by solving explicitly for ϵ as a function of α and ρ, fixing $\alpha = 10^{-4}$ and graphing the resulting function. (b) Draw a graph of the corresponding surface over the α-ρ plane restricted to the regions where all three parameters are positive.

Exercise 4.20. (a) Use a computer to verify the validity of the claims made about the Hopf bifurcation for the two-dimensional system (4.39) as a curve in the parameter space crosses the Hopf curve depicted in Fig. 4.4. Take, for example, the horizontal line in the parameter space $\epsilon = 0.01$ and show that a limit cycle is born as ρ increases from 0 and the point $(\rho, 0.01)$ moves across the Hopf curve. Also show that the corresponding family of limit cycles dies as ρ is increased further and passes out of the bounded region. (b) Draw graphs of the amplitudes (with respect to variation in x) and periods of the limit cycles along the parameter curve of part (a). Describe the qualitative changes in these quantities. (c) Find the value of ρ such that the limit cycle corresponding to a parameter value $(\rho, 0.1)$ has maximum amplitude with respect to x.

Exercise 4.21. How many units of time t are required to reproduce the left-hand panel of Fig. 4.6 using the slow-time system (4.42)? Make the graph to verify your result.

Exercise 4.22. Determine the invariance relation, given by Eq. (4.49), by computing the normal to the surface and taking its inner product with the vector field corresponding to the fast system.

Exercise 4.23. (a) Suppose the initial concentration of species A is increased. How does the concentration oscillation period of the chemical species corresponding to x change as a prediction from the two-dimensional system (4.39)? Discuss. (b) Suppose the initial concentration of B is increased. How does the oscillation amplitude change as a prediction from the two-dimensional system (4.39)? Discuss.

Exercise 4.24. The Brusselator model for a hypothetical oscillating chemical reaction is

$$\begin{aligned} A &\to X, \\ B + X &\to Y + D, \\ 2X + Y &\to 3X, \\ X &\to E. \end{aligned}$$

Assume that the concentrations of A and B are constant. (a) Write the rate equations for the concentrations of the species X and Y. (b) Determine a change of variables (including time) for your model to obtain its dimensionless form

$$\begin{aligned} \dot{x} &= a - bx + x^2 y - x, \\ \dot{y} &= bx - x^2 y. \end{aligned}$$

(c) Determine the steady state(s) of the dimensionless system and their stability types. (d) Show that oscillations can occur for some parameter values. (e) Draw a bifurcation diagram for oscillatory solutions in the (a, b) parameter space.

Exercise 4.25. Consider the system

$$\begin{aligned} \dot{x} &= \alpha x - \beta y - \gamma x z + \epsilon x z^2, \\ \dot{y} &= \beta x + \alpha y - \gamma y z + \epsilon y z^2, \\ \epsilon \dot{z} &= -z + x^2 + y^2 + 2\epsilon(x^2 + y^2)(\alpha - \gamma(x^2 + y^2)), \end{aligned}$$

where α, $\beta > 0$, $\gamma > 0$ are parameters and ϵ is a small parameter. (a) Use singular perturbation theory to show that this family of ODEs has periodic solutions for some values of these parameters where $\epsilon \neq 0$. (b) The system has a slow manifold given as the graph of a function B defined on the (x, y) plane. It may be expressed as a series in ϵ of the form

$$B(x, y) = B_0(x, y) + B_1(x, y)\epsilon + B_2(x, y)\epsilon^2 + O(\epsilon^3).$$

Determine the coefficients B_0, B_1, and B_2. (c) Is it possible to determine a function whose graph is exactly the slow manifold for small $|\epsilon|$? (d) Prove that limit cycles in this family always lie in planes that are parallel to the (x, y) plane. How does the position of this plane depend on the parameters in the system? Find a formula for the limit cycles that depends on the parameter values.

CHAPTER 5

Reaction, Diffusion, and Convection

Conservation of mass is a fundamental physical law used to model many physical processes.

5.1 FUNDAMENTAL AND CONSTITUTIVE MODEL EQUATIONS

Suppose that some substance is distributed in $\mathbb{R}^n$ and let Ω denote a bounded region in $\mathbb{R}^n$ with boundary $\partial\Omega$ and *outer* normal η. The density of the substance is represented by a function $u : \mathbb{R}^n \times \mathbb{R} \to \mathbb{R}$, where $u(x,t)$ is the numerical value of the density in some units of measurement at the site with coordinate $x \in \mathbb{R}^n$ at time t. Usually $n \in \{1,2,3\}$, but the following model is valid for an arbitrary dimension. To avoid using the names for geometric objects in each dimension, the derivation is given with respect to the geometry of $n = 3$ where area and volume have their usual meanings. Also, recall from vector calculus the following concepts: gradient of a function (grad f or ∇f), divergence of a vector field (div X or $\nabla \cdot X$), and Laplacian of a function (Δf or $\nabla \cdot \nabla f$ or $\nabla^2 f$).

The time rate of change of the amount of the substance in Ω is given by the negative flux of the substance through the boundary of Ω plus the amount of the substance generated in Ω; that is,

$$\frac{d}{dt}\int_{\Omega} u \, d\mathcal{V} = -\int_{\partial\Omega} X \cdot \eta \, d\mathcal{S} + \int_{\Omega} f \, d\mathcal{V},$$

where X is the vector field on $\mathbb{R}^n$ (sometimes called the diffusion flux) of the substance pointing in the direction in which the substance is moving and with the magnitude of the amount of substance per area per time passing through the plane perpendicular to this direction; $d\mathcal{V}$ is the volume element; $d\mathcal{S}$ is the surface element; the vector field η is the outer unit normal field on the boundary of Ω; and f (a function of density, position, and time) is the amount of the substance generated in Ω per volume per time. The minus sign on the flux term is required because we are measuring the rate of change of the amount of substance *in* Ω. For example, when the flow is all out of Ω, the inner product $X \cdot \eta$ is not negative and the minus sign is required because the rate of change of the amount of substance in Ω is negative.

An Invitation to Applied Mathematics: Differential Equations, Modeling, and Computation.
http://dx.doi.org/10.1016/B978-0-12-804153-6.50005-1, Copyright © 2017 Elsevier Inc. All rights reserved.

Using the divergence theorem (also called Gauss's theorem) to rewrite the flux term and by interchanging the time derivative with the integral of the density, we have the relation

$$\int_\Omega u_t \, d\mathcal{V} = -\int_\Omega \operatorname{div} X \, d\mathcal{V} + \int_\Omega f \, d\mathcal{V}.$$

Moreover, because the region Ω is arbitrary in this integral identity, it follows that

$$u_t = -\operatorname{div} X + f. \tag{5.1}$$

To obtain a useful dynamical equation for u from Eq. (5.1), we need a constitutive relation between the density u of the substance and the flow field X. In most applications, it is not at all clear how to derive this relationship from the fundamental laws of physics. Thus, we have an excellent example of a class of important modeling problems where physical intuition must be used to propose a constitutive law whose validity can only be tested by comparing the results of experiments with the predictions of the corresponding model. Problems of this type lie at the heart of applied mathematics and physics.

For Eq. (5.1), the classic constitutive relation—called Darcy's, Fick's, or Fourier's law depending on the physical context—is

$$X = -K \operatorname{grad} u + \mu V \tag{5.2}$$

where $K \geq 0$ and μ are functions of density, position, and time; and V denotes the flow field for the medium in which our substance is moving. The minus sign on the gradient term represents the assumption that the substance diffuses from higher to lower concentrations.

By inserting the relation (5.2) into the balance law [Eq. (5.1)], we obtain the dynamical equation

$$u_t = \operatorname{div}(K \operatorname{grad} u) - \operatorname{div}(\mu V) + f. \tag{5.3}$$

Also, by assuming that the diffusion coefficient K is equal to k^2 for some constant k, the function μ is given by $\mu(u, x, t) = \gamma u$ where $0 \leq \gamma \leq 1$ is a constant that determines the amount of u that moves with the velocity field V, and V is an incompressible vector field ($\operatorname{div} V = 0$) such as the velocity field for the motion of water, we obtain the most often used reaction-diffusion-convection model equation

$$u_t + \gamma \operatorname{grad} u \cdot V = k^2 \Delta u + f. \tag{5.4}$$

The quantity $\gamma \operatorname{grad} u \cdot V$ is called the *convection term,* $k^2 \Delta u$ is called the *diffusion term,* and f is the *source term.*

In case no diffusion in involved in the motion of the substance ($K = 0$ in Eq. (5.3)), all of the substance moves with the velocity field V (that is, $\gamma = 1$), the source function f vanishes, and the velocity V is not necessarily incompressible, the balance law reduces to the differential form of the law of conservation of mass, also called the *continuity equation*, given by

$$u_t + \operatorname{div}(uV) = 0. \tag{5.5}$$

Because Eq. (5.4) is derived from physical laws and generally accepted constitutive laws, this partial differential equation (PDE) is used to model many physical processes where reaction, diffusion, or convection is involved.

In case the substrate medium is stationary (that is, $V = 0$), the model [Eq. (5.4)] is the diffusion (or heat) equation with a source

$$u_t = k^2 \Delta u + f. \tag{5.6}$$

This equation, where u is interpreted as temperature, is the standard model for heat flow. In this application, the conserved substance may be viewed as the average kinetic energy of the particles under consideration (molecules in a metal, molecules in a gas, electrons, and so on). Temperature is a measure of the density of this kinetic energy. A deeper understanding of heat and temperature requires a study of thermodynamics. Here, an intuitive understanding of these concepts is sufficient to apply the model equation for temperature to many heat flow scenarios.

An understanding of the derivation of the model equation [Eq. (5.4)] should also explain the widespread appearance of the Laplacian in applied mathematics: It is the divergence of the gradient vector field whose flow is supposed to carry some substance from regions of its higher concentration toward regions of lower concentration.

Exercise 5.1. Show that the gradient of a function evaluated at a point p points in the direction of maximum increase of the function at p. Also, the gradient is orthogonal to each level set of the function at each point on a level set.

Exercise 5.2. Discuss the meaning of the divergence of a vector field. In particular, discuss positive divergence, negative divergence, and zero divergence. Give examples. Hint: You may wish to consider the equation

$$\operatorname{div} X(p) = \lim_{\Omega \to \{p\}} \frac{1}{\operatorname{vol}(\Omega)} \int_{\partial\Omega} X \cdot \eta \, dS,$$

where the limit is taken over all bounded open sets with smooth boundaries that contain the point p whose diameters shrink to zero.

5.2 REACTION-DIFFUSION IN ONE SPATIAL DIMENSION: HEAT, GENETIC MUTATIONS, AND TRAVELING WAVES

5.2.1 One-Dimensional Diffusion

Imagine the diffusion of heat in an insulted bar with insulated ends (that is, zero heat flux through the surface of the bar) under the further assumptions that the temperature is the same over each cross section perpendicular to the bar's axis and there are no heat sources or sinks along the bar. Given the initial temperature distribution along the bar, the basic problem is to determine the temperature at each point of the bar as time increases.

Under the assumptions, we need only consider the spatial distribution of heat along the axis of the bar that we idealize as an interval of real numbers. For a bar of length L, we thus let x denote the spatial coordinate in the open interval $(0, L)$. The model equation [Eq. (5.6)] is

$$u_t = \kappa u_{xx}, \tag{5.7}$$

where $u(x, t)$ is the temperature at position x along the bar at time $t \geq 0$ and $\kappa > 0$ is a constant (called the diffusivity) that depends on the material used to construct the bar. The value of κ must be determined by experiment. (How would you set up and conduct such an experiment?) Eq. (5.7) is often called the heat equation.

A function $f : (0, L) \to \mathbb{R}$ representing the initial temperature along the bar gives the *initial condition*

$$u(x, 0) = f(x). \tag{5.8}$$

Zero flux conditions at each end of the bar provide the *boundary conditions*

$$u_x(0,t) = 0, \qquad u_x(L,t) = 0. \tag{5.9}$$

These boundary conditions are also called the zero Neumann boundary conditions.

The problem is to find a function u that satisfies the heat equation [Eq. (5.7)], the initial condition [Eq. (5.8)], and the boundary conditions [Eqs. (5.9)]. This is a classic problem first solved by Joseph Fourier in 1822. His basic ideas, which have far-reaching consequences, are milestones in the history of science. Fourier's law of heat conduction—heat flows from regions of high temperature to regions of lower temperature—is used in the derivation of the PDE (5.7). His mathematical solution of the heat equation with initial and boundary conditions introduced Fourier series, one of the main tools of mathematical analysis.

A fundamental technique that often works to solve linear PDEs with rectangular spatial domains is separation of variables: Look for solutions of the form $u(x,t) = X(x)T(t)$, where X and T are unknown functions. Inserting this guess into the heat equation [Eq. (5.7)] yields the formula

$$X(x)T'(t) = \kappa X''(x)T(t),$$

which must hold if $(x,t) \mapsto X(x)T(t)$ is a solution. Assume for the moment that $X(x)$ and $T(t)$ do not vanish and rearrange the last formula to the form

$$\frac{T'(t)}{\kappa T(t)} = \frac{X''(x)}{X(x)}.$$

If $X(x)T(t)$ is a solution that does not vanish, then the left-hand side of the equation is a function of t alone and the right-hand side a function of x alone. It follows that the left-hand side and the right-hand side of the equation are equal to the same constant c. In other words, there must be a constant c such that

$$T'(t) = c\kappa T(t), \quad X''(x) = cX(x);$$

that is, X and T must be solutions of the given ordinary differential equations (ODEs). Whether or not solutions X and T of these ODEs vanish at some points, u defined by $u(x,t) = X(x)T(t)$ satisfies PDE (5.7).

The ODE $X''(x) = cX(x)$ is easily solved for each of the usual cases: $c > 0$, $c = 0$, and $c < 0$. Indeed, an essential requirement of the method is to find *all* solutions of this ODE that satisfy the boundary conditions. For $c = \lambda^2 > 0$, the general solution (expressed with arbitrary constants a and b) is

$$X(x) = ae^{\lambda x} + be^{-\lambda x};$$

for $c = 0$,

$$X(x) = ax + b;$$

and, for $c = -\lambda^2 < 0$,

$$X(x) = a\cos\lambda x + b\sin\lambda x.$$

Solutions of the PDE are required to satisfy the boundary conditions. This is possible only for $c = 0$ and $X(x) = b$ or $c < 0$ and

$$X(x) = a\cos\frac{n\pi}{L}x,$$

where n is an integer. As cosine is an even function, all of the solutions for $c < 0$ are obtained with n ranging over the nonnegative integers.

For $c = -\lambda^2 = -\left(\frac{n\pi}{L}\right)^2$, the corresponding solution of $T'(t) = c\kappa T(t)$ is

$$T(t) = e^{-(\kappa n^2\pi^2/L^2)t}.$$

An infinite number of solutions have been constructed:

$$u_n(x,t) = X_n(x)T_n(t) := e^{-(\kappa n^2\pi^2/L^2)t}\cos\frac{n\pi}{L}x, \qquad n = 0, 1, 2, \dots, \infty.$$

The principle of superposition for the PDE with the zero Neumann boundary conditions is valid and easy to prove. It states that if u and v are solutions of the PDE that also satisfy the zero Neumann boundary conditions, then so is every linear combination $au + bv$ of these solutions, where a and b are scalars. As a corollary, every finite sum

$$u(x,t) = b_0 + \sum_{n=1}^{N} b_n e^{-(\kappa n^2\pi^2/L^2)t}\cos\frac{n\pi}{L}x \tag{5.10}$$

is a solution of the PDE and the boundary conditions. Warning: The superposition principle is not valid for nonzero Neumann boundary conditions (for instance, $u_x(0,t) = a$ and $u_x(L,t) = b$ where a and b are constants and at least one of them is not zero).

What about the initial condition? One fact is clear: If

$$f(x) = b_0 + \sum_{n=1}^{N} b_n \cos \frac{n\pi}{L} x$$

for some choice of $b_0, b_1, b_2, \ldots, b_N$, then the function u in Eq. (5.10) is a solution of the PDE, boundary conditions, and initial conditions. This result suggests the question: Which functions f can be written as a sum of cosines? The surprising answer, first given by Fourier, is that most functions defined on the interval $[0, L]$ can be written as an infinite sum of cosines (or sines). We will discuss this result in more detail in subsequent sections. A more precise (but not the most general) fact is that *every piecewise continuously differentiable function f defined on $[0, L]$, with at most a finite number of jump discontinuities, can be represented by a (pointwise) convergent Fourier cosine series*; that is,

$$f(x) = b_0 + \sum_{n=1}^{\infty} b_n \cos \frac{n\pi}{L} x. \tag{5.11}$$

If in addition f is continuous (no jump discontinuities), then the partial sums of the Fourier series converge uniformly to f. This is a powerful result. By applying it, we know that the PDE model with zero Neumann boundary conditions has a solution for all initial conditions that are likely to be encountered. In fact, once the initial condition f is expressed as a Fourier series, the function u defined by

$$u(x,t) = b_0 + \sum_{n=1}^{\infty} b_n e^{-(\kappa n^2 \pi^2/L^2)t} \cos \frac{n\pi}{L} x \tag{5.12}$$

is a solution of our PDE that satisfies the boundary and initial conditions.

It turns out that there is a simple method to determine the Fourier coefficients b_n of a function f defined on $[0, L]$. Using the convergence theorem and assuming for simplicity that f is continuously differentiable, this function may be expressed as in Eq. (5.11) and the sum on the right-hand side of the equation may be integrated term-by-term. Thus,

$$\int_0^L f(x)\,dx = \int_0^L b_0\,dx + \sum_{n=1}^{\infty} \int_0^L b_n \cos\frac{n\pi}{L}x\,dx.$$

Every integral in the infinite summation vanishes. Hence,

$$b_0 = \frac{1}{L}\int_0^L f(x)\,dx.$$

For each positive integer m, we have that

$$\int_0^L f(x)\cos\frac{m\pi}{L}x\,dx = \int_0^L b_0\cos\frac{m\pi}{L}x\,dx + \sum_{n=1}^{\infty}\int_0^L b_n\cos\frac{n\pi}{L}x\cos\frac{m\pi}{L}x\,dx.$$

As before, the first integral on the right-hand side of the equation vanishes. The integrals of products of cosines behave in the best possible way: they all vanish except for the product where $n = m$. (Check this statement carefully; it is a basic result that makes Fourier series useful.) By an application of this fact,

$$\int_0^L f(x)\cos\frac{m\pi}{L}x\,dx = \int_0^L b_m\cos^2\frac{m\pi}{L}x\,dx = b_m\frac{L}{2},$$

and we have the general formula

$$b_m = \frac{2}{L}\int_0^L f(x)\cos\frac{m\pi}{L}x\,dx.$$

All the Fourier coefficients may be computed simply by integrating the product of the given function and an appropriate cosine. A similar result holds for Fourier sine series. Indeed, a function f in the same class of functions may be represented as the Fourier sine series

$$f(x) = \sum_{n=1}^{\infty} a_n \sin\frac{n\pi}{L}x.$$

A solution for the heat equation with zero Neumann boundary conditions and arbitrary piecewise continuously differentiable initial value has been constructed. This is a wonderful result, but using the PDE with the given boundary and initial data as a model would be useless if there were other solutions not detected by the construction of the solution using separation of variables and Fourier series. Which solution would we choose? To be predictive, a proposed model should have a *unique* solution. Suppose there were two solutions u and v in the class of piecewise continuously differentiable functions. By superposition, $w := u - v$ is also a solution of the same boundary value problem (BVP) but with zero initial condition; that is,

$$w_t = \kappa w_{xx}, \quad w(x,0) = 0, \quad w_x(0,t) = 0, \quad w_x(L,t) = 0.$$

Note that

$$\int_0^L w_t w \, dx = \int_0^L \kappa w_{xx} w \, dx.$$

By an application of integration by parts on the right-hand integral and taking the time derivative outside the left-hand integral, we have the equality

$$\frac{d}{dt} \int_0^L \frac{1}{2} w^2 \, dx = \kappa w w_x \Big|_0^L - \kappa \int_0^L w_x^2 \, dx.$$

Using the boundary conditions,

$$\frac{d}{dt} \int_0^L \frac{1}{2} w^2 \, dx = -\kappa \int_0^L w_x^2 \, dx \leq 0.$$

It follows that the function

$$t \mapsto \int_0^L \frac{1}{2} w^2 \, dx$$

is nonnegative and it does not increase as t increases. But, at $t = 0$, the initial condition is $w = 0$. Hence, w must be the zero function for all $t > 0$ for which the solution exists; therefore, $u = v$ as desired. The proof method used here is called the energy method.

The reader may ask: Could a solution exist that is not piecewise continuously differentiable? If so, then perhaps solutions of the model problem are not unique. A full answer to the question goes beyond the mathematics developed here. At least the assumptions used in the proof of uniqueness are reasonable for the basic heat flow model. But, the desire to answer such questions is a very good reason to pursue more advanced mathematics.

The solution of our diffusion model equation can be used to answer questions and make predictions. For example, the solution [Eq. (5.12)] predicts that the temperature distribution on the rod will go to a constant

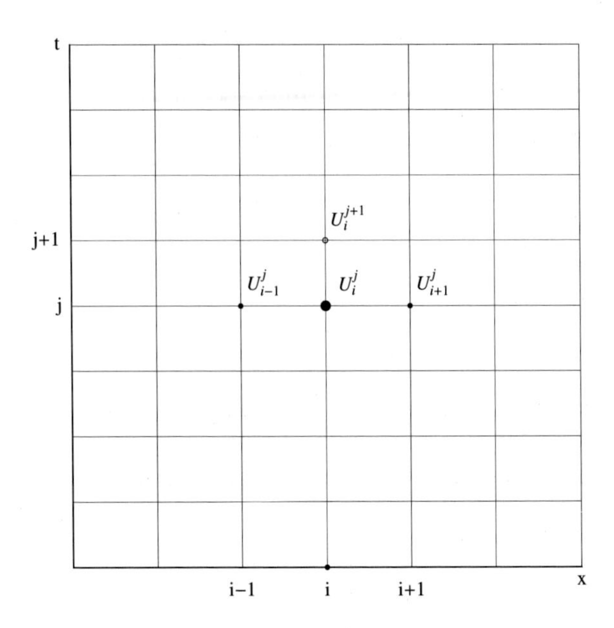

Fig. 5.1 A schematic discretization of space and time is depicted for numerical solutions of the diffusion equation. The discrete solution values are $U_i^j := u(i\Delta x, j\Delta t)$. *The unknown interior value* U_i^{j+1} *is computed using the previously computed values* U_{i-1}^j, U_i^j, *and* U_{i+1}^j.

steady state as time increases and this steady state is the average value of the initial heat distribution.

In case heat sources and sinks are included in the model or other physical phenomena are taken into account, especially those that produce nonlinear terms, Fourier solution methods will likely fail. In fact, there are no methods that will produce exact solutions for all such models. Thus, to make predictions from PDE model equations, approximation methods are often used. The most important approximation methods are called numerical methods; they use arithmetic to approximate solutions of partial differential equations.

New methods are best applied to problems where the exact answer is known. Here, the diffusion model serves as a simple example to illustrate a numerical method, called a finite difference method, that is often used for approximating solutions of PDEs.

To make a numerical approximation of a solution of the heat equation, the diffusivity k, the length of the spatial domain L, and the initial data f must be known. The first step of a finite difference method is to discretize space and time. This may be done by choosing a positive integer M with corresponding spatial increment $\Delta x := L/M$, a temporal increment Δt, and by agreeing to consider the value of u only at the interior spatial domain points with coordinates $x_i := i\Delta x = iL/M$ for $i = 1, 2 \ldots M-1$ and temporal coordinates $j\Delta t$ for $j = 0, 1, 2, \ldots$, where we leave unspecified the (finite) number of time steps that might be computed. In other words, the basic idea is to determine approximate values of u at the interior gridpoints $U_i^j = u(i\Delta x, j\Delta t)$ as depicted in Fig. 5.1.

Having discretized space and time, the second step is to discretize the time and space derivatives that appear in the PDE. For this, we use Taylor's theorem. Recall that if a function f is $(n+1)$ times continuously differentiable at a point a, then the value of f at a nearby point x is given by

$$\begin{aligned} f(x) = f(a) + f'(a)(x-a) + \frac{1}{2!}f''(a)(x-a)^2 + \frac{1}{3!}f''(a)(x-a)^3 \\ + \cdots + \frac{1}{n!}f^n(a)(x-a)^n + \frac{1}{(n+1)!}f^{n+1}(c)(x-a)^{n+1} \end{aligned}$$

for some number c that lies between a and x. We often say that the right-hand side of the last formula is the Taylor expansion of the function f at a.

Consider a point (x, t) and the nearby point $(x, t+\Delta t)$. By an application of Taylor's formula to the function $\Delta t \mapsto u(x, t+\Delta t)$ at $\Delta t = 0$, we have the expansion

$$u(x, t+\Delta t) = u(x,t) + u_t(x,t)\Delta t + \frac{1}{2!}u_{tt}(x,c)\Delta t^2.$$

A rearrangement yields the equation

$$u_t(x,t) = \frac{u(x,t+\Delta t) - u(x,t)}{\Delta t} - \frac{1}{2!}u_{tt}(x,c)\Delta t.$$

and the approximation

$$u_t(x,t) \approx \frac{u(x,t+\Delta t) - u(x,t)}{\Delta t}$$

with an error of order Δt. The left-hand side of our PDE may be approximated by

$$u_t(i\Delta x, j\Delta t) \approx \frac{U_i^{j+1} - U_i^j}{\Delta t}. \tag{5.13}$$

Using a similar procedure,

$$u(x-\Delta x,t)=u(x,t)-u_x(x,t)\Delta x+\frac{1}{2}u_{xx}(x,t)\Delta x^2-\frac{1}{3!}u_{xxx}(x,t)\Delta x^3,$$
$$u(x+\Delta x,t)=u(x,t)+u_x(x,t)\Delta x+\frac{1}{2}u_{xx}(x,t)\Delta x^2+\frac{1}{3!}u_{xxx}(x,t)\Delta x^3,$$

up to an error of fourth-order in Δx. These approximations may be added and rearranged to obtain the formula

$$u_{xx}(x,t)\approx\frac{u(x-\Delta x,t)-2u(x,t)+u(x+\Delta x,t)}{\Delta x^2}$$

with an error of order Δx^2, which is one order more accurate than the approximation chosen for the time derivative. This discrepancy will be addressed further on; finite difference methods that are second order in both space and time are certainly desirable and will be constructed. Here, the right-hand side of our PDE is approximated by

$$u_{xx}(i\Delta x,j\Delta t)\approx\frac{U_{i-1}^j-2U_i^j+U_{i+1}^j}{\Delta x^2}. \tag{5.14}$$

Equating the last two approximations [Eqs. (5.13) and (5.14)] and rearranging the result, a discrete approximation of the PDE is

$$U_i^{j+1}=U_i^j+\frac{\kappa\Delta t}{\Delta x^2}(U_{i-1}^j-2U_i^j+U_{i+1}^j). \tag{5.15}$$

The left-hand side of Eq. (5.15) is the value of u at the $(j+1)$st time step; all values of u on the right-hand side are evaluated at the jth time step. As the values of u at the zeroth time step ($j=0$) are given by the U_i^0 via the initial condition, the approximate values of u at the first time step ($j=1$) are given by U_i^1 at the nodes on the grid by Eq. (5.15). The values U_i^{j+1}, corresponding to the time step $(j+1)$, are determined using the previously computed values at the jth time step. Well, almost.... There is a problem: The values of U_i^0 for $i=0,1,2,3\ldots M$ are given by the initial data; thus, all is well when computing U_i^1 at the interior nodes ($i=1,2,3,\ldots,M-1$). But, the boundary values U_0^1 at the left end and U_M^1 at the right end, which will be needed in the next time step, must be determined from the boundary conditions. (The subject of PDEs would be simple absent boundary conditions, but there is no reprieve; boundary conditions are essential in modeling and in numerical approximations. Be careful.)

For an approximation of the zero Neumann boundary conditions (which require the partial derivative u_x to vanish at each end of the computational domain), one possibility is to insist that for all time steps

$$U_0^j = U_1^j, \qquad U_M^j = U_{M-1}^j. \tag{5.16}$$

These conditions impose an approximation of the zero first derivatives at the ends of the bar, whose spatial coordinates are x_0 and x_M, by viewing the approximation of u to be constant over the intervals $[x_0, x_1]$ and $[x_{M-1}, x_M]$. This is not the only possible approximation of the zero Neumann boundary conditions, but this approximation is consistent with the desired condition and has the virtue of being simple to implement in computer code. For the case $j = 1$, simply set U_0^1 equal to the already computed (interior) value U_1^1 and set U_M^1 equal to the already computed interior value U_{M-1}^1. The same procedure is used for all subsequent time steps.

All the ingredients are now in place to approximate solutions of the heat equation [Eq. (5.7)] with Neumann boundary conditions [Eqs. (5.9)] and the initial condition [Eq. (5.8)]. Determine the initial data U_i^0, for $i = 1, 2, 3, \ldots, M-1$, from the initial condition. Set $j = 1$ and compute U_i^1 for $i = 1, 2, 3, \ldots, M-1$ using Eq. (5.15) and impose the end conditions [Eqs. (5.16)]. Repeat the process to compute U_i^2 using the previously determined values U_i^1, and continue to compute in turn U_i^{j+1}, for $j \geq 2$, over $i = 1, 2, 3, \ldots, M-1$ using the previously computed values U_i^j. The process is stopped when j reaches some preassigned integer value N. The size of the increment Δt can be adjusted so that after a finite number of steps $j = 1, 2, 3, \ldots, N$, the time $T = N\Delta t$ is equal to a preassigned final value.

The numerical scheme may also be viewed in vector form. Define the $(M-1)$ vector W^j to be the transpose of the row vector $(U_1^j, U_2^j, U_3^j, \ldots, U_{M-1}^j)$, let $\alpha := \kappa\Delta t/\Delta x^2$, and define the $(M-1) \times (M-1)$ matrix A whose main diagonal is $(1-\alpha, 1-2\alpha, 1-2\alpha, 1-2\alpha, \ldots, 1-2\alpha, 1-\alpha)$ (that is, the first and last components are $1-\alpha$ and the other components are all $1-2\alpha$), whose first superdiagonal and first subdiagonal elements are all α, and all remaining elements are set to zero.

In case $M = 5$, the matrix is

$$A = \begin{pmatrix} 1-\alpha & \alpha & 0 & 0 \\ \alpha & 1-2\alpha & \alpha & 0 \\ 0 & \alpha & 1-2\alpha & \alpha \\ 0 & 0 & \alpha & 1-\alpha \end{pmatrix}.$$

The iteration scheme [Eq. (5.15)] (including the boundary conditions) takes the vector form

$$W^{j+1} = AW^j.$$

In other words, the iteration scheme is simply matrix multiplication by A. The special first and last rows of A are due to the Neumann boundary conditions.

The initial vector W^0 is determined by the initial condition for the PDE. Subsequent iterates are $W^1 = AW^0$, $W^2 = AW^1$, and so on. Or, in a more compact form,

$$W^{j+1} = A^j W^0,$$

where A^j denotes the jth power of the matrix A and W^j is the jth element in the sequence of iterates whose first three elements are W^1, W^2, and W^3. This type of iteration scheme, iterating a function (which is a linear transformation here), is ubiquitous in applied mathematics. Learning the corresponding theory is certainly worthwhile.

There are at least two reasons to expect difficulties: (1) The discrete first-order approximation of the time derivative u_t is less accurate than the second-order approximation of the spatial derivative u_{xx}; and (2) perhaps the approximate values U_i^j begin to grow or oscillate due to discretization errors and thus the approximation does not remain close to the solution of the continuous model.

Alternative numerical algorithms that overcome difficulty (1) are available. In fact, a viable method that makes the time discretization second-order is discussed in Section 5.5.4.

Numerical instability will occur (see Exercise 5.9) unless the space and time increments chosen to discretize the model satisfy the Courant–

Friedrichs–Lewy (CFL) condition

$$\kappa \frac{\Delta t}{\Delta x^2} \leq \frac{1}{2}, \tag{5.17}$$

a requirement that is revisited and explained more fully in Section 5.4.4. In practice, the CFL condition determines the maximum allowable time-step size for our numerical method after a spatial discretization is set.

To appreciate the CFL condition for numerical approximations of the diffusion model, suppose that a roundoff error is introduced in the computation at the first time step. Instead of computing the exact value W^1 from W^0, the machine computes $W^1 + \epsilon$ (where ϵ is an $(M - 1)$ vector representing the error). Of course, further errors might be introduced at subsequent steps. But, for simplicity, consider only the propagation of the first error, and assume that the computed results are exact after the first error occurs. Under these assumptions, the computed values are

$$W^2 = AW^1 + A\epsilon, \quad W^3 = A^2W^1 + A^2\epsilon, \quad W^4 = A^3W^1 + A^3\epsilon, \ldots.$$

Note that the vector $A^j\epsilon$ represents the error at each step. The algorithm will produce useless results if the norm (which may be taken to be the Euclidean length) of the vector $A^j\epsilon$ grows as j increases.

What happens in case the matrix A were the diagonal 2×2 matrix

$$A = \begin{pmatrix} a & 0 \\ 0 & b \end{pmatrix}?$$

Clearly

$$A^j = \begin{pmatrix} a^j & 0 \\ 0 & b^j \end{pmatrix}.$$

If either $|a| > 1$ or $|b| > 1$ and there is a corresponding nonzero element of the vector ϵ, then the size of the propagated error will grow. For example, if $|b| > 1$ and ϵ is the transpose of the vector $(0.01, 0.035)$ the number $b^j 0.035$ will grow to infinity as j goes to infinity. In this case, roundoff errors are amplified under iteration and the numerical approximation of the PDE becomes increasingly less accurate as the number of time steps increases. On the other hand, if both $|a|$ or $|b|$ are less than or equal to 1, then the propagated error will remain bounded as j goes to infinity.

The matrix A that appears in our numerical scheme is not diagonal. But, it has a special form: A is symmetric; that is, A is equal to its transpose. *Every symmetric matrix is diagonalizable; in other words, if A is symmetric, then there is an invertible matrix B such that $B^{-1}AB$ is diagonal. Also, every eigenvalue of a symmetric matrix is real.* Using these facts, let us suppose that every eigenvalue of our matrix A lies in the closed interval $[-1, 1]$. The matrix $C := B^{-1}AB$ has the same eigenvalues as A. (Why?) Iterations of a vector v by C remain bounded because C is diagonal and all its eigenvalues have absolute value less than or equal to 1. In fact,

$$|C^j v| \le |v|$$

for all vectors v and all positive nonzero integers j. Iteration of v by the matrix A also remains bounded because

$$|A^j v| = |(BCB^{-1})^j v| \le |BC^j B^{-1} v| \le \|B\||B^{-1}v| \le \|B\|\|B^{-1}\||v|.$$

The CFL condition [Eq. (5.17)] implies that all eigenvalues of A are in the closed interval $[-1, 1]$ (see Section 5.4.4).

The heat equation is well established as a model and is widely used. But, as for all mathematical models that rely on constitutive laws, it is not in complete agreement with nature. To reveal a basic flaw, reconsider the Fourier series solution [Eq. (5.12)]

$$u(x,t) := b_0 + \sum_{n=1}^{\infty} b_n e^{-(\kappa n^2\pi^2/L^2)t} \cos\frac{n\pi}{L}x$$

of this model [Eq. (5.7)]. The Fourier coefficients are determined by the initial data. Imagine the initial temperature is zero everywhere along the bar except near some place $x = \xi$ along the rod. In fact, the temperature far away from this point will remain zero for some finite time interval: the effects of the motions of the molecules near ξ cannot be transferred instantly to some distant location. In more prosaic language, there is no action at a distance. The model does not agree with this physical reality. Indeed, every Fourier coefficients $b_n e^{-(\kappa n^2\pi^2/L^2)t}$ in the series representation of u is changed from its initial value by a different exponential factor for arbitrarily small $t > 0$. At some prespecified distance along the rod from ξ and for arbitrarily small $t > 0$, the model predicts nonzero temperatures at points at least this distance from ξ. The influence of the nonzero temperature at ξ is predicted to be felt instantly at every point along the rod. Of course, the predicted influence is so small in practical situations that it is not measurable. In most practical applications, predicted temperatures agree with measured temperatures. Thus, although the heat equation model predicts a violation of a physical law at a fundamental level, it is an excellent model for applications to heat transfer phenomena. The violation is due to Fourier's constitutive law $X = -K \operatorname{grad} u$, not the conservation of mass.

Exercise 5.3. (1) Write the Fourier sine series for the function $f(x) \equiv 1$ on the interval $[0, 2]$ and make graphs of a few of the partial sums to indicate its convergence. (2) Repeat the exercise for the Fourier cosine series. (3) Write the Fourier sine series for the unit step function that is zero on the interval $[0, 1)$ and one on the interval $[1, 2]$. (3) Show that the series evaluated at $x = 1$ converges to $1/2$. Also, draw graphs to indicate the convergence of the partial sums. Describe the behavior of the partial sums on a small interval $[1, 1 + \epsilon)$. The (perhaps strange) behavior is called the Gibbs phenomenon.

Exercise 5.4. Let h be a smooth function defined on the interval $[0, L]$ and consider the function f given by

$$f(x) = \sum_{n=1}^{N} a_n \sin(\frac{2\pi n}{L} x).$$

Define the error in approximating h by f to be

$$\Lambda = \int_0^L |f(x) - h(x)|^2 \, dx.$$

(1) Find the numbers $\{a_1, a_2, a_3, \ldots, a_N\}$ that minimizes the error. Compare with the Fourier coefficients. (2) Approximate the least error in case $N = 4$, $L = 2$, and $h(x) = 1$. (3) What is the minimum N so that the error is less that 0.01?

Exercise 5.5. (a) Solve the diffusion equation on the spatial domain $[0, L]$ with initial condition and zero Dirichlet boundary conditions:

$$u_t = \kappa u_{xx}, \quad u(x, 0) = f(x), \quad u(0, t) = 0, \quad u(L, t) = 0.$$

(b) Show that your solution is unique.

Exercise 5.6. Solve the diffusion equation on the spatial domain $[0, L]$ with initial condition and nonzero Neumann boundary conditions:

$$u_x(0, t) = a, \qquad u_x(L, t) = b,$$

where a and b are real numbers. Hint: Look for a solution $u = v + w$, where v is a function that satisfies the boundary conditions and w satisfies the PDE with zero boundary conditions.

Exercise 5.7. Solve the diffusion equation on the spatial domain $[0, L]$ with initial condition and nonzero Dirichlet boundary conditions:

$$u(0, t) = a, \qquad u(L, t) = b,$$

where a and b are real numbers.

Exercise 5.8. The zero flux boundary condition for the diffusion equation has the physical interpretation that no substance is lost as time increases. Prove this fact by showing that the time derivative of the total amount of substance (its integral over the spatial domain) vanishes.

Exercise 5.9. (a) Write computer code to implement the numerical method described in this section to approximate the solution of the diffusion equation in one space dimension on a finite interval with zero Neumann boundary conditions and given initial condition. As a test case, consider the spatial domain to be one unit in length, the diffusivity $\kappa = 1$, and the initial data given by $f(x) = 1 + \cos \pi x$. Compare your numerical results with the analytic solution. (b) Test the CFL number with at least two discretizations, one such that $\Delta t/\Delta x^2 = 0.4$ and the other with $\Delta t/\Delta x^2 = 0.6$. Discuss your results.

Exercise 5.10. (a) Modify your code written for Exercise 5.9 to approximate the solution of the diffusion equation in one space dimension on a finite interval with mixed boundary conditions and given initial condition. As a test case, consider the spatial domain to be one unit in length, the diffusivity $\kappa = 10^{-3}$, the initial data given by $f(x) = 1$, and the boundary conditions

$$u_x(0,t) = 0.01, \qquad u(1,t) = 1.$$

Draw a graph of the initial heat profile, its steady state profile, and five profiles at equally spaced times between these extremes. (b) Determine the value of u at the midpoint of the space interval (that is, at $x = 0.5$ at $t = 10$). The engineers who need this result want it correct to 4 decimal places. Can you assure them that your value meets this requirement? (c) Draw a graph of the function $t \mapsto u(0.5,t)$ on the time interval $0 \le t \le 20$.

Exercise 5.11. [Ink Diffusion] The following is a constitutive model for the ink diffusion experiment in Section 1.3: Let $u(x,t)$ denote the concentration of ink at position x at time t and L the trough length. Also, let a and b be positive constants. The model equation of motion for u is the initial BVP

$$u_t = ku_{xx}, \qquad u_x(0,t) = u_x(L,t) = 0, \quad u(x,0) = \begin{cases} a, & 0 \le x \le b, \\ 0, & x > b. \end{cases}$$

(a) What constitutive law is used to construct this model? (b) Show that the model predicts the presence of ink at every position along the trough for every positive time. Does this fact invalidate the model? (c) Show that the model predicts that the total amount of ink remains constant in time. (d) Define the diffusion front to be the largest distance from the origin where the ink concentration is 1% of a. Use the model to determine the diffusion front. If necessary, choose values for the parameters in the problem. (e) Is the distance of the diffusion front from the origin a linear function of time? If not, what type is this function? (f) Calibrate the model to the data given in Section 1.3 and discuss the model prediction in view of the experimental data. (g) Construct a model that takes into account two (or three) space dimensions. Compare the new front speed with the front speed obtained for the one-dimensional model. (h) Can you refine the model to give a more accurate representation of the experiment? You may wish to perform your own experiment. Note: Perhaps diffusion is not the dominant mechanism that causes the ink to disperse. The water was left undisturbed for a long period of time to minimize residual fluid motions, but perhaps the effects of temperature cause convection currents that drive the ink movement. Could the proposed model be based on incorrect physics for the process under consideration? Discuss.

Exercise 5.12. Imagine a chemical element that dissolves in water and a mixture of 50 grams per liter of this element and water poured into the bottom of a round cylindrical flask. Pure water is added on top of the fluid mixture in a very careful manner so that no stirring takes place. The pure water has a depth of 10 centimeters above the surface Σ of the mixture. The flask has a small hole located exactly 2 centimeters above Σ that allows small samples to be extracted and tested for their chemical concentrations every 24 hours for 10 days. Here are the experimental results given in the form (days, grams/liter of the chemical element):

days	*concentration*	*days*	*concentration*
1	*0.0*	*6*	*2.598*
2	*0.0*	*7*	*3.806*
3	*0.096*	*8*	*5.031*
4	*0.594*	*9*	*6.23*
5	*1.484*	*10*	*7.382*

What is the diffusion coefficient for this chemical element in water?

Exercise 5.13. (1) Consider a manufactured rod with uniform cross sections. The rod is 1 meter long and the thermal diffusivity of the construction material has been previously measured to be 10^{-4} square meters per second. Also, the rod is in an insulating jacket so that no heat can escape to the ambient environment. The initial temperatures along the rod, measured in degrees celsius at 21 positions along the rod, are given in the following table where the first coordinate is the position along the rod measured in meters and the second is the corresponding measured celsius temperature.

position	*temperature*	*position*	*temperature*
0.00	*200.00*	*0.55*	*254.37*
0.05	*200.25*	*0.60*	*264.33*
0.10	*201.00*	*0.65*	*274.29*
0.15	*202.25*	*0.70*	*284.31*
0.20	*204.00*	*0.75*	*294.00*
0.25	*206.27*	*0.80*	*279.50*
0.30	*209.00*	*0.85*	*264.18*
0.35	*214.40*	*0.90*	*249.32*
0.40	*224.21*	*0.95*	*234.49*
0.45	*234.38*	*1.00*	*219.05*
0.50	*244.88*		

Heat sources are removed and the heat in the rod is allowed to diffuse. Determine a good approximation for the temperature of the rod at its midpoint 10 minutes later. Consider a numerical approximation and an approximation using Fourier series. Compare your results. (2) Assume the same conditions and data as in part (1) except that the rod is no longer completely insulated; only its ends are insulated. Heat is transferred from positions along the rod to the environment such that the rate of change of temperature is proportional to their temperature differences with the constant ambient temperature of 20 degrees celsius. The constant of proportionality is 0.05 per minute. Determine a good approximation for the temperature of the rod at its midpoint 10 minutes later. (3) An early version of part (2) stated a constant of proportionality of 0.5 per minute. A fall of 1/2 a degree per minute does not seem unreasonable. But, this reasoning is not a correct interpretation of the influence often constant of proportionality. What is a correct interpretation?

Exercise 5.14. [Oscillations Carried by Diffusion] Consider the diffusion model defined on the interval $[0, 1]$:

$$u_t = k^2 u_{xx}, \qquad u(0,t) = f(t), \quad u_x(1,t) = 0, \quad u(x,0) = g(x),$$

where g gives the initial data and $f(t)$ is a periodic function (perhaps $f(t) = A \sin \omega t$). Determine $h(t) := u(1,t)$. Is h periodic? If so, what is its period? Could the amplitude of h ever exceed the amplitude of f? Hint: Use analytic and numerical methods. The best method to determine a Fourier series solution is the expansion method described in [103].

5.2.2 Propagation of a Mutant Gene

Consider a population with a mutant gene whose concentration is u. Its allele (the parent gene) has concentration $1 - u$. The individuals in the population diffuse along a one-dimensional spatial domain (for example a shoreline) and they interact with each other to produce offspring. A simple model (introduced in 1937 independently by R. A. Fisher [38] and A.

N. Kolmogorov, I. Petrovskii, and N. Piscounov [58]) is

$$u_t = \kappa u_{xx} + au(1 - u). \tag{5.18}$$

The choice of the model interaction term, given by $f(u) = au(1 - u)$, is akin to the principle of mass action and may be more fully justified using probability theory (see [58]). Of course, this term may also be viewed as perhaps the simplest model of interaction that agrees (qualitatively) with experiments. The constant a is meant to model the utility of the gene for the organism to survive: $a > 0$ for an advantageous mutation; $a < 0$ for a disadvantageous mutation. When the concentration of the mutant gene is zero—no mutant genes in the population—the reaction term $au(1 - u)$ has value zero and $u = 0$ is a solution of the model equation; and, when the entire population has the mutant gene—the concentration is one—the function $u = 1$ is a solution of the model equation.

There is a natural scientific question: How will an advantageous mutation spread if it occurs in some individual or group of individuals at a specified spatial location?

The quantity $u(x, t)$ is the number of individuals at time t at shoreline position x with the mutant gene divided by the total number of individuals in the population at x. Of course this is an idealization. In reality, we would measure this ratio over some area (a fixed width times a length of shoreline). The concentration is more precisely defined to be the limit of this ratio as the area of the region shrinks to zero at x for the fixed time t. Thus, u—the ratio of two tallies—is a dimensionless quantity. The time derivative u_t has the dimensions of inverse time. In symbols, we write $[u_t] = 1/T$, where in this formula the square brackets denote units of the enclosed expression and T denotes the unit of time (perhaps T is years). With the choice of a unit L of length, the diffusion term has units $[\kappa u_{xx}] = [\kappa]/L^2$. These must agree with the units of the time derivative. Thus, $[\kappa]/L^2 = 1/T$ and $[\kappa] = L^2/T$. Likewise, the units of the interaction term are carried by a and $[a] = 1/T$; thus, a has the units of a rate.

We are unlikely to know good values of the diffusivity κ (a measure of how fast the organisms carrying the mutant gene spread along the shoreline) or a (which is the growth rate of the population with the mutant gene). Thus, we should not expect to make reliable quantitative predictions; rather, we should use the model to predict the qualitative behavior of the spread of a mutant gene.

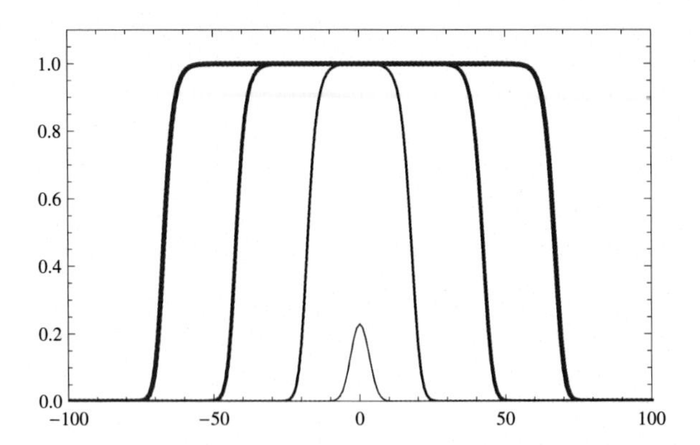

Fig. 5.2 Graphs of the spatial distribution of the mutant gene modeled by PDE (5.20) *at times* $t = 5, 15, 30,$ *and* 45 *are depicted for the initial condition* (5.21) *with* $a = -2$, $b = 2$, *and* $\mu = 0.01$. *The plotted curves are thicker as time increases. The numerical method is forward Euler (for the temporal variable) with time step 0.25, with Neumann boundary conditions, and a spatial grid of 200 interior points on the spatial domain* $[-100, 100]$.

The qualitative behavior of the solutions of our model are independent of the (positive) parameters κ and a. In fact, we can simply eliminate the parameters by a change of variables. Let $t = \tau s$, where $[\tau] = T$ (that is, τ has the dimensions of time), and $x = \ell\xi$, where $[\ell] = L$. The change of variables is accomplished by applying the chain rule. Ignoring the temporal variable for this computation, note that

$$\frac{du}{dt} = \frac{du}{ds}\frac{ds}{dt} = \frac{1}{\tau}u_s,$$
$$\frac{du}{dx} = \frac{du}{d\xi}\frac{d\xi}{dx} = \frac{1}{\ell}u_\xi,$$
$$\frac{d^2u}{dx^2} = \frac{1}{\ell^2}u_{\xi\xi}.$$

By substitution into the model PDE [Eq. (5.18)]

$$\frac{1}{\tau}u_s = \kappa\frac{1}{\ell^2}u_{\xi\xi} + au(1-u)$$

and, by rearranging, we have the equation

$$u_s = \kappa\frac{\tau}{\ell^2}u_{\xi\xi} + \tau au(1-u).$$

We may choose

$$\tau = \frac{1}{a}, \qquad \ell = \sqrt{\frac{\kappa}{a}}, \tag{5.19}$$

to obtain the desired dimensionless model

$$u_s = u_{\xi\xi} + u(1-u).$$

Reverting to the usual notation, the PDE

$$u_t = u_{xx} + u(1-u) \tag{5.20}$$

is discussed here.

For the biological application, the model PDE [Eq. (5.20)] may be considered for mathematical convenience with the spatial variable x defined on the whole real line, or we may impose boundary conditions at the ends of some interval. For instance, the zero Dirichlet boundary condition (u vanishes at an end of the portion of the shoreline under consideration) or the zero Neumann boundary condition (u_x vanishes) may be used. In the present context, the Dirichlet condition means that individuals at the end of the shoreline where it is imposed never carry the mutant gene; the Neumann condition means that no individuals with the mutant gene leave or enter the population through the end of the shoreline where it is imposed. More generally, nonzero constant or time-dependent boundary conditions might also have plausible physical meanings.

The model [Eq. (5.20)] is a nonlinear PDE; the superposition of solutions is generally not a solution. There is no known general explicit solution (cf. Exercise 5.15). But, it is possible to prove that unique solutions exist for appropriate initial conditions $u(x,0) = f(x)$ (see, for example, [89]). Lct us assume these results.

The simplest model has no diffusion. In this case, the PDE reduces to the ODE $\dot{u} = u(1-u)$ (see Exercise 2.4). There are two steady states $u = 0$ and $u = 1$. If $0 < u(x,0) < 1$ (that is, some individuals at the position x have the mutation), then the corresponding solution $u(x,t)$ grows monotonically to $u = 1$ as t goes without bound. Thus, with the passage of time, the mutant gene eventually is established in the entire population at each location along the shoreline where there were some individuals with the mutant gene.

Does the model predict that an advantageous mutant gene will spread to the entire population when diffusion (and thus spatial dependence) is taken into account?

To begin the analysis of the PDE with diffusion, suppose that the initial population with the mutant gene is found only in one location along the

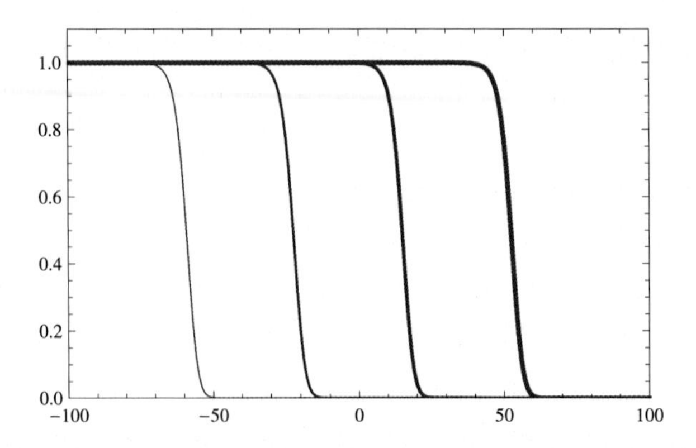

Fig. 5.3 Graphs of the spatial distribution of the mutant gene for Fisher's model [Eq. (5.20)] at times $t = 20, 40, 60$, *and* 80 *are depicted for the initial condition (5.21) with* $\alpha = -100$, $\beta = -90$, *and* $\mu = 1$. *The plotted curves are thicker as time increases. The numerical method is forward Euler on the spatial domain* $[-100, 100]$ *with Neumann boundary conditions, time step 0.1, and a spatial grid of 200 interior points.*

shoreline. This situation may be modeled by the initial function

$$u(x,0) = f(x) = \begin{cases} 0; & x < \alpha, \\ \mu; & a \le x \le \beta, \\ 0; & x > \beta, \end{cases} \tag{5.21}$$

where $\alpha < \beta$ and $0 < \mu \le 1$. Numerical experiments are used to obtain the population profiles depicted in Fig. 5.2. The spread of the mutant gene seems to be a wave spreading in both directions from the spatial location of the initial population that carried the mutation. The wave speed for this simulation is approximately 10.

To help determine the wave speed in general, let us note that in our scaling [Eq. (5.19)], the characteristic velocity is length divided by time. Our length scale is $\ell = \sqrt{\kappa/a}$ and our time scale is $\tau = 1/a$. There is a unique characteristic velocity given by

$$\text{characteristic velocity} = \sqrt{\kappa a}. \tag{5.22}$$

The wave speed should be a function of this characteristic velocity. Thus the dependence of the wave speed of a wave solution of the original model [Eq. (5.20)] should be $h(\sqrt{\kappa a})$ for some scalar function h (see Exercise 5.19). This argument is a simple example of an application of dimensional analysis, which is often useful to help determine the functional dependence of some phenomenon on the parameters in a model. Dimensioned characteristic quantities (such as wave speed) should be functions of terms that are ratios of monomials in the scales (such as τ and ℓ), which are used to make the model dimensionless.

Numerical experiments, using the PDE model for the spread of a mutant gene from a location where the mutation arises, suggest a basic prediction: The spread of the mutant gene has a wave front moving in both directions away from the initial location with the concentration of the mutant gene and the mutant gene saturates the population as the wave passes each remote location.

A related physical phenomenon is the spread of a mutant gene that is already dominant on the left side of a location along our beach but not present on the right side. Some of the results of numerical experiments are reported in Fig. 5.3; they suggest the local concentration of the mutant gene rises to dominate the population where the mutation is already present and it spreads to the right at a constant speed with each location being fully saturated as the wave front passes. A mathematical idealization of this situation leads to the question: Is there a solution u of the dimensionless model [Eq. (5.20)] such that, for each fixed position x,

$$0 \leq u(x,t) \leq 1, \qquad \lim_{t\to-\infty} u(x,t) = 0, \qquad \lim_{t\to\infty} u(x,t) = 1? \quad (5.23)$$

The existence of a solution of this type would show that some solutions of the PDE have the same qualitative behavior as a solution of $\dot{u} = u(1-u)$ (that is, the model differential equation without diffusion) with initial value $u(0)$ restricted to $0 < u(0) < 1$.

Inspection of Fig. 5.3 (or better yet an animation of the wave) suggests that the spatial concentration quickly approaches a wave that maintains its profile while moving to the right with constant velocity. Ideally, there is some function $\phi : \mathbb{R} \to \mathbb{R}$ such that

$$u(x,t) = \phi(x - ct), \quad (5.24)$$

where $c > 0$ is the wave speed. A solution of this form is called a traveling wave with wave form ϕ.

By substitution of Eq. (5.24) into PDE (5.20), we obtain the differential equation

$$-c\phi'(x-ct) = \phi''(x-ct) + \phi(x-ct)(1 - \phi(x-ct)), \quad (5.25)$$

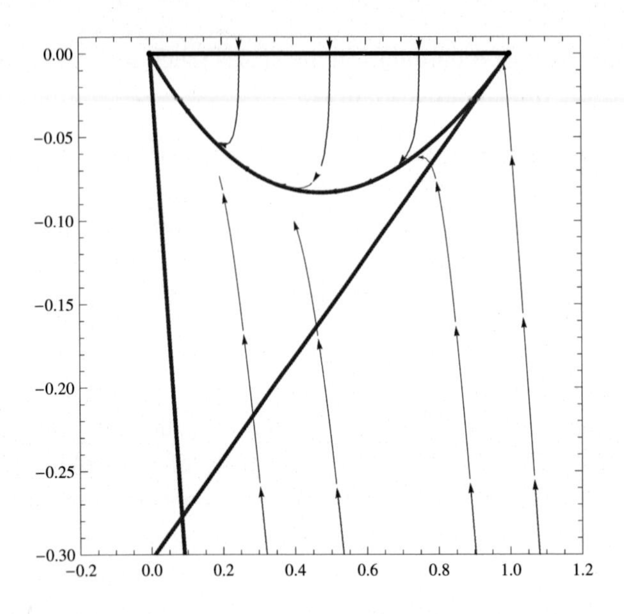

Fig. 5.4 A portion of the phase portrait for the ODE system (5.28) with $c = 3$ is depicted. The flow crosses each thick line segment in the same direction. The thick curve is an approximate trajectory connecting the rest points at $(0, 0)$ *and* $(1, 0)$.

which we may view as an ODE for the unknown wave profile ϕ with auxiliary conditions

$$0 \leq \phi \leq 1, \qquad \lim_{s \to -\infty} \phi(s) = 1, \qquad \lim_{s \to \infty} \phi(s) = 0. \tag{5.26}$$

A basic fact is that *if $c \geq 2$, then there is a solution of the ODE*

$$\ddot{\phi} + c\dot{\phi} + \phi(1 - \phi) = 0 \tag{5.27}$$

that satisfies the auxiliary conditions (5.26). This result implies that Fisher's model [Eq. (5.20)] has a traveling wave solution that satisfies the original auxiliary conditions (5.23).

The second-order ODE [Eq. (5.27)] is equivalent to the first-order system of ODEs

$$\begin{aligned} \dot{\phi} &= v, \\ \dot{v} &= -cv - \phi(1 - \phi) \end{aligned} \tag{5.28}$$

in the phase plane.

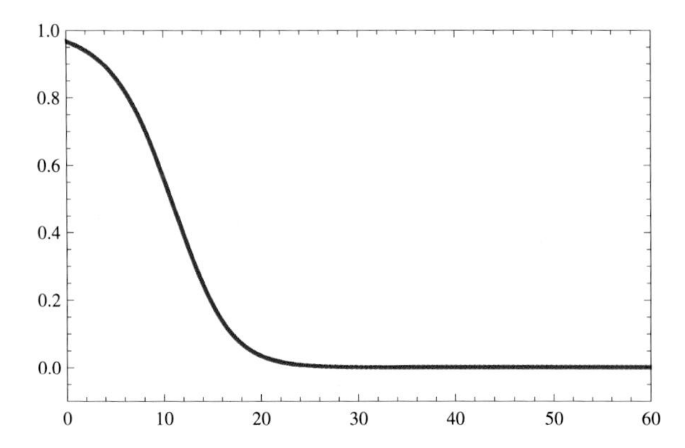

Fig. 5.5 The figure depicts an approximation of the graph of the function ϕ that is the first coordinate function of the solution of the ODE system (5.28) *with $c = 3$ along the unstable manifold of the saddle point at $(1, 0)$ that connects this point to the sink at the origin.*

First-order system (5.28) has two rest points at coordinates $(0, 0)$ and $(1, 0)$ in the phase plane. The rest point at the origin is asymptotically stable and the rest point at $(1, 0)$ is a saddle point. If $c \geq 2$, then the triangle depicted by thick line segments is positively invariant. The horizontal segment connects the rest points, the ray with negative slope is in the direction of the eigenspace corresponding to the negative eigenvalue of the linearized system matrix at $(0, 0)$ with the largest absolute value, and the ray with positive slope is in the direction of the eigenspace corresponding to the unstable manifold of the linearized system matrix at $(1, 0)$. For $c < 2$, the eigenvalues of the linearized system at the origin are complex. Thus, the incoming unstable trajectory from the saddle point winds around the origin and ϕ has negative values, violating the condition that $0 \leq \phi \leq 1$ (see [20] for a detailed proof and Exercise 5.22).

The connecting orbit in Fig. 5.4 corresponds to the desired function ϕ that defines a traveling wave solution. Of course, ϕ is the first coordinate function of the corresponding solution of the ODE system (5.28). Its graph, which has the profile of the expected traveling wave, is depicted in Fig. 5.5. Kolmogorov, Petrovskii, and Piscounov [58] proved that solutions of PDE (5.20) with initial data such that $0 \leq u(x, 0) \leq 1$, $u(x, 0) = 1$ for $x \leq \alpha$, and $u(x, 0) = 0$ for $x \geq \beta > \alpha$ approach a traveling wave solution with wave speed $c = 2$ (see Exercise 5.21).

Exercise 5.15. Show that $u(x,t) = \phi(x - ct)$ is a solution of PDE (5.20) in case $c = 5/\sqrt{6}$, K is a constant, and

$$\phi(z) = (1 + Ke^{z/\sqrt{6}})^{-2}.$$

Exercise 5.16. (a) Repeat the experiment reported in Fig. 5.2. (b) Repeat the experiment with Dirichlet boundary conditions and compare the results.

Exercise 5.17. How does the speed of the wave front(s) of solutions of PDE (5.20) depend on the amplitude of the initial population with the mutant gene?

Exercise 5.18. How does the speed of the wave front(s) of solutions of PDE (5.20) depend on the length of the spatial interval occupied by the initial population with the mutant gene?

Exercise 5.19. Use numerical experiments to test the characteristic velocity approximation in Eq. (5.22). Set an initial mutant gene concentration and vary the parameters κ and a in PDE (5.18).

Exercise 5.20. Reproduce Figs. 5.3–5.5.

Exercise 5.21. Use numerical experiments to verify the theorem of Kolmogorov, Petrovskii, and Piscounov [58] that solutions of Fisher's model [Eq. (5.20)] with initial data such that $0 \le u(x,0) \le 1$, $u(x,0) = 1$ for $x \le \alpha$, and $u(x,0) = 0$ for $x \ge \beta > \alpha$ approach a traveling wave solution with wave speed $c = 2$.

Exercise 5.22. (a) Find the system matrix at each rest point of system 5.28, find the eigenvalues and eigenvectors, and determine the stability types of the rest points. Show the triangle as in Fig. 5.4 is positively invariant by proving the vector field corresponding to the system of differential equations points into the region bounded by the triangle along the boundary of the region.

Exercise 5.23. Consider Fisher's equation with a disadvantageous gene ($a < 0$) in Eq. (5.18). (a) What happens when the disadvantageous gene is carried by all individuals counted in a finite interval of the spatial domain? (b) Does the disadvantageous gene always disappear as time increases? Discuss.

Exercise 5.24. Consider the PDE

$$u_t = u_{xx} + u(1 - u^2)$$

and restrict attention to solutions u such that $|u| \le 1$. Note that $u = 0$ and $u = \pm 1$ are steady states. (1) Which pairs of these steady states have solutions connecting them in the space of solutions of the ODE $u_t = u(1-u^2)$ obtained by ignoring the diffusion term u_{xx}? (2) Which pairs of these steady states have traveling wave solutions connecting them in the space of solutions of the PDE? Hint: Gather evidence using numerical experiments before attempting a pencil and paper solution.

Exercise 5.25. Consider Jin-ichi Nagumo's equation

$$u_t = u_{xx} + u(1 - u)(u - a),$$

where $0 < a < 1$ is a parameter. A traveling wave $f(x - ct)$ is called a front if its profile f has finite distinct limits at $\pm\infty$ and a pulse if these limits are equal. Does the Nagumo equation have front or pulse type traveling waves for some parameter values? Discuss using analysis and numerical experiments. If such solutions exist, graph typical front and pulse profiles and determine the corresponding wave speeds.

Exercise 5.26. Consider a population where the concentration of individuals with an advantageous gene u (which is the number with the gene divided by total number at a given spatial position) is modeled by Fisher's equation $u_t = \kappa u_{xx} + au(1 - u)$, where time is measured in months and distance in kilometers. The population resides on a shoreline 0.8 kilometers long where at the left end of the shore no individuals can enter or leave the population, but at the right end of the shoreline, individuals without the advantageous gene enter cyclically so that during the summer months of June, July, August, and September the concentration u at the right end of the beach is always measured to be approximately 20% less than the concentration at the left end. During the winter months (December, January, February, and March) the concentration at the right end of the beach is unaffected by migration. Spring and fall are transitional periods where the concentration builds up in fall and tapers off during spring to the mentioned levels. Suppose that the system constants are $\kappa = 0.002$ square kilometers per month and $a = 0.05$ per month. The initial population with the advantageous gene is measured in June to be 10% over three-quarters of the beach measured from left to right and 8% over the remainder of the beach. What is the percent of the population with the gene halfway along the beach after 5 years? Hint: There could be more than one viable model for the boundary conditions. Discuss your choice(s).

5.3 REACTION-DIFFUSION SYSTEMS: THE GRAY–SCOTT MODEL AND PATTERN FORMATION

Consider two concentrations u and v of two substances in some process involving diffusion and interaction. Imagine, for example, interacting populations (perhaps a predator and its prey) or a chemical reaction.

Absent diffusion, the interaction of two species is often modeled by a (nonlinear) system of ODEs

$$\dot{u} = f(u, v), \qquad \dot{v} = g(u, v). \tag{5.29}$$

For example, the basic interaction between a predator concentration u and its prey concentration v might be modeled by

$$\dot{u} = -au + buv, \qquad \dot{v} = cv(1 - \frac{1}{k}v) - buv. \tag{5.30}$$

Here, a is the death rate of the predator, c is the growth rate of the prey, k is the carrying capacity of the prey's environment, and b is the success rate of the predator. By taking into account the (spatial) diffusion of the two species, we obtain the reaction-diffusion model

$$u_t = \lambda \Delta u - au + buv, \qquad v_t = \mu \Delta v + cv(1 - \frac{1}{k}v) - buv, \tag{5.31}$$

where now u and v are functions of space and time.

At a practical level, the derivation of phenomenological reaction-diffusion models of the form

$$u_t = \lambda \Delta u + f(u, v), \quad v_t = \mu \Delta v + g(u, v) \tag{5.32}$$

can be as simple as our derivation of a predator-prey model. The derivation of more accurate models of course requires a detailed understanding of the underlying reaction [79, 84].

A famous and influential paper [113] by Alan Turing suggests reaction-diffusion models to explain pattern formation in biological systems (morphogenesis). Indeed, the patterns we see in nature (for example, mammalian skin spots and stripes, fish skin patterns, snow flakes, and many others) can all be generated approximately from graphs of solutions of reaction diffusion equations. The wider application of his idea suggests a broad form of Turing's principle: Reaction and diffusion are the underlying mechanisms for pattern formation in the natural world. Although Turing's principle is controversial, the application of mathematics to understand reaction and diffusion has proved enormously successful.

One of the models Turing considered is

$$u_t = \lambda \Delta u + r(\alpha - uv), \qquad v_t = \mu \Delta v + r(uv - v - \beta). \tag{5.33}$$

Another widely studied reaction-diffusion model for pattern formation is the dimensionless Gray–Scott model [31, 45, 78, 83, 123]

$$\begin{aligned} u_t &= \lambda \Delta u + F(1 - u) - uv^2, \\ v_t &= \mu \Delta v + uv^2 - (F + \kappa)v. \end{aligned} \tag{5.34}$$

It is derived from a hypothetical chemical reaction of the form

$$u + 2v \to 3v, \qquad v \to P,$$

where the second reaction creates an inert product; the coefficient F is the dimensionless feed/drain rate (which feeds u and drains u, v, and P), and

κ is the dimensionless rate of conversion in the second reaction $v \to P$. A standard theory in chemical kinetics, called the law of mass action, assumes that the rate of a reaction is proportional to the product of the concentrations of the chemicals involved in the reaction. For a reaction of the form $nA + mB \to C$ the reaction rate is $r = k[A]^n[B]^m$, where the square brackets denote concentration, k is the constant of proportionality called the rate constant, and n and m are the stoichiometric coefficients that specify the number of molecules of the corresponding chemical species that combine in the reaction. Thus, according to the first reaction equation $u + 2v \to 3v$, the rate at which v is increased by the reaction is proportional to uvv—one molecule of u plus two molecules of v combine to form three molecules of v.

For a system of reaction-diffusion equations

$$U_t = \lambda \Delta U + H(U),$$

Turing's fundamental idea—which has had a profound influence on developmental biology—is that spatial patterns can form even for the case of small diffusion for a reaction ODE

$$\dot{U} = H(U)$$

that has an attracting steady state. In effect, the diffusion can act against the tendency of the process to proceed to the steady state of the reaction. The implication is that a pattern (for example, the spots and stripes of animal skins) might arise from a chemical process involving reaction and diffusion. Thus, there is an ongoing field of research devoted to describing pattern formation and understanding the underlying mechanisms that produce patterns. The mathematical models are systems of reaction-diffusion equations.

Initial and boundary conditions must be specified to obtain a unique solution of a reaction-diffusion PDE.

Consider a two-dimensional spatial domain Ω and the concentration vector $U = (u, v)$, which is a function $U : \Omega \times [0, T] \to \mathbb{R}^2$. The initial condition simply gives the initial spatial distribution of the concentration vector; that is,

$$U(x, 0) = U_0(x) \tag{5.35}$$

for some function $U_0 : \Omega \to \mathbb{R}^2$.

The boundary conditions depend on the underlying physical problem. A Dirichlet boundary condition is used for the case where the concentration vector at the boundary of Ω is a known (vector) function $g : \partial\Omega \times [0, T] \to \mathbb{R}^2$; that is,

$$U(x, t) = g(x, t) \tag{5.36}$$

whenever $x \in \partial\Omega$ and $t \in [0, T]$. A Neumann boundary condition is used when the reactants are known to penetrate the boundary at some specified rate; that is, the flux through the boundary is known. In this case, let η denote the outer unit length normal on $\partial\Omega$. The Neumann boundary condition states that the normal derivative of the concentration vector is a known function on the boundary; that is,

$$\eta \cdot \nabla U(x, t) = g(x, t) \tag{5.37}$$

whenever $x \in \partial\Omega$ and $t \geq 0$.

A computationally convenient (but perhaps less physically realistic) boundary condition is the periodic boundary condition

$$U(x + a, y) = U(x, y), \qquad U(x, y + b) = U(x, y),$$

where a and b are fixed positive constants. In other words, the concentration vector is supposed to be defined on a torus given by a rectangle with opposite sides identified with the same orientations.

A reaction-diffusion system defined on a bounded domain Ω with a piecewise smooth boundary, either Dirichlet, Neumann, or periodic boundary conditions, and an initial condition, has a unique solution that exists for some finite time interval $0 \leq t < \tau$ (see [84, 99]). This is our hunting license. In the next section we will consider qualitative and numerical methods that can be used to understand some of the evolution of species concentrations predicted by the Gray–Scott model, which is used here as a mathematically rich prototype for reaction-diffusion models and as a specific example where numerical methods can be used to approximate solutions.

5.4 ANALYSIS OF REACTION-DIFFUSION MODELS: QUALITATIVE AND NUMERICAL METHODS

The mechanism for pattern formation is not well understood. At least it is very difficult to define the properties of the patterns that arise and prove their existence. To show that patterns arise begins with numerical approximations. But, as we will see, qualitative methods for understanding the solutions of differential equations arise naturally in this investigation.

Numerical methods and numerical analysis for differential equations are vast subjects, which are essential tools for applied mathematics. This section is meant as a glimpse into a few numerical methods together with some discussion of the issues that are encountered in practice. Our context will be an analysis of the Gray–Scott model [Eq. (5.34)]. What is the fate of the concentrations of the reactants for this hypothetical chemical reaction?

5.4.1 Euler's Method

To begin the adventure into the world of reaction-diffusion, recall Euler's method. It is the prototypical method for solving ODEs; for example, the system of ODEs

$$\dot{u} = f(u, v), \qquad \dot{v} = g(u, v) \tag{5.38}$$

obtained by ignoring the spatial dependence in our reaction-diffusion system [Eq. (5.32)]. The idea is to discretize the time derivatives and thus approximate the continuous flow of time by a series of discrete steps that can be easily computed.

To accomplish the desired discretization, recall from calculus that if $F : \mathbb{R} \to \mathbb{R}$ is a sufficiently smooth function of one variable, then the functions $h \to F(x + h)$ and $h \to F(x - h)$ each have Taylor series expansions at $h = 0$:

$$\begin{aligned} F(x+h) &= F(x) + F'(x)h + \frac{1}{2!}F''(x)h^2 + \frac{1}{3!}F'''(x)h^3 + O(h^4), \\ F(x-h) &= F(x) - F'(x)h + \frac{1}{2!}F''(x)h^2 - \frac{1}{3!}F'''(x)h^3 + O(h^4), \end{aligned} \tag{5.39}$$

where the big O notation is explained in Appendix A.8 and Taylor's formula in Appendix A.9. Using the first equation of the display, it follows that

$$F'(x) = \frac{F(x+h) - F(x)}{h} + O(h).$$

Thus, by replacing $F'(x)$ with the (forward) difference quotient $F(x+h) - F(x)/h$, we make an error of order h.

The numerical *method* is to make this replacement; the numerical *analysis* is to prove the error has order h.

The replacement of the derivative of F with the forward difference quotient leads to Euler's method for approximating solutions of the system of ODEs [Eqs. (5.38)]: For an initial time t_0 and initial state $(u, v) = (u_0, v_0)$, future times at the regular time increment Δt and the corresponding future states are given recursively by

$$\begin{aligned} u_n &= u_{n-1} + \Delta t f(u_{n-1}, v_{n-1}), \\ v_n &= v_{n-1} + \Delta t g(u_{n-1}, v_{n-1}), \\ t_n &= t_{n-1} + \Delta t. \end{aligned} \tag{5.40}$$

The discretization error is $O(h)$; it is proportional to the step size h. The local truncation error, or the error per step, is the norm of the difference between the solution and the approximation after one step. In vector form ($\dot{U} = H(U)$) and using Taylor's formula, this error is

$$\begin{aligned} |U(t_0 + \Delta t) - U_1| &= |U(t_0) + \Delta t U'(t_0) + \frac{\Delta t}{2!} U''(\tau) - (U_0 + \Delta t H(U_0))| \\ &= |\frac{\Delta t}{2!} U''(\tau)| \\ &= O(\Delta t^2), \end{aligned}$$

where τ is a number between t_0 and $t_0 + \Delta t$.

5.4.2 The Reaction Equations

Euler's method can be applied to the Gray–Scott reaction model [Eq. (5.34)]; that is, the system of ODEs

$$\begin{aligned} \dot{u} &= F(1 - u) - uv^2, \\ \dot{v} &= uv^2 - (F + \kappa)v. \end{aligned} \tag{5.41}$$

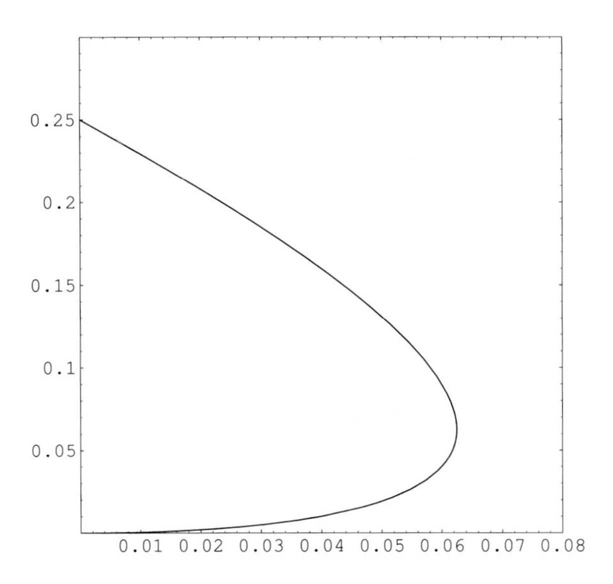

Fig. 5.6 A plot of the parabola $F = 4(F + \kappa)^2$ with horizontal axis κ and vertical axis F is depicted. Outside the curve, system (5.41) *has one rest point at $(u, v) = (1, 0)$. Inside the curve it has three rest points.*

To compute an approximate solution, we must set values for the parameters $\kappa > 0$ and $F > 0$, an initial condition $(u(0), v(0))$, and a time step Δt. How should these choices be made? This question does not have a simple answer. The choices depend on the intended application. For the Gray–Scott model, we are interested in the fate of a typical solution. Thus, the choice of initial condition is not essential as long as it is generic. On the other hand, the reaction model has a two-dimensional parameter space. We will be computing for a very long time if we wish to exhaust all the possible parameter values. A wiser course of action is to rely on a fundamental principle of applied mathematics: *Think before you compute.*

Our current goal is to understand the general behavior of system (5.41). How does this behavior depend on the parameter values?

The state variables u and v are supposed to represent chemical concentrations; thus, their values should be nonnegative. Moreover, the evolution of the system from a physically realistic state ($u(0) \geq 0$ and $v(0) \geq 0$) should produce only physically realistic states. In other words, the closed first quadrant in the state space should remain invariant under the flow. This fact is easily checked. Simply note that the u-axis is invariant (because $\dot{v} = 0$

whenever $v = 0$), and the vector field points into the first quadrant along the positive v-axis (because $\dot{u} > 0$ whenever $u = 0$).

The simplest dynamics is given by rest points; that is, constant solutions of the system of ODEs given by solutions of the algebraic equations

$$F(1-u) - uv^2 = 0, \qquad uv^2 - (F+\kappa)v = 0.$$

Note that $(u, v) = (1, 0)$ is a solution for all values of κ and F. Also, $(1/2, F/(2(F+\kappa)))$ is a double root whenever $F = 4(F+\kappa)^2$ and $F > 0$. Fig. 5.6 depicts this curve (a parabola) in the parameter space. It meets the vertical axis at $(\kappa, F) = (0, 0)$ and $(\kappa, F) = (0, 1/4)$; the point $(\kappa, F) = (1/16, 1/16)$ is the point on the curve with the largest first coordinate. Our ODE has three rest points for the parameter vector (κ, F) in the region bounded by the curve and the vertical axis and one rest point outside of this region. For parameter values inside the region, the two new rest points have coordinates

$$u = \frac{1}{2}\left(1 \pm \sqrt{1 - 4(F+\kappa)^2/F}\right), \qquad v = \frac{F}{F+\kappa}(1-u). \tag{5.42}$$

The system matrix for the linearization of the system of ODEs at $(1, 0)$ is the diagonal matrix whose main diagonal has the components $-F$ and $-(F+\kappa)$ (see Appendix A.17 for a discussion of linearization and stability). These negative numbers are the eigenvalues of this matrix; hence, $(1, 0)$ is asymptotically stable independent of the parameter values.

The new rest points appear as the parameters cross the curve from outside to inside; on the curve a double root appears "out of the blue." This is called a saddle-node bifurcation or a blue sky catastrophe. The prototype for this bifurcation (called a normal form) is given by

$$\dot{x} = b - x^2; \qquad \dot{y} = \pm y,$$

where b is the bifurcation parameter. Note that for $b < 0$ (which corresponds in our case to being outside the parabola $F = 4(F+\kappa)^2$), there are no rest points. The system with $b = 0$ has a semistable rest point called a saddle-node. For $b > 0$, there are two rest points: one saddle and one sink in the $+$ sign case, and one saddle and one source in the $-$ sign case. This scenario is exactly what happens for the reaction model [Eq. (5.41)]. (The rest point at $(1, 0)$ plays no role in this bifurcation.)

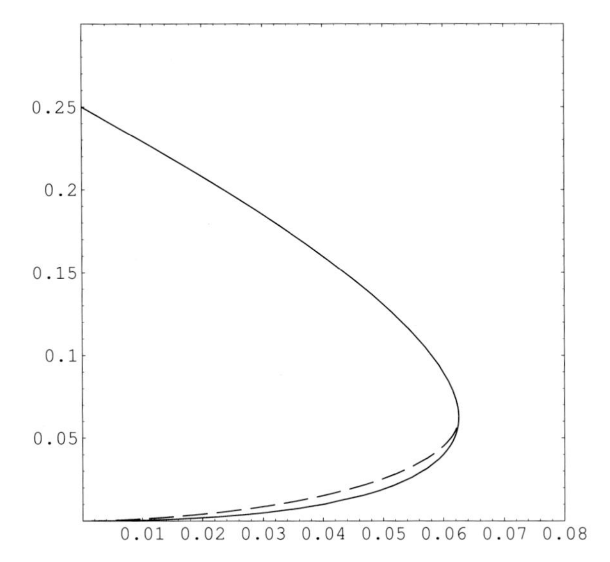

Fig. 5.7 The figure (with horizontal axis κ and vertical axis F) depicts with a solid curve the parabola $F = 4(F + \kappa)^2$ (the saddle-node bifurcation curve) and with dashed curve $F = 1/2(\sqrt{\kappa} - 2\kappa - \sqrt{\kappa - 4\kappa^{3/2}})$ (the Hopf bifurcation curve). Outside the saddle-node curve, the system (5.41) has one stable rest point at $(u, v) = (1, 0)$. Inside this curve the system has three rest points. Above the Hopf bifurcation curve and inside the saddle-node curve there are three rest points, two sinks, and one saddle. Below the Hopf curve and above the saddle-node curve there is one sink, one source, and one saddle. For F decreasing and $\kappa < 0.0325$ (approximately), the Hopf bifurcation is supercritical; for $\kappa > 0.0325$, the Hopf bifurcation is subcritical.

The system matrix at the rest points [Eqs. (5.42)] is

$$A = \begin{pmatrix} -F - v^2 & -2uv \\ v^2 & 2uv - (F + \kappa) \end{pmatrix}. \tag{5.43}$$

At these rest points $uv = F + \kappa$; therefore, we have the formulas

$$\text{Trace}\,(A) = \kappa - v^2, \qquad \text{Determinant}\,(A) = (F + \kappa)(v^2 - F). \tag{5.44}$$

The eigenvalues of A are the roots of the characteristic equation

$$\text{Determinant}\,(A - \lambda I) = 0,$$

where I denotes the 2×2 identity matrix. By an easy computation, the characteristic equation is seen to have the general form

$$\lambda^2 - \text{Trace}\,(A)\lambda + \text{Determinant}\,(A) = 0$$

and the roots

$$\lambda = \frac{\text{Trace}(A) \pm \sqrt{\text{Trace}(A)^2 - 4\text{Determinant}(A)}}{2}.$$

The stability types of rest points are determined from the signs of Trace(A) and Determinant(A). For example, if Trace$(A) < 0$ and Determinant$(A) > 0$, then the radicand is either complex or, in case it is real, less than Trace(A) < 0. Hence, both roots have positive real parts and the corresponding rest point is a source.

Consider Trace(A) on the parabola $F = 4(F + \kappa)^2$. Using the formulas for the corresponding rest points [Eqs. (5.42)], it follows that

$$v = \frac{F}{2(F + \kappa)};$$

hence,

$$\begin{aligned} \text{Trace}(A) &= \kappa - v^2 \\ &= \kappa - F\Big(\frac{F}{4(F + \kappa)^2}\Big) \\ &= \kappa - F \\ &= \frac{\sqrt{F}}{2} - 2F. \end{aligned}$$

By simple analysis of the function $F \mapsto \frac{\sqrt{F}}{2} - 2F$, we see that Trace$(A)$ is positive for $0 < F < 1/16$, zero at $F = 1/16$, and positive for $1/16 < F < 1/4$. Also, by the continuity of Trace(A) as a function of κ and F, it maintains its sign along curves in the parameter space that cross the parabola, except those that cross at the point $(\kappa, F) = (1/16, 1/16)$.

The quantity Determinant(A) as a function of κ and F vanishes on the parabola. To determine its sign along a curve in the parameter space that crosses into the region bounded by the parabola and the coordinate axes, note that $F > 4(F+\kappa)^2$ in this region; thus, in the bounded region and near the boundary parabola, there is some (small) $\epsilon > 0$ such that

$$\frac{4(F+\kappa)^2}{F} = 1 - \epsilon.$$

The value of v at the corresponding rest points is

$$\begin{aligned} v_\pm^2 &= \frac{F}{F+\kappa}\Big(\frac{1}{2} \mp \frac{1}{2}\sqrt{1-(1-\epsilon)}\Big) \\ &= \frac{F}{2(F+\kappa)}(1 \mp \eta), \end{aligned}$$

where $\eta := \sqrt{\epsilon}$. The sign of Determinant(A) is determined by the sign of $v^2 - F$, which is given by

$$\begin{aligned} v_\pm - F &= \frac{F^2}{4(F+\kappa)^2}(1 \mp 2\eta + \eta^2) - F \\ &= F\Big(\frac{F}{4(F+\kappa)^2}(1 \mp 2\eta + \eta^2) - 1\Big) \\ &= F\Big(\frac{1}{1-\eta^2}(1 \mp 2\eta + \eta^2) - 1\Big) \\ &= \frac{2F\eta}{1-\eta^2}(\mp 1 + \eta). \end{aligned}$$

From this computation, it is clear that Determinant(A) is positive (respectively, negative) at the rest point whose second component is v_- (respectively, v_+). Also, the value of this determinant goes to zero as $\epsilon > 0$ approaches zero.

In summary, a source and a saddle appear in a saddle-node bifurcation upon crossing the parabola into the bounded region at points on the parabola corresponding to $0 < F < 1/16$; a sink and a saddle appear in a saddle-node bifurcation upon crossing the parabola into the bounded region at points on the parabola corresponding to $1/16 < F < 1/4$.

Another type of bifurcation, called Hopf bifurcation, occurs in the regime where A has complex eigenvalues; it occurs whenever, along a curve in the parameter space, a pair of complex conjugate eigenvalues crosses the imaginary axis in the complex plane with nonzero speed. A normal form for this important bifurcation is most easily understood in polar coordinates (r, θ); it is given by

$$\dot{r} = br \pm r^3, \qquad \dot{\theta} = \omega + ar^2.$$

the corresponding system in Cartesian coordinates has the linear part

$$\dot{x} = bx - \omega y, \qquad \dot{y} = \omega x + by. \tag{5.45}$$

With $\omega \neq 0$ and $a \neq 0$ fixed, a pair of complex conjugate eigenvalues $(b \pm i\omega)$ crosses the imaginary axis as the bifurcation parameter b passes through zero in the positive direction. The rest point at the origin thus changes from a sink to a source. For the plus sign and $b < 0$, there is an unstable limit cycle (that is, an isolated periodic orbit), which is a circle of radius $\sqrt{-b}$ that disappears into the rest point at $b = 0$. The bifurcation in this case is called a subcritical Hopf bifurcation. For the minus sign, a stable limit cycle (which in this special case is the circle with radius $\sqrt{b}$) emerges out of the rest point as b increases from zero.

Hopf bifurcations are detected by finding the curve(s) in the parameter space where the characteristic equation of the linearization at a rest point has pure imaginary eigenvalues. Generically, Hopf bifurcations occur whenever some other curve in the parameter space crosses this "Hopf curve."

The eigenvalues of the matrix A are pure imaginary whenever its trace vanishes and its determinant is positive. For parameter values in the region of parameter space bounded by the saddle-node bifurcation curve, the system matrix of the linearization at the rest point v_+ has positive determinant. From its formula [Eq. (5.44)], the trace of this matrix at v_+ is given by $\kappa - v_+^2$. Using the rest points [Eqs. (5.42)] and

$$G := 1 - \frac{4(F + \kappa)^2}{F},$$

the trace vanishes if and only if (κ, F) lie on the Hopf curve

$$\kappa = \frac{F^2}{4(F+\kappa)^2}(1+\sqrt{G})^2.$$

After some algebra, the Hopf curve in the first quadrant is also given by

$$(\kappa + F)^2 = F\sqrt{\kappa},$$

a quadratic equation for F that has solutions

$$F = \frac{\sqrt{\kappa} - 2\kappa \pm \sqrt{\kappa - 4\kappa^{3/2}}}{2}. \tag{5.46}$$

Because the trace of A (given by $\kappa - v^2$) vanishes and the determinant of A (given by $(F+\kappa)(v^2 - F)$) is positive on the Hopf curve, $F < \kappa$ on this curve. By inspection of the graphs of the solutions [Eqs. (5.46)], it follows that this condition is satisfied only with the minus sign on the interval $0 < \kappa < 1/16 = 0.0625$. The corresponding graph, depicted in Fig. 5.7, is the Hopf curve. It meets the saddle-node curve at $(\kappa, F) = (0,0)$ and $(\kappa, F) = (1/16, 1/16)$.

Outside the saddle-node curve, our first-order system (5.41) has one stable rest point at $(u,v) = (1,0)$. Inside this curve the system has three rest points. Above the Hopf bifurcation curve and inside the saddle-node curve there are three rest points, two sinks, and one saddle. Below the Hopf curve and above the saddle-node curve there is one sink, one source, and one saddle. There is a critical value $\kappa_* \approx 0.0325$ of κ such that for F decreasing and $\kappa < \kappa_*$, the Hopf bifurcation is supercritical; for $\kappa > \kappa_*$, the Hopf bifurcation is subcritical.

To prove the statements about the subcritical and supercritical Hopf bifurcations is not trivial. An idea for a proof is to translate the corresponding rest point to the origin of a new coordinate system and then transform (by a linear transformation) the resulting first-order system of ODEs so that its linearization has the normal form for the Hopf bifurcation given by first-order system (5.45). After this process is complete, the resulting second- and third-order terms of the right-hand side of the system of differential equations determine the stability index, which gives the direction of the Hopf bifurcation (see [20]). For our differential equation in the form

$$\dot{x} = \alpha x - y + p(x,y), \qquad \dot{y} = x + \alpha y + q(x,y)$$

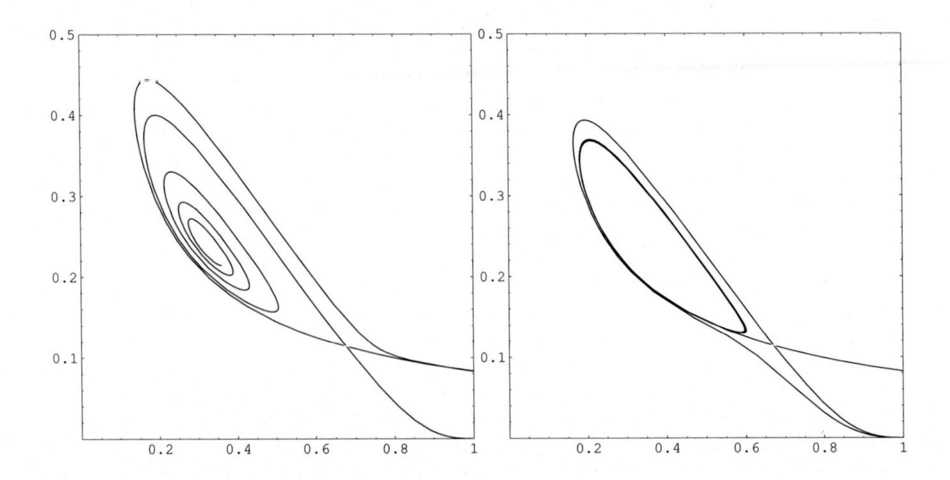

Fig. 5.8 The left panel depicts the phase portrait for system (5.41) *for the parameter values* $\kappa = 0.05$ *and* $F = 0.02725$. *An orbit in the unstable manifold of the saddle point is asymptotic to a spiral sink. In the opposite direction on the stable manifold, an orbit is asymptotic to the sink at* $(1, 0)$. *The right panel depicts the phase portrait for the parameter values* $\kappa = 0.05$ *and* $F = 0.0265$. *The positions of the stable and unstable manifolds have crossed. An orbit in the stable manifold of the saddle point is asymptotic to an unstable limit cycle, which surrounds a spiral sink (not depicted).*

where $p = \sum_{j=1}^{\infty} p_j(x, y)$ and $q = \sum_{j=1}^{\infty} q_j(x, y)$ with

$$p_j(x, y) := \sum_{i=0}^{j} a_{j-i,i} x^{j-i} y^i, \qquad q_j(x, y) := \sum_{i=0}^{j} b_{j-i,i} x^{j-i} y^i,$$

the sign of the stability index is given by the sign of the expression

$$\begin{aligned} L_4 = \frac{1}{8}(&a_{20}a_{11} + b_{21} + 3a_{30} - b_{02}b_{11} \\ &+ 3b_{03} + 2b_{02}a_{02} - 2a_{20}b_{20} - b_{20}b_{11} + a_{12} + a_{02}a_{11}). \end{aligned} \tag{5.47}$$

If the quantity $L_4 > 0$ (respectively, $L_4 < 0$), then our Hopf bifurcation is subcritical (respectively, supercritical). The bifurcation diagram in Fig. 5.7 is not complete. For example, the existence of the subcritical Hopf bifurcation for $\kappa > \kappa_*$ as F decreases through the Hopf curve means that an unstable limit cycle disappears into the corresponding rest point at this bifurcation. Where did the limit cycle come into existence for some larger value of F? The answer is that there is a third bifurcation in our system called a homoclinic loop bifurcation.

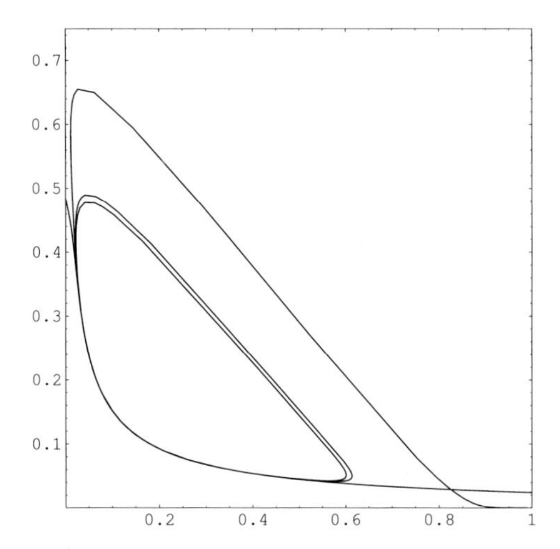

Fig. 5.9 The phase portrait of system (5.41) *is depicted for the parameter values* $\kappa = 0.02$ *and* $F = 0.0039847$. *An orbit in the unstable manifold of the saddle point is asymptotic to a stable limit cycle. In the opposite direction on the stable manifold, an orbit is asymptotic to the sink at* $(1, 0)$.

Two phase portraits (corresponding to a value of the parameter F slightly larger than its critical value and slightly smaller than its critical value) are depicted in Fig. 5.8. Pay attention to the portion of the unstable manifold of the saddle point (which is at the point where two curves appear to cross) that lies above and to the left of the saddle; the other part of the unstable manifold goes to the sink at $(u, v) = (1, 0)$. In the left panel, the unstable manifold lies below the stable manifold; in the right panel, it lies above. Between the parameter values corresponding to these phase portraits, there is a value of F where the stable and unstable manifolds meet to form a (homoclinic) loop. During this process, there is a sink surrounded by the loop. In the second panel the stable manifold winds around this sink. But, it can't be asymptotic to the sink. So, (by the Poincaré–Bendixson theorem) there must be an unstable periodic orbit surrounding the sink and surrounded by the spiral formed by the stable manifold (see [20]). This is an example of a homoclinic loop bifurcation.

A homoclinic loop bifurcation also occurs for $\kappa < 0.035$ as F decreases to a critical value below the Hopf curve. In this case the homoclinic loop bifurcation absorbs a stable limit cycle. An example of this cycle near the homoclinic loop bifurcation is depicted in Fig. 5.9.

The conjectural position of the homoclinic loop curve in the bifurcation diagram is between the saddle-node and Hopf curve for $\kappa < 0.035$ and above the Hopf curve for $\kappa > 0.035$. Its end points are the same as the end points of the Hopf curve (see Exercise 5.28 and [3] for more details of the bifurcation analysis of this system).

Exercise 5.27. Show that there is a critical value $\kappa_* \approx 0.0325$ of κ such that, for $\kappa < \kappa_*$, system (5.41) has a supercritical Hopf bifurcation as F decreases through Hopf curve (5.46).

Exercise 5.28. Make a numerical study of the homoclinic loop curve for system (5.41) and plot it together with the saddle-node and Hopf curves. Hint: Public domain software such as AUTO or XPPAUTO can approximate bifurcation curves. Even better, write your own code.

Exercise 5.29. Add a periodic forcing term to system (5.41) and compute the stroboscopic Poincaré map (that is, start at some initial condition, integrate forward for one period of the forcing, plot the final point, and iterate this process). Vary the amplitude and frequency of the forcing and plot some typical phase portraits of the Poincaré map.

Exercise 5.30. Add a periodic parametric forcing term to system (5.41) and compute the stroboscopic Poincaré map (that is, start at some initial condition, integrate forward for one period of the forcing, plot the final point, and iterate this process). Here the most interesting scenario is to make F or κ periodically time dependent so that these parameters sweep through some of their bifurcation values (cf. [53]). For example, fix $\kappa = 0.02$ and replace F by

$$F = 0.004 + 0.00005 \sin(\omega t),$$

where ω is small (perhaps $\omega = 0.0001$) so that F changes relatively slowly. See if you can obtain a long transient that exhibits beats. Our system is not well-suited to dynamic bifurcation because the basin of attraction of the sink at $(1, 0)$ is likely to capture our orbit.

5.4.3 Diffusion and Spatial Discretization

We have succeeded in understanding the bifurcation diagram for the Gray–Scott model without diffusion. What happens when diffusion is introduced?

It is reasonable to assume that adding sufficiently small diffusion (λ and μ) and choosing parameters κ and F in the region where the unique steady state $(u, v) = (1, 0)$ resides will produce a spatially constant steady state for the Gray–Scott model (5.34). On the other hand, it is not at all clear what happens in this model for parameter values close to the bifurcations of the reaction ODEs [Eqs. (5.41)]. But, the dynamical behavior of this PDE near these values may be explored using numerical approximations.

By adding the Taylor series (5.39) and rearranging terms, note that

$$F''(x) = \frac{F(x+h) - 2F(x) + F(x-h)}{h^2} + O(h^2). \tag{5.48}$$

This approximation is used to discretize the second derivatives in our PDE.

The first step of our numerical procedure is to choose a spatial discretization of a spatial domain. For the Gray–Scott model there is no natural spatial domain; it is just some region in two-dimensional space. Following Pearson [83], let us consider the spatial domain to be the square $[0, L] \times [0, L]$ in the (x, y) plane. A lattice (or grid) in this square is defined by the points $(i\Delta x, j\Delta y)$ (called nodes), for $i = 0, 1, 2, \ldots, m$ and $j = 0, 1, 2, \ldots, n$, where $\Delta x := L/m$ and $\Delta y := L/n$. As in Euler's method for ODEs, let us also choose a time step Δt. The concentration $u(i\Delta x, j\Delta y, k\Delta t)$ is denoted by $U_{i,j}^k$ and $v(i\Delta x, j\Delta y, \Delta t)$ by $V_{i,j}^k$.

The next step is to approximate the PDE by difference quotients; this idea leads to many possible alternatives. Perhaps the simplest viable scheme is the *forward Euler method* given by

$$\begin{aligned} U_{i,j}^{k+1} &= U_{i,j}^k + \Delta t\Big[\frac{\lambda}{\Delta x^2}(U_{i+1,j}^k - 2U_{i,j}^k + U_{i-1,j}^k) \\ &\quad + \frac{\lambda}{\Delta y^2}(U_{i,j+1}^k - 2U_{i,j}^k + U_{i,j-1}^k) + F(1 - U_{i,j}^k) - U_{i,j}^k(V_{i,j}^k)^2\Big], \\ V_{i,j}^{k+1} &= V_{i,j}^k + \Delta t\Big[\frac{\mu}{\Delta x^2}(V_{i+1,j}^k - 2V_{i,j}^k + V_{i-1,j}^k) \\ &\quad + \frac{\mu}{\Delta y^2}(V_{i,j+1}^k - 2V_{i,j}^k + V_{i,j-1}^k) + U_{i,j}^k(V_{i,j}^k)^2 - (F + \kappa)V_{i,j}^k\Big]; \end{aligned} \tag{5.49}$$

it is exactly Euler's method applied to PDEs.

The third step is to impose the boundary conditions. For periodic boundary conditions, the concentrations are updated at (what are called) the interior nodes corresponding to $i = 0, 1, 2, \ldots, m-1$ and $i = 0, 1, 2, \ldots, m-1$. Updates are required also at the remaining fictitious nodes with (i, j) coordinates $(-1, j)$, for $j = 0, 1, 2, \ldots, n-1$ and $(i, -1)$, for

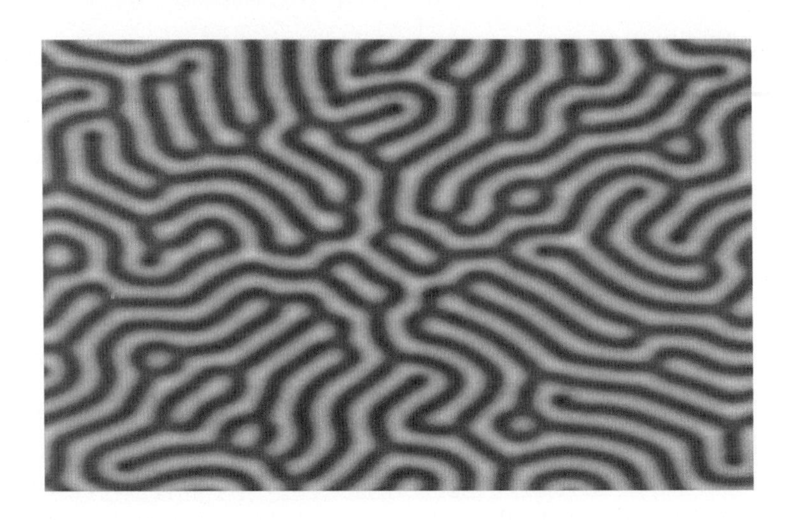

Fig. 5.10 The figure depicts the u concentration for a computer-generated approximate state of the Gray–Scott model, with periodic boundary conditions, for the parameter values $\kappa = 0.06$, $F = 0.038$, $\lambda = 2\times10^{-5}$, and $\mu = 10^{-5}$. The system evolves from the depicted state to states with similar configurations for at least 200,000 time steps of unit length.

$i = 0, 1, 2, \ldots, m - 1$. These are given by

$$\begin{aligned} U^k_{-1,j} &= U^k_{m-1,j}, \quad V^k_{-1,j} = V^k_{m-1,j}, \\ U^k_{i,-1} &= U^k_{i,n-1}, \quad V^k_{i,-1} = V^k_{i,n-1}. \end{aligned} \tag{5.50}$$

Likewise, values at the unassigned portion of the boundary of the square are

$$\begin{aligned} U^k_{m,j} &= U^k_{0,j}, \quad V^k_{m,j} = V^k_{0,j}, \\ U^k_{i,n} &= U^k_{i,0}, \quad V^k_{i,n} = V^k_{i,0}. \end{aligned} \tag{5.51}$$

To complete the grid (for graphics), define $U^k_{m,n} = U^k_{0,0}$.

To implement this procedure, start at $k = 0$ and assign (initial) values of the concentrations at all the interior nodes. The periodic boundary conditions are used to assign values to the unassigned portion of the boundary of the square and at the fictitious nodes. Update the values of the concentrations (to U^1_{ij} and V^1_{ij}) at all the interior nodes using Eqs. (5.49), and update the values of the concentrations at the reminder of the boundary and the fictitious nodes using Eqs. (5.50) and (5.51). This process is continued until some preassigned final time is reached, or until some other test applied to the array of concentrations is met.

To reproduce the numerical experiments in [83] (which are reported with color graphics), use the assignments

$$L = 2.5, \quad m = n = 256, \quad \lambda = 2 \times 10^{-5}, \quad \mu = 10^{-5}, \quad \Delta t = 1.0$$

with (κ, F) chosen near the bifurcation curves in Fig. 5.7 and impose the initial data as follows: Set the initial concentrations to the spatial steady state value $(u, v) = (1, 0)$, reset the values in the 20×20 central square of nodes to $(u, v) = (0.5, 0.25)$, and then perturb every grid point value by a random number that changes the originally assigned value by no more than 1%. Integrate forward for some number of time steps, usually chosen in the range of 100,000–200,000, and plot the final approximate concentrations.

The result of a typical numerical experiment, for the parameter values $\kappa = 0.06$ and $F = 0.038$, 100,000 time steps of unit length, and a rendering of the value of the function $(x, y) \mapsto u(100000, x, y)$ into a gray scale on the interval $[0, 1]$, is depicted in Fig. 5.10. This pattern seems to evolve over a long time interval and eventually it reaches a (time-dependent) steady state that retains the basic qualitative features in the figure (see Exercise 5.32).

5.4.4 Numerical Stability

A virtue of the forward Euler method is that it is easy to program. Its error per step is $O(\Delta x^2 + \Delta y^2)$ in space and $O(\Delta t)$ in time. Unfortunately, there is a hidden danger: the discretization can introduce instabilities that are not present in the PDE.

To understand how a numerical instability might occur, imagine that the values of the concentrations at the interior grid points are written as a column vector W (of length $2mn$). The update given by Eqs. (5.49) can be recast in the matrix form

$$W^{k+1} = AW^k + H(W^k),$$

where A is the matrix corresponding to the discretization of the second derivatives and H is the nonlinear function whose components are the reaction terms. Is this process stable?

The linear part of our update equation is $W^{k+1} = AW^k$. At some step of our numerical computation a vector error ϵ might be introduced into the computed value of W^k. Let us suppose the error occurs in the computation of W^1 so that the computed value is $W^1 + \epsilon$ instead of W^1. This error will propagate in future steps as follows:

$$W^2 = AW^1 + A\epsilon, \quad W^3 = A^2W^1 + A^2\epsilon, \ldots, W^{k+2} = A^{k+1}W^1 + A^{k+1}\epsilon.$$

The error will not cause too much trouble if its propagation by the iteration process remains bounded; that is, if there is some constant M such that $|A^k\epsilon| \leq M|\epsilon|$ for all $k > 1$. A sufficient condition for the desired stability is that all eigenvalues of A are in the open unit disk in the complex plane, or equivalently, the spectral radius of A is strictly less than one. If at least one eigenvalue of A lies outside the closed unit disk, then a generic error will grow without bound. (The error will not grow in the unlikely situation that it remains in the eigenspace of an eigenvalue that is in the open unit disk, but all other errors will grow.) In case there are eigenvalues on the unit disk, a stability analysis requires additional information. Consider two simple examples:

$$A = \begin{pmatrix} 1 & 0 \\ 0 & 1 \end{pmatrix}, \qquad B = \begin{pmatrix} 1 & 1 \\ 0 & 1 \end{pmatrix}.$$

In this example, A is the 2×2 identity matrix; its eigenvalues are both 1. Because $A^k v = v$ for every vector v and every integer k, it follows that $|A^k v| = |v|$. Thus, the propagation of v remains bounded. The eigenvalues of B are also both equal to 1; but, for this matrix

$$\begin{pmatrix} 1 & 1 \\ 0 & 1 \end{pmatrix}^k \begin{pmatrix} v_1 \\ v_2 \end{pmatrix} = \begin{pmatrix} v_1 + kv_2 \\ v_2 \end{pmatrix}.$$

Thus, if $v_2 \neq 0$, the propagation of v by iteration of B grows without bound. These examples are indicative of the general situation. To fully analyze the general case requires some concepts from linear algebra, especially the Jordan canonical form. A simple result, adequate for our purposes, states that if all the eigenvalues of a matrix A lie in the closed unit disk and the matrix is diagonalizable (that is, there is an invertible matrix B such that $B^{-1}AB$ is a diagonal matrix), then errors remain bounded under iteration by A.

To apply these observations to the stability of our numerical methods, consider the A matrix for the forward Euler method in case there is only one spatial dimension. The forward Euler update equation is given by

$$U_i^{k+1} = U_i^k + \lambda(U_{i+1}^k - 2U_i^k + U_{i-1}^k),$$

for $i = 1, 2, \dots m$, where $\lambda := \mu \Delta t/\Delta x^2$ and $\mu \geq 0$. Using the rearrangement

$$U_i^{k+1} = (1 - 2\lambda)U_i^k + \lambda U_{i+1}^k + \lambda U_{i-1}^k$$

and assuming that $U_{-1}^k = 0$ and $U_{m+1}^k = 0$, it follows that

$$A = \begin{pmatrix} 1-2\lambda & \lambda & 0 & 0 & 0 & 0 & \cdots & \lambda \\ \lambda & 1-2\lambda & \lambda & 0 & 0 & 0 & \cdots & 0 \\ 0 & \lambda & 1-2\lambda & \lambda & 0 & 0 & \cdots & 0 \\ \vdots & & & & & & & \vdots \\ 0 & 0 & \cdots & 0 & 0 & \lambda & 1-2\lambda & \lambda \\ \lambda & 0 & 0 & \cdots & 0 & 0 & \lambda & 1-2\lambda \end{pmatrix}. \tag{5.52}$$

The matrix A is symmetric (A is equal to its transpose). By a result from operator theory, a symmetric matrix has real eigenvalues. Thus, the propagation of errors will remain bounded if all eigenvalues lie in the closed interval $[-1, 1]$. By an application of Gerschgorin's theorem (A.5), the eigenvalues of A all lie in the interval $[1 - 4\lambda, 1]$. Thus, there is the possibility of an eigenvalue outside the unit disk if $1 - 4\lambda < -1$; that is, if $\lambda \geq 1/2$. To eliminate this possibility our stability criterion is

$$\mu \frac{\Delta t}{\Delta x^2} \leq \frac{1}{2}, \tag{5.53}$$

the Courant–Friedrichs–Lewy (CFL) condition. Instabilities that lead to meaningless numerical results might occur if this condition is not met (see Exercise 5.34). The forward Euler method is called conditionally stable because inequaltiy (5.53) must be met to avoid instabilities. Unfortunately, the number 1 is an eigenvalue of A (see Exercise 5.31). Thus, although the propagation of errors will remain bounded, the errors will not be reduced under iteration by A.

Exercise 5.31. Show that 1 is an eigenvalue of the matrix (5.52).

Exercise 5.32. (a) Perform numerical experiments (using the forward Euler method with periodic boundary conditions) that reproduces a pattern for the Gray–Scott model similar to the pattern in Fig. 5.10. (b) Demonstrate that the qualitative results for part (a) do not change with the choice of the choice of step sizes in space and time. For example, repeat the experiment(s) in part (a) with a spatial grid containing twice the number of nodes. (b) Use the code written for part (a) to find a pattern generated by the Gray–Scott model that is clearly different from the pattern in Fig. 5.10 and does not correspond to a constant value of u.

Exercise 5.33. Explore the parameter space for the Gray–Scott model (using the forward Euler method with no flux boundary conditions).

Exercise 5.34. Write a forward Euler code for the one dimensional heat equation $u_t = \lambda u_{xx}$ with Neumann boundary conditions on the interval $[0, 1]$. Taking into account the stability condition [Eq. (5.17)], demonstrate with carefully designed numerical experiments that numerical instability occurs when the stability condition is not met.

Exercise 5.35. Determine a stability criterion analogous to inequality (5.17) for the forward Euler scheme with periodic boundary conditions for PDE (5.34).

We have explored some of the qualitative long-term behavior of the Gray–Scott model. The forward Euler method converges to the solution of the PDE as the discretization step sizes in space and time approach zero. But, how do we know that the qualitative behavior—for example, the qualitative behavior depicted in Fig. 5.10—is present in the solution of the PDE? This is a difficult question. The correct answer: We don't know! The only way to be sure is to prove a mathematical theorem. On the other hand, confidence in the correctness of the results of a numerical experiment can be gained in several ways that will be discussed. For example, the answer might be computed with different step sizes in (space and time) to confirm that the qualitative features of the computed values do not change (see Exercise 5.32), or the same solution might be approximated with a different numerical scheme.

Good practice suggests starting with a moderately small step size (perhaps the step size equal to 1) that is decreased systematically until the result of the experiment does not appear to change with the choice of the step size. A decease in step size theoretically decreases the truncation error. On the other hand, a decrease in step size tends to increase the number of numerical operations (additions, subtractions, multiplications, and divisions) thus increasing the effects of roundoff error due to the finite number of decimal digits stored in the computer. Another source of error is called condition error. One manifestation of this type of error is due to the big O estimates that determine the order of the numerical method. Recall that these estimates state that the error in some approximation is proportional to a power of the step size. But, the size of the constant of proportionality is ignored. It might happen that these constants are very large for a particular application, in which case it is called ill-conditioned. The errors due to roundoff and condition tend to accumulate with subsequent iterations of our numerical method. Hence, the global error will generally decrease with step size to some minimum value as the order estimates dominate, but it will increase (perhaps not monotonically) as the step size is decreased farther due to the accumulation of roundoff and condition errors.

5.4.5 Quantitative PDE: A Computational Challenge

The interesting qualitative behavior of the Gray–Scott model is pattern formation. Because it is difficult to quantify a particular pattern, we will consider a simpler quantitative problem as a vehicle for discussing the fundamental problem of numerical computation: *Do we obtain a good approximation of the correct answer*?

Let us suppose we are interested in an accurate and efficient computation of the concentrations u and v over some finite time interval on the spatial domain $[0, L] \times [0, L]$ with (zero) Neumann boundary conditions and given initial data. We must define (or have our scientific collaborator with expertise in chemistry define) what is meant by "accurate." For an applied problem, the desired accuracy will often be determined by the accuracy of the measurements in some experiment or the accuracy of the measurements of some physical parameters. The relative error (percent error) for approximation of numbers larger than 1 is usually more meaningful than the absolute error. Recall that for numbers a and b, where b is viewed as an approximation of a, the absolute error is $|a - b|$; and, in case $a \neq 0$, the relative error is

$$\text{relative error} := \frac{|a-b|}{|a|}.$$

For the concentrations in the Gray–Scott model, which will all be in the interval $[0, 1]$, the absolute error is the better choice. In practice, we do not know the number a that we are trying to approximate. But, it is sometimes possible to obtain a theoretical error estimate.

Suppose that an approximation of the solution of the Gray–Scott PDE is desired with an absolute error of less than 10^{-2}.

To meet the expectations and desires of our imaginary collaborator, let us set the following mathematical problems:

Problem 5.1. For the Gray–Scott model (5.34) on the spatial domain $[0, L] \times [0, L]$ with (zero) Neumann boundary conditions, with $L = 2.5$, the diffusions

$$\lambda = 2 \times 10^{-5}, \qquad \mu = 10^{-5},$$

the parameters

$$F = 0.0225, \qquad \kappa = 0.05,$$

and initial data (at time $t = 0$)

$$\begin{aligned} u(x,y) &= 1 - 0.5\frac{2^8}{L^8}(x^2y^2(x-L)^2(y-L)^2), \\ v(x,y) &= 0.25\frac{2^8}{L^8}(x^2y^2(x-L)^2(y-L)^2), \end{aligned} \tag{5.54}$$

approximate the average values of the concentrations u and v over the spatial domain at time $T = 1024$ with an absolute error of less than 10^{-2}.

Problem 5.2. Using the data of Problem 5.1, approximate the average values of the concentrations u and v at time $T = 10$ with an error of less than 10^{-2}.

The initial state in Problem 5.1 satisfies the boundary conditions and is chosen to lie between the steady states of the system. (No technical meaning is intended for the word "between.")

In practice, an "efficient" method is one that can be used to obtain the desired accuracy in a short time. This definition is not precise; the idea is that efficient methods are fast.

Our theoretical estimate of the global error for the forward Euler method is $O(\Delta t)$ and $O(\Delta x^2 + \Delta y^2)$. This means that we can expect the absolute error, which is defined to be the norm of the difference between the solution and its approximation, to be proportional to a quantity controlled by the sizes of $|\Delta t|$ and $|(\Delta x, \Delta y)|^2$. There are many possible choices for norms. In a finite-dimensional space all norms are equivalent, so the choices are not of fundamental importance. On the other hand, the choices of norms might be dictated by the physical problem.

A practical choice for the norm might lead to the representation of the theoretical error estimate in the form

$$|u - u_{\text{appx}}| + |v - v_{\text{appx}}| \leq M(|\Delta t| + \Delta x^2 + \Delta y^2),$$

where M is some unknown constant and the norms on the left-hand side of the inequality are Euclidean norms, maximum norms, or ℓ_1 norms. The maximum norm would be appropriate if we desire the computed concentrations at all points in the spatial grid to differ from the exact values by no more than some specified amount; that is, we would hope to achieve the result

$$\max_{i,j,k}\{|u(k\Delta t, i\Delta x, j\Delta y) - u_{ij}^k|\} < 10^{-2},$$
$$\max_{i,j,k}\{|v(k\Delta t, i\Delta x, j\Delta y) - v_{ij}^k|\} < 10^{-2},$$

where $i = 0, 1, 2, \ldots, m-1$, $j = 0, 1, 2, \ldots, n-1$, and $k = T/\Delta t$.

For Problems 5.1 and 5.2, we are challenged to approximate the average values of the states u and v over the spatial domain. Our numerical method produces approximations of the state variables on a grid of points covering the square $[0, L] \times [0, L]$. The desired averages are the integrals of the states over this domain divided by its area. Hence, we must approximate the integrals. For definiteness and simplicity, the trapezoidal rule is a viable choice. With kend $\Delta t = T$ (that is, kend corresponds to the final time T), we wish to achieve the error estimates

$$|\frac{1}{L^2}\int_0^L\int_0^L u(T,x,y)\,dxdy - \frac{1}{L^2}\mathrm{Trap}(u)| < 10^{-2},$$
$$|\frac{1}{L^2}\int_0^L\int_0^L v(T,x,y)\,dxdy - \frac{1}{L^2}\mathrm{Trap}(v)| < 10^{-2}, \tag{5.55}$$

where

$$\mathrm{Trap}(u) := \frac{\Delta x \Delta y}{4}\sum_{i,j}(u_{i,j}^{\mathrm{kend}} + u_{i+1,j}^{\mathrm{kend}} + u_{i+1,j+1}^{\mathrm{kend}} + u_{i,j+1}^{\mathrm{kend}}).$$

There is no simple method known to *prove* that the desired error is achieved. Of course, if we know the error exactly, we know the solution. So we cannot expect to prove that a numerical approximation achieves a desired error bound except in some special cases. On the other hand, we can test our numerical results in various ways to achieve a high confidence level that they are correct (cf. [10]).

5.4.6 Use a Convergent Algorithm

The numerical algorithms (for instance, the forward Euler method) used in this book are known to converge to the corresponding exact solutions. The proofs for their convergence are given in books on theoretical numerical analysis. The issue of convergence is fundamental. But, the mathematics required to present the proofs is beyond the scope of this book.

5.4.7 Use a Numerically Stable Algorithm

Algorithms for numerical computation must be either unconditionally stable or satisfy appropriate conditional stability criteria. For example, to successfully employ the forward Euler method in one space dimension, its stability criterion (5.17) must be respected. The appropriate stability criterion for two space dimensions will be discussed in the next subsection (see Eq. (5.57)).

5.4.8 Test Code against Known Solutions

The first confidence-building step is to test the numerical code that implements an algorithm against an example where the exact solution is known. The preferred choice for an example is a special case of the problem under consideration. For the Gray–Scott model with Neumann boundary data, it seems that no explicit solution is known. The next best choice is to use an exact solution for a model with similar features.

Let us test the forward Euler method with Neumann boundary data against the exact solution

$$v(t,x,y) = e^{(-F-2\lambda\pi^2/L^2)t}\cos\frac{\pi}{L}x\cos\frac{\pi}{L}y, \qquad u(t,x,y) = 1 + v(t,x,y)$$

of the linear system

$$\begin{aligned} u_t &= \lambda\Delta u + F(1-u), \\ v_t &= \lambda\Delta v - Fv, \\ u_x(t,0,y) &= u_x(t,L,y) = u_y(t,x,0) = u_y(t,x,L) = 0, \\ v_x(t,0,y) &= v_x(t,L,y) = v_y(t,x,0) = v_y(t,x,L) = 0 \end{aligned} \tag{5.56}$$

defined on the square $[0,L]\times[0,L]$.

The Neumann boundary condition, which in our case states that the normal derivative vanishes on the boundary, is implemented using the central difference approximation of the first derivative. For example, the vanishing of the partial derivative $u_x(t,0,y)$ (along the left-hand boundary of the spatial domain) leads to the numerical approximation

$$u(k\Delta t, -\Delta x, j\Delta y) = u(k\Delta t, \Delta x, j\Delta y).$$

After the boundary data is imposed, the forward Euler difference equations are applied to all grid points including the grid points on the boundary of the rectangular domain. The required values of the concentrations outside the spatial domain (which are required in the difference equations applied to grid points on the boundary of the spatial domain) are assigned by the Neumann boundary data to values at grid points in the spatial domain.

How should we make a numerical experiment to test against the known solution of the linear problem? The simplest idea would be to simply choose a grid size, compute over some time interval, and compare the results of the numerical experiment with the exact answer. How do we judge the success of our experiment? Which time-step size and grid size should we choose? How small an error is acceptable?

To answer the questions posed in the last paragraph, let us design an experiment that uses some numerical analysis.

Recall the stability criterion for the forward Euler method: For the diffusion equation in the case of one space dimension, the method is stable if

$$\lambda \frac{\Delta t}{\Delta x^2} \leq \frac{1}{2}.$$

For two space dimensions and $\Delta x = \Delta y$, the stability criterion is

$$\mathrm{CFL} := \lambda \frac{\Delta t}{\Delta x^2} \leq \frac{1}{4}. \tag{5.57}$$

The denominator on the right-hand side of this inequality comes from the factor 4 in the forward Euler difference equations written in the form

$$\begin{aligned}
U_{i,j}^{k+1} &= (1 - 4\,\mathrm{CFL})U_{i,j}^{k} + \mathrm{CFL}\big[(U_{i+1,j}^{k} + U_{i-1,j}^{k}) + U_{i,j+1}^{k} + U_{i,j-1}^{k}\big],\\
V_{i,j}^{k+1} &= (1 - 4\,\mathrm{CFL})U_{i,j}^{k} + \mathrm{CFL}\big[(U_{i+1,j}^{k} + U_{i-1,j}^{k}) + U_{i,j+1}^{k} + U_{i,j-1}^{k}\big].
\end{aligned}$$

The quantity $1 - 4\,\mathrm{CFL}$ (which replaces $1 - 2\lambda$ in our stability analysis of the one-dimensional forward Euler scheme) appears along the diagonal of the matrix representation of these difference equations. The abbreviation CFL stands for Courant–Friedrichs–Lewy. Indeed, this quantity is often called the Courant–Friedrichs–Lewy number (see [25]).

A prosaic way to describe the meaning of the CFL number is to say that once the spatial discretization is set, the time step must be set small enough so that nothing important happens over the space of one computational cell during one time step. The CFL number answers the question: How small? In truth, there is a subtlety that is not fully addressed here: The CFL condition as treated in this book is related to stability of numerical methods. But, a numerical method applied to a PDE might be stable and yet not give a good approximation of the desired solution. To be sure the solution does give a good approximation; exact discretized solutions should approach the solution of the PDE as the discretization sizes in space and time approach zero. A necessary condition to capture a reasonable approximation to the solution without passing to the limit—as we hope to do in every numerical simulation—is to at least use in the numerical scheme the part of the initial data that determines the solution at the desired point in space at some desired finite time. The necessity of this condition more accurately describes the CFL condition; it is not directly related to the numerical stability of the discrete time-marching scheme. A slightly more precise way to describe the prosaic meaning of CFL is to assume some measure V of the velocity of change of the solution in the spatial domain; perhaps the velocity of a passing wave of the velocity of a fluid. Also, consider the maximum number num in the normal direction to a face of a grid cell that are used to discretize. For example, this number for the centered difference, approximation of the second derivative is num $= 1$, as only one neighboring cell in each direction is used. The corresponding CFL condition is

$$\Delta t \leq \frac{\text{num}\Delta x}{V}.$$

The trouble is in determining V. Note that for our diffusion processes, the given CFL condition involves the diffusion speed.

According to theory, the global error for the forward Euler method is $O(\Delta t)$ and $O(\Delta x^2)$. If we fix CFL $\leq 1/4$ and choose the time step

$$\Delta t = \text{CFL}\frac{\Delta x^2}{\lambda},$$

then the global error should be $O(\Delta x^2)$. This observation is the basis for a useful test: Put $\lambda = 2 \times 10^{-5}$, $\mu = \lambda$, $F = 10^{-3}$, and $L = 2.5$; for $\ell = 3, 4, \ldots, 8$, set $m = 2^\ell$, $\Delta x = \Delta y = L/m$, and choose the grid $(i\Delta x, j\Delta y)$ with $i = 0, 2, \ldots, m$ and $j = 0, 2, \ldots, m$; define

$$\text{CFL} = \lambda\frac{256^2}{L^2} \approx 0.2097,$$

$$\Delta t = \text{CFL}\frac{\Delta x^2}{\lambda} = 2^{16-2\ell};$$

and apply the forward Euler method for $2^{2\ell-6}$ time steps $\ell = 3, 4, \ldots, 8$ so that the final time is always $T = 1024$. For each choice of ℓ compute

Interior Grid	Δt	error	pe/ce
8×8	1024	0.362346	
16×16	256	0.05383	6.7313
32×32	64	0.0126742	4.24721
64×64	16	0.00312647	4.05384
128×128	4	0.000779079	4.01304
256×256	1	0.000194612	4.00323

Table 5.1 Results of a numerical experiment are listed for system (5.56) with $\lambda = 10^{-3}$, $F = 10^{-1}$, and total integration time $T = 1024$. The last column is the previous maximum absolute error (over all the nodes in the spatial domain) divided by the current absolute error.

the error e_ℓ to be the maximum norm of the difference of the exact and computed values at the final time over all interior grid points; and, for $\ell > 3$, also compute the quantity $e_{\ell-1}/e_\ell$. Note that if

$$e_{\ell-1} = K\Delta x^2 \text{ and } e_\ell = K\Big(\frac{\Delta x}{2}\Big)^2,$$

then

$$\frac{e_{\ell-1}}{e_\ell} = 4;$$

in other words, our code is performing properly if the computed quotients of the errors approaches 4 as the number of nodes in the grid increases. On the other hand, note that the time-step size decreases as the grid size increases. For very large grids, roundoff errors are likely to accumulate and destroy the expected second-order global error.

The results of a numerical experiment are given in Table 5.1. It seems that the code is performing as expected with a second-order global error relative to the spatial discretization provided that the time step is chosen so that the CFL condition is met. Moreover, with a grid size of 256×256 and $\Delta t = 1$, the computed value agrees with the exact value to three decimal places; the error is less than 10^{-3}.

Warnings for the novice programmer: Implementing algorithms into computer code for solving partial differential equations is complicated. Standard mistakes occur when indexing arrays. Be careful when coding arrays with indices starting at zero (or some other number not equal to one). Remember that in most computer systems, the equality sign is used to mean replacement; for example, the statement $a = a + 1$ replaces the current value of a with the value $a + 1$. When updating arrays while solving difference equations, it might be tempting to replace an array U_{ij} with its updated value as defined perhaps by the forward Euler formulas. But, using the same symbol U_{ij} could lead to trouble: the original value U_{ij} might be needed later. Be sure to incorporate comments on every block within your code. Without thoughtful comments, your code will be incomprehensible when you return to check, debug, or modify its content. Write code using subroutines that can be checked independently. This method helps to preserve modularity. Also, well-tested subroutines can be used without modification in other projects. Declare the types of all variables used in your code. *The only way to debug a code is by printing out computed values at strategic locations within your code.*

5.4.9 Test Approximation Order Using Richardson Extrapolation

How can we test a code to see if it performs as expected with respect to the theoretical global error in case the exact solution is not available? Our answer to this question uses a gem of numerical analysis: Richardson extrapolation. We will discuss this method and apply it to the Gray–Scott model.

By adding Eqs. (5.39) and rearranging the resulting expression, the centered difference approximation of the second derivative is easily seen to be given by

$$U^0(h) := \frac{F(x+h) - 2F(x) + F(x-h)}{h^2} = F''(x) + 2\sum_{j=1}^{\infty} \frac{F^{(2j)}(x)h^{2j}}{(2j)!}. \tag{5.58}$$

Take note of the form of the error, which is a Taylor series in h^2. Richardson extrapolation requires knowledge of the form of the error written as a series in powers of some parameter, which is usually the size of the discretization as in this example. We will proceed under the assumption that the form of the error is a power series in even powers of the discretization size h, but this requirement is not essential (see Exercise 5.37 and compare [10, Appendix A]). Let us assume $U^0(h)$ is used to approximate some quantity U such that

$$U^0(h) = U + \sum_{j=1}^{\infty} K_{2j}^0 h^{2j}, \tag{5.59}$$

h	U^0	U^1	U^2	pe/ce
$h = 1$	-0.5972732058359690			
$h = 1/2$	-0.6362151182684900	-0.649195755745997		
$h = 1/4$	-0.6462604545058728	-0.649635179283789	-0.649636442863156	
$h = 1/8$	-0.6487914980893103	-0.649636828977658	-0.649636931241486	63.30447
$h = 1/16$	-0.6494254962555712	-0.649636932196862	-0.649636938957249	63.8253
$h = 1/32$	-0.6495840732115392	-0.649636938649834	-0.649636939078142	63.96
$h = 1/64$	-0.6496237222902605	-0.649636939053173	-0.649636939080032	64
$h = 1/128$	-0.6496336348624447	-0.649636939078382	-0.649636939080062	64

Table 5.2 A Richardson table for centered difference approximation of the second derivative of $\sin x$ *at* $x = 1/\sqrt{2}$ *is listed. The last column is the previous absolute error divided by the current absolute error, where the error is computed from the value* $\sin \frac{1}{\sqrt{2}} \approx -0.6496369390800624$. *This column indicated that the error in the third column is* $O(h^6)$.

where the K_{2j} are constants. This is exactly the case for the centered difference formula used to approximate the second derivative as soon as we fix x.

Because Eq. (5.59) is supposed to hold for all h, it holds for $h/2$, for which we have

$$U^0(\frac{h}{2}) = U + \sum_{j=1}^{\infty} K_{2j}^0 \frac{h^{2j}}{2^{2j}}. \tag{5.60}$$

The fundamental observation of Richardson extrapolation is simple: The formulas $U^1(h)$ and $U^1(h/2)$ both approximate U with a $O(h^2)$ error. By multiplying the second formula by 4, subtracting the first formula from the second, and dividing through by 3, we obtain the formula

$$U^1(h) := \frac{4U^0(\frac{h}{2}) - U^0(h)}{3} = U + \sum_{j=2}^{\infty} K_{2j}^1 h^{2j}, \tag{5.61}$$

where the K_{2j}^1 are the new constants obtained from the algebraic manipulations. *The new formula* $U^1(h)$ *approximates* U *with a* $O(h^4)$ *error!* Moreover, the new approximation formula is again a formula to which the same procedure can be applied. Doing so results in a new formula with a $O(h^6)$ error, and so on. Proceeding inductively, the Richardson extrapolation formula U^ℓ, for $\ell > 0$, is given by

$$U^\ell(h) = \frac{2^{2\ell} U^{\ell-1}(\frac{h}{2}) - U^{\ell-1}(h)}{2^{2\ell} - 1}; \tag{5.62}$$

it approximates U with a $O(h^{2(\ell+1)})$ error.

Discretization	Approximation	Error
$h = 1/2^2$	-0.646260455	0.00337648457
$h = 1/2^4$	-0.649425496	0.000211442825
$h = 1/2^6$	-0.649623722	1.321679E-05
$h = 1/2^8$	-0.649636113	8.26053417E-07
$h = 1/2^{10}$	-0.649636887	5.17169042E-08
$h = 1/2^{12}$	-0.649636935	3.63737629E-09
$h = 1/2^{14}$	-0.649636924	1.48132472E-08
$h = 1/2^{16}$	-0.649636745	1.93627182E-07
$h = 1/2^{18}$	-0.649635315	1.62413866E-06
$h = 1/2^{20}$	-0.649536133	0.000100806268

Table 5.3 The centered difference approximation of the second derivative of $\sin x$ *at* $x = 1/\sqrt{2}$ *and the approximation error are listed.*

Richardson extrapolation is used in several different ways. A direct application yieds high-order approximation formulas from low-order formulas. For instance, by an application of Eq. (5.62) with $\ell = 1$ to the centered difference [Eq. (5.58)], we produce the approximation

$$F''(x) = -\frac{1}{3h^2}[F(x+h) - 16F(x+\frac{h}{2}) + 30F(x) - 16F(x-\frac{h}{2}) + F(x-h)] + O(h^4). \tag{5.63}$$

Note that the higher-order accuracy requires two additional function evaluations. Alternatively, a sequence of approximations using the original centered difference formula for step sizes

$$h, \quad \frac{h}{2}, \quad \frac{h}{2^2}, \quad \ldots, \quad \frac{h}{2^j}, \quad \ldots$$

can be calculated and Richardson extrapolation can be applied to this sequence of numbers. An example of this procedure for the centered difference formula is tabulated in Table. 5.2. Note that accuracy to over 12 digits is obtained using $h = \Delta x = 1/128$ at the bottom of the third column of the Richardson table.

The centered difference approximation of the second derivative (of a sufficiently smooth) function certainly converges to the second derivative of the function. This fact is consistent with the data in Table 5.2. On the other hand, in a numerical computation with finite decimal approximations of numbers, the roundoff error and the condition error (which arises here from the error made in the function evaluations) will destroy the expected convergence as the discretization size decreases. The typical situation for a numerical approximation is for the error of the computation to decrease with the discretization size until it reaches some optimal value and then the error begins to increase as the roundoff and condition errors dominate. This phenomenon is illustrated by the data in Table 5.3. Using 10-digit arithmetic, the optimal error for approximating the second derivative of sin at $1/\sqrt{2}$ is on the order of 10^9 (which is perhaps not too surprising) with a discretization size of $1/2^{12} \approx 0.000244141$ (which might be surprising).

Let us return to our original question on assessing the order of the error of a numerical procedure with respect to a discretization parameter applied to approximate a quantity that is not known. Good estimates are possible using a reinterpretation of Richardson's extrapolation.

Suppose the theoretical error for an order-two procedure $U^0(h)$ is known to be as in Eq. (5.59), and we wish to verify that our implementation is performing with the expected $O(h^2)$ error. We again compute $U^0(h/2)$ and subtract the error formulas to obtain

$$U^0(h) - U^0(\frac{h}{2}) = \frac{3}{4}K_2h^2 + O(h^4).$$

Next, we substitute

$$K_2h^2 = U^0(h) - U + O(h^4)$$

to obtain the formula

$$U^0(h) - U^0(\frac{h}{2}) = \frac{3}{4}(U^0(h) - U) + O(h^4),$$

or its rearrangement

$$U^0(h) - U = \frac{4}{3}(U^0(h) - U^0(\frac{h}{2})) + O(h^4). \tag{5.64}$$

The quantity

$$\text{RE}(h) := \frac{4}{3}(U^0(h) - U^0(\frac{h}{2})), \tag{5.65}$$

which can be computed without knowing U, is a $O(h^4)$ approximation of the error $U^0(h) - U$. Under our assumption that the error is $O(h^2)$, we can substitute $U^0(h) - U = O(h^2)$ in Eq. (5.64) and conclude that

$$\text{RE}(h) = O(h^2).$$

h	32×32	64×64	128×128	256×256
$\Delta t = 1/4$	0.955307	0.955303	0.955301	0.955300
$\Delta t = 1/8$	0.977365	0.977362	0.977361	0.977361
$\Delta t = 1/16$	0.988587	0.988585	0.988585	0.988585
$\Delta t = 1/32$	0.994266	0.994266	0.994266	0.994265
$\Delta t = 1/64$	0.997126	0.997126	0.997126	0.997126
$\Delta t = 1/128$	0.998561	0.998561	0.998561	0.998561

Table 5.4 The order estimates in this table are for an implementation of the forward Euler method using the data of Problem 5.2. The floating point numbers are computed for the average of the concentrations $(u + v)/2$ using Eq. (5.67).

h	32×32	64×64	128×128	256×256
$\Delta t = 1/4$	0.0068953	0.006898	0.00690026	0.00690022
$\Delta t = 1/8$	0.00353214	0.00353455	0.00353486	0.00353496
$\Delta t = 1/16$	0.00178909	0.00178955	0.00178951	0.00178949
$\Delta t = 1/32$	0.000900304	0.000900353	0.000900337	0.000900331
$\Delta t = 1/64$	0.000451592	0.00045157	0.000451573	0.00045157
$\Delta t = 1/128$	0.000226156	0.000226133	0.000226138	0.000226139

Table 5.5 The error estimates in this table are for an implementation of the forward Euler method using the data of Problem 5.2. The floating point numbers are computed by averaging the absolute error estimates [Eq. (5.69)*] corresponding to the two concentrations.*

The desired test of our procedure is obtained by computing RE(h). In fact, we can be confident that our procedure is performing at the theoretical second-order if

$$\frac{\text{RE}(h)}{\text{RE}(\frac{h}{2})} \approx 2^2.$$

The factor $4/3$ cancels in the quotient; it does not play a role in this test.

Using the same idea in the general case, where the error is given by an order estimate whose first nonzero term is $K_j h^j$, the corresponding ratio is

$$\frac{U^0(2h) - U^0(h)}{U^0(h) - U^0(\frac{h}{2})} = 2^j, \tag{5.66}$$

where j is the order of the method (see Exercise 5.37). In other words, the order of the method is given (approximately) by

$$j = \log_2 \frac{U^0(2h) - U^0(h)}{U^0(h) - U^0(\frac{h}{2})}. \tag{5.67}$$

Exercise 5.36. Apply the test given in Eq. (5.66) to confirm that Euler's method is order one for the Gray–Scott system with no diffusion.

Exercise 5.37. Show that the Richardson extrapolation formula for an approximation scheme $U^0(h)$, where

$$U^0(h) = U + \sum_{j=1}^{\infty} K_j h^j,$$

is given inductively by

$$U^j(h) = \frac{2^j U^{j-1}(\frac{h}{2}) - U^{j-1}(h)}{2^j - 1} \tag{5.68}$$

and the error estimate (corresponding to Eq. (5.65)) is

$$\mathrm{RE}^1(h) := 2(U^0(h) - U^0(h/2)). \tag{5.69}$$

In Problem 5.1, the PDE includes diffusion and nonlinear reaction. The forward Euler method is a second-order scheme for approximating the diffusion process provided that we choose the time step so that the CFL number is fixed and less than $\frac{1}{4}$. On the other hand, the forward Euler method applied to the reaction ODE is first-order by Exercise 5.36. Thus, we can expect the forward Euler method to be first-order when applied to the Gray–Scott PDE, which combines reaction and diffusion. To confirm this estimate by a numerical experiment, we can compute (with the data in Problem 5.2) using a fixed spatial grid and Eq. (5.67). The results of an experiment are given in Table 5.4; they indicate that the forward Euler approximations are indeed order-one. Richardson absolute error estimates, computed using Eq. (5.69) and given in Table 5.8 suggest that the absolute error for step sizes smaller than 1/64 and for grid sizes at least 64×64 is less than 10^{-3}. The Richardson tables strongly suggest that the average over both concentrations is approximately 0.4530.

A conservative analysis suggests that we can be confident that the computed values are correct with an error of less than 10^{-2} for a spatial-grid size of 128×128 and a time step of about 1/16. A similar accuracy seems to be achieved with larger time steps (perhaps $1/2$, $1/4$, and $1/8$) together with one Richardson extrapolation. To reach time $T = 10$, a time step of $1/128$ requires 1280 steps. On the other hand, the proposed Richardson extrapolation with time steps $1/2$ and $1/4$ requires 60 time steps. Clearly, the Richardson extrapolation offers a more efficient procedure.

h	U^0	U^1	U^2	U^3	U^4	U^5
$\Delta t = 1/4$	0.453113					
$\Delta t = 1/8$	0.453069	0.453036				
$\Delta t = 1/16$	0.453046	0.453027	0.453084			
$\Delta t = 1/32$	0.453035	0.453024	0.453054	0.453062		
$\Delta t = 1/64$	0.453029	0.453024	0.453039	0.453043	0.453045	
$\Delta t = 1/128$	0.453026	0.453023	0.453031	0.453033	0.453034	0.453034
$\Delta t = 1/4$	0.453112					
$\Delta t = 1/8$	0.453067	0.453035				
$\Delta t = 1/16$	0.453045	0.453025	0.453083			
$\Delta t = 1/32$	0.453033	0.453023	0.453052	0.453061		
$\Delta t = 1/64$	0.453028	0.453022	0.453037	0.453042	0.453043	
$\Delta t = 1/128$	0.453025	0.453022	0.45303	0.453032	0.453033	0.453033
$\Delta t = 1/4$	0.453111					
$\Delta t = 1/8$	0.453067	0.453034				
$\Delta t = 1/16$	0.453044	0.453025	0.453082			
$\Delta t = 1/32$	0.453033	0.453022	0.453052	0.45306		
$\Delta t = 1/64$	0.453027	0.453022	0.453037	0.453041	0.453043	
$\Delta t = 1/128$	0.453024	0.453022	0.453029	0.453031	0.453032	0.453033
$\Delta t = 1/4$	0.453111					
$\Delta t = 1/8$	0.453067	0.453034				
$\Delta t = 1/16$	0.453044	0.453025	0.453082			
$\Delta t = 1/32$	0.453033	0.453022	0.453052	0.45306		
$\Delta t = 1/64$	0.453027	0.453022	0.453037	0.453041	0.453043	
$\Delta t = 1/128$	0.453024	0.453021	0.453029	0.453031	0.453032	0.453032

Table 5.6 Four Richardson extrapolation tables are listed for the average concentration $(u+v)/2$ *using Eq.* (5.68) *applied to an implementation of the forward Euler method for the data of Problem 5.2. The spatial grid sizes for the tables (top to bottom) are* 32×32, 64×64, 128×128, *and* 256×256.

h	32×32	64×64	128×128	256×256
$\Delta t = 1/4$	1.83847	1.07313	0.983787	1.01402
$\Delta t = 1/8$	1.49828	1.51129	0.982416	1.00673
$\Delta t = 1/16$	1.26883	1.50625	0.989738	1.00323
$\Delta t = 1/32$	1.14148	1.33578	0.994583	1.00157
$\Delta t = 1/64$	1.0734	0.997854	0.997225	1.00077
$\Delta t = 1/128$	0.904951	0.867657	0.998597	1.00038

Table 5.7 The order estimates in this table are for an implementation of the forward Euler method using the data of Problem 5.1. The floating point numbers are computed for the average of the concentrations $(u+v)/2$ *using Eq.* (5.67).

h	32×32	64×64	128×128	256×256
$\Delta t = 1/4$	2.96827×10^{-8}	0.0482585	0.0856009	0.180438
$\Delta t = 1/8$	1.08615×10^{-8}	0.0308683	0.0435401	0.0936266
$\Delta t = 1/16$	4.62043×10^{-9}	0.0168982	0.0221594	0.0459164
$\Delta t = 1/32$	2.12758×10^{-9}	0.00875027	0.0111671	0.0232914
$\Delta t = 1/64$	1.02037×10^{-9}	0.00444146	0.00560387	0.0117545
$\Delta t = 1/128$	4.99665×10^{-10}	0.00223628	0.0028068	0.00590156

Table 5.8 The error estimates in this table are for an implementation of the forward Euler method using the data of Problem 5.1. The floating point numbers are computed by averaging the absolute error estimates [Eq. (5.69)*] corresponding to the two concentrations.*

h	U^0	U^1	U^2	U^3	U^4	U^5
$\Delta t = 1/4$	0.485018					
$\Delta t = 1/8$	0.485018	0.485018				
$\Delta t = 1/16$	0.485018	0.485018	0.485018			
$\Delta t = 1/32$	0.485018	0.485018	0.485018	0.485018		
$\Delta t = 1/64$	0.485018	0.485018	0.485018	0.485018	0.485018	
$\Delta t = 1/128$	0.485018	0.485018	0.485018	0.485018	0.485018	0.485018
$\Delta t = 1/4$	0.463847					
$\Delta t = 1/8$	0.463847	0.463850				
$\Delta t = 1/16$	0.463844	0.463840	0.463847	0.463847		
$\Delta t = 1/32$	0.463838	0.463841	0.463843	0.463843		
$\Delta t = 1/64$	0.463836	0.463837	0.463839	0.463840	0.463840	
$\Delta t = 1/128$	0.463836	0.463835	0.463837	0.463838	0.463838	0.463838
$\Delta t = 1/4$	0.352699					
$\Delta t = 1/8$	0.352471	0.352276				
$\Delta t = 1/16$	0.352357	0.352248	0.352549			
$\Delta t = 1/32$	0.352300	0.352243	0.352395	0.352439		
$\Delta t = 1/64$	0.352271	0.352242	0.352319	0.352340	0.352349	
$\Delta t = 1/128$	0.352257	0.352242	0.352280	0.352291	0.352296	0.352298
$\Delta t = 1/4$	0.336485					
$\Delta t = 1/8$	0.336403	0.336317				
$\Delta t = 1/16$	0.336362	0.336320	0.336430			
$\Delta t = 1/32$	0.336342	0.336321	0.336376	0.336392		
$\Delta t = 1/64$	0.336332	0.336321	0.336349	0.336357	0.336360	
$\Delta t = 1/128$	0.336327	0.336322	0.336335	0.336339	0.336341	0.336341

Table 5.9 Four Richardson extrapolation tables for the average concentration $(u + v)/2$ *using Eq.* (5.68) *applied to an implementation of the forward Euler method for the data of Problem 5.1 are listed. The spatial grid sizes for the tables (top to bottom) are* 32×32, 64×64, 128×128, *and* 256×256.

For a less conservative and more efficient procedure, the computation can be made with a 64×64 spatial grid and a time step of $1/4$. Computed results, with a high degree of confidence, should meet the 10^{-2} error tolerance.

Note that Richardson extrapolation is used only on the computed values at the final time in the experiments reported in this section. Clearly, the accuracy of the code can be improved by incorporating Richardson extrapolation at each time step (see Exercise 5.39).

Although some appreciation of the meaning of the word "efficiency" for a computational algorithm can be gained by reading books, its true meaning is best understood by direct experience of the time required to perform computations. The choice between codes that produce a result in a few seconds or a few hours may not be of fundamental importance if the code is only run once to solve a textbook problem. But, an improvement of just a few percent might be essential for the success of a production code.

Because we are dealing with a system of equations in Problem 5.2, it is natural to consider both concentrations u and v in our error analysis. Thus, the corresponding averaged values are reported in this section. Taking into account our analysis and the computed values of u and v not listed but used to obtain the averaged concentrations, the result of our forward Euler approximations strongly suggest that (at time $T = 10$ and with an error of less than 10^{-2}) the average u concentration is 0.8084 and the average v concentration is 0.09768.

Computational results for Problem 5.1 are listed in Tables 5.7, 5.8, and 5.9. They suggest that the forward Euler scheme is operating as an order-one method for the grid sizes 128×128 and 256×256. The computed error is on the order of 10^{-2}, but this is not assured. In this case, the average of the concentrations in Table 5.9 do not seem to converge beyond the first decimal point. Taking the corresponding extrapolated values for the averaged concentrations u and v, reasonable approximations for these averages at $T = 1024$ are 0.5 for u and 0.1 for v. The accuracy of the computation seems to degrade as the time interval of computation increases. This is probably due to the accumulation of truncation error. An error of 10^{-3} per step might accumulate to an error of 1.0 after 1000 steps.

Problem 5.1 is not a typical textbook exercise; it is a genuine computational challenge.

Our reaction-diffusion model has an important complicating feature that has not yet been mentioned: The system evolves on two time scales. The diffusion evolves on a slow time scale of order 10^{-5} due to the diffusion coefficients; the reaction evolves on a fast time scale of order one. How can we take this into account to obtain a more accurate numerical method?

At this point, we should have a high level of confidence in our computed averages at $T = 10$ and a low level of confidence in our computation at $T = 1024$. Certainly, more testing is warranted.

Exercise 5.38. Discuss which is better (taking into account efficiency and accuracy): Richardson extrapolation applied to three runs of forward Euler (with the data of Problem 5.1) for the step sizes $1/4$, $1/8$, and $1/16$, and a 256×256 spatial grid or one run with step size $1/128$ and a 128×128 spatial grid.

Exercise 5.39. Repeat the numerical work for Problems 5.1 and 5.2 with a forward Euler integration that incorporates Richardson extrapolation after the second time step: compute two time steps, extrapolate, use the extrapolated value to be the computed value, compute another step, and so on. Compare the results with the results reported in this section.

Exercise 5.40. Repeat the numerical work for Problems 5.1 and 5.2 with one change: use Dirichlet boundary conditions such that the concentrations u and v on the boundary are held constant at $u = v = 0.5$.

5.5 BEYOND EULER'S METHOD FOR REACTION-DIFFUSION PDE: DIFFUSION OF GAS IN A TUNNEL, GAS IN POROUS MEDIA, SECOND-ORDER IN TIME METHODS, AND UNCONDITIONAL STABILITY

The forward Euler method is easy to program, it runs fast, and with acceleration (extrapolation applied to increase accuracy) it can be used to produce reasonably efficient and accurate results. On the other hand, it has at least two weaknesses: (1) the ratio of the time-step size and the square of the spatial discretization size must be small to avoid numerical instability (see Exercise 5.34) and (2) the method is first-order in time. Both of these deficiencies have been overcome with the development of discretization methods that are unconditionally stable and second-order in time. One such method that is often used is due to John Crank and Phyllis Nicolson [28]; its derivation, order estimates, numerical stability, and implementation are discussed in the following sections.

Second-order methods are introduced after development of two model problems: gas diffusion in air and gas motion in porous media. These models have independent interest, but their main function here is to demonstrate via realistic applications that numerical simulations using Euler's method are not efficient due to the CFL restriction on step size. Of course, to appreciate this fact, the reader is encouraged to perform the suggested exercises. Further on, after second-order methods are introduced, the exercises can be repeated to compare and contrast the efficacy of first- and second-order numerical algorithms in practice.

5.5.1 Gas Diffusion in a Tunnel

A portion of a cylindrical tunnel with radius $a = 4\,\mathrm{m}$ is $\ell = 25\,\mathrm{m}$ long. A gas with molecular weight of $21.3\,\mathrm{g}\,/\,\mathrm{mol}$ leaks near the entrance of the tunnel at the rate of $2\,\mathrm{L}\,/\,\mathrm{s}$. A sensor, installed in the tunnel $22\,\mathrm{m}$ from the entrance, records the concentration of this gas in $\mathrm{mg}\,/\,\mathrm{L}$ once per minute. The gas is absorbed in a filter at the far end of the tunnel; it absorbs 90% of the concentration of the leaked gas from the $1.5\,\mathrm{m}^3\,/\,\mathrm{s}$ of air passing through the filter. The problem is to simulate the sensor output as a function of time from the instant the gas is released until the sensor reads $2\,\mathrm{mg}\,/\,\mathrm{L}$. Assume the following: The diffusivity of the gas in air is $0.5 \times 10^{-4}\,\mathrm{m}^2\,/\,\mathrm{s}$, $1\,\mathrm{mol}$ of gas occupies a volume of $22.4\,\mathrm{L}$, and $1\,\mathrm{L} = 0.001\,\mathrm{m}^3$.

Let u denote the concentration of gas in $\mathrm{mol}\,/m^3$ and assume the gas concentration is constant in each cross section of the tunnel so that u is a function of the distance x measured in meters from the entrance and time t measured in seconds. In this case, the diffusion in the tunnel is modeled by the usual PDE

$$u_t = \kappa u_{xx} \tag{5.70}$$

for $0 < x < \ell$, $\kappa = 0.6 \times 10^{-4}\,\mathrm{m}^2\,/\,\mathrm{s}$, and initial data is $u(x, 0) = 0$ on the same spatial domain.

At the end of the tunnel, gas is filtered out. Thus, there is a net loss of gas corresponding to outflux across the cross section at $x = \ell$. The outflux is

$$\pi a^2 \kappa u_x,$$

a quantity measured in $\mathrm{mol}\,/\,\mathrm{s}$. This is due to the filter processing air at the rate $R = 1.5\,\mathrm{m}^3\,/\,\mathrm{s}$ with an efficiency $E = 0.9$ times the gas concentration of $u(\ell, t)$ mol / s. Thus, the boundary condition at the end of the tunnel is

$$\pi a^2 \kappa u_x(\ell, t) = -ERu(\ell, t). \tag{5.71}$$

The negative sign corresponds to gas leaving the tunnel. In fact, $u(\ell - \Delta x, t)$ must be larger than $u(\ell, t)$ because gas is removed from the air at the boundary. This means the approximate derivative

$$u_x(\ell, t) \approx \frac{u(\ell - \Delta x, t) - u(\ell, t)}{-\Delta x}$$

is negative as in the boundary condition.

At the entrance of the tunnel, there seems to be more than one choice for a viable boundary condition. In case the gas is injected into the tunnel, it is natural to consider its influx across the cross section at $x = 0$, which is the rate of injection

$$r = 2\frac{\mathrm{L}}{\mathrm{s}} \times \frac{1}{22.4}\frac{\mathrm{mol}}{\mathrm{L}} = \frac{1}{11.2}\frac{\mathrm{mol}}{\mathrm{s}}.$$

The corresponding boundary condition is

$$\pi a^2 \kappa u_x(0, t) = -r(t). \tag{5.72}$$

The minus sign indicates that the concentration of the gas is larger outside the tunnel.

An alternative boundary condition, corresponding to pure diffusion across the boundary, is obtained by setting the concentration at the entrance to be the concentration of gas in an adjoining chamber or part of the tunnel. A typical choice for the adjoining volume might be $\mathcal{V} = 1.0\,\mathrm{m} \times \pi a^2$. In this case, the concentration at $x = 0$ is set equal to the concentration C (measured in mol / m^3) in the chamber that is given by

$$C(t) = \frac{t}{11.2\pi a^2}.$$

The boundary condition is

$$u(0, t) = C(t) \tag{5.73}$$

(see Exercise 5.41).

Building a mathematical model for a real application is not a series of precise mathematical deductions. Assumptions are made as in the models of the two boundary conditions at the entrance of the contaminated tunnel. The best choice of boundary condition would be determined by experiments in controlled conditions for a scaled physical model of the tunnel in a laboratory or perhaps in some real tunnel. The model derived here has several shortcomings: The filtering process will cause the air in the tunnel to move. Convection is expected to transport gas (or heat) at a much faster rate than diffusion. The gas from the leak might increase the gas pressure at the entrance of the tunnel. This force would cause the mixture of gas and air to move away from the entrance. Perhaps you will produce a more accurate model after reading further in this book (see Exercise 19.10).

The contaminated tunnel model for the concentration u of the contaminant gas (in moles per cubic meter) discussed in subsequent sections is

$$\begin{aligned} u_t &= \kappa u_{xx}, \\ u(x,0) &= 0, \\ \pi a^2 \kappa u_x(0,t) &= -r(t), \\ \pi a^2 \kappa u_x(\ell,t) &= -ERu(\ell,t), \end{aligned} \tag{5.74}$$

where

$$\kappa = 0.5 \times 10^{-4}, \qquad a = 4.0, \qquad \ell = 25,$$
$$E = 0.9, \qquad R = 1.5, \qquad r = 1/11.2.$$

The desired output stt (the sensor time trace) is the concentration at $x = 22$ in $\mathrm{mg}\,/\,\mathrm{L}$ with time τ measured in minutes given by the formula

$$\mathrm{stt}(\tau) = 21.3u(22, 60\tau). \tag{5.75}$$

Exercise 5.41. (a) Approximate the sensor output for the contaminated tunnel model (5.74). In particular, determine the length of time in hours until the sensor output is $2.0\,\mathrm{mg}\,/\,\mathrm{L}$. (b) Approximate the sensor output with the boundary condition (5.73) and determine the length of time in hours until the sensor output is $2.0\,\mathrm{mg}\,/\,\mathrm{L}$. (c) Compare and contrast the simulated sensor output for the two models, and discuss the viability of the model and the suggested boundary conditions. (d) Suppose the gas leak persists for 24 hours after which it ceases. Determine the time in hours when the concentration at the sensor reaches its maximum value and specify this maximum. (e) Consider Exercise 5.78 after reading the following sections on the trapezoidal and Crank–Nicolson methods.

5.5.2 Gas in a Porous Medium

For the motion of a gas in a porous medium, diffusion due to the concentration gradient of the gas is generally so slow (due to the obstruction of the porous material) that it is ignored in the modeling process in favor of motion due to gas pressure.

To make a model, reconsider the conservation laws (5.1) and (5.2) that lead to the dynamical equation

$$u_t + \operatorname{div}(-K \operatorname{grad} u + \mu V) = f,$$

where u is the amount of the substance under consideration (the gas), V is the velocity of the medium moving the substance, and f is the function representing a model of the sources or sinks of the substance. For simplicity, assume that there are no sources or sinks, so that $f = 0$, and ignore the diffusion by setting the diffusivity K to zero. These assumptions reduce the conservation law to the equation

$$u_t + \operatorname{div}(\mu V) = 0.$$

For gas flow in a porous medium, the usual state variable is the density ρ measured in some average void in the porous material. In the derivation of the conservation law, u is the density of some substance in the space that it occupies, which in this case would include the porous media. More precisely, the density of the substance in the conservation law is to be determined by taking the limit of the total amount of the substance in an open set divided by the volume of the open set as the size of the set shrinks to a point. This viewpoint in the case of a gas in a porous medium would produce a density that would be some fraction of the density ρ of the fluid itself, a quantity given by $\epsilon\rho$, where $0 < \epsilon < 1$ is the ratio of the volume of the voids to the total volume of the material. The dimensionless number ϵ is called the porosity of the material. Using this dimensionless parameter, the conservation law for the density of a gas in a porous medium is

$$\epsilon\rho_t + \operatorname{div}(\rho V) = 0, \tag{5.76}$$

where V is the velocity of the gas and the function μ in the conservation law here equals ρ because all the gas is moving with velocity V.

Reusing the symbol μ for the viscosity of the gas and k for the permeability of the medium, Darcy's (constitutive) law relates the velocity of the fluid to the pressure p (that is, force per unit area) acting on the fluid:

$$\mu V = -k \operatorname{grad} p. \tag{5.77}$$

The minus sign is there because a positive pressure gradient drives the gas toward a region of lower pressure.

Using Darcy's law and the conservation of mass, the governing equation takes the form

$$\epsilon \rho_t = \operatorname{div}(\frac{k}{\mu}\rho \operatorname{grad} p).$$

The model is closed by specifying the pressure or relating the pressure to the gas density via an equation of state (a relation between state variables), which in this case relates pressure to density. In symbols, equations of state are functional relations $F(\rho, p) = 0$ determined by the modeling process. For gases, the usual choice (which can be partially justified by using generally accepted theories of thermodynamics) is a power law

$$p = p_0 \Big(\frac{\rho}{\rho_0}\Big)^{\gamma}, \tag{5.78}$$

where p_0 is a reference pressure, ρ_0 is some reference density, and the exponent γ is in the range $\gamma \geq 1$. Note that the presence of the reference pressure and density in the formula ensure that p has the units of pressure (force per area).

The (constitutive) equation of state combined with the governing equation is the dynamical equation for gas flow in a porous medium:

$$\epsilon \rho_t = \operatorname{div}(\frac{k}{\mu}\rho \operatorname{grad}(p_0\Big(\frac{\rho}{\rho_0}\Big)^{\gamma})) = \operatorname{div}(\frac{p_0 k \gamma}{\mu}\rho\Big(\frac{\rho}{\rho_0}\Big)^{\gamma-1} \operatorname{grad} \frac{\rho}{\rho_0}). \tag{5.79}$$

By defining the dimensionless and positive state variable

$$u = \frac{\rho}{\rho_0},$$

the dimensionless time

$$s = \frac{\gamma p_0}{\epsilon \mu} t,$$

and dimensionless length

$$\xi = \frac{x}{\sqrt{k}},$$

the dynamical equation [Eq. (5.79)] is recast into the dimensionless porous medium equation

$$u_s = \nabla \cdot (u^\gamma \nabla u). \tag{5.80}$$

It is a nonlinear PDE whose dimensionless form is suitable for theoretical work.

As a specific test case, consider a rectangular block of sandstone (located underground) with length $2.0\,\mathrm{m}$, width and height 1.0 m. Its porosity is 0.1 and its permeability is $10^{-8}\,\mathrm{cm}^2$. A gas (perhaps methane) exists in a reservoir adjacent to one end of the block so that a face with area $1.0\,m^2$ is exposed to the gas reservoir. The other end of the block is open to the atmosphere, and the lateral surface is bounded by impermeable rock. The gas has viscosity $12.119 \times 10^{-6}\,\mathrm{Pa\,s}$ and pressure $6000\,\mathrm{kPa}$. Assume the given equation of state [Eq. (5.78)] has exponent $\gamma = 2$; the density of methane is approximately 16.04 grams per mole; the universal gas constant R is approximately 8.3 joules per mole per kelvin, and the ideal gas law is $PV = nRT$, where P is pressure, V is volume, n is number of moles, and T is absolute temperature (kelvins). The problem is to determine the transient to a steady state and the gas flux through the downstream end of the block at steady state (see Exercise 5.42).

The change of variables used to derive the porous medium equation is not suitable for numerical approximations of solutions for the gas-sandstone application. Why? Instead, let ℓ be the characteristic length of $1.0\,\mathrm{m}$, ρ_0 be a reference density, and use the scaling

$$x = \ell\xi, \qquad t = \frac{\ell^2\mu}{kp_0}s, \qquad \rho = \rho_0 U \tag{5.81}$$

to obtain the dimensionless porous medium model

$$U_s = \frac{\gamma}{\epsilon}\nabla \cdot (U^\gamma \nabla U), \tag{5.82}$$

where it is essential to note that ∇ now denotes differentiation with respect to the dimensionless variable ξ. In fact, u is to be viewed as a function

$U = U(\xi, s)$, where the corresponding function in the original variables is

$$u(x,t) := U(\frac{x}{\ell}, \frac{kp_0}{\ell^2\mu}t).$$

Boundary conditions are problematic as there is no obvious choice. One approach is to relate the gas flux through a boundary to the pressure difference inside and outside the medium.

Under the assumption of a homogeneous and isotropic medium, so that the geometry of the physical domain is idealized to one spatial dimension, the gas flux through a cross section is then (in dimensioned variables) area times density times velocity

$$\ell^2 \frac{p_0 \rho_0 k\gamma}{\mu} u^\gamma u_x$$

at the cross section. It has units of mass per time. This quantity is related to the (dimensionless) pressure difference

$$c(\Big(\frac{\rho}{\rho_0}\Big)^\gamma - \frac{p_a}{p_0})$$

using the ambient pressure p_a (of the gas) and a new constant c that has units of mass per time. The corresponding boundary condition is

$$\ell^2 \frac{p_0 \rho_0 k\gamma}{\mu} u^\gamma u_x = \pm c(u^\gamma - \frac{p_a}{p_0}); \tag{5.83}$$

or, in dimensionless variables,

$$U^\gamma U_\xi = \pm\alpha(U^\gamma - \frac{p_a}{p_0}), \tag{5.84}$$

where

$$\alpha := \frac{c\mu}{\ell p_0 \rho_0 k\gamma}.$$

Note that pressure refers to pressure of the substance with density ρ and, for example, high pressure in one vicinity relative to another means the presence of *less* of the substance in the latter vicinity. Using this fact and taking into account the direction of the outer normal, the plus sign is taken at the left end boundary (with respect to the direction of the spatial coordinate axis); the minus sign is taken at the right end boundary.

By choosing the reference pressure of $p_0 = 6000\,\text{kPa}$, computing an approximate reference density $\rho_0 = 1.7\,\text{kg}\,/\,\text{m}^3$ via the ideal gas law, and (absent experimental data) taking $c = 0.01\,\text{kg}\,/\,\text{s}$, the dimensionless constant α is computed to be approximately $\alpha = 0.00594$.

Although all the ingredients are in place to make a numerical computation using Euler's method for time stepping and the second-order central difference approximation for second derivatives, there are several issues that need to be addressed in writing code.

The porous medium PDE written in dimensionless form [Eq. (5.82)] may not reveal the many options for discretization of the spatial derivative $\nabla \cdot (U^\gamma \nabla U)$. One possibility is to expand the derivative to

$$\nabla \cdot (U^\gamma \nabla U) = \gamma U^{\gamma-1} \nabla U \cdot \nabla U + U^\gamma \Delta U$$

and discretize (in one space dimension) as follows

$$\gamma U_i^{\gamma-1} \Big(\frac{U_{i+1} - U_{i-1}}{2\Delta x} \Big)^2 + U_i^\gamma \frac{U_{i+1} - 2U_i + U_{i-1}}{\Delta x^2}. \tag{5.85}$$

The central difference approximation of the first derivative seems natural, but it could be replaced by a forward or backward difference. An alternative method is obtained by compressing the spatial derivative to

$$\nabla \cdot (U^\gamma \nabla U) = \frac{1}{\gamma + 1} \Delta (U^{\gamma+1})$$

and using the discretization

$$\frac{1}{\gamma+1} \frac{U_{i+1}^{\gamma+1} - 2U_i^{\gamma+1} + U_{i-1}^{\gamma+1}}{\Delta x^2}. \tag{5.86}$$

The boundary conditions [Eqs. (5.84)] may also be discretized in several ways. A natural choice at the left boundary, using the interior nodes from $i = 1$ to $i = n - 1$ as the computational domain and $i = 0$ as the left boundary, is

$$U_0^\gamma \frac{U_1 - U_0}{\Delta x} = \alpha (U_0^\gamma - \frac{p_a}{p_0}). \tag{5.87}$$

At each time step, U_1 is known; the required value of U_0 can be obtained by approximating the solution of the nonlinear equation. A similar approach can be used at the right boundary.

Exercise 5.42. (a) For the case of one spatial variable, use the porous medium equation, the boundary conditions, zero density in the porous medium at the initial time, and the other data given in this section to approximate the gas flux at the middle of the porous block as a function of time until 30 seconds after the initial time taken to be $t = 0$. Report the flux at the middle cross-section of the block in kg / s and the time in seconds.
(b) For the case of one spatial variable, use pencil and paper to determine the general solution of the porous medium equation at steady state. Impose the boundary conditions and use a numerical computation to approximate the solution(s) of the steady state BVP. Use your result to determine the gas flux at the downstream boundary when the flow is in steady state.
(c) How long is the transient in minutes measured from the instant the flow starts until the root mean square distance of the density profile in the porous block is within 1% of the steady state density profile.
Hints: One good way to debug a numerical code is to use a stable steady state as initial data. Stepping forward in time should leave the steady state unchanged. For part (a) it might be wise to use a second-order accurate numerical method to approximate the velocity at the cross section. This can be achieved by using a centered difference across the section; that is, by approximating the first derivative of an appropriate function f via $(f(x + \Delta x) - f(x - \Delta x))/(2\Delta x)$. The discretizations mentioned in Eqs. (5.85) and (5.86) are both viable. Which is better? Be careful with the former discretization when starting with zero initial data. The computed solution will remain at zero if nothing is done to coax it from this state. The CFL condition is not obvious for the nonlinear PDE under consideration, but something like it must be respected to avoid numerical instability. A small step size in time might be required. What is the CFL condition (approximately) for the two proposed numerical methods? The right-hand boundary condition does not require solution of a nonlinear equation: the ambient gas pressure can be taken to be zero. For part (b), think before making simplifications of the ODE for the steady state. The ODE is very easy to solve. The root mean square distance between two functions is the square root of the integral over the spatial domain of the square of the difference between the two functions divided by the length of the interval; that is, $\sqrt{1/(b-a)\int_a^b |f(x) - g(x)|^2\,dx}$. Using a laptop computer programmed with noncompiled and nonoptimized code at the time of writing of this book might require hours of computer time to obtain good results. In case the computation time seems excessive for your code, change problems (a) and (c) to make the computation time shorter. For example, consider the flux through a cross section one centimeter from the left boundary after 10 seconds of time and, instead of reaching within 1% of the steady state profile, consider integration from zero density until reaching within 30% of the steady state profile.
(d) Consider Exercise 5.79 after reading the following sections on the trapezoidal and Crank–Nicolson methods.

5.5.3 The Trapezoidal Method and Crank–Nicolson in One Spatial Dimension

Reconsider a solution $t \mapsto u(t)$ of the ODE $\dot{u} = f(u)$; it satisfies the equation

$$u'(t) = f(u(t)).$$

By integrating both sides of this identity with respect to the independent variable, we obtain the integral equation

$$u(t + \Delta t) - u(t) = \int_t^{t+\Delta t} f(u(s))\, ds.$$

The left-hand rectangle rule approximation of the integral (that is, the value $\Delta t f(u(t))$) yields the forward Euler method; the trapezoidal rule approximation

$$u(t + \Delta t) \approx u(t) + \frac{\Delta t}{2}(f(u(t + \Delta t)) + f(u(t)))$$

yields the *trapezoidal method* also called the *implicit improved Euler method*

$$U^{k+1} = U^k + \frac{\Delta t}{2}(f(U^{k+1}) + f(U^k)) \tag{5.88}$$

which, for the nonautonomous case, is

$$U^{k+1} = U^k + \frac{\Delta t}{2}(f(U^{k+1}, t + \Delta t) + f(U^k, \Delta t)). \tag{5.89}$$

As might be expected, the trapezoidal method is more accurate than Euler's method; in fact, it is a second-order method. To prove this fact, we simply assume that the solution and its approximation agree at time t and estimate the norm of the difference between the true and approximate solutions at $t + \Delta t$. Using Taylor's theorem, the solution at $t + \Delta t$ is

$$\begin{aligned} u(t + \Delta t) &= u(t) + u'(t)\Delta t + \frac{1}{2}u''(t)\Delta t^2 + O(\Delta t^3) \\ &= U^0 + f(U^0)\Delta t + \frac{1}{2}f(U^0)f'(U^0)\Delta t^2 + O(\Delta t^3). \end{aligned}$$

The trapezoidal approximation

$$T := U^0 + \frac{\Delta t}{2}(f(U^0) + f(U^1)),$$

can be expanded in the Taylor series

$$
\begin{aligned}
T &= U^0 + \frac{\Delta t}{2}(f(U^0) + f(U^0 + \frac{\Delta t}{2}(f(U^0) + f(U^1))) \\
&= U^0 + \frac{\Delta t}{2}(f(U^0) + f(U^0) + \frac{\Delta t}{2} f'(U^0)(f(U^0) + f(U^1))) + O(\Delta t^3) \\
&= U^0 + f(U^0)\Delta t + \frac{\Delta t^2}{4} f'(U^0)(f(U^0) + f(U^0)) + O(\Delta t^3) \\
&= U^0 + f(U^0)\Delta t + f'(U^0)f(U^0)\frac{\Delta t}{2} + O(\Delta t^3).
\end{aligned}
$$

Thus, the local truncation error is

$$|u(t + \Delta t) - T| = O(\Delta t^3).$$

Because the local truncation error is $O(\Delta t^3)$, the method is second-order in time (see Exercise 5.47).

Numerical stability, when carefully considered, leads to some difficult problems. At the minimum, a stable method should produce a good approximation when applied to a stable solution of an ODE. More precisely, suppose the ODE $\dot{u} = f(u)$ has a rest point at u_0 so that $f(u_0) = 0$ and this rest point is asymptotically stable; that is, for every open ball Ω centered at u_0 there is a smaller concentric ball B such that for each initial state ν in B, the initial value problem $\dot{u} = f(u)$ and $u(0) = \nu$ has a solution that stays in Ω and converges to u_0 as time grows without bound. A stable numerical method should at least produce good approximations for these initial value problems. Recall a basic theorem: If u_0 is a rest point of the ODE $\dot{u} = f(u)$ and all eigenvalues of the system matrix $Df(u_0)$ of the linearized system have negative real parts, then u_0 is asymptotically stable (see Appendix A.17). In view of this result, the numerical method should also be stable when applied to the linear system $\dot{w} = Df(u_0)w$ (which has the rest point $w = 0$) for arbitrary initial condition $w(0) = \omega$. The matrix $Df(u_0)$ is generically diagonalizable. In this case, the linear system reduces to a decoupled system of ODEs all of the form $\dot{x} = \lambda x$, where λ has negative real part. Thus, a reasonable (but not definitive) test of numerical stability of a numerical algorithm is to apply it to the ODE

$$\dot{x} = -\lambda x, \tag{5.90}$$

with $x(0) = 1$ and $\lambda > 0$ and determine the condition (if any) on the step size that ensures stability for this special case. For example, Euler's method produces the iteration scheme

$$x^{j+1} = x^j + \Delta t(-\lambda x^j) = (1 - \lambda\Delta t)x^j = (1 - \lambda\Delta t)^j x^0.$$

The iterates x^j converge to zero if and only if $|1 - \lambda\Delta t| < 1$. The method is numerically stable provided the time step is restricted so that $0 < \Delta t < 2/\lambda$. A time step that exceeds $2/\lambda$ will result in a sequence of iterates whose absolute values grow without bound. (What about $\Delta t = 2/\lambda$?)

The result for Euler's method is closely related to the CFL condition (5.17) for numerical stability when Euler's method [Eq. (5.15)] is applied to approximate solutions of the heat equation. Recall the numerical method used previously, when applied to $u_t = \kappa u_{xx}$. is

$$U_i^{j+1} = U_i^j + \Delta t \frac{\kappa}{\Delta x^2}(U_{i-1}^j - 2U_i^j + U_{i+1}^j). \tag{5.91}$$

It has exactly the form of Euler's method applied to an ODE except that the function f in the ODE $\dot{x} = f(x)$ is replaced by an operator $F(u) := \kappa u_{xx}$, which is discretized in the scheme (5.91).

Similarly, the trapezoidal method applied to the test ODE [Eq. (5.90)] is the iteration scheme

$$x^{j+1} = x^j - \lambda \frac{\Delta t}{2}(x^{j+1} - x^j).$$

By a simple rearrangement, the iterates are given by

$$x^{j+1} = \frac{1 - \lambda \frac{\Delta t}{2}}{1 + \lambda \frac{\Delta t}{2}} x^j. \tag{5.92}$$

Thus, the trapezoidal method produces a scheme that is *unconditionally stable*. Indeed, the coefficient of x^j has absolute value less than 1 for every positive Δt. No restriction on the step size is required to maintain numerical stability.

The Crank–Nicolson method, which will be discussed in detail in this section, is the trapezoidal scheme applied to PDEs. It is unconditionally stable for many PDEs; in fact, the method is numerically stable with no restriction on the size chosen for the positive time steps. Warning: Stable does not imply accurate. A small step size might still be required to achieve some desired global error. Of course, one can hope that step sizes can be taken larger for a stable second-order method than for a first-order method.

For the heat equation $u_t = \kappa u_{xx}$ in one spatial dimension, the Crank–Nicolson method with $U_i^j := u(i\Delta x, j\Delta t)$ is given by the difference scheme

$$U_i^{j+1} = U_i^j + \frac{\kappa \Delta t}{2\Delta x^2}(U_{i-1}^j - 2U_i^j + U_{i+1}^j + U_{i-1}^{j+1} - 2U_i^{j+1} + U_{i+1}^{j+1}), \quad (5.93)$$

where, for example, $i = 1, 2, 3, \ldots n$, and zero Dirichlet boundary conditions are assumed: $U_0 = U_{n+1} = 0$. Let

$$\alpha := \frac{\kappa \Delta t}{2\Delta x^2}$$

and define W^j to be the n-dimensional vector with components U_i^j for $i = 1, 2, 3, \ldots n$. Also, let A be the tridiagonal matrix with all components on the main diagonal equal to -2, all elements on the first super and subdiagonals equal to 1, and all other components equal to zero. Using these notations and an algebraic rearrangement, the difference scheme (5.93) is given in the vector form

$$(I - \alpha A)W^{j+1} = (I + \alpha A)W^j, \quad (5.94)$$

where $\alpha > 0$.

Using the Gerschgorin theorem (see Appendix A.5) and the symmetry of A, every eigenvalue of this matrix is a real number in the closed interval $[-4, 0]$. An easy calculation shows that μ is an eigenvalue of $I - \alpha A$ (that is, $(I - \alpha A)v = \mu v$ for some nonzero vector v) if and only if $\mu = 1 - \alpha\lambda$ for some eigenvalue λ of A. Because every eigenvalue of A is nonpositive, every eigenvalue of $I - \alpha A$ is positive. In particular, this matrix has no zero eigenvalue; it is invertible. More precise results are available for the eigenvalues of A; in fact, they can be computed explicitly (see Appendix A.19). This matrix or its negative is often called the discrete Laplacian.

Because $I - \alpha A$ is invertible, the Crank–Nicolson scheme for the heat equation reduces to the iteration process

$$W^{j+1} = (I - \alpha A)^{-1}(I + \alpha A)W^j \quad (5.95)$$

To prove that *the Crank–Nicolson method for the heat equation is unconditionally numerically stable* it suffices to show that every eigenvalue of the matrix $(I - \alpha A)^{-1}(I + \alpha A)$ is in the closed unit disk in the complex plane. Again, this fact follows by relating the eigenvalues of this matrix to the eigenvalues of A.

The number γ is an eigenvalue of $(I - \alpha A)^{-1}(I + \alpha A)$ if there is a nonzero vector v such that

$$(I + \alpha A)v = \gamma(I - \alpha A)v.$$

By an easy calculation, this eigenvalue must be given by

$$\gamma = \frac{1 + \alpha\lambda}{1 - \alpha\lambda},$$

where λ is an eigenvalue of A. Because $\alpha > 0$ and $-4 \leq \lambda \leq 0$, the eigenvalue γ is such that $-1 < \gamma \leq 1$ (compare to Eq. (5.92)). In particular, every eigenvalue of the matrix $(I - \alpha A)^{-1}(I + \alpha A)$ lies in the closed unit disk. In fact, $-1 < \gamma < 1$ (see Exercise 5.43); therefore, the Crank–Nicolson method for the heat equation is unconditionally numerically stable.

The numerical stability result does not take into account boundary conditions that alter the matrix A. Dirichlet and Neumann boundary conditions are considered in this context in Exercise 5.67.

Unconditional stability comes with a price: the trapezoidal method [Eq. (5.89)] is implicit. At each time step or (equivalently) each iteration of the method, applied to approximate a solution of the ODE $\dot{x} = f(x)$, the equation

$$U^{k+1} - \frac{\Delta t}{2} f(U^{k+1}) = U^k + \frac{\Delta t}{2} f(U^k)$$

must be solved for the unknown U^{k+1}. This equation is nonlinear whenever the ODE is nonlinear.

The premier method for approximating the solutions of nonlinear equations is Newton's method; it is one of the most important algorithms in analysis. To find a root a of a function $g : \mathbb{R}^n \to \mathbb{R}^n$, suppose that x is close to a, Using Taylor's theorem,

$$g(a) = g(x) + Dg(x)(a - x) + O((x - a)^2).$$

Because $g(a) = 0$, the solution b of the equation $g(x) + Dg(x)(b - x) = 0$ should yield a good approximation of the desired root a. By some simple matrix algebra,

$$b = x - Dg(x)^{-1}g(x).$$

This observation is the motivation for Newton's method: Choose an approximation x^0 of the expected root a and compute the sequence $\{x^k\}_{k=0}^{\infty}$ according to the iteration process

$$x^{k+1} = x^k - Dg(x^k)^{-1}g(x^k). \tag{5.96}$$

Under the three assumptions (1) *the starting value x^0 is sufficiently close to a*, (2) the function g is continuously differentiable, and (3) *$Dg(a)$ is invertible*, the sequence of iterates defined by Newton's method converges to the desired root a. Moreover, the rate of convergence is quadratic; that is, for $r = 2$ and some *positive* number C,

$$|x^{k+1} - a| \le C|x^k - a|^r \tag{5.97}$$

(see Appendix A.14 for more information).

To approximate a solution $x = a$ of the equation $g(x) = 0$ for $g : \mathbb{R} \to \mathbb{R}$ by Newton's method (the scalar case), choose some initial approximation x^0 of a and iterate using the formula

$$x^{k+1} = x^k - \frac{g(x^k)}{g'(x^k)}. \tag{5.98}$$

In the vector case, the inverse of the matrix $Dg(x^k)$ is not computed. (Why?) Instead, Newton's method is implemented in an alternate form: Solve the linear system

$$Dg(x^k)y = -g(x^k) \tag{5.99}$$

for y and define $x^{k+1} = y + x^k$ (see page 174 for a discussion of some numerical methods for approximating solutions of systems of linear equations).

For an arbitrary convergent sequence $\{x^k\}_{k=1}^{\infty}$ with limit a, we say that the order of convergence is r and the asymptotic error is C whenever inequality (5.97) is satisfied. In other words, the error at index $k + 1$ (which we will denote by $\text{err}(k + 1)$) is approximately proportional (with a fixed constant C of proportionality) to $\text{err}(k)^r$; in particular, the error at index

iterate	Newton	asymptotic error	iteration	asymptotic error
0	2.0		2	
1	1.4		0.875	
2	1.1	0.4	0.929932	0.125
3	1.0087	0.625	0.962554	0.560547
4	1.00007	0.869565	0.980582	0.534420
5	1.0	0.987124	0.990104	0.518548
6	1.0	0.999888	0.995003	0.509662

Table 5.10 The table lists iterates and asymptotic errors, using $|x^{k+1} - a|/|x^k - a|^r$ with $r = 2$ for Newton's method and $r = 1$ for composition, to approximate the real root $x = 1$ of the polynomial $x^3 + x^2 - x - 1$.

$k+1$ is $O(\text{err}(C)^r)$. The quadratic convergence of Newton's method makes it a pillar of numerical (and theoretical) analysis (cf. [52]).

As a simple example of the convergence properties of Newton's method, consider the real root, $x = 1$, of the cubic polynomial $g(x) = x^3 + x^2 - x - 1$. The results of an implementation of Newton's method with initial guess $x = 2$ are given in the left-half of Table 5.10. The asymptotic error computed with $r = 2$ as in Eq. (5.97) seems to converge to $C = 1$. Thus, the experiment confirms that Newton's method is quadratically convergent in this case (see Exercise 5.48 for a proof).

For comparison, note that the root $x = 1$ can also be found as a fixed point of the function

$$G(x) = x - \frac{1}{8}(x^3 + x^2 - x - 1). \tag{5.100}$$

Starting from an initial guess $x = x_0$, the composition method (also called fixed point iteration) is to compute the compositional iterates of the function G; that is, to proceed inductively using the formula $x^{k+1} = G(x^k)$. In other words, we view G as a dynamical system on the real line and seek the real root of the polynomial as a stable fixed point of this dynamical system. The factor $1/8$ is used to ensure that G has a stable fixed point at $x = 1$ (see Exercise 5.56). The data in Table 5.10 suggests that the iterates of G are linearly convergent to the fixed point.

Newton's method gives the correct root with an error of less than 10^{-2} after three iterates; the composition method requires five. To achieve an error less than 10^{-3}, Newton's method requires four iterates; the composition method requires nine. Let us also note that the computational cost of the two methods is comparable. It should be clear that Newton's method is superior.

iterate	Newton	Aitken	Steffensen
0	2.0		
1	1.4		
2	1.1	0.92737	0.92737
3	1.0087	1.01026	1.00308
4	1.00007	1.00285	1.00000
5	1.00000	1.00076	1.00000
6	1.00000	1.00020	1.00000

Table 5.11 The table lists iterates for Newton's, Aitken's, and Steffensen's methods applied to find the root ($x = 1$) of the polynomial $x^3 + x^2 - x - 1$.

Acceleration by Richardson extrapolation of certain low-order methods with discretization errors has been discussed. It is also possible to accelerate some linearly convergent sequences by a staple of numerical analysis called Aitken's Δ^2 method. The idea is simple. A linearly convergent sequence $\{x_k\}_{k=1}^{\infty}$ with limit x^∞ should satisfy

$$|x^{k+1} - x^\infty| \approx C|x^k - x^\infty|,$$

for some constant C and all large k. In addition, assume that *the errors all have the same signs*. By taking one more iterate, there are two approximations

$$x^{k+1} - x^\infty \approx C(x^k - x^\infty), \qquad x^{k+2} - x^\infty \approx C(x^{k+1} - x^\infty).$$

Eliminating C, we have the relation

$$(x^{k+1} - x^\infty)^2 \approx (x^{k+2} - x^\infty)(x^k - x^\infty),$$

and solving for x^∞ yields

$$x^\infty \approx \frac{x^k x^{k+2} - x^{k+1}}{x^{k+2} - 2x^{k+1} + x^k},$$

or equivalently,

$$x^\infty \approx x^k - \frac{(x^k - x^{k+1})^2}{x^{k+2} - 2x^{k+1} + x^k}.$$

Of course, this last relation is turned into a numerical method by replacing the unknown x^∞ by the kth Aitken Δ^2 approximation; that is,

$$\text{Aitken}(k) := \frac{x^k x^{k+2} - x^{k+1}}{x^{k+2} - 2x^{k+1} + x^k}. \tag{5.101}$$

Table 5.11 lists the Aitken approximations obtained by applying Eq. (5.101) to the sequence of iterates of the function G defined in Eq. (5.100). The convergence is faster than the linear convergence of the simple iterates of G, but the new sequence does not converge quadratically (see Exercise 5.57). This deficiency can be remedied: Instead of applying Aitken's method directly to the sequence of iterates, compute two iterations, apply Eq. (5.101) to x_0, x_1 and x_2 to obtain Aiken(0), use this value—which should be a better approximation of the fixed point than x_3—as the new x_0, and repeat this process. This algorithm is called Steffensen's method. The Steffensen sequence for our test example is listed in Table 5.11. For this example, it performs as well as Newton's method (see Exercise 5.57).

The assumption that the errors $x^k - x^\infty$ in our sequence all have the same signs might not be satisfied; for example, the sequence given by $x^0 = 1$ and $x^{k+1} = g(x^k)$ (where $g(x) := -x/2$) converges to zero but alternates in sign. For a situation like this where the iterates alternate in sign, Steffensen's method can be applied to every second iterate, or in other words, it can be applied to the sequence generated by $g^2(x) := g(g(x))$.

Why arc Aitken's and Steffensen's methods used when Newton's method is available? Answer: Aitken's method applies to general linearly convergent sequences; hence, it can be applied in situations where no quadratically convergent method is known. Steffensen's method is valuable in cases where the derivative of the function being iterated is difficult to obtain or expensive to evaluate (cf. Exercise 5.58). As might be expected, Richardson extrapolation, Aitken's Δ^2, and Steffensen's arc only a few of the important methods that have been developed to accelerate convergence of low-order methods. Of course, extrapolation methods, like all numerical methods, do not always work. The complete story is beyond the scope of this book (cf. [10]).

Exercise 5.43. Prove that the Crank–Nicolson method is unconditionally stable. Hint: See Appendix A.19.

Exercise 5.44. [Newton's Method in Mountain Terrain] Via satellite (for example) elevations are mapped over a region of the Earth's surface and the function

$$z = F(x, y) := a^3x^3 - 3ab^2xy^2 - 0.1(a^4x^4 + b^4y^4) + 1500, \qquad a = 0.01, b = 0.01$$

is fit to this data to produce an approximation to the elevation of the terrain over an imaginary plane with coordinates (x, y). (How do you suppose the fitting is done?) An observer resides at the point with (x, y, z) coordinates $(-550, 10, 1242.284)$ on

the mapped terrain, where all distances are measured in meters. The first coordinate measures the east-west direction with the positive direction pointing east. Likewise, the second coordinate measures north-south with the positive direction pointing north.
(1) Determine the curve of points that are 700 meters away via laser shots in three dimensions from this observation point and lie farther north. Note: A more difficult project is to determine the curve of points that is 700 meters away when distance is measured along the terrain.
(2) Is there a point on the terrain equidistant via laser shots from the first observation point, the second observation point $(-900, 990, 1800.574)$, and third observation point $(-300, 250, 1517.244)$? If so, give the coordinates of such a point.
(3) How many solutions of (2) exist within the mapped terrain?
(4) What happens for (2) in case the third observation point is $(1000, 0, 1501)$?
(5) More difficult projects: What is the distance from observation point 1 to observation point 2 along the terrain? Solve part (2) with distances measured along the terrain.

Exercise 5.45. [Newton's Method for an ODE Model] For the model equation $\ddot{x}+x = x^3$ (where x might measure the deflection of a beam) suppose the beam is pulled down a units and released from rest (that is, $x(0) = a$, $\dot{x}(0) = 0$). Is there a choice of a such that $x(8) = a$ and $x(t) \neq a$ for $0 < t < 8$? If so, determine a correct to three decimal places.

5.5.4 The Crank–Nicolson Method in Two Spatial Dimensions

The Crank–Nicolson method is simply the trapezoidal method adapted to the context of parabolic PDEs by viewing a parabolic PDE as an abstract evolution equation $\dot{u} = f(u)$ (which has the form of an ODE) where f is a differential operator.

For our basic reaction-diffusion PDE [Eq. (5.32)], the Crank–Nicolson method is given by the scheme

$$\begin{aligned}
U_{i,j}^{k+1} &= U_{i,j}^{k} + \frac{\Delta t}{2}\Big[\frac{\lambda}{\Delta x^2}(U_{i+1,j}^{k+1} - 2U_{i,j}^{k+1} + U_{i-1,j}^{k+1} + U_{i+1,j}^{k} - 2U_{i,j}^{k} + U_{i-1,j}^{k}) \\
&\quad + \frac{\lambda}{\Delta y^2}(U_{i,j+1}^{k+1} - 2U_{i,j}^{k+1} + U_{i,j-1}^{k+1} + U_{i,j+1}^{k} - 2U_{i,j}^{k} + U_{i,j-1}^{k}) \\
&\quad + (f(U_{i,j}^{k+1}, V_{i,j}^{k+1}) + f(U_{i,j}^{k}, V_{i,j}^{k}))\Big], \\
V_{i,j}^{k+1} &= V_{i,j}^{k} + \frac{\Delta t}{2}\Big[\frac{\lambda}{\Delta y^2}(V_{i+1,j}^{k+1} - 2V_{i,j}^{k+1} + V_{i-1,j}^{k+1} + V_{i+1,j}^{k} - 2V_{i,j}^{k} + V_{i-1,j}^{k}) \\
&\quad + \frac{\lambda}{\Delta y^2}(V_{i,j+1}^{k+1} - 2V_{i,j}^{k+1} + V_{i,j-1}^{k+1} + V_{i,j+1}^{k} - 2V_{i,j}^{k} + V_{i,j-1}^{k}) \\
&\quad + (g(U_{i,j}^{k+1}, V_{i,j}^{k+1}) + g(U_{i,j}^{k}, V_{i,j}^{k}))\Big],
\end{aligned} \tag{5.102}$$

where the symbols f and g are now redefined to denote the reaction terms in the PDE.

To implement the Crank–Nicolson scheme directly, we must solve a nonlinear system of equations to compute each update. This can be done using Newton's method or Steffensen's method (see Exercise 5.55).

Another possibility is to modify the trapezoidal method into an explicit second-order scheme (which may no longer be unconditionally stable). One idea for doing this is simple: Compute the implicit update using Euler's method. This results in the explicit second-order scheme

$$U^{k+1} = U^k + \frac{\Delta t}{2}(f(U^k + \Delta t f(U^k)) + f(U^k)), \tag{5.103}$$

which is often called the improved Euler method (see Exercise 5.59).

A third possibility is to modify the method to make it explicit (using the Euler approximation as in Eq. (5.103)) for the reaction terms and leave the update equations implicit for the diffusion terms. To complete each time step for this algorithm, a nonlinear solver is avoided as the modified scheme requires only the solution of a system of *linear* equations. This method is theoretically second-order with respect to both the space and time discretization sizes; in other words, the method is $O(\Delta t^2)$ and $O(\Delta x^2 + \Delta y^2)$.

Although Newton's method is the premier choice for approximating the solution of a set of nonlinear equations, there are superior methods for approximating the solution of large systems of linear equations. The subject of solving systems of linear equations is itself an important branch of numerical analysis (and pure mathematics); a glimpse of a few topics in this area is provided here.

The simplest PDEs that lead to systems of linear equations are of course those with one spatial dimension. New concepts are required to treat the reaction diffusion model [Eq. (5.32)]

$$u_t = \lambda \Delta u + f(u, v), \quad v_t = \mu \Delta v + g(u, v) \tag{5.104}$$

for a two-dimensional rectangular spatial domain with side lengths L_1 and L_2. This example is used to illustrate some of the challenges that arise in multidimensional problems.

For a spatial discretization, we may choose the spatial increments $\Delta x = L_1/m$ and $\Delta y = L_2/n$ and the nodes $((i-2)\Delta x, (j-2)\Delta y)$ with $i = 2, 3 \ldots, m{+}2$ and $j = 2, 3 \ldots, n{+}2$. The corners of our spatial rectangle are labeled by the indices $(2,2)$ (lower left), $(2, n{+}2)$ (upper left), $(m{+}2, n{+}2)$ (upper right), and $(m+2, 2)$ (lower right). Each of the $(m+1) \times (n+1)$ nodes in the closed rectangle will correspond to two linear equations, one for U_{ij}^{k+1} and one for V_{ij}^{k+1}. The resulting systems of linear equations can and will be solved separately.

The equations for the approximate state variables at the nodes must be ordered in some convenient manner; that is, we need a bijective function defined on the set of nodes whose range is the set of integers $\{1, 2, 3, \ldots, (m+1) \times (n+1)\}$. We will use the bijection that corresponds to the ordering of the nodes on the spatial grid from top left to bottom right along rows from left to right. The required function, here called nodes, is given by

$$\text{nodes}(i,j) = i - (m+1)j + (n+1) + m(n+2) \tag{5.105}$$

(see Exercise 5.60).

Using our bijection, the system of linear equations (for the unknown updates of the state variables at the nodes) can be written in matrix form

$$AU^{k+1} = b, \tag{5.106}$$

where the $(m{+}1)(n{+}1) \times (m{+}1)(n{+}1)$ matrix A and the $(m{+}1)(n{+}1)$ vector b are defined using the Crank–Nicolson scheme. This procedure is accomplished in three steps:
(1) Use the Neumann boundary condition to set the values of U and V at the fictitious nodes that lie outside the spatial domain, which are needed to compute the discretized spatial second derivatives; for example, $U_{1,j}^k := U_{3,j}^k$ along the left-hand edge of the spatial domain.
(2) Define the matrix A using the form of the equations at each node; for example, the generic $\text{nodes}(i,j)$ row of the matrix A for the U variables is given by the assignments

$$\begin{aligned}
A(\text{nodes}(i,j), \text{nodes}(i,j)) &:= 1 + \lambda\Big(\frac{\Delta t}{\Delta x^2} + \frac{\Delta t}{\Delta y^2}\Big), \\
A(\text{nodes}(i,j), \text{nodes}(i+1,j)) &:= -\lambda\frac{\Delta t}{2\Delta x^2}, \\
A(\text{nodes}(i,j), \text{nodes}(i-1,j)) &:= -\lambda\frac{\Delta t}{2\Delta x^2},
\end{aligned}$$

$$A(\text{nodes}(i,j), \text{nodes}(i,j-1)) := -\lambda\frac{\Delta t}{2\Delta y^2},$$
$$A(\text{nodes}(i,j), \text{nodes}(i,j+1)) := -\lambda\frac{\Delta t}{2\Delta y^2}, \tag{5.107}$$

and all other components are set to zero in this row. Unfortunately, the matrix contains many nongeneric rows. Perhaps an example will help to illustrate the extra complications due to the Neumann boundary conditions. For the case $m = n = 2$ with $\beta := \lambda\Delta t/(2\Delta x^2)$, the matrix A is the 9×9 matrix such that all components on the main diagonal are $1 + 4\beta$, the first super diagonal is

$$(-2\beta, -\beta, -\beta, -\beta, -\beta, -\beta, -\beta, -\beta, -\beta),$$

the third super diagonal is

$$(-2\beta, -2\beta, -2\beta, -\beta, -\beta, -\beta),$$

the first subdiagonal is

$$(-\beta, -\beta, -\beta, -\beta, -\beta, -\beta, -\beta, -\beta, -2\beta),$$

the third subdiagonal is

$$(-\beta, -\beta, -\beta, -2\beta, -2\beta, -2\beta),$$

and all other components are zero.

The factor two appears due to the boundary conditions. Note that the matrix structure is naturally separated into five strips from top to bottom: The top strip is the first row, strip two is rows 2 through $m + 1$, strip three (the generic strip) is rows $m + 2$ through $(m + 1)(n + 1) - (m + 1)$, strip four is rows $(m+1)(n+1) - m$ to $(m+1)(n+1) - 1$, and strip five is the last row. The upper nonzero diagonals are the first and the $(m + 1)$ upper diagonals; the lower nonzero diagonals are the first and the $(m + 1)$ lower diagonals.

(3) The vector b is defined accordingly:

$$b(\text{nodes}(i,j)) := (1 - \lambda\Big(\frac{\Delta t}{\Delta x^2} + \frac{\Delta t}{\Delta y^2}\Big))U_{i,j}^k$$
$$+ \lambda\frac{\Delta t}{2\Delta x^2}(U_{i+1,j}^k + U_{i-1,j}^k)$$

$$
\begin{aligned}
&+\lambda\frac{\Delta t}{2\Delta y^2}(U_{i,j+1}^k+U_{i,j-1}^k)\\
&+\frac{\Delta t}{2}(f(\mathrm{FEU}_{i,j}^k,\mathrm{FEV}_{i,j}^k)+f(U_{i,j}^k,V_{i,j}^k)),
\end{aligned}
$$

where $(\mathrm{FEU}_{i,j}^k,\mathrm{FEV}_{i,j}^k)$ is the forward Euler approximation *of the full reaction-diffusion PDE with boundary conditions imposed* of the approximate states $(U_{i,j}^{k+1},V_{i,j}^{k+1})$ computed from the known approximation $(U_{i,j}^k,V_{i,j}^k)$ and the same Δt, Δx and Δy used for the Crank–Nicolson scheme.

Our choice of ordering determines the structure of the matrix A and vector b, but in a computer code to implement the Crank–Nicolson method, the matrix A need not be stored—there is no need for reserve storage of its zero elements. Because the matrices involved are large (for instance, a 129×129 grid requires solving a system of 16641 equations), it is natural to take advantage of the structure of our (sparse and banded) matrices, which have only five diagonals with nonzero components. We do not want the computer to waste our time computing values that are known to be zero.

At this juncture, we could enter the world of numerical linear algebra. This is a vast subject (see [12, 112]), which has certainly been influenced by the necessity of dealing with the matrix systems that arise in solving PDEs. This book is mainly about differential equations in applied mathematics, so we will not develop the theory in detail here. On the other hand, we need a viable method to approximate the solutions of the (large) linear systems that arise in implementations of implicit schemes such as the Crank–Nicolson algorithm. Hence, we must at least have a working knowledge of some fundamental results of this subject.

There are two basic methods for the numerical solution of linear systems: Gaussian elimination (which should be a familiar method from linear algebra) and fixed point iteration.

A viable implementation of Gaussian elimination for our (large and sparse) matrices demands adaptations that take into account the matrix structures. In the special case of tridiagonal matrices (which arise for PDEs with one spatial dimension), there is an adaptation of Gaussian elimination that is simple to program, fast, and practical (see [12]). For banded matrices there are similar but more complicated methods. Also, in applications to PDEs (and many other places), numerical algorithms require solving $AW =$

b repeatedly for the same system matrix A but with different column vectors b. Efficient numerical schemes compute and store a factorization of A, for example a factorization $PA = LU$, where P is a permutation matrix, L is a lower triangular matrix, and U is upper triangular. The basic idea, for the case where $P = I$, is to multiply A by a sequence of elementary matrices E_i that encode row reductions so that $E_n E_{n-1} \cdots E_1 A = U$, where U is upper triangular. This is simply Gauss elimination in case every pivot is nonzero. The P matrix is used to make sure every pivot is nonzero and more generally to permute rows so that the largest available pivot is used, a process called partial pivoting. The reason to use the large pivots is to avoid multiplying rows by large numbers—which might cause large numerical errors for some matrices that have very small nonzero eigenvalues. For the simple case where $P = I$, the inverse L of the product of elementary matrices is lower triangular. Thus, $A = LU$. The advantage of this decomposition is that it stores most of the work in solving the original system $AW = b$. The factored matrix system $LUW = b$ is solved in two steps: $LY = b$ and $UW = y$. Each of the two linear systems are solved immediately by simple recursive substitution. No elimination is necessary. In case $PA = LU$ is used, simply multiply b by P and then use the LU decomposition. There are refinements of this method for special types of matrices; especially, symmetric matrices, banded matrices, and sparse matrices.

Iterative methods used to solve linear systems are also called methods of successive relaxation. In its simplest form, the basic method is a generalization of the idea used (recall Eq. (5.100)) to approximate solutions of the nonlinear equation $g(x) = 0$ by iteration of the function $G(x) = x - \omega g(x)$ for a choice of ω that makes the desired solution a stable fixed point of the dynamical system defined by G. Indeed, the solution of a linear system $Az = b$ might be solved by iterating the linear transformation T given by

$$Tz = z - \omega(Az - b), \tag{5.108}$$

where ω is a *nonzero* real number. There are at least three important questions related to this procedure: (1) What is a sufficient condition to guarantee that the process converges? (2) What is the best value to choose for the parameter ω? (3) What is a good choice for z to start the process?

Question (3) does not have a definite answer. Surely, the optimal choice of the initial point is the solution vector for the equation $Az = b$. As this is the value we wish to find, we will have to settle for a less than optimal choice. One useful idea for our problem is to note that we plan to solve our system many times as we step along in time; hence, we can use current values of the concentrations at the nodes as the starting value to obtain the updated concentrations.

Questions (1) and (2) are answered by doing some numerical analysis. Recall that a fixed point of a map is asymptotically stable whenever all eigenvalues of the derivative of the map at the fixed point lie inside the open unit disk in the complex plane (see Exercise 5.56). Fortunately, the derivative of a linear transformation is itself and therefore does not depend on the point at which it is evaluated. The derivative of our proposed function [Eq. (5.108)], which defines our linear dynamical system, is

$$\mathrm{DT} = I - \omega A,$$

where I is the identity matrix. The eigenvalues of DT are in correspondence with the eigenvalues of A; in fact, μ is an eigenvalue of DT if and only if $(1-\mu)/\omega$ is an eigenvalue of A.

For the matrix A given by the assignments in Eqs. (5.107) where no boundary condition is taken into account, all the elements on the main diagonal of A are equal. In case $\Delta x = \Delta y$ and for computational convenience $\alpha := 2\beta$, these diagonal elements are all equal to $1 + 2\alpha$. The sum of the absolute values of the off-diagonal elements in each row of A is at most 2α. Moreover, A is a symmetric matrix; therefore, it has real eigenvalues. By Gerschgorin's theorem (see Appendix A.5), the eigenvalues of A lie in the interval $[1, 1 + 4\alpha]$, which consists of only positive real numbers.

Let $\sigma(A)$ denote the set of eigenvalues (the spectrum) of A and note that $\xi \in \sigma(A)$ corresponds to the eigenvalue $\mu = 1 - \xi\omega$ in $\sigma(\mathrm{DT})$. The maximum absolute value of the eigenvalues of DT (which is its spectral radius $\rho(DT)$) is given by

$$\rho(DT) = \min_{\omega\in\mathbb{R}} \max_{\xi\in\sigma(A)} |1 - \xi\omega|. \tag{5.109}$$

Under the assumption that $\rho(DT) < 1$, the optimal value of ω is the value at which the minimum occurs. By Exercise 5.69,

$$\rho(DT) = \frac{2\alpha}{1+2\alpha}, \qquad \omega = \frac{1}{1+2\alpha}.$$

Thus, in this case, we have that $\rho(DT) < 1$ as desired, and the optimal choice is $\omega = 1/(1 + 2\alpha)$. The sequence of iterates of the transformation T will be linearly convergent to the solution of $Az = b$ for every starting vector.

The implementation of our iterative method $z^{\ell+1} = Tz^\ell$ should take advantage of the structure of the matrix A. In particular, there is no reason to store this matrix. The update from z^ℓ to $z^{\ell+1} = Tz^\ell$ is given, for $\gamma = 1, 2, 3, \ldots (m+1)(n+1)$, by

$$\begin{aligned} z_\gamma^{\ell+1} &= (1 - \omega(1 + \lambda\Big(\frac{\Delta t}{\Delta x^2} + \frac{\Delta t}{\Delta y^2}\Big))z_\gamma^\ell \\ &\quad - \omega\lambda\frac{\Delta t}{2\Delta x^2}z_{\gamma+1}^\ell - \omega\lambda\frac{\Delta t}{2\Delta x^2}z_{\gamma-1}^\ell \\ &\quad - \omega\lambda\frac{\Delta t}{2\Delta y^2}z_{\gamma+m+1}^\ell - \omega\lambda\frac{\Delta t}{2\Delta x^2}z_{\gamma-(m+1)}^\ell, \end{aligned} \tag{5.110}$$

where z_p^ℓ is set to zero if the index

$$p \in \{\gamma+1, \gamma-1, \gamma+m+1, \gamma-(m+1)\}$$

is not in the range of γ. We will have to encode a stopping procedure for our iteration process; for example, the procedure can be stopped as soon as an iterative update does not change by more than some prespecified tolerance from its previous value. After gaining some experience by monitoring the performance of a code, we might fix the number of iterations to decrease its execution time. The final iterate Z_γ is used to update the matrix U^{k+1} in the code used to solve the PDE setting

$$U_{i,j}^{k+1} = Z_{\text{nodes}(i,j)}.$$

The iteration process can be accelerated by an application of Steffensen's method. For applications of this method to systems, there are at least two natural questions: (1) Can Steffensen's method be applied componentwise? (2) Can the signs of the errors be determined to check that the hypothesis for the convergence of Steffensen's method is satisfied?

The idea underlying the answers to our questions can be illustrated by examining the iteration of linear transformation of the plane. Let us suppose that T is a linear transformation of the plane whose spectrum is inside the open unit disk. In this case, the sequence of vectors $\{x^k\}_{k=0}^\infty$ defined by the iteration process $x^{k+1} = Tx^k$ (for an arbitrary choice of x^0) converges to zero. Under the assumption that T is symmetric and positive definite, the eigenvalues of T are positive real numbers $\lambda_1 \leq \lambda_2$ and there is a basis of $\mathbb{R}^2$ consisting of eigenvectors of T. In this basis, T has the matrix representation

$$\begin{pmatrix} \lambda_1 & 0 \\ 0 & \lambda_2 \end{pmatrix}.$$

The eigenspaces (that is, the subspaces spanned by the eigenvectors) in this basis correspond to the coordinate axes. Pick a vector in the plane, say the vector with components x_0 and y_0, and iterate the process. The kth iterate has components $\lambda_1^k x_0$ and $\lambda_2^k y_0$. Note that if $\lambda_1 < \lambda_2$, then the kth iterate becomes nearly parallel to the vertical coordinate axis. In other words, the dot product of the kth iterate and the usual basis vector e_1 is small compared with the dot product of the kth iterate and e_2. The first two vectors are nearly orthogonal. After a few iterations, the iterates align with the eigenspace corresponding to the largest eigenvalue. Thus, up to a small error, the iterates converge monotonically to zero along this one-dimensional subspace. Also, all the components of the iterates converge linearly to zero and the errors with respect to each component have the same signs.

Exactly the same behavior occurs in general as long as the matrix that determines the iteration process is positive definite and symmetric. More generally, it suffices to have a positive real eigenvalue in the open unit disk that is larger than the absolute values of the real parts of all the other eigenvalues.

In practice, several iterates are computed so that the expected alignment takes place, Steffensen's method is applied, several more iterations are computed so that the expected alignment takes place, Steffensen's method is applied, and this process is continued until the iterations are no longer changing up to some preassigned tolerance or a maximum preset number of iterations is exceeded. It is possible to check internally that the alignment has occurred; for example, the computed differences $x_i^{k+1} - x_i^k$ for the components can be tested to see if they are all of the same sign and approximately the same magnitude (see Exercise 5.73).

Although the iterative method just described is viable (especially when it is accelerated via Aitken extrapolation), this method is seldom used because there are superior alternative methods due to Jacobi, Gauss, Seidel and others. All of these methods are based on a simple idea. To solve $Az = b$ for z, write the system matrix A as a sum $A = P + Q$, separate the product Az

accordingly into a sum $Pz + Qz$, move one summand to the right-hand side of the equation $Pz = b - Qz$, and multiply both sides by the inverse of the remaining matrix on the left-hand side $z = P^{-1}(b - Qz)$ to set up a fixed point iteration scheme $z^{k+1} = P^{-1}(b - Qz^k)$. With an appropriate choice of P and Q, new iteration schemes can be constructed that (usually) converge more rapidly than iterations of the function T defined by Eq. (5.108).

Suppose A is an $n \times n$ matrix and ω is a nonzero scalar. To solve for z in the linear system $Az = b$, precondition the equation by multiplication with a scalar variable ω (or, equivalently, the matrix ωI) to obtain the equivalent system

$$0 = -\omega(Az - b), \tag{5.111}$$

and split A into the sum of three $n \times n$ matrices: D, whose main diagonal is the main diagonal of A and all its other components are zero; L, the lower triangular part of A (that is, L consists of the components of A below its main diagonal and all of its other components are zero); and U, similarly, the upper triangular part of A. Using this notation and with the intention of defining a transformation whose fixed point is a solution of the linear system, recast Eq. (5.111) into the form

$$0 = Dz - Dz - \omega(Dz + Lz + Uz - b),$$

which may be rearranged to obtain the matrix equation

$$(D + \omega L)z = ((1 - \omega)D - \omega U)z + \omega b.$$

If $D + \omega L$ is invertible, the unknown vector z is a fixed point of the linear transformation

$$\zeta \mapsto (D + \omega L)^{-1}(((1 - \omega)D - \omega U)\zeta + \omega b). \tag{5.112}$$

Iteration of this transformation starting from an arbitrary initial guess z_0 defines the numerical method. This iteration process is sometimes called successive overrelaxation (SOR). More precisely, the process is called an overrelaxation method if $\omega > 1$, an underrelaxation method if $\omega < 1$, and the Gauss–Seidel method if $\omega = 1$.

Eq. (5.112) requires the matrix

$$S := D + \omega L$$

to be invertible, which is the case if and only if every element on its main diagonal is not zero. In general, the inversion of matrices should be avoided in numerical computation because the number of operations required to invert a matrix increases rapidly with its size. In fact, the number of operations for inversion is on the order of n^3 for an $n \times n$ matrix. In contrast, the SOR method is practical because the required inversion of a lower triangular matrix is accomplished simply and efficiently by back substitution. Inspection of the matrix S reveals that the components of z in the linear system

$$Sz = v,$$

where v is a given n-dimensional vector, are

$$\begin{aligned} z_1 &= \frac{1}{S_{ii}} v_1, \\ z_i &= \frac{1}{S_{ii}} (v_i - \sum_{j=1}^{i-1} S_{ij} z_j). \end{aligned} \tag{5.113}$$

In this scheme, z_1 is used to solve for z_2, z_1 and z_2 are used to solve for z_3, and so on; that is, the components of the solution are used to solve for subsequent values as soon as they are obtained. For this reason, SOR is expected to converge more rapidly than the naive iteration method defined in Eq. (5.108). In the SOR iteration scheme [Eq. (5.113)],

$$S_{ii} = a_{ii}, \qquad S_{ij} = \omega a_{ij}.$$

To implement SOR, choose an initial approximation z^0 and compute successive approximations z^k in two steps:

$$\begin{aligned} v^k &:= [(1 - \omega)D - \omega U] z^k + \omega b, \\ (D + \omega L) z^{k+1} &= v^k. \end{aligned} \tag{5.114}$$

The components v_i^k, $i = 1, 2, 3, \ldots n$, of v^k are

$$v_i^k = (1 - \omega) a_{ii} z_i^k - \omega \sum_{j=i+1}^{n} a_{ij} z_j^k + \omega b_i, \tag{5.115}$$

and the second step is completed using the back substitution formulas of the SOR iteration [Eq. (5.113)].

h	32×32	64×64	128×128	256×256
$\Delta t = 1/4$	1.93461	1.93451	1.93497	1.92266
$\Delta t = 1/8$	1.96882	1.96848	1.96688	2.00855
$\Delta t = 1/16$	1.98589	1.98510	1.98407	1.83655
$\Delta t = 1/32$	1.99551	1.99388	1.99383	2.07241
$\Delta t = 1/64$	2.0028	1.99951	1.99816	2.01279
$\Delta t = 1/128$	2.01155	2.00488	2.00157	2.00958

Table 5.12 The order estimates in this table are for an implementation of the Crank–Nicolson method using the data of Problem 5.2 and the Gauss–Seidel iteration with the stopping criterion: successive iterations that differ in Euclidean norm by less than 10^{-6}. The floating point numbers are computed for the average of the concentrations $(u+v)/2$ using Eq. (5.67).

The SOR algorithm will converge if the eigenvalues of the matrix

$$(D + \omega L)^{-1}((1 - \omega)D - \omega U)$$

all lie inside the unit circle in the complex plane (see Exercise 5.64). To obtain faster convergence, the parameter ω may be adjusted to make these eigenvalues as close to zero as possible. An analysis of the structure of the decomposition $A = D + U + L$ or numerical computation of the required eigenvalues in special cases might lead to a theorem that would ensure convergence. Unfortunately, SOR does not converge for every $n \times n$ matrix A.

The SOR method for the five-diagonal matrix A arising in the Crank–Nicolson scheme (5.106) can be programmed so that only the nonzero elements of the matrix are used (compare to Eq. (5.110)).

If A is a positive-definite and symmetric matrix and $0 < \omega < 2$, then the spectrum of the matrix

$$(D + \omega L)^{-1}((1 - \omega)D - \omega U)$$

lies in the open unit disk in the complex plane; hence, SOR converges whenever A is a positive-definite symmetric matrix and ω is in this range (see [114] and Exercise 5.66). In general, it is a difficult problem to determine the optimal value of ω.

Computational results for Problem 5.2 using the Crank–Nicolson algorithm, where the solutions of the corresponding matrix systems are approximated using the Gauss–Seidel iteration (SOR with $\omega = 1$ and with the stopping criterion: successive iterations that differ in Euclidean norm by less than 10^{-6}) are reported in Tables 5.12 and 5.13. The order

h	32×32	64×64	128×128	256×256
$\Delta t = 1/4$	0.00022356	0.00022363	0.00022355	0.00022420
$\Delta t = 1/8$	0.00005730	0.00005731	0.000057347	0.00005776
$\Delta t = 1/16$	0.00001450	0.00001451	0.000014569	0.00001497
$\Delta t = 1/32$	3.65×10^{-6}	3.65×10^{-6}	3.664×10^{-6}	4.14×10^{-6}
$\Delta t = 1/64$	9.1×10^{-7}	9.1×10^{-7}	9.2×10^{-7}	9.8×10^{-7}
$\Delta t = 1/128$	2.3×10^{-7}	2.3×10^{-7}	2.3×10^{-7}	2.4×10^{-7}

Table 5.13 The error estimates in this table are for an implementation of the Crank–Nicolson method using the data of Problem 5.2 and Gauss–Seidel iteration with the stopping criterion: successive iterations that differ in Euclidean norm by less than 10^{-6}. The floating point numbers are computed by averaging the absolute error estimates [Eq. (5.69)] corresponding to the two concentrations.

h	32×32	64×64	128×128	256×256
$\Delta t = 1/4$	1.20052	2.97511	1.97269	2.06121
$\Delta t = 1/8$	1.42807	4.57200	2.17178	-0.457879
$\Delta t = 1/16$	1.65060	0.41385	1.94150	0.03870
$\Delta t = 1/32$	0.99641	1.79209	1.96255	3.38090
$\Delta t = 1/64$	4.77933	1.94427	1.98211	3.95762
$\Delta t = 1/128$	-4.55519	1.97100	1.99247	1.69587

Table 5.14 The order estimates in this table are for an implementation of the Crank–Nicolson method using the data of Problem 5.1 and Gauss–Seidel iteration with the stopping criterion: successive iterations that differ in Euclidean norm by less than 10^{-6}. The floating point numbers are computed for the average of the concentrations $(u + v)/2$ using Eq. (5.67).

estimates in Table 5.12 suggest that this implementation of the algorithm is performing as expected at order two in time. The computed value of the average over both concentrations is consistently computed (over the viable grid and step sizes) to be 0.453021. The averaged u concentration is consistently 0.808316 and the averaged v concentration is 0.0977254. These values agree with the values obtained using the forward Euler method reported on page 150. We can have a high level of confidence that these values are within 1% of the exact corresponding values.

Computational results for Problem 5.1, using the Crank–Nicolson algorithm and Gauss–Seidel iteration, are reported in Tables 5.14 and 5.15. The order estimates in Table 5.14 suggest that this implementation of the algorithm is performing at the expected order two in time. The computed values of the averages over both concentrations converge, and these values are in good agreement with the values obtained by the forward Euler method. Perhaps the most trustworthy results are for the 128×128 grid with $\Delta t = 1/4$. The corresponding computed value for the averaged u concentration is 0.531848 and the averaged v concentration is 0.139837.

h	32×32	64×64	128×128	256×256
$\Delta t = 1/4$	3.42×10^{-9}	1.91×10^{-3}	2.42×10^{-3}	2.08×10^{-3}
$\Delta t = 1/8$	1.70×10^{-9}	4.80×10^{-4}	5.63×10^{-4}	1.00×10^{-3}
$\Delta t = 1/16$	8.49×10^{-10}	1.20×10^{-4}	1.43×10^{-4}	8.21×10^{-4}
$\Delta t = 1/32$	4.24×10^{-10}	3.01×10^{-5}	3.60×10^{-5}	1.06×10^{-4}
$\Delta t = 1/64$	2.11×10^{-10}	7.54×10^{-6}	9.05×10^{-6}	3.00×10^{-5}
$\Delta t = 1/128$	1.07×10^{-10}	1.89×10^{-6}	2.97×10^{-6}	1.40×10^{-5}

Table 5.15 The error estimates in this table are for an implementation of the Crank–Nicolson method using the data of Problem 5.1 and Gauss–Seidel iteration with the stopping criterion: successive iterations that differ in Euclidean norm by less than 10^{-6}. *The floating point numbers are computed by averaging the absolute error estimates [Eq.* (5.69)*] corresponding to the two concentrations.*

h	32×32	64×64	128×128	256×256
$\Delta t = 1/4$	0.485018	0.463835	0.352109	0.335842
$\Delta t = 1/8$	0.485018	0.463835	0.352000	0.335848
$\Delta t = 1/16$	0.485018	0.463835	0.352098	0.335842
$\Delta t = 1/32$	0.485018	0.463835	0.352097	0.335842
$\Delta t = 1/64$	0.485018	0.463835	0.352097	0.335842
$\Delta t = 1/128$	0.485018	0.463835	0.352097	0.335842

Table 5.16 The average concentrations in this table are for an implementation of the Crank–Nicolson method using the data of Problem 5.1 and Gauss–Seidel iteration with the stopping criterion: successive iterations that differ in Euclidean norm by less than 10^{-6}. *The floating point numbers are the computed averages of the concentrations* $(u + v)/2$.

These values also agree with the values obtained using the forward Euler method reported on page 150. The confidence level for these results extends to two or three decimal places at best.

How can more accurate results be obtained for Problem 5.1?

As already mentioned, a viable alternative is to use the trapezoidal method together with Newton's method to approximate the reaction terms.

A direct method (some adapted form of Gaussian elimination) could be used to solve the large linear systems that arise in the Crank–Nicolson algorithm to try to avoid errors due to iteration. The immediate difficultly (which can be overcome) is to write code that incorporates the banded matrix structure into the method. Recall that a banded $n \times n$ matrix has all components a_{ij} equal to zero whenever $|i - j| > m$ for some $m < n$. In other words, the nonzero components are confined to a diagonal band parallel to the main diagonal.

More accurate approximations for the boundary conditions could be employed; for example, an approximation based on the formula

$$F'(a) = \frac{1}{12\Delta x}(F(a-2\Delta x) - 8F(a-\Delta x) + 8F(a+\Delta x) - F(a+2\Delta x)) \\ + O(\Delta x^2). \tag{5.116}$$

Another idea is to use a more accurate spatial discretization instead of centered differences to approximate second derivatives. For instance, there are higher-order approximations based on the formula

$$\begin{aligned} F''(a) = &-\frac{1}{12\Delta x^2}(F(a-2\Delta x) - 16F(a-\Delta x) + 30F(x), \\ &- 16F(a+\Delta x) + F(a+2\Delta x)) + O(\Delta x^4), \end{aligned} \tag{5.117}$$

which is an alternate form of Eq. (5.63) (see Exercise 5.70).

Of course, there are many other methods to try. Can you devise a method that produces more accurate results for Problem 5.1?

Exercise 5.46. Suppose that $f : \mathbb{R}^n \to \mathbb{R}^n$ is a smooth function. Prove that the initial value problem $\dot{u} = f(u)$, $u(0) = u_0$ is equivalent to the integral equation

$$u(t) = u_0 + \int_0^t f(u(s))\, ds.$$

Exercise 5.47. Suppose that the local truncation error with discretization step size h for some method is $O(h^{n+1})$. Assume that the global error for the computation is the number of steps times the error per step. In case the computation is carried out over an interval of length L, show that the global error is proportional to Lh^n; that is, the global error is $O(h^n)$.

Exercise 5.48. Prove that Newton's method is quadratically convergent. Hint: Write the Newton iteration scheme as a function so that $x^{k+1} = F(x^k)$ and use the Taylor series approximation for F at the fixed point of F.

Exercise 5.49. (a) Derive the vector form of Newton's method and use it to write a numerical code formulated to approximate the rest points of the system of ODEs given by

$$\begin{aligned} \dot{U}_{ij} &= \epsilon(U_{i+1,j} - 2U_{i,j} + U_{i-1,j} + U_{i,j+1} - 2U_{i,j} + U_{i,j-1}) \\ &\quad - U_{i,j}V_{i,j}^2 + F(1 - U_{ij}), \\ \dot{V}_{ij} &= \mu(V_{i+1,j} - 2V_{i,j} + V_{i-1,j} + V_{i,j+1} - 2V_{i,j} + V_{i,j-1}) \\ &\quad + U_{i,j}V_{i,j}^2 - (F + \kappa)V_{ij}, \end{aligned}$$

for $i = 1, 2, 3, \ldots m$ and $j = 1, 2, 3, \ldots n$. Check that your implementation of Newton's method is quadratically convergent. (b) Find all the rest points in case $\lambda = 2.0 \times 10^{-5}$, $\mu = 10^{-5}$, $\kappa = 0.05$, $F = 0.02725$, $m = 32$, and $n = 32$. (c) Repeat part (b) for $m = n = 64$, 128, and, 256. Check for convergence of the rest points to values independent of the choice of the grid size.

Exercise 5.50. Recall Exercise 2.17 concerning the ODE

$$\dot{x} = 1, \qquad \dot{y} = axy,$$

where a is a parameter. The solution of the initial value problem with data $x(0) = y(0) = -1$ has the exact value $(x(2), y(2)) = (1, -1)$ at $t = 2$ independent of a. Apply forward Euler, backward Euler, improved Euler, and the trapezoidal method to this ODE, and determine the largest value of a for which each of your numerical codes returns the correct answer.

Exercise 5.51. An alternative to Newton's method for vector functions is "Newton's method one variable at a time (NOVAT)." Suppose we wish to apply Newton's method to find a zero of the function $f : \mathbb{R}^n \to \mathbb{R}^n$ and this function is given in coordinates by

$$y_1 = f_1(x_1, x_2, \ldots, x_n), \quad y_2 = f_2(x_1, x_2, \ldots, x_n), \quad \ldots, \quad y_n = f_n(x_1, x_2, \ldots, x_n).$$

The alternative iteration scheme is

$$\begin{aligned}
x_1^{k+1} &= x_1^k - \frac{f_1(x_1^k, x_2^k, \ldots, x_n^k)}{\frac{\partial f_1}{\partial x_1}(x_1^k, x_2^k, \ldots, x_n^k)}, \\
x_2^{k+1} &= x_2^k - \frac{f_2(x_1^{k+1}, x_2^k, \ldots, x_n^k)}{\frac{\partial f_2}{\partial x_2}(x_1^{k+1}, x_2^k, \ldots, x_n^k)}, \\
&\vdots \\
x_n^{k+1} &= x_n^k - \frac{f_2(x_1^{k+1}, x_2^{k+1}, \ldots, x_{n-1}^{k+1}, x_n^k)}{\frac{\partial f_n}{\partial x_n}(x_1^{k+1}, x_2^{k+1}, \ldots, x_{n-1}^{k+1}, x_n^k)}.
\end{aligned}$$

(a) Write code to implement NOVAT and test its convergence rate on several examples where the root is known in advance. Does it converge quadratically? Report your results. (b) Code an ODE solver using the trapezoidal method and NOVAT to solve the implicit equation for the updated state variable. Apply your code to several ODE systems and report your result. (c) Show that this version of Newton's method applied to a linear system $Ax = b$ produces Gauss–Seidel iteration. (d) Construct an example where Newton's method converges but NOVAT does not. (e) Is there an example where NOVAT converges but Newton's method does not? (f) Determine the criterion for convergence of NOVAT.

Exercise 5.52. Test the code for Exercise 5.51 on the ODE of Exercise 2.17. Compare the results obtained with the improved Euler method.

Exercise 5.53. Consider a test example for root finding: Let n be an even integer and consider the system of n equations given, for $i = 1, 2, 3, \ldots, n$, by

$$(x_i - 1)(x_{n-i+1}^2 + x_{n-i+1} + 3) = 0. \tag{5.118}$$

The system has real root $x_i = 1$. (a) Write a numerical code to implement Newton's method for this system of equations. Make a graph showing the number of iterations your code takes (starting at $x = 0$) to converge to the root (with absolute error less than 10^{-5}) as a function of n. How large can you make n so that your algorithm running on your computer finishes in less than one minute of CPU time? (b) Write a NOVAT code, as in Exercise 5.51 and repeat the test in part (a). (c) Approximate the Jacobian matrix using function evaluations (via the definition of the derivative as in Eq. (19.34)) and repeat the test in part (a). (d) Read Appendix A.14, write a code implementing Broyden's method, and repeat the test in part (a). This part requires more advanced coding experience. As an alternative, use the open source codes LAPACK and BLAS. These are state-of-the-art standards for linear algebra computations and solving systems of linear equations. (e) Discuss your findings. (f) Modify the system of equations or the other conditions in some interesting manner, which you must explain, and repeat parts (a)–(e). For example, what happens if the leading factor in each equation is changed to $(x_i - 1)^2$? (g) What happens if n is odd?

Exercise 5.54. (a) Working definition: An ODE with solutions that converge (or diverge) exponentially fast to (or from) other solutions is called stiff. There is no universally accepted definition of stiffness. An alternative definition might be the following: An ODE is stiff if some numerical methods require small step sizes to make accurate approximations of some of its solutions. The simplest example of a stiff ODE is $\dot{x} = \lambda x$, where $\lambda \neq 0$. If λ is negative, then all solutions approach the zero solution exponentially fast. Write the formulas and prove this fact. We may gain some insight on the performance of ODE solvers on stiff equations by applying them to this equation. For the analysis, it suffices to consider the solution of the ODE with initial condition $x(0) = 1$. Show that Euler's method applied to the test ODE gives the iteration scheme $x^{k+1} = (1 + \lambda\Delta t)x^k$. In case $\lambda < 0$, we know the solution decays. Show that the Euler approximation will decay if and only if $|1 + \lambda\Delta t| < 1$ and that this puts a restriction on the step size Δt: the step size must be smaller than $2/|\lambda|$. Thus, if λ is large, the step size must be small to have a chance of making an accurate approximation. Show that the same restriction applies in case $\lambda > 0$. An alternative view is the restriction $|\lambda\Delta t| < 2$. The set of points z in the complex plane (which must be considered for some more complicated examples) with $|z| < 2$ is called the region of absolute stability. (b) Determine the region of absolute stability for the improved Euler method. (c) Show that the region of absolute stability for the trapezoidal method is *the entire complex plane*; that is, there is no restriction on the step size. This result tells us that the trapezoidal method is useful to approximate solutions of stiff ODEs. It should perform well with larger step sizes than will be required for the one-step methods we have discussed. The trapezoidal method is a viable method for many stiff ODEs if the accuracy requirements are not too restrictive. Higher-order methods have been devised for stiff equations. To learn more, consult any book on the numerical analysis of ODE solvers. (d) Is the ODE in Exercise 2.17 stiff? Discuss your answer. (e) Consider the ODE

$$\dot{x} = \lambda x - y - \lambda(x^2 + y^2)x, \qquad \dot{y} = x + \lambda y - \lambda(x^2 + y^2)y$$

with initial data $(x(0), y(0)) = (\xi, 0)$. Derive the exact solution

$$x(t) = \frac{\xi e^{\lambda t}}{\sqrt{1 + \xi^2(e^{2\lambda t} - 1)}} \cos t, \quad y(t) = \frac{\xi e^{\lambda t}}{\sqrt{1 + \xi^2(e^{2\lambda t} - 1)}} \sin t.$$

Hint: Change to polar coordinates and recall Bernoulli's differential equation. (f) Show that the ODE of part (e) is stiff. (g) Test codes for improved Euler and trapezoidal integration with λ small (perhaps $\lambda = 1$) and big (perhaps $\lambda = 10$) and integration from $t = 0$ until $t = 6$. Do both codes return the correct answer? What is the maximum allowable step size for your codes to return an answer with relative error less than 1%?

Exercise 5.55. Write code for solving the Gray–Scott model using the Crank–Nicolson method. Use Newton's method to solve the nonlinear equations for the updates. Apply the code to Problems 5.1 and 5.2.

Exercise 5.56. [Discrete Dynamical Systems] Consider the function $G : \mathbb{R}^n \to \mathbb{R}^n$ as a dynamical system; that is, the initial state $x = x_0$ in $\mathbb{R}^n$ evolves according to the rule $x^{k+1} = G(x^k)$. A fixed point of G is a state x_0 such that $G(x_0) = x_0$. A fixed point is called stable if for each $\epsilon > 0$ there is a $\delta > 0$ such that all (forward) iterates of every initial state in the (open) ball of radius δ centered at the fixed point are in the ball of radius ϵ. A fixed point is called asymptotically stable if it is stable and, in addition, if the sequence of iterates of every state starting in some ball of radius δ centered at the fixed point converges to the fixed point. (a) Prove that a fixed point x_0 is asymptotically stable if $n = 1$ and $|G'(x_0)| < 1$. (b) The function $G(x) = x - \alpha(x^3 + x^2 - x - 1)$ has a fixed point at $x = 1$ for every $\alpha \in \mathbb{R}$. For which α is this fixed point asymptotically stable? (c) Prove that a fixed point x_0 is asymptotically stable if all the eigenvalues of the derivative $DG(x_0) : \mathbb{R}^n \to \mathbb{R}^n$ are in the open unit disk in the complex plane. (d) Consider the two-dimensional iteration scheme for the function G defined by

$$G(\phi, \theta) = (\phi + a \sin \phi, \theta + \phi + a \sin \phi)$$

defined on the (ϕ, θ) plane modulo 2π. This means a point starting in the square $[0, 2\pi] \times [0, 2\pi]$ maps back into this set by reducing each component of the image to the remainder when it is divided by 2π. This iteration scheme has been widely studied. It is called the standard map. It has the physical interpretation as a model of the motion of a kicked rotor, where θ is the rotation angle of the rotor and ϕ is the momentum variable. The origin is a fixed point. Is it asymptotically stable for some values of the parameter a? Plot 1000 iterates of the map on the same figure starting at several different points chosen at random in the square and taken as initial values where a is fixed at some point in the range $[0.5, 1]$. (e) Consider the second-order differential equation model for the displacement x of the free end of a clamped beam under sinusoidal excitation:

$$\ddot{x} + \epsilon \dot{x} - x + x^3 = a \sin(2\pi t),$$

where ϵ is the (positive) damping parameter and a is the positive amplitude of the sinusoidal forcing. Rewrite the differential equation as a first-order system and set up a code to compute the stroboscopic Poincaré map (see Exercise 5.29 for the definition). Start with $\epsilon = 0.1$ and $a = 0.0$. Does iteration of the Poincaré map converge to a fixed point for some starting value? Same question for $\epsilon = 0.1$ and $a = 0.5$. Hint: Perhaps some graphs of orbits of the Poincaré map would give insight into the behavior of the iteration scheme. A few hundred iterations of each initial state should be sufficient. (The dynamics of this map has a rich structure that is worth exploring. Try some other choices of the parameters.)

Exercise 5.57. (a) Show that Aitken's Δ^2 method applied to the sequence of iterates of the function G defined by Eq. (5.100) with starting value $x_0 = 2$ is not quadratically convergent. (b) Show that Steffensen's method applied to this sequence is quadratically convergent. (c) It is possible to prove that Aitken's Δ^2 method applied to a sequence of iterates of a function is superlinearally convergent; that is,

$$\lim_{k\to\infty} \frac{|\text{Aitken}(k+1) - x^\infty|}{|\text{Aitken}(k) - x^\infty|} = 0.$$

Give some (convincing) numerical evidence for this fact.

Exercise 5.58. [ODE Nonlinear BVPs] Consider the following BVP: Find a solution $t \mapsto \phi(t)$ of the differential equation $\ddot{x} + x - x^3 = 0$ that satisfies the conditions $\phi(0) = 0$, $\phi(2) = 0$, and $\phi(t) > 0$ for $0 < t < 2$. (a) Approximate the solution (is there only one?) by the shooting method; that is, let $t \mapsto x(t,\xi)$ denote the solution of the ODE such that $x(0,\xi) = 0$ and $\dot{x}(0,\xi) = \xi$. Use a numerical method to find a root of the equation $x(2,\xi) = 0$. Suggestion: Draw the phase portrait of the ODE and use it to determine a reasonable starting value ξ_0. Perform some numerical experiments to determine a value for λ so that the iterates of $G(\xi) = \xi - \lambda x(2,\xi)$ converge. Accelerate the convergence using Steffensen's method. Your code might be tested using instead the ODE $\ddot{x} + \dot{x}/10 + x = 1$ so that exact solutions are available. Why is the right-hand side of this test equation not set to zero? (b) Shoot with Newton's method. Hint: Find the required derivative by solving (numerically) a variational equation. (c) The suggested methods in parts (a) and (b) may not be the best. Don't shoot! Approximate the solution of the BVP by discretizing in time with say N equally spaced subintervals on $[0,2]$, write ϕ^k for the value of the desired solution at the kth node, use the discretization to write one equation in these N variables at each node, and approximate a solution of the resulting nonlinear system of N equations in N variables. (d) Discuss the accuracy and efficiency of various methods for solving BVPs for ODEs. Perhaps you can find a new method that is better than any known method.

Exercise 5.59. Show that the improved Euler method [Eq. (5.103)] is second order.

Exercise 5.60. (a) Prove that the nodes function defined by Eq. (5.105) is bijective. (b) Show how to determine the formula for the nodes function by supposing the function is affine and solving a system of linear equations.

Exercise 5.61. Consider the (x, y) data

$$(0, 0.09), (1/4, 0.12), (1/2, 0.16), (3/4, 0.20), (1, 0.26), (5/4, 0.32),$$

$$(3/2, 0.39), (7/4, 0.46), (2, 0.54), (9/4, 0.61), (5/2, 0.68), (11/4, 0.75),$$
$$(3, 0.80), (13/4, 0.85), (7/2, 0.88), (15/4, 0.91), (4, 0.93), (17/4, 0.95),$$
$$(9/2, 0.96), (19/4, 0.97), (5, 0.98).$$

The theory underlying the experiment that produced the data implies that the data should be on the graph of a function of the form

$$f(x, a, b) = \frac{e^{a+bx}}{10 + e^{a+bx}},$$

where a and b are parameters. Use nonlinear regression to find the best constants a and b. To do this, denote the data points with (x_i, y_i), for $i = 1, \ldots, 21$, and define a new function by

$$\Gamma(a, b) = \sum_{i=1}^{21} (y_i - f(x_i, a, b))^2.$$

The desired (a, b) is the minimum of Γ. To find this point, use calculus. Take the derivative and set it equal to zero. Use Newton's method to solve the resulting nonlinear equations.

Exercise 5.62. Consider the (x, y) data

$$(250, 0.0967), (500, 0.159), (750, 0.201), (1000, 0.233), (1250, 0.257),$$
$$(1500, 0.276), (1750, 0.291), (2000, 0.305), (2250, 0.315), (2500, 0.325),$$
$$(2750, 0.333), (3000, 0.340), (3250, 0.346), (3500, 0.351), (3750, 0.356),$$
$$(4000, 0.360).$$

The theory (Michaelis–Menten kinetics) underlying the experiment that produced the data implies that the data should be on the graph of a function of the form

$$f(x, a, b) = \frac{ax}{b + x},$$

where a and b are parameters. Use nonlinear regression, as in Exercise 5.61, to find the best constants a and b.

Exercise 5.63. Find a nonsingular matrix A and a nonzero vector b such that there is no choice of ω such that iteration of T, defined to be $Tz = z - \omega(Az - b)$, converges to the solution of the linear system $Az = b$.

Exercise 5.64. (a) The iteration schemes mentioned in this chapter for approximating solutions of $Ax = b$ all have the general form $x^{k+1} = Mx^k + a$, where M is an $(n \times n)$ matrix and a is a given n-dimensional vector. Show that if the sequence x^k produced by the method converges, then the limit of the sequence satisfies the equation $x = Mx + a$; in other words, the limit is a fixed point of the map $x \mapsto Mx + a$.
(b) Show: If M is a diagonal matrix and every eigenvalue of M has absolute value less

than unity, then the iteration scheme converges for every starting point x^0.
(c) Show: If M is a diagonalizable matrix and every eigenvalue of M has absolute value less than unity, then the iteration scheme converges for every starting point x^0.
(d) Show: If M is in Jordan canonical form and every eigenvalue of M has absolute value less than unity, then the iteration scheme converges for every starting point x^0.
(e) Show: If every eigenvalue of M has absolute value less than unity, then the iteration scheme converges for every starting point x^0.
(f) Show: In all cases, the fixed point is unique; that is, the same fixed point is obtained independent of the starting point.
Hint: Define the error $e^k := x - x^k$, where x is the fixed point. Show that the desired convergence follows if the sequence produced by $e^{k+1} = Me^k$, independent of the starting point e^0, converges to zero. Also, the same is true if the sequence of matrices M^k converges to zero.

Exercise 5.65. Is Newton's method, which is quadratically convergent, a useful choice for solving *linear* systems of equations?

Exercise 5.66. Many special properties of symmetric matrices can be used in numerical methods. Consider solving $Ax = b$ for a general matrix A. (a) Show that the matrix A^TA, where A^T denotes the transpose of A, is symmetric. (b) Show that if A is invertible, then A^TA is positive-definite. (c) There are good results concerning convergence of SOR for symmetric positive-definite matrices. Discuss the possible utility of replacing the linear system $Ax = b$ with $A^TAx = A^Tb$ in theory and using numerical experiments.

Exercise 5.67. (a) Consider the Crank–Nicolson scheme applied to the heat equation $u_t = \kappa u_{xx}$ on a finite interval with zero Dirichlet boundary conditions at the end of the interval. Prove that this numerical method is unconditionally numerically stable. (b) Repeat part (a) for zero Neumann boundary conditions.

Exercise 5.68. (a) Derive the backward Euler method for $\dot{x} = f(x)$:

$$x^{k+1} = x^k + \Delta t f(x^{k+1}).$$

(b) Show this method is first order. (c)Write a numerical code to implement backward Euler that incorporates Newton's method to solve for the state variable update at each time step. (d) Repeat the computations for Problems 5.1 and 5.2 using the backward Euler method and compare results obtained with the forward Euler method. (e) Apply the backward Euler scheme to obtain a finite difference method for approximating solutions of the heat equation $u_t = \kappa u_{xx}$. Is the method unconditionally numerically stable?

Exercise 5.69. Show that

$$\rho(DT) = \frac{2\alpha}{1+2\alpha}, \qquad \omega = \frac{1}{1+2\alpha}$$

in the context of Eq. (5.109). Hint: The maximum occurs at an end point of the interval $[1-\alpha, 1+3\alpha]$. Draw graphs of the functions $\omega \mapsto |1-(1-\alpha)\omega|$ and $\omega \mapsto |1-(1+3\alpha)\omega|$.

Exercise 5.70. (a) Derive Eq. (5.116) directly from Taylor series approximations of F. (b) Derive Eq. (5.117) directly from Taylor series approximations of F.

Exercise 5.71. Write a code to approximate solutions of the Gray–Scott model using the Crank–Nicolson method with periodic boundary conditions, and compare the results with those obtained using the forward Euler method.

Exercise 5.72. [Method of Lines] The method of lines is a useful numerical method for some PDEs. For the case of reaction-diffusion equations, the idea is very simple: Discretize in space but not in time and treat the resulting equations (one for each spatial node) as a (perhaps large) system of ODEs (with time as the independent variable). The forward Euler method, discussed previously in this text, is an example of the method of lines where Euler's method is used to solve the system of ODEs. More sophisticated methods can be used to solve the ODEs. (a) Solve the BVP given by Eq. (5.56) by the method of lines using the improved Euler (explicit) method to solve the resulting systems of ODEs. (b) Solve Problem 5.2 by the method of lines using the improved Euler (explicit) method to solve the system of ODEs. The next part of this problem requires standard numerical methods for approximating the solutions of ODEs that are not explained in this book, but are found in standard textbooks (see, for example, [12]). (c) Solve the BVP given by Eqs. (5.56) by the method of lines using fourth-order Runge–Kutta. (d) Solve the BVP given by Eqs. (5.56) by the method of lines using Runge–Kutta–Fehlberg, a method that includes adaptive step size control. (e) Solve the BVP given by Eqs. (5.56) by the method of lines using a fourth-order (Adams) predictor-corrector, a multistep method that uses several previously computed steps. (f) Solve the BVP given by Eqs. (5.56) by the method of lines using your favorite ODE integration method. In all cases, compare numerical results (accuracy, stability, and efficiency) relative to previously coded methods.

Exercise 5.73. Implement Steffensen's method (incorporating the ideas discussed on page 177) to solve five-diagonal $(m+1)^2 \times (m+1)^2$ matrix systems $Ax = b$ whose nonzero diagonals are the main diagonal of A, the lower (m+1)st diagonal, the lower first diagonal, the upper first diagonal, and the upper (m+1)st diagonal. (a) Use your code to solve the system $Ax = b$, where the components of the vector b are all equal to 1, all components on the main diagonal of A are equal to 5 and all elements on the lower (m+1)st diagonal, the lower first diagonal, the upper first diagonal, and the upper (m+1)st diagonal are equal to 1. (b) Incorporate Steffensen's method into a Crank–Nicolson code and use it to solve the BVP given by Eqs. (5.56).

Exercise 5.74. Jacobi's method for approximating solutions of $Ax = b$ is obtained by splitting the system matrix as $A = D + (L + U)$, where D is the main diagonal and $L+U$ is the off diagonal part of the matrix (which for lack of a better notation, is written here as the sum of the strict lower L and upper U triangular parts of A). The iterative method is simply obtained by moving the off diagonal part to the right-hand side:

$$Dx^{k+1} = b - (L + U)x^k.$$

(a) Show that this method is given in components by

$$x_i^{k+1} = \frac{1}{a_{ii}}(b - \sum_{j \neq i} a_{ij} x_j^k).$$

(b) Write the Gauss–Seidel method in components.
(c) Discuss the statement: "The Jacobi method programed badly is Gauss–Seidel." Hint: Think about mathematical = and computer code =, which often means replacement.
(d) Set up and run tests to compare the number of iterations required to achieve some prespecified accuracy for Jacobi and Gauss–Seidel iteration. Report on your results.

Exercise 5.75. Recall the two-dimensional Oregonator reaction model [Eqs. (4.39)]. Add diffusion and explore spatial pattern formation for this model using the methods developed in this chapter.

Exercise 5.76. Suppose that the current dimensionless concentration of a substance u in space is $u(w, t)$ at the dimensionless spatial position $w = (x, y, z)$ and dimensionless time t. Suppose the measured current concentration is approximately

$$\frac{z(1 - e^{-(1-x^2-y^2)})}{(1 + z^2)^2}$$

for w in the half cylinder $\{(x, y, z) : x^2 + y^2 \leq 1, z \geq 0\}$ and zero everywhere else. The process that led to the current state is believed to be modeled by the three-dimensional diffusion equation $u_t = \frac{1}{2}\Delta u$. Determine the dimensionless concentration at the point $w = (1, 0, 2)$ at time $t = -1$. Hint: Solutions of the diffusion equation do not make sense for negative time. To treat the problem, consider the process beginning at time $t = -1$ and moving forward to time zero. Alternatively, reverse the direction of time in the dynamical equation; that is, work with the PDE $u_t = -\frac{1}{2}\Delta u$. Although it is possible to use the numerical methods developed in this chapter to obtain an approximation of the desired value, alternative methods are more efficient. The key result states that the solution of $u_t = k\Delta u$ with bounded continuous initial data $u(w, 0) = f(w)$ is

$$u(w, t) = \int_{\mathbb{R}^3} K(w - \omega) f(\omega)\, d\omega,$$

where K, called the heat kernel (or diffusion kernel), is given in n-dimensional space by

$$K(p) = \frac{1}{(4\pi kt)^{n/2}} e^{-|p|^2/(4kt)}.$$

The desired concentration value may now be determined by approximating an integral over three-dimensional space. A numerical method that is widely used due to its simplicity is Monte Carlo integration. Suppose we desire the value of the integral $\int_a^b g(x)\, dx$. We may use a random number generator to generate a finite sequence $\{x_j\}_{j=1}^N$ of (uniformly distributed) random numbers in the interval $[a, b]$. An approximate value of the integral is given by

$$\int_a^b g(x)\,dx \approx \frac{b-a}{N}\sum_{j=1}^{N} g(x_j).$$

Try this method on a few test cases to convince yourself that it gives reasonable approximations and discuss why the method should work. The integral of the heat kernel (for all k and t such that $kt > 0$) over all of space is exactly one. Check this by Monte Carlo integration. (A better way to apply Monte Carlo integration in this case is to notice the relation between the heat kernel and the normal distribution function from probability theory.) Use Monte Carlo integration to solve the original problem. Check the result by using an alternative method to approximate the solution of the heat equation.

Exercise 5.77. [Modeling Project] According to folklore, starting a fire in a fireplace warms the room in which the fireplace resides and cools the outlying rooms. The mechanism for this process is the movement of air required by the fire. Air from the outlying rooms used by the fire is replaced by cold outside air. Develop a mathematical model to describe this physical situation and determine to what extent this bit of folklore is true.

Exercise 5.78. (a) Revisit Exercise 5.41 using a second-order in time method. Approximate the sensor output for the model given by Eqs. (5.74). In particular, determine the length of time in hours until the sensor output is 2.0 mg / L. (b) Compare and contrast simulated sensor output and the efficiency of your code relative to Euler's method for time stepping. In particular, discuss the time-step sizes that may be used in these simulations.

Exercise 5.79. Revisit Exercise 5.42 using a second-order in time method. (a) How long is the transient in hours measured from the instant the flow starts until the root mean square distance of the density profile in the porous block is within 1% of the steady state density profile. Compare and contrast simulated sensor output and efficiency of your code relative to Euler's method for time stepping. In particular, discuss the time-step sizes that may be used in these simulations.

CHAPTER 6

Excitable Media: Transport of Electrical Signals on Neurons

One of the most important discoveries of the 20th century in biophysics is the understanding of how nerves transmit information: the transport of ions of sodium and potassium (also sodium and calcium) across the outer membrane of a nerve cell is responsible for electrical signals that may propagate in traveling waves along the membrane after an appropriate stimulation. Alan Hodgkin and Andrew Huxley (working in the early 1950s) described the biological basis of the ion transport, created a mathematical model, and explained experimental data on electrical signals excited in squid giant axons; they were awarded the Nobel Prize in Physiology or Medicine in 1963.

The original Hodgkin–Huxley model is a system of four ordinary differential equations (ODEs). It was not meant as a predictive model as it does not include the details of the ion transport. The utility of the model lies in its aid to understanding the qualitative behavior of signals on neurons. Simplifications of the basic model, modifications for other excitable media (for example, muscle cells), and spatial dependence have been extensively investigated. One of the most influential simplifications of the Hodgkin–Huxley model was introduced by Richard FitzHugh, who also pioneered its mathematical and numerical analysis. An electric circuit analog for a similar model was constructed by Jin-Ichi Nagumo. Their two-state model—which is still widely used in the study of neural networks—describes the qualitative electrical behavior of stimulated nerve cells. We will investigate this model.

Excitable media in biology are far from being completely understood. Much contemporary work is focused on ion transport. Living membranes contain a variety of ion channels (across the membrane) that are selective to specific ions. The transport mechanisms and the switches that open and close ion channels are fundamental to the function of many biological processes. Also, networks of nerve cells and other excitable media are ubiquitous in biology. The study of such networks may lead to an understanding of how the brain works. Mathematics is playing an increasingly important role in this area of interdisciplinary research.

An Invitation to Applied Mathematics: Differential Equations, Modeling, and Computation.
http://dx.doi.org/10.1016/B978-0-12-804153-6.50006-3, Copyright © 2017 Elsevier Inc. All rights reserved.

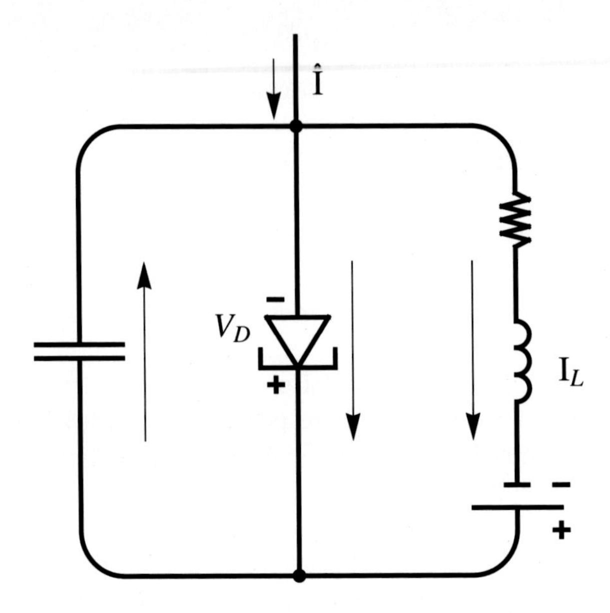

Fig. 6.1 The FitzHugh–Nagumo circuit.

6.1 THE FITZHUGH–NAGUMO MODEL

The FitzHugh–Nagumo model treats the nerve membrane as the electric circuit depicted in Fig. 6.1. The differential equations for the important states, the voltage V_D across the diode, and the current through the inductor I_L are obtained using standard circuit theory.

An electric circuit is a network of electrical components (capacitors, resistors, inductors, diodes, transistors, batteries, and so on) connected by wires. In this context, the basic physics of electromagnetism, which is encoded in Maxwell's laws and the Lorentz force law, may be simplified to a few basic rules that are most often used to approximate the currents and voltages in an electrical circuit.

Current I is defined to be the rate of change of charge Q with respect to time:

$$\frac{dQ}{dt} = I.$$

The magnitude of a current is measured in amperes; the sign of the current determines the direction of the flow of electrons. For historical reasons—

Benjamin Franklin being responsible—the direction of a current in a circuit is usually taken to be opposite to the direction of the flow of electrons.

An electron carries one unit of negative electric charge. A proton carries a unit of positive charge. Two charged particles attract or repel according to Coulomb's law: the magnitude of the force on a charged particle due to a second charged particle is proportional to the product of the charges and inversely proportional to the square of the distance between them. The direction of the force is from the charged particle toward the second charge; that is, the force on charge q_1 at position r_1 due to the charge q_2 at position r_2 is

$$\text{Coulomb force on } q_1 = k\frac{q_1 q_2}{|r_1 - r_2|^3}(r_1 - r_2),$$

where k is Coulomb's constant. The force on q_2 due to the presence of q_1 is the negative of the force on q_1 due to q_2. Charged particles with the same signs repel; charged particles with opposite signs attract.

A static charge (no motion relative to an inertial coordinate system) produces an electric field

$$E = \nabla \phi$$

where

$$\phi = k\frac{q}{r}$$

is called the electric potential and r is the distance from the charge that produces the field. The field of a collection of charges is the sum of the (vector) fields produced by each charge in the collection. A test charge (a particle with small charge) interacts with an electric field. In case the electric field E is static, negative test charges (electrons) move from positions of high potential to positions with lower potential. Positive charges move in the opposite direction. The equation of motion (when the particle is moving slowly relative to the speed of light) is given by the force law

$$m\frac{dv}{dt} = qE,$$

where m is the mass of the particle, v its velocity, and q its charge.

Voltage is a scalar (measured in volts) defined to be the potential difference at two positions in an electric field. Note that, because we differentiate the potential to obtain the electric field, the electric potential is defined up to the addition of a constant.

The electric field produced by moving charges is complicated. In this case electricity and magnetism are entwined, and an electromagnetic field is produced that consists of an electric field E and a magnetic field B. A test particle moves according to the Lorentz force law

$$m\frac{dv}{dt} = q(E + v \times B).$$

Well almost.... When a charge moves in an electric field, the field that the charge produces also acts on the particle. This leads to serious complications, which are not completely understood. These facts are mentioned to give an impression of what might be involved in fundamental models of electric circuits and circuit elements.

Fortunately, for most practical applications, a useful approximation of the fundamental field theory of electromagnetism can be used to analyze electric circuits. In fact, much of the electromagnetic theory for currents and voltages in wires connected into circuits is reduced to a basic assumption and two simple rules due to Gustav Kirchhoff:

Basic Assumption The current in a wire is the same at all points along the wire and through each two-terminal circuit element connected in series along the wire.

Kirchhoff's Current Rule The sum of the currents at a node in an electric circuit (where two or more wires are joined) is zero.

Kirchhoff's Voltage Rule The sum of the potential differences (voltages) across circuit elements around every loop in a circuit is zero.

The basic assumption is reasonable provided the circuit elements are small compared with the wave length of the electromagnetic waves produced by the motion of the electrons in the circuit. More precisely, the speed of light divided by the frequency of operation (which is the wave length) must be much larger than the size of the circuit. The rule of thumb is to assume the basic assumption is true when the wave length is at least 10 times the size of the circuit. The current rule is essentially a statement of conservation of charge; the voltage rule is a statement of conservation of energy.

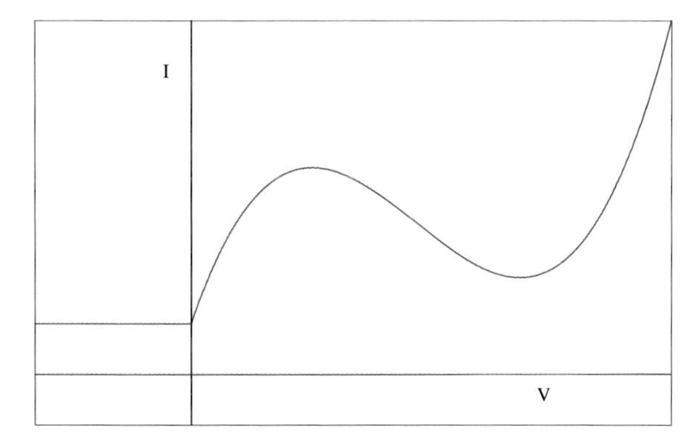

Fig. 6.2 The figure depicts a schematic I-V relation for the diode in the FitzHugh–Nagumo circuit. The axes are offset to show zero current through the diode for voltages across the device less than the threshold voltage at the position of the vertical axis.

We are now ready to consider the currents and voltages in the FitzHugh–Nagumo circuit. Although there are systematic ways to solve circuits, the process for simple circuits is an application of Kirchhoff's laws for enough nodes and loops to obtain a closed system of differential equations. We begin by assigning positive directions to the currents. The assigned directions are not important as long as we remember that currents can be negative (which simply means that the flow in the circuit might be opposite to some of our choices). Circuit elements affect the current flowing though them. The effect of each element in the FitzHugh–Nagumo circuit, a capacitor, a tunnel diode, a resistor, an inductor, and a battery, is given mathematically by a function that relates the current I flowing through the element and the voltage V across it. These I-V relations are the building blocks of circuits. The relations for the most basic elements are as follows:

Resistor The I-V relation is Ohm's law $V = IR$, where R is a factor of proportionality called the resistance.

Capacitor A capacitor is a storage device, which at a basic level stores charge. The total charge Q produces a field such that the voltage across the capacitor is $V = Q/C$, where the factor C is called the capacitance. Using the definition of current as the rate of change of charge, the I-V relation for a capacitor is

$$\frac{dV}{dt} = \frac{I}{C}.$$

Inductor An inductor is a coil of wire. Its effect is a result of Faraday's law of induction, which states that a nonzero current produces a magnetic field. A magnetic field is produced in the space bounded by the coil that opposes changes in the current through the coil. The I-V relation in an inductor is

$$V = L\frac{dI}{dt},$$

where the factor L is called the inductance. (Note: Faraday's law is symmetric: the motion of a magnet in the space bounded by the coil produces a current in the coil. This is the basic principle underlying the electric motor.)

Battery A battery is a storage device that produces a constant voltage across its terminals.

Diode Diodes are circuit elements that cause currents to flow in a specified direction. In modern circuits, diodes are semiconductor devices that come in several varieties. Generally their I-V relations are *nonlinear*. The diode in the FitzHugh–Nagumo circuit has an I-V relation in the form of a cubic, as depicted in Fig. 6.2. The current through the diode is zero for voltages across the diode that are less than the threshold voltage indicated by the position of the vertical axis in the figure.

The top node of the FitzHugh–Nagumo circuit, with the current directions as in Fig. 6.1, is a junction of four wires that carry the currents I_C of the capacitor, I_D of the diode, the induced current $\hat{I}$, and the current through the circuit branch containing the resistor, inductor, and battery. By Kirchhoff's current rule, the current through each of the latter elements is the same. Thus, we may choose the current through one of these elements (for example, the current I_L through the inductor) to represent the current in the branch. With these choices, an application of the current rule yields the relation

$$\hat{I} + I_C = I_D + I_L.$$

Using the I-V relation for a capacitor and the relation $I_D = F(V_D)$ for the diode (where F is the function whose graph is depicted in Fig. 6.2), we have the equation

$$\hat{I} + C\frac{dV_C}{dt} = F(V_D) + I_L.$$

Kirchhoff's voltage rule applied to the leftmost loop in the circuit implies that $V_C = -V_D$; therefore,

$$\hat{I} - C\frac{dV_D}{dt} = F(V_D) + I_L. \tag{6.1}$$

The cubic shape of the I-V curve for the diode may be approximated by shifting and lifting the cubic function given by $g(z) = Az^3 - Bz$; that is,

$$F(V_D) = g(V_D - \lambda) + \mu,$$

where A, B, λ, and μ are positive constants and V_D exceeds the threshold voltage for the diode. In the analysis to follow, the operating conditions will always be assumed to exceed the threshold value.

We may now replace $F(V_D)$ in Eq. (6.1) to obtain

$$\hat{I} - C\frac{dV_D}{dt} = g(V_D - \lambda) + \mu + I_L$$

or

$$C\frac{dV_D}{dt} = -g(V_D - \lambda) - (\mu + I_L) + \hat{I}. \tag{6.2}$$

The voltage rule applied to the rightmost loop in the circuit yields the relation

$$V_D = V_R + V_L + V_B.$$

The voltage across the resistor is RI_R and the voltage across the inductor is $L(dI_L/dt)$. Moreover, the current through the resistor is the same as the current through the inductor. Thus we have the differential equation

$$L\frac{dI_L}{dt} = V_D - RI_L - V_B$$

or

$$L\frac{dI_L}{dt} = V_D - R(\mu + I_L) + R\mu - V_B. \tag{6.3}$$

With the change of variables $\hat{V} = V_D - \lambda$ and $\hat{W} = I_L + \mu$, we have the basic form of the circuit equations:

$$C\frac{d\hat{V}}{dt} = B\hat{V} - A\hat{V}^3 - \hat{W} + \hat{I},$$
$$L\frac{d\hat{W}}{dt} = \hat{V} - R\hat{W} + (\lambda + R\mu - V_B). \tag{6.4}$$

By scaling $\hat{V}$, $\hat{W}$, and t to V, W, and s, respectively, this system can be converted to the dimensionless form

$$\frac{dV}{ds} = V - \frac{1}{3}V^3 - W + I,$$
$$\frac{dW}{ds} = aV - bW + c, \tag{6.5}$$

where a, b, and c are nonnegative constants (see Exercise 6.1).

The FitzHugh–Nagumo circuit is a model of the electrical activity at a point on a neuron. The process of opening and closing ion channels is modeled by diffusion of the voltage (corresponding to the dimensionless state V) along the neuron. For the FitzHugh–Nagumo model, the spatial dependence is considered to be one-dimensional with respect to a measure of distance $\hat{x}$ in the axial direction of the neuron. As in heat conduction, the mathematical model of diffusion is $\delta\partial^2 V_D/\partial\hat{x}^2$, where δ is the diffusivity. By adding this term to the right-hand side of the circuit model and scaling of the spatial variable, we obtain the dimensionless form of the FitzHugh–Nagumo equations:

$$\frac{\partial V}{\partial s} = \frac{\partial^2 V}{\partial x^2} + V - \frac{1}{3}V^3 - W + I,$$
$$\frac{\partial W}{\partial s} = aV - bW + c \tag{6.6}$$

(see Exercise 6.2). The state variable V is a representation of the voltage; it is also called the action or membrane potential, W is called the recovery variable, and I is the stimulus. In keeping with tradition, we may forget that the model is a dimensionless version of the original model and revert to the usual time variable; that is, we will consider this PDE model in the form

$$\frac{\partial V}{\partial t} = \frac{\partial^2 V}{\partial x^2} + V - \frac{1}{3}V^3 - W + I,$$

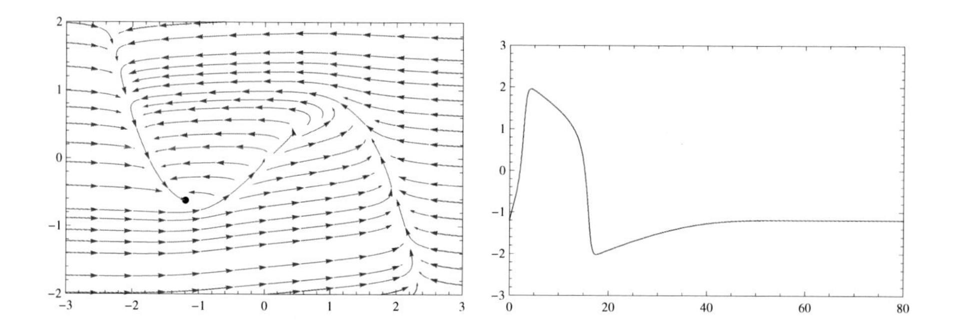

Fig. 6.3 The left panel depicts a computer-generated phase portrait of ODE system (6.9) *for the parameter values* (6.8) *and stimulus* $I = 0$ *with the spiral sink at* $(V, W) \approx (-1.199, -0.624)$ *marked with a disk. The right panel is a* V *profile for a trajectory starting half a unit below the rest point.*

$$\frac{\partial W}{\partial t} = aV - bW + c, \tag{6.7}$$

where we may imagine that t is the temporal independent variable and x is the spatial variable.

The FitzHugh–Nagumo model is not meant to be predictive; rather, its purpose is to capture the main qualitative features of the electrical activity along a neuron. The most important prediction of the model (which agrees with experiments) is the existence of a threshold stimulus impulse that produces traveling voltage (and recovery) waves, which propagate away from the spatial position of the stimulus. The membrane potential traveling wave is the mechanism responsible for carrying information along the neuron.

For definiteness, let us consider the case first explored by FitzHugh [39], which for our version of the model has parameter values

$$a = 0.08, \qquad b = (0.08)(0.8), \qquad c = (0.08)(0.7). \tag{6.8}$$

In this case,

$$a > b > c > 0$$

and each of these parameters is much smaller than 1.

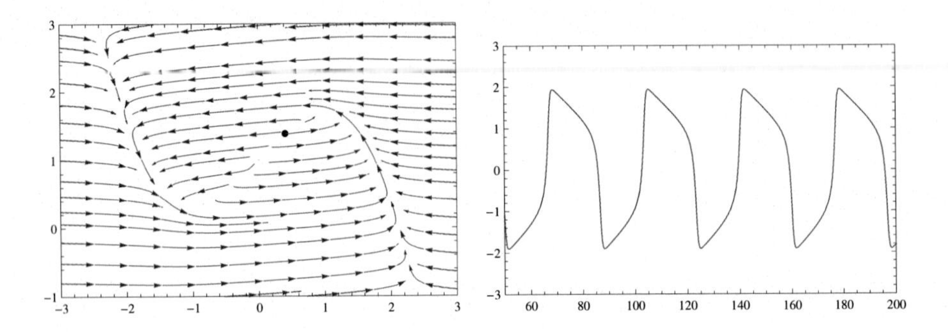

Fig. 6.4 The left panel depicts a computer-generated phase portrait of the ODE system (6.9) *for the parameter values* (6.8) *and stimulus* $I = 1$ *with the source at* $(V, W) \approx (0.409, 1.386)$ *marked with a disk. The right panel is a* V *profile for the periodic trajectory surrounding the rest point.*

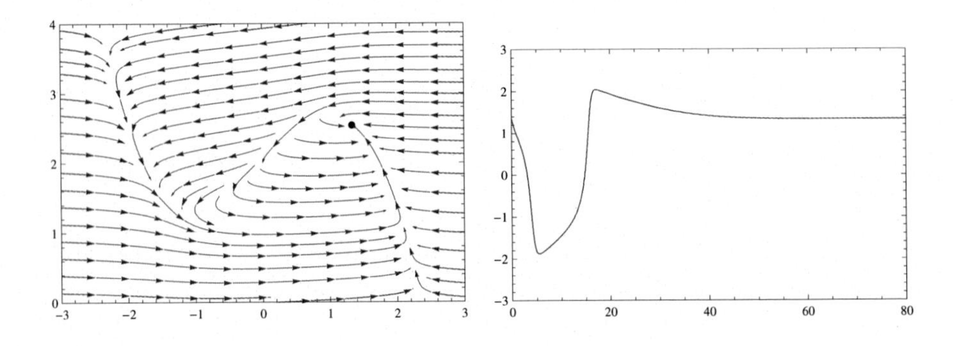

Fig. 6.5 The left panel depicts a computer-generated phase portrait of the ODE system (6.9) *for the parameter values* (6.8) *and stimulus* $I = 2$ *with the source at* $(V, W) \approx (1.334, 2.543)$ *marked with a disk. The right panel is a* V *profile for a trajectory starting half a unit above the rest point.*

Some insight is gained by analyzing the model without diffusion

$$
\begin{aligned}
\dot{V} &= V - \frac{1}{3}V^3 - W + I, \\
\dot{W} &= aV - bW + c,
\end{aligned}
\tag{6.9}
$$

which may be viewed as a model for the action potential and recovery variable restricted to a single point along a neuron.

Fig. 6.3 depicts the phase portrait of ODE model (6.9) with stimulus $I = 0$. Note the presence of the vertical isocline $W = V - V^3/3$ (the curve in the phase plane where $\dot{V} = 0$). Away from this curve, $\dot{V}$ is at least an order of magnitude larger than $\dot{W}$ because the parameters a, b, and c are at least an order of magnitude smaller than 1 (which may be viewed as the coefficient of the V component of the vector field). Thus, the flow below the vertical isocline moves rapidly toward the right and the flow above it moves rapidly toward the left. Because this isocline can be crossed in only one direction in each of the half planes bounded by the horizontal isocline $W = (aV + c)/b$, there is a layer near the horizontal isocline where the flow moves relatively slowly and stays near the isocline. For example, a solution starting on the negative W-axis moves right and rapidly approaches the vertical isocline. Similarly, solutions starting to the right of this isocline, from about the same distance below the V-axis, move rapidly toward the isocline in the opposite direction (right to left). As the flow is attracted from both directions toward the isocline in this region, once a trajectory in the phase plane is near the isocline it must stay near the isocline (where it moves slowly) until it reaches the vicinity of the local maximum of this cubic curve. Above this point in the phase plane, the trajectory moves rapidly left until it reaches the vicinity of the vertical isochrone in the left half-plane. The trajectory follows the isochrone downward until it reaches the vicinity of the local minimum and then moves rapidly to the right again *above* its starting point on the negative vertical axis. In other words, such an orbit is spiraling inward. Solutions that start near and below the rest point (which is a sink for the given parameter values), traverse a nearly closed loop as described and are eventually confined to a small neighborhood of the rest point. Thus, the rest point is their forward limit as the temporal parameter goes to infinity. The behavior of the action potential (V-component) along such a typical trajectory is also shown in the figure. Note the fast and slow motions are reflected in the figure. Clearly, this system has two timescales: one for solutions while they are near the vertical isocline and one when they are away from this curve.

Fig. 6.4 depicts the phase portrait with stimulus $I = 1$. As the stimulus is increased through a critical value ($I \approx 0.3313$) at which the rest point changes stability from a sink to a source (as a pair of complex conjugate eigenvalues passing through the imaginary axis in the complex plane), a stable limit cycle is produced via a Hopf bifurcation. The action potential V corresponding to the limit cycle has multiple spikes. A second critical value is reached as I is increased further and the limit cycle disappears in a Hopf bifurcation.

Fig. 6.5 depicts the phase portrait with stimulus $I = 2$ (which is larger than the second critical value [see Exercise 6.4]), and a typical action potential profile. The pulses (spikes) in voltage observed in the figures may be interpreted as voltage changes induced by a stimulus at a site along a neuron.

The basic scenario just described for changes in the phase portrait of the ODE system (6.9) with respect to the parameter I can be proved without assigning values to a, b, and c. Using the hypothesis $b > a$, simple arguments show that there is exactly one rest point and the first component of its coordinates increases with I (see Exercise 6.5). Examination of the signs of the trace and determinate of the system matrix of the linearization as functions of I shows that the determinant is always positive and thus the sign of the trace determines the signs of the real parts of its eigenvalues. For sufficiently negative I, the rest point is a sink. As I increases past a critical value (exactly at the first value of I that makes the trace vanish), a pair of complex conjugate eigenvalues cross the imaginary axis and a stable limit cycle is born in a (supercritical) Hopf bifurcation. These eigenvalues move back to the negative half of the complex plane as I increases past a second, larger critical value where the trace vanishes. At this second critical value the stable limit cycle dies in a (subcritical) Hopf bifurcation.

Although the existence of a limit cycle in the spatially independent model is not related directly to the biological application of the electrical activity of neurons, the bifurcation analysis predicts that the nerve fires—which might be taken to mean that the system has oscillatory behavior—over a closed and bounded range of stimuli. Too small or too large stimuli do not cause the nerve to fire; it relaxes to a rest state. This result gives a correct indication of the predictions implied by the FitzHugh–Nagumo model, which includes diffusion.

The FitzHugh–Nagumo equations (with diffusion) model the spatial coupling among ion channels along a neuron. A stimulus with magnitude larger than a threshold value (whose magnitude is not too large) produces an action potential that moves as time increases, as depicted in Fig. 6.6. The voltage pulse caused by the stimulus splits into two parts that move left and right away from the spatial position of the stimulus. This wave of voltage (the action potential) represents a bit of information carried along a neuron. The resting neuron transmits 0; the fired neuron transmits 1. At a fundamental level the confirmation of this observation in experiments

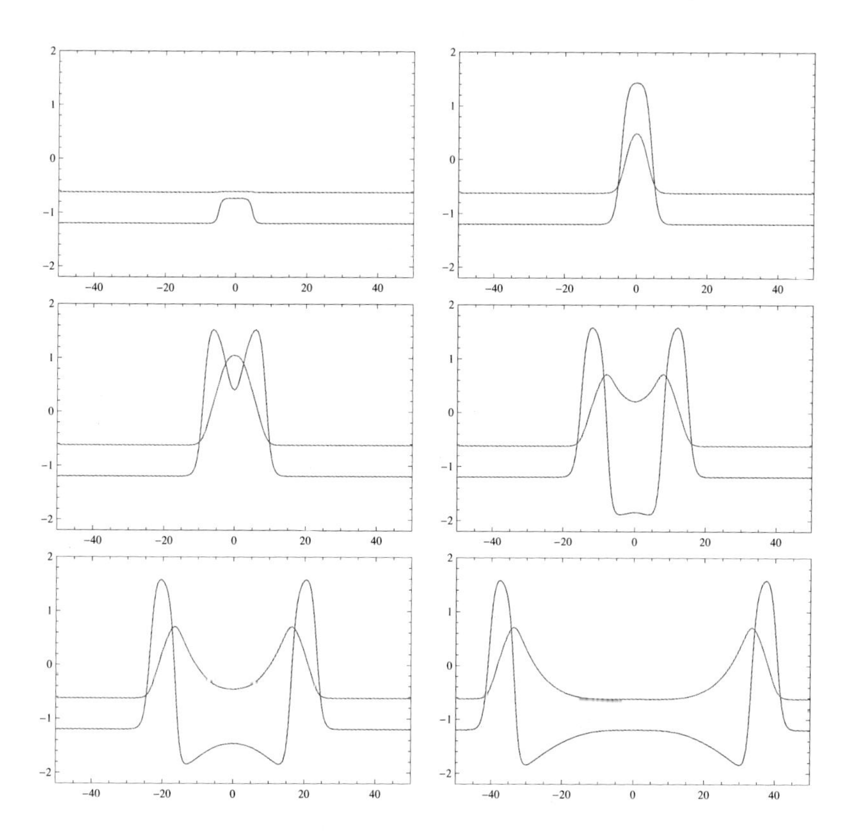

Fig. 6.6 The upper left panel depicts an approximation of the action potential (lower graph) and the recovery variable (upper graph) at $t = 0.84$ *for the FitzHugh–Nagumo model* (6.7) *with the parameter values* (6.8). *The stimulus* I *is given by* $I(x, t) = 0.69$ *for* $|x| < 5$ *and* $0 \le t \le 1$ *and by* $I(x, t) = 0$ *otherwise, the initial* V *and* W *are set equal to their rest state values (approximately* $(-1.199, -0.624)$*), and Dirichlet boundary conditions are set at* ± 50 *equal to these rest state values. The panels left to right and top to bottom depict the action potential and recovery variable at* $t = 0.84, 10.5, 16.8, 24.5, 35,$ *and* 56.

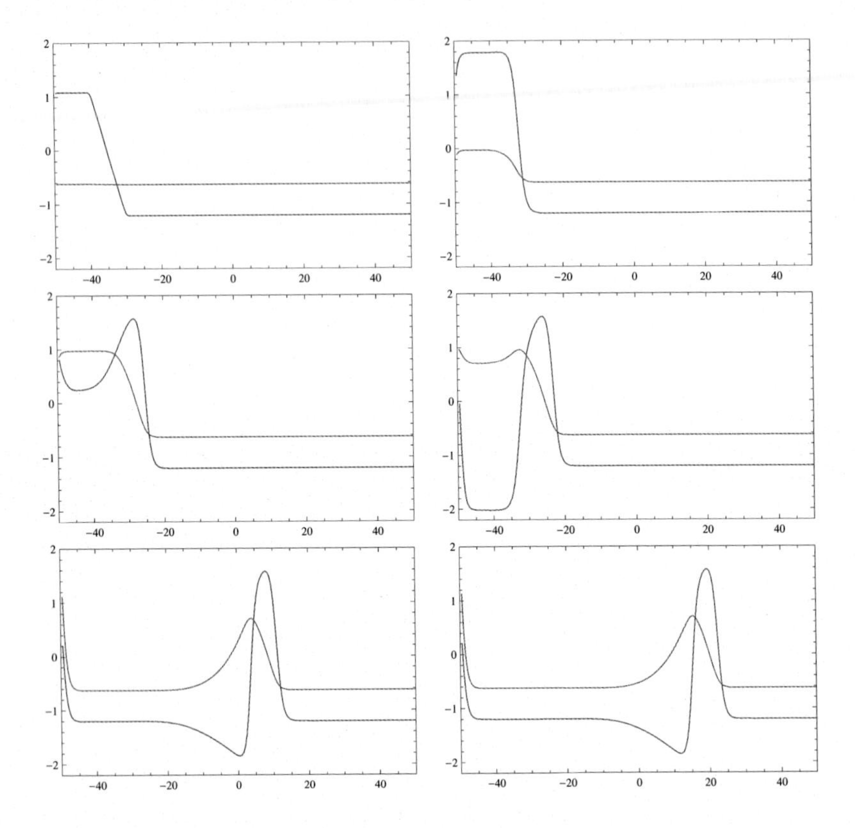

Fig. 6.7 The upper left panel depicts an approximation of the action potential (lower graph) and the recovery variable (upper graph) at t $= 0.07$ *for the FitzHugh–Nagumo model* (6.7) *with parameter values* (6.8). *The stimulus is* $I = 0$; *the initial* V *is given by* $V = 1$ *for* $x < -40$, $(V_0 - 1.0)(x + 30)/10 + V_0$ *for* $-40 \leq x \leq -30$ *with* $V_0 \approx -1.199$ *(the corresponding ODE steady state first component), and* $V = V_0$ *for* $x > -30$; *the initial* W *is set to its ODE* (6.9) *rest state value (approximately* -0.624*); and Dirichlet boundary conditions are set at* ± 50, *which are equal to the ODE rest state values. The panels left to right and top to bottom depict the action potential and recovery variable at t* $= 0.84, 2.8, 11.2, 14, 56,$ *and* 70.

with living neurons explains how electrical activity carries information in organisms with nerves.

As in the example depicted in Fig. 6.7, a sufficiently strong stimulus that spreads to one end of the nerve axon produces an action potential front that moves away from the stimulated region. The front in the figure seems to converge to a fixed profile; that is, it seems to converge to a traveling wave solution of the partial differential equation (PDE).

Recall that a solution (V, W) of the PDE model (6.7) is called a traveling wave if there are two functions f and g and a number $\gamma \neq 0$, called the wave speed, such that $V(x, t) = f(x - \gamma t)$ and $W(x, t) = g(x - \gamma t)$.

The pair (f, g) is called the wave profile. We will assume that $\gamma > 0$ so that the solution is in the form of a wave traveling to the right along the x-axis with speed γ. The expected physical wave is in the form of a pulse, which is defined as a traveling wave solution such that V converges to the same fixed value as $|x - \gamma t|$ grows to infinity, and W behaves the same way except that the values at infinity may be different from the values of V. For our biological application, these values at infinity are the rest state values (V_0, W_0) given by the coordinates of the rest point of ODE (6.9); they correspond to the rest state of the nerve, which is the corresponding spatially constant solution of the PDE. The state of the nerve is resting before stimulus, a traveling pulse after stimulation, and relaxation to rest after firing.

The natural setting for the existence of a traveling wave solution (which must be a solution of the PDE defined on the whole real line) is the PDE model (6.7) with zero stimulus and the boundary conditions $(V(x,t), W(x,t)) \to (V_0, W_0)$ as $|t| \to \infty$.

To seek a traveling wave solution, substitute $V(x,t) = f(x - \gamma t)$ and $W(x,t) = g(x - \gamma t)$ into the PDE, and seek solutions for f, g, and γ. This substitution results in a family of ODEs parameterized by γ:

$$\begin{aligned} f'' + \gamma f' + f - \frac{1}{3} f^3 - g &= 0, \\ \gamma g' + af - bg + c &= 0 \end{aligned} \tag{6.10}$$

with independent variable $s = x - ct$. To satisfy the boundary conditions, a solution is desired such that

$$\lim_{|s| \to \infty} (f(s), g(s)) = (V_0, W_0), \tag{6.11}$$

where (V_0, W_0) is the rest point of the ODE model (6.9) with the same parameter values. In this case, the wave profile is called a pulse; each component f and g has the same values at $s = \pm\infty$. The new ODE (6.10) is second-order; it is equivalent to the first-order system

$$\begin{aligned} \dot{u} &= v, \\ \dot{v} &= -\gamma v - u + \frac{1}{3} u^3 + w, \\ \dot{w} &= \frac{1}{\gamma}(-au + bw - c). \end{aligned} \tag{6.12}$$

The point (V_0, W_0) corresponds to the steady state $(u, v, w) = (V_0, 0, W_0)$ for this system of ODEs. Thus, we seek a homoclinic orbit (a trajectory that is asymptotic in forward and backward time to the same steady state) of this three-dimensional system of ODEs.

Proof of the existence of traveling wave solutions for the three-dimensional system (6.12) is much more complicated than for Fisher's equation: the new system of ODEs is three-dimensional instead of two-dimensional and it is no longer true that there is a continuum of parameters γ for which a traveling wave with the appropriate boundary conditions exists. The reason is that homoclinic orbits are unstable to perturbations; in contrast, heteroclinic orbits from a saddle to a sink are stable. Another way to see the difficulty is simply by inspection of the differential equations. There are three equations, but four unknowns: u, v, w, and γ. Thus, the problem is underdetermined. When a new equation is added to make four equations in four unknowns, solutions are expected to be isolated; that is, there is likely to be a ball in four-dimensional space associated with each solution such that no other solution is in the same ball. A viable solution method is discussed in the next section.

Although traveling wave solutions for the FitzHugh–Nagumo model illustrate the motion of action potentials on neurons, this model does not explain the underlying mechanism of ion flow that creates the electrical currents and voltage differences responsible for these waves. Modeling and mathematics have important roles to play in understanding ion channel flows at a fundamental level where electrodynamics, fluid flow, particle motions, and the behavior of living tissue must be taken into account. At present, theories of ion channel flow are not well developed.

Exercise 6.1. Determine a scaling of the state variables and time that converts system (6.4) to the dimensionless form (6.5).

Exercise 6.2. Show that an appropriate rescaling after the addition of a diffusion term to the first equation in system (6.4) yields the dimensionless equations (6.6).

Exercise 6.3. Reproduce Figs. (6.3)–(6.5).

Exercise 6.4. (a) Determine the values of the stimulus I, for system (6.9) with parameter values (6.8), for which Hopf bifurcations occur. (b) Using numerical experiments, find the value of I for which the area bounded by the limit cycle is maximal.

Exercise 6.5. (a) Write a detailed proof of the Hopf bifurcation scenario described on page 206. Assume that a Hopf bifurcation occurs if a pair of complex conjugate eigenvalues cross the imaginary axis. (b) Study a formulation (and proof) of the Hopf bifurcation theorem (see, for example, [20]) and apply this theorem to prove that the claimed Hopf bifurcations occur.

Exercise 6.6. Discuss via numerical experiments the dependence of the wave speed on initial data for the FitzHugh–Nagumo model for waves determined as in Fig. 6.7 and compare with Fisher's model.

Exercise 6.7. Design and perform a numerical experiment to determine homoclinic solutions of the system (6.12) that does not use the methodology of Section 6.2. Hint: One possibility is to locate a saddle point and try to adjust γ so that, for some fixed value of this parameter, a solution on the unstable manifold of the saddle rest point returns to this rest point on its stable manifold.

6.2 NUMERICAL TRAVELING WAVE PROFILES

A beautiful approach to the existence problem for traveling waves, which will be briefly discussed here, has resulted in a far-reaching theory for the existence and stability of several types of nonlinear waves in many different models (see [13] for a useful introduction).

The first idea is to seek the profile of a traveling wave solution as a stable steady state of a PDE. We may hope that solutions of such a PDE converge as time goes to infinity to the desired steady state. In principle, we may then choose initial data as we please and let the corresponding solution of the PDE lead us to the traveling wave profile. This is a powerful idea in many different contexts.

For the dimensionless FitzHugh–Nagumo model (6.7), the idea may be implemented by introducing the family (depending on the parameter γ) of new variables

$$\xi := x - \gamma t, \qquad \tau = t.$$

With the new functions

$$\hat{V}(\xi,\tau) := V(\xi + \gamma\tau, \tau), \qquad \hat{W}(\xi,\tau) := W(\xi + \gamma\tau, \tau),$$

which are simply V and W in the new variables, PDE (6.7) (without the stimulus because we are interested in long-term behavior only) is recast in the form

$$\frac{\partial \hat{V}}{\partial \tau} = \frac{\partial^2 \hat{V}}{\partial \xi^2} + \hat{V} - \frac{1}{3}\hat{V}^3 - \hat{W} + \gamma \frac{\partial \hat{V}}{\partial \xi},$$

$$\frac{\partial \hat{W}}{\partial \tau} = a\hat{V} - b\hat{W} + c + \gamma \frac{\partial \hat{W}}{\partial \xi}. \tag{6.13}$$

Note that, for a traveling wave solution of the FitzHugh–Nagumo equation, the state variables have the form

$$V(x,t) = f(x - \gamma t), \qquad W(x,t) = g(x - \gamma t),$$

where f and g are the wave profiles for V and W, respectively. In view of the new variable $\hat{V}$, for example, a traveling wave solution is given by

$$f(x - \gamma t) = \hat{V}(x - \gamma t, t),$$

or equivalently,

$$f(\xi) = \hat{V}(\xi, \tau)$$

with the function $\xi \mapsto \hat{V}(\xi, \tau)$ independent of τ. In other words, a steady state solution of system (6.13), which has appropriate limits as ξ approaches $\pm\infty$ as in Eq. (6.11), is the profile of a traveling wave solution of the dimensionless FitzHugh–Nagumo model (6.7) with wave speed γ. Indeed, the steady state equation is exactly the system (6.10). Thus, after the change of variables, steady states of the new system (6.13) are traveling wave solutions of the original system (6.7).

We have not yet accomplished a full implementation of the first idea because system (6.13) contains the free parameter γ. A remaining problem is to determine values of this parameter so that there is a steady state corresponding to a profile for a traveling wave solution of the original FitzHugh–Nagumo model.

The next idea is natural once we view the evolution of system (6.13) in an infinite-dimensional space of functions and consider what we are trying to achieve. The (wave speed) parameter γ has to be allowed to evolve with time so that its value might approach the fixed value that would correspond to an appropriate steady state. The idea is to assign $\gamma(\tau)$, for each τ, so that the square of the length of the vector field (the right-hand side of PDE (6.13)) is minimized. By minimizing this length, the velocity vector of the solution would be as close to zero as possible at each τ as this temporal variable increases toward infinity. Thus, the imposition of this constraint should push solutions toward a steady state; that is, to a point in the function space where the vector field is zero.

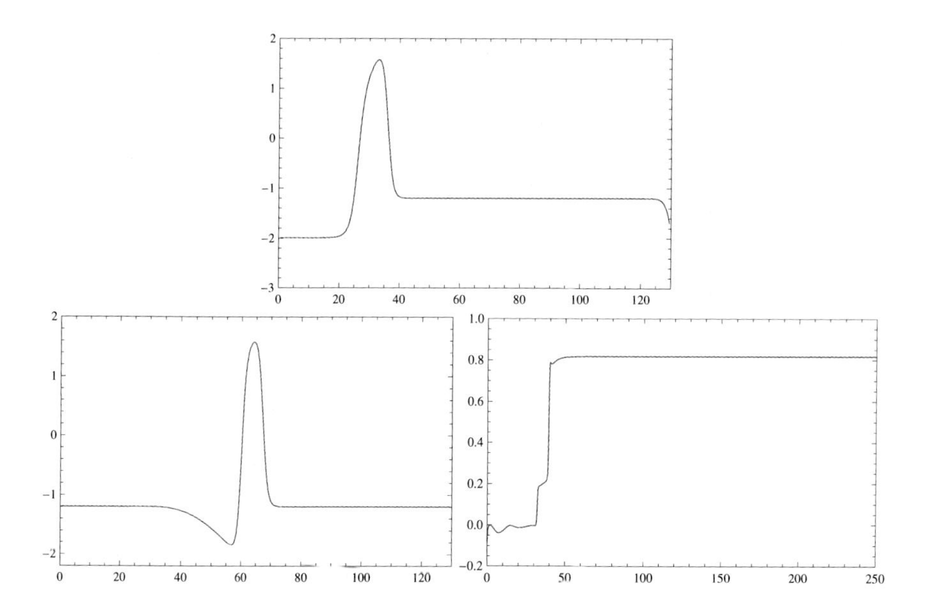

Fig. 6.8 The lower left panel depicts a numerical approximation of the graph of a steady state for $\hat{V}$ for the differential algebraic system (6.13) *and* (6.15) *for the finite interval* $(0, 130)$, *the parameters given in Eqs.* (6.8), *and zero Neumann boundary conditions. It is an approximation of the desired traveling wave profile. The right panel depicts a numerical approximation of γ versus τ for the same system and initial data. The top panel depicts the initial $\hat{V}$ used for the numerical experiment. The initial $\hat{W}$ is set to the constant steady state value -0.62426.*

A vector field is a special type of function. What is meant by the length of a function? One possibility (among many) is to define the length (more often called the norm) of a function ϕ to be

$$\|\phi\| := \Big(\int_{-\infty}^{\infty} (\phi(\xi))^2\, d\xi\Big)^{1/2}.$$

The function ϕ is called square integrable if the integral exists so that this norm is finite. Mathematicians call this length the L^2 norm; and, they call a function, which has finite L^2 norm, a square integrable function or an L^2 function.

Although there are many possible norms, a major advantage of the L^2 norm is that it is defined by an inner product on the vector space of all square integrable (real valued) functions. Indeed, the L^2 inner product for two such functions ϕ and ψ is defined by

$$\langle \phi, \psi \rangle := \int_{-\infty}^{\infty} \phi(\xi)\psi(\xi)\, d\xi$$

and the norm of ϕ is

$$\|\phi\| = \langle \phi, \phi \rangle^{1/2}.$$

We may extend the L^2 norm to vectors of functions in several ways. Here, we will define the norm for the two-dimensional vector (ϕ, ψ) of L^2 functions by

$$(\|\phi\|^2 + \|\psi\|^2)^{1/2}.$$

Returning to our traveling wave problem, the idea is to add the condition that the L^2 norm of the vector field given by the right-hand side of PDE (6.13) (viewed as an infinite-dimensional ODE) is minimized at each τ with respect to γ. Equivalently, it suffices to minimize the square of the norm. By taking advantage of the L^2 inner product, the square of the norm is given by the scalar expression

$$\begin{aligned}\langle \frac{\partial^2 \hat{V}}{\partial \xi^2} + \hat{V} - \frac{1}{3}\hat{V}^3 - \hat{W} + \gamma\frac{\partial \hat{V}}{\partial \xi}, \frac{\partial^2 \hat{V}}{\partial \xi^2} + \hat{V} - \frac{1}{3}\hat{V}^3 - \hat{W} + \gamma\frac{\partial \hat{V}}{\partial \xi}\rangle \\ + \langle a\hat{V} - b\hat{W} + c + \gamma\frac{\partial \hat{W}}{\partial \xi}, a\hat{V} - b\hat{W} + c + \gamma\frac{\partial \hat{W}}{\partial \xi}\rangle.\end{aligned}$$

The appropriate condition for an extreme point is obtained in the usual manner: differentiate with respect to γ and set the derivative equal to zero. The resulting equation is

$$0 = \langle \frac{\partial \hat{V}}{\partial \xi}, \frac{\partial^2 \hat{V}}{\partial \xi^2} + \hat{V} - \frac{1}{3}\hat{V}^3 - \hat{W} + \gamma \frac{\partial \hat{V}}{\partial \xi} \rangle + \langle \frac{\partial \hat{W}}{\partial \xi}, a\hat{V} - b\hat{W} + c + \gamma \frac{\partial \hat{W}}{\partial \xi} \rangle. \tag{6.14}$$

We must impose boundary conditions at $\xi = \pm\infty$ to obtain unique solutions of PDE (6.13) for given initial $\hat{V}$ and $\hat{W}$ and fixed γ. Because the method relies on state variables being L^2 functions as τ increases, improper integrals must converge on the whole real line. Inspection of the critical point Eq. (6.14) reveals that space derivatives of the state variables must also be square integrable. At a minimum, this requirement suggests that the state variables $\hat{V}$ and $\hat{W}$ together with their first partials with respect to ξ must vanish in both directions with respect to ξ at infinity. The second partial derivative of $\hat{V}$ with respect to ξ also appears. Unfortunately, the requirement that the state variables vanish at infinity is incompatible with the properties of the desired steady state, a pulse traveling wave two-component profile that has limiting value (V_0, W_0) at $\xi = \pm\infty$.

By ignoring the desired zero Dirichlet boundary conditions and simply imposing zero Neumann boundary conditions at infinity, the critical point Eq. (6.14) can be simplified. In fact, one term obtained by expanding the inner products vanishes:

$$\begin{aligned}
\langle \frac{\partial \hat{V}}{\partial \xi}, \frac{\partial^2 \hat{V}}{\partial \xi^2} \rangle &= \int_{-\infty}^{\infty} \frac{\partial \hat{V}}{\partial \xi} \frac{\partial^2 \hat{V}}{\partial \xi^2}\, dx \\
&= \lim_{\xi \to \infty} \frac{1}{2}\Big(\frac{\partial \hat{V}}{\partial \xi}\Big)^2 - \lim_{\xi \to -\infty} \frac{1}{2}\Big(\frac{\partial \hat{V}}{\partial \xi}\Big)^2 \\
&= 0.
\end{aligned}$$

For theoretical work, careful definitions of new function spaces is required to study the existence, uniqueness, and asymptotic behavior of solutions $\tau \mapsto (\hat{V}(\tau), \hat{W}(\tau), \gamma(\tau))$ of the system of equations consisting of the PDEs (6.13), the constraint (6.14), and the zero Neumann boundary conditions. In particular, the problem of finite limits of $\hat{V}(\tau)$ and $\hat{W}(\tau)$ at infinity must be resolved.

For numerical experiments, the dynamical system must be restricted to a finite interval, for example $(0, L)$ for some $L > 0$. This restriction removes the requirement that solutions vanish at $\pm\infty$. With zero Neumann boundary conditions imposed at the ends of this interval, the L^2 inner products $\langle \frac{\partial \hat{V}}{\partial \xi}, \frac{\partial^2 \hat{V}}{\partial \xi^2} \rangle$ defined now so that integration is over the interval $(0, L)$, vanish just as they do on the infinite interval. Thus, the minimization constraint (with some rearrangement) reduces to

$$
\begin{aligned}
0 = \gamma \int_0^L \Big(\frac{\partial \hat{V}}{\partial \xi}\Big)^2 + \Big(\frac{\partial \hat{W}}{\partial \xi}\Big)^2 \, d\xi + \int_0^L (\hat{V} - \frac{1}{3}\hat{V}^3 - \hat{W}) \frac{\partial \hat{V}}{\partial \xi} \, d\xi \\
+ \int_0^L (a\hat{V} - b\hat{W} + c) \frac{\partial \hat{W}}{\partial \xi} \, d\xi. \qquad (6.15)
\end{aligned}
$$

Fig. 6.8 shows some of the results of a numerical experiment to approximate the solution of the differential algebraic equation (DAE) system (6.13) and (6.15) for the finite interval $(0, 130)$, the parameters in display (6.8), and zero Neumann boundary conditions. The initial configuration depicted in the top panel evolves to a steady state in variables $\hat{V}$, $\hat{W}$, and γ. An approximation of the graph of $\hat{V}$ at this steady state is shown. It is an approximation to a traveling wave profile for the dimensionless FitzHugh–Nagumo system, where the wave speed for the corresponding traveling wave solution is the steady state value of $\gamma \approx 0.8168481086$ (compare [13, p. 101]). The initial configuration for this experiment was obtained by evolving forward (to $t = 13$) the dimensionless FitzHugh–Nagumo solution depicted in Fig. 6.7. The result of the experiment suggests that this initial configuration is indeed close to a traveling wave solution with the computed wave speed. The profile at $t = 13$ is not the same as the traveling wave profile, but the profile of the long-time evolution of this initial configuration seems to approach the traveling wave profile.

An interesting aspect of the effectiveness of the numerical method is that the desired profile for the pulse traveling wave, which has the correct (Dirichlet) end values, is obtained using Neumann boundary conditions; the desired limits at $\xi = \pm\infty$ are not used. Note that the zero Neumann conditions specify that the profile approaches a point where its derivative vanishes at each end point. The steady state profile is the solution of an autonomous ODE. Derivatives of solutions have zero limits when they approach rest points. Thus, a solution with zero Neumann conditions may be interpreted as the solution of an ODE that approaches a rest point in forward and backward time. In case there is only one rest point and the solution is not constant, this behavior specifies a homoclinic orbit. This is precisely the desired solution.

Unfortunately, the DAE system (6.13) and (6.15) has multiple (stable) steady states. For example, in addition to the desired steady state approximated as in Fig. 6.8, the system has the constant steady state given by the solutions of the algebraic equations

$$\hat{V} - \frac{1}{3}\hat{V}^3 - \hat{W} = 0, \qquad a\hat{V} - b\hat{W} + c = 0,$$

and γ arbitrary. At the parameter values specified in display (6.8), the constant state is $\hat{V} \approx -1.19941$ and $\hat{W} \approx -0.62426$. For the modification of the numerical experiment reported in Fig. 6.8, where the initial configuration is obtained in the same manner for $t < 10$, the system evolves to this constant steady state. The upshot is that to obtain the desired traveling wave profile, the initial profile cannot be too far away, or in other words, the initial data must be in the basin of attraction of the desired steady state and this basin of attraction is not the entire state space.

The approach to traveling wave profiles and wave speeds discussed here is a powerful tool for theoretical and numerical work. But, at present, no method is known for choosing initial profiles that ensures the DAE system converges to a steady state corresponding to a desired unknown traveling wave profile. It is wise to realize that a method can be useful without being universally applicable.

Some basic numerical methods for differential algebraic systems are suggested in the exercises.

Exercise 6.8. Consider differential algebraic systems of the form

$$\dot{x} = f(x, y), \qquad 0 = g(x, y)$$

with initial conditions $x(0) = \xi$ and $y(0) = \eta$ restricted so that $g(\xi, \eta) = 0$. Euler's method is easily modified to be a numerical method for such a system. Indeed, we may approximate the evolution of the system with the discrete time process

$$x^{n+1} = x^n + \Delta t f(x^n, y^n), \qquad 0 = g(x^n, y^{n+1})$$

where $x^0 = \xi$ and $y^0 = \eta$. (a) Analyze the proposed numerical method and write code to implement the method. (b) Solve the DAE

$$\dot{x} = y, \qquad 0 = x + 2y$$

with the initial condition $x(0) = -2$ and $y(0) = 1$. (c) Test your code against the exact solution of the system in part (b). Is your code accurate to first order in Δt? Hint: Accuracy to first order means in effect that halving the step size halves the error. (d) Compare the accuracy of the alternative numerical algorithm (the implicit trapezoidal method)

$$x^{n+1} = x^n + \frac{\Delta t}{2}(f(x^n, y^n) + f(x^{n+1}, y^{n+1})), \qquad 0 = g(x^{n+1}, y^{n+1})$$

to the accuracy of Euler's method. (e) There are many other possible numerical methods. An obvious alternative is to replace $0 = g(x^n, y^{n+1})$ by $0 = g(x^{n+1}, y^{n+1})$ in part (a). Implement this method and discuss its accuracy and efficiency.

Exercise 6.9. (a) Repeat the numerical experiment reported in Fig. 6.8. Hint: The numerical method of Exercise 6.8 is adequate provided the discretization in space is second order. The integrals that appear in the algebraic constraint may be approximated with sufficient accuracy by the trapezoidal rule. (b) Modify the experiment, by choosing the initial voltage profile so that the system evolves to a constant steady state. Determine the asymptotic value of γ as $\tau \to \infty$. (c) Repeat the experiment as in part (a) with the usual second-order discretization of the second-order derivatives with respect to ξ, but only first-order accurate discretizations of the first-order spatial derivatives. Does the numerical simulation reach a steady state?

Exercise 6.10. Consider the differential equation $e^{\dot{x}} + \dot{x} + x = 0$. (a) Assume the initial condition $x(0) = 1$. Find (numerically) $x(1)$ correct to four decimal places. Hint: Recast the ODE as a DAE. (b) Using numerical experiments, describe the long-term behavior of the solution starting at $x(0) = 0$. (c) Prove (using pencil and paper) that the solution of the ODE behaves as you predict in part (b). (d) Does your answer to part (c) depend on the zero initial condition?

Exercise 6.11. Consider the DAE

$$\dot{x} = -x + y - xz, \qquad \dot{y} = x + y - yz, \qquad z^3 - (x^2 + y^2)z - 1 = 0.$$

(a) Given the initial data $x(0) = 0.5$, $y(0) = 0$, and $z(0) = 1$, approximate z at time 6π. (b) What can you say about the general behavior of solutions of this system?

Exercise 6.12. [Nonlinear Eigenvalue Problem via DAEs] The eigenvalue problem for an $n \times n$ matrix A is to solve the nonlinear equation $(A - \gamma I)v = 0$ for γ and $v \neq 0$. Eigenvector-eigenvalue pairs are steady states of the ODE $\dot{v} = (A - \gamma I)v$. We must solve for both γ and v. An appropriate constraint is obtained by minimizing the square of the length of $\dot{v}$ over γ. This leads to the constraint $\langle v, (A - \gamma I)v\rangle = 0$ and suggests integrating the differential algebraic system

$$\dot{v} = (A - \gamma I)v, \qquad 0 = \langle v, (A - \gamma I)v\rangle$$

with state variables v and γ. If the solution approaches a steady state whose v component does not vanish, then this steady state corresponds to an eigenvector-eigenvalue pair for A. Analyze this approach with mathematical analysis and numerical experiments. Notes: Start with 2×2 matrices. Eigenvalues of real matrices can be complex numbers. Perhaps, for this case, an augmentation of the dynamical system to include complex state variables would be helpful.

Exercise 6.13. [Stability of Numerical ODE Solvers] One of the methods for testing the stability of an ODE numerical method is to test it on the differential equation $\dot{y} = -\lambda y$ to determine how errors might grow. (a) Consider Euler's method and show that an error ϵ in computing one time step grows (or decays) according to the formula $z_0 = \epsilon$ and $z_{n+1} = (1 - \Delta t\lambda)z_n$, which is exactly the Euler method applied to the ODE. (b) Show that the method is stable on the test equation in case $\lambda > 0$ and $\Delta t < 2/\lambda$. (c) In textbooks on numerical analysis, λ is taken to be a complex number and the stability criterion is $|1 + \Delta t\lambda| < 1$, where the left-hand side is the modulus of the complex number $1 + \Delta t\lambda$. Why are complex numbers considered? Hint: Consider the origin of the numerical test. In the simplest case, the numerical method is given by $y_{n+1} = F(y_n)$ where F is some function. For Euler applied to the ODE $\dot{y} = f(y)$, the function F is defined by $F(y) := y + \Delta t f(y)$. Imagine that f is a vector function of a vector variable—which is the case for a system of first-order ODEs—and note that the stability criterion is related to the eigenvalues of the Jacobian matrix DF evaluated at an appropriate point. The eigenvalues of a matrix can be complex. The test equation will arise if DF were diagonal. A good answer to the question would consider the hints in writing a few paragraphs describing in detail why the test equation is used. (d) Consider the stability of the Euler method and the trapezoidal method using the test equation for a complex λ.

Exercise 6.14. [Numerical Pulse Type Traveling Waves] (a) Reconsider the numerical experiment reported in Fig. 6.8 and Exercise 6.9. Write and use a numerical code to approximate solutions of the corresponding DAE with nonzero Dirichlet boundary conditions given by the desired constant steady state values V_0 and W_0 at the end points of the interval $(0, L)$ instead of zero Neumann boundary conditions. Ignore the use of the zero Neumann boundary conditions to simplify the critical point equation. (b) Repeat part (a) but recompute the critical point equation using only Dirichlet boundary conditions. (c) The goal of the project is to determine the effectiveness of the DAE with Dirichlet boundary conditions in approximating the profile of the desired pulse type traveling wave profile. Write a report on the details of your code and your numerical experiments.

CHAPTER 7

Splitting Methods

This chapter introduces an interesting and not completely understood algorithm for approximating solutions of evolution equations based on the idea of separating differential equations into sums of simpler equations.

7.1 A PRODUCT FORMULA

Let A be a matrix and define the matrix exponential of A by

$$e^{A} = I + \sum_{j=1}^{\infty} \frac{A^j}{j!}, \tag{7.1}$$

where I is the identity matrix. Also, let $\mathcal{L}(E)$ denote the finite-dimensional vector space of all $N \times N$ matrices. The infinite sum in Eq. (7.1) converges (absolutely with respect to every norm on $\mathcal{L}(E)$) and the matrix exponential satisfies the usual rules of exponents except that

$$e^{A+B} = e^{A}e^{B}$$

if and only if $AB = BA$ (that is, the matrices A and B commute). Also,

$$\frac{d}{dt}e^{tA} = Ae^{tA} = e^{tA}A,$$

or in other words, the function $t \mapsto e^{tA}$ is the matrix solution of the initial value problem

$$\dot{x} = Ax, \qquad x(0) = I.$$

These results are proved in textbooks on ordinary differential equations (ODEs) (see, for example, [20]).

The next theorem is a special case of the Lie–Trotter product formula for the exponential of a sum of two $N \times N$ matrices when the matrices do not necessarily commute (see [20] for a detailed proof).

Theorem 7.1. If $\gamma : \mathbb{R} \to \mathcal{L}(E)$ is a continuously differentiable function with $\gamma(0) = I$ and $\dot{\gamma}(0) = A$, then the sequence $\{\gamma^n(t/n)\}_{n=1}^{\infty}$ converges

An Invitation to Applied Mathematics: Differential Equations, Modeling, and Computation.
http://dx.doi.org/10.1016/B978-0-12-804153-6.50007-5, Copyright © 2017 Elsevier Inc. All rights reserved.

to $\exp(tA)$. In particular, if A and B are $k \times k$ matrices and $\gamma(t) := e^{tA}e^{tB}$, then

$$e^{t(A+B)} = \lim_{n\to\infty} \left(e^{\frac{t}{n}A}e^{\frac{t}{n}B}\right)^n.$$

The solution of the initial value problem

$$\dot{x} = Ax + Bx, \qquad x(0) = x_0 \tag{7.2}$$

is

$$x(t) = e^{t(A+B)}x_0.$$

But, this compact notation hides the infinite sum that is used to define the matrix exponential. To obtain numerical values, this sum must be approximated. A natural and viable way to proceed is to approximate the exponential with a partial sum of its Taylor series, but this is certainly not the only way (see [74]).

Suppose good approximation schemes are known for computing $\exp tA$ and $\exp tB$. How can they be used to compute $\exp(t(A+B))$? There are many possible answers. Here, methods are discussed for computing the latter quantity based on the product formula. The results lead to numerical methods that will be used later in this book for solving some partial differential equations (PDEs) (see the next section and Section 22.8).

Take $n = 1$ in the product formula to obtain the approximation

$$e^{t(A+B)} \approx e^{tA}e^{tB}.$$

How much error does this introduce? To answer this question, simply expand in Taylor series (with respect to t at $t = 0$) and subtract:

$$\begin{aligned}\|e^{t(A+B)} - e^{tA}e^{tB}\| = \|&I + t(A+B) + \frac{t}{2!}(A+B)^2 \\ &- (I + t(A+B) + \frac{t^2}{2!}(A^2 + 2AB + B^2))\| + O(t^3).\end{aligned}$$

In the generic case where A and B do not commute, the second-order terms do not cancel and the estimate reduces to

$$\|e^{t(A+B)} - e^{tA}e^{tB}\| = O(t^2).$$

Viewing this as a numerical algorithm to solve the initial value problem (IVP) (7.2), we have a first-order method given by

$$x^0 = x_0, \qquad x^{k+1} = e^{\Delta t A} e^{\Delta t B} x^k.$$

To obtain a second-order method, the idea is to cancel the second-order terms in the Taylor series that appears in the error estimates. Note that for

$$\gamma(t) = e^{t/2A} e^{tB} e^{t/2A}$$

we have that $\gamma(0) = I$ and $\dot{\gamma}(0) = A + B$. Thus, this choice satisfies the hypothesis of Theorem 7.1; moreover,

$$\ddot{\gamma}(0) = A^2 + AB + BA + B^2 = (A + B)^2.$$

Thus, with this choice for γ, we have the estimate

$$\|e^{t(A+B)} - e^{t/2A} e^{tB} e^{t/2A}\| = O(t^3)$$

and the second-order numerical method

$$x^0 = x_0, \qquad x^{k+1} = e^{\Delta t/2A} e^{\Delta t B} e^{\Delta t/2A} x^k. \tag{7.3}$$

A fourth-order method can be obtained uisng Richardson extrapolation. To accomplish this, note that one step of our second-order procedure is given by

$$U(h) = e^{h/2A} e^{hB} e^{h/2A} x_0; \tag{7.4}$$

it is the approximation of the solution of our ODE at $t = h$. With half the step size, the solution is

$$U(\frac{h}{2}) = e^{h/4A} e^{h/2B} e^{h/4A} e^{h/4A} e^{h/2B} e^{h/4A} x_0.$$

Be careful here; the notation can be confusing. To use Richardson extrapolation, we must compare approximations of the same value (for instance, the solution of our initial value problem at time $t = h$). Here, $U(h)$ denotes the value at $t = h$ of our procedure, not a function with name U defined by Eq. (7.4). An application of Richardson extrapolation yields the fourth-order method

$$U^1(h) = \frac{1}{3}(4U(\frac{h}{2}) - U(h)). \tag{7.5}$$

Exercise 7.1. (a) Implement the numerical method (7.3) for the matrices

$$A := \begin{pmatrix} 0 & -1 \\ 1 & 0 \end{pmatrix} \qquad B := \begin{pmatrix} -1 & 1 \\ 0 & -1 \end{pmatrix}$$

(b) Show by numerical experiments for this example that the method is second order.
(c) Show by numerical experiments that the Richarson Eq. (7.5) leads to a fourth-order method for this example.

Exercise 7.2. (a) Consider taking several steps with the second-order method (7.3) and note that the result looks like the first-order method except for the first and last factors; for example, taking three steps yields

$$x^3 = e^{\Delta t/2A} e^{\Delta tB} e^{\Delta tA} e^{\Delta tB} e^{\Delta tA} e^{\Delta tB} e^{\Delta t/2A} x^0.$$

As we have proved, this scheme has higher-order accuracy than the approximation

$$y^3 = e^{\Delta tB} e^{\Delta tA} e^{\Delta tB} e^{\Delta tA} e^{\Delta tB} e^{\Delta A} x^0$$

produced by the first-order method of simply alternating the application of the two operators. But, it would seem that because the two methods are so similar when compared over many steps, the performance of the first-order method should be better than might be expected. Do and report on numerical experiments to further this discussion. (b) When do the two methods produce the same result? (c) What is the worst case scenario (for some choice of A, B, and the initial data) where the second method performs much worse than the first or the first performs much worse than the second? This question does not seem to have a simple answer. What can you say?

7.2 PRODUCTS FOR NONLINEAR SYSTEMS

To generalize the numerical method (7.3) to nonlinear systems we will take several leaps of faith, which is not an uncommon practice in applied mathematics. The idea is simple: replace the matrix exponential solution $t \mapsto \exp(tA)$ of $\dot{x} = Ax$ by the flow of the ODE $\dot{x} = f(x)$.

Recall that the flow of an ODE is defined to be the function $\phi(t, \xi)$ such that $t \mapsto \phi(t, \xi)$ is the solution of the initial value problem

$$\dot{x} = f(x), \qquad x(0) = \xi.$$

Solutions of differential equations may not be defined for all time, but it is easy to prove that

$$\phi(0,\xi)=\xi, \qquad \phi(t,\phi(t,\xi))=\phi(t+s,\xi)$$

whenever both sides are defined (see [20]). Of course, the flow of the linear system $\dot{x}=Ax$ is given by

$$\phi(t,\xi)=e^{tA}\xi.$$

It is convenient to view flows as one-parameter groups of transformations of the state space. To emphasize this interpretation, let us write $\phi_t(x)=\phi(t,x)$ so that

$$\phi_0=I, \qquad \phi_t\circ\phi_s=\phi_{t+s}.$$

We can approximate the solutions of the system $\dot{x}=f(x)+g(x)$, where we know the flow ϕ of $\dot{x}=f(x)$ and the flow ψ of $\dot{x}=g(x)$ with the numerical method

$$x^0=\xi, \qquad x^{k+1}=\phi_{\Delta t/2}\circ\psi_{\Delta t}\circ\phi_{\Delta t/2}(x^k), \tag{7.6}$$

where compositions of maps replaces the products in the iteration scheme (7.3).

More generally, the solution of a well posed parabolic PDE is given by a semiflow; that is, a semigroup of transformations of the underlying infinite-dimensional state space. Although flows are defined for positive and negative time (perhaps restricted to some interval of zero), semiflows are defined only for some interval of the form $[0,T)$ on the real line for $T>0$ or $T=\infty$. Fortunately, the formula in Eq. (7.6) makes sense for semiflows.

Recall the Gray–Scott model:

$$\begin{aligned}\dot{u}&=\lambda(u_{xx}+u_{yy})+F(1-u)-uv^2,\\ \dot{v}&=\mu(v_{xx}+v_{yy})-(F+\kappa)v+uv^2\end{aligned} \tag{7.7}$$

with zero Neumann boundary conditions. The obvious splitting corresponds to the reaction and diffusion systems

$$\begin{aligned}\dot{u}&=F(1-u)-uv^2,\\ \dot{v}&=-(F+\kappa)v+uv^2\end{aligned} \tag{7.8}$$

table $\Delta t = 1/4$	4.03697	3.97303	3.88698	3.28067
$\Delta t = 1/8$	3.72054	3.44124	3.21635	2.98286
$\Delta t = 1/16$	2.86821	2.45000	2.17921	2.56180
$\Delta t = 1/32$	1.82492	1.51948	1.36262	2.00486
$\Delta t = 1/64$	1.25103	1.08322	1.08139	1.48712
$\Delta t = 1/128$	1.82523	1.04201	0.987641	1.09860

Table 7.1 The order estimates in this table are for an implementation of method defined by the iteration Eq. (7.10) using the data of Problem 5.2. The floating point numbers are computed for the average of the concentrations $(u+v)/2$ using Eq. (5.67).

and

$$\begin{aligned}\dot{u} &= \lambda(u_{xx} + u_{yy}),\\ \dot{v} &= \mu(v_{xx} + v_{yy}).\end{aligned} \tag{7.9}$$

The system of linear PDEs (7.9) with (almost any choice of) boundary data has a semiflow L_t. It is the operator on initial data (concentrations u and v defined on the spatial domain $[0, L] \times [0, L]$ and satisfying the boundary conditions) that evolves the data forward from time zero to time t by solving the system of PDEs (7.9) including boundary conditions. Also, let ϕ_t denote the flow of the system of ODEs (7.8). A useful numerical method (which incorporates Richardson extrapolation as in Section 7.1) is given schematically by

$$\begin{aligned}w^0 &= w_0\\ w^{k+1} &= \frac{1}{3}\big[4L_{\frac{\Delta t}{4}} \circ \phi_{\frac{\Delta t}{2}} \circ L_{\frac{\Delta t}{2}} \circ \phi_{\frac{\Delta t}{2}} \circ L_{\frac{\Delta t}{4}}(w^k)\\ &\quad - L_{\frac{\Delta t}{2}} \circ \phi_{\Delta t} \circ L_{\frac{\Delta t}{2}}(w^k)\big],\end{aligned} \tag{7.10}$$

where $w = (u, v)$.

There is a hidden danger in the iteration scheme (7.10): Perhaps the numerical method does not preserve the boundary data. Although, by definition, the semiflow L_t preserves the boundary conditions, this is not true in general for the flow ϕ_t (see Exercise 7.3). Fortunately, for applications to Problem 5.1, the flow does preserve the (no flux) zero Neumann boundary conditions. Indeed, suppose that $u(x, y)$ and $v(x, y)$ are concentrations that satisfy the Neumann boundary conditions. These concentrations are changed by the flow ϕ to new concentrations given by the first and second components of the function

$$(x, y) \mapsto \phi_t(u(x, y), v(x, y)).$$

By the chain rule, we have that

$$D\phi_t(u(x, y), v(x, y)) \begin{pmatrix} u_x(x, y) & u_y(x, y) \\ v_x(x, y) & v_y(x, y) \end{pmatrix}.$$

The partial derivatives with respect to the first variable of the new concentrations vanish at $x = 0$ and $x = L$ because the corresponding partial derivatives of u and v *both* vanish on these lines. Similarly, the partial derivatives with respect to the second variable of the new concentrations vanish at $y = 0$ and $y = L$.

Implementation of the algorithm (7.10) leaves open many choices for numerical approximations of the semiflow L and the flow ϕ.

A viable method is to use the Crank-Nicolson scheme to approximate L and the trapezoidal method to approximate ϕ. These are, as mentioned previously, essentially the same second-order methods. The order estimates, for a code written for this algorithm and applied to Problem 5.2, are listed in Table 7.1. Note that this implementation performs at nearly the theoretical fourth-order estimate using the larger step sizes. Accuracy degrades for smaller step sizes, where there is a more likely accumulation of roundoff errors. The computed averages $((u+v)/2$ are consistently 0.453021 over the full range of step and grid sizes. The average u concentration is consistently 0.808316 and the average v concentration is consistently 0.097725 in agreement with forward Euler and Crank–Nicolson. It is important to note that this new code returns the correct values at much larger step sizes, even for unit steps.

For Problem 5.1, the computed order estimates degrade for smaller step sizes to approximately order one. The computed value for the average values of u decrease with the grid size (from 0.9597 for the 32×32 grid to 0.532 for the 256×256 grid, and the computed average value for v increases from 0.0135 to 0.1398. These results are consistent with the values 0.5, for u, and 0.1, for v, using forward Euler and Crank–Nicolson. Tables 7.2 and 5.16 show almost exact agreement. Is there now enough data to state that the required average is 0.3 correct to one decimal place?

h	32×32	64×64	128×128	256×256
$\Delta t = 1/4$	0.485018	0.463834	0.352093	0.335836
$\Delta t = 1/8$	0.485018	0.463835	0.352096	0.335841
$\Delta t = 1/16$	0.485018	0.463835	0.352097	0.335842
$\Delta t = 1/32$	0.485018	0.463835	0.352097	0.335842
$\Delta t = 1/64$	0.485018	0.463835	0.352097	0.335842
$\Delta t = 1/128$	0.485018	0.463835	0.352097	0.335842

Table 7.2 The average concentrations in this table are for the product method using the data of Problem 5.1. Gauss–Seidel iteration is used for the diffusion terms with the stopping criterion: successive iterations that differ in Euclidean norm by less than 10^{-6}. Newton's method is used in conjunction with the trapezoidal method for the reaction terms. The floating point numbers are the computed averages of the concentrations $(u + v)/2$.

Exercise 7.3. (a) Show that zero Dirichlet boundary conditions are usually not preserved by the numerical scheme (7.10). (b) Show that periodic boundary conditions are preserved.

Exercise 7.4. An alternative splitting that might be used to approximate solutions of the Gray–Scott model is obtained by incorporating the linear part of the reaction equations into the linear diffusion system. The remaining "reaction" system (that is, the two equations $\dot{u} = -uv^2$ and $\dot{v} = uv^2$) might be solved analytically. Implement algorithm (7.10) for the Gray-Scott model (5.1) using this new splitting. Compare your results with the original splitting.

Exercise 7.5. Use extrapolation across rows in Table 7.2 to estimate $(u + v)/2$ at $T = 1024$. Discuss the result.

Exercise 7.6. Apply splitting to produce a numerical scheme for approximating the solution of the PDE $u_t = u_{xx} + u(1 - u)$ with zero Neumann boundary conditions on the interval $0 \le x \le 1$ and some initial data localized near $x = 1/2$, perhaps of the form suggested in Eq (5.21). Compare your results with those obtained with Euler's method and the usual second-order central difference approximation of the second derivative for the spatial discretization.

CHAPTER 8

Feedback Control

8.1 A MATHEMATICAL MODEL FOR HEAT CONTROL OF A CHAMBER

The problem of maintaining a desired temperature in a chamber equipped with a controlled heater/cooler is discussed in this section (see Fig. 8.1). Basic modeling with heat transfer and control are considered. This control problem is encountered in industrial processes (for example, in furnaces and tank reactors) and in laboratory work where it is desirable to control the temperature of a sample. Similar control systems are ubiquitous in biology, automobile cruise controls, steering of ships, and many other applications.

The heater/cooler (henceforth simply called the heater) and chamber are modeled as open subsets $\hat{\Omega}$ and Ω of $\mathbb{R}^3$, respectively, and the wall separating them is represented by $\Sigma := \partial\hat{\Omega}\cap\partial\Omega \subseteq \mathbb{R}^2$. The temperature in the heater at $x \in \hat{\Omega}$ and time t is denoted by $\hat{u}(x,t)$, and the temperature in the chamber at $x \in \Omega$ and time t is denoted by $u(x,t)$.

Heat transport through the interface wall Σ will be modeled under the assumption that the wall is thin and the heat flux through the wall is proportional to the difference in temperature on the two sides of the wall (Newton's law of cooling). The temperature of the heater is actuated by a controller whose input is the continuous measurement of temperature by a thermometer placed at a fixed point x_0 in the interior of the chamber and whose action is to change the temperature in some region (for example, a heating element coil) in the heater box. For simplicity, assume that the temperature may be raised or lowered by the heater instantaneously. Convection is ignored and the heater and chamber are assumed to be insulated at all exterior walls. Heat loss due to all causes (radiation, conduction or leakage through walls, opening and closing access to the chamber, etc.) not specifically modeled is included in a simplistic manner by modeling all such causes as if the chamber as a whole was subject to Newton's law of cooling. The cases where the temperature can be changed in only one direction or where there is a delay in the time required to obtain the heat change called for by the controller are left to the reader in the form of exercises.

An Invitation to Applied Mathematics: Differential Equations, Modeling, and Computation.
http://dx.doi.org/10.1016/B978-0-12-804153-6.50008-7, Copyright © 2017 Elsevier Inc. All rights reserved.

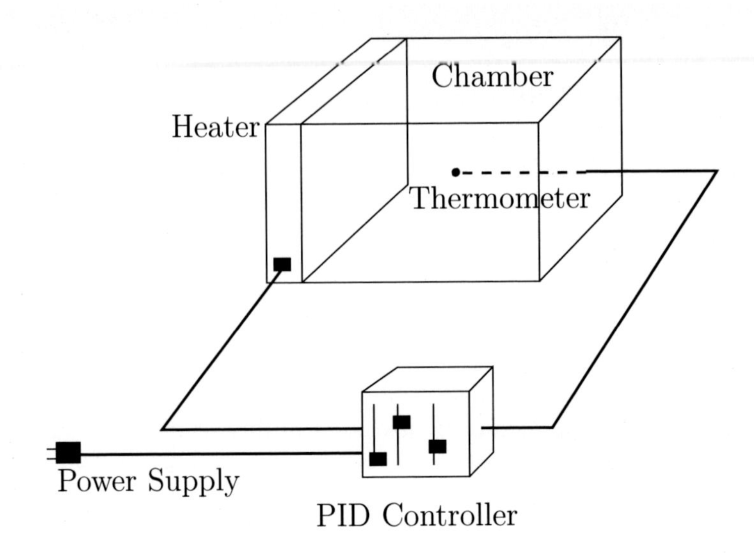

Fig. 8.1 Schematic of control system.

The physical parameters are the heat transfer coefficient λ across the wall, the heat transfer rate α from all causes that are not specifically modeled, the desired (set point) temperature T_s, the diffusion constants $\hat{K}$ and K, the initial heater temperature $\hat{u}_0$ and chamber temperature u_0, and the ambient temperature T_a.

The model equations, under our assumptions and with the notation just described, is given by a system of partial differential equations (PDEs), boundary conditions, and initial conditions. The PDEs are

$$\begin{aligned}\frac{\partial \hat{u}}{\partial t} &= \hat{K}\Delta\hat{u} + h(x)\mathrm{PID}(T_s - u(x_0,t)), \quad (x,t) \in \hat{\Omega} \times (0,\infty),\\ \frac{\partial u}{\partial t} &= K\Delta u + \alpha(T_a - u), \quad (x,t) \in \Omega \times (0,\infty),\end{aligned} \tag{8.1}$$

where the function h is nonzero at points in the heating element and zero otherwise, and PID denotes the output of the controller. A model for heat flow through the interface Σ between the heater and the chamber is determined from Newton's law of cooling: *The heat flux through a surface is proportional to the difference in the temperatures across the surface*; and Fourier's law: *The vector field that determines the heat flow is proportional to the negative gradient of the temperature*. (Note: At a slightly deeper level Newton's law is a special case of Fourier's law.) There are related boundary conditions for $\hat{u}$ and u on Σ. For the outer unit normal $\hat{\eta}$ on Σ, which points into the chamber, the boundary condition is

$$\int_{\Sigma} -\hat{K} \operatorname{grad} \hat{u} \cdot \hat{\eta} \, dS = \lambda(\hat{u} - u),$$

where λ is the (positive) constant of proportionality in Newton's law and $\hat{K}$ is the (positive) constant of proportionality in Fourier's law (the thermal diffusivity). For example, suppose that $\hat{u} > u$. In this case, the gradient of $\hat{u}$ points into the heater, $-K \operatorname{grad} \hat{u}$ points into the chamber, the dot product is positive, and the positive integral agrees in sign with the difference $\hat{u} - u$. The constant λ depends on the material properties of the interface. For the boundary of the chamber,

$$\int_{\Sigma} -K \operatorname{grad} u \cdot \eta \, dS = \lambda(u - \hat{u})$$

with the same λ because Σ is the boundary for the heater and the chamber.

The zero flux assumption for the insulated exterior walls reduces to several conditions, one for each wall $\mathcal{W}$, of the form

$$\int_{\mathcal{W}} \operatorname{grad} v \cdot N \, dS = 0$$

where v is either $\hat{u}$ or u and N is the outer normal on a exterior boundary wall of the heater (corresponding to the boundary conditions for $\hat{u}$) or the outer normal on a boundary of the chamber (corresponding to u). The initial conditions are

$$\begin{aligned} \hat{u}(x,0) &= \hat{u}_0(x), \quad x \in \hat{\Omega}, \\ u(x,0) &= u_0(x), \quad x \in \Omega, \end{aligned} \tag{8.2}$$

for $\hat{u}_0$ a given temperature distribution in the heater and u_0 a temperature distribution in the chamber.

The proportional–integral–derivative (PID) feedback control—which is written in the PDEs with an abuse of notation—is given by

$$\mathrm{PID}(f)(t) = K_p f(t) + K_i \int_0^t f(\sigma) \, d\sigma + K_d f'(t), \tag{8.3}$$

where K_p, K_i, and K_d are the respective parameters that determine the proportional, integral, and derivative controller gains. The control input in the model equations may be rewritten in the explicit form

$$\begin{aligned}\mathrm{PID}(T_s - u(x_0,t)) = K_p(T_s - u(x_0,t)) + K_i \int_0^t (T_s - u(x_0,\sigma))\, d\sigma \\ - K_d \frac{\partial u}{\partial t}(x_0,t).\end{aligned}$$

Note that the integral part of the PID control depends on the entire past history of the control process. In practice, the controller should be influenced by only part of the past history; that is, integration should be over some interval $[t-\tau, t]$, where τ is some appropriately chosen positive number.

8.2 A ONE-DIMENSIONAL HEATED CHAMBER WITH PID CONTROL

The geometry of the controlled heater-chamber model may be collapsed to one spatial dimension under the assumption that the axial direction from the heater to the thermometer carries the essential information. In other words, we may assume that the temperature in each slice perpendicular to this axis is constant. The heater reduces to a rod of length $\hat{L}$ parameterized by the interval $\hat{\Omega} := (0, \hat{L})$ (which represents the interior of the heater) and a rod of length L parameterized on the interval $\Omega := (\hat{L}, \hat{L}+L)$ (which represents the interior of the chamber). With this change of notation, $\Sigma := \{\hat{L}\}$, $\partial\Omega_1 \setminus \Sigma = \{0\}$, and $\partial\Omega \setminus \Sigma = \{\hat{L}+L\}$. The original model [Eqs. (8.1) and (8.2)] reduces to the coupled system of boundary value problems

$$\begin{aligned}\hat{u}_t &= \hat{K}\hat{u}_{xx} + h(x)(K_p(T_s - u(x_0,t)) + K_i \int_0^t (T_s - u(x_0,\sigma))\, d\sigma \\ &\quad - K_d u_t(x_0,t)), \quad (x,t) \in \Omega_1 \times (0,\infty), \\ \hat{u}_x(0,t) &= 0, \\ -A\hat{K}\hat{u}_x(\hat{L},t) &= \lambda(\hat{u}(\hat{L},t) - u(\hat{L},t))\end{aligned} \tag{8.4}$$

and

$$\begin{aligned}u_t &= K u_{xx} + \alpha(T_a - u), \quad (x,t) \in \Omega \times (0,\infty), \\ u_x(\hat{L}+L,t) &= 0, \\ AK u_x(\hat{L},t) &= \lambda(u(\hat{L},t) - \hat{u}(\hat{L},t))\end{aligned} \tag{8.5}$$

with the initial conditions

$$\begin{aligned}\hat{u}(x,0) &= \hat{u}_0(x), \quad x \in \hat{\Omega},\\ u(x,0) &= u_0(x), \quad x \in \Omega.\end{aligned} \tag{8.6}$$

The change in sign of the left-hand side of the last interface boundary condition is due to the use of the partial derivative $\partial u/\partial x$ to replace the dot product of the gradient and the normal. The outer normal on the boundary of the chamber points in the negative direction of the spatial coordinate.

The parameters in our model are $\hat{K}$, K, λ, T_s, $\hat{L}$, L, and A (the area of the interface Σ). More specifically, $\hat{K}$ and K are thermal diffusivities with units of area per time (with values in SI units of square meters/sec), λ is diffusivity per length, T_s is temperature (measured in SI units in kelvins), and the length and areas (which are measured in meters and square meters respectively). There are many thermal quantities defined in the literature; thus care is required to make sure named measured quantities have units compatible with a given model. For example, standard thermal diffusivities for various materials are available; they are sometimes given by using thermal diffusivity equal to $\kappa/(\rho c_s)$, where κ is the thermal conductivity, ρ is the material density, and c_s is the specific heat. Similarly, λ is thermal diffusivity times length, where the appropriate length for our heat control problem is the thickness of the interface wall.

The natural temperature scale for our model is T_s, a natural length scale is $\hat{L}$, and a natural time scale is $A\hat{L}/\lambda$. There are many other possibilities. Using these scales, define the dimensionless variables

$$\xi := \frac{x}{\hat{L}}, \quad \tau = \frac{\lambda t}{A\hat{L}}, \quad T_s\hat{U}(\xi,\tau) := \hat{u}(x,t), \quad T_sU(\xi,\tau) := u(x,t)$$

and dimensionless groups

$$\ell := \frac{L}{\hat{L}}, \quad T_r := \frac{T_a}{T_s},$$
$$\hat{k} := \frac{\hat{K}A}{\lambda\hat{L}}, \quad k := \frac{KA}{\lambda\hat{L}}, \quad \hat{\mu} := \frac{\lambda\hat{L}}{A\hat{K}}, \quad \mu := \frac{\lambda\hat{L}}{AK}, \quad a := \frac{\hat{L}A\alpha}{\lambda},$$
$$k_p := \frac{\hat{L}AK_p}{\lambda}, \quad k_i := \frac{\hat{L}AK_i}{\lambda}, \quad k_d := \frac{\hat{L}AK_d}{\lambda}$$

to obtain the dimensionless model

$$
\begin{aligned}
\hat{U}_\tau &= \hat{k}\hat{U}_{\xi\xi} + H(\xi)(k_p(1 - U(\xi_0,\tau)) + k_i \int_0^\tau (1 - U(\xi_0,\sigma))\,d\sigma \\
&\quad - k_d U_\tau(\xi_0,\tau)), \\
U_\tau &= kU_{\xi\xi} + a(T_r - U), \\
\hat{U}_\xi(0,\tau) &= 0, \\
-\hat{U}_\xi(1,\tau) &= \hat{\mu}(\hat{U}(1,\tau) - U(1,\tau)), \\
U_\xi(\ell+1,\tau) &= 0, \\
U_\xi(1,\tau) &= \mu(U(1,\tau) - \hat{U}(1,\tau)), \\
\hat{U}(\xi,0) &= \hat{U}_0(\xi), \\
U(\xi,0) &= U_0(\xi),
\end{aligned}
\tag{8.7}
$$

where $H(\xi) = h(\hat{L}\xi)$, $\hat{U}_0(\xi) = \hat{u}_0(\hat{L}\xi)/T_s$, and $U_0(\xi) = u_0(\hat{L}\xi)/T_s$.

In an ideal world we would like to look in tables in the literature to find values for the physical parameters $\hat{K}$, K, and λ, substitute these into our model with appropriate lengths and areas for the problem at hand, and use the model to predict the behavior of the system in open loop (without the control) and in closed loop (with the control). This might be possible if the model were a close representation of reality. But, the assumptions used to make the model are crude approximations. Thus, a model with table-book values for physical parameters may not produce accurate results. Perhaps the model is too simplistic or simply wrong. In case the model is physically reasonable (like the model Eqs. (8.4)–(8.6)), a better approach is to try to calibrate the model by modifying its coefficients to fit experimental data gathered from the physical problem at hand (in this case, the open loop heater-chamber system).

Problem 8.1. Suppose the physical control system discussed in this section has a heating/cooling element located in the center of an air-filled cubic box with axial length $\hat{L} = 20$ centimeters; the function h in the model has value $h = 1$ on this set and zero otherwise; the interface is a steel plate with area 400 square centimeters and thickness 0.5 centimeters (which in our model is a portion of a plane with zero thickness); the chamber is a rectangular box with axial length 60 centimeters filled with air; and the thermometer is placed in the center of the chamber. In reality, the order of magnitude of the thermal diffusivity of air is 10^{-5} square meters per second and the same for (some types of) steel. For definiteness, use SI units (meters, seconds, kilograms, and kelvins) and assume the physical model parameters are

$$\hat{K} = K = 10^{-5}, \quad \lambda = 5 \times 10^{-8}, \quad \hat{L} = 0.2, \quad L = 0.6, \quad A = 0.04.$$

The corresponding dimensionless groups are approximated by

$$\hat{k} = k = 40.0, \qquad \hat{\mu} = \mu = 0.025.$$

Also, assume that the rate of heat loss α due to causes not specifically modeled is 6.86×10^{-6}, which is about a drop of 5 degrees kelvin in 1 hour when the chamber temperature is 500 kelvins and the ambient temperature is 295 kelvins. The corresponding dimensionless variable has value $a \approx 1.097$.

(a) Discuss tuning the control gains to achieve the shortest time startup from system temperature 295 kelvins to the set-point temperature 500 kelvins without overshoot that would exceed 550 kelvins as the system continues to run with the same control gains. To make the problem somewhat realistic, limit the dimensionless control gains (which would be related to the limits of the heater/cooler) to be in the range $[0, 1000]$.
(b) Discuss tuning the controller gains to find those that keep the chamber temperature within 5 kelvins of the set point when subjected to a fluctuation of ambient temperature of at most 15 kelvins.

Simulation of the dimensionless heat control problem may be accomplished by seeking an analytic solution, approximation of our model, or the use of a numerical method to approximate solutions of the model equations.

The method of lines (recall, Exercise 5.72) is well suited to numerical simulation for our model equations. The idea is simple: discretize in space (but not in time) to obtain a system of ordinary differential equations (ODEs) that incorporates the boundary conditions. Then approximate solutions using a numerical method for ODEs. In principle, Euler's method applied to PDEs is a special case of the method of lines. Thus, the new method is simply a change in point of view. To be clear, consider the heat equation in one space dimension with zero Neumann boundary conditions and initial data given by the initial boundary value problem

$$u_t = u_{xx}, \quad u_x(0,t) = 0, \quad u_x(L,t) = 0, \quad u(x,0) = f(x). \tag{8.8}$$

Discretize the spatial domain (0,L) into n parts of length $\Delta x = L/n$ and define $n - 1$ variables $\{u_i\}_{i=1}^{n-1}$. The spatial derivative may be discretized according to the desired accuracy. For simplicity and in keeping with usual practice for a first approximation, the second-order centered difference approximation (with $u_i := u(i\Delta x, t)$)

$$u_{xx}(i\Delta x, t) \approx \frac{u_{i-1} - 2u_i + u_{i+1}}{\Delta x^2}$$

is a reasonable choice. The Neumann boundary conditions may be approximated by taking $u_0 := u_1$ and $u_n := u_{n-1}$. This choice is simple to program, but not second-order accurate. Higher-order accuracy for the boundary conditions can be achieved by using multipoint approximations of the first derivative. The choices made here produce the system of $n - 1$ ODEs given by

$$\begin{aligned}
\dot{u}_1 &= \frac{1}{\Delta x^2}(-u_1 + u_2), \\
\dot{u}_i &= \frac{1}{\Delta x^2}(u_{i-1} - 2u_i + u_{i+1}), \qquad i = 2, 3, 4, \ldots, n-2 \\
\dot{u}_{n-1} &= \frac{1}{\Delta x^2}(u_{n-2} - u_{n-1}).
\end{aligned} \tag{8.9}$$

Initial data for the system is defined in the obvious manner from the initial data for the PDE. Note the presence of the factor $1/\Delta x^2$, which will be large whenever Δx is small. This could cause numerical difficulties. One possible remedy is the change of timescale $t = \Delta x^2 s$. It removes the undesirable factor, but forces the computation of a large number of time steps. These considerations should remind you of the Courant–Lewy–Friedrichs condition discussed in Section 5.4. There is no obvious resolution of these difficulties, which are inherent in most numerical methods. They may be partially resolved by further numerical analysis that is beyond the scope of this book.

For a simple implementation of the method of lines to approximate solutions of the dimensionless model (8.7), we may use second-order accurate discretizations of the second-order spatial derivatives and first-order accurate discretizations of the zero Neumann boundary conditions. To describe a viable discretization of the boundary conditions at the interface ($\xi = 1$), consider the discretization of the interval $(0, \ell + 1)$ given by choosing positive integers m and n, defining $\Delta\hat{\xi} := 1/m$ and $\Delta\xi = \ell/n$, and using $\hat{u}_i := \hat{U}(i\Delta\hat{\xi}, \tau)$ and $u_i := U(i\Delta\xi, \tau)$. The temperature at the thermometer (placed in the middle of the chamber) may be approximated by $(u_{[n/2]_-} + u_{[n/2]_+})/2$, where $[j]_+$ is the smallest integer larger than or equal to j (often called ceiling(j)) and $[j]_-$ is the largest integer less than or equal to j (called floor(j)). This approximation may be used to code the proportional control. The integral control may be implemented by adding the differential equation and initial value

$$\frac{dz}{d\tau} = 1 - U(\xi_0, \tau) \approx 1 - \frac{u_{[n/2]_-} + u_{[n/2]_+}}{2}, \qquad z(0) = 0$$

to the method-of-lines system of ODEs. The new variable z is the desired integral. Finally, the derivative control may be added by simply using the right-hand sides of the method-of-lines ODEs at $[n/2]_-$ and $[n/2]_+$ as these are equal to the appropriate time derivatives. The boundary conditions at the interface couple the system. Using first-order approximations (which are adequate, but not very accurate) and the proposed numbering scheme, the boundary conditions are

$$\begin{aligned} \frac{\hat{u}_m - \hat{u}_{m-1}}{\Delta\hat{\xi}} &= \hat{\mu}(u_0 - \hat{u}_m), \\ \frac{u_1 - u_0}{\Delta\xi} &= \mu(u_0 - \hat{u}_m). \end{aligned} \tag{8.10}$$

The state variables for the method of lines are $\hat{u}_i$ for $i = 1, 2, 3, \ldots, m-1$ and u_i for $i = 1, 2, 3, \ldots, n-1$; the system of boundary conditions (8.10) is two equations for the two unknowns $\hat{u}_m$ and u_0; that is, the (scaled) temperatures of the heater and the chamber at the interface. By solving this system, we have that

$$\begin{aligned} \hat{u}_m &= \frac{1}{1 + \hat{\mu}\Delta\hat{\xi} + \mu\Delta\xi}((1 + \mu\Delta\xi)\hat{u}_{m-1} + \hat{\mu}\Delta\hat{\xi}u_1), \\ u_0 &= \frac{1}{1 + \hat{\mu}\Delta\hat{\xi} + \mu\Delta\xi}(\mu\Delta\xi\hat{u}_{m-1} + (1 + \hat{\mu}\Delta\hat{\xi})u_1). \end{aligned} \tag{8.11}$$

These formulas are used whenever $\hat{u}_m$ or u_0 are needed for computation of the discretized second derivatives.

All the ingredients are now in place to simulate the system using the method of lines. The system of $m + n - 2$ ODEs may be approximated with a variety of methods. Euler's method is a possibility. To maintain the second-order accuracy for the spatial discretizations of second derivatives, implicit or explicit trapezoidal time stepping is preferable.

Exercise 8.1. Implement the method of lines to approximate the PDE initial BVP (8.8) using ODEs (8.9). Discuss and compare the efficiency and accuracy of the method for the ODEs in their given form and after scaling time with the change of variables $t = \Delta x^2 s$. Compare your numerical results with the exact solution for your choice of initial data.

Exercise 8.2. Implement the method of lines to approximate the PDE initial BVP (8.8) using ODEs (8.9). Discuss the accuracy of the numerics for the ODEs and for their modification using a second-order accurate approximation for the boundary conditions. State explicitly your second-order accurate discrete approximation of the second derivative. Compare your numerical results with the exact solution for your choice of initial data.

Exercise 8.3. Consider the following dimensionless model for heat diffusion with proportional boundary control:

$$u_t = u_{xx}, \quad u_x(0,t) = u(0,t) - k_p(1 - u(\frac{1}{2},t)), \quad u(0,1) = 3, \quad u(x,0) = 0.$$

(a) Show that it is not possible to choose the control gain k_p so that the system will maintain the set point $u(1/2,t) = 1$. (b) What is the behavior of the function $t \to u(1/2,t)$ as t increases—that is, when the system is allowed to run—for a fixed control gain? (c) Show that the set point can be maintained using a PI control.

Exercise 8.4. Problem 8.1 is the main problem for this section. (1) Implement a numerical method to approximate the solution of part (a) of this problem. (2) Approximate the solution of part (b).

Exercise 8.5. Modify Problem 8.1 for the case where no active cooling is available; that is, a heater is available subject to PID control but there is no active cooling. The heater may be turned off. Using the given parameters solve part (a).

Exercise 8.6. [Heat Fluctuations in a Bar] (a) Suppose one end of an insulated bar is insulated and the temperature at the other end is changed (instantaneously) with a preassigned function of time. How does the temperature at the insulated end respond to the input heating protocol? For example, suppose the input is a periodic function of time. Discuss the temperature recorded at the insulted end. Could this temperature ever exceed the input temperature? Perform numerical experiments and interpret your results with general statements. Can you prove your statements using Fourier series or other methods? Discuss.

(b) Suppose that one end of an insulted bar is fixed to a material of unknown thermal properties. Heat flows through this end according to Newton's law of cooling, where the unknown properties are encoded in the constant of proportionality. Can this constant be measured by heating the bar at the opposite end and recording the heat flux through this end?

(c) Suppose an insulated bar is subjected to unknown temperature fluctuations within some specified range of temperatures at one of its ends and is insulated at the other end. An actuator is installed that allows the temperature to be changed (instantaneously) at the midpoint of the bar. Set up and simulate a control system designed to maintain a constant temperature at the insulated end of the bar.

Exercise 8.7. [Modeling and Control Project] Consider a cylindrical tank with base radius a and height h. Suppose water flows into the tank intermittently. The tank has a circular drain in the center of its base with radius r and a controllable valve that is designed to change the radius of the drain from zero (fully closed) to r fully open. A sensor measures the depth of the water in the tank and is connected to a PID controller and a servo mechanism that can continuously activate the valve control. Suppose that the inflow never exceeds the outflow capacity with the valve fully open and the flow velocity in the drain is $\sqrt{2gz}$, where g is the acceleration due to gravity and z is the depth of water in the tank (see Exercise 13.2).

(a) Make a mathematical model of the open loop system (no control); that is, set a fixed drain radius.

For the following parts of this problem assume that the system parameters are

$$a = 3\,\text{m}, \quad h = 6\,\text{m}, \quad r = 0.12\,\text{m}, \quad g = 9.8\,\text{m}\,/\,\text{sec}^2;$$

the desired depth is $h_{\text{set}} = 5\,\text{m}$, and the density of water is $\rho = 10^3\,\text{kg}\,/\,\text{m}^3$.

(b) Discuss controller gain tuning for P control used to maintain the depth of the water in the tank at the set point for a constant inflow of 300 kg / sec.

(c) Discuss controller gain tuning for PI control with the set point depth as in part (b) for the system startup from an empty tank.

(d) Discuss controller gain tuning for PID control with the set point depth as in part (b) for the system startup from an empty tank.

(e) Discuss controller gain tuning in case the inflow rate fluctuates between 80 and 120 kg / sec on a 1 minute cycle. Note: There is usually no reason to tune control gains to meet all possible situations. For example, a good tuning for startup might not be suitable to maintain a desired set point during later operation. Setting different gains for different regimes is called gain scheduling.

CHAPTER 9

Random Walks and Diffusion

Processes that might be modeled by random motions are ubiquitous in physics, engineering, finance, biochemistry, and biology. The classic example is Browning motion: a particle of pollen immersed in water moves due to molecular collisions. In biology and the life sciences, cell motion, particles in cells, molecular motion, and many other such phenomena are fruitfully modeled by random processes. This is a vast subject with many deep results. Here, the simplest part of the theory will be introduced and its relation to diffusion processes will be discussed. How do random processes lead to differential equations? That is the question!

9.1 BASIC PROBABILITY THEORY

The language of probability theory is used to describe random processes. No prior knowledge of the subject is required here.

Probability theory starts by agreeing on a procedure for assigning a set of values to the outcomes of some type of experiment or trial. The set S of all possible outcomes of an experiment (or trial) is called the *sample space*. Perhaps the most important example is the most familiar: flipping a coin. Each trial has two possible outcomes: H or T, heads or tails. The sample space is $S = \{H, T\}$. Probabilities are assigned to the possible events corresponding to the sample space. An *event* is a subsets of S: one of the sets $\varnothing$, $\{H, T\}$, $\{H\}$, and $\{T\}$; and, the collection of all subsets, denoted here by 2^S, is called the *power set* of S. Probabilities are assigned to events with the intuitive idea that the *probability of an event* is the fraction of at least one of its elements occurring (by a random choice) from all possibilities. Exactly what is meant by a random choice is not defined. By definition the probability of an event is a number in the interval $[0, 1]$. A coin flip (one trial) produces exactly one outcome, heads H or tails T. A fair coin is defined to produce equal probabilities $1/2$ for each outcome, or in formal language, the probability of the event $\{H\}$ is 1/2 and the probability of the event $\{T\}$ is 1/2. This is supposed to agree with our intuition. To make a precise definition, a *probability distribution* is a function $P : 2^S \to [0, 1]$. Its domain is a collection of sets and its range is contained in the interval of real numbers $[0, 1]$. For the coin toss, the probability distribution is given

An Invitation to Applied Mathematics: Differential Equations, Modeling, and Computation.
http://dx.doi.org/10.1016/B978-0-12-804153-6.50009-9, Copyright © 2017 Elsevier Inc. All rights reserved.

by $P(\varnothing) = 0$, $P(S) = 1$, $P(\{H\}) = 1/2$, and $P(\{T\}) = 1/2$. Sample spaces and probability distributions are the fundamental building blocks of the theory.

When two events E and F are disjoint (that is, $E \cap F = \varnothing$), the probability of their union is the sum of their probabilities

$$P(E \cup F) = P(E) + P(F). \tag{9.1}$$

This statement should be intuitively clear, but in any case, it is taken as an axiom about probabilities. The two events are called *independent* if

$$P(E \cap F) = P(E)P(F). \tag{9.2}$$

This notion is more difficult to understand. The idea is that the occurrence of event E does not affect the occurrence of the event F. For the coin toss example, the events $\{H\}$ and $\{T\}$ are *not* independent. If heads comes up, tails does not. The events $\{H, T\}$ and $\{T\}$ are independent. The occurrence of heads or tails puts no restriction on which alternative comes up. It does not effect the event that tails comes up.

In many different situations, an experiment with quantifiable outcomes depends on some underlying random process. As an example, suppose a clinical trial involves two groups where blood pressure is under study. A coin is flipped. In case the outcome is H, the average blood pressure of the individuals in group one is recorded; in case the outcome is T, the average blood pressure of group 2 is recorded. This scenario defines a function $f : S \to \mathbb{R}$ by the rule $f(H)$ is the average blood pressure of group 1 and $f(T)$ is the average blood pressure of group 2. Formally, a function from the sample space to the real numbers is called a *random variable*. The interplay between random variables, sample spaces, and probability distributions lies at the heart of probability theory.

A fundamental notion related to random variables is expected value. In case S is a finite set, the *expected value* is a function $\mathbb{E}$ from the set of random variables to the real numbers given by

$$\mathbb{E}(f) = \sum_{s \in S} f(s)P(\{s\}). \tag{9.3}$$

It is an average of the random variable over the sample space weighted by the probability distribution. For the clinical trial example,

$$\mathbb{E}(f) = \frac{1}{2}(f(H) + f(T)).$$

Random variables that take on a finite (or countably infinite) number of values are called *discrete random variables*. Clearly, every random variable on a finite sample space is discrete.

When the sample space S is an infinite set, random variables can take on a continuum of values (for example, all values in some interval of real numbers). They are called *continuous random variables*. The sum defining the expected value of a discrete random variable is replaced by an integral. To appreciate this in full generality requires some knowledge of real analysis. But, informally, a function P that assigns a unique number in the interval $[0, 1]$ to each element of some collection Σ of subsets of a sample space S in such a way that (1) $\varnothing$ and S are in Σ, $P(\varnothing) = 0$, and $P(S) = 1$; and (2) for every sequence of elements $\{E_i\}_{i=1}^{\infty}$ in Σ such that $\cup_{i=1}^{\infty} E_i$ is in Σ and $E_i \cap E_j = \varnothing$ for every $i \neq j$, the function P is additive; that is,

$$P(\cup_{i=1}^{\infty} E_i) = \sum_{i=1}^{\infty} P(E_i)$$

is called a *probability measure*. The *expected value of a random variable* with respect to a probability measure P is the integral of the random variable with respect to this measure:

$$\mathbb{E}(f) = \int_S f(s)dP. \tag{9.4}$$

A *probability space* is a triple (S, Σ, P) consisting of a sample space S, a collection Σ of subsets of the sample space (events), and a probability measure $P : \Sigma \to [0, 1]$. The simplest infinite sample spaces can be viewed as subsets of the real line $\mathbb{R}$ where the measure P on S is given by some function $\rho : S \to \mathbb{R}$, called its (probability) *density*, with two properties:

$$\int_S \rho\, dx = 1 \tag{9.5}$$

and, for every interval (a, b),

$$P((a, b)) := \int_a^b \rho\, dx. \tag{9.6}$$

In this case, the expected value of a random variable $f : S \to \mathbb{R}$ is defined by

$$\mathbb{E}(f) = \int_S f(x)\rho(x)\, dx, \tag{9.7}$$

where the integral is defined and computed in the usual manner.

The mathematical notion of a measure works equally well for finite sets where the definition of expected value reduces to Eq. (9.3). Here, for simplicity, the finite and infinite cases are treated separately. In the general case, the functions on S allowed to be random variables must at least be nice enough so that the integral in Eq. (9.7) is defined. In most practical situations this restriction is met automatically.

Using the language of measures, the probability of an event E is simply the integral of the probability density over the event

$$P(E) = \int_E \rho\, dx.$$

An additional important idea is to make precise a notion of how much a random variable (discrete or continuous) deviates from its expected value. The most important quantification of this deviation is called the *variance* of the random variable; it is a function var from the set of random variables to the real numbers defined by

$$\text{var}(f) = \mathbb{E}((f - \mathbb{E}(f))^2). \tag{9.8}$$

In view of the definition of expected value [Eq. (9.7)] and the properties of the probability density [Eq. (9.5)],

$$\begin{aligned} \text{var}(f) &= \int_S (f - \mathbb{E}(f))^2 \rho\, dx \\ &= \int_S (f^2 - 2\mathbb{E}(f)f + \mathbb{E}(f)^2)\rho\, dx \\ &= \mathbb{E}(f^2) - 2\mathbb{E}(f)\mathbb{E}(f) + \mathbb{E}(f)^2 \qquad (9.9) \\ &= \mathbb{E}(f^2) - \mathbb{E}(f)^2. \qquad (9.10) \end{aligned}$$

The *standard deviation* of a random variable is the square root of its variance: $\sigma(f) := \sqrt{\mathrm{var}(f)}$.

In general, two random variables f and g on a sample space S are called *independent random variables* if for every pair of subsets A and B of S the events

$$F := \{s \in S : f(s) \in A\}, \qquad G := \{s \in S : g(s) \in B\}$$

are independent. This concept—whatever its intuitive meaning—is important as evidenced by a useful theorem: If the random variables f and g are independent, then $\mathbb{E}(fg) = \mathbb{E}(f)\mathbb{E}(g)$ and $\mathrm{var}(f + g) = \mathrm{var}(f) + \mathrm{var}(g)$. The proofs of these statements are left to the reader.

9.2 RANDOM WALK

A classic problem is the motion of a person (perhaps a forgetful mathematician) who starts walking from a signpost along a street. After a short time and a short walk, the person either continues walking in the same direction again for the same short time and distance, or turns around and walks for the same short time and distance in the opposite direction. After every equal time interval a random decision is made to continue in the same direction or to turn around and walk in the opposite direction always walking the same distance in the same short time. Where is the walker after some fixed finite number of steps? The answer to this question depends on the sequence of random decisions on which direction to walk. Clearly, there is no definite answer unless each choice is known. A prediction of where the walker will be at some future time can only be given as the probability of the walker being at some location at some specified time. This is the quintessential random process. But, there is a surprising and powerful result (which will be discussed later): The probability of being at some location at some specified time in the limit as the time steps and the length of each walk become infinitesimally small is given as the solution of a *deterministic* process, in fact a partial differential equation (PDE).

To analyze the random walk problem mathematically, set the distance ℓ and the duration τ of each step of the random walk; that is, during a step the walker goes a distance ℓ (in one of two possible directions) in time τ. After each step in the process, a walker who starts at the origin ends up at a point on the real line in the lattice $\mathcal{L} := \{k\ell : k \in \mathbb{Z}\}$, where $\mathbb{Z}$ denotes the set of all integers (positive, negative, and zero).

The decision on which direction to walk (in the positive or negative direction along the real line) is made randomly, perhaps by tossing a coin. A fair coin toss could be used, but to be a bit more general, consider instead choosing at random (whatever that means) a point in the interval $[0, 1]$ and assigning the letter L (for left) if the chosen point is in the subinterval $[0, p]$ and the letter R (right) in case the chosen point is in the interval $(p, 1]$. Note that the left subinterval has length p and the right subinterval has length $q := 1 - p$. For each step, we may define the sample space $\widehat{S} := \{L, R\}$ and the probability distribution $\widehat{P}$ by $\widehat{P}(\varnothing) = 0$, $\widehat{P}(\widehat{S}) = 1$, $\widehat{P}(\{L\}) = p$, and $\widehat{P}(\{R\}) = q$. In this interpretation, the probability is the length of the interval corresponding to the outcome of a random experiment: choosing a point in the unit interval. Note that for each event, P is given as the integral over the corresponding interval with probability density $\rho(x) \equiv 1$. For example,

$$\widehat{P}(\{L\}) = \int_0^p 1\,dx.$$

Define a random variable $\hat{f} : \widehat{S} \to \mathbb{R}$ by the rule

$$\hat{f}(L) = -\ell, \qquad \hat{f}(R) = \ell, \tag{9.11}$$

and the *probability space* $(\widehat{S}, \Sigma, \widehat{P})$, where Σ is the collection of sets generated by taking unions and complements of open intervals in $[0, 1]$. The random variable corresponds to the action of moving either left or right a distance ℓ.

To define the probability distribution for the random walk after n steps, define the sample space S_n whose elements are all words of length n composed of the letters L and R. Note this is simply an alternative way of looking at the product set $\times_{i=1}^{n} S$ whose elements are ordered n-tuples of elements of S. At step $n = 3$, for example,

$$S_3 = \{LLL, LLR, LRR, LRL, RLL, RLR, RRL, RRR\}.$$

The probability distribution P_n on S_n, which defines the random walk probabilities, is given by assigning to each word the probability of obtaining the corresponding sequence of events by n independent trials of choosing a random point in the unit interval and assigning L or R according to the probability distribution $\widehat{P}$. For example, $P_3(\{LRL\}) = pqp = p^2q$, $P_3(\{LRR\}) = pqq = pq^2$, and $P(\{RRR\}) = q^3$. The probabilities of the events containing more than one word are determined by the general rule (9.1): for disjoint events, the probability of their union is the sum of their probabilities. For instance, $P_3(\{LRL, LRR\}) = p^2q + pq^2$. The probability of the event S_n must be unity. Is it? This fact is proved by noticing that the probability of each event corresponding to a single word has the form p^iq^j with $i + j = n$. The sum of all terms of this form is

$$\sum_{i+j=n} p^iq^j = (p+q)^n = (p+(1-p))^n = 1$$

by the binomial formula.

A word s in S_n may be represented in the form

$$s = s_1s_2s_3\cdots s_n,$$

where s_i is either L or R. Recall the random variable $\hat{f}$ defined in Eq. (9.11) defined on the sample space on $\widehat{S}$, where the probability distribution is defined to be $\widehat{P}$. For each step $i = 1, 2, 3, \ldots n$, define the random variable f^i on S_n with probability distribution P_n by the rule

$$f^i(s_1s_2s_3\cdots s_n) = \hat{f}(s_i); \tag{9.12}$$

and, the random variable f_n on the same probability space by

$$f_n(s) = f^1(s_1s_2s_3\cdots s_n) + f^2(s_1s_2s_3\cdots s_n) + f^3(s_1s_2s_3\cdots s_n) + \cdots + f^n(s_1s_2s_3\cdots s_n).$$

The value of $f_n(s)$ is exactly the position of the walker on the lattice $\mathcal{L}$ in case the random choices at each step were $s_1, s_2, s_3, \ldots, s_n$. For example,

$$f_3(LRL) = -\ell + \ell - \ell = -\ell;$$

that is, with the choices L, R, L, the walker is at the position $-\ell$ in the lattice.

After three steps, there is only one way to reach 3ℓ or -3ℓ, no way to reach $\pm 2\ell$ or 0, and three ways to reach each of the points ℓ and $-\ell$. The number of ways to end on a reachable lattice point is given by binomial coefficients defined for nonnegative integers $a \geq b$ by

$$\binom{a}{b} = \frac{a!}{b!(a-b)!}$$

where $0! := 1$. For step three, the numbers are exactly $\binom{3}{j}$ for $j = 0, 1, 2, 3$. Careful inspection of the cases $n = 3$ and $n = 4$ should suggest (and you should prove) that the exact formula for the probability of the event defined by being at the lattice point $m\ell$ at time $n\tau$ is

$$\begin{aligned} u(m\ell, n\tau) &:= P_n(\{s \in S_n : f_n(s) = m\ell\}) \\ &= \begin{cases} \binom{n}{n-j} p^j q^{n-j}, & m = n - 2j \ \& \ j \in \{0, 1, 2, \dots n\}; \\ 0, & \text{otherwise} \end{cases} \\ &= \begin{cases} \binom{n}{\frac{n-m}{2}} p^{\frac{(n-m)}{2}} q^{\frac{(n+m)}{2}}, & m = n - 2j \ \& \ j \in \{0, 1, 2, \dots n\}; \\ 0, & \text{otherwise.} \end{cases} \end{aligned} \tag{9.13}$$

This is the solution of the one-dimensional random walk problem.

Similar scenarios are ubiquitous in probability theory where repeated trials of some experiment are under investigation. A process consisting of a series of experiments (such as hitting a target, choosing a point in $[0, 1]$, or asking a yes-no question), each with only two possible outcomes (which might be interpreted as success or failure, moving left or right, or answering yes or no), leads to (1) a sample space that is the Cartesian product of a number of copies of the two-element sample space and (2) a random variable that is the sum of random variables each defined so that it depends on only one factor of the product. On this factor, the random variable is the same as all the other random variables in the sum. This is exactly the situation in the random walk example where each factor is the sample space S=$\{L, R\}$ and the random variables are the f^i. Each summand has the same probability distribution function and each pair of distinct summands are independent. In this case the summands are said to form an *independent set of identically distributed random variables.* Such a set is often called *idd*. Many elementary texts in probability theory are confusing, to say the least, on this subject when they seem to imply that a idd sequence is a set of random variables defined on S instead of on a product of several copies of S. The problem is that the random variables defined in the former way fail to be independent by the general definition of independence.

For the case just described, consider the sequence whose nth element is the random variable

$$\frac{f^1 + f^2 + f^3 + \cdots f^n}{n}.$$

What is the value of this random variable on the infinite sequence $s = s_1 s_2 s_3 \cdots$? The answer surely depends on the first n factors of this product. But thinking probabilistically when n is large and s is chosen at random, it seems reasonable to expect that the number of occurrences of L and R would be dictated by the probabilities p and q. In fact, the fraction of the number of occurrences of the letter L should be p and of R should be q; that is,

$$\frac{f^1 + f^2 + f^3 + \cdots f^n}{n} \approx -p\ell + q\ell.$$

This is not true for all s, but it is true most of the time. A fancy way to express this observation is to say that

$$\lim_{n \to \infty} \frac{f^1 + f^2 + f^3 + \cdots f^n}{n} = g, \tag{9.14}$$

where g is the random variable with constant value $-p\ell + q\ell$ (on the infinite product of S with itself) except for a small exceptional set (called a set of measure zero). This constant value is exactly the expected value of each f^i. This fact is a special case of an important theorem called the *law of large numbers*: For every sequence $\{X_i\}_{i=1}^{\infty}$ of identically distributed random variables all with the same expected value μ,

$$\lim_{n \to \infty} \frac{X^1 + X^2 + X^3 + \cdots X^n}{n} = g,$$

where (except for a set of measure zero) g is the constant random variable whose value is μ. This result is one of the pillars of probability theory. As the number of trials of an experiment grows, the average outcome is the expected value of the experiment.

Exercise 9.1. Show that the set of random variables f^i, $i = 1, 2, 3, \ldots, n$ defined by Eq. (9.12) are independent and identically distributed.

Exercise 9.2. Prove Eq. (9.14).

9.3 CONTINUUM LIMIT OF THE RANDOM WALK

Although the analysis of the last section is fundamental, it is just the beginning of a wonderful story. An underlying theme is the physics of matter at different scales. Roughly speaking, motion at the quantum level is probabilistic, at the microscopic level motion is deterministic but very complicated when many particles are involved, and at the macroscopic level motion is deterministic (following Newton's laws). Classical modeling at the microscopic level is usually probabilistic to simplify the dynamics of a very large number of bodies interacting according to Newton's laws. In an appropriate limit, the probability distribution is defined on a collection of subsets of a continuum (perhaps the real line) and its time-dependent evolution is the solution of a deterministic evolution equation, which is in fact a PDE. Another model of microscopic particle motions results in an ordinary differential equation containing a term that has random values. Molecular dynamics becomes more approachable with numerical approximations as computer speed increases. This important subject—how to write efficient codes to evolve many body systems—is beyond the scope of this book (but see Exercise 10.10).

To proceed, consider the following alternate approach to solving the random walk problem, where as before $P_n(m\ell)$ is the probability of the event defined by being at the lattice point $m\ell$ at the nth step of the random walk. What is the probability of being at the same point after taking another step? Answer:

$$P_{n+1}(m\ell) = P_n(m\ell - \ell)q + P_n(m\ell + \ell)p, \tag{9.15}$$

or in words, the desired probability is given by the probability of being at the lattice point $m\ell - \ell$ and on the next step moving to the right (which has probability q) plus the probability of being at the lattice point $m\ell + \ell$ and on the next step moving to the left (which has probability p). The statement requires that being at a point and moving left (respectively moving right) on the next step are independent events. Thus, these probabilities are multiplied. Also the event defined by being at a point and then moving left is disjoint from the event defined by being at a different point and then moving right in the next step. Thus, these probabilities are added.

The difference equation [Eq. (9.15)] is called the *master equation* for the random process. As might be expected, the binomial probability distribution given by Eq. (9.13) satisfies the master equation (see Exercise 9.3).

In most scientific applications, random walks are used to model processes where a large number of short distance steps occur. For instance, imagine a pollen grain in water. The water molecules (or many of them) impact the pollen grain a large number of times over a very short time. The exact number of collisions per unit time and the length scale of the motion per collision are not usually known with precision. The underlying physical motion is deterministic, but extremely complicated. To avoid modeling the motion by taking into account Newton's law and the forces due to the collisions, the motion is modeled as a random process. The exact probability for the position of the particle after some finite number of steps (which for the pollen grain model would have to be computed for a random walk in three-dimensional space) is usually impossible to determine. Instead, a good approximation of this probability is sought that is valid in the limit as the number of steps approaches infinity and the length scale approaches zero. This idea has proven to be useful in understanding many different physical motions. To implement it, the correct limit process must be employed that takes into account the number of collisions per unit of time and the length scale of the motion per collision. Different motions (for instance a particle of dust moving in a room filled with air due to collisions with air molecules) would be modeled by the same random walk, but with a different probability distribution, a different number of collisions per unit time and a different length scale. A useful model should take into account the physical characteristics of these different motions: the velocity of the molecules, the mass of the particles being tracked, and perhaps external forces acting on the particles.

Our task is to set the probability p and the correct limit process to take into account the physical characteristics of the motion being modeled.

A clue to determining the correct model is apparent by imagining what might be observed in the random motion of a particle. The actual motion of the particle might be observed over several units of time. The position of the particle at each point in time is (as before) given by a random variable. Although we might wish to model the exact motion, the idea is to assume it is not deterministic and use a random process as the model. Thus, we seek to determine the probability of the particle being at some place at a given time. Perhaps the mean (also called the average) distance to some origin might be recorded in an experiment. This quantity is modeled by the expected value of the random variable measuring the position of the particle. Thus the expected value of the motion under investigation provides one constraint on the random process; its variance is an obvious additional parameter that is sure to play a role. In fact, the expected value of the motion and its variance turn out to be the natural parameters in random walk problems.

The expected value for the discrete random variable f_n (as in the previous section) that measures the position of the random walker after n steps is computable and given by

$$\mathbb{E}(f_n) = \sum_{s\in S_n} f_n(s)P_n(\{s\}) = \sum_{j=0}^{n}(n-2j)\ell p^j q^{n-j} = n\ell(q-p). \quad (9.16)$$

The combinatorial sum can be evaluated directly (out of context) to obtain the formula

$$\mathbb{E}(f_n) = n\ell(q-p)(p+q)^{n-1}, \quad (9.17)$$

which reduces to sum (9.16) because $p+q=1$. An easier way to obtain the desired result is to use the properties of the expected value. By its definition, the expected value function $\mathbb{E}$ is linear (that is, $\mathbb{E}(f+g) = \mathbb{E}(f)+\mathbb{E}(g)$). Note that (from Eq. (9.11))

$$\mathbb{E}(\hat{f}) = -\ell p + \ell q = \ell(q-p).$$

Using the linearity,

$$\mathbb{E}(f_n) = \sum_{i=1}^{n}\mathbb{E}(f^i).$$

The expected value of the ith summand is

$$\begin{aligned}\mathbb{E}(f^i) &= \sum_{s\in\{s'\in S_n: s_1=L\}} f^i(s)P(\{s\}) + \sum_{s\in\{s'\in S_n: s_1=R\}} f^i(s)P(\{s\})\\ &= -\ell p\sum_{s\in S_{n-1}} P(\{s\}) + \ell q\sum_{s\in S_{n-1}} P(\{s\})\\ &= (q-p)\ell,\end{aligned}$$

which is (as it should be) the expected value of $\hat{f}$ on the probability space $(\widehat{S}, \widehat{P})$. Also, the answer is independent of i. Therefore,

$$\mathbb{E}(f_n) = (q-p)n\ell. \quad (9.18)$$

In case $p=q$, the mean is zero.

The simplest method for computing the variance uses the independence of the random variables summed to obtain f_n (see Exercise 9.4). But, it is certainly instructive when first introduced to probability theory to compute the variance from the definition using the formula $\text{var}(f_n) = \mathbb{E}(f_n^2) -$

$\mathbb{E}(f_n)^2$. The expected value $\mathbb{E}(f_n)$ is already computed. For the expected value of the square of the random variable, we have that

$$\mathbb{E}(f_n^2) = \mathbb{E}((f^1 + f^2 + f^3 + \cdots + f^n)^2).$$

The square on the right-hand side multiplied out yields n^2 terms, each of which is of the form $f^i f^j$. There are n terms where $i = j$ and $n^2 - n$ where $i \neq j$. In case $i \neq j$,

$$\begin{aligned}
\mathbb{E}(f^i f^j) &= \sum_{s \in S_n} f^i(s) f^j(s) P(\{s\}) \\
&= \sum_{s \in S_n} \hat{f}(s_i)\hat{f}(s_j) P(\{s\}) \\
&= \sum_{s \in \{s' \in S_n : s_i = L, s_j = L\}} \ell^2 P(\{s\}) + \sum_{s \in \{s' \in S_n : s_i = L, s_j = R\}} \ell^2 P(\{s\}) \\
&\quad + \sum_{s \in \{s' \in S_n : s_i = R, s_j = L\}} \ell^2 P(\{s\}) + \sum_{s \in \{s' \in S_n : s_i = R, s_j = R\}} \ell^2 P(\{s\}) \\
&= \sum_{s \in S_{n-2}} p^2 \ell^2 P(\{s\}) - 2 \sum_{s \in S_{n-2}} pq\ell^2 P(\{s\}) + \sum_{s \in S_{n-2}} q^2 \ell^2 P(\{s\}) \\
&= \ell^2 (p - q)^2.
\end{aligned}$$

The result is independent of i and j. For $i = j$,

$$\mathbb{E}((f^i)^2) = \sum_{s \in S_{n-1}} p\ell^2 P(\{s\}) + \sum_{s \in S_{n-1}} q\ell^2 P(\{s\}) = \ell^2.$$

Note that the last equality uses $p + q = 1$ and that p and q appear as factors for this case, not p^2 and q^2. (Why?) The result is again independent of i. Putting the two cases together, we have (using $(p + q)^2 = 1$) that

$$\mathbb{E}(f_n^2) = n\ell^2 + (n^2 - n)\ell^2 (p - q)^2 = n^2 \ell^2 (p - q)^2 + 4n\ell^2 pq. \tag{9.19}$$

The variance of the random variable f_n is

$$\mathrm{var}(f_n) = \mathbb{E}(f_n^2) - \mathbb{E}(f_n)^2 = 4n\ell^2 pq. \tag{9.20}$$

One interpretation of the expected value $\mathbb{E}(f_n^2)$ is interesting: it is the expected value of the square of the distance traveled by the random walker after n steps because positive and negative final distances from the origin are taken to be the same for this quantity. For the case $p = 1/2$, the expected value is $\ell^2 n$. What does this really mean? One way to interpret this fact is to rely on the law of large numbers. It says that if the experiment (of taking a random walk of n steps) is repeated many times, then the average outcome is this expected value. In (perhaps nonrigorous) discussions of an application that uses a random walk with equal probabilities, one of the most used facts is

$$\text{the distance from the origin is expected to be proportional to } \sqrt{n}. \tag{9.21}$$

The constant of proportionality is known but not obvious (see Exercise 9.8).

A great idea is to approximate the random walk probability distribution by a continuous probability density by passing to a continuum limit as the length ℓ of each step and its duration τ approach zero. To carry out this procedure rigorously requires some advanced mathematics that is beyond the scope of this discussion. Fortunately, the main ideas that lie at the heart of a rigorous derivation are simple; they are presented here.

To pass from the discrete probability distributions P_n defined at each step of the random walk to a limiting probability density requires that each point x on the real line be assigned a density, which is defined by integrating the probability over intervals containing x, dividing by the length of the corresponding interval, and taking the limit as the lengths of the intervals approach zero. A probability density on the line must have units of probability per length. At the nth step, the event $\{s \in S_n : f_n(s) = m\ell\}$ corresponding to the walker being at the lattice point $m\ell$ is assigned a probability by P_n. The idea for passing to the continuum limit is to change the sample space from S_n to the real line and extend accordingly the discrete random variable f_n to the piecewise constant function defined on the real line such that f_n is constant with value $m\ell$ in the half-open interval $[(m-1)\ell, (m+1)\ell)$ whenever $P_n(f_n = m\ell)$ is not zero. Each interval has length 2ℓ. Also, this piecewise constant random variable approximates the continuous random variable $X_{n\tau}$ defined on the real line by $X_{n\tau}(x) = x$; and the piecewise constant probability density function $P_n(f_n = x)$ is used to approximate the probability density function $x \mapsto u(x, n\tau)$ at time $n\tau$ by the formula

$$u(x, n\tau) = \frac{P_n(f_n = x)}{2\ell}. \tag{9.22}$$

The factor 2 is appropriate given the extension to piecewise constant functions, but this is not essential. An arbitrary nonzero factor will work just as well. The important feature is that the denominator be proportional to some length scale associated with the lattice.

The next idea is to use the master equation to derive a difference equation for the limiting probability density function $x \mapsto u(x,t)$. With $x = m\ell$ and $t = n\ell$, the master equation for the density u is

$$u(x, t+\tau) = u(x-\ell, k\tau)q + u(x+\ell, k\tau)p. \tag{9.23}$$

Proceeding with the intention of obtaining an equivalent difference equation that might be associated with a differential equation in the continuum limit, the left-hand side suggests a partial derivative with respect to t. To approximate this partial, subtract $u(x,t)$ and divide by τ. The right-hand side suggests a partial derivative with respect to x. Because there are terms with $x+\ell$ and $x-\ell$, perhaps a second derivative with respect to x is involved. Using these ideas and keeping in mind the identity $p+q=1$, check the following algebraic manipulations:

$$\begin{aligned}
\frac{u(x,t+\tau)-u(x,\tau)}{\tau} &= \frac{1}{\tau}(u(x-\ell,t)q + u(x+\ell,t)p - u(x,t)) \\
&= \frac{1}{\tau}(u(x-\ell,t) - 2u(x,t) + u(x+\ell,t))q \\
&\quad + \frac{1}{\tau}(u(x-\ell,t) - 2u(x,t) + u(x+\ell,t))p \\
&\quad + \frac{1}{\tau}[(2u(x,t) - u(x+\ell,t)q \\
&\quad + (2u(x,t) - u(x-\ell,t)p - u(x,t)] \\
&= \frac{1}{\tau}(u(x-\ell,t) - 2u(x,t) + u(x+\ell,t)) \\
&\quad + \frac{1}{\tau}[(u(x,t) - u(x+\ell,t)q + (u(x,t) - u(x-\ell,t)p] \\
&= \frac{1}{\tau}(u(x-\ell,t) - 2u(x,t) + u(x+\ell,t)) \\
&\quad + \frac{1}{\tau}[(u(x,t) - u(x+\ell,t)q + (u(x,t) - u(x-\ell,t)p] \\
&= \frac{\ell^2}{\tau}\frac{u(x-\ell,t) - 2u(x,t) + u(x+\ell,t)}{\ell^2} \\
&\quad - \frac{\ell q}{\tau}\frac{u(x+\ell,t) - u(x,t)}{\ell} + \frac{\ell p}{\tau}\frac{u(x-\ell,t) - u(x,t)}{-\ell}
\end{aligned}$$

Passing to the limit as ℓ and τ go to zero and *assuming* that these limits exist produces a partial derivative with respect to t, two partials with respect to x, and one second partial with respect to x. There is only one problem: What happens to the coefficients ℓ^2/τ, $p\ell/\tau$, and $q\ell/\tau$? Maybe one or more of these expressions does not have a limit. Suppose for example that $\ell = 1/j$ and $\tau = 1/j$ with $j \to \infty$. In this case, the second two expressions have finite limits; the first does not.

To obtain a PDE, we must have that ℓ^2/τ converges to a finite limit. Note that, in the limit process, the expression

$$-\frac{\ell q}{\tau}\frac{u(x+\ell,t)-u(x,t)}{\ell}+\frac{\ell p}{\tau}\frac{u(x-\ell,t)-u(x,t)}{-\ell}$$

approaches the form

$$-\frac{\ell(p-q)}{\tau}u_x(x,t).$$

Thus, we must also have that $\ell(p-q)/\tau$ converges to a finite limit.

There are three possibilities: (1) $\ell/\tau \to 0$ and $q-p \to \infty$, (2) ℓ/τ and $q-p$ converge to finite limits, or (3) $\ell/\tau \to \infty$ and $q-p \to 0$.

Possibility (1) cannot occur because p and q are probabilities confined to the unit interval $[0,1]$.

As $\ell \to 0$, in case (2) the quantity ℓ^2/τ converges to zero. This is theoretically possible, but highly unlikely in a truly random process because, in view of the variance for the underlying random process (9.20) and the equation $t = n\tau$, it follows that the variance of the underlying process goes to zero. In this case, the limit process produces the PDE model

$$u_t(x,t) = -\alpha u_x(x,t). \tag{9.24}$$

Case (3) is expected for most applications of the theory. The limits of the quantities ℓ^2/τ and $\ell(p-q)/\tau$ can both be specified.

As $q-p \to 0$ and $p+q=1$, either $p=1/2$ and $q=1/2$ for all the processes considered, or these values are approached in the limit. Something nice happens: The coefficient $(q-p)\ell/\tau$ is exactly the expected value of the random variable f_n for the underlying random walk. Thus, it is natural to arrange the limit to approach a value α given by the observed variance of

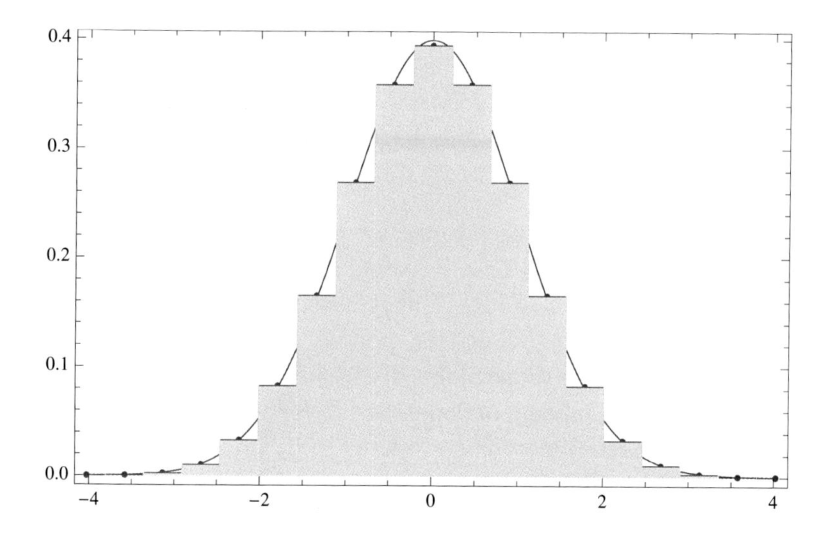

Fig. 9.1 The piecewise continuous approximation of the probability density function of the continuous random walk random variable X_t at $t = 1$ is depicted for the case $n = 20$. The continuous curve is given by the exact continuous probability density $x \to u(x, 1)$ as defined in Eq. (9.29).

the process being modeled. Likewise, the limit process may be arranged so that ℓ^2/τ converges to $\kappa/2$, which is the variance of the underlying process divided by two (see Exercise 9.7). In this case, the limit process produces the PDE, often called the Fokker–Planck (Adrian Fokker 1887–1972 and Max Planck 1858–1947) equation

$$u_t(x,t) = -\alpha u_x(x,t) + \frac{\kappa}{2} u_{xx} \tag{9.25}$$

for the probability density $x \mapsto u(x, t)$ at time t, where α is the observed expected value and κ is the variance of the corresponding random variable X_t, which is used to determine the position of the continuous time random walker at time t. Formally, this random variable on the real line is given by $X_t(x) = x$ and has probability density function $x \mapsto u(x, t)$.

For some finite number of particles N undergoing random walks, the Fokker–Planck model gives (for each particle) the probability density for finding it at x at time t. Thus, it is also true that $C(x, t) := Nu(x, t)$ is the proportion of the total number of particles at x at time t, or equivalently, the concentration of particles at x at time t. Because N is constant in this scenario, the concentration C also satisfies Eq. (9.25), a model previously obtained simply from the conservation of mass. Thus the random walk analysis is one way to pass from the microscopic (think molecules moving in a liquid) to the macroscopic (think concentrations of some substance dissolved in the liquid). This basic idea was introduced by Albert Einstein in 1905. His analysis helped to convince the scientific community that molecules exist. Since his seminal work, randomness has played a fundamental role in modeling of many different physical phenomena.

There is an alternate approach to the limit process that passes from the sample spaces S_n to the real line. It should be clear that this involves an infinite product of copies of $\{L, R\}$. This is correct, but the analysis required to understand the class of sets Σ and the properly defined probability density on this space is beyond the scope of this discussion. Instead, simply note that elements in the infinite product of copies of $\{L, R\}$ are infinite strings of the letters L and R. This set may be considered to be infinite strings of zeros and ones. Put a decimal point before such a string and consider it to be the binary number written $s = .s_1 s_2 s_3 \cdots$, where $s_i = 0$ or $s_i = 1$. What does this mean? The corresponding number (base ten) in the unit interval is

$$\sum_{i=1}^{\infty} \frac{s_i}{2^i}$$

(see Exercise 9.11). In this way, the infinite product of copies of $\{L, R\}$ may be viewed as the unit interval on which all the random variables may be defined.

To visualize the limit process, reconsider the probability density function for the piecewise constant, extended, random variable f_n defined on the real line as in Eq. (9.22). Also note that this function is the corresponding piecewise constant extension of the binomial distribution [Eq. (9.13)] with each probability at step n divided by the length 2ℓ. A graph of the piecewise constant probability density function for the case $n = 20$ is depicted in Fig. 9.1, where for the graphical representation $\tau = 1/n$ and $\ell = \sqrt{\kappa\tau}$. The smooth curve in the figure that is so well approximated by the density function is called a normal distribution. It is defined later in this section following a short discussion of general continuous random variables and their probability density functions.

The function $x \to u(x, t)$ discussed above may be taken as the probability density function of the random variable X_t given by $X_t(x) = x$; which might be interpreted as the experiment of looking at the position x.

In this case, the probability of the event

$$\{\xi \in \mathbb{R} : a \le X_t(x) \le b\}$$

is

$$P(\{x \in \mathbb{R} : a \le X_t(x) \le b\}) = \int_a^b u(\xi, t)\, d\xi,$$

which is interpreted to be the probability of the random walker being in the interval (a, b) at time t. Cumbersome expressions, for example those involving the definitions of sets, are often replaced with simpler notation in the probability literature where one might read

$$P(a \le X_t \le b) = \int_a^b u(x, t)\, dx$$

for the same probability. Although the interpretation of such expressions in view of the technical definitions of the subject can be confusing, familiarity with the language eventually makes such expressions seem natural. In any case, the probability of the event represented by the interval is essentially the sum (integral) of the probabilities of the events represented by each point of the interval. Here the random variable X_t is said to have the probability density $\xi \to u(\xi, t)$.

When the sample space is the entire real line as in the limit of the random walks, a general continuous random variable X (for example X_t) is, of course, a function defined on the real line. The class of random variables is restricted to contain those functions X such that there is another function g, called its *probability density function*, associated with X such that (1) g is nonnegative, (2) g is integrable with

$$\int_{-\infty}^{\infty} g(x)\, dx = 1,$$

and, (3) for every (measurable) subset A of the real line

$$P(\{x \in \mathbb{R} : X(x) \in A\}) = \int_A g(x)\, dx.$$

The last statement is usually written

$$P(X \in A) = \int_A g(x)\, dx. \tag{9.26}$$

It should be clear from this notation that X is thought of as a variable—thus the name random variable—taking on real values. The interpretation is simple: the probability that X takes values in A is given by integrating (summing) the associated probability density over this set. Note that the probability of the event $\{X = r\}$ where X takes on one particular real value r is zero; indeed,

$$P(X = r) = P(\{x \in \mathbb{R} : X(x) = r\}) = \int_r^r g(x)\,dx = 0.$$

Changing the random variable on a set of zero measure has no effect; for example, $P(X < r) = P(X \leq r)$. In the random walk example, the coordinate function x is viewed as a random variable with probability density $\xi \to u(\xi, t)$.

The function G defined by

$$G(t) = P(X \leq t) = \int_{-\infty}^t g(s)\,ds$$

is called the probability distribution function. Of course, $G' = g$.

The probability density function relates the random variable to the underlying probability measure. Of course, the same random variable (function on a sample space) can have different probability density functions according to the probability density chosen on the class of subsets of the sample space under consideration: for discrete measures this subset is usually all subsets but for the continuous case the set of subsets is restricted to those subsets on which the probability measure is defined. The precise statement requires knowledge of measure theory. A typical example of such a restricted class is the set of all Borel subsets of the real line: all open intervals together with the sets obtained from these by countable intersections and by taking complements.

Canonical examples (called uniform random variables) have sample space $S = \mathbb{R}$ and probability density ρ supported on some interval $[a, b]$ given by

$$\rho(x) = \begin{cases} \frac{1}{b-a}, & a \leq x \leq b; \\ 0, & x < a \text{ or } x > b. \end{cases} \tag{9.27}$$

When the random variable X is given by $X(x) = x$ with corresponding probability density function ρ, note that

$$P(\{x \in S : X(s) \in A\}) = P(X \in A) = \int_A \rho(x)\, dx,$$

the expected value of X is

$$\mathbb{E}(X) = \int_S X(x)\rho(x)\, dx = \int_{-\infty}^{\infty} x\rho(x)\, dx = \frac{1}{b-a}\int_a^b x\, dx = \frac{1}{2}(b-a),$$

and its variance is

$$\text{var}(X) = \mathbb{E}(X^2) - \mathbb{E}(X)^2 = \frac{1}{12}(b-a)^2.$$

Uniform random variables are designed to convey a meaning: the probability of an event in the interval $[a, b]$ is the length of the interval, or all points in the interval are equally likely to occur.

For a continuous random variable X with probability density function f, the expected value is defined, as it should be, as the average of the function X over the sample space with measure P; that is,

$$\mathbb{E}(X) := \int_S X(x)\, dP.$$

But, often the expectation is defined by

$$\mathbb{E}(X) := \int_{-\infty}^{\infty} xg(x)\, dx, \tag{9.28}$$

where g is the probability density function of X. This is exactly the formula obtained for the expected value of a uniform random variable. As might be expected, these definitions are equivalent—the same number $\mathbb{E}(X)$ is produced by both definitions. The idea of the proof of this fact is instructive; the proof requires some measure theory. To see why the statement should be true, start with the probability density function property (3) and note that

$$P(X^{-1}(A)) = P(X \in A) = \int_A g(x)\, dx.$$

Assume the probability P is itself given by the probability density function ρ so that

$$P(B) = \int_B \rho(x)\,dx.$$

On a set A where the random variable X is defined, has range B, is continuously differentiable, and invertible, the change of variables formula states that

$$\int_B g(x)\,dx = P(X^{-1}(B)) = \int_{X^{-1}(B)} \rho(x)\,dx = \int_B \rho(X^{-1}(x))(X^{-1})'(x)\,dx.$$

Because this holds for all such sets B, we should have

$$g(x) = \rho(X^{-1}(x))(X^{-1})'(x).$$

For $X : \mathbb{R} \to \mathbb{R}$, we have $X^{-1}(\mathbb{R}) = \mathbb{R}$. So,

$$\begin{aligned}\mathbb{E}(X) &= \int_{\mathbb{R}} X(x)\rho(x)\,dx \\ &= \int_{X^{-1}(\mathbb{R})} X(x)\rho(x)\,dx \\ &= \int_{\mathbb{R}} X(X^{-1}(x))\rho(X^{-1}(x))(X^{-1})'(x)\,dx \\ &= \int_{\mathbb{R}} xg(x)\,dx.\end{aligned}$$

This last result points out that working with the sample space and its probability distribution is essentially the same as working with the image of the random variable (which is in the real line) and its probability density on the line. For this reason, in practical probability theory where the sample space is itself identified as a subset of the real line and x denotes position, the usual practice is to view all random variables as the inclusion function $X(x) = x$ but with different probability densities on the real line. For example, consider the experiment of determining the speed (measured in feet per second) of a particle moving moving through a thin tube. The outcome of the experiment is a positive real number (which defines the sample space of all positive real numbers). The obvious random variable is simply $X(x) = x$, where the image is considered on the real line. This variable X has a certain probability density on the real line, which is likely to be the unknown that the experiment is designed to determine. In this scenario, the usual practice is to ignore formality and simply view X as the position variable along the real line with a (as yet undetermined) probability density on the real line. In case the speed were measured in meters per second, the random variable Y is still the position variable along the real line ($Y(x) = x$), but its probability density function is different. After a while, the formally correct notion of viewing the random variable as a map from the sample space to the real numbers is replaced by simply working with the probability density function of the random variable.

The initial data (for the random walk) is concentrated at the origin $x = 0$ at time $t = 0$ because the random walker starts at this position. In the limit, the probability distribution at time $t = 0$ should be concentrated at $x = 0$. As the integral of the probability over the whole sample space must be unity [Eq. (9.5)], the initial probability distribution would have to have the property that it is zero everywhere on the real line except at zero and

$$\int_{-\infty}^{\infty} u(x,0)\,dx = 1.$$

No function has this property. But, viewed as the limit of functions defined over decreasing short intervals centered at the origin all with integral equal to unity over the line, the limit is represented by the Dirac delta function. With this initial condition, the corresponding solution of PDE (9.25) exists and is unique; it is the prediction of the model for the probability of finding the walker at position x at time t.

The solution of the initial value problem for PDE (9.25) is given by

$$u(x,t) = \frac{1}{\sqrt{2\pi\kappa t}} \exp\left(\frac{-(x-\alpha t)^2}{2\kappa t}\right). \tag{9.29}$$

This formula is derived in Exercise 23.5. For each $t > 0$, the density $\xi \to u(\xi, t)$ is called normal (or Gaussian). The expected value of X_t is αt with variance κt. At $t = 1$, the unit used to define $n\tau = 1$. Of course the expected value of x is α and its variance is κ. This is the reason for the factor $\kappa/2$ in the Fokker–Planck PDE: with this coefficient, the solution has variance κ as a probability density. In this context, for a general random variable X with probability density given by the right-hand side of Eq. (9.29) with $t = 1$, we say it has a *normal (or Gaussian) distribution and call it a normal (or Gaussian) random variable*. The graph of each normal distribution is the famous bell-shaped curve, the icon of probability and statistics (depicted as the continuous curve in Fig. 9.1).

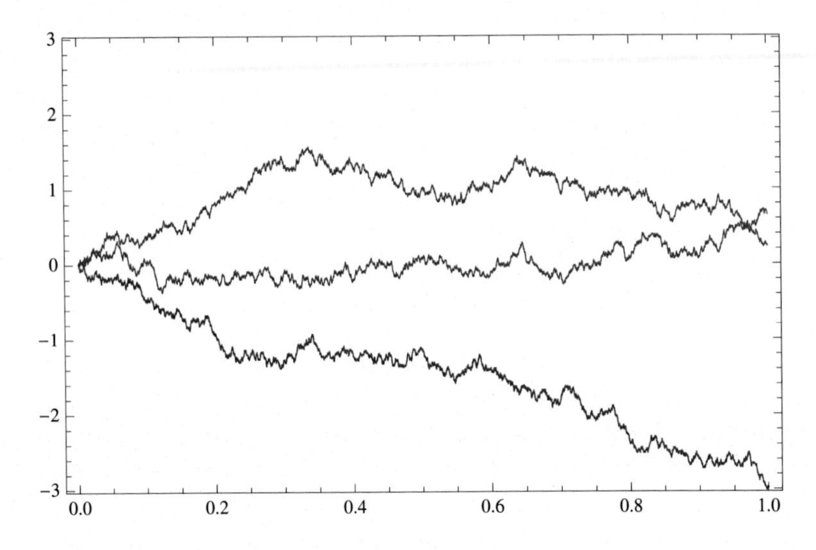

Fig. 9.2 Three random walks are depicted that approximate the Browning motion in the continuum limit. The depicted random walks are for lattice spacing $\ell = 10^{-2}$ and length scale $\tau = 10^{-4}$ random steps. The vertical axis represents the line on which the random walk takes place: the horizontal line is the time with 10^4 steps per one unit of time.

Inspection of the solution [Eq. (9.29)] reveals that the probability of finding the random walker is largest for events containing the point $x = \alpha t$ and goes to zero rapidly for events that are bounded away from zero as their lower bounds increase to infinity or their upper bounds decrease to minus infinity. This is exactly what is expected. The PDE $u_t = \alpha u_x$ contributes the drift of the maximum probability events away from $x = 0$ while the PDE $u_t = \frac{\kappa}{2} u_{xx}$ contributes diffusion around the maximum probability.

The position of a random walker remains uncertain (by definition) because the process is assumed to be random, but the probability density of the corresponding random variable (in the limit described in this section) is given by a deterministic process, which is governed by the Fokker–Planck PDE.

Random walkers can of course be simulated by taking small grid spacing ℓ and time step τ and using a random number generator instead of a coin flip. Fig. 9.2 depicts three random walks all starting at the origin for the same length and timescales. These approximate the limit behavior of the random walk as ℓ and τ go to zero in the manner discussed in this section. This is the mathematical model of Brownian motion. The paths are all continuous and nowhere differentiable. They have many other fascinating properties that are special cases of random processes called Wiener processes (Norbert Wiener

1894–1964, who was born at a location that is now part of the campus of the University of Missouri near the current location of the Tiger Fountain).

For completeness, one more result must be mentioned: If $\{X_i\}_{i=1}^{\infty}$ is a sequence of identically distributed independent random variables each with expected value μ and variance σ^2, then the probability density functions of the random variables $(X_1 + X_2 + X_3 + \cdots + X_n)/\sqrt{n}$ converge to a Gaussian with expected value zero and variance σ^2. This second pillar of probability theory is called the *central limit theorem.* It ensures the prevalence of the Gaussian distribution when dealing with a large number of *independent* random variables. Note also that the averages in the law of large numbers (the sum divided by n) are associated with the random variables in the central limit theorem via

$$\frac{n}{\sqrt{n}}\frac{X_1 + X_2 + X_3 + \cdots + X_n}{n} = \frac{X_1 + X_2 + X_3 + \cdots + X_n}{\sqrt{n}}.$$

Thus, by the central limit theorem, these averages are well approximated by a random variable whose density function is a Gaussian with expected value zero and variance σ^2/n. In particular, the associated density functions of the averages approach a Gaussian no matter what the density function of all the summands. This powerful result has many applications.

Exercise 9.3. Show by substitution that the formula for u given in Eq. (9.13) satisfies difference equation (9.23).

Exercise 9.4. Show that the random variables f^i defined in Eq. (9.12) are independent and use the linearity of the variance for independent random variables to show the variance of f_n agrees with Eq. (9.20).

Exercise 9.5. Show that condition (3) on page 259 implies condition (2).

Exercise 9.6. (a) Suppose the probability measure on the real line is the uniform probability measure on the set $[0, 1]$ with density given by Eq. (9.27). Compute the probability density function, with respect to the uniform density, for the random variable $X : \mathbb{R} \to \mathbb{R}$ given by $X(x) = x^2$. (b) Repeat the exercise for an arbitrary interval $[a, b]$.

Exercise 9.7. Suppose $\kappa/2$ and α are given positive numbers. Construct functions ℓ, τ, p, and q all depending on the same variable so that in the limit as this variable approaches infinity $\ell \to 0$, $\tau \to 0$, $p \to 1/2$, $q \to 1/2$, $\ell^2/\tau \to \kappa/2$, and $(p - q)\ell/\tau \to \alpha$.

Exercise 9.8. (a) Make a simulation for the random walk with the distance of each step fixed at 0.1 and equal probabilities of changing direction. Let the random walker take 1000 steps. According to Eq. (9.19), the square of the distance from the origin of the random walker (at the end of the walk) is expected to be 10. Gather experimental evidence from your simulation that this average distance from the origin after many trials (random walks) approaches the stated value. Discuss the expected value and the averages using the law of large numbers. (b) You might expect that the random walker's distance from the origin in part (a) is $\ell\sqrt{n}$. Gather and report on evidence that this is not true. (c) Explain why the expected value for the distance is not necessarily the square root of the expected value for the square of the distance. (d) Gather evidence (by considering different numbers of steps for the random walk) that the expected value is indeed proportional to $\sqrt{n}$ and determine the constant of proportionality. Hint: You can write the constant of proportionality (which is an irrational number) in the form $2^\alpha\pi^\beta$, where α and β are real numbers.

Exercise 9.9. In the context of Fig. 9.2, make a simulation to show that the end point of the random walker at time $t = 1$ is distributed on the vertical axis according to a normal distribution. Which normal distribution?

Exercise 9.10. (a) Suppose the probability measure on the real line is the uniform probability measure on the set $[0, 1]$ with density ρ defined in Eq. (9.27). The random variable $X(x) = x$ has probability density ρ. There are many continuous random variables with the same probability density. For example, change the value of X at a finite number of points. Construct a continuous random variable Y that is a *continuous function* on $\mathbb{R}$, has the same probability density, and is such that there is a point x in every open interval (a, b) contained in $[0, 1]$ with $X(x) \neq Y(x)$. Note: Such random variables are called identically distributed; that is, they have the same probability density functions.

Exercise 9.11. (a) What real number corresponds to the binary $.11111\cdots$? (b) What real number corresponds to the binary $.101010\cdots$? (c) Determine the binary representation of the number 1/3. (d) An interesting function ϕ on the set of binary decimals, called the shift map, is defined by

$$\phi(.s_1s_2s_3\cdots) = .s_2s_3s_4\cdots ;$$

that is, the map shifts the decimal one place to the right and drops the binary digit before the decimal point. Show that this map has periodic orbits of all periods. For example find a binary decimal s so that $\phi(s) = t$ and $\phi(t) = s$, which would be a periodic point of period 2. This function corresponds to a map of the interval into itself given by $h(x) = 2x$ if $x \leq 1/2$ and $h(x) = 2 - 2x$ if $x > 1/2$. Write a simple computer code to iterate h. Start, for example, with the point 1/3, iterate some finite number of times (say 50 times), and plot the result. Does the plot have any interesting features? The map is called the tent map and its properties are discussed in most books on dynamical systems (for example, [106]).

Exercise 9.12. Consider the following experiment akin to a coin toss described in three steps: (1) Pick a number in the interval $[0, 1]$ by some deterministic process; for example, use a clock (in a computer) to find the current time in hours, minutes, and seconds since the previous midnight, record the number of seconds s, and use $a = s/60$. (2) Define the function $f : [0, 1] \to [0, 1]$ by $f(x) = 4x(1 - x)$ and iterate 100 times starting with $x = a$ (that is, compute $f(a)$, $f(f(a))$, and so on). (3) Define the outcome of the experiment to be 0 if the final iterate is in the interval $[0, 1/2]$ and 1 otherwise.

Compare this experiment with using a random number generator (or flipping a coin) to generate 0 or 1 randomly. When simulating the iteration process using a computer to make several trials of the experiment, the computer may be so fast that the seed value a is not changed when it is passed to the function f. To avoid this problem iterate $100 + i$ times where i is the trial number. This is similar to flipping a fair coin. The number of revolutions of the coin must change with each flip to avoid having the coin always come up heads! A skilled human could bias the coin by controlling the number of revolutions before catching the coin. As a test, consider a run of some number of trials of the experiment, say 50. For each run, assign the value of the random variable that counts the number of times the outcome is 1. Approximate the expected value and the variance of this random variable by conducting many runs (say 100). Compare the results of your simulations with similar tests using a random number generator (or a coin) instead of the function iterations. Discuss your results.

Exercise 9.13. Suppose that X and Y are continuous random variables with probability distribution functions G and H. Show that X and Y are independent if and only if their joint probability distribution function K defined by

$$K(t) = P(t : X(t) \leq t \text{ and } Y(t) \leq t) = P(X \leq t, Y \leq t) = G(t)H(t).$$

Thus, this property may be taken (as it often is) as the definition of independent random variables.

9.4 RANDOM WALK GENERALIZATIONS AND APPLICATIONS

The one-dimensional random walk is fundamental. It can be generalized to several dimensions and more general random walks, and the generalizations can be used to model many physical processes.

Key ingredients for modeling are appropriate generalizations of the master equation [Eq. (9.15)] for the one-dimensional classical random walk repeated here for the convenience of the reader:

$$P_{n+1}(m\ell) = P_n(m\ell - \ell)q + P_n(m\ell + \ell)p.$$

This master equation will be recast into a form that can be easily generalized.

Recall that, for passage to a continuum limit, the probability P_n can be viewed as the probability distribution on the one-dimensional lattice (of possible positions of the random walker) for the random variable f_n given by $f_n(m\ell) = m\ell$ on the lattice. Or, in other words, $P_n(m\ell)$ is the probability that, at step n, the random walker is at the lattice point $m\ell$. For the one-dimensional case considered so far, p and q are the probabilities of moving left or right one position on the lattice. Consider a new random variable defined at each step, which in the spirit of modern probability theory is named $\Delta \mathrm{f}_n$ to connote displacement or increment. It is given by $\Delta \mathrm{f}_n(m\ell) = m\ell$ on the lattice with probability distribution $\Delta \mathrm{P}_n$ defined on the lattice by

$$\Delta \mathrm{P}_n(m\ell) = \begin{cases} p, & m = 1; \\ q, & m = -1; \\ 0, & \text{otherwise.} \end{cases} \tag{9.30}$$

The probability distribution $\Delta \mathrm{P}_n$ does not depend on n; it depends only on the ordering of the lattice points. This fact is important as it ensures that the events corresponding to the increments at each step of the process are independent. (A process where this distribution does depend on n might be interesting, but is beyond this discussion.) Thus, for the rest of the discussion, the subscript is eliminated with the understanding that $\Delta \mathrm{P}$ does not depend on n. Using this new probability distribution, the master equation can be rewritten as a sum over all lattice points:

$$P_{n+1}(m\ell) = \sum_{k\ell} P_n(k\ell)\, \Delta \mathrm{P}(m\ell - k\ell). \tag{9.31}$$

Here, as usual, $P_n(k\ell)$ is shorthand for $P_n(f_n = k\ell)$ or for the probability of the event that the random variable is equal to the lattice point $k\ell$. Likewise, $m\ell - k\ell$ is a lattice point obtained by subtracting two numbers, and $\Delta \mathrm{P}(m\ell - k\ell)$ is the probability of the event that the random variable $\Delta \mathrm{f}_n$ has value equal to the lattice point $m\ell - k\ell$. This latter probability vanishes except when $m - k = \pm 1$; therefore, Eq. (9.31) makes the same statement as the original master equation.

The right-hand side of Eq. (9.31) is a discrete convolution. Using the change of variables $j\ell = m\ell - k\ell$, the equation can be recast in the form

$$P_{n+1}(m\ell) = \sum_{j\ell} P_n(m\ell - j\ell)\, \Delta \mathrm{P}(j\ell), \tag{9.32}$$

which is more convenient for analysis.

To generalize, note that by simply redefining P_n and $\Delta \mathrm{P}$ to be other probability distributions for position and displacement random variables on an arbitrary lattice, perhaps one in two- or three-dimensional space, the same form of the master equation [Eq. (9.32)] with an appropriate change in the indices to denote points on the new lattice remains valid: the probability of being at a point in the lattice at step $n+1$ is the sum over all points in the lattice of the products of the probabilities of being at a point in the lattice and the probability of the displacement from the current point to the given lattice point. The increment might allow a nonzero probability for points anywhere on the lattice to move to a given point. Of course, the more general random walks in one dimension can be treated; perhaps the walker is allowed to move two places or one place according to some probability distribution.

Assume that the lattice under consideration lies in the d-dimensional space $\mathbb{R}^d$ and a typical point z in the lattice $\mathcal{L}$ has Cartesian coordinates

$$z = (i_1\ell_1, i_2\ell_2, i_3\ell_3, \ldots, i_n\ell_d),$$

where the i_j are integers and ℓ_j are positive real numbers. In this case, the master equation is

$$P_{n+1}(z) = \sum_{\zeta \in \mathcal{L}} P_n(z-\zeta)\, \Delta\mathrm{P}(\zeta). \tag{9.33}$$

Define the density $z \to u(z, \tau)$ on the lattice at step n by

$$u(z, n\tau) = \frac{P_n(z)}{\ell_1\ell_2\ell_3\cdots\ell_d}. \tag{9.34}$$

In one-dimensional space (the line), each pair of adjacent lattice points bounds an interval, in two dimensions the lattice is associated with a grid of rectangles, in three dimensions with a grid of boxes, and so on. For simplicity, agree to call these intervals, rectangles, and boxes by the same name: boxes. Choose a consistent way to assign a box to each grid point; for example, given a grid point one possibility would be to move the origin of the coordinate system to this grid point by a translation of $\mathbb{R}^d$ and assign the box at this grid point that has all positive coordinates and corner at the origin. Use this choice to extend $z \to u(z, n\tau)$ to a piecewise constant function by assigning to every point in the box associated with the grid point z the value $u(z, n\tau)$. The extended function is piecewise constant over the entire space containing the grid.

The master equation in this generality does not lend itself (without further assumptions about the probability densities associated with the underlying random walk) to the algebraic manipulations that produced a difference equation including approximations of partial derivatives, which led to the Fokker–Planck PDE. In general it is not clear how to take the same approach.

A new idea is to treat the increment probability density the same way as the probability density $z \to u(z, \tau)$ by defining

$$\rho(z) = \frac{\Delta \mathrm{P}(z)}{\ell_1 \ell_2 \ell_3 \cdots \ell_d}. \tag{9.35}$$

The extended densities satisfy the equation

$$u(x, n\tau + \tau) = \sum_{\zeta \in \mathcal{L}} u(x - \zeta, n\tau)\rho(\zeta)\ell_1 \ell_2 \ell_3 \cdots \ell_d \tag{9.36}$$

for every $x \in \mathbb{R}^d$. And, because the functions u and ρ are piecewise constant, the sum can be replaced by an integral; in fact,

$$u(x, (n+1)\tau) = \int_{\mathbb{R}^d} u(x - \xi, n\tau)\rho(\xi)\, d\xi. \tag{9.37}$$

For large n the piecewise constant functions u and ρ should be closely approximated by smooth functions with as many continuous partial derivatives as desired. Warning: This is not a rigorous statement. But, it is true for most situations that are encountered in applications. To make a rigorous statement, the probability distributions P_n and $\Delta \mathrm{P}$ must be known. Assume that u and ρ can be replaced by smooth functions so that (up to an acceptable approximation)

$$u(x, n\tau + \tau) = \int_{\mathbb{R}^d} u(x - \xi, n\tau)\rho(\xi)\, d\xi. \tag{9.38}$$

To determine the Fokker–Planck PDE, the idea is to expand $\xi \mapsto u(x - \xi, n\tau)$ in a Taylor series at $\xi = 0$ to obtain the partial derivatives that will appear in the PDE. For this approach to succeed (without too much analysis), only the first few terms of the series would be retained. In fact, the plan is to expand to second order and ignore the contributions of terms with order at least $O(|\xi|^3)$. The validity of such an approximation requires $|\xi|$ to be small. In general, this is certainly not the case; indeed, the integration in the master equation [Eq. (9.38)] is over all of space where $|\xi|$ takes on values that are certainly not small. To go further, some assumption must be made to justify the desired approximation. The simplest idea is to impose a restriction on the probability distribution of the increment $\Delta \mathrm{P}$; that is, some bound is placed on how far a walker can move in the lattice. For the classical one-dimensional random walk, this quantity [Eq. (9.30)] vanishes outside the closed ball (closed interval in one dimension) of radius ℓ centered at the origin of the lattice. For the more general case, it suffices to suppose there is some positive integer ω such that $\Delta \mathrm{P}(m\ell)$ vanishes whenever $|m| > \omega$. This restriction imposes the length scale $\omega\ell$ on the limit process.

The multivariable Taylor series is a direct generalization of the one-dimensional case:

$$u(x-\xi, n\tau) = u(x, n\tau) - \nabla u(x, n\tau)\xi + \frac{1}{2}\xi^T \,\mathrm{Hess}\, u(x, n\tau)\xi + O(|\xi|^3),$$

where Hess (the Hessian) is the matrix of second partial derivatives obtained by differentiating the vector valued function $x \to \nabla u(x, n\tau)$ and T denotes the transpose. All terms in the Taylor series are scalars. By substitution into the master equation [Eq. (9.37)] and some simple reductions, we have that

$$\begin{aligned} u(x,(n+1)\tau) &= \int_{\mathbb{R}^d} [(u(x,n\tau) - \nabla u(x,n\tau)\xi \\ &\quad + \frac{1}{2}\xi^T \,\mathrm{Hess}\, u(x,n\tau)\xi)\rho(\xi)]\, d\xi + O(|\xi|^3) \\ &= u(x,n\tau)\int_{\mathbb{R}^d} \rho(\xi)\, d\xi - \nabla u(x,n\tau)\int_{\mathbb{R}^d} \xi\rho(\xi)\, d\xi \\ &\quad + \int_{\mathbb{R}^d} \frac{1}{2}\xi^T \,\mathrm{Hess}\, u(x,n\tau)\xi\rho(\xi)\, d\xi + O(|\xi|^3) \\ &= u(x,n\tau) - \nabla u(x,n\tau)\int_{\mathbb{R}^d} \xi\rho(\xi)\, d\xi \\ &\quad + \frac{1}{2}\sum_{i,j=1}^{d} \int_{\mathbb{R}^d} \xi_i\xi_j\rho(\xi)\, d\xi\, u_{x_i x_j}(x,n\tau) + O(|\xi|^3). \end{aligned} \tag{9.39}$$

Here, because ρ is a probability density,

$$\int_{\mathbb{R}^d} \rho(\xi)\, d\xi = 1.$$

The integrals

$$\int_{\mathbb{R}^d} \xi_i\rho(\xi)\, d\xi, \qquad \int_{\mathbb{R}^d} \xi_i\xi_j\rho(\xi)\, d\xi,$$

called first and second *moments* of the probability density ρ, come equipped with the length scale imposed by the restriction on the increment density. This arises because the integrals in the formula over all of $\mathbb{R}^d$ can be replaced with integrals over the ball $B_{\omega\ell}$ centered at the origin of radius $\omega\ell$; for example,

$$\int_{\mathbb{R}^d} \xi_i \rho(\xi)\, d\xi = \int_{B_{\omega\ell}} \xi \rho(\xi)\, d\xi.$$

Rearrange Eq. (9.39) and divide by the time increment τ to obtain the new equation

$$\frac{u(x, n\tau+\tau) - u(x, n\tau)}{\tau} = -\frac{1}{\tau}\nabla u(x, n\tau) \int_{\mathbb{R}^d} \xi \rho(\xi)\, d\xi$$
$$+ \frac{1}{2\tau} \sum_{i,j=1}^{d} \int_{\mathbb{R}^d} \xi_i \xi_j \rho(\xi)\, d\xi\, u_{x_i x_j}(x, n\tau) + O(|\xi|^3).$$

Without going further into details, appropriate choices of timescale τ and length scale $\omega\ell$, allow passage to the limit as before to obtain the Fokker–Planck PDE

$$u_t = -\alpha \nabla u + \frac{1}{2}\mathcal{D}u \tag{9.40}$$

where α is a vector of constants and $\mathcal{D}$ is the second-order differential operator given by the sum of second moments and partial derivatives. Here, the unknown probability density depends on the vector variable x and time t, the ∇u term contributes drift, and the differential operator (which is similar to the Laplace operator as they both belong to the well-studied class of elliptic operators) contributes diffusion. In specific applications where further assumptions are made, the second moment coefficients of the operator $\mathcal{D}$ are known and the PDE can be solved in all of $\mathbb{R}^d$ with appropriate initial data.

A word is in order about boundary data. This topic is not covered here to keep the discussion short, but boundary data arises naturally from random walk problems where the random walker is, for example, absorbed by a wall or reflected from a wall. Of course, this is an excellent topic for further exploration.

The classic application that introduced the random walk and probabilistic modeling (especially in the life sciences) is the question Karl Pearson

(1857–1936) asked during 1905 : “A man starts from a point O and walks l yards in a straight line; he then turns through any angle whatever and walks another l yards in a second straight line. He repeats this process n times. I require the probability that after these n stretches he is at a distance between r and $r + dr$ from his starting point, O." An approximate answer given by Lord Rayleigh (J. W. Strutt, 1842–1919) in the same year is “If n be very great, the probability sought is

$$\frac{2}{n} e^{-r^2/n} r dr.\text{”}$$

Both brief communications are published in *Nature* July 1905 (p. 294 and p. 318, respectively). Rayleigh’s answer will be obtained from the Fokker–Planck model [Eq. (9.40)]. Pearson was thinking about the spread of insect infestations in forests, but he wisely framed his question to avoid extraneous context.

In Pearson’s problem, the random walk takes place in two dimensions. Consider a two-dimensional rectangular lattice whose nodes have coordinates $(i\ell, j\ell)$ for integers i and j and box side length ℓ. The random walker moves l yards during each step. By taking ℓ much smaller than l, a new random walker confined to horizontal and vertical moves on the two-dimensional lattice will reach nodes that approximate every point (corresponding to “any angle whatsoever") on the circle centered at the origin with radius l. In other words, the new random walker approximates Pearson’s walker when the new random walker takes enough steps to reach points close to the circle (say less than or equal to a distance ℓ from the circle). By moving the origin to one of these points, the process can be continued. This approach is of course compatible with the derivation of the Fokker–Planck PDE [Eq. (9.40)].

To obtain the Fokker–Planck PDE for a two-dimensional random walk designed to model the Pearson problem (or any other applied problem of this type), the main issue is choosing an appropriate probability density ρ for one step. In Pearson’s problem, every direction is equally likely. Starting at the origin, there are four available nodes on the lattice with spacing ℓ, each equally likely to be reached in one step by the new random walker. The limiting probability density for the increment is approximated by the piecewise constant function, again named ρ, that is constant on the closed disk of radius ℓ with value $1/(\pi\ell^2)$, and zero in the complement of the disk.

In more compact form,

$$\rho(x, y) := \begin{cases} \frac{1}{\pi \ell^2}, & x^2 + y^2 \leq \ell^2; \\ 0, & x^2 + y^2 > \ell^2. \end{cases} \tag{9.41}$$

The constant value of ρ makes all four directions equally likely and the integral of ρ over all of $\mathbb{R}^2$ is equal to one, as it must be for a probability density.

Using the definition of ρ, an easy computation using a change to polar coordinates (see Exercise 9.14) can be used to show that

$$\int_{\mathbb{R}^2} x\rho(x, y)\, dxdy = \int_{\mathbb{R}^2} y\rho(x, y)\, dxdy = \int_{\mathbb{R}^2} xy\rho(x, y)\, dxdy = 0$$

and

$$\int_{\mathbb{R}^2} x^2 \rho(x, y)\, dxdy = \int_{\mathbb{R}^2} y^2 \rho(x, y)\, dxdy = \frac{\ell^2}{4}.$$

All first moments vanish—the expected value for the position of the random walker is the origin. The cross second moments vanish. The second moments in the directions of the grid are equal. In particular, the expected value of the square of the distance of the walker from the origin after one step is

$$\int_{\mathbb{R}^2} (x^2 + y^2)\rho(x, y)\, dxdy = \frac{\ell^2}{2}.$$

Using these results, the Fokker–Planck equation reduces to

$$u_t = \frac{\ell^2}{8\tau}(u_{xx} + u_{yy}). \tag{9.42}$$

The theory tells us that in the limit process (as ℓ and τ approach zero) ℓ must be proportional to $\sqrt{\tau}$, but the constant of proportionality is left undetermined. It must be obtained from the physical process being modeled. In Pearson's walk, no timescale is given. Thus, it makes sense to choose $\tau = 1/k$, for some positive integer k that will increase without bound during the limit process as finer and finer meshes are considered. Using this choice, one Pearson step is modeled by k steps for the new walker on the lattice. One unit of physical time, whichever units are chosen, should be $1 = k\tau$, and n Pearson steps are completed at time $t = n$. Take

$$\ell = \sqrt{4\kappa\tau}$$

with κ to be determined. This choice ensures

$$\frac{\ell^2}{8\tau} = \frac{\kappa}{2},$$

so that κ corresponds to the variance of the one-dimensional normal probability distribution expected to be related to the solution of the new Fokker–Planck equation. But, this choice is not too important; the key result is to determine κ. The Fokker–Planck PDE takes the form

$$u_t = \frac{\kappa}{2}(u_{xx} + u_{yy}). \tag{9.43}$$

The solution of this equation for the initial distribution modeled by the delta function at the origin (where the walker is defined to start with probability one) is

$$u(x, y, t) = \frac{1}{2\pi\kappa t} e^{-(x^2+y^2)/(2\kappa t)}. \tag{9.44}$$

Note that the leading factor has denominator $2\pi\kappa t$, not $\sqrt{2\pi\kappa t}$. For three-dimensions, the corresponding PDE

$$u_t = \frac{\kappa}{2}(u_{xx} + u_{yy} + u_{zz})$$

has a solution of the same form but with denominator $(2\pi\kappa t)^{3/2}$; in d-dimensions the exponent is $d/2$. Also, there is special terminology for multidimensional normal distributions that is not used here.

What value should be assigned to κ? One answer to this question is to assign κ so that the expected value of the *square* of the position X_t^2 of the random walker in the continuum limit at time $t = 1$ is l^2, the distance from the origin after one Pearson step. The expected value is

$$\mathbb{E}(X_1^2) = \int_{\mathbb{R}^2} (x^2 + y^2)\, u(x, y, 1)\, dx dy = 2\kappa \tag{9.45}$$

and

$$\kappa = \frac{l^2}{2}.$$

By substituting this value for κ into Eq. (9.44) the desired probability density $(x, y) \mapsto u(x, y, t)$ is determined at time t. After n Pearson steps, this probability density is given by $(x, y) \mapsto u(x, y, n)$.

Pearson asked for the probability of being at "a distance between r and $r + dr$ from his starting point, O." To estimate the answer to his question, we must compute the probability of the event that X_n is in the annulus A with radii r and $r + dr$ and center at the origin. This assumes the increment dr is positive. We know the probability density. The probability of this event is

$$\int_A u(x, y, n)\, dxdy \approx \frac{1}{\kappa n} e^{-r^2/(2\kappa n)} = \frac{2}{l^2 n} e^{-r^2/(l^2 n)}. \tag{9.46}$$

The approximation is due to treating dr as an infinitesimal—as Pearson surely intended—by dropping terms of order dr^2. Rayleigh's result is obtained for Pearson length taken to be $l = 1$. This estimate is expected to be better "If n be very great." Why?

Exercise 9.14. A rotationally invariant function $\rho : \mathbb{R}^2 \to \mathbb{R}$ has the form $\rho(x, y) = f(x^2 + y^2)$, for some function $f : [0, \infty) \to \mathbb{R}$. Suppose that $\rho : \mathbb{R} \to \mathbb{R}$ is a continuous function with finite first and second moments. Show that

$$\int_{\mathbb{R}^2} x f(x^2 + y^2)\, dxdy = \int_{\mathbb{R}^2} y f(x^2 + y^2)\, dxdy = \int_{\mathbb{R}^2} xy f(x^2 + y^2)\, dxdy = 0$$

and

$$\int_{\mathbb{R}^2} x^2 f(x^2 + y^2)\, dxdy = \int_{\mathbb{R}^2} y^2 f(x^2 + y^2)\, dxdy = \pi \int_0^\infty r^2 f(r^2)\, dr.$$

Exercise 9.15. Make detailed computations to verify approximation (9.46)

Exercise 9.16. Simulate Pearson's random walk and make a comparison with the approximation (9.46). Discuss your methodology.

Exercise 9.17. Repeat the derivation of the approximation (9.46) for the case of three-dimensional space. The result would be relevant to modeling a dust particle moving in a room filled with air or a molecule of some substance moving in a bath of water due to collisions with water molecules.

CHAPTER 10

Problems and Projects: Concentration Gradients, Convection, Chemotaxis, Cruise Control, Constrained Control, Pearson's Random Walk, Molecular Dynamics, Pattern Formation

Exercise 10.1. [Concentration Gradients] (a) Make a mathematical model to describe the diffusion of a solute (call it's concentration u) in water through a permeable membrane idealized as a cross section of a tube (perhaps an idealized blood vessel) with one closed end. Suppose the tube radius is a, its length is L , and the membrane is placed at a distance pL (for some $0 < p < 1$) from the open end of the tube. The tube is filled with pure water and its open end is connected to a reservoir with a large supply of solution containing the mentioned solute with concentration c. This solute diffuses in the tube. The solute moves across the membrane according to a form of Fick's law: the solute flux across the membrane is proportional to the concentration difference across the membrane and in the direction from higher to lower concentration. Let k denote the diffusion constant for the solute in water and λ the diffusivity constant (the constant of proportionality across the membrane).
(b) Using the same geometry, make a model for the situation where the solute is secreted through the closed end of the tube and leaves the open end of the tube that is immersed in pure water. Assume that the flux across the closed end of the tube is known. (c) Choose numerical values for all the parameters in this problem and determine the corresponding steady state solute concentration in the tube as a function of position along the tube.

Exercise 10.2. [Chemotaxis] Let u denote the density of a population of bacteria, cells, insects, or other organism whose motion is influenced by the presence of a chemical in their environment with concentration c and recall the basic conservation equation [Eq. (5.1)]

$$u_t = -\operatorname{div}(X) + f,$$

where X is the diffusive flux of u and f is the amount of substance generated per volume per time, which is here taken to be zero. Our organisms are assumed to diffuse from higher concentration to lower concentration independent of the presence of the chemical. Use the usual constitutive law $X = -K \operatorname{grad} u$ to model this process and note that X is measured in mass per area per time. In the present case mass is essentially the number of organisms at a point (the units u times volume). The orientation of organisms with respect to the presence of the chemical (the chemotaxis) is modeled by modifying the

An Invitation to Applied Mathematics: Differential Equations, Modeling, and Computation.
http://dx.doi.org/10.1016/B978-0-12-804153-6.50010-5, Copyright © 2017 Elsevier Inc. All rights reserved.

flux term to read

$$X = -K \operatorname{grad} u + u\chi(c) \operatorname{grad}(c),$$

where χ is the chemotactic sensitivity measured in units of inverse chemical concentration times length per time. This assumption leads to the chemotaxis equation

$$u_t = \operatorname{div}(K \operatorname{grad} u - u\chi(c) \operatorname{grad}(c)).$$

In case the chemical concentration is not affected by the presence of the organisms, it may be specified. Or its concentration may be modeled by a reaction-diffusion equation

$$c_t = k\Delta c + g(c, u),$$

where g models the creation and consumption of the chemical in the environment. (a) Suppose that the organisms do not diffuse ($K = 0$), the chemical concentration in a two-dimensional environment is given by a function of the form $e^{-x^2-y^2}$, and the chemotactic sensitivity is constant (either $+1$ or -1). Discuss the change in concentration of the organisms with respect to time under the assumption that the organisms are initially uniformly distributed over the environment. (b) Repeat part (a) with diffusion ($K > 0$). (c) Suppose the organisms secrete the chemical at a rate inversely proportional to the presence of the chemical. Write the model under this assumption and discuss the effect on the concentration of organisms. Use a two-dimensional environment on a finite part of the plane and discuss your choice of boundary conditions.

Exercise 10.3. [Chemotaxis for Individuals and Agent-Based Modeling] Imagine a hypothetical insect that has three antennae: one on each side of its head and one pointing forward along the insect's axis of symmetry. The insect can sense the concentration of a chemical with each of its antennae. Also, the insect has short-term memory: It can remember the output of its sensors over some fixed period of time. The insect feeds on substances that excrete the chemical it can sense. (a) Design a controller (that would mimic the controller that would have been designed through evolution for the insect) so that the insect will follow a gradient of increasing concentration of the chemical to its source, where the food is likely to reside. Your model might be continuous or discrete and the controller might not be proportional–integral–deriviative (PID). (b) Test your control strategy by simulation using chemical trails of your own design. (c) Suppose there are many insects. Augment your control rules so that two insects cannot occupy the same space at the same time. Simulate the motion of several insects using your control strategy. Your model is likely to be an example of an agent-based model: a set of agents, an environment in which they reside, and a set of rules that determine how each agent behaves.

Exercise 10.4. [Cruise Control] A basic model for an automobile cruise control starts with a model of the motion of an automobile over a road. For simplicity, assume that the controller will be tested on a straight road over hilly terrain. Choose an inertial coordinate system—in this case, a coordinate system fixed to the Earth is a reasonable approximation—such that the positive horizontal axis is in the direction of the road and the vertical axis points away from the surface of the Earth. The road may then be idealized as the graph of a function f that gives the elevation y over each horizontal position x via $y = f(x)$. This reduces the problem to two dimensions. As usual, let the position of the automobile as a function of time t be denoted by $(x(t), y(t))$. There are several forces acting on the automobile: the forward force provided by the automobile's engine, gravity, aerodynamic drag, rolling resistance due to deformation of the tires, and the force that keeps the automobile from dropping through the road. Assume that the latter force F_N is everywhere normal to the road and lump the sum of all other forces except gravity into one force denoted by G. According to Newton's second law and assuming the mass m of the automobile does not change with time due to fuel consumption or other reasons) the equation of motion is

$$m\begin{pmatrix} \ddot{x} \\ \ddot{y} \end{pmatrix} = -mg\begin{pmatrix} 0 \\ 1 \end{pmatrix} + F_N + G. \tag{10.1}$$

The unit tangent to the road is the vector

$$T := \frac{1}{\sqrt{1 + (f'(x))^2}}\begin{pmatrix} 1 \\ f'(x) \end{pmatrix}.$$

By computing the inner product of both sides of differential equation (10.1) with T, the model is reduced to

$$\frac{m}{\sqrt{1 + (f'(x))^2}}(\ddot{x} + f'(x)\ddot{y}) = -mg\frac{f'(x)}{\sqrt{1 + (f'(x))^2}} + G \cdot T.$$

The system is reduced to one dimension by using the relation $y = f(x)$ to obtain

$$m\Big(\ddot{x}\sqrt{1 + (f'(x))^2} + \frac{f'(x)f''(x)}{\sqrt{1 + (f'(x))^2}}\Big) = -mg\frac{f'(x)}{\sqrt{1 + (f'(x))^2}} + G \cdot T. \tag{10.2}$$

Although the equation of motion [Eq. (10.2)] with G specified is a model of the automobile motion, it is not well suited to designing and simulating a cruise control in case the sensor determines the vehicle speed along the road; that is, with respect to arc length along the graph of f, which is given by

$$\ell := \int_0^x \sqrt{1 + (f'(\xi))^2}\, d\xi.$$

For this reason, it is advantageous to recast the model [Eq. (10.2)] with the dependent variable being speed along the road; that is,

$$s := \frac{d\ell}{dt} = \frac{d\ell}{dx}\frac{dx}{dt} = \dot{x}\sqrt{1 + (f'(x))^2}.$$

Note that

$$\dot{s} = \ddot{x}\sqrt{1+(f'(x))^2} + \frac{f'(x)f''(x)}{\sqrt{1+(f'(x))^2}}.$$

Thus, we may consider the model equation as the system

$$\begin{aligned} m\dot{s} &= -mg\frac{f'(x)}{\sqrt{1+(f'(x))^2}} + G\cdot T, \\ \dot{x} &= \frac{s}{\sqrt{1+(f'(x))^2}}. \end{aligned} \tag{10.3}$$

The drag force acts in the direction opposite to the motion and its magnitude may be approximated by

$$\frac{1}{2}\rho C_d A s^2,$$

where ρ is the density of the air, C_d is the dimensionless drag coefficient for the particular automobile under consideration, and A is the cross-sectional area presented by the projection of the automobile onto a plane perpendicular to the direction of its motion.

A good model for rolling resistance requires some knowledge of the physics of tires. A crude approximation is made by simply regarding this force as a resistance to the motion proportional to the speed of the vehicle; that is,

$$F_{rr} := C_{rr}s,$$

where $C_r r$ is a constant with units of mass per time.

The force due to the automobile engine may be modeled in several different ways. A simple approach requires the tire radius r and the angle θ (measured in radians) that the vector from axis of rotation to a point on the tread of a tire makes with the horizontal. In the ideal situation where there are no forces acting, the speed s of forward motion is given by $s = 2\pi r\dot{\theta}$. The automobile mass times its acceleration is

$$m\dot{s} = 2\pi r\ddot{\theta} = \frac{2\pi r m}{I}|\tau|,$$

where I is the moment of inertia of the wheel and τ is the torque on the wheel provided by the engine.

By inserting the forces into model system (10.3), the automobile motion model is

$$m\dot{s} = -mg\frac{f'(x)}{\sqrt{1+(f'(x))^2}} - \frac{1}{2}\rho C_d A s^2 - C_{rr}s + \frac{2\pi r m}{I}|\tau|,$$

$$\dot{x} = \frac{s}{\sqrt{1+(f'(x))^2}}. \tag{10.4}$$

The cruise control actuator may be taken to be a linkage to a variable valve that changes the amount of fuel flowing to the engine. In turn, the fuel supply to the engine determines the torque supplied to the wheel through the drive train of the automobile. The valve, fuel pump, fuel line, injection system, transmission, differential, and so on combine into a complicated mechanical system. Is it necessary to have a precise model of this system to design a useful cruise control? This question can only be answered by building and testing cruise control systems. The purpose of the model is preliminary design. It seems that a detailed model of the fuel supply to the engine and the drive train would lead to unnecessary complication. A reasonable way to proceed is to simply assume the torque ($|\tau|$) supplied to the wheel is proportional to the fuel supply. For example, we may assume that

$$|\tau| = \lambda Q,$$

where Q is the volumetric flow rate (dV/dt) of fuel to the engine. In this case, the constant of proportionality λ has units of mass per length per time. Perhaps the constant of proportionality would be easier to measure by experiments with the automobile if this coefficient were dimensionless. To achieve this we might redefine our relationship to be

$$|\tau| = \lambda \frac{m}{r} Q,$$

where m is the mass of the automobile and r is the tire radius so that the new λ is a dimensionless coefficient of proportionality between volumetric fuel consumption of the engine and the torque supplied to the wheels. The volumetric flow rate Q may be subjected to a PID control with desired (set point) speed s_d. In this application, the volumetric fuel consumption is limited by the design of the automobile engine. There are two positive rates α and ω such that

$$\alpha \le Q \le \omega.$$

The lower limit is positive to keep the engine running even when the speed exceeds the set point. Thus, the PID control must be composed with an appropriate function H to impose the saturation limits; that is, the controlled volumetric flow rate is

$$Q = H(k_P(s_d - s) + k_I \int_{t-a}^{t} (s_d - s(\sigma))\, d\sigma - k_D \dot{s}),$$

where $a \ge 0$ is the amount of time kept in the controller memory.

A closed loop control model is

$$m\dot{s} = -mg\frac{f'(x)}{\sqrt{1+(f'(x))^2}} - \frac{1}{2}\rho C_d A s^2 - C_{rr} s$$

$$+\frac{2\pi m^2}{I}H(k_P(s_d - s) + k_I\int_{t-a}^{t}(s_d - s(\sigma))\,d\sigma - k_D\dot{s}),$$

$$\dot{x} = \frac{s}{\sqrt{1+(f'(x))^2}}. \tag{10.5}$$

(a) Criticize the closed loop control model.
(b) Set all system parameters equal to unity. Suppose that the road elevation is given by $f(x) = 2\exp(-(x-3)^2)$ and the automobile starts at position $x = 0$ at time $t = 0$. Optimize the control parameters k_P, k_I, k_D, and a over the interval $0 \le x \le 5$.
(c) Using the parameters for part (b), define a typical road terrain f that might be used to test cruise control systems, defend your choice, and optimize the control parameters over a finite distance on your terrain. Define two new terrains (one of which might have some atypical feature) and test your optimized control system over the new terrains. Write a report on your findings.
(d) Consult the literature on automotive engineering for approximations of the system parameters for a particular automobile. Define a typical terrain f and optimize the control parameters for a set-point speed of 70 miles per hour over a finite distance on your terrain. Define two new terrains (which perhaps have some atypical but physically reasonable features) and test your optimized control system over the new terrains. Write a report on your findings.

Exercise 10.5. [Unicycle Control] A wheel (imagine a unicycle or a rolling disk) with control mechanisms can roll on a flat horizontal surface and rotate about the axis through its axle and the point where it contacts the surface. The wheel is not allowed to skid; that is, it is not allowed to move normal to its present heading (which is defined to be the unit vector in the direction of the projection of its velocity vector in the plane). Let (x, y) denote the coordinates in the plane and θ the angle between the heading of the wheel and the positive x axis. A model for the motion of the wheel is

$$\begin{aligned}
\dot{x} &= f(x, y, \theta, u(t))\cos\theta,\\
\dot{y} &= f(x, y, \theta, u(t))\sin\theta,\\
\dot{\theta} &= g(x, y, \theta, v(t)),
\end{aligned}$$

where u and v are control inputs; that is, functions of time to be specified. (a) Describe in detail how the proposed model reflects the description of the controlled motion. (b) Suppose that $f(x, y, \theta, u(t)) = u(t)$ and $g(x, y, \theta, v(t)) = v(t)$. Is it possible to specify u and v so that the wheel will move on a circular path about the origin of the coordinate system with radius r? What about an arbitrary smooth path?

Exercise 10.6. [Math Project: Iteration and Eigenvalues] Give a complete proof that the iteration scheme $z_{n+1} = Az_n + b$ for an $n \times n$ matrix A and an n vector b converges whenever every eigenvalue of A has absolute value less than unity. Hint: You may need to use the Jordan canonical form and at least one important result from the theory of matrix norms: Given an $n \times n$ matrix A, there is a norm on n-dimensional space such that the associated matrix norm of A is as close as desired to the spectral radius of A (that is, the absolute value of the largest eigenvalue of A).

Exercise 10.7. [Exact Solution of Pearson's Random Walk] Recall the formula [Eq. (9.46)] for an estimate of the probability of Pearson's random walker, who takes steps of length l, to be in the annulus centered at the origin with radii r and $r + dr$ after n steps in random directions:

$$\frac{2}{l^2 n} e^{-r^2/(l^2 n)}.$$

In 1906, J. C. Kluyver and later Rayleigh (1919) discovered exact formulas for Pearson's random walk and extended this to random walks where the distances of the steps can be taken to be different lengths and the walker can move in d dimensions (see [117, p. 419] for an early account). The basic results for the two-dimensional case uses Bessel functions of order zero and one to express the probability of the walker being at a distance less than $r > 0$ from the origin after n steps of varying step lengths l_i, for $i = 1, 2, 3, \ldots, n$,

$$P = r \int_0^\infty J_1(rs) \prod_{j=1}^{n} J_0(sl_i)\, ds.$$

The probability of being between distances r and R is $(R - r)dP/dr$. Give a short description of why Bessel functions are important. Compare Rayleigh's estimate to the exact two-dimensional solution of Pearson's problem with equal step distances, and include an exposition of the methodology used to obtain the exact result. Your discussion should also include appropriate simulations, which involve a numerical challenge of evaluating improper integrals of products of Bessel functions. A discussion including mathematical estimates to compare the exact and approximate solutions would be best. Why is the exact solution of this problem ignored in most of the probability literature in favor of estimates involving the normal distribution? A more ambitious project would be to include unequal probabilities of moving in different directions or some bias in the choice of direction in Pearson's original problem. Give good estimates of finding the random walker between r and $r + dr$. Is there an exact solution?

Exercise 10.8. [Stability of Numerical ODE Solvers] Consider the (vector) initial value problem

$$\dot{x} = f(x, t), \qquad x(t_0) = x_0$$

where all partial derivatives of f are continuously differentiable. The midpoint method (also called the leapfrog method) for approximating solutions is defined by

$$x_{n+1} = x_{n-1} + 2\Delta t f(x_n, t_n), \qquad t_{n+1} = t_n + \Delta t,$$

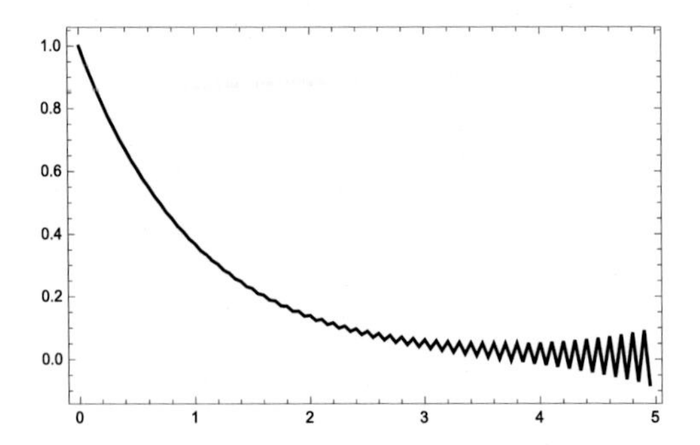

Fig. 10.1 Graph of numerical approximation to the solution of the initial value problem $\dot{x} = -x$ and $x(0) = 1$.

where Δt is the length of one time step.
(a) Derive the midpoint method. Hint: Write, as usual,

$$x(t + \Delta t) = x(t) + \int_t^{t+2\Delta t} f(x(s), s)\, ds$$

and approximate the integral by the value of the integrand at the midpoint of the time interval $t \le s \le t + 2\Delta t$.
(b) Show that the midpoint method is second order.
(c) Note that implementation of the method requires two values of the state variable x: x_0 and x_1. Thus the midpoint method is a multistep method where previously computed values are used to make the approximation at each new step. The standard way to start the midpoint method is to use Euler's method

$$x_1 = x_0 + \Delta t f(x_0, t_0).$$

Recall that Euler's method is a first-order method. Nonetheless, show that starting the midpoint method with Euler's method does not destroy its second-order accuracy. Suppose instead a small step size is chosen and x_1 is simply set equal to x_0 to start. Is second-order accuracy preserved? Hint: How is the order of a method defined?
(d) Write code that implements the midpoint method using Euler's method to start the multistep procedure and verify numerically that your code produces second-order accuracy to the exact solution for the test equation $\dot{x} = x$ with initial condition $x(0) = 1$.
(e) Use the code written in part (d) to approximate the solution of the initial value problem $\dot{x} = -x$ with initial condition $x(0) = 1$.
For a step size $\Delta t = 0.05$, your approximation should look like the graph in Fig. 10.1. Show with a pencil and paper argument that the oscillatory behavior is to be expected. Hint: Look for explicit solutions of the difference equation $x_{n+1} = x_{n-1} - 2\Delta t x_n$ of the form $x_n = \lambda^n$ for some (real or complex) number λ and write the general form of the solution of this difference equation as a linear combination of a pair of such solutions.
(f) There are at least three possible cures for the oscillatory behavior (numerical instability): (1) restart the midpoint method after every N steps, where N is an even integer (perhaps $N = 20$); (2) compute on some interval $[t_0, t_0 + T]$ (perhaps with

T some unit of time so that the desired approximation is at time $t_0 + KT$ for some integer K) with step size Δt, recompute with step size $\Delta t/2$, and use Richardson extrapolation to improve the approximation at time T before restarting the midpoint method; and (3) smooth the approximation before restarting from x_N by computing one additional time step, averaging over x_{N-1}, x_N, and x_{N+1} using (for example) $(x_{N-1} + 2x_N + x_{N+1})/4$, and restating with the averaged value as the initial state. Test these ideas on two or three differential equations with known solutions and discuss your results. The original source for this methodology is [44] (see also [100]).

(g) Imagine a modification of the limited growth model $u_t = u(1 - u)$, where u represents the concentration of some substance or life-form growing in a limited environment. Suppose the substance is distributed in a one-dimensional environment where growth depends on the total concentration in some neighborhood. A model of this scenario is given by

$$u_t = U(1 - U/K), \qquad U(x,t) := \frac{\lambda\sigma}{\sqrt{\pi}} \int_{-\infty}^{\infty} u(\xi,t)e^{-\sigma^2(\xi - x)^2}\, d\xi,$$

where λ has dimensions of the reciprocal of length times time, K has the same dimensions as U (that is, concentration per time), σ the dimension of reciprocal length, and a the dimension of length. The exponential factor makes the concentration at the field point x more significant than the surrounding concentrations. This model has constant steady state solutions at $u = 0$ and $u = K/\lambda$. Prove this fact. Question: Given an initial distribution of concentration on the line, say $u(x, 0) = f(x)$ for some specified function f with $0 \le f(x) \le K/\lambda$, does the concentration evolve to a steady state[1]? Explore this model seeking to gain insight into answering this question using numerical experiments that employ the midpoint method. Hint: The differential equation is an ordinary differential equation (ODE) in a function space: for each time t the state is a function $x \mapsto u(x,t)$. Make it dimensionless. To discretize, choose some finite interval (to approximate) the whole real line, split it into equal sized subintervals whose endpoints are called nodes and approximate functions by the vector of function values at these nodes in the same way that functions are approximated when discretizing a partial differential equation (PDE) such as the heat equation. Also, approximate the integral using, for example, the trapezoidal rule or Simpson's rule with respect to the nodes in the discretization. This leads to a system of coupled ODEs, one for each node, whose solutions may be approximated by the midpoint method. You should see why this method might be used: function evaluations are expensive and at least second-order accuracy is desirable.

Exercise 10.9. [Numerical Integration of Newton's Equation] The fundamental equation of motion (Newton's second law) reduces to the mathematical form $\ddot{x} = f(x,t)$ called Newton's equation. One method of integration is to turn the equation into a first-order system, say $\dot{x} = y$ and $\dot{y} = f(x,t)$ and use one of the numerical methods already described (Euler, improved Euler, trapezoidal, and so on). Another, more direct method, is to simply approximate the second derivative directly and make a numerical method

[1] At this writing, the author does not know the answer.

from this approximation. The basic example is to take the usual centered difference approximation

$$\frac{x(t+\Delta t)-2x(t)+x(t-\Delta t)}{\Delta t}=f(x(t),t)$$

and rearrange it keeping the forward step $x(t+\Delta t)$ on the left-hand side and moving the other terms to the right-hand side. (a) Show that this leads to the numerical method (sometimes called the Störmer-Verlet method)

$$t_{n+1}=t_n+\Delta t, \qquad x_{n+1}=2x_n-x_{n-1}+\Delta t f(x_n,t_n).$$

(b) To solve the initial value problem for Newton's equation requires the initial data $x(0)$ and $\dot{x}(0)$, two numbers. To start the suggested numerical scheme requires knowing x_0 and x_1. Clearly $x_0=x(0)$. What is the correct choice for x_1? Hint: Expand x at $t=0$ in a Taylor series.
(c) Write code to implement the method and use it to approximate the solution of $\ddot{x}+x=0$ with $x(0)=1$ and $\dot{x}(0)=0$. Check the accuracy of the method using the exact solution after integration on the interval $0\le t\le 2\pi$. What is the order or the method? Guess the order by numerical experiment and then prove your guess is correct.
(d) Compare the efficiency of the method against the improved Euler method or the trapezoidal method.
(e) Suppose that in a physical problem, the velocity of some particle is more important than its position. What is the best way to obtain the velocity from Störmer–Verlet integration. How does your method compare with obtaining the velocity from improved Euler or the trapezoidal method?
(f) Consider the interaction of three particles in space. Each of them produces a potential that affects the other particles. The equation of motion, according to Newton's second law and after some rearrangement, is

$$\ddot{x}^i=-\sum_{j\neq i}\nabla V_j(x^i), \quad i=1,2,3,$$

where V_j is the potential produced by the jth particle. The classic potential of Newtonian gravitation or Coulomb electrostatics has the form $V_j(x^i)=a_{ij}/|x^i-x^j|$ for $i\neq j$, where the constant a_{ij} measures the strength and direction of the interaction between the ith an jth particles. To focus on a specific problem, suppose that the initial state of the particles has positions

$$x^1=(1,0,0), \qquad x^2=(-1,0,0), \qquad x^3=(0,0,1)$$

and the velocities

$$v^1=(0,0.2,0), \qquad v^2=(0,-0.2,0), \qquad v^3=(0,0,c),$$

where c is a parameter. Moreover, suppose that $a_{i1}=a_{i2}=-0.15$ (that is, the first two particles attract each of the other particles with the same strength), $a_{i3}=0$ (that is, the third particle does not attract either of the first two particles), and $c=0$ (that is, the third particle starts from rest). In the gravitational interpretation, think of two

massive bodies and a third body of small mass. With this data, plot the motion of the particles over at least an interval 40 units of time. Describe the motions of the particles. To compare results with other experimenters, specify (with three correct decimal places) the positions and velocities of the particles after 10 units of time has elapsed. Is it possible to solve the equations of motion exactly for the special case?
(g) Experiment with different interaction strengths and initial velocities for the third particle. Report on your results. In particular, seek to determine what happens when the third particle is initially not at rest.
(h) What would the a_{ij} be for the Earth-Moon-Sun system? How about the initial data? Make a simulation and check against the known orbital periods and orbital geometries.
(i) Choose a different initial state for the three particles that might be physically relevant for gravity or electromagnetism. Determine the motion of the particles and report your results. Does the motion agree with your physical intuition?
(j) Experiment with different potentials. In particular, try the Lennard–Jones potential. Report on at least one case that you find physically relevant and interesting.

Exercise 10.10. [Molecular Dynamics I] The subject of molecular dynamics in the general sense refers to the internal motions of molecules and their motions as particles. In full generality, bonds between atoms in a molecule are considered along with van der Waal forces and other electromagnetic forces between molecules. This project concerns an idealized model for a gas: the molecules are hard spheres with no internal structure undergoing elastic collisions. Also, the model is reduced to motions in two-dimensional space.
(a) Consider the two-dimensional box with corners $(1, 1)$, $(-1, 1)$, $(-1, -1)$, and $(1, -1)$. Suppose that the box contains particles modeled by nonoverlapping disks all with the same radius r (for r much smaller than one) and the same mass m. No external forces act on the particles. They move according to Newton's laws with the additional assumption that collisions with the boundary of the box or between disks are perfectly elastic. Start with just one particle. Simulate its motion taking into account bounces off box walls according to the collision law: angle of incidence equals angle of reflection. Consider two fixed metric disks in the box both with the same radius. Does the particle spend the same amount of time in both of these sets?
(b) Upgrade your program to allow several disks undergoing elastic collisions among themselves and the box walls. Suppose all the disks are initially confined in the left half of the box. Gather evidence that the disks spread out so their density is the same everywhere. Also consider how long it takes for this state to occur as a function of their initial average velocity. Hint: As the number of particles increases so will the duration of your simulation. A parallel code will run (much) faster than a serial code. Programming based on determining the first collision time by considering collisions between particles while ignoring the box boundaries and taking into account collisions with the walls of the box separately can lead to a more efficient code than one that follows disks through wall collisions. Perfectly elastic collisions are discussed in elementary books on classical mechanics (for example, [62]). The basic principle is that kinetic energy and momentum are conserved.
(c) One additional disk with the same radius, called the dust particle, has mass much smaller than m. It is placed in the box at the position with coordinates $(0, 0)$ and with zero velocity. It also moves according to Newton's laws and perfectly elastic collisions.

Is the dust particle's motion well approximated by a two-dimensional random walk as the number of (nondust) particles is increased?
(d) Conduct the same basic experiment with different interactions. Possibilities for research are endless; for example, introduce various repulsive potentials between the particles or at the walls, remove the walls and consider the motion on the surface of a torus—a particle that penetrates a wall shows up on the opposite wall with the same velocity and relative position, or consider a three-dimensional box.

Exercise 10.11. [Molecular Dynamics II] The random walk is one way to view a physical process like Brownian motion; the pollen grain's erratic motions is modeled by a random walk. Such a model might reproduce the motion of the pollen grain, but the underlying modeling is not designed to explain the underlying physics of pollen grain motion. Think about the situation from a molecular perspective: The pollen grain is a ship in a sea of water molecules, which are much smaller and lighter than the grain and moving (on average) perhaps much more rapidly due to thermal forces. Remember that temperature is essentially a measure of the average speed of molecular motion. Newton's laws govern the motion of the pollen grain with mass m:

$$m\frac{dv}{dt} = F.$$

But what is the force? To do things correctly would require a system of such equations for the pollen grain and *all* the molecules in the water. The pollen grain motion would be determined after its initial position and velocity were specified along with the system of equations, and this system were solved. Of course, to determine such a model and analyze it is beyond current understanding. All the forces between the molecules and the impact forces with the pollen grain would have to be specified. What would we do with such an equation if we had it? One mole is about 10^{23} molecules, thus about this many equations. To gain some insight, a good idea is to use Newtonian forces derived from physical reasoning about the average behavior of the molecules. For example, we know from common experience and a lot of macroscopic modeling that an object moving through a fluid experiences viscous drag. More will be said about this force later in the book. An often used model for viscous drag is simply $-\gamma v$, where γ is some constant that could be determined by experiment, or with a bit more theory, it can be determined by the shape of the pollen grain. Using this force, the model equation for the motion of the pollen grain becomes

$$m\frac{dv}{dt} + \gamma v = 0,$$

which is essentially the ubiquitous exponential growth law. You know how to solve this equation; the solution is

$$v(t) = v(0)e^{-\gamma t}.$$

Also, the position of the pollen grain can be determined by substituting $\dot{x} = v$ and integrating one more time. Easy. But, this model cannot be viable: The velocity of the grain goes exponentially fast to zero, which is not what is observed. Molecules in the fluid are bombarding the pollen grain in all directions. What is this sum of these forces on the grain? In truth, we don't know. But, we might expect some random fluctuations due to the thermal properties of the fluid. To strengthen this weak statement requires some serious thermodynamics and statistical mechanics. But, accepting this rather vague statement, there should be some *random* force R (which is rapidly fluctuating relative to the speed of the pollen grain) that acts on the pollen grain. The addition of this force produces the simplest version of Langevin's model (Paul Langevin in 1908)

$$\begin{aligned} \frac{dx}{dt} &= v, \\ m\frac{dv}{dt} + \gamma v &= R(t). \end{aligned} \tag{10.6}$$

It is an important example of a stochastic differential equation. To be a bit more complete, there could be a body force (due, for example, to the presence of some other deterministic force such as the Lorentz force from an electric field). Under the further assumption that this force is conservative (the gradient of a potential U), Langevin's equation is damped Newtonian motion with a random force:

$$\begin{aligned} \frac{dx}{dt} &= v, \\ m\frac{dv}{dt} + \gamma v + U_x(x,t) &= R(t). \end{aligned} \tag{10.7}$$

In the presence of many pollen grains there would be a (perhaps coupled) system of Langevin equations, one for each grain. Exactly how to model the random variable $R(t)$ is not obvious. This force is to be chosen from physical intuition. Predictions of the resulting model are tested against experiments. When the model fits the experimental data, there is strong evidence for the correctness of the physical intuition that introduced a choice of the random force. For such a model to be consistent with thermodynamics, the expected value of R must be zero. This random variable must have some other properties (for example, no correlation between the random effects at distinct times), which might be the subject of further research for this project. In practice , the basic problem is to determine the statistical properties of the random state x or velocity v (the expected value, variance, and so on) from the known statistics of R. In general, these are nontrivial problems, which are well worth exploring beyond this introductory project. The subject of the properties of the solutions of Langevin equations and stochastic differential equations has an illustrious history that is still under development today. The theory and applications are important to current understanding of some basic physics, some biological processes, and finance.

(a) A random walk on the line is a (discrete) model of Brownian motion with the units of length per time. The stochastic term R should have the units of mass times length per time squared. In short, R should not be the random walk; rather, it should be the derivative of the random walk—whatever that means. One way to view the situation is to recall that the one-dimensional random walk is given by iterating the process $x_{n+1} = x_n + \ell g^{n+1}$, where g is the random variable akin to $\hat{f}$ in Eq. (9.11) with values ± 1 and g_n is the value of g on the nth trial. Or, if you like, g^n is simply the result of the nth coin flip. Using the recursion, prove that

$$x_n = \ell(g^1 + g^2 + g^3 + \cdots + g^n).$$

The discrete derivative should be a difference. This suggests using the left-hand side of the equality $x_n - x_{n-1} = \ell g^n$. Thus, the stochastic term should be proportional to the random variable g that has values ± 1. For definiteness, take $R(t) = \sigma g$ for a constant σ that determines the strength of the random force. This random variable has expected value zero. Gather evidence by numerical simulation that the probability density function for the random variable v at some fixed time $T > 0$ in the Langevin equation is a normal distribution with expected value zero. Hint: Don't take T too small.
(b) Prove: If X is a random variable with mean (= expected value) zero, then the expected value of v in the Langevin equation converges to a random variable with zero mean as $t \to \infty$. Hint: Compute the expected value of both sides of the equation and derive a differential equation for $\mathbb{E}(v)$.
(c) What can you say about the variance of v?
(d) Does the random variable x in the Langevin equation have a normal distribution?
(e) In some sense, the Langevin equation can be solved exactly. (Though by now you may be wondering if there is a theory that ensures it has a solution. After all it is not an ODE in the usual sense. The usual existence and uniqueness theory does not apply.) Working formally, treat the Langevin equation for velocity as an ODE and apply variation of constants to write a formula for v. A very good idea is to forget the original differential equation and consider the new integral equation (which you have just obtained) as the model equation. In other words, simply take the expression for v given by variation of constants as the solution of the Langevin equation. In cases where there is a nonlinear force, this trick will not solve the differential equation, but it can be used to replace the stochastic differential equation with an integral equation that is often more amenable to analysis. In even more generality, the stochastic differential equation

$$\dot{x} = f(x) + R(t)$$

may be considered to have a solution if x is a random variable that solves the integral equation

$$x(t) = x(0) + \int_0^t f(x(s))\,ds + \int_0^t R(s)\,ds.$$

To solve this equation, seek a random variable x, which when substituted into the right-hand side of the equation, produces an expression equal to itself. (You should be aware that some authors write this equation more briefly as

$$dx = f(x)dt + R(t)dt,$$

which might be called the differential form of the equation.) As a too simple example, consider the integral equation corresponding to $\dot{x} = -\gamma x$; substitute a guess for the solution with initial condition $x(0) = 1$ (for example, the function $x(t) = 1$); take the result and substitute it again, and so on. Does this sequence of functions produced by this process converge to a function that satisfies the integral equation? Hint: This is one

idea used to prove the general existence and uniqueness theorem for ODEs.
(f) The Langevin equation and the Fokker–Planck equations are obviously related. Discuss how they are related.
(g) The particular Langevin equation discussed here is closely related to the Ornstein–Uhlenbeck stochastic process. Find out what this means and discuss it in the context of this project.
(h) Find out what are reasonable values for m, γ, and σ for a pollen grain in still water. Simulate the motion of a pollen particle using Langevin's equation and report on your findings relative to the motion of real pollen grains. Perhaps look through a microscope yourself!
The subject of stochastic processes is vast. A mere glimpse is provided by the questions posed here. Thus, this project is open ended to say the least. A wise educational choice is to pursue this subject more deeply.

Exercise 10.12. [Pattern Formation] The method of lines (Exercise 5.72) is a framework for conducting numerical experiments that can provide a glimpse into pattern formation for the Gray–Scott model. Theoretically, the dynamics of a system of ODEs produced by spatial discretization may be considered to be a model for pattern formation. A basic question: What can be said about the dynamics of such a system of ODEs? One idea is to consider the ODE system as a perturbation problem where the unperturbed system is the reaction at each node and the discretized diffusion is the couplings between the reactions. The persistence or bifurcation from structures in the reaction equations might be used to account for some aspects of the pattern formation that is approximated by the system of ODEs. (a) Show that spatial pattern formation requires distinct diffusion coefficients. Hint: Use the system of PDEs and subtract the two equations. (b) Show that the coupled systems given by the method of lines have periodic solutions for sufficiently small diffusion coefficients. (c) What else can you say? Hint: There are many unanswered questions in this context. An early paper in this direction by Stephen Smale, "A mathematical model of two cells via Turing's equation" appears on pages 354–367 in [70].

CHAPTER 11

Equations of Fluid Motion

The mathematical description of the motion of fluids plays a fundamental role in applied mathematics. The basic model is a system of partial differential equations of evolution type derived from conservation of mass and Newton's second law of motion.

Fluids consist of molecules; thus, on a microscopic level, a fluid is a discrete material. To obtain a useful approximation, we will describe fluid motion on the macroscopic level by taking into account forces that act on a parcel of fluid, which we assume to be a collection of sufficiently many molecules of the fluid so that the *continuity assumption* is valid. More precisely, we will assume the fluid to be a continuous medium contained in three-dimensional space $\mathbb{R}^3$ such that every parcel of the fluid, no matter how small in comparison with the whole body of fluid, can be viewed as a continuous material.

Mathematically, a parcel of fluid (at every moment of time) is a bounded open subset A of the fluid that is assumed to be in an open set $\mathcal{R}$ whose closure is the region containing the fluid. To ensure correctness of the mathematical operations that follow, we assume that each parcel A has a C^1 boundary. At every given moment of time, a *particle* of fluid is identified with a *point* in $\mathcal{R}$.

Let $\rho = \rho(x,t)$ denote the density and $u = u(x,t)$ the velocity of the fluid at position $x \in \mathcal{R}$ and time $t \in \mathbb{R}$. The motion of the fluid is modeled by differential equations for ρ and u determined by the fluid's internal material properties, its container, and the external forces acting on the fluid.

The components of the velocity u with respect to the usual coordinates of three-dimensional space are denoted by (u_1, u_2, u_3). Moreover, the function u is assumed to be twice continuously differentiable and to satisfy other properties—which will be mentioned when needed—that are necessary for the correctness of the mathematical operations used in the following discussion.

An Invitation to Applied Mathematics: Differential Equations, Modeling, and Computation.
http://dx.doi.org/10.1016/B978-0-12-804153-6.50011-7, Copyright © 2017 Elsevier Inc. All rights reserved.

The changing position of the particle of fluid with initial position $x_0 \in \mathcal{R}$ at time $t = 0$ is given by the curve $t \mapsto \gamma(x_0, t)$ in $\mathcal{R}$ that is the solution of the initial value problem (IVP)

$$\dot{\xi} = u(\xi, t), \qquad \xi(0) = x_0 \tag{11.1}$$

(see Appendix A.3 for a theorem on existence of solutions of ordinary differential equations). When γ is viewed as a function of two variables $\gamma : \mathbb{R}^3 \times \mathbb{R} \to \mathbb{R}^3$ it is called the flow of u. For a subset A of $\mathbb{R}^3$ and a slight abuse of notation, let $\gamma(A, t)$ denote the set of all points in $\mathbb{R}^3$ obtained by solving to time t the differential equation with initial condition $\xi(0) = x_0$ for x_0 in A.

Let x_1, x_2, and x_3 denote the Cartesian coordinates in $\mathbb{R}^3$ and e_1, e_2, e_3 the usual unit direction vectors. Using this notation, the velocity vector u is $u = u_1 e_1 + u_2 e_2 + u_3 e_3$.

The gradient operator in Cartesian coordinates (also called nabla or del) is

$$\nabla := e_1 \frac{\partial}{\partial x_1} + e_2 \frac{\partial}{\partial x_2} + e_3 \frac{\partial}{\partial x_3},$$

or equivalently,

$$\nabla := \begin{pmatrix} \frac{\partial}{\partial x_1} \\ \frac{\partial}{\partial x_2} \\ \frac{\partial}{\partial x_3} \end{pmatrix}.$$

This operator applied to a function $f : \mathbb{R}^3 \to \mathbb{R}$ gives its gradient in Cartesian coordinates

$$\nabla f = \begin{pmatrix} \frac{\partial f}{\partial x_1} \\ \frac{\partial f}{\partial x_2} \\ \frac{\partial f}{\partial x_3} \end{pmatrix}.$$

Applied to a vector field u (with the notation $\nabla \cdot u$, where $\cdot$ denotes the usual inner product in Euclidean space), the gradient operator gives the divergence of u in Cartesian coordinates

$$\nabla \cdot u = \frac{\partial u_1}{\partial x_1} + \frac{\partial u_2}{\partial x_2} + \frac{\partial u_3}{\partial x_3}.$$

The expression $(u \cdot \nabla)u$, often written $u \cdot \nabla u$, is the vector

$$\begin{pmatrix} u_1 \frac{\partial u_1}{\partial x_1} + u_2 \frac{\partial u_1}{\partial x_2} + u_3 \frac{\partial u_1}{\partial x_3} \\ u_1 \frac{\partial u_2}{\partial x_1} + u_2 \frac{\partial u_2}{\partial x_2} + u_3 \frac{\partial u_2}{\partial x_3} \\ u_1 \frac{\partial u_3}{\partial x_1} + u_2 \frac{\partial u_3}{\partial x_2} + u_3 \frac{\partial u_3}{\partial x_3} \end{pmatrix}.$$

Here, $u \cdot \nabla$ is to be viewed as if it were the dot product of two vectors. The result is a scalar differential operator. The expression $(u \cdot \nabla)u$ is used to denote the vector field resulting from this operator acting on each component of the vector field u.

A parcel A of fluid identified at time zero is moved to $\gamma(A, t)$ at time t. The time rate of change of the total mass of fluid in the parcel is, by Reynolds's transport theorem,

$$\frac{d}{dt} \int_{\gamma(A,t)} \rho(x, t)\, dx = \int_{\gamma(A,t)} \rho_t(x, t) + \operatorname{div}(\rho u)(x, t)\, dx \tag{11.2}$$

(see A.11). By conservation of mass, the rate of change of the total mass in A does not change as the parcel is transported by the flow; therefore, the left-hand side of Eq. (11.2) vanishes. Because A may be taken arbitrarily small (for example, a ball with arbitrarily small radius) and the velocity field and density are assumed to be continuously differentiable, it follows that

$$\rho_t + \operatorname{div}(\rho u) = 0, \tag{11.3}$$

or equivalently,

$$\rho_t + \nabla \cdot (\rho u) = 0. \tag{11.4}$$

Eq. (11.4) (or (11.3)) is called the equation of continuity; it states that the mass of a fluid parcel is conserved by the fluid motion. Eq. (11.2) is a general statement of the rate of change of total mass that holds as long as u is an arbitrary (smooth) vector field with flow γ.

A differential equation for the velocity field u is obtained from Newton's second law of motion: the total momentum of a body is conserved unless it is acted on by forces; when forces act, the time rate of change of the momentum of the body is equal to the sum of these forces. There are two types of forces acting on a body of fluid: body forces and internal forces. A fluid has mass; it might also be charged. Thus, a fluid is subjected to body forces, which by definition are forces that act per unit of mass or per unit of charge. The most important body force is the gravitational force, which acts on every fluid simply because fluids have mass. Electromagnetic interactions are also important, but not discussed in this chapter. Unlike the motion of particles or rigid bodies, fluids (by definition) have internal stress (force per area) that is caused by the action of the fluid on itself. Stress is modeled by a function σ that assigns a vector in $\mathbb{R}^3$ to each pair consisting of a point (x, t) in space-time and a unit-length vector η in $\mathbb{R}^3$ at this point. The value of the stress function at this pair is called the stress vector at the point x; it represents the force per area exerted by the fluid on each imaginary surface passing through the point x and with outer normal η at time t. Using conservation of momentum and angular momentum, it is possible to prove that the stress function at each point in space-time is a symmetric linear transformation of space (see [21] and the discussion following Eq. (18.7)). This fact is simply incorporated here as an assumption. Thus, instead of writing $\sigma(x, t, \eta)$ for the stress vector at (x, t) on the plane with outer normal η, this expression is written $\sigma(x, t)\eta$. The function $(x, t) \to \sigma(x, t)$ assigns to each point in space-time a linear transformation of three-dimensional space.

A linear and symmetric transformation on vectors may be represented by a (diagonalizable) matrix in the usual Cartesian coordinates. Thus, for each point (x, t) in space-time, $\sigma(x, t)$ may be viewed as a symmetric matrix, which is thus defined by six numbers (the elements on and above the main diagonal of the matrix). This matrix may also be viewed as defining a bilinear form S at each point of space-time that acts on tangent vectors at this point: $S(x, t)(v, w) := v^T\sigma(x, t)w$, where v and w are (tangent) vectors at (x, t) and the superscript T denotes transpose. The assignment of a bilinear form at each point in space-time defines a rank-two tensor, which in this case is called the stress tensor.[1] Although the precise definition of tensors requires more mathematical structure, the intuitive definition has just been presented: a tensor is the (smooth) assignment of a (multi) linear map, which may not be symmetric, at each point in space or space-time.

Perhaps the most familiar example of a tensor is the inner product: It assigns the same multilinear map to each point in space. Indeed, the map $(X, Y) \mapsto X \cdot Y$ is multilinear; that is, the function $Y \mapsto X \cdot Y$ is a linear transformation for each fixed vector X and the function $X \mapsto X \cdot Y$ is a linear transformation for each fixed vector Y. This tensor is symmetric. It

[1]The word tensor seems to be derived from the latin *tendere*, which means "to stretch."

may be viewed as the assignment of a symmetric matrix (in this case the identity matrix) to each point of space.

Using the inner product, $(X, Y) \mapsto \langle X, Y\rangle$, every rank-two tensor R on space-time is given by

$$R(x,t)(X,Y) = \langle X, A(x,t)Y\rangle,$$

where $(x,t) \mapsto A(x,t)$ is a (smooth) matrix valued function. The tensor is symmetric if the matrix $A(x,t)$ is symmetric for every (x,t).

Total stress over the current position of parcel A at time t is given by

$$\text{TS} := \int_{\partial\gamma(A,t)} \sigma(x,t)\eta(x)\, d\mathcal{S},$$

where η is the outer unit normal on the boundary of $\gamma(A,t)$ and $d\mathcal{S}$ is the element of surface area. Using the body force b per unit of mass, the total body force on $\gamma(A,t)$ is

$$\text{TB} := \int_{\gamma(A,t)} \rho(x,t)b(x,t)\, d\mathcal{V},$$

where $d\mathcal{V}$ is the element of volume.

The total momentum of $\gamma(A,t)$ is

$$\int_{\gamma(A,t)} \rho(x,t)u(x,t)\, d\mathcal{V}.$$

By Newton's second law of motion (the time rate of change of momentum on a body is equal to the sum of the forces acting on the body), the mathematical expression for momentum-force balance is

$$\frac{d}{dt}\int_{\gamma(A,t)} \rho(x,t)u(x,t)\, d\mathcal{V} = \int_{\partial\gamma(A,t)} \sigma(x,t)\eta(x)\, d\mathcal{S} + \int_{\gamma(A,t)} \rho(x,t)b(x,t)\, d\mathcal{V}. \tag{11.5}$$

The region $\gamma(A,t)$ in space is moving with time. Using the equation of continuity (11.3), the transport theorem (A.11), and some algebra, it follows that the left-hand side of the momentum balance (11.5) is given by

$$\frac{d}{dt}\int_{\gamma(A,t)} \rho(x,t)u(x,t)\,d\mathcal{V} = \int_{\gamma(A,t)} \rho(x,t)(u_t + (u\cdot\nabla)u)(x,t)\,d\mathcal{V}, \tag{11.6}$$

where u_t denotes the partial derivative of u with respect to t.

The expression $u_t + (u \cdot \nabla)u$ that appears in the right-hand side of Eq. (11.6) is the *material derivative* of the velocity field u, which is often denoted by $\frac{Du}{Dt}(x,t)$; its definition takes into account the motion of the fluid particles with time:

$$\begin{aligned}\frac{d}{dt}u(\gamma(t,x_0),t) &= u_t(\gamma(t,x_0),t) + Du(\gamma(t,x_0),t)\dot\gamma(t,x_0)\\ &= u_t(\gamma(t,x_0),t) + Du(\gamma(t,x_0),t)u(\gamma(t,x_0),t)\\ &= (u_t + (u\cdot\nabla)u)(\gamma(t,x_0),t)\\ &= \frac{Du}{Dt}(\gamma(t,x_0),t).\end{aligned}$$

Note that the material derivative operator is denoted by D/Dt. The symbol D denotes differentiation with rcspect to spatial variables. For example, Du denotes the derivative of the function $(x,y,z) \mapsto u(x,y,z,t)$. In Cartesian coordinates Du is given by the Jacobian matrix of partial derivatives.

The total stress may be viewed in components. Note that $\sigma \cdot \eta$ (at each point on the boundary of $\gamma(A,t)$ is a vector whose first component is the dot product of the first row σ_1 of σ and the unit normal η. The second and third components are dot products of each of these rows with the same η. Thus, the ith component (with $i \in \{1,2,3\}$) is

$$\int_{\partial\gamma(A,t)} \sigma_i(x,t)\cdot\eta(x)\,d\mathcal{A}.$$

By the divergence theorem,

$$\int_{\partial\gamma(A,t)} \sigma(x,t)_i\cdot\eta(x)\,d\mathcal{S} = \int_{\gamma(A,t)} \operatorname{div}\sigma_i(x,t)\,d\mathcal{V}.$$

Using nabla notation, the vector whose ith component is $\operatorname{div} \sigma_i$ is $\nabla \cdot \sigma$. Written in full,

$$\nabla \cdot \sigma = \begin{pmatrix} \frac{\partial \sigma_{11}}{\partial x_1} + \frac{\partial \sigma_{12}}{\partial x_2} + \frac{\partial \sigma_{13}}{\partial x_3} \\ \frac{\partial \sigma_{21}}{\partial x_1} + \frac{\partial \sigma_{22}}{\partial x_2} + \frac{\partial \sigma_{23}}{\partial x_3} \\ \frac{\partial \sigma_{31}}{\partial x_1} + \frac{\partial \sigma_{32}}{\partial x_2} + \frac{\partial \sigma_{33}}{\partial x_3} \end{pmatrix}. \tag{11.7}$$

Using these facts, the total stress is

$$\mathrm{TS} = \int_{\partial \gamma(A,t)} \sigma(x,t)\eta(x)\, d\mathcal{S} = \int_{\gamma(A,t)} \nabla \cdot \sigma(x,t)\, d\mathcal{V}. \tag{11.8}$$

By substitution of Eqs. (11.6) and (11.8) into the momentum balance [Eq. (11.5)], we have the equivalent integral expression

$$\int_{\gamma(A,t)} \big(\rho \frac{Du}{Dt} - \nabla \cdot \sigma - \rho b\big)\, d\mathcal{V} = 0. \tag{11.9}$$

This equation holds for every parcel of fluid. Under the assumption that the integrand is continuous (which might be a strong assumption in some circumstances), the corresponding differential equation is

$$\rho \frac{Du}{Dt} = \nabla \cdot \sigma + \rho b, \tag{11.10}$$

or equivalently, Cauchy's equation (Augustin-Louis Cauchy 1789–1857)

$$\rho(u_t + (u \cdot \nabla)u) = \nabla \cdot \sigma + \rho b. \tag{11.11}$$

There is an additional conservation law: the conservation of energy. It is required, for example, in case temperature changes are important (see, for example, [60]). But, for simplicity, we will treat only situations where this law can reasonably be ignored.

An alternate form of the momentum balance [Eq. (11.11)] is often useful. Simply add the continuity equation $\rho_t + \nabla(\rho u) = 0$ to the original momentum balance, and rewrite the left-hand side of the result to obtain the equation

$$\frac{\partial}{\partial t}(\rho u) + \rho u \cdot \nabla u + \nabla \cdot (\rho u)u = -\nabla \cdot \sigma + \rho b. \tag{11.12}$$

To simplify the sum of the second and third terms on the left-hand side of the equation, consider one more step into tensor calculus: we may view u as a one-tensor. Recall that a tensor is a multilinear map of vectors to the real numbers. The vector (field) u determines a tensor (field) T via the inner product:

$$T(v) = \langle v, u \rangle, \tag{11.13}$$

which is a tensor field as soon as u and v are considered as vector fields that depend on space and time so that T is a function of space and time whose range is in the vector space of multilinear maps. The tensor field assigns a tensor to each point (p,t) in space-time and thus acts on a vector field v by

$$T(p,t)(v) = \langle v(p,t), u(p,t) \rangle.$$

Consider a second one-tensor S, perhaps $S(v) = \langle v, w \rangle$. A pair of tensors (not necessarily of the same rank) can be multiplied to form a new tensor $T \otimes S$ by the formula

$$T \otimes S(v, y) = T(v)S(y), \tag{11.14}$$

where the rank of the new tensor (called the tensor product) is the sum of the ranks of the two tensors used to form the product. For the rank-one tensors just defined, the tensor product is a rank-two tensor. The same construction defines a new tensor field simply by including the functional dependence on space and time. An important note is that $T \otimes S$ is generally not the same as $S \otimes T$.

The reason for this digression is to introduce the tensor $u \otimes u$, which is defined via an abuse of proper notation to be the tensor $T \otimes T$; that is, in this tensor product, u is to be viewed as the tensor T determined by u as defined in Eq. (11.13). Being a rank-two tensor, $u \otimes u$ is given by a matrix A in Cartesian coordinates in the same manner that the stress tensor is determined by a matrix. Indeed, there is a 3×3 matrix A such that

$$u \otimes u(v, y) = \langle Av, y \rangle,$$

where the (i, j) component of A is given by

$$A_{ij} = \langle Ae_i, e_j \rangle = u \oplus u(e_i, e_j) = u_i u_j.$$

In coordinate notation, the tensor would be called A and its components are $A_{ij} = u_i u_j$. A clear understanding of the construction for this example should be helpful in understanding tensor notation and the tensor product.

The new tensor $u \otimes u$ appears in the desired formula

$$\nabla \cdot (\rho u \otimes u) = \rho u \cdot \nabla u + \nabla \cdot (\rho u)u, \tag{11.15}$$

which is easy to prove by simply using the definitions to express the fields in coordinates. Of course, a function multiple of a tensor is again a tensor of the same rank defined on the argument of the tensor to be the value of the tensor multiplied by the scalar function. Using the identity (11.15), the desired alternate form of the momentum balance is

$$\frac{\partial(\rho u)}{\partial t} + \nabla \cdot (\rho u \otimes u - \sigma) = \rho b. \tag{11.16}$$

An important special case is steady flow (no dependence on t) with no body force where the momentum balance reduces to

$$\nabla \cdot (\rho u \otimes u - \sigma) = 0. \tag{11.17}$$

In the special case where σ is a scalar function p times the identity matrix (which will play an important role in the discussion of fluid motion), the corresponding result is

$$\nabla \cdot (\rho u \otimes u + pI) = 0, \tag{11.18}$$

where of course pI is shorthand for the rank-two tensor given by the inner product: $(v, y) \mapsto p\langle v, y\rangle$.

Eq. (11.11) holds for arbitrary elastic media, not just fluids. This model would lead to fundamental models if internal stress could be described exactly by a symmetric tensor that was derived without additional assumptions from electromagnetism. Although a fundamental representation of internal stress might be possible in principle, it would likely be so complex as to be useless for making predictions. Instead, an approximation—based on physical principles and physical intuition—called a constitutive law *for the type of fluid (or elastic material) being modeled* is used to define the stress tensor as a function of the other state variables. The choice of constitutive stress law is the most important ingredient in a viable model. Because constitutive laws are approximations, predictions derived from corresponding models must be validated by physical experiments.

Perhaps the simplest model for internal stress arises from the constitutive assumption that the stress is the same in all directions (no shear stress). In this case, there is a scalar function p on space-time, called the pressure, such that the stress tensor is given by

$$\sigma(x, t) = -p(x, t)I, \tag{11.19}$$

where I denotes the identity transformation on space. The minus sign is taken because the stress on the boundary of a fluid parcel is $\sigma\eta$, where η is the outer unit normal on the boundary of the parcel and a positive pressure (from outside the parcel) is assumed to act in the direction of the inner normal. Indeed, pressure is correctly defined to be the normal component of force per area. A fluid that satisfies constitutive law (11.19) is called an *ideal fluid*. Using Eq. (11.7), the divergence of the stress for an ideal fluid is

$$\nabla \cdot \sigma = -\nabla p,$$

and the corresponding equations of motion are

$$\begin{aligned} \rho(u_t + (u \cdot \nabla)u) &= -\nabla p + \rho b, \\ \rho_t + \nabla \cdot (\rho u) &= 0. \end{aligned} \tag{11.20}$$

The momentum balance equation is called Euler's equation.

The partial differential equations (PDEs) (11.20) constitute four scalar PDEs, counting the three components of the vector equation for momentum balance, for five unknowns: the three components of the velocity u, the density ρ, and the pressure p. One more equation is needed to close the system. The missing ingredient is conservation of energy. A physically realistic approach to conservation of energy requires a digression into thermodynamics. Although the kinetic energy $\frac{1}{2}\rho\|u\|^2$ is a quantity from classical mechanics, models of internal energy require a deeper analysis that is not completely understood. Thus, the most general form of the equations of fluid dynamics remain somewhat controversial. Fortunately, for some practical applications, internal energy considerations can be bypassed or modeled with relatively simple constitutive laws. Another approach to closing the system is to make a simplifying assumption on the nature of the fluid. Two standard (closely related) possibilities are the assumptions that the fluid is incompressible (that is, the flow preserves volume) or that the density of the fluid is constant.

Incompressibility (via Reynolds's transport theorem) is equivalent to the velocity field being divergence free: $\operatorname{div} u = 0$. In this case, by using the identity

$$\nabla \cdot (\rho u) = \nabla \rho \cdot u + \rho \nabla \cdot u,$$

the equations of motion

$$\begin{aligned} \rho(u_t + (u \cdot \nabla)u) &= -\nabla p + \rho b, \\ \rho_t + \nabla \rho \cdot u &= 0, \\ \nabla \cdot u &= 0 \end{aligned} \tag{11.21}$$

are called Euler's equations for an ideal fluid.

A more restrictive assumption is constant density; it leads to the closed system of (four) equations

$$\begin{aligned} \rho(u_t + (u \cdot \nabla)u) &= -\nabla p + \rho b, \\ \nabla \cdot u &= 0 \end{aligned} \tag{11.22}$$

for the three unknown components of u and the pressure.

Because our fluid is confined to a region $\mathcal{R}$ of space, boundary conditions must be imposed. Physical experiments demonstrate that the correct boundary condition is $u \equiv 0$ on the (stationary) solid boundaries of $\mathcal{R}$. This is called the no-slip boundary condition. To perform a simple experiment, consider cleaning a metal plate by using a hose to spray it with water; for example, try cleaning a dirty automobile. As the pressure of the water increases, the size of the particles of dirt that can be removed decreases. But, it is very difficult to remove all the dirt by spraying alone. This can be checked by polishing with a clean cloth. In fact, the speed of the spray decreases rapidly in the boundary layer near the plate. Dirt particles with sufficiently small diameter are not subjected to flow speeds that are high enough to dislodge them.

For moving boundaries, the no-slip boundary condition requires the fluid velocity at each point of a solid boundary to be the same as the velocity of that point. When moving boundaries are modeled, the no-slip condition reads $u = f$ on the boundary of the region where f is the vector function that assigns the velocity at each boundary point.

Euler's equations are not well posed with the no-slip boundary condition; that is, under this boundary condition, the equations of motion do not always have unique solutions depending continuously on the initial position of the fluid. The mathematically correct boundary condition *for Euler's equations* is that the fluid does not penetrate the boundary, or equivalently, the fluid velocity is everywhere tangent to the boundary. The no-slip condition is allowed, but not required. Although this fact implies that Euler's equations cannot be the correct model for physical fluids, Euler's model gives experimentally verifiable predictions as long as measurements are taken far from the fluid's boundary. Indeed, this observation is fundamental in many applications (for example, flow over an airplane wing) where there is a thin layer of fluid near the boundary that must obey a more realistic stress constitutive law to satisfy the no-slip boundary condition, but away from this layer, the motion of the fluid is well-approximated by Euler's equations.

To obtain more realistic models of fluid motion, the viscosity of the fluid must be taken into account. The modeling process requires several assumptions, which are different for different types of fluids. For fluids similar to water and air, called Newtonian fluids, the main assumptions are the isotropy of the fluid (that is, the fluid is the same in all directions) and a linear relation between stress and velocity. For some fluids, for example blood, accurate models require nonlinear stress-velocity relations.

The modeling process to obtain the stress-velocity relation for Newtonian fluids—for which stress is proportional to strain rate—is fundamental in fluid mechanics (see, for example, [21, 60]). A related discussion and derivation of this stress-strain relation in the context of the equation of motion for elastic media is presented in Chapter 18. The same underlying ideas lead to models of stress for fluids. Some of the stresses are given by pressure (due to molecular motion). This part of the stress is modeled by the Eulerian fluid where the stress tensor is given by the pressure; that is, $\sigma = -pI$. Stresses in moving Newtonian fluids are determined by constitutive laws, formulated as linear relations between stress and strain (via Hooke's law).

Recall that stress is defined as a force per area on a surface. At a point q in a fluid at time t_0, a stress is defined for each pair of vectors v and w at q via the stress tensor. In fact, $\sigma(q, t_0)(v, w)$ is the force on surfaces normal to v in the direction given by the projection of w on this surface. A constitutive model for the stresses is built from a basic assumption: the

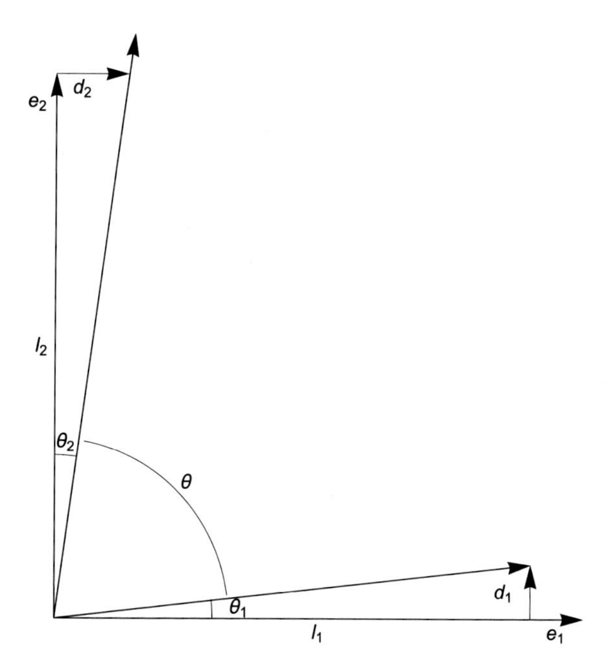

Fig. 11.1 The figure is a schematic diagram of a fluid cube face distorted by the flow. Angles and lengths are depicted. The shear corresponding to the motion of e_1 *with respect to the direction* e_2 *is* d_1/ℓ_1.

stresses are proportional to strains at q and these strains are related to the spatial derivative of the velocity of the fluid at this point. Although there are many ways to describe the relationship between these quantities, the approach taken here is to define a tensor (which has a clear geometric meaning) and show that its values give the strain rates.

Strains are relative changes in length. They are measured by the (infinitesimal) changes in positions of material points of the moving fluid. Strain is dimensionless. Stresses in Newtonian fluids are proportional to strain *rates*, which have units of inverse time. Strains come in several varieties: shear strains, normal strains, and volumetric strains.

Imagine a unit cube of fluid surrounding a point q at time t_0. Suppose for simplicity that the cube is situated so that its faces are parallel to the Cartesian coordinate planes. In this case, the cube can be viewed as the unit cube in the first octant of three-dimensional space. Shear strains can be defined from the shape distortions of the cube as it moves with the fluid.

Consider, for instance, the face of the cube corresponding to the first two basis vectors e_1 and e_2 at q at time t_0 and the unit square determined by these vectors. The shear strain on this face is determined from the parallelogram formed as the square is distorted by the flow as depicted in Fig. 11.1. The imagined cube will of course not remain a cube under the flow except in some extraordinary cases. When the cube is subjected to the fluid motion for a sufficiently short time and first-order approximations (that is, linearizations) are used, a rigid motion of the distorted cube may be performed so that the corner originally at the origin is moved back to the origin, and the distorted face is moved back into the plane of the original face spanned by e_1 and e_2. Physical shear strain is then measured by the ratios d_1/ℓ_1 and d_2/ℓ_2 of the signed lengths as depicted in the figure. The signs of the lengths d_1 and d_2 are determined by the directions of the fluid motion: If the points along the vector e_2 move in the direction of e_1, as in the figure, the sign is positive; if the motion is in the direction of $-e_1$ the sign is negative. The shear strain rate is defined to be the time rate of change of the strain; for example, $\frac{d}{dt}(d_1/\ell_1)$. Shear strain is the ratio of two lengths, a dimensionless quantity; thus, the shear strain rate has units of inverse time.

Note that for the angle θ_1 depicted in Fig. 11.1, $\tan\theta_1 = d_1/\ell_1$, and

$$\dot{\theta}_1 \sec^2\theta_1 = \frac{d}{dt}\tan\theta_1 = \frac{d}{dt}\Big(\frac{d_1}{\ell_1}\Big).$$

For θ_1 near zero (which will be the case after a sufficiently short flow time), we have that $\sec^2\theta_1 \approx 1$. Thus, as a first-order approximation, $\dot{\theta}_1$ is the shear strain rate. Similarly for $\dot{\theta}_2$. The total shear strain rate for the face spanned by e_1 and e_2 is the sum of these two derivatives.

The angle θ between the distorted vectors (given as a function of time) is

$$\theta = \frac{\pi}{2} - (\theta_1 + \theta_2);$$

thus, the shear strain rate is $-\dot{\theta}$ evaluated at the initial time $t = t_0$. This quantity will be determined as a function of the fluid velocity field.

Although vectors are not transported in an obvious way by the flow, curves of fluid particles are transported by moving the particles on the curve by the flow. Every vector is the velocity vector of some curve. Thus, the correct way to transport a vector is to choose a (parameterized) curve such that the vector under consideration is the velocity of this curve at one of its points, move the curve (of fluid particles) by the fluid flow, and define the transported vector to be the velocity vector of the transported curve at the transported point. This definition of transport of a vector by a flow is fundamental to understanding continuum mechanics.

To determine the motion of the curve by the flow, recall that the motion of the point q, starting at time $t = t_0$ in the moving fluid is given by $t \mapsto \gamma(q, t_0, t)$, where

$$\frac{\partial \gamma}{\partial t}(q, t_0, t) = u(\gamma(q, t_0, t), t) \tag{11.23}$$

with initial condition $\gamma(q, t_0, t_0) = q$. Here, the temporal variable t measures time relative to some prespecified origin and the fluid is observed at time $t = t_0$. For simplicity, suppress the second argument of the flow, which records this observation time, and write $\gamma(q, t)$ for the position of q at time t. In this notation, $\gamma(q, t_0) = q$.

Let α be a curve at q parameterized by s with tangent vector v; that is, $s \mapsto \alpha(s)$ is a curve such that $\alpha(0) = q$ and $\alpha'(0) = v$. The infinitesimal distortion of the curve at time t is the directional derivative

$$\frac{d}{ds}\gamma(\alpha(s), t)\Big|_{s=0} = D\gamma(q, t)v,$$

where D denotes the derivative of the transformation $q \mapsto \gamma(q, t)$ for fixed t. Thus, $D\gamma(q, t_0) = I$ and the vector v is transported by the flow to the vector $D\gamma(q, t)v$ at the point $\gamma(q, t)$.

Writing the shear strain using the time derivative of the angle θ is useful because the angle is given by a dot product: In fact, the cosine of the angle θ between an arbitrary pair of (nonzero) vectors v and w at q at time t_0 moved by the flow γ to $D\gamma(q, t)v$ and $D\gamma(q, t)w$ is

$$\cos\theta(q, t) = \frac{\langle D\gamma(q, t)v, D\gamma(q, t)w\rangle}{|D\gamma(q, t)v||D\gamma(q, t)w|}. \tag{11.24}$$

For simplicity, assume that v is orthogonal to w and both have unit length. Applications of the quotient rule and Leibniz's rule together with the formula $D\gamma(q, t_0) = I$ and the equality $\theta(q, t_0) = \pi/2$ can be used to show that

$$\begin{aligned} -\dot{\theta}(q, t_0) &= \frac{d}{dt}\langle D\gamma(q, t)v, D\gamma(q, t)w\rangle\Big|_{t=t_0} \\ &= \big(\langle v, \frac{d}{dt}D\gamma(q, t)w\rangle + \langle \frac{d}{dt}D\gamma(q, t)v, w\rangle\big)\Big|_{t=t_0}. \end{aligned} \tag{11.25}$$

Differentiate both sides of differential equation (11.23) with respect to the spatial variable in the direction v and use the independence of partial differentiation from the order of taking partial derivatives to show that

$$\frac{d}{dt}D\gamma(q,t)v = Du(\gamma(q,t),t)D\gamma(q,t)v.$$

In effect, d/dt and D commute. At $t = t_0$, this identity reduces to

$$\frac{d}{dt}D\gamma(q,t)v\big|_{t=t_0} = Du(q,t)v.$$

Thus,

$$\begin{aligned} -\dot{\theta}(q,t_0) &= \langle Du(q,t_0)v, w\rangle + \langle v, Du(q,t_0)w\rangle \\ &= \langle (Du(q,t_0)^T + Du(q,t_0))v, w\rangle. \end{aligned} \tag{11.26}$$

In view of this result, define the shear strain rate tensor ε by

$$\varepsilon(q,t_0)(v,w) = \langle (Du(q,t_0)^T + Du(q,t_0))v, w\rangle. \tag{11.27}$$

This function is linear in each of its second two arguments; that is, $v \mapsto \varepsilon(q,t)(v,w)$ for fixed q, t_0, and w and $w \mapsto \varepsilon(q,t)(v,w)$ for fixed q, t_0, and w are linear transformations from three-dimensional space to the real numbers. Thus, ε defines a rank-two tensor, which is called the strain rate tensor[2], and its Cartesian coordinate representation is given by the components of the matrix $Du(q,t_0)^T + Du(q,t_0)$; that is, in coordinate notation,

$$\varepsilon_{ij} := \varepsilon(q,t_0)(e_i,e_j) = \frac{\partial u_i}{\partial x_j}(q,t_0) + \frac{\partial u_j}{\partial x_i}(q,t_0).$$

By Hooke's law, the stress $\sigma(q,t_0)(e_1,e_2)$ is proportional to the shear strain rate ε_{12}, as this is the only strain rate produced by the internal fluid force in the direction e_2 on the imaginary surface normal to e_1. The proportionality factor μ is called the dynamic viscosity. It could, in principle, depend on the state variables, temperature, or electromagnetic radiation. Likewise all the stresses given by orthogonal unit basis vectors e_i and e_j

[2]The strain rate tensor is often defined to be $\varepsilon/2$ when the factor $1/2$ is natural due to a different approach to modeling stress as in Chapter 18.

(with $i \neq j$) are related to the velocity of the fluid by the constitutive law

$$\sigma_{ij} = \sigma(q, t_0)(e_i, e_j) = \mu(\frac{\partial u_i}{\partial x_j}(q, t_0) + \frac{\partial u_j}{\partial x_i}(q, t_0)) = \mu\varepsilon_{ij}. \quad (11.28)$$

To determine a constitutive law for the (diagonal) stresses σ_{ii}, recall that the fluid pressure at a point is the same in all directions and has already been modeled by the tensor $-pI$. This pressure arises from thermodynamic properties of the fluid and is present when the fluid is at rest or in motion. For a moving fluid there are stresses in the normal directions to a parcel's surface due to the forces of the fluid on itself, which are not due to the fluid pressure. These stresses do not cause shears.

From the definition of the strain rate tensor [Eq. (11.27)], each diagonal element of the strain rate tensor gives the time rate of change of the square of the length of a unit basis vector:

$$\varepsilon(q, t_0)(e_i, e_i) = 2\frac{\partial u_i}{\partial x_i}(q, t_0). \quad (11.29)$$

These strains (called normal strains) should be present and caused by stresses in the normal directions to the faces of the hypothetical cube of fluid at q at time t_0. They can of course be recognized as physical strains by relating them to relative changes in lengths corresponding to the positions of the cube faces as they move with the flow.

By Hooke's (constitutive) law, stress is a linear function strain. For a moving fluid, stresses are linear functions of strain rates. The simplest model is obtained by taking the stress to be a scalar multiple of the strain rates. Thus, a model that maintains the fluid pressure and takes into account the shear and normal stresses is

$$\sigma = -pI + \mu\varepsilon,$$

where μ is called the dynamic viscosity of the fluid.

In case a viscous fluid is not undergoing shear or normal stress, σ should reduce to $\sigma = -pI$. In particular, the trace of σ in this case is $-3p$. The trace of σ is

$$-3p + 2\mu \operatorname{div} u.$$

Thus, for the model to be consistent with the possibility that the moving fluid is not undergoing shear or normal stress, either $\operatorname{div} u = 0$ in such regimes or $\mu = 0$. By definition μ is not zero for a viscous flow, but taking $\operatorname{div} u = 0$ would not allow for compressible flow. This model is not compatible with the motion of compressible viscous flow.

To remedy the model for compressible flow, strain due to change of volume of the moving fluid must be taken into account. The volumetric strain rate is defined to be the limit, as a volume shrinks to the point q, of the infinitesimal rate of change of the relative change in volume. Using the transport theorem, this scalar quantity is

$$\lim_{\Omega \to \{q\}} \frac{d}{dt} \frac{\int_{\gamma(\Omega,t)} d\mathcal{V}}{\int_{\Omega} d\mathcal{V}}\bigg|_{t=t_0} = \operatorname{div} u(q, t_0). \tag{11.30}$$

As for the fluid pressure, to have the volumetric stress part of the integrand $\sigma \cdot \eta$ in the integral for total stress [Eq. (11.8)] be the same in all directions, the corresponding volumetric strain rate tensor is taken to be $\operatorname{div} uI$.

Again using Hooke's law, a model that takes into account all the stresses is

$$\sigma = -pI + \mu\varepsilon + \lambda \operatorname{div} uI,$$

where μ is the dynamic viscosity and λ, another factor of proportionality, is called the second viscosity of the fluid.

In case a viscous fluid is not undergoing normal, volumetric, or shear stress, the stress tensor σ should reduce to $\sigma = -pI$. In particular, the trace of σ in this case is $-3p$. The trace of σ is

$$-3p + 2\mu \operatorname{div} u + 3\lambda \operatorname{div} u = -3p + (2\mu + 3\lambda) \operatorname{div} u.$$

Thus, to be consistent with the possibility that the fluid is not undergoing normal, volumetric, or shear stress, either $\operatorname{div} u = 0$ in such regimes or $2\mu + 3\lambda = 0$. To allow for compressible flow, the two viscosities are related by $\lambda = -2/3\mu$, and the stress tensor is taken to be

$$\sigma = -pI + \mu\varepsilon - \frac{2}{3}\mu \operatorname{div} uI. \tag{11.31}$$

The derivation just presented is somewhat fanciful. Although it contains the usual ingredients and correctly determines the strains, it lacks a compelling case on physical grounds for the exact form of the stress-velocity constitutive law. A perhaps better approach would start with the (discrete) dynamics of the individual molecules that constitute the fluid and seek to determine the continuous equations of fluid motion via averaging over the molecular motions. This approach has been partially successful in reproducing expressions similar to the suggested stress-velocity constitutive law, but these results are not definitive. From an applied point of view, the constitutive law derived here has been enormously successful in predicting fluid motion. It agrees with experiments in cases where the physical fluid flow is not too complicated (turbulent). The universal validity of the model is not known.

Using $\delta_{ij} = 0$ if $i \neq j$ and $\delta_{ij} = 1$ if $i = j$, the components of σ are given by

$$\sigma_{ij} = -p\delta_{ij} + \mu\Big(\frac{\partial u_i}{\partial x_j} + \frac{\partial u_j}{\partial x_i}\Big) - \frac{2}{3}\mu(\nabla \cdot u)\delta_{ij}. \tag{11.32}$$

The function p is the pressure, μ is the viscosity, and $\partial u_i/\partial x_j + \partial u_j/\partial x_i$ is the (i, j) component of the strain rate tensor. With this choice of the stress tensor σ, the corresponding partial differential equations (PDEs)

$$\begin{aligned} \rho(u_t + (u \cdot \nabla)u) &= -\nabla p + \mu\Delta u + \frac{\mu}{3}\nabla(\nabla \cdot u) + \rho b, \\ \rho_t + \nabla \cdot (\rho u) &= 0 \end{aligned} \tag{11.33}$$

(which are derived with an easy computation from Cauchy's equation and the continuity equation) are called the Navier–Stokes equations for a Newtonian fluid. Here Δu denotes the Laplacian applied componentwise to the vector u; that is,

$$\Delta u = \begin{pmatrix} \frac{\partial^2 u_1}{\partial x_1^2} + \frac{\partial^2 u_1}{\partial x_2^2} + \frac{\partial^2 u_1}{\partial x_3^2} \\ \frac{\partial^2 u_2}{\partial x_1^2} + \frac{\partial^2 u_2}{\partial x_2^2} + \frac{\partial^2 u_2}{\partial x_3^2} \\ \frac{\partial^2 u_3}{\partial x_1^2} + \frac{\partial^2 u_3}{\partial x_2^2} + \frac{\partial^2 u_3}{\partial x_3^2} \end{pmatrix}.$$

The no-slip boundary condition is enforced at solid boundaries.

In some realistic situations (for example, the flow of water at constant temperature) it is reasonable to make two further assumptions: the viscosity μ and density ρ are constant. Using the equation of continuity, the constant density assumption implies that the vector field u is divergence free; that is, the fluid is incompressible (see Appendix A.11). Under these assumptions, the Navier–Stokes equations (11.33) simplify to

$$\begin{aligned} \rho(u_t + (u \cdot \nabla)u) &= -\nabla p + \mu\Delta u + \rho b, \\ \nabla \cdot u &= 0. \end{aligned} \tag{11.34}$$

Exercise 11.1. (a) Prove that if u is a sufficiently smooth vector field, then

$$\nabla \cdot \Delta u = \Delta(\nabla \cdot u).$$

Conclude that if u is divergence free, then $\nabla \cdot \Delta u = 0$. (b) Use part (a) to show that for incompressible, constant density, flow with body force b satisfying $\nabla \cdot b = 0$ (such a body force is called *conservative*), the pressure is given by

$$\Delta p = -\rho\nabla \cdot ((u \cdot \nabla)u).$$

Exercise 11.2. Prove Eq. (11.15). Hint: Change from the coordinate-free form of the vector equation to the coordinate form.

Exercise 11.3. Consider incompressible flow with no body force. Let ϕ be a harmonic function ($\Delta\phi = 0$) on three-dimensional space and f an arbitrary real function of one real variable. (a) Show that the velocity field $u(x, t) = f(t)\nabla\phi$ solves the Navier–Stokes equations, ignoring the boundary condition. Find the exact expression for the pressure. (b) Is it possible to find nontrivial solutions as in part (a) on a bounded domain such that the velocity field vanishes at the boundary? Discuss.

Exercise 11.4. Consider the vector field (on $\mathbb{R}^2$) given by $(2y, 0)$. (a) Determine the flow of this vector field. (b) Consider the distortion of the parallelogram determined by the usual Cartesian unit vectors e_1 and e_2 as it is moved by the flow. Determine the time rate of change at time $t = 0$ (when the parallelogram is a square) of the negative of the angle between the moving legs of the parallelogram corresponding to e_1 and e_2. (c) Consider the vector field $(y, (1 - x^2)y - x)$. It has a flow $\gamma = \gamma((x, y), t_0, t)$. Approximate the vector

$$D\gamma((1, 0), 0, 2)e_1.$$

Hint: Use a numerical computation for part (c).

Exercise 11.5. Show that it is not necessary to assume the density is constant (everywhere) to obtain Eqs. (11.34) from Eqs. (11.33). More precisely, prove that it suffices to have $D\rho/Dt = 0$ (that is, the density is constant along fluid particle paths).

Exercise 11.6. Define

$$\phi(x) = \begin{cases} 1, & x \le 0; \\ 1 - 3x^2 + 2x^3, & 0 < x < 1; \\ 0, & x \ge 1. \end{cases}$$

(a) Show that ϕ is continuously differentiable and write out the formula for ϕ'. (b) Recall the conservation of mass formula (the continuity equation) $\rho_t + \operatorname{div}(\rho u) = 0$. Suppose that u is the constant vector field $(2, 0, 0)$ and the density is distributed in space so that the density is the same on each plane that is orthogonal to the x-axis. This means $\rho(x, y, z, t)$ does not depend on y or z; it can be viewed as a function of x and t only. Using these simplifications, consider the initial value problem

$$\rho_t + \operatorname{div}(\rho u) = 0, \qquad \rho(x, 0) = \phi(x).$$

Determine the value of $\rho(3, 5/4)$. Hint: Look for a traveling wave solution.

Exercise 11.7. The total kinetic energy of a fluid with density ρ and velocity u over the parcel A moving with the fluid via the flow γ is

$$\text{KE} := \int_{\gamma(A,t)} \frac{1}{2}\rho(x, t)u(x, t) \cdot u(x, t)\, d\mathcal{V}.$$

Use the transport theorem and a computation in coordinates to show that

$$\frac{d\text{KE}}{dt} = \int_{\gamma(A,t)} \rho u \cdot (u_t + (u \cdot \nabla)u)\, d\mathcal{V}.$$

Exercise 11.8. [Stratified flow]. Consider a steady, incompressible, inviscid fluid modeled by Euler's equation in three-dimensional space. Assume no body force is acting on the fluid but it has nonconstant density ρ and pressure P so that the equations of motion for the fluid velocity u and pressure P are

$$\begin{aligned} \rho(u \cdot \nabla)u + \nabla P &= 0, \\ (u \cdot \nabla)\rho &= 0, \\ \nabla \cdot u &= 0. \end{aligned}$$

Show that $w := \sqrt{\rho}\, u$ is a solution of the inviscid incompressible Euler equations

$$\begin{aligned} (w \cdot \nabla)w + \nabla P &= 0, \\ \nabla \cdot w &= 0, \end{aligned}$$

that model flow with constant unit density. (This observation is mentioned in [122].) How do you interpret this fact?

11.1 SCALING: THE REYNOLDS NUMBER AND FROUDE NUMBER

For simplicity, the Navier–Stokes equations (11.34) for a constant density (thus incompressible) flow is considered in this section. By introducing a characteristic length L and a characteristic velocity V—more correctly a characteristic speed—for the flow in question together with the corresponding natural timescale $\tau = L/V$, the state variables are rendered dimensionless via the assignments $X = x/L$, $s = t/\tau$, $U(X,s) = u(x,t)/V$, and $P(X,s) = p(x,t)/(V^2\rho)$. Using the kinematic viscosity $\nu := \mu/\rho$, the Reynolds (dimensionless) number $\mathrm{Re} := LV/\nu$, and the Froude number $\mathrm{Fr} := V/\sqrt{L|b|}$, which depends on the magnitude of the body force, the system of equations for the velocity and pressure can be rescaled to a dimensionless form of the Navier–Stokes equations

$$\begin{aligned} U_s + (U \cdot \nabla)U &= \frac{1}{\mathrm{Re}}\Delta U - \nabla P + \frac{1}{\mathrm{Fr}^2}\frac{b}{|b|}, \\ \nabla \cdot U &= 0, \end{aligned} \tag{11.35}$$

where the equations hold on the region $\mathcal{R}^*$ defined to be the image of the region $\mathcal{R}$ under the change of coordinates. The no-slip boundary condition is to be enforced at solid boundaries of $\mathcal{R}^*$.

The existence of this scaling is important: If two flows have the same Reynolds and Froude numbers, then the flows have the same dynamics. For example, flow around a *scaled model* of an airplane in a wind tunnel might be tested at the same Reynolds and Froude numbers expected for the airplane under certain flight conditions. Perhaps the same Reynolds number can be obtained for testing by increasing the velocity in the wind tunnel to compensate for the smaller length scale of the scaled model. In principle, the behavior of the flow around the scaled model is the same as for the full-sized aircraft.

A standard question is how to choose the characteristic length and velocity for a given application. There is no simple answer, but there are conventions. For flow in a pipe, the inlet flow velocity is taken to be the characteristic velocity and the characteristic length is the diameter of the pipe. Another example is flow over a blunt body (for example, a truck moving on a highway) where the front-view height of the truck is usually taken to be the characteristic length and the truck speed is the characteristic velocity. Conventions are often used in specialized fields of applied science to compare experimental and theoretical results, but for precision, the chosen characteristic length and velocity must be explicitly specified. As long as this is done, the choice of characteristic length and velocity can be made arbitrarily. To illustrate, suppose convention is not followed and some phenomenon in pipe flow is observed at Reynolds number 8,888 by someone who prefers to take the radius of their round pipe as characteristic length. The result is easily translated to the conventional value $\mathrm{Re} = 17,776$ because the conventional Reynolds number is simply twice the reported Reynolds number.

11.2 THE ZERO VISCOSITY LIMIT

The scaled Euler equations defined in the region $\mathcal{R}^*$ as in display (11.35) are

$$\begin{aligned} U_s + (U \cdot \nabla)U &= -\nabla P + B, \\ \nabla \cdot U &= 0 \end{aligned} \tag{11.36}$$

can be viewed as an idealization of the scaled Navier–Stokes equations (11.35) for a fluid with zero viscosity. The no-slip boundary condition for the Navier–Stokes equations is replaced by the no-penetration condition: There is no fluid passing through solid boundaries.

A naive expectation is that the limit of a family of solutions of the Navier–Stokes equations, as the Reynolds number increases without bound, is a solution of Euler's equations. After all, the term $\Delta U/\mathrm{Re}$ would seem to approach zero as $\mathrm{Re} \to \infty$. Note, however, the possibility that the second derivatives of the velocity field are unbounded in the limit and the different boundary conditions for the Navier–Stokes and Euler equations. For these and other reasons, the limiting behavior of the Navier–Stokes equations for large values of the Reynolds number is not yet completely understood.

The necessity of different boundary conditions in passing from the Navier–Stokes to the Euler equations is the starting point for one of the most important aspects of fluid dynamics, which was introduced by L. Prandtl in 1904 (see [88, 93]), called boundary layer theory. The fundamental idea is that for flows of interest in aerodynamics, for instance, the viscous effects are only important in a thin layer near the boundary where the no-slip condition must hold; away from the boundary, the flow is well-approximated by Euler's equations. Boundary layer theory will be discussed more fully in Section 17.3.

The possibility of large second derivatives of the velocity field is important in another area of fluid dynamics: the study of turbulence, a subject that is beyond the scope of this book.

11.3 THE LOW REYNOLDS NUMBER LIMIT

In the zero viscosity limit, the pressure is scaled by $P = p/(V^2\rho)$. Physically, the pressure is comparable to the momentum per unit volume. For flows with low Reynolds's numbers, the pressure is comparable to the forces due to viscosity. For this reason, the correct scaling in this regime is $X = x/L$, $s = t/\tau$, $U(X, s) = u(x, t)/V$, and $P(X, s) = p(x, t)L/(\mu V)$, which leads to the equation

$$\frac{L}{V}U_s + U \cdot \nabla U = -\frac{1}{\mathrm{Re}}\nabla P + \frac{1}{\mathrm{Re}}\Delta U.$$

After multiplying by the Reynolds number and passing to the limit as $\mathrm{Re} \to 0$, we obtain the dimensionless Stokes equations

$$\begin{aligned} \nabla P &= \Delta U, \\ \nabla \cdot U &= 0. \end{aligned} \tag{11.37}$$

This approximation of the Navier–Stokes equations is a useful model in many different flow regimes where the velocity is small, the viscosity is large, or the size of some body immersed in the fluid is small. There are numerous applications in the field of developmental biology [16], lubrication theory [92], and other areas of science and engineering (see Section 19.6).

Note that the temporal variable does not appear in the low Reynolds number limit; Stokes flow is always in a steady state.

In the low Reynolds number limit, the model retains the assumption that the fluid has nonzero viscosity (due to the presence of the Laplacian of the velocity). Thus, the no-slip boundary condition remains appropriate at solid boundaries.

The Stokes equations are linear and much simpler than the Navier–Stokes model. For example, although existence and uniqueness for the Navier–Stokes equations is a highly nontrivial problem, these issues are much easier to resolve for the Stokes model [Eqs. (11.37)]. Indeed, consider the question of uniqueness. Suppose the velocity is specified everywhere on the boundary of $\mathcal{R}^*$; that is, there is a given function F such that $U = F$ on the boundary. Then, the Stokes model for the fluid motion has at most one solution. More precisely, a solution consists of a velocity field U and a pressure function P. Solutions are never unique in the usual sense, because a constant may be added to the pressure to obtain a new solution. In fluid mechanics, a solution pair (U, P) is understood to mean that the pressure is defined up to the addition of a constant.

One elementary proof of uniqueness requires an indirect approach using the stress tensor components [Eq. (11.32)], which of course are the fundamental quantities that underly the Navier–Stokes equations. Name the strain rate tensor components

$$e_{ij} := \frac{\partial U_i}{\partial X_j} + \frac{\partial U_j}{\partial X_i}$$

and use the equation $\nabla \cdot U = 0$ to write the components of the stress tensor (still named σ) as follows:

$$\sigma_{ij} = -P\delta_{ij} + e_{ij}. \tag{11.38}$$

The important formula (akin to the one used in the derivation of the first equation in display (11.33)) to be employed in the uniqueness proof is

$$\nabla \cdot \sigma = -\nabla P + \Delta U. \tag{11.39}$$

For Stokes flow, the divergence of the stress tensor vanishes in agreement with the first equation in display (11.37). Also, Eq. (11.7) for the divergence of the stress tensor in components will be used; in particular, each component of this vector vanishes for Stokes flow.

Suppose there are two solutions (U_1, P_1) and (U_2, P_2) of the Stokes equations both satisfying the same boundary condition: the (same) velocity of the boundary is specified at each point of the boundary of some *bounded* domain $\mathcal{R}^*$ in the scaled coordinates. There are corresponding stress tensors σ_1 and σ_2 and corresponding strain rate tensors e_1 and e_2. Set $U = U_1 - U_2$, $P = P_1 - P_2$, $\sigma = \sigma_1 - \sigma_2$, and $e = e_1 - e_2$. Note that $U = 0$ everywhere on the boundary of the domain. The strategy of the proof is to sum over the squares of the components of e, integrate this quantity over the region $\mathcal{R}^*$, and prove that the value of this integral is zero. Once this is accomplished, each squared quantity must be zero; that is, $e_{ij} = 0$ for each choice of i and j. In particular, $e_{ii} = 0$. Using the definition of e, this implies

$$2\frac{\partial U_i}{\partial X_i} = 0.$$

Choose a point $a = (a_1, a_2, a_3)$ in the domain and, for example, let $i = 1$. By the vanishing of the partial derivative with respect to X_1, the function

$s \mapsto U_1(a_1 + s, a_2, a_3)$ is a constant. As s increases from zero, the path $s \mapsto (a_1 + s, a_2, a_3)$ must cross the boundary of the bounded domain $\mathcal{R}^*$. Thus, the function U_1 has the same value at a and at the crossing point on the boundary. Because $U_1 = 0$ on the boundary, U_1 is zero at a. By the same argument for the other choices of i, the vector field U is proved to be zero everywhere in the domain. Thus, $U_1 = U_2$. Using Stokes' equation, $\nabla P_1 = \nabla P_2$. It follows that $P_1 - P_2$ is a constant. Thus, the two solutions are the same, as claimed.

For notational convenience in showing that the integral of the sum of the squares of the components of e over the domain $\mathcal{R}^*$ vanishes, use the Einstein convention: summation of repeated indices is implied over the specified range(s). For example, under this convention and the specification that all indices range over $\{1, 2, 3\}$, the quantity a_{ii} appearing in a formula is an abbreviation for $a_{11} + a_{22} + a_{33}$. The quantity $e_{ij}e_{ij}$ in the remaining part of the proof is the double sum over the same range.

The objective of the remaining part of the proof is to show that

$$J := \int_{\mathcal{R}^*} e_{ij}e_{ij}\, d\mathcal{V} = 0.$$

From the Stokes equations, $\nabla \cdot U = 0$. Using this assumption and the symmetry of the stress tensor,

$$\begin{aligned} J &:= \int_{\mathcal{R}^*} \sigma_{ij}e_{ij}\, d\mathcal{V} + \int_{\mathcal{R}^*} Pe_{ii}\, d\mathcal{V} \\ &= \int_{\mathcal{R}^*} \sigma_{ij}\Big(\frac{\partial U_i}{\partial X_j} + \frac{\partial U_j}{\partial X_i}\Big)\, d\mathcal{V} + 2\int_{\mathcal{R}^*} P\frac{\partial U_i}{\partial X_i}\, d\mathcal{V} \\ &= 2\int_{\mathcal{R}^*} \sigma_{ij}\frac{\partial U_i}{\partial X_j}\, d\mathcal{V}. \end{aligned}$$

Using the integration by parts formula (A.1), which is an easy consequence of Green's theorem,

$$J = 2\Big(-\int_{\mathcal{R}^*} \frac{\partial \sigma_{ij}}{\partial X_j} U_i + \int_{\partial\mathcal{R}^*} \sigma_{ij}U_i\eta_j\, dS\Big),$$

where η is the outer unit normal on the boundary. The second integral vanishes because $U = 0$ on the boundary; the first integral vanishes because $\nabla \cdot \sigma = 0$. To show the latter fact, use Eq. (11.7). This completes the proof.

As an example where the uniqueness result has physical meaning, suppose a fluid is enclosed in a solid container that is moving with zero velocity. Stokes' model has zero Dirichlet boundary condition $U = 0$. Clearly the flow velocity $U = 0$ and $P = 0$ is a solution of this boundary value problem. By the uniqueness result, it is the only solution. Thus, Stokes flow in a bounded domain surrounded by a solid boundary is complete stagnation unless some portion of the solid boundary has nonzero velocity. See Section 19.7 for more on Stokes flows surrounded by solid boundaries.

Exercise 11.9. Prove that Poisson's equation $\Delta u = f$ on a bounded domain with Dirichlet boundary condition $u = g$ on the boundary for given f and g has at most one solution. Hint: Suppose there are two solutions, define their difference to be U, and consider the integral of $\nabla U \cdot \nabla U$ over the domain. What regularity (smoothness) assumptions do you need to make your argument rigorous. Comment: The proof that solutions exist requires some new concepts; but, it is part of the subject matter in the basic study of PDEs (see, for example, [103]).

Exercise 11.10. (a) Show that the pressure in Stokes' flow is a harmonic function; that is, $\Delta P = 0$. (b) Show that the Stokes velocity field satisfies the homogeneous biharmonic equation; that is, $\Delta\Delta U = 0$. In particular, each component of U satisfies the biharmonic equation $\nabla^4\zeta := \Delta\Delta\zeta = 0$. (c) Suppose that c is a constant vector, ζ is a function, and set

$$U_i = (\frac{\partial\zeta}{\partial X_i \partial X_j} - \delta_{ij}\zeta)c_j, \qquad P = \frac{\partial\Delta\zeta}{\partial X_j}c_j.$$

Show that (U, P) is a solution of Stokes' equations if and only if ζ satisfies the biharmonic equation. As a consequence, this result shows how to construct solutions of Stokes' equations from the solution of the scalar biharmonic equation.

CHAPTER 12

Flow in a Pipe

As an example of the solution of a fluid flow problem, let us consider perhaps the most basic example of the subject: a special case of flow in a round pipe.[1]

Consider cylindrical coordinates r, θ, and z where the z-axis is the axis of symmetry of a round pipe with radius a. More precisely, we consider the coordinate transformation

$$x_1 = r\cos\theta, \quad x_2 = r\sin\theta, \quad x_3 = z.$$

The basis vector fields for cylindrical coordinates are defined in terms of the usual basis of Euclidean space by

$$e_r := (\cos\theta, \sin\theta, 0), \quad e_\theta := (-\sin\theta, \cos\theta, 0), \quad e_z := (0, 0, 1).$$

We denote the coordinates of a vector field F on the cylinder $\{(r, \theta, z) : r \leq a\}$ with respect to the basis vector fields e_r, e_θ, and e_z by F_r, F_θ, and F_z, respectively. *The use of subscripts to denote coordinates in this section must not be confused with partial derivatives of F.*

It is natural to expect that there are *some* flow regimes for which the velocity field has its only nonzero component in the axial direction of the pipe; that is, the velocity field has the form

$$u(r, \theta, z, t) = (0, 0, u_z(r, \theta, z, t)), \tag{12.1}$$

where the components of this vector field are taken with respect to the basis vector fields e_r, e_θ, e_z, and u_z denotes the z component of the field (not the partial derivative with respect to z).

We will express the Euler and the Navier–Stokes equations in cylindrical coordinates. For a function f and a vector field $F = F_r e_r + F_\theta e_\theta + F_z e_z$ on

[1]The general nature of flow in a round pipe is not completely understood (see, for example, B. Eckhardt (2008), Turbulence transition in pipe flow: some open questions. *Nonlinearity*. **21**, T1–T11.)

An Invitation to Applied Mathematics: Differential Equations, Modeling, and Computation.
http://dx.doi.org/10.1016/B978-0-12-804153-6.50012-9, Copyright © 2017 Elsevier Inc. All rights reserved.

Euclidean space, the basic operators are given in cylindrical coordinates by

$$\begin{aligned}\nabla f &= \frac{\partial f}{\partial r}e_r + \frac{1}{r}\frac{\partial f}{\partial \theta}e_\theta + \frac{\partial f}{\partial z}e_z,\\ \nabla \cdot F &= \frac{1}{r}\frac{\partial}{\partial r}(rF_r) + \frac{1}{r}\frac{\partial F_\theta}{\partial \theta} + \frac{\partial F_z}{\partial z},\\ \Delta f &= \frac{1}{r}\frac{\partial}{\partial r}\big(r\frac{\partial f}{\partial r}\big) + \frac{1}{r^2}\frac{\partial^2 f}{\partial \theta^2} + \frac{\partial^2 f}{\partial z^2}\end{aligned} \tag{12.2}$$

(see Exercise 12.2). To obtain the incompressible Navier–Stokes equations in cylindrical coordinates, consider the unknown velocity field $u = u_r e_r + u_\theta e_\theta + u_z e_z$. Write this vector field in the usual Cartesian components by using the definitions of the direction fields given above, insert the result into the Navier–Stokes equations, and then compute the space derivatives using the operators given in display (12.2). After multiplication of the vector consisting of the first two of the resulting component equations (that is, the equations in the directions e_x and e_y) by the matrix

$$\begin{pmatrix} \cos\theta & \sin\theta \\ -\sin\theta & \cos\theta \end{pmatrix},$$

we obtain the equivalent system

$$\begin{aligned}\frac{\partial u_r}{\partial t} + (u\cdot\nabla)u_r - \frac{1}{r}u_\theta^2 &= \frac{1}{\mathrm{Re}}\big(\Delta u_r - \frac{1}{r^2}(u_r + 2\frac{\partial u_\theta}{\partial \theta})\big) - \frac{\partial p}{\partial r},\\ \frac{\partial u_\theta}{\partial t} + (u\cdot\nabla)u_\theta + \frac{1}{r}u_r u_\theta &= \frac{1}{\mathrm{Re}}\big(\Delta u_\theta - \frac{1}{r^2}(u_\theta - 2\frac{\partial u_r}{\partial \theta})\big) - \frac{1}{r}\frac{\partial p}{\partial \theta},\\ \frac{\partial u_z}{\partial t} + (u\cdot\nabla)u_z &= \frac{1}{\mathrm{Re}}\Delta u_z - \frac{\partial p}{\partial z},\\ \nabla\cdot u &= 0,\end{aligned} \tag{12.3}$$

where the operators ∇ and Δ are represented in cylindrical coordinates.

The Euler equations in cylindrical coordinates for the fluid motion in the pipe are obtained from system (12.3) by deleting the terms that are divided by the Reynolds number. If the velocity field u has the form given in Eq. (12.1), then u automatically satisfies the appropriate boundary condition for the incompressible Euler equation; that is, the Neumann boundary condition $\frac{\partial u}{\partial n} = 0$, where n is a unit normal on the cylinder that models the wall of the pipe. Thus, the Euler equations for the (scaled) velocity and pressure fields u and p reduce to the system

$$\frac{\partial p}{\partial r} = 0, \quad \frac{\partial p}{\partial \theta} = 0, \quad \frac{\partial u_z}{\partial t} + u_z \frac{\partial u_z}{\partial z} = -\frac{\partial p}{\partial z}, \quad \frac{\partial u_z}{\partial z} = 0.$$

The first two equations imply that p is a function of z and t only; the second two equations imply that

$$\frac{\partial u_z}{\partial t} = -\frac{\partial p}{\partial z}. \tag{12.4}$$

After differentiation of Eq. (12.4) with respect to z, it follows that $\partial^2 p/\partial z^2 = 0$. Therefore, $p = \alpha + \beta z$ for some functions α and β that depend only on t. Using Eq. (12.4), $u_z = v(r, \theta) - \int_0^t \beta(s)\, ds$ for an arbitrary choice of initial velocity v that may depend on r and θ, but not on z. The general solution for the class of velocities that have zero first and second components is

$$\begin{aligned} u(x, y, z, t) &= (0, 0, v(r, \theta) - \int_0^t \beta(s)\, ds), \\ p(x, y, z, t) &= \alpha(t) + \beta(t) z. \end{aligned}$$

The no-penetration boundary condition is always satisfied under the assumption that the first two components of u vanish. The no-slip boundary condition can also be satisfied by taking $\beta = 0$ and $v(a, \theta) = 0$ for all θ, where a is the radius of the pipe.

The general solution is for a pipe with infinite length. A more realistic scenario requires a pressure differential to push the flow. Suppose, for example, the pressure has the constant values p_0 at $z = 0$ and p_1 at $z = 1$, and $p_0 > p_1$. The model should predict a nonzero flow velocity with positive u_z component. In this case, the pressure must be

$$p = p_0 + (p_1 - p_0) z$$

and the third component of velocity is

$$u_z = v(r, \theta) + (p_0 - p_1) t.$$

The flow moves in the expected direction, but the velocity increases without bound as t grows to infinity. Thus, the model predicts an unrealistic flow independent of the choice of initial velocity v.

Note that for arbitrary α, the pair of functions $u_z = 0$ and $p = \alpha(t)$ is a solution of the Euler model with appropriate boundary condition. At first sight it might seem that this simple example shows that solutions of Euler's equations, with zero initial velocity and the no-penetration boundary condition, do not have unique solutions, but this is not the case: the nonuniqueness of pressure is allowed. The reason is simple: the gradient of the pressure appears in the equations of motion. Thus, the pressure is never unique in a fluid flow problem; it is required to be unique up to the addition of a constant function of time.

There is one useful physically reasonable (but not physically realistic) Euler pipe flow. For $\beta = 0$, the pressure does not depend on the position in the pipe and the fluid velocity field is constant with respect to z (along the flow direction in the pipe). This idealization is called *plug flow.* Because of its mathematical simplicity, plug flow is often used as a model. For example, plug flow is often used to model flow in tubular reactors studied in chemical engineering. At least this flow stays bounded.

What about Navier–Stokes flow?

By considering the same pipe, the same coordinate system, and the same hypothesis about the direction of the velocity field, the Navier–Stokes equations reduce to

$$\frac{\partial p}{\partial r} = 0, \quad \frac{\partial p}{\partial \theta} = 0, \quad \frac{\partial u_z}{\partial t} + u_z \frac{\partial u_z}{\partial z} = \frac{1}{\mathrm{Re}} \Delta u_z - \frac{\partial p}{\partial z}, \quad \frac{\partial u_z}{\partial z} = 0,$$

with the no-slip boundary condition at the wall of the pipe given by

$$u_z(a, \theta, z, t) \equiv 0.$$

This apparently simple system of fluid equations is difficult to solve, but we can obtain a solution under two additional assumptions: The velocity field is in steady state and it is symmetric with respect to rotations about the central axis of the pipe. In other words, the partial derivatives of u with respect to t and θ vanish. With these assumptions and taking into account that $\partial u_z / \partial z = 0$, it suffices to solve the single equation

$$\frac{1}{\mathrm{Re}} \Big(\frac{1}{r} \frac{\partial}{\partial r} \big(r \frac{\partial u_z}{\partial r} \big) \Big) = \frac{\partial p}{\partial z}.$$

Because $\partial p/\partial r = 0$ and $\partial p/\partial \theta = 0$, $\partial p/\partial z$ depends only on z and the left-hand side of the last equation depends only on r. Thus, the functions on both sides of the equation must have the same constant value, say β.

Under the assumptions just imposed, $p = \alpha + \beta z$ and

$$\frac{d(ru_z'(r))}{dr} = (\beta \,\mathrm{Re})r$$

with the initial condition $u_z(a) = 0$. The general solution of this ordinary differential equation (ODE) has a free parameter because there is only one initial condition given for a second-order differential equation. The term with the free parameter contains the factor $\ln r$, which blows up as r approaches zero (that is, at the center of the pipe). Solutions with this property are discarded because they do not agree with experience. With the free parameter set to zero, the solution

$$u = (0, 0, u_z(r)) = (0, 0, \frac{1}{4}\beta \,\mathrm{Re}\,(r^2 - a^2))$$

is continuous in the pipe and physically realistic. This steady state velocity field, called *Poiseuille flow*, predicted from the Navier–Stokes model is parabolic with respect to the radial coordinate with the fastest flow at the center of the pipe and flow velocity zero at the pipe wall.

Poiseuille flow is a close approximation to physical flow in a pipe for small Reynolds numbers. As the Reynolds number is increased, a critical value is reached at which the radial symmetry hypothesis used to obtain the Poiseuille flow is violated. For a Reynolds number above this critical value, the steady state physical flow measured in experiments and the velocity field given as a solution of the Navier–Stokes model become more complex. The flow regime changes from laminar flow to turbulent flow (the regime where, among other properties, eddies are found at all length scales). The causes of the onset of turbulent flow, the transition from laminar to turbulent flow, and the nature of fully turbulent flow are not fully understood; they are important unsolved problems in physics and applied mathematics.

Exercise 12.1. Consider Poiseuille flow in a section of length L of an infinite round pipe with radius a. Suppose that the pressure is p_{in} at the inlet of the section and the flow speed at the center of the pipe is v_{in}. Determine the pressure at the outlet. What happens in the limit as the Reynolds number grows without bound? Compare with the prediction of plug flow.

Exercise 12.2. (a) Write a detailed derivation of the gradient, divergence, and Laplacian in cylindrical coordinates (see display (12.2)). There are several methods that can be employed. One method begins by defining $f^c(r, \theta, z) = f(r\cos\theta, r\sin\theta, z)$ so that $f(x, y, z) = f^c(\sqrt{x^2 + y^2}, \arctan(y/x), z)$ and continues by differentiating the latter equality with respect to x, y, and z. (b) Derive the expressions for the gradient, divergence, and Laplacian in spherical coordinates.

Exercise 12.3. [Hagen–Poiseuille law] Return to unscaled variables and reconsider steady state Poiseuille flow in a round pipe of radius a. With the z coordinate in the direction of the pipe axis, the equations of motion are

$$p_x = 0, \qquad p_y = 0, \qquad p_z = \mu(v_{xx} + v_{yy}), \qquad v_z = 0$$

and the boundary condition is no-slip at the pipe wall. (a) Write out the full physically realistic general solution for the pressure and velocity. (b) Suppose the pressure at $z = 0$ is p_0 and the pressure at some $z = \ell > 0$ is p_1 with $p_0 > p_1$. The pressure drop on this section of pipe is $p_0 - p_1$. Determine the relation between the pressure drop and the volumetric flow rate in the pipe (volume per time of fluid passing a given cross section of the pipe). This relation is called the Hagen–Poiseuille law. (c) Determine the average velocity through a cross section.

Exercise 12.4. Determine the profile for Stokes flow in a round pipe of radius a under the assumption that the flow is invariant with respect to rotations around the central axis of the pipe.

Exercise 12.5. Polar coordinates were used in this section to show their use in fluid problems, but they are not necessary to determine the Euler or Poiseuille flow. Write out the derivation of these results using Cartesian coordinates.

CHAPTER 13

Eulerian Flow

13.1 BERNOULLI'S FORM OF EULER'S EQUATIONS

A flow is called isentropic if there is some function q called the enthalpy, such that[1]

$$\operatorname{grad} q = \frac{1}{\rho} \operatorname{grad} p. \tag{13.1}$$

More precisely, q is the enthalpy per unit of mass; it has units of energy/mass usually measured in Joules/kilogram or square meter/second/second. In case the density is constant, the flow is isentropic with $q := p/\rho$.

Using the vector identity

$$\frac{1}{2} \operatorname{grad}(u \cdot u) = u \times \operatorname{curl} u + (u \cdot \nabla)u,$$

Euler's equation of motion [Eq. (11.20)] may be recast in the form

$$u_t - u \times \operatorname{curl} u = b + \operatorname{grad}(-\frac{1}{2}(u \cdot u) - q).$$

With

$$\alpha := -\frac{1}{2}|u|^2 - q\,,$$

Bernoulli's form of Euler's equations for incompressible flow is

$$\begin{aligned} u_t &= u \times \operatorname{curl} u + \operatorname{grad} \alpha + b, \\ \operatorname{div} u &= 0, \\ u \cdot \eta &= 0 \quad \text{in } \partial\mathcal{R}. \end{aligned} \tag{13.2}$$

The curl of the flow velocity field u appears in Bernoulli's equations and plays an important role in fluid motion. This quantity $\operatorname{curl} u$ (or $\nabla \times u$ in

[1]Nabla notation has been used in this book for the discussion of fluid motion. Here the notation is switched to div, grad, and curl so the reader will gain experience in reading literature that uses this alternate notation.

An Invitation to Applied Mathematics: Differential Equations, Modeling, and Computation.
http://dx.doi.org/10.1016/B978-0-12-804153-6.50013-0, Copyright © 2017 Elsevier Inc. All rights reserved.

nabla notation) is called the vorticity field of the flow; and, a flow is called irrotational (sometimes curl free) if its vorticity vanishes everywhere.

In case the flow velocity is steady (does not depend on time), *irrotational*, and the body force is conservative ($b = \operatorname{grad} \beta$), the first equation in display (13.2) may be recast in the form

$$\operatorname{grad}(\frac{1}{2}(u \cdot u) + p/\rho + \beta) = 0.$$

Thus, in this case, there must be a constant C such that

$$\frac{1}{2}(u \cdot u) + p/\rho + \beta = C.$$

The most important example, when the body force is gravity and z is the vertical spatial coordinate, is Benoulli's equation

$$\frac{1}{2}(u \cdot u) + p/\rho + gz = C.$$

Fluid particle motion for the velocity field $u = (u_1, u_2, u_3)$ is governed by the ordinary differential equations (ODEs)

$$\dot{x} = u_1(x, y, z, t), \quad \dot{y} = u_2(x, y, z, t), \quad \dot{z} = u_3(x, y, z, t).$$

More precisely, the path of a fluid particle starting at time t_0 at position (x_0, y_0, z_0) (called its path line) is the parametric curve $t \mapsto (x(t), y(t), z(t))$ that satisfies these ODEs and the initial data $(x(0), y(0), z(0)) = (x_0, y_0, z_0)$. For each fixed t, the curves in space that are traced out by solutions $s \mapsto (x(s), y(s), z(s))$ of the ODEs

$$\frac{dx}{ds} = u_1(x, y, z, t), \quad \frac{dy}{ds} = u_2(x, y, z, t), \quad \frac{dz}{ds} = u_3(x, y, z, t)$$

are called streamlines. Thus, the pattern of streamlines changes at each instant of time. For steady flows (no dependence on t), streamlines and the curves traced out by the path lines coincide.

For steady flow, a conservative body force, and with the Bernoulli function

$$B := \frac{1}{2}(u \cdot u) + p/\rho + \beta,$$

Bernoulli's form of Euler's equations yields the vector equation

$$\operatorname{grad} B = u \times \operatorname{curl} u.$$

Consider the time derivative of B along a streamline and note that (by the properties of cross product)

$$\frac{d}{dt}B(x(t), y(t), z(t)) = \operatorname{grad} B \cdot u = (u \times \operatorname{curl} u) \cdot u = 0. \tag{13.3}$$

This is an important fact: For steady flow with conservative body force, the Bernoulli function is constant along streamlines.

13.2 POTENTIAL FLOW

An irrotational Eulerian (no viscosity) flow is called a potential flow.

From vector analysis, an irrotational field is locally the gradient of some function. More precisely, at each point in a region throughout which the curl of a vector field vanishes, there is an open disk containing the point and a function defined on this disk whose gradient is the curl free vector field. Recall that a set is called simply connected if every loop can be shrunk to a point without leaving the set. Otherwise the set is called multiply connected. For example, a disk in the plane is simply connected and an annulus is multiply connected. Also, from vector calculus, an irrotational vector field on a simply connected domain is the gradient of a function defined on the entire domain; the vector field is globally a gradient. Such a function is called a potential for the vector field and thus the name potential flow. In a multiply connected domain an irrotational vector field may not be the gradient of a function defined on the entire domain.

Bernoulli's form of the incompressible Euler's equations [Eqs. (13.2)] for the velocity field u of a potential flow in a region $\mathcal{R}$ are given by

$$\begin{aligned} u_t &= -\operatorname{grad}(\frac{1}{2}|u|^2 + \frac{p}{\rho}) + b, \\ \operatorname{div} u &= 0, \\ u \cdot \eta &= 0 \quad \text{in } \partial\mathcal{R}. \end{aligned} \tag{13.4}$$

Suppose that $u = \operatorname{grad} \phi$ on all of $\mathcal{R}$ and the body force is the gradient of a potential β so that $b = \operatorname{grad} \beta$. By substitution into these partial differential equations (PDEs) and some rearrangement, the equations of motion can be recast into the form

$$\operatorname{grad}\Big(\frac{\partial \phi}{\partial t} + \frac{1}{2}|\operatorname{grad}\phi|^2 + \frac{p}{\rho} - \beta\Big) = 0, \qquad \Delta\phi = 0. \tag{13.5}$$

As a result, there is a number C, constant with respect to the space variable, such that

$$\frac{\partial \phi}{\partial t} + \frac{1}{2}|\operatorname{grad}\phi|^2 + \frac{p}{\rho} - \beta = C. \tag{13.6}$$

When the body force is gravity and the flow is steady (the velocity field is not changing with time), Bernoulli's law states that (at a height h above a reference surface)

$$p + \frac{1}{2}\rho|u|^2 + g\rho h = C. \tag{13.7}$$

Suppose the density ρ is constant and the body force b is given by a potential β so that $\operatorname{grad}\beta = b$. In this case, every (sufficiently smooth) time independent and irrotational vector field u (that is, a field with $u_t = 0$ and $\operatorname{curl} u = 0$) satisfies the incompressible Euler equations

$$u \cdot \nabla u = -\nabla(p/\rho) + \nabla\beta, \qquad \nabla \cdot u = 0$$

with the pressure p defined by

$$p = -\frac{1}{2}\rho|u|^2 + \rho\beta. \tag{13.8}$$

An important consequence of Bernoulli's law is Bernoulli's principle: At constant height, if the velocity of an incompressible fluid increases, then its pressure decreases.

Bernoulli's principle is often used as an explanation of lift caused by flow over an airplane wing. The usual argument is that a wing produces lift because the velocity of air relative to the wing is greater for flow over the top of the wing compared to flow over its bottom; thus, the pressure is less on top and the pressure difference between top and bottom produces lift. Exactly why the flow is faster over the top of an airplane wing can be explained to some extent, but not simply (see, for example, [60]). Also, the Bernoulli principle's argument is somewhat undercut by the observation that airplanes can fly upside down—though not efficiently—when the pitch is adjusted appropriately. An alternate explanation of why airplanes fly includes the observation that the wings and the hull are pitched (angle of attack) in flight to direct more airflow downward than upward. This is supposed to produce a (lift) force in the up direction by Newton's third law of motion. It is not clear, however, that this is a correct application of Newton's law. Because a simple correct physical argument for why airplanes can fly is not available, there is no shortage of discussion and controversy on this topic.

From an applied mathematics point of view, the explanation of lift does not require a simple physical explanation; the phenomenon is explained via the modeling process. Basic physics is used to derive the Navier–Stokes model for fluid motion. Application of this model for aerodynamic airflow (that is, the identification of an appropriate domain around an airplane wing or the entire airplane and the imposition of the no-slip boundary condition) will produce a solution such that the pressure difference is a close approximation to the measured lift. Experimental evidence to the contrary (which does not exist at present) would require a new explanation. Either the basic physics used to construct the Navier–Stokes model is wrong, the approximations used to construct the model are too crude, or the model does not incorporate all the relevant physics. Thus, a full explanation lies in the physics of fluid motion, not the particular phenomenon of lift. Unfortunately, the Navier–Stokes equations are not easy to solve. Thus, there is no simple way to determine the predictions of the model. This fact fuels the desire for a more direct intuitive explanation of lift, which (due to the complexity of fluid flow) is probably impossible. Why not embrace the modeling process instead? The correct approach to understanding (from this point of view) is to explore the Navier–Stokes model until its predictions do not agree with experiment or observation. Should this happen, abandon or modify the model and test its predictions against the experiments. Perhaps life is not this easy, but this is the correct point of view.

Exercise 13.1. Can a fluid particle undergo acceleration in a steady flow? Hint: The motion of a particle is governed by the ODE $\dot{x} = u(x, t)$. It's velocity is $\dot{x}$.

Exercise 13.2. Consider an open tank containing water with a round hole in the bottom of the tank. Show that the velocity of the fluid in the drain is (approximately) proportional to the square root of the depth of the fluid in the tank. More precisely, this velocity is approximately $\sqrt{2gh}$, where h is the depth of the water in the tank and g is the acceleration due to gravity. What assumptions are required to make the velocity exactly $\sqrt{2gh}$.

Exercise 13.3. Consider flow with density ρ and viscosity μ in a river with a flat bottom whose slope is $-\tan\theta$ with respect to the horizontal where θ is a small angle. Ignore the influence of the banks of the river on the flow, assume that *the flow velocity is all in the downstream direction*, and the body force driving the flow is gravity. In addition, assume—as an approximation—that the flow near the surface of the river forgets that it is viscous and behaves as if it is Eulerian. (a) Show that the velocity at z units above the river bottom is $u = \frac{1}{2\mu}\rho g z(2h-z)\sin\theta$. (b) The Mississippi River drops 2.5 inches per mile. Suppose its average depth is 30 feet and width is 1 mile. Compute the surface velocity of the river. Does the velocity computed using these assumptions agree with observation? Which, if any, of the assumptions is not realistic? Discuss.

Exercise 13.4. Assume that the curl of the body force on a fluid vanishes. (a) Show that the vorticity $\omega = \operatorname{curl} u$ satisfies the ODE

$$\dot{\omega} = \operatorname{curl}(u \times \omega).$$

(b) Show that the solution of the initial value problem

$$\dot{\omega} = \operatorname{curl}(u \times \omega), \qquad \omega(0) = 0$$

is $\omega = 0$. Interpretation: Euler flow with constant body force (such as gravity) that is initially irrotational does not develop vorticity in the future. (c) Show that (in two space dimensions) the vorticity satisfies the differential equation

$$\omega_t + (u \cdot \nabla)\omega = 0.$$

Is this equation the same as the one in part (a)? What about three dimensions?
(d) Suppose γ is a solution of the ODE system

$$\dot{x} = u_1(x, y, t), \qquad \dot{y} = u_2(x, y, t)$$

where $u = (u_1, u_2)$ is the fluid velocity field. Show that the vorticity is constant along γ. What is the interpretation of this result?

13.3 POTENTIAL FLOW IN TWO DIMENSIONS

Imagine, as an example of two-dimensional flow that occurs in applications, an invariant two-dimensional plane of fluid in a three-dimensional fluid flow (that is, fluid particles starting on this plane stay on the plane). The flow equations restricted to a two-dimensional plane are more amenable to analysis than the full three-dimensional model. Some of the properties of two-dimensional potential flow are discussed in this section. A few results from the theory of complex variables are used. Accept these results without proof if you have not studied functions of complex variables.

Consider a *steady* fluid flowing on a plane with Cartesian coordinates (x, y) whose velocity field u is given by a potential ϕ; that is, $u = \nabla\phi$. In view of the second equation of system (13.5) (that is, $\Delta\phi = 0$), the potential is a harmonic function. By some basic theory of functions of complex variables, ϕ is *locally* the real part of a holomorphic function, say $f = \phi + i\psi$, and the pair ϕ, ψ satisfies the Cauchy–Riemann equations

$$\frac{\partial\phi}{\partial x} = \frac{\partial\psi}{\partial y}, \qquad \frac{\partial\phi}{\partial y} = -\frac{\partial\psi}{\partial x}.$$

Thus, the assumption that $u = \operatorname{grad}\phi$ implies that the fluid motions are solutions of an ODE that can be viewed in two different ways: as the gradient system

$$\dot{x} = \frac{\partial\phi}{\partial x}, \qquad \dot{y} = \frac{\partial\phi}{\partial y}; \tag{13.9}$$

or the Hamiltonian system

$$\dot{x} = \frac{\partial\psi}{\partial y}, \qquad \dot{y} = -\frac{\partial\psi}{\partial x}. \tag{13.10}$$

The function ψ, a Hamiltonian function for system (13.10), is the *stream function;* but, this is not the definition of the stream function (see Exercise 13.6). The orbits of system (13.10), called *streamlines,* all lie on level sets of ψ.

Because streamlines are also orbits of the gradient system (13.9), there are no (nonconstant) periodic motions of fluid particles for steady potential flow. In fact, an autonomous gradient system (such as the gradient system (13.9)) cannot have (nonconstant) periodic solutions. (Why?)

It should be clear that function theory can be used to study planar potential flow. For example, if ψ is a harmonic function defined in a simply connected region of the complex plane such that the boundary of the region is a level set of ψ, then ψ is the imaginary part of a holomorphic function defined in the region, and therefore ψ is the stream function of a steady state flow. This fact can be used to find steady state solutions of Euler's equations.

Consider plug flow in a round pipe with radius a and notice that every planar slice containing the axis of the pipe is invariant under the flow. Thus, it seems reasonable to consider two-dimensional flow on the strip

$$\mathcal{S} := \{(x, y) : 0 < y < 2a\},$$

which is viewed as such a slice where x as the axial direction and the center of the pipe lies on the line with equation $y = a$. The plug flow solution of Euler's equations in $\mathcal{S}$ is given by the velocity field $u = (c, 0)$ and pressure $p = p_0$, where c and p_0 are constants. It is a potential flow, with potential $\phi(x, y) = cx$, stream function $\psi(x, y) = cy$, and *complex potential* $f(x, y) = cz = cx + icy$.

Suppose that Q is an invertible holomorphic function defined on $\mathcal{S}$ and that $\mathcal{R}$ is the image of $\mathcal{S}$ under Q, then $w \mapsto f(Q^{-1}(w))$ for $w \in \mathcal{R}$ is a holomorphic function on $\mathcal{R}$ with real part $w \mapsto \phi(Q^{-1}(w))$ and imaginary part $w \mapsto \psi(Q^{-1}(w))$. Thus, the function given by $w \mapsto \psi(Q^{-1}(w))$ is a stream function for a steady state potential flow in $\mathcal{R}$ with potential $w \mapsto \phi(Q^{-1}(w))$. In particular, streamlines of ψ map to streamlines of $w \mapsto \psi(Q^{-1}(w))$. And this flow is a solution of Euler's equation in the domain $\mathcal{R}$.

For example, note that $w := Q(z) = \sqrt{z}$ has a holomorphic branch defined on the strip $\mathcal{S}$ such that this holomorphic function maps $\mathcal{S}$ into the region $\mathcal{R}$ in the first quadrant of the complex plane bounded above by the hyperbola $\{(\sigma, \tau) : \sigma\tau = a\}$. In fact, $Q^{-1}(w) = w^2$ so that

$$x = \sigma^2 - \tau^2, \qquad y = 2\sigma\tau.$$

The velocity field of the new (corner) flow is

$$(2c\sigma, -2c\tau),$$

and the corresponding pressure is found from Bernoulli's equation (13.7). In fact, there is a constant p_1 such that

$$p = p_1 - 2c^2(\sigma^2 + \tau^2). \tag{13.11}$$

Streamlines for the flow at a corner are all hyperbolas.

The flow near a wall is essentially plug flow. In fact, if we consider, for example, the velocity field on a vertical line orthogonal to the σ-axis, say the line with equation $\sigma = \sigma_0$, then the velocity field near the wall, where $\tau \approx 0$, is closely approximated by the constant vector field $(2c\sigma_0, 0)$. In other words, the velocity profile is nearly linear.

Exercise 13.5. Consider the plug flow vector field $u = (c, 0)$ defined in a horizontal strip in the upper half plane of width $2a$. Find the push forward of u into the first quadrant with respect to the map $Q(z) = \sqrt{z}$ with inverse $Q^{-1}(w) = w^2$. Is this vector field a steady state solution of Euler's equations at the corner for an incompressible fluid? Explain.

Exercise 13.6. (a) Suppose that (u, v) are components of a (smooth) vector field in the plane whose divergence vanishes. Show that there is a function ψ, defined up to an additive constant, such that

$$u = \psi_y, \qquad v = -\psi_x.$$

This function ψ is called the stream function. (b) Show that the Laplacian of the stream function is the vorticity of the vector field. (c) Can a potential flow have nonzero vorticity?

13.4 CIRCULATION, LIFT, AND DRAG

A fundamental problem in applied fluid dynamics is to describe the motion of solid bodies moving in fluids as in aerodynamics, ballistics, and the motion of ships and submarines. Although numerical computation with the Navier–Stokes equations has superseded theoretical analysis for most realistic applications, the problem is certainly illuminated by making predictions from basic theory. This subject matter uses foundational applied mathematics, which is useful in many other contexts.

Imagine a three-dimensional body B immersed in a fluid. Although in reality the fluid is confined to some container, for the subject at hand it is natural to assume that the outer boundary of the fluid does not affect the fluid motion of the body as long as the container is much larger than the body. A notable exception is a scaled model in a wind tunnel where the fluid motion near the wall of the wind tunnel may have to be taken into account. As a consequence of Newton's law, the motion of the body is determined by the sum of the forces acting on it. Ignoring forces due to gravity or electromagnetism, which may later be included as summands of force, the forces acting on the body are due to the surrounding fluid. Two of the most important of these forces are the fluid pressure and the shear stress due to the no-slip boundary condition. Shear stress in the boundary layer is discussed briefly in Section 17.3. The force on B due to pressure is discussed here. In the aerodynamics literature, this force is called pressure drag. Drag due to shear stress is called viscous (or frictional) drag. In case the drag is mainly due to viscosity, the body is called streamlined; it is called blunt (or bluff) in case the drag is dominated by pressure. The purpose of streamlining by design is to reduce pressure drag.

Recall that pressure is force per area; thus, to obtain the total force due to pressure, the fluid pressure is integrated over the surface of the body. More precisely, let η denote the outer unit normal on B and p the pressure. The total pressure is

$$F := -\int_{\partial B} p\eta \, dS. \tag{13.12}$$

The minus sign is included to ensure that a positive pressure acts toward the interior of the body. Of course, F is a vector.

Although the more physically realistic Navier–Stokes equations should be used to determine the pressure force F, these equations are probably too complicated to analyze by elementary analysis; thus, Euler's equations are used instead. More precisely, the assumptions for this section are

(a) The density of the fluid is constant; that is, ρ is constant.
(b) The fluid velocity is irrotational; that is $\nabla \times u = 0$.
(c) No body forces act on the fluid or the body.
(d) The fluid is in steady state.
(e) The viscosity of the fluid is zero.
(f) The body B immersed in the flow is a finite solid (not necessarily round) cylinder whose axis is taken to be vertical in the usual Cartesian coordinates (see Fig. 13.1 for a cross sectional view); that is, in the usual Cartesian coordinates, when $(x, y, 0)$ is in the body so is (x, y, z) for all z in some closed interval $[c, d]$.
(g) The fluid velocity u has zero component in the direction of the axis of the cylinder; that is, $u = (u_1, u_2, 0)$ in components, the velocity does not depend on the axial coordinate, and it satisfies the no-penetration condition on the boundary of B.
(h) Far away from B, the velocity of the fluid is nearly constant with zero component in the direction of the axis of the cylinder; that is, $u = U + v$, where $U = (U_1, U_2, 0)$ is a constant vector field and the function $(x, y) \mapsto (x^2 + y^2)^{1/2}|v(x, y)|^2$ approaches zero uniformly as $x^2 + y^2$ grows without bound.

Consider a portion of the cylindrical body B of unit axial length, and consider the solid right circular cylinder $\{(x, y, z) : x^2 + y^2 \le a^2,\ 0 \le z \le 1\}$ with closed ends such that the radius of its base a is so large that this circular cylinder contains B. The boundary of the circular cylinder consists of its lateral boundary $\Sigma := \{(x, y, z) : x^2 + y^2 = a^2,\ 0 < z < 1\}$ and two end disks $D_0 := \{(x, y, z) : x^2 + y^2 = a^2,\ z = 0\}$ and $D_1 := \{(x, y, z) : x^2 + y^2 = a^2,\ z = 1\}$. Let Ω denote the complement of the set B in the solid circular cylinder. Also, let Γ denote the curve given by the intersection of B and the plane with equation $z = 0$; it is a curve on the boundary of a cross section of B.

Drag on the body B is defined to be the total pressure F in Eq. (13.12) in the direction of the constant field U. The mathematical expression for the drag is

$$\text{drag} = \frac{F \cdot U}{|U|} \frac{U}{|U|}.$$

Lift on B is defined to be the component of F in the direction, perpendicular to the constant field U in the cross sectional plane, given by $U^\perp := (-U_2, U_1, 0)/|U|$. Thus, the lift is

$$\text{lift} = (F \cdot U^\perp)U^\perp.$$

Because the orientation of the cylinder is taken for notational beauty so that the coordinates in its two-dimensional cross sections are (x, y), a lift force would move the cylinder side-to-side instead of up or down according to drawings of the usual Cartesian coordinates. Although this picture is perhaps not in concert with the usage of the word lift, the position of the cylinder in space is obviously of no importance for mathematical analysis as long at the coordinates are chosen with respect to this position.

One other quantity plays a fundamental role: the circulation of the flow velocity on closed curves. For Υ a closed oriented curve and u a vector field, the circulation of u on (or around) Υ is defined to be

$$\text{circulation} = \int_\Upsilon u \cdot ds,$$

where ds is the differential of arc length taken in the positive direction on the curve. Of course, the integrand is the dot product of the unit tangent vector to the curve in the positive direction and the vector field u. When the curve resides in a plane, which has a standard orientation (for example, counterclockwise in the usual Cartesian plane), positive orientation of the curve is assumed. In case the curve is the boundary of a surface Ξ, we have

Stokes's theorem:

$$\int_{\Upsilon} u \cdot ds = \int_{\Xi} (\nabla \times u) \cdot \eta \, d\mathcal{S}.$$

If the surface Ξ lies entirely in the fluid and the fluid is irrotational, then the circulation around $\Upsilon = \partial\Xi$ vanishes. Although there is a definite relationship between circulation and rotation, they are not the same. Imagine a body immersed in a fluid and note that the fluid vector field u is not defined inside the body where there is no fluid. For this reason, Stokes's theorem does not apply. Thus, it is possible to have nonzero circulation around a curve with respect to an irrotational vector field. Vorticity measures a local property of a flow; circulation is a global property.

One of the most important results concerning circulation, drag, and lift is the Kutta–Zhukovsky theorem (Martin Wilhelm Kutta and Nikolay Yegorovich Zhukovsky): Under the assumptions (a)–(h), with B the cylindrical body defined above, and Γ the closed curve surrounding B, the drag on B is zero and the lift is given by

$$\text{lift} = -\rho|U| \int_{\Gamma} u \cdot ds \, U^{\perp}. \tag{13.13}$$

In particular, nonzero lift requires nonzero circulation.

The proof given here uses two previously established facts about the pressure p and steady state fluid velocity u: the alternate form of the momentum balance [Eq. (11.18)], which is

$$\rho\nabla \cdot (u \otimes u + \frac{p}{\rho} I) = 0,$$

and the existence of a constant p_0 such that Bernoulli's law [Eq. (13.7)] holds in the form

$$p = -\frac{1}{2}\rho u \cdot u + p_0.$$

By the momentum balance and the divergence theorem,

$$0 = \int_{\Omega} \rho\nabla \cdot (u \otimes u + \frac{p}{\rho} I) \, d\mathcal{V} = \int_{\partial\Omega} \rho(u \otimes u \cdot \eta + \frac{p}{\rho}\eta) \, d\mathcal{S},$$

where η is the outer normal on the boundary of Ω, which is the union of the lateral boundary, the end disks, and the lateral boundary of B. The integral of $u \otimes u \cdot \eta$ over the lateral boundary of B vanishes because the fluid velocity (which satisfies the no-penetration boundary condition on B) is perpendicular to the normal on this boundary. The sum of the integrals over the end disks vanish for one of two reasons: the fluid velocity is perpendicular to the outer normals on these disks, and for the pressure and constant terms, the integrals are equal and of opposite signs on these disks. Thus, the vanishing of the boundary integral implies

$$-\int_{\partial B} p\eta \, d\mathcal{S} = \int_{\Sigma} \rho(u \otimes u \cdot \eta + \frac{p}{\rho}\eta) \, d\mathcal{S}, \tag{13.14}$$

where η is the inner normal on B. In particular, the desired total pressure F is given by

$$F = -\int_{\Sigma} \rho(u \otimes u \cdot \eta + \frac{p}{\rho}\eta) \, d\mathcal{S}.$$

Using Bernoulli's law,

$$F = -\int_{\Sigma} \rho(u \otimes u \cdot \eta - \frac{1}{2}u \cdot u\eta - \frac{p_0}{\rho}\eta) \, d\mathcal{S}.$$

Substitute $u = U + v$ and expand terms using linearity to obtain

$$\begin{aligned} F = -\rho \int_{\Sigma} & U \otimes U \cdot \eta + U \otimes v \cdot \eta + v \otimes U \cdot \eta + v \otimes v \cdot \eta - \frac{1}{2}U \cdot U\eta \\ & - U \cdot v\eta - \frac{1}{2}v \cdot v\eta - \frac{p_0}{\rho}\eta \, d\mathcal{S}. \end{aligned} \tag{13.15}$$

The right circular cylinder Σ is parameterized by $\gamma(\theta, s) = (a\cos\theta, a\sin\theta, s)$ with $0 \le \theta \le 2\pi$ and $0 < s < 1$. Recall that, for a continuous function g,

$$\begin{aligned} \int_{\Sigma} g \, d\mathcal{S} &= \int_0^1 \int_0^{2\pi} g(a\cos\theta, a\sin\theta, s) D\gamma(\theta, s)e_1 \times D\gamma(\theta, s)e_2 \cdot \eta \, d\theta ds \\ &= a \int_0^1 \int_0^{2\pi} g(a\cos\theta, a\sin\theta, s) \, d\theta ds. \end{aligned}$$

In the application of this formula to the terms in the integrand of the integral in Eq. (13.15), the hypotheses on the flow imply there is no functional dependence on s. Thus,

$$F = -\rho \int_0^{2\pi} U \otimes U \cdot \eta + U \otimes v \cdot \eta + v \otimes U \cdot \eta + v \otimes v \cdot \eta - \frac{1}{2} U \cdot U \eta - U \cdot v \eta - \frac{1}{2} v \cdot v \eta - \frac{p_0}{\rho} \eta \, d\theta. \quad (13.16)$$

Constant terms all integrate to zero as each of them reduces to integration of a sine or cosine over the interval $[0, 2\pi]$. The remaining integral, expressed in coordinates, is

$$\begin{aligned} F = & -a\rho \int_0^{2\pi} (v_1(\theta) \cos\theta + v_2(\theta) \sin\theta) \begin{pmatrix} U_1 \\ U_2 \end{pmatrix} \\ & + (U_1 \cos\theta + U_2 \sin\theta) \begin{pmatrix} v_1(\theta) \\ v_2(\theta) \end{pmatrix} \\ & - (U_1 v_1(\theta) + U_2 v_2(\theta)) \begin{pmatrix} \cos\theta \\ \sin\theta \end{pmatrix} \\ & - \frac{1}{2}(v_1(\theta)^2 + v_2(\theta)^2) \begin{pmatrix} \cos\theta \\ \sin\theta \end{pmatrix} d\theta. \end{aligned}$$

To determine the drag in the direction $U/|U|$, compute the dot product of F with U, expand all products, and make the obvious cancellations. The result, in vector form, is

$$\text{drag} = -a\rho(|U|^2 \int_0^{2\pi} v \cdot \eta \, d\theta + \frac{1}{2} \int_0^{2\pi} |v|^2 U \cdot \eta \, d\theta) \frac{U}{|U|^2}.$$

Because U is constant,

$$\int_0^{2\pi} U \cdot \eta \, d\theta = 0;$$

therefore

$$\int_0^{2\pi} v \cdot \eta \, d\theta = \int_0^{2\pi} u \cdot \eta \, d\theta,$$

where η is the outer normal on Σ. But by the divergence theorem and the assumption $\nabla \cdot u = 0$,

$$\begin{aligned} 0 &= \int_\Omega \nabla \cdot u \, d\mathcal{V} \\ &= \int_{\partial B} u \cdot \eta \, dS + \int_\Sigma u \cdot \eta \, dS \end{aligned}$$

$$= \int_{\Sigma} u \cdot \eta \, dS$$
$$= \int_{0}^{2\pi} u \cdot \eta \, d\theta$$
$$= \int_{0}^{2\pi} v \cdot \eta \, d\theta.$$

It follows that

$$\text{drag} = -\frac{\rho}{2} \int_{0}^{2\pi} (a|v|^2) U \cdot \eta \, d\theta \frac{U}{|U|^2}.$$

In view of assumption (h) and passing to the limit as a increases without bound, the drag vanishes.

To determine the lift in the direction $U^{\perp}$, compute the dot product of F with $U^{\perp}$, expand, and cancel to obtain

$$\text{lift} = \left(-\rho|U| \int_{0}^{2\pi} a(v_2(\theta)\cos\theta - v_1(\theta)\sin\theta)\, d\theta + \frac{\rho}{2} \int_{0}^{2\pi} a|v|^2 U^{\perp} \cdot \eta \, d\theta\right) U^{\perp}.$$

As for the calculation for drag, the second integral vanishes in the limit; the first integral (including the factor a in its integrand) is equal to the circulation

$$\int_{C} v \cdot ds,$$

where C is the circle centered at the origin with radius a. As before, because U is constant,

$$\int_{C} v \cdot ds = \int_{C} u \cdot ds. \tag{13.17}$$

Using Stokes's theorem,

$$\int_{A} \nabla \times u \cdot \eta \, dS = \int_{C} u \cdot ds + \int_{\Gamma} u \cdot ds,$$

where A is the annular surface bounded by the circle C centered at the origin with radius a, and Γ is the previously defined cross sectional boundary of B. Because $\nabla \times u = 0$ for the potential flow u, the circulation of u around B is the same as its circulation around C, which is the same as the circulation of v around C by Eq. (13.17). Hence,

$$\text{lift} = -\rho|U| \int_\Gamma u \cdot ds\, U^\perp. \tag{13.18}$$

This completes the proof of the Kutta–Zhukovsky theorem.

The proof given here is an excellent application of vector calculus. An alternative approach is to recognize and consider the two-dimensional nature of the result from the start and use complex function theory instead of vector analysis (see, for example, [1]).

The Kutta–Zhukovsky theorem should be surprising: it states that the fluid flow does not cause drag, and unless there is circulation, there is also no lift. This is often called d'Alembert's paradox; the result is contrary to experience. As we now know, the paradox arises by *ignoring viscosity* and compressibility, but this fact was not clear at the birth of aerodynamics. A better understanding was obtained much later with the advent of boundary layer theory. In this more realistic model, the viscosity is still not important and can be ignored *except* in a thin layer near the skin of the body where the fluid remembers it has viscosity and therefore sticks to the skin; that is, the fluid velocity vanishes on this boundary instead of merely satisfying the no-penetration condition. The no-slip boundary condition produces shear stress (skin friction), which is one of the effects that must be taken into account to model the drag force. For streamlined bodies, skin friction as explained by boundary layer theory is the dominant contribution to drag, at least for flows whose Reynolds' numbers are not too high. Drag due to pressure differences between the leading and trailing edges of the body (called pressure or form drag) is dominant for nonstreamlined bodies (called bluff or blunt bodies).

The Kutta–Zhukovsky lift formula (13.18) is a good approximation even for viscous flow; it can be used to predict the lift force for bodies of various shapes. To state the obvious, maximization of lift and minimization of drag—within other design constraints—are basic goals of aerodynamics.

Continuing with the classical theory, the simplest and natural example of a body that might be subjected to a flow is a right circular cylinder B. The cross section is a disk in the Cartesian plane of some radius $a > 0$. In the unbounded region Ω exterior to this disk with outer normal η on its finite circular boundary, the existence of a potential flow given by a velocity potential ϕ defined in Ω such that this velocity field has nearly constant velocity U far from the body (to model the body subjected to a constant freestream flow) is equivalent to the existence of a solution of the exterior Neumann BVP

$$\begin{aligned} \Delta\phi &= 0 \quad \text{in } \Omega, \\ \nabla\phi\cdot\eta &= 0 \quad \text{in } \partial\Omega, \\ \nabla\phi &\to U \quad \text{as } x^2+y^2\to\infty \end{aligned} \tag{13.19}$$

for the unknown velocity potential ϕ.

The circular geometry of the exterior Neumann problem (13.19) leads to a pencil and paper solution using polar coordinates.

The Laplacian Δ is easily converted to polar coordinates. Simply consider a scalar valued function f on the Cartesian plane so that the value of f at the point with coordinates (x, y) is $f(x, y)$. The same function in polar coordinates, F, is given by $F(r,\theta) = f(r\cos\theta, r\sin\theta)$ and $f(x,y) = F(\sqrt{x^2+y^2}), \arctan(y/x))$, ignoring for the moment division by zero when $x = 0$. Away from this value of x, simply compute the Laplacian of the right-hand side of the latter equation using the chain rule and substitute r for $\sqrt{x^2+y^2}$ and θ for $\arctan(y/x)$ in the result to obtain

$$\Delta F = \frac{1}{r}\frac{\partial}{\partial r}(r\frac{\partial F}{\partial r}) + \frac{1}{r^2}\frac{\partial^2 F}{\partial\theta^2}. \tag{13.20}$$

This formula is singular at $r = 0$ due to the singularity of polar coordinates at the origin; the underlying problem, which is often ignored when using polar coordinates, is that there is no global coordinate system on the circle. This is the source of the problem using the arctangent function to define the angular coordinate θ. Indeed, the formula $\theta = \arctan(y/x)$ is only valid for $x > 0$. Another formula, $\theta = \pi/2 - \arctan(x/y)$, gives the angular coordinate for $y > 0$. There are infinitely many such formulas. The proper way to treat the circle and the related polar coordinates on the punctured plane (the subset of the plane with the origin removed) requires some ideas from the theory of differentiable manifolds, but this level of sophistication is not necessary here. All that is required is careful attention to the use of the polar coordinates. When in doubt, all results can be checked by simply transforming back to Cartesian coordinates. A detailed discussion of polar coordinates from the correct point of view is in [20, p. 65].

A good idea to try when seeking an explicit solution of a partial differential equation (PDE) is separation of variables. Fortunately, this method can be used to find a solution of the BVP (13.19). Look for a solution of the

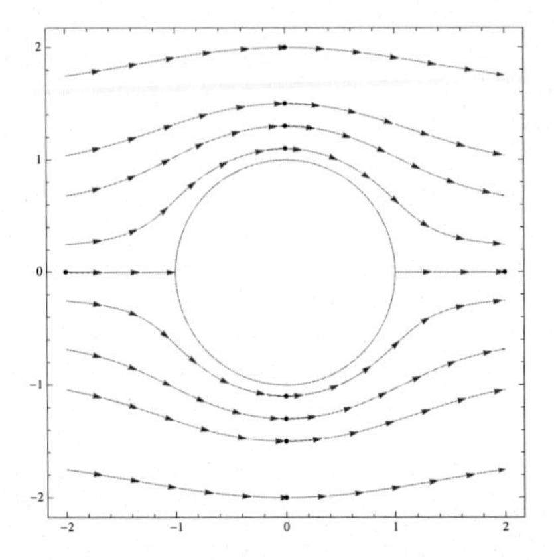

Fig. 13.1 The figure depicts a numerical approximation of potential flow around a cylinder of radius one such that the fluid velocity at infinity is $(1, 0)$.

form $\Phi(r,\theta) = R(r)\Theta(\theta)$, where of course $\Phi(r,\theta) := \phi(r\cos\theta, r\sin\theta)$. Note that for such a solution to be defined on the punctured plane viewed as the Cartesian product of the line and the unit circle, the function Θ must be 2π-periodic. After computing the Laplacian of Φ, setting the result to zero, and using subscripts for partial derivatives, the variables are separated via the formula

$$r\frac{(rR_r)_r}{R} = -\frac{\Theta_{\theta\theta}}{\Theta}.$$

Because the left-hand side of this equation does not depend on θ and the right-hand side does not depend on r, both sides of the equation must be equal to some constant λ. The problem is solved if there is a choice of λ such that the product of the corresponding R and Θ also satisfies the boundary conditions.

For an arbitrary choice of λ, the function Θ must satisfy the ODE

$$\Theta'' + \lambda\Theta = 0$$

and the solution Θ must be 2π-periodic. Thus, λ must be nonnegative. In case $\lambda = 0$, the corresponding periodic solution is constant. For $\lambda > 0$, the general solution is a linear combination of sinusoids:

$$\Theta(\theta) = K\sin\sqrt{\lambda}\,\theta + L\cos\sqrt{\lambda}\,\theta.$$

For Θ to be 2π-periodic, the constant λ must be the square of an integer, say $\lambda = n^2$ and $n > 0$.

The corresponding equation for R, after expanding the indicated derivatives, is

$$r^2R'' + rR' - n^2R = 0,$$

a second-order linear ODE with nonconstant coefficients. It is an Euler type second-order equation. From the elementary theory of ODEs, the usual solution method is to try a function $R = r^\alpha$ for some real or complex exponent α. Implementation of this procedure yields the indicial equation

$$\alpha(\alpha - 1) + \alpha - n^2 = 0$$

that has roots $\alpha = \pm n$. Thus, the general solution of the second-order equation for $n > 0$ is

$$R(r) = Cr^n + Dr^{-n},$$

for constants C and D. For $n = 0$, the solution (for $r > 0$) is

$$R(r) = C \ln r + D.$$

There are two types of separable solutions of $\Delta\phi = 0$.

For $\lambda = 0$ (in Cartesian coordinates) the candidate solution is

$$\phi(x, y) = \frac{C}{2}\ln(x^2 + y^2) + D.$$

It does not satisfy the boundary condition on the circle with radius a (the boundary of Ω) unless $C = 0$. In this case the gradient of ϕ vanishes, and the boundary condition at infinity is not satisfied unless $U = 0$, a case of no physical interest.

With $\lambda = n^2 > 0$,

$$\Phi(r, \theta) = (Cr^n + \frac{D}{r^n})(K \sin n\theta + L \cos n\theta).$$

To check the boundary conditions, the gradient of Φ is best computed in polar coordinates. Using the same method as for computing the Laplacian [Eq. (13.20)], the polar form of the gradient in Cartesian components is found by computing the partial derivatives with respect to x and y of $f(x, y) = F(\sqrt{x^2 + y^2}), \arctan(y/x))$. The result is

$$\text{grad}\,\Phi(r, \theta) = (F_r \cos\theta - \frac{1}{r}F_\theta \sin\theta)e_1 + (F_r \sin\theta + \frac{1}{r}F_\theta \cos\theta)e_2.$$

This vector field is expressed in the polar basis

$$e_r = \begin{pmatrix} \cos\theta \\ \sin\theta \end{pmatrix}, \qquad e_\theta = \begin{pmatrix} -\sin\theta \\ \cos\theta \end{pmatrix}$$

by

$$\operatorname{grad}\Phi(r,\theta) = F_r e_r + \frac{1}{r} F_\theta e_\theta.$$

The fluid velocity field is

$$\begin{aligned}\operatorname{grad}\Phi(r,\theta) =& (nCr^{n-1} - n\frac{D}{r^{n+1}})(K\sin n\theta + L\cos n\theta)e_r \\ &+ \frac{1}{r}(Cr^n + \frac{D}{r^n})(nK\cos n\theta - L\sin n\theta)e_\theta.\end{aligned}$$

Its normal derivative vanishes on the circle centered at the origin with radius a when $D = Ca^{2n}$. At infinity, the gradient blows up unless $n = 1$. In case $n = 1$, the potential in Cartesian coordinates is given by

$$\phi(x,y) = (1 + \frac{a^2}{r^2})(Kx + Ly),$$

where the constant C has been subsumed into K and L, and the fluid velocity field $u = \operatorname{grad}\phi$, given in Cartesian components by

$$\begin{aligned} u(x,y) = (&-\frac{2a^2x(Kx+Ly)}{(x^2+y^2)^2} + K(1 + \frac{a^2}{x^2+y^2}), \\ &-\frac{2a^2y(Kx+Ly)}{(x^2+y^2)^2} + L(1 + \frac{a^2}{x^2+y^2})), \end{aligned} \tag{13.21}$$

has constant limit $U = (K, L)$ as $x^2 + y^2 \to \infty$, as required. A phase portrait of the flow in case $U = (1, 0)$ and $a = 1$ is depicted in Fig. 13.1.

The gradient flow velocity [Eq. (13.21)], which is a solution of BVP (13.19), produces zero lift and drag on the cylinder. Thus, it might seem that this is the end of the story for potential flow. Indeed it would be if the BVP had a unique solution for a given U. With the extra condition that only gradient flows are to be considered, the solution is unique. But, for our exterior Neumann problem, there are other potential flows that have nonzero circulation and hence produce lift. The key observation is directly related to the aforementioned difficulties with division by zero in the arctangent function, the nonexistence of a global coordinate system for the circle, and

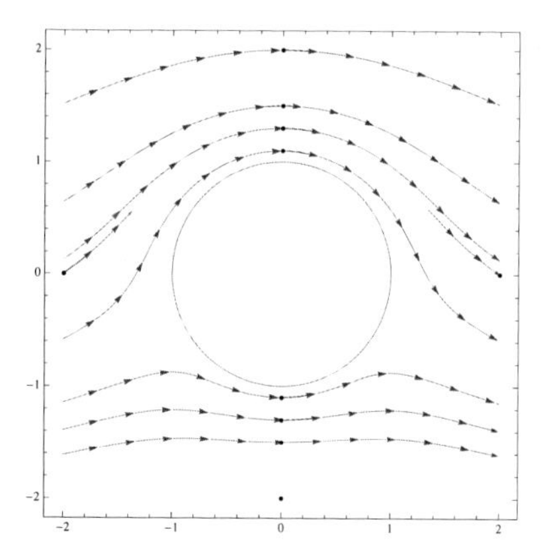

Fig. 13.2 The figure depicts a numerical approximation of the flow for the vector field $w = u - v$ where u is defined in Eq. (13.21) *with $K = 1$ and $L = 0$ and v is defined in Eq.* (13.22).

other manifestations of the annular geometry. The exterior problem is posed on a multiply connected subset of the plane. Every loop in a disk can be shrunk to a point without going outside of the disk, but some loops in the exterior of the disk—those that go around the disk such as a circular loop with radius larger than the disk radius—cannot be shrunk to a point within the exterior region. This nontrivial topology allows for additional solutions of the BVP that are potential flows but not gradient flows with respect to a potential defined in the *entire* exterior domain.

The angle θ that might be assigned to each point (x, y) in the punctured plane by its polar representation $(r\cos\theta, r\sin\theta)$ is not a continuous function on this domain. For example, starting with the value $\theta = 0$ at the point $(1, 0)$ in the plane, a continuous assignment of θ around the unit circle requires that $\theta = 2\pi$ also be assigned to $(1, 0)$. But by ignoring this fact, the function $(x, y) \mapsto \arctan(y/x)$ defined for $x \neq 0$, which is often defined by $\theta = \arctan(y/x)$, has gradient

$$v = (\frac{-y}{x^2 + y^2}, \frac{x}{x^2 + y^2}), \tag{13.22}$$

which is defined on the punctured plane. This vector field is locally gradient. For example, it is the gradient of the function $\pi/2 - \arctan(x/y)$ in the upper half-plane that includes some of the points where $x = 0$. In fact, given an arbitrary point (x, y) in the punctured plane, there is an open disk containing this point and a function (which is an assignment of angle between rays from the origin to points in the set and the positive polar axis) such that its gradient is v. Because these functions all assign angles, it is natural to write $v = \text{grad}\, \theta$, but this is an abuse of notation because there is no global function θ on the punctured plane whose gradient is v. Nonetheless, v is smooth on the punctured plane. It is irrotational and divergence free; that is,

$$\nabla \times v = 0, \qquad \nabla \cdot v = 0,$$

it satisfies the zero Neumann boundary condition on the exterior of the disk centered at the origin with radius a (that is, $v \cdot \eta = 0$ where η is the unit normal on the complement of this disk in the plane), and $v \to (0,0)$ as $x^2 + y^2 \to \infty$. This vector field is a solution of the Neumann boundary value problem

$$\begin{aligned} \nabla \times v &= 0 \quad \text{in } \Omega, \\ \nabla \cdot v &= 0 \quad \text{in } \Omega, \\ v \cdot \eta &= 0 \quad \text{in } \partial\Omega, \\ v &\to (0,0) \quad \text{as } x^2 + y^2 \to \infty. \end{aligned} \tag{13.23}$$

Thus, with the pressure as in Eq. (13.8) given by

$$p = -\frac{1}{2}\rho|u|^2 + \rho B$$

and zero body force, v is a solution of the incompressible Euler's equations; indeed, it is the velocity field of a potential flow.

For u defined in Eq. (13.21), v defined in Eq. (13.22), and every real number $\mathcal{L}$, the flow velocity field $w = u + \mathcal{L}v$ is a potential flow that satisfies the incompressible Euler's equations with constant density for the pressure as in Eq. (13.8) and no body force in the exterior region Ω with the Neumann boundary condition and limit $U = (K, L)$ at infinity. Moreover, w produces lift in the direction $U^\perp$

$$\text{lift} = -\rho|U| \int_\Gamma w \cdot ds U^\perp = -2\pi a \mathcal{L} \rho |U| U^\perp \tag{13.24}$$

on the cylinder with radius a. To obtain positive lift for the freestream velocity $U = (1,0)$, the factor $\mathcal{L}$ must be negative. An example is depicted in Fig. 13.2 and further analysis is suggested in Exercise 13.9.

The boundary value problem

$$\nabla \times w = 0 \quad \text{in } \Omega,$$

$$\begin{aligned} \nabla \cdot w &= 0 \quad \text{in } \Omega, \\ w \cdot \eta &= 0 \quad \text{in } \partial\Omega, \\ w &\to U \quad \text{as } x^2 + y^2 \to \infty \end{aligned} \tag{13.25}$$

is ill-posed; solutions are not unique because the previously defined $w = u + \mathcal{L}v$ is a family of solutions parameterized by $\mathcal{L}$. A similar nonuniqueness result is expected for cylindrical bodies B whose cross sections are not circles. For a general cross section, there will certainly be no explicit solution. To determine the flow, we might resort to numerical approximations. But, there is a serious problem: Which solution would be produced by a numerical approximation? For that matter, which solution is physically correct when a body is subjected to a constant freestream flow? These questions do not arise when a model is well posed; in particular, solutions exist and are unique. Martin Kutta (1902) argued that the flow velocity must have a zero at the trailing edge of an airfoil. Using this hypothesis, a unique solution can be determined. The rest of the story is told in many textbooks on aerodynamics.

Another important result, called the Kelvin circulation theorem (William Thomson, 1869), states that the circulation around a loop remains constant as the loop is moved by an isentropic and Eulerian flow that is allowed to include a body force given by a potential B.

To prove Kelvin's theorem, consider a simple closed curve C, the flow velocity field u, and its flow γ. Without loss of generality, let α be the arc-length parameterization of C (that is, $\alpha : [0, \ell] \to C$, where ℓ is the length of C and $|\dot{\alpha}(s)| = 1$ for every $s \in [0, \ell)$), and define

$$C(t) = \{\gamma(\alpha(s), t) : s \in [0, \ell)\};$$

it is the loop obtained by moving C by the flow for t time units. There is no loss of generality, because the circulation is independent of the parameterization of C (see Exercise 13.10). We must show that

$$\frac{d}{dt} \int_{C(t)} u \cdot ds = 0. \tag{13.26}$$

By definition,

$$\frac{d}{dt}\gamma(\alpha(s), t) = u(\gamma(\alpha(s), t), t). \tag{13.27}$$

We also have,

$$\frac{d}{ds}\gamma(\alpha(s),t) = D\gamma(\alpha(s),t))\dot{\alpha}(s),$$

where D denotes the derivative with respect to the space variable, and differentiating both sides of the ODE (13.27) with respect to the space variable,

$$\frac{d}{dt}D\gamma(\alpha(s),t) = Du(\gamma(\alpha(s),t),t)D\gamma(\alpha(s),t)).$$

Using the arc length parameterization, the differential equations in the last paragraph, the enthalpy q (defined by Eq. (13.1)), and the body force $\rho\nabla B$,

$$\begin{aligned}\frac{d}{dt}\int_{C(t)} u\cdot ds &= \frac{d}{dt}\int_0^\ell u(\gamma(\alpha(s),t),t)D\gamma(\alpha(s),t))\dot{\alpha}(s)\,ds\\ &= \int_0^\ell (Du(\gamma(\alpha(s),t),t)u(\gamma(\alpha(s),t),t)+u_t)D\gamma(\alpha(s),t))\dot{\alpha}(s)\\ &\quad + u(\gamma(\alpha(s),t),t)Du(\gamma(\alpha(s),t),t)D\gamma(\alpha(s),t))\dot{\alpha}(s)\,ds\\ &= \int_0^\ell (u_t + u\cdot\nabla u)D\gamma\dot{\alpha} + uDuD\gamma\dot{\alpha}\,ds\\ &= \int_0^\ell (-\nabla q + \nabla B)D\gamma\dot{\alpha} + uDuD\gamma\dot{\alpha}\,ds\\ &= \int_0^\ell \frac{d}{ds}(q+B) + \frac{1}{2}\frac{d}{ds}|u|^2\,ds.\end{aligned}$$

The last integral vanishes because the antiderivative of its integrand has the same values at zero and ℓ. This completes the proof of the Kelvin circulation theorem.

The circulation theorem leads to another difficulty with the Eulerian theory applied to bodies, for example airplanes, moving in a fluid. Imagine an airplane waiting to take off at the end of a runway in still air. There is zero circulation around its wings. The Kutta–Zhukovsky theorem says that there must be circulation for the wings to be subjected to lift and the airplane to fly. The circulation theorem tells us that the circulation would remain zero as the airplane accelerates down the runway. Where does the circulation necessary for flight come from? The answer is of course that the Eulerian fluid flow is simply not the correct model: it does not account for the viscosity of air. In reality, the viscous flow creates a vortex, often called the starting vortex, which introduces rotation into the flow. As this vortex is shed off the wing, the flow around the wing has nonzero circulation; hence it is subjected to lift. This subject was considered in great detail well into the 20th century before the advent of electronic computers. It is an important chapter in the history of applied mathematics as much progress in understanding was achieved. As mentioned previously, boundary layer theory completed the picture with a model that was sufficiently accurate to make predictions and good approximations. The basic result of this theory is that viscosity must be taken into account only in a thin layer (the boundary layer) near the surface of the wings; the Eulerian model is viable outside this layer. An approximation of the Navier–Stokes equations (called the boundary layer equations and more fully described in Section 17.3) are solved in the boundary layer and matched at the boundary of the boundary layer to an Eulerian flow. Although far from easy to carry out in realistic applications, this scenario produced good approximations that were used effectively for understanding lift and drag in aerodynamics. Of course, the mathematical theory of this model and the basic understanding it provides of fluid flow around bodies remains valid, but for practical applications, analysis based on the boundary layer theory has been replaced by numerical computation using the full Navier–Stokes model.

Another approximation, which was widely used in aerodynamics throughout the 20th century, takes into account the compressibility of air. The main assumptions are no body force, zero viscosity (inviscid flow), no rotation (irrotational flow), an equation of state that produces an enthalpy q (isentropic flow), and the existence of a potential whose gradient is the fluid velocity (potential flow). For simplicity, the approximation is carried out here for steady flow.

For steady, irrotational, and compressible flow with density ρ, velocity u, and enthalpy q, the continuity equation is

$$0 = \nabla \cdot \rho u = \nabla \rho u + \rho \nabla \cdot u \tag{13.28}$$

and Bernoulli's form of Euler's equation is

$$\nabla(\frac{1}{2}|u|^2 + q) = 0. \tag{13.29}$$

An equation of state—a constitutive law relating pressure and density—is used to determine q. The simplest choice is derived from the thermodynamics of gases and defined for pressure p, reference density ρ_0, sound speed c_0 at the reference density, and exponent γ by

$$p = \frac{\rho_0 c_0^2}{\gamma}\Big(\big(\frac{\rho}{\rho_0}\big)^\gamma - 1\Big). \tag{13.30}$$

The exponent γ can be estimated using thermodynamics or fit to experiments. For simplicity, and because thermodynamics is not discussed in this text, a reasonable choice of the dimensionless constant γ in numerical computations for air is a value that satisfies $1 < \gamma \leq 2$; for example, $\gamma = 1.5$. Using the equation of state, the corresponding enthalpy q (defined so that $\nabla p = \rho \nabla q$) is

$$q = \frac{c_0^2}{\gamma - 1}\big(\frac{\rho}{\rho_0}\big)^{\gamma-1}. \tag{13.31}$$

Moreover, at the reference density $dp/d\rho = c_0^2$, as it should be for an ideal gas.

Substitute the enthalpy [Eq. (13.31)] into the momentum balance [Eq. (13.29)], compute gradient of the enthalpy, and solve for the gradient of the density to obtain

$$\nabla \rho = -\frac{\rho_0}{2c_0^2}\big(\frac{\rho_0}{\rho}\big)^{\gamma-2}\nabla |u|^2.$$

Again, substitute this value of $\nabla \rho$ into the conservation of mass [Eq. (13.28)], divide by ρ, and rearrange the resulting equation into the form

$$\nabla \cdot u = \frac{1}{2c_0^2}\big(\frac{\rho_0}{\rho}\big)^{\gamma-1} u \cdot \nabla |u|^2. \tag{13.32}$$

This nonlinear scalar relation for velocity and density encodes the conservation laws and the equation of state for steady, irrotational, and compressible flow, but it is clearly not closed. There are four unknown scalar quantities: three components of velocity and the density.

A closed equation is obtained from the relation (13.32) by making one further assumption and an approximation: the velocity field u is assumed to be the gradient of a potential ϕ and the relation is approximated by its linearization about the freestream velocity and reference pressure.

What is linearization? The simple answer is the processes of approximating a nonlinear expression by the first-order term in its Taylor expansion. More precisely, consider an equation $G(u,\rho) = 0$; for example, the relation (13.32) viewed in this manner. Suppose that U and ρ_0 satisfies the equation; that is, $G(U,\rho_0) = 0$, as do the constant freestream velocity and density in the example. Consider the Taylor expansion of G at (U,ρ_0) to obtain

$$0 = G(u,\rho) = G(U,\rho_0) + DG(U,\rho_0)(u-U,\rho-\rho_0) + \text{higher order terms},$$

where D denotes the derivative of the function G. The linearization of the original nonlinear equation is, by definition, the new equation

$$0 = DG(U,\rho_0)(w,\delta)$$

for the new unknowns w and δ. A solution (w,δ) of the linearized equation is expected to be a good approximation of the deviations $u-U$ and $\rho-\rho_0$—also called perturbations—from a desired solution (u,ρ) of original equation provided that $|u-U|$ and $|\rho-\rho_0|$ are small.

A useful way to carry out the linearization procedure is to set $w = u-U$ and $\delta = \rho-\rho_0$, substitute to obtain

$$0 = G(u,\rho) = G(w+U,\delta+\rho_0),$$

expand in Taylor series about $(w,\delta) = 0$, and discard all nonlinear terms in the variables w and δ.

Returning to relation (13.32), substitute $w = u-U$ and $\delta = \rho-\rho_0$ to obtain

$$\nabla\cdot(U+w) = \frac{1}{2c_0^2}\Big(\frac{\rho_0}{\delta+\rho_0}\Big)^{\gamma-1}(U+w)\cdot\nabla((U+w)\cdot(U+w)).$$

As U is constant,

$$\nabla\cdot w = \frac{1}{2c_0^2}\Big(\frac{\rho_0}{\delta+\rho_0}\Big)^{\gamma-1}(U+w)\cdot\nabla(2U\cdot w+w\cdot w).$$

Using $\mathcal{N}$ for the current sum of nonlinear combinations of the variables w and δ and their derivatives, the relation reduces to

$$\nabla\cdot w = \frac{1}{2c_0^2}\Big(\frac{\rho_0}{\delta+\rho_0}\Big)^{\gamma-1}U\cdot\nabla(2U\cdot w)+\mathcal{N}.$$

By Taylor expansion at $\delta = 0$, note that

$$\left(\frac{\rho_0}{\delta+\rho_0}\right)^{\gamma-1} = 1 - \frac{\gamma-1}{\rho_0}\delta + \mathcal{N}.$$

Use this expansion to simplify the previous relation to

$$\nabla \cdot w = \frac{1}{2c_0^2} U \cdot \nabla(2U \cdot w) + \mathcal{N}.$$

Thus, we have the linearization

$$\nabla \cdot w = \frac{1}{c_0^2} U \cdot \nabla(U \cdot w). \tag{13.33}$$

To close Eq. (13.33), use the assumption that w is a potential flow with potential v. In case $u = \operatorname{grad}\phi$ and $U = \operatorname{grad}(U_1x + U_2y + U_3z)$, the potential v approximates $\phi - (U_1x + U_2y + U_3z)$. The desired linearized system for the potential v is, after a simple rearrangement of Eq. (13.33) with ∇v substituted for w, the scalar equation

$$\Delta v = \frac{1}{c_0^2}(U_1^2 v_{xx} + U_2^2 v_{yy} + U_3^2 v_{zz} + 2U_1U_2v_{xy} + 2U_1U_3v_{xz} + 2U_2U_3v_{yz}). \tag{13.34}$$

By reassigning the original coordinate system so that $U = (U_1, 0, 0)$ and $U_1 > 0$, and using the Mach number $M_0 := |U_1|/c_0$—named after Ernst Mach (1880s) and defined to be the magnitude of the velocity of interest divided by the speed of sound—the linearized equation reduces to the Prandtl–Glauert equation (Ludwig Prandtl and Hermann Glauert, 1928)

$$(1 - M_0^2)v_{xx} + v_{yy} + v_{zz} = 0; \tag{13.35}$$

it is the equation that was used extensively, before the possibility of computing with the full Navier–Stokes equations, to determine the flow around bodies (such as wings) in aerodynamics (see, for example, [34]).

An immediate and important observation signaled by the Prandtl–Glauert equation is that something changes when the freestream velocity passes through the speed of sound. In fact, for speeds less than the speed of sound ($M_0 < 1$), the equation is elliptic and has the same character as Laplace's equation; but, when $M_0 > 1$, the equation is akin to the wave equation where x plays the role usually assigned to time:

$$v_{xx} = \frac{1}{M_0^2 - 1}(v_{yy} + v_{zz}) \text{ may be compared to } v_{tt} = C^2(v_{yy} + v_{zz}).$$

Supersonic flow is associated with shock waves.

Exercise 13.7. The Laplacian and gradient have been determined in polar coordinates. What are their corresponding expressions in three dimensions in cylindrical coordinates?

Exercise 13.8. Suppose a solid sphere of radius a made of a material with density d is immersed in water so that its center is at a depth h. Compute the magnitude of the sum of the forces on the sphere due to gravity and fluid pressure. Is this force related to lift? Explain.

Exercise 13.9. (a) Make a more detailed drawing of the flow depicted in Fig. 13.2 that includes the flow on the circle of radius $a = 1$. (b) This flow has rest points. Determine them and their stability types. (c) Determine the pressure at each point on the circle. (d) Where is the maximum and the minimum of the pressure? (e) Discuss your results in comparison to the Kutta–Zhukovsky theorem.

Exercise 13.10. Prove that the circulation of a vector field around a curve is independent of the parameterization of the curve.

Exercise 13.11. Reconsider the nonlinear equation (13.32), linearize ρ but not u, assume that $u = \nabla\phi$, and assume two-dimensional flow. (a) Show that these assumptions lead to the full potential equation

$$(1 - \frac{\phi_x^2}{c_0^2})\phi_{xx} + (1 - \frac{\phi_y^2}{c_0^2})\phi_{yy} = \frac{2\phi_x\phi_y}{c_0^2}\phi_{xy}.$$

(b) Derive the full potential equation for three-dimensional flow. (c) State appropriate further assumptions and use them to derive the Prandtl–Glauert equation from the result of part (b).

Exercise 13.12. Determine a change of variables that transforms the Prandtl-Glauert equation to the Laplace equation in case the Mach number is less than unity.

CHAPTER 14

Equations of Motion in Moving Coordinate Systems

Moving coordinate systems are discussed in this chapter in the context of general mechanical systems. The theory is applied to the equations of fluid dynamics and particle mechanics.

14.1 MOVING COORDINATE SYSTEMS

An inertial coordinate system is a coordinate system in which Newton's laws of motion are valid; in particular, a free particle moves along a (straight) line.

Imagine an inertial coordinate system in Euclidean space, with rectangular coordinates (ξ, η, ζ), and the motion of a particle whose position vector in this coordinate system is R. Suppose there is a second (moving) rectangular coordinate system (with coordinates (x, y, z)) such that, after the translation in space that moves its origin to the origin of the inertial coordinate system, its translated frame (ordered basis) of coordinate unit direction vectors $[e_x, e_y, e_z]$ can be rigidly rotated in space to coincide with the inertial frame $[e_\xi, e_\eta, e_\zeta]$. In other words, assume that the moving frame is given at each instant of time by a translation and rotation of the inertial frame.

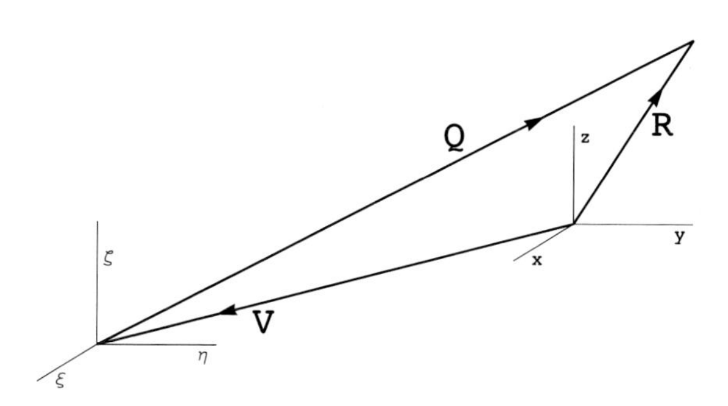

Fig. 14.1 A position vector R is depicted in a coordinate system that has been parallel transported from an inertial coordinate system with coordinates (x, y, z). The vector $Q = R - \mathcal{V}$ is this position vector in the inertial coordinates.

An Invitation to Applied Mathematics: Differential Equations, Modeling, and Computation.
http://dx.doi.org/10.1016/B978-0-12-804153-6.50014-2, Copyright © 2017 Elsevier Inc. All rights reserved.

Translation in space, which more properly should be called parallel translation or parallel transport, is given by vector addition. For example, let $\mathcal{V}$ denote the vector from the origin of the moving frame to the origin of the inertial frame. The vector $Q := R - \mathcal{V}$, where the addition is with respect to the inertial vector space, represents the parallel translation of R along the straight line joining the origins to the position of the origin of the moving frame (see Fig. 14.1). Likewise, the inertial frame $[e_\xi, e_\eta, e_\zeta]$ is parallel transported to a frame $[\hat{e}_\xi, \hat{e}_\eta, \hat{e}_\zeta]$. Note that the basis vectors in this frame are not necessarily parallel to the moving frame in the case where this motion includes rotation.

The vector connecting the origin of the moving frame to the point Q in space is expressed in the coordinates of the moving frame $[e_x, e_y, e_z]$ by

$$Q = \begin{pmatrix} a \\ b \\ c \end{pmatrix}.$$

In other words,

$$Q = ae_x + be_y + ce_z.$$

The basis vectors for the moving frame have coordinate representations in the parallel transported inertial coordinates:

$$e_x = \begin{pmatrix} a_{11} \\ a_{21} \\ a_{31} \end{pmatrix}, \qquad e_y = \begin{pmatrix} a_{12} \\ a_{22} \\ a_{32} \end{pmatrix}, \qquad e_x = \begin{pmatrix} a_{13} \\ a_{23} \\ a_{33} \end{pmatrix}.$$

Thus, the coordinate representation of Q in the parallel transported inertial coordinates is

$$a \begin{pmatrix} a_{11} \\ a_{21} \\ a_{31} \end{pmatrix} + b \begin{pmatrix} a_{12} \\ a_{22} \\ a_{32} \end{pmatrix} + c \begin{pmatrix} a_{13} \\ a_{23} \\ a_{33} \end{pmatrix}.$$

Or equivalently,

$$\begin{pmatrix} \alpha \\ \beta \\ \gamma \end{pmatrix} := \begin{pmatrix} a_{11} & a_{12} & a_{13} \\ a_{21} & a_{22} & a_{23} \\ a_{31} & a_{32} & a_{33} \end{pmatrix} \begin{pmatrix} a \\ b \\ c \end{pmatrix} \tag{14.1}$$

are the coordinates of R in the parallel transported inertial frame. In a more compact form,

$$R = \mathcal{V} + AQ, \tag{14.2}$$

where A is the matrix with components a_{ij}. This matrix changes the coordinates relative to the moving frame $[e_x, e_y, e_z]$ to the coordinates relative to the parallel transported inertial frame $[\hat{e}_\xi, \hat{e}_\eta, \hat{e}_\zeta]$. To see this more explicitly, note that the representation of e_x in the transported inertial coordinates means

$$e_x = a_{11}\hat{e}_\xi + a_{21}\hat{e}_\eta + a_{31}\hat{e}_\zeta,$$

with similar expressions for e_y and e_z. Thus,

$$\begin{aligned} ae_x + be_y + ce_z = a(a_{11}\hat{e}_\xi + a_{21}\hat{e}_\eta + a_{31}\hat{e}_\zeta) + b(a_{12}\hat{e}_\xi + a_{22}\hat{e}_\eta + a_{32}\hat{e}_\zeta) \\ + c(a_{13}\hat{e}_\xi + a_{23}\hat{e}_\eta + a_{33}\hat{e}_\zeta). \end{aligned}$$

When the right-hand side of the last equality is collected with respect to the frame $[\hat{e}_\xi, \hat{e}_\eta, \hat{e}_\zeta]$, the result is exactly

$$ae_x + be_y + ce_z = \alpha\hat{e}_\xi + \beta\hat{e}_\eta + \gamma\hat{e}_\zeta.$$

This change from the coordinates (a, b, c) with respect to the rotating frame to the coordinates (α, β, γ) in the transported inertial frame is exactly what is accomplished by the matrix multiplication in Eq. (14.1). When viewed as a linear transformation, A takes the moving frame to the transported inertial frame; that is,

$$Ae_x = \hat{e}_\xi, \quad Ae_y = \hat{e}_\eta, \quad Ae_z = \hat{e}_\zeta.$$

The matrix A is an orthogonal transformation with respect to the usual inner product, which will be denoted by angled brackets $\langle \; \rangle$; that is, $\langle Au, Av\rangle = \langle u, v\rangle$ for every pair of vectors u, v. Equivalently, $A^{-1} = A^T$; that is,

$$AA^T = A^T A = I.$$

Using this fact,

$$Q = A^T(\mathcal{V} + R). \tag{14.3}$$

By Newton's second law and assuming constant mass, the equation of motion of the particle with position Q (with measurements made in the inertial frame) is

$$m\ddot{Q} = F, \tag{14.4}$$

where m is the mass of the particle and F is the (vector) sum of the forces acting on the particle.

The next goal is to express the equation of motion of the particle in the moving coordinate system where its position vector is R. To do this, it is notationally convenient to define

$$B = A^T, \qquad \mathcal{U} = A^T\mathcal{V} = B\mathcal{V}$$

and use the three quantities

$$r := BR, \qquad v := B\dot{R}, \qquad a := B\ddot{R}.$$

These latter three quantities are the position, velocity, and acceleration of the particle in the moving frame. Also note that B^T is the inverse of B.

In view of Eq. (14.2), the velocity of the particle can be expressed in the form

$$\begin{aligned} \dot{Q} &= \dot{\mathcal{U}} + \dot{B}R + B\dot{R}, \\ &= \dot{\mathcal{U}} + \dot{B}B^T(BR) + v \\ &= \dot{\mathcal{U}} + \dot{B}B^T r + v. \end{aligned} \tag{14.5}$$

To make further progress, note that the transformation $\Omega := \dot{B}B^T$ has a useful property: it is skew-symmetric; that is,

$$\Omega^T = -\Omega.$$

This fact is easily proved by differentiation with respect to time in the identity $BB^T = I$. In fact,

$$\dot{B}B^T + B\dot{B}^T = 0; \tag{14.6}$$

therefore,

$$\Omega^T = (\dot{B}B^T)^T = B\dot{B}^T = -\dot{B}B^T = -\Omega.$$

By using the definition of skew-symmetry, the matrix representation of Ω must have the form

$$\begin{pmatrix} 0 & -\omega_3 & \omega_2 \\ \omega_3 & 0 & -\omega_1 \\ -\omega_2 & \omega_1 & 0 \end{pmatrix} \tag{14.7}$$

with respect to every orthonormal basis. Moreover, the action of Ω on a vector W (given by the multiplication ΩW) can also be represented using the vector cross product with respect to the orthonormal basis used to obtain the matrix representation of Ω. Suppose the basis is $[e_\xi, e_\eta, e_\zeta]$. For vectors $u = (u_1, u_2, u_3)$ and $v = (v_1, v_2, v_3)$ expressed in components with respect to this orthonormal basis, the cross product is given by

$$u \times v = (u_2 v_3 - u_3 v_2)e_\xi - (u_1 v_3 - u_3 v_1)e_\eta + (u_1 v_2 - u_2 v_1)e_\zeta.$$

Using the vector $\omega := (\omega_1, \omega_2, \omega_3)$,

$$\Omega W = \omega \times W. \tag{14.8}$$

Thus, the velocity of the particle may be expressed in two ways:

$$\begin{aligned} \dot{R} &= \dot{\mathcal{U}} + \Omega r + v, \\ \dot{R} &= \dot{\mathcal{U}} + \omega \times r + v. \end{aligned} \tag{14.9}$$

Using the first equation in display (14.9), the acceleration of the particle is

$$\ddot{R} = \ddot{\mathcal{U}} + \dot{v} + \dot{\Omega} r + \Omega \dot{r}. \tag{14.10}$$

Because $r = BR$ and $v = B\dot{R}$, their time derivatives are

$$\begin{aligned} \dot{r} &= \dot{B}B^T(BR) + B\dot{R} \\ &= \Omega r + v, \end{aligned} \tag{14.11}$$

and

$$\begin{aligned} \dot{v} &= \dot{B}B^T(B\dot{R}) + B\ddot{R} \\ &= \Omega v + a. \end{aligned} \tag{14.12}$$

These formulas for $\dot{r}$ and $\dot{v}$ are substituted into Eq. (14.10) to obtain the particle's acceleration in the form

$$\ddot{R} = \ddot{\mathcal{U}} + a + 2\Omega v + \dot{\Omega} r + \Omega^2 r, \tag{14.13}$$

or equivalently,

$$\ddot{R} = \ddot{\mathcal{U}} + a + 2\omega \times v + \dot{\omega} \times r + \omega \times (\omega \times r). \tag{14.14}$$

The equation of motion (14.4) may be recast in the forms

$$ma = -m(\ddot{\mathcal{U}} + 2\Omega v + \dot{\Omega} r + \Omega^2 r) + F \tag{14.15}$$

or

$$ma = -m(\ddot{\mathcal{U}} + 2\omega \times v + \dot{\omega} \times r + \omega \times (\omega \times r)) + F, \tag{14.16}$$

where the vectors a, v, and r give the acceleration, velocity and position of the particle, moving under the influence of the force F, with respect to an observer in the moving coordinate system but referred to the frame $[\hat{e}_\xi, \hat{e}_\eta, \hat{e}_\zeta]$. These equations in the moving frame $[e_x, e_y, e_z]$ are obtained by multiplication on the left by the orthogonal matrix B^T. For example, using Eq. (14.15), the equation of motion is given by

$$mB\ddot{R} = -m(B\ddot{\mathcal{V}} + 2\Omega B\dot{R} + \dot{\Omega} BR + \Omega^2 BR) + F, \tag{14.17}$$

and in the moving coordinate frame $[e_x, e_y, e_z]$, it is

$$m\ddot{R} = -m(\ddot{\mathcal{V}} + 2B^T\Omega B\dot{R} + B^T\dot{\Omega} BR + B^T\Omega^2 BR) + B^T F. \tag{14.18}$$

The last equation can also be written in the form

$$m\ddot{R} = -m(B^T\ddot{\mathcal{U}} + 2\Gamma\dot{R} + \dot{\Gamma} R + \Gamma^2 R) + B^T F, \tag{14.19}$$

where $\Gamma := B^T\Omega B$; that is, Γ and Ω are names for the same linear transformation expressed in different bases.

Because the equation of motion (14.19) is a valid form of Newton's equation of motion for the particle moving under the influence of the force F, from the point of view of a noninertial observer sitting at the moving origin and measuring with respect to coordinates in the moving frame, this observer treats the terms on the right-hand side of the equation of motion as additional forces. Put another way, the Newton in this observer's world would discover the laws of motion with the additional forces always included.

The vector $\ddot{\mathcal{V}}$ is the acceleration of the origin of the noninertial frame, $\dot{\Omega}r = \dot{\omega} \times r$ is the acceleration due to the rotation of the moving frame, $2m\Omega v = 2m\,\omega \times v$ is the Coriolis force, and $m\Omega^2 r = m\,\omega \times (\omega \times r)$ is the centrifugal force. Of course, from Newtonian mechanics, F is the only force acting on the particle. The remaining fictitious forces are artifacts of reference to a noninertial coordinate system.

Exercise 14.1. (a) Prove: The linear transformation A is an orthogonal transformation with respect to the usual inner product; that is, $\langle Au, Av\rangle = \langle u, v\rangle$ for every pair of vectors u, v if and only if $A^{-1} = A^T$. (b) Prove: The linear transformation A in Eq. (14.2) is orthogonal. (c) Prove: The product of two orthogonal transformations is orthogonal. (d) Note that I is orthogonal. (e) Let $GL(n, \mathbb{R})$ denote the set of all invertible $n \times n$ matrices with real components. Suppose (a, b) is an interval of real numbers containing zero and that there is a smooth curve $\gamma : (a, b) \to GL(n, \mathbb{R})$ such that $\gamma(0) = I$ and $\gamma(t)$, for $a < t < b$, is an orthogonal matrix. Prove that $\dot{\gamma}(0)$ is a skew symmetric matrix. (f) Construct a smooth curve $\gamma : (a, b) \to GL(n, \mathbb{R})$ such that $\gamma(0) = I$ and $\gamma(t)$, for $a < t < b$, is an orthogonal matrix. Hint: Consider the exponential function on matrices $e^A := I + A + \frac{1}{2!}A^2 + \cdots$. (g) In fancy language, the tangent space at the identity of the orthogonal group is the set of all skew-symmetric matrices. Explain all the concepts in the last statement. (h) (For readers who have studied topology.) The orthogonal group of dimension n inherits a topology by virtue of the fact that each orthogonal matrix may be considered to be a point in Euclidian n^2 dimensional space. Just string out the rows. Is the orthogonal group a connected set in this topology? Hint: Consider $n = 1, 2, 3$ to gain insight. (i) Using the identification of an $n \times n$ matrix with a point in n^2 dimensional space, determine the dimension of the set of orthogonal matrices as a subset of this space.

14.2 PURE ROTATION

To consider pure rotation, imagine a rotating disk whose motion is measured with respect to a rectangular rotating coordinate system (with coordinates (x, y, z)), whose third coordinate axis (say the axis with tangent vector e_z) coincides with the axis of rotation of the disk, whose origin is at the intersection of the disk with this axis, and whose first two coordinate axes are fixed in the disk. In addition, suppose without loss of generality, that an inertial frame with coordinates (ξ, η, ζ) has the same origin.

The inertial coordinates would most likely be chosen so that the ζ axis is the axis of rotation, but to illustrate an important feature of three-

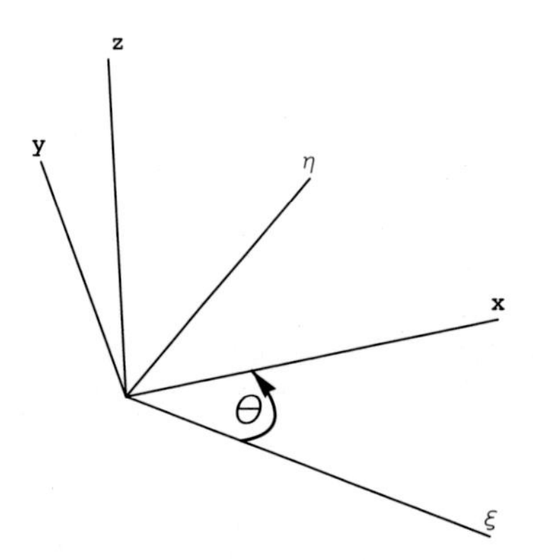

Fig. 14.2 A rotating frame (labeled x, y, and z) is depicted that has rotated through the positive angle θ around its coordinate vector e_z with respect to a fixed inertial frame labeled ξ and η, where the third positive coordinate axis of the inertial frame coincides with the positive coordinate axis labeled z.

dimensional space, consider (for the moment) the general case where inertial coordinates are chosen arbitrarily. Rotation in three dimensions is determined by four numbers: the components of a three-dimensional vector (specifying the axis of rotation) and the rotation angle about this direction (compare Exercise 14.1).

Under our assumption of a fixed axis of rotation, there is (as in Section 14.1) an orthogonal transformation A such that

$$Ae_\xi = e_x, \quad Ae_\eta = e_y, \quad Ae_\zeta = e_z.$$

The determinant of A is ± 1 (see Exercise 14.2). In case the determinant is $+1$, the two frames are said to have the same orientation. They have opposite orientation when the determinant is -1. For definiteness, suppose A has determinant $+1$ so that the inertial and moving frames have the same orientation (compare Exercise 14.1). Also, recall that $R = AQ$ for the position vectors of the same point measured with respect to the inertial frame (where the vector is Q) and the moving frame (where it is R).

Recall that two matrices G and H are similar if there is an invertible matrix C such that $G = C^{-1}HC$. In this case C is called the similarity transformation. A basic fact is that every orthogonal matrix in three-dimensional space with determinant $+1$ is similar to the orthogonal matrix

$$\mathcal{R} := \begin{pmatrix} \cos\theta & \sin\theta & 0 \\ -\sin\theta & \cos\theta & 0 \\ 0 & 0 & 1 \end{pmatrix} \tag{14.20}$$

for some choice of θ and the similarity transformation is also orthogonal (see Exercise 14.3 and compare Exercise 14.1). Geometrically, $\mathcal{R}$ corresponds to negative rotation about the vector e_ζ. Its inverse $\mathcal{R}^T$ corresponds to positive rotation. Using this fact, there is an orthogonal matrix C such that $C^T\mathcal{R}C = A$. Thus, there is a reasonable way to transform an arbitrary inertial frame to the moving frame attached to the rotating disk. Doing so requires consideration of a time-dependent similarity transformation.

Return to the standard choice of inertial coordinates in the present scenario, where the frame $[e_\xi, e_\eta, e_\zeta]$ is positively oriented, $e_\zeta = e_z$, and e_ξ and e_η are in the plane of the rotating disk. In this case, $A = \mathcal{R}$ where θ is changing as the disk rotates. For simplicity assume also that $\dot{\theta} > 0$ with the passage of time. So, the disk is rotating in the positive angular direction with respect to e_ζ (see Fig. 14.2).

By a direct computation, it is easy to check that for $B = A^T$,

$$\Omega = \dot{B}B^T = \begin{pmatrix} 0 & -\dot{\theta} & 0 \\ \dot{\theta} & 0 & 0 \\ 0 & 0 & 0 \end{pmatrix}, \tag{14.21}$$

which is exactly the angular speed in matrix form; that is, the action of this matrix on a vector W is

$$\beta \times W,$$

where β is the vector with components $(0, 0, \dot{\theta})$.

For rotations in three-dimensional space of the form $\mathcal{R}$ as in display (14.20), note the useful identity

$$\dot{B}B^T = B^T\dot{B}, \tag{14.22}$$

which is recorded here for later use (see Exercise 14.5).

For pure rotation, where the inertial and rotating coordinates are chosen so that $A = \mathcal{R}$ and $R = AQ$, the equation of motion [Eq. (14.17)] with $B := A^T$ reduces to

$$mB\ddot{R} = -m(2\Omega B\dot{Q} + \dot{\Omega}BR + \Omega^2 BR) + F, \tag{14.23}$$

and in case the angular frequency is constant,

$$mB\ddot{R} = -m(2\Omega B\dot{R} + \Omega^2 BR) + F. \tag{14.24}$$

In this case, only the centrifugal and Coriolis forces appear.

Exercise 14.2. (a) Prove: The determinant of an orthogonal matrix is ± 1. Hint: As always a proof depends on what is assumed. One way to proceed is to first show the determinant of a matrix is the same as the determinant of its transpose. Try $n = 2$ and $n = 3$ before constructing a general proof. (b) Show that two frames in three-dimensional space both satisfying the right-hand rule have the same orientation.

Exercise 14.3. (a) Show that every (three-dimensional) orthogonal matrix with determinant $+1$ is similar to matrix (14.20) for some choice of θ by a similarity transformation that is orthogonal. (b) What happens for orthogonal matrices with determinant -1.

Exercise 14.4. Revisit Exercise 2.13 and derive the equation of motion referred to a coordinate system rotating with the hoop using the abstract equation of motion (14.24).

Exercise 14.5. (a) Prove identity (14.22) for two- and three-dimensional rotations transformations. (b) Is the same identity true for a time-dependent family of orthogonal matrices? If it is not true in general, is there a subset of families of orthogonal matrices strictly larger than the rotation matrices for which the result is true? (c) What about dimensions larger than three?

Exercise 14.6. (a) Solve the initial value problem

$$\begin{aligned}
\ddot{u} &= 2\dot{v} + u, \\
\ddot{v} &= -2\dot{u} + v, \\
\ddot{w} &= 2\dot{w}, \\
u(0) = 1, &\quad u'(0) = 2, \\
v(0) = 2, &\quad v'(0) = -1, \\
w(0) = 3, &\quad w'(0) = 0.
\end{aligned}$$

(b) Imagine an observer riding at the center of a rotating disk with unit angular speed ($\dot{\theta} = 1$) in an inertial reference frame. This scenario might be used for a simple model of an observer on Earth who views distant objects. Suppose a star is fixed in the sky relative to the inertial reference frame with inertial coordinates $(1, 2, 3)$. Write the equation of motion of the star with respect to the coordinates of the rotating observer. (c) What does the rotating observer observe?

Exercise 14.7. Suppose an observer riding on a rotating disk with angular speed α sets up a mass-spring system with spring constant k and mass m by attaching the massless spring to the axis of rotation so that the spring attachment may rotate freely around the peg used to attach the spring. This observer stretches the spring so that the mass is held fixed at unit distance along the positive x-axis of an orthonormal coordinate system attached to the disk with e_z in the direction of the axis of rotation. The observer then releases the spring from rest in the observers coordinates. What motion of the mass will the observer observe? What will someone in an inertial laboratory frame observe. You may ignore friction and assume that the inertial frame is defined in the usual manner relative to the rotating disk.

14.3 FLUID MOTION IN ROTATING COORDINATES

The equation of motion of a particle in a moving coordinate system [Eq. (14.18)] is of course valid for a particle of fluid. The goal of this section is to explain the transformation of the equations of fluid motion, which are written with respect to an inertial frame, to a rotating coordinate system.

Recall the differential form

$$\begin{aligned} \rho(u_t + u \cdot \nabla u) &= -\nabla p + \mu \Delta u + \rho g, \\ \rho_t + u \cdot \nabla \rho &= -\rho \nabla \cdot u \end{aligned} \tag{14.25}$$

of Newton's second law and the conservation of mass for a fluid in a gravitational field. Because these equations use Newton's law, they can be written in coordinates relative to an inertial coordinate system. This fact was tacitly assumed in their derivation. Fix an inertial frame and consider a rotating frame whose origin is fixed at the origin of the inertial frame. Recall the transformation from the inertial to the moving coordinates given by

$$R = AQ, \tag{14.26}$$

where as before A is the orthogonal matrix taking the inertial frame coordinates to the inertial frame coordinates.

The position of a moving fluid particle is determined by solving the ordinary differential equation (ODE)

$$\dot{Q} = u(Q, t). \tag{14.27}$$

Density is a scalar field; that is, at each point of space and time, this field assigns a real number. It is clear that this number is invariant under a arbitrary change of coordinates. How does a scalar field change coordinates? This question is easy to answer, simply compose the scalar field with the change of coordinates. The given ρ is a function of Q and t; thus, there is a function ϱ such that

$$\varrho(R,t) = \rho(A^T R, t).$$

Or, using previous notation where $B := A^T$, this new function ϱ is given by

$$\varrho(R,t) = \rho(BR, t).$$

The velocity field u is a vector field; its change of coordinates is more complicated. If u were simply a function from the product of three-dimensional space and time to three-dimensional space, the change of coordinates would be exactly the same as for a scalar field: there would be a function U such that $U(R,t) = u(BR,t)$. But this is not the case here because particles move according to ODE (14.27); thus, an ODE for the position R is expected in the rotating coordinates. In other words, u is a (time-dependent) vector field defined to be the velocity vector of the fluid at each point in the space-time region occupied by the fluid.

To perform the change of coordinates, start with ODE (14.27) and definition (14.26) to obtain $\dot{R} = A\dot{Q} + \dot{A}Q$ or

$$\dot{R} = Au(A^T R, t) + \dot{A}A^T R.$$

Thus, the vector field u is transformed by the change of coordinates $R = AQ$ to the vector field U given by

$$U(R,t) = Au(A^T R, t) + \dot{A}A^T R. \tag{14.28}$$

In the moving coordinates, the motion of a fluid particle satisfies the ODE

$$\dot{R} = U(R,t), \tag{14.29}$$

where U is defined in Eq. (14.28).

For a fluid particle with inertial position Q at time t, conservation of mass is expressed by the differential equation

$$\frac{d}{dt}\rho(Q,t) = \rho_t(Q,t) + \nabla\rho(Q,t)u(Q,t) = -\rho(Q,t)\nabla\cdot u(Q,t). \tag{14.30}$$

Because

$$\rho(Q,t) = \varrho(R,t),$$

where $R = AQ$,

$$\begin{aligned}\frac{d}{dt}\rho(Q,t) &= \frac{d}{dt}\varrho(R,t)\\ &= \varrho_t(R,t) + \nabla\varrho(R,t)\dot{R}\\ &= \varrho_t(R,t) + \nabla\varrho(R,t)U(R,t).\end{aligned} \tag{14.31}$$

In these equations, ∇ is defined with respect to the moving coordinate frame.

Using Eqs. (14.30) and (14.31),

$$\varrho_t(R,t) + \nabla\varrho\cdot U(R,t) = -\varrho(R,t)\nabla\cdot u(Q,t).$$

The desired result is obtained by transforming the divergence operator to the moving coordinate frame.

Divergence of a vector field is equal to the trace of its spatial derivative; that is,

$$\nabla\cdot u(Q,t) = \operatorname{tr} Du(Q,t),$$

where as always D denotes the derivative of the function $Q \mapsto u(Q,t)$ for fixed t. In (orthonormal) coordinates Du is the Jacobian matrix of partial derivatives of u with respect to the space variables ξ, η, and ζ. Using the transformed vector field (14.28),

$$\operatorname{tr} DU(R,t) = \operatorname{tr} D(Au(A^TR,t) + \dot{A}A^TR) = \operatorname{tr}(ADu(Q,t)A^T + \dot{A}A^T).$$

The matrix $\Gamma := \dot{A}A^T$ is skew symmetric; that is, $\Gamma^T = -\Gamma$. This fact follows as in Section 14.1 for the matrix

$$\Omega = \dot{B}B^T$$

after differentiating both sides of the identity $AA^T = I$ with respect to t. Taking the trace of both sides of the equation for Γ and using the obvious fact that the trace of a matrix is the same as the trace of its transpose, conclude that $\operatorname{tr}(\dot{A}A^T) = 0$. Because of this fact and the properties of tr (in particular, it is a linear transformation that is invariant under linear changes of coordinates as in Exercise 14.9),

$$\operatorname{tr} DU(R,t) = \operatorname{tr} Du(Q,t).$$

In other words, the divergence changes coordinates according to

$$\nabla \cdot U(R,t) = \nabla \cdot u(A^T R, t), \tag{14.32}$$

where ∇ operates as usual in (orthonormal) coordinates on both sides of the equation. On the left-hand side, ∇ is defined with respect to the moving coordinate frame $[e_x, e_y, e_z]$; on the right-hand side it operates in coordinates measured with respect to the inertial frame $[e_\xi, e_\eta, e_\zeta]$. The final result is the invariance of the scalar differential equation for the conservation of mass:

$$\varrho_t + U \cdot \nabla \varrho = \frac{d}{dt}\varrho(R,t) = -\varrho(R,t)\nabla \cdot U(R,t), \tag{14.33}$$

or

$$\varrho_t + U \cdot \nabla \varrho + \varrho(R,t)\nabla \cdot U(R,t) = 0.$$

A more sophisticated proof relies on the derivation of this conservation law by simply observing that the derivation does not depend on the choice of orthonormal coordinates.

To change coordinates in the differential equation for the momentum balance, acceleration due to the rotation will appear. To see this, use

$$B := A^T$$

and rearrange Eq. (14.28) to the form

$$u(Q,t) = BU(R,t) - B\dot{B}^T BR.$$

By the usual argument (differentiation with respect to t of the identity $BB^T = I$), replace $-B\dot{B}^T$ with $\Omega = \dot{B}B^T$ (the skew-symmetric matrix that appeared in Secton 14.1) so that

$$u(Q,t) = BU(AQ,t) + \Omega Q. \tag{14.34}$$

The time derivative of the function $u(Q,t)$ is

$$\frac{d}{dt}u(Q,t) = B\frac{d}{dt}U(AQ,t) + \dot{B}U(AQ,t) + \dot{\Omega}Q + \Omega\dot{Q}$$

$$
\begin{aligned}
&= B\frac{d}{dt}U(AQ,t) + \dot{B}B^T BU(AQ,t) + \dot{\Omega}BR \\
&\quad + \Omega(BU(AQ,t) + \Omega BR) \\
&= B(U_t(R,t) + (U \cdot \nabla)U) \\
&\quad + \dot{\Omega}BR + 2\Omega BU(R,t) + \Omega^2 BR, \qquad (14.35)
\end{aligned}
$$

where again ∇ in the last expression is with respect to the moving frame coordinates. The right-hand side of the last equation replaces

$$u_t(Q,t) + (u(Q,t) \cdot \nabla u)(Q,t) = \frac{d}{dt}u(Q,t)$$

in the partial differential equation (PDE) for conservation of momentum.

To complete the change of coordinates for the momentum balance, we must determine the gradient of the pressure and the Laplacian of the velocity field in the moving coordinate system.

As mentioned for the transformation of the conservation of mass to rotating coordinates, a scalar field f (for example, the pressure field) changes coordinates via the relation $f(Q,t) = F(AQ,t)$, where as usual the capitalized function is defined on the rotating coordinate space. By the chain rule,

$$Df(Q,t) = DF(AQ,t)A.$$

And by the definitions of D and ∇,

$$Df^T = \nabla f.$$

Therefore, the gradient changes coordinates according to the formula

$$\nabla f(Q,t) = A^T \nabla F(AQ,t) = B\nabla F(R,t). \qquad (14.36)$$

In particular,

$$\nabla p(Q,t) = B\nabla P(R,t). \qquad (14.37)$$

The Laplacian of a vector field v with respect to a frame is defined to be a new vector field whose components in this frame are the Laplacians of the components of v in the same frame. The next objective is to prove the formula

$$\Delta u(Q,t) = B\Delta U(R,t). \qquad (14.38)$$

Here the Laplacian on the left-hand side is the Laplacian with respect to the inertial frame and the one on the right-hand side is with respect to the moving frame.

A proof is constructed using four basic facts: The formula for the change of coordinates of a vector field (14.34), the formula for the change of coordinates of the divergence of a vector field [Eq. (14.32)], the change of coordinates for the gradient [Eq. (14.36)], and a new result on change of coordinates of the Laplacian. Suppose that G is the function given by $G(Q) = AQ$, where (as always in this section) A is an orthogonal transformation, and F is an arbitrary (sufficiently smooth) scalar function on $\mathbb{R}^3$, then

$$\Delta(F \circ G)(Q) = (\Delta F) \circ G(Q). \tag{14.39}$$

The same result is true for functions on $\mathbb{R}^n$, not just $n = 3$.

To begin the proof, note that $\nabla(F \circ G)$ is a vector field. Call it v; that is,

$$v(Q) = \nabla(F \circ G)(Q).$$

Using this new name,

$$\Delta(F \circ G)(Q) = (\nabla \cdot \nabla(F \circ G))(Q) = \nabla \cdot v(Q)$$

By a rearrangement of the change of variables formula for vector fields, the vector field v is transformed to V via

$$V(R) = B^T(v(BR) - \Omega BR).$$

And by the change of coordinates for the divergence,

$$\nabla \cdot v(Q) = \nabla \cdot V(R)$$

whenever $R = AQ$. Using this fact,

$$\Delta(F \circ G)(Q) = \nabla \cdot (B^T(v(BR) - \Omega BR)),$$

where the right-hand side is an abuse of notation: ∇ is meant to act on the function $R \mapsto B^T(v(BR) - \Omega BR)$ and is then evaluated at R.

The vector field $R \mapsto B^T \Omega BR$ has zero divergence. Indeed, the divergence is computed as the trace of the derivative of the field. Because this function is linear, its derivative is the skew-symmetric operator $B^T \Omega B$, and the trace of a skew-symmetric operator is zero. (Why?)

Using this result and $\nabla f = Df^T$,

$$\Delta(F \circ G)(Q) = \nabla \cdot (B^T v(BR)) = \nabla \cdot (B^T (D(F \circ G)(BR))^T).$$

By the chain rule and the identities $B = A^T$ and $B^T B = I$,

$$\Delta(F \circ G)(Q) = \nabla \cdot (B^T (DF(R)A)^T) = \nabla \cdot (DF(R)^T).$$

The right-hand side is exactly

$$(\nabla \cdot \nabla F)(AQ) = (\Delta F) \circ G(Q).$$

This completes the proof of Eq. (14.39).

To prove Eq. (14.38), make new notations for the inertial and moving frames:

$$[e_1, e_2, e_3] := [e_\xi, e_\eta, e_\zeta], \qquad [\epsilon_1, \epsilon_2, \epsilon_3] := [e_x, e_y, e_z].$$

Define, as before, the function G given by $G(Q) = AQ$ and the scalar function F by

$$F(R) = e_i^T (BU(R, t) + \Omega BR),$$

where i is one of the integers 1, 2, or 3.

In components,

$$u = u_1 e_1 + u_2 e_2 + u_3 e_3, \qquad U = U_1 \epsilon_1 + U_2 \epsilon_2 + U_3 \epsilon_3.$$

Because the frame is orthonormal,

$$u_i = e_i^T u, \qquad U_i = \epsilon_i^T U.$$

Thus, using Eq. (14.39),

$$\Delta(e_i^T u)(Q) = \Delta(F \circ G)(Q) = \Delta F \circ G(Q).$$

Because $R \mapsto e_i^T \Omega BR$ is a linear function, its Laplacian vanishes. (Why?) Also, because the Laplace operator is linear ($\Delta(f + g) = \Delta f + \Delta g$),

$$\Delta(e_i^T u)(Q) = \Delta(e_i^T BU) \circ G(Q).$$

Simple manipulation of the right-hand side using properties of transpose produces

$$\Delta(e_i^T u)(Q) = \Delta((U^T B^T e_i)^T) \circ G(Q)$$

$$
\begin{aligned}
&= \Delta((U^T \epsilon_i)^T) \circ G(Q) \\
&= \Delta(\epsilon_i^T U) \circ G(Q).
\end{aligned}
$$

The last formula states that $\Delta u_i(Q) = \Delta U_i(AQ)$ for each $i \in \{1, 2, 3\}$; whence,

$$
\begin{aligned}
\Delta u(Q) &= \Delta u_1(Q) e_1 + \Delta u_2(Q) e_2 + \Delta u_3(Q) e_3 \\
&= \Delta U_1(AQ) e_1 + \Delta U_2(AQ) e_2 + \Delta U_3(AQ) e_3 \\
&= \Delta U_1(AQ) B\epsilon_1 + \Delta U_2(AQ) B\epsilon_2 + \Delta U_3(AQ) B\epsilon_3 \\
&= B(\Delta U_1(AQ)\epsilon_1 + \Delta U_2(AQ)\epsilon_2 + \Delta U_3(AQ)\epsilon_3) \\
&= B\Delta U(AQ).
\end{aligned}
$$

This completes the proof.

By putting together the results of this section, the momentum balance [Eq. (14.25)] in the moving coordinates $R = AQ$ (where A is an orthogonal transformation, $B := A^T$, $\Omega := \dot{B}B^T$, and U is the transformation of u under the change of coordinates) is

$$
\begin{aligned}
\varrho \frac{dU}{dt} = &- \nabla P + \mu \Delta U \\
&+ \varrho B^T g - \varrho(B^T \dot{\Omega} B R + 2B^T \Omega B U + B^T \Omega^2 B R).
\end{aligned}
\tag{14.40}
$$

As in Eq. (14.8), there is a vector ψ such that $B^T \Omega B W = \psi \times W$ for every vector W on $\mathbb{R}^3$. Using this representation and the identity

$$
\dot{\psi} := B^T \dot{\Omega} B \tag{14.41}
$$

(see Exercise 14.10), an alternative form of the conservation of momentum is

$$
\begin{aligned}
\varrho \frac{dU}{dt} = &- \nabla P + \mu \Delta U \\
&+ \varrho B^T g - \varrho(\dot{\psi} \times R + 2\psi \times U + \psi \times (\psi \times R)).
\end{aligned}
\tag{14.42}
$$

Recall that $\dot{\psi} \times R$ is the acceleration due to the rotation of the moving frame, $\varrho\psi \times U$ is the Coriolis force per volume, and $\varrho\psi \times (\psi \times R)$ is the centrifugal force per volume.

Exercise 14.8. A time-dependent vector field changes coordinates according to Eq. (14.28) for the time-dependent linear change of coordinates $R = AQ$. How does a time-dependent vector field change coordinates under a general smooth coordinate change $R = F(Q, t)$?

Exercise 14.9. Prove that tr is a linear transformation that is invariant under linear changes of coordinates. Hint: The trace of a matrix is related to its eigenvalues.

Exercise 14.10. Prove identity (14.41). Hint: Write both sides as an expression in Ω.

14.4 WATER DRAINING IN SINKS VERSUS HURRICANES

Hurricanes always rotate counterclockwise. Large-scale low-pressure systems in the Southern Hemisphere always rotate clockwise. Water (usually) drains counterclockwise in a bathtub or sink in the Northern Hemisphere and clockwise in the Southern Hemisphere. The model equations for fluid dynamics will be used to discuss these phenomena.

For fluid motion on the surface of the Earth, as viewed by an observer rotating with the Earth, the apparent motion is governed by Eq. (14.42) (or (14.40)) and the corresponding equation of continuity (14.33). By custom, we take our (right-hand orthonormal) inertial coordinate system (ξ, η, ζ) with the positive ζ-axis passing through the North Pole, the origin at the center of the Earth, and the other two coordinates fixed relative to distant stars. (There is certainly a serious question about how to choose an inertial frame, but it is assumed in this section that such a frame exists.) The rotation is counterclockwise (west to east) looking down from above the North Pole. We will consider a (right-hand rectangular) rotating coordinate system (x, y, z) fixed to the Earth whose origin is at the center of the Earth and whose positive z-axis passes through the North Pole. The orthogonal transformation B from the rotating to the inertial coordinates ($BR = Q$ in the notation of this chapter) is

$$B = \begin{pmatrix} \cos\theta & -\sin\theta & 0 \\ \sin\theta & \cos\theta & 0 \\ 0 & 0 & 1 \end{pmatrix} \tag{14.43}$$

(compare Fig. 14.2). As in Eq. (14.40),

$$\Omega := \dot{B}B^T = \begin{pmatrix} 0 & -\dot{\theta} & 0 \\ \dot{\theta} & 0 & 0 \\ 0 & 0 & 0 \end{pmatrix} \tag{14.44}$$

and by simple calculation for this special case,

$$B^T \Omega B = \Omega, \qquad \psi = \begin{pmatrix} 0 \\ 0 \\ \dot{\theta} \end{pmatrix}. \tag{14.45}$$

Assume that the angular velocity of the Earth has *constant* value

$$\dot{\theta} := \frac{2\pi}{24}\ \mathrm{hr}^{-1} \approx 7.3 \times 10^{-5}\ \mathrm{sec}^{-1}. \tag{14.46}$$

The equation of motion (14.42) applied to the Earth in the rotating coordinates takes the form

$$\frac{d}{dt}U + \frac{1}{\varrho}\nabla P = \frac{\mu}{\varrho}\Delta U + g - 2\psi \times U - \psi \times (\psi \times R). \tag{14.47}$$

To determine the effect of the Coriolis acceleration $-2\psi \times U$ for air moving with velocity U, which is measured by the rotating observer relative to what is perceived as the stationary surface of the Earth, imagine a position Q in the Northern Hemisphere on the surface of the Earth and note that the vector ψ at Q can be decomposed into a vector tangent to the Earth at Q and a vector normal to the Earth at this point. In reality, the velocity U is a vector in space that may not be tangent to the Earth, but for a large-scale weather system (like a hurricane), wind velocity at this scale is measured as if it were tangent to the surface of the Earth.

The formation of a hurricane (in an idealized scenario) starts when there is a large low-pressure air mass surrounded by relatively high-pressure air. The pressure difference causes high-pressure air to move toward the low-pressure region (see Exercise 14.16). Imagine a circle parallel to the Earth whose center is at the center of the low-pressure region. In an ideal situation, the velocity vectors of the airflow at each point on the circle point toward its center.

Recall that the cross product $\alpha \times \beta$ of two vectors α and β has length $|\alpha||\beta| \sin\varphi$, where φ is the angle between the vectors and it points in the direction, say n, such that $[\alpha, \beta, n]$ is a right-hand system of vectors; or equivalently, the matrix $[\alpha, \beta, n]$ partitioned by columns has positive determinant. In other words, the basis $\{\alpha, \beta, n\}$ is positively oriented.

The cross product of a vector that is tangent to the surface of the Earth and the velocity field U is a vector that is normal to the Earth; it does not affect rotation. On the other hand, the cross product of U and a vector normal to the Earth is tangent to the Earth. In particular, consider the normal component ψ_N of ψ taken in the direction of the outer normal on the surface of the Earth and the velocity vector U, which points toward the center of the circle with its center at the center of the low-pressure air. By the right-hand rule, the cross product of these vectors is tangent to the circle and points in the clockwise direction on the circle viewed from above. The Coriolis acceleration is twice the negative of this cross product. Hence, the Coriolis acceleration is counterclockwise in the Northern Hemisphere. In the Southern Hemisphere, the analysis is the same except that the normal component of ψ points toward the center of the Earth instead of away from the center of the Earth. If follows that the Coriolis acceleration is clockwise.

Exactly the same analysis that shows the Coriolis acceleration is counterclockwise in the Northern Hemisphere applies to water moving toward a drain. By observation, hurricanes always rotate counterclockwise in the Northern Hemisphere but water does not always drain in this direction. What is the difference? It must be that the size of the Coriolis effect is different. How can we determine the size of this effect?

The determination of relative size is not obvious in fluid mechanics. The usual comparison is via scaling as in the determination of the Reynolds number in the scaled fluid model equations (11.35) (see [21]). We will consider a length scale L (a characteristic length) and a velocity scale V (a characteristic velocity) together with the natural induced timescale $T := L/V$. Notice that in Eq. (14.47) all the terms have units of acceleration L/T^2. Using our scaling, the fluid acceleration dU/dt can be viewed as having units $L/T^2 = V^2/L$. Its relative size is viewed as the square of the characteristic velocity divided by the characteristic length. The relative size of the normal component of Coriolis acceleration is twice the characteristic velocity times the angular velocity times the sine of the latitude (see Exercise 14.12). Where, of course, the latitude is the angle of elevation above the equator measured from the center of Earth; it is 0 at the equator and 90 degrees at the poles. By custom, the relative sizes of fluid acceleration and Coriolis acceleration is measured by the (dimensionless)

Rossby number (see [42, 60])

$$\text{Rossby number} := \text{fluid acceleration/Coriolis acceleration} = \frac{V}{2|\psi|L}. \tag{14.48}$$

The Coriolis acceleration is important whenever this number is *smaller* than 1. Thus, Coriolis acceleration is significant when

$$V < 2|\psi|L$$

At a latitude of 23 degrees,

$$|\psi| = \frac{2\pi}{24}\sin(23\pi/180)\,\text{hour}^{-1}.$$

For hurricane formation, consider the characteristic length to be the diameter of the low-pressure region at the given latitude. Perhaps $L = 100\,\text{miles}$. In this case, a wind velocity less than or equal to about $20.5\,\text{miles}\,/\,\text{hour}$ would lead to counterclockwise rotation in the Northern Hemisphere. Of course, hurricane formation is a complex phenomenon. But, the Coriolis effect does prevail. Hurricanes always rotate counterclockwise in the Northern Hemisphere.

For water draining in a kitchen sink at the same latitude, the characteristic length might be the diameter of the sink (say 1 meter). In this case, the Coriolis force is important when

$$V < 5.7 \times 10^{-4}\,\text{cm}\,/\,\text{sec}.$$

Velocities of this magnitude might be present due to residual vorticity from filling the sink. If so, the residual fluid motion might overwhelm the rotational effect due to the Coriolis force that requires the fluid velocity to be toward the drain. This is easily checked by moving a sink filled with water to create strong clockwise rotation before opening its drain. On the other hand, for initially still water the velocity induced by the drain should be sufficiently small for the rotational effect to occur. Does it (see [96])?

Our discussion so far leaves out consideration of the centrifugal acceleration $-\psi \times (\psi \times R)$ given in components by

$$-\psi \times (\psi \times R) = -\dot{\theta}^2 \begin{pmatrix} R_1 \\ R_2 \\ 0 \end{pmatrix}. \tag{14.49}$$

It has magnitude $\dot{\theta}^2\sqrt{R_1^2 + R_2^2}$ and, for a point R on the surface of the Earth, it points toward the axis of rotation and is parallel to the equatorial plane. The magnitude of centrifugal acceleration is largest at the equator and vanishes at the poles. Because of its direction, the centrifugal acceleration is naturally added to the gravitational acceleration thus defining an effective gravitational force acting on a fluid at the surface of the Earth. The effective gravity does not point toward the center of mass of the Earth; it also has a nonzero component that points toward the poles. In practice, this term is neglected because it is small relative to the gravitational acceleration (see Exercise 14.18).

Exercise 14.11. Suppose a particle moves radially outward with constant velocity on a disk that is stationary with respect to an inertial coordinate system. Determine the motion of the particle with respect to a rotating coordinate system that is rotating with angular velocity ϑ. Assume that the first two coordinate axes of the rotating frame rotate in the plane of the disk and its third axis is perpendicular to the disk at the center of the disk. The inertial frame has the same configuration but is not rotating.

Exercise 14.12. Show that the normal component of the Coriolis acceleration is twice the characteristic velocity times the angular velocity times the sine of the latitude.

Exercise 14.13. Do hurricanes form at the equator? Check the data and discuss this issue using the Navier–Stokes model.

14.5 A COUNTERINTUITIVE RESULT: THE PROUDMAN–TAYLOR THEOREM

Imagine a cylindrical tank partially filled with water that is being rotated rapidly at a constant angular velocity about its axis of symmetry. The Rossby number, for a given characteristic fluid velocity and characteristic length, will approach zero as the rotation speed increases without bound. Thus, it is (physically) reasonable to ignore the inertial acceleration given by the material derivative of the velocity of the fluid. The resulting equations of motion are called the geostrophic equations. Likewise, because the viscosity of water is small, it seems reasonable to ignore the Laplacian term in the equations of motion. By doing so, we are tacitly expecting that solutions with small nonzero viscosity will be near the corresponding solutions with zero viscosity. Although this expectation may be true in special cases, results in this setting—called singular perturbation—where the highest-order derivatives are multiplied by a small parameter, are mathematically challenging. Undaunted, we will also assume that the density of the fluid remains constant and that the fluid is incompressible (that is, $\nabla \cdot U = 0$). With all these assumptions and in view of Eq. (14.47), the geostrophic equations for inviscid incompressible flow are

$$\frac{1}{\varrho}\nabla P = g - 2\psi \times U - \psi \times (\psi \times R), \tag{14.50}$$

$$\nabla \cdot U = 0. \tag{14.51}$$

The Proudman–Taylor theorem states that every (smooth) solution of the inviscid incompressible geostrophic equations is constant in the direction of the rotation axis; that is,

$$\frac{\partial U_i}{\partial R_3} = 0 \tag{14.52}$$

for $i = 1, 2, 3$.

To prove this theorem, apply the curl operator $(\nabla\times)$ to both sides of the equation of motion. It is not difficult to simply write out all the terms in components and compute, but it is perhaps more elegant (and certainly more instructive) to use some results from vector analysis; in particular, for vector fields X and Y,

$$\begin{aligned}\nabla \times (\nabla X) &= \operatorname{curl}\operatorname{grad} X = 0,\\ \nabla \times (X \times Y) &= (\nabla \cdot Y)X - (\nabla \cdot X)Y + (Y \cdot \nabla)X - (X \cdot \nabla)Y\\ &= (\operatorname{div} Y)X - (\operatorname{div} X)Y\\ &\quad + \langle Y, \operatorname{grad} X\rangle - \langle X, \operatorname{grad} Y\rangle.\end{aligned} \tag{14.53}$$

Using the first identity in display (14.53) and the constant density, it follows that the curl of the pressure term vanishes. The gravity term is constant; hence, it is clearly curl free. The vector ψ is constant and U is divergence free; therefore, by an application of the second identity, we have that

$$\operatorname{curl}(2\psi \times U) = -2(\psi \cdot \nabla)U. \tag{14.54}$$

By Exercise 14.19,

$$\psi \times (\psi \times Q) = -\nabla(\frac{1}{2}|\psi \times Q|^2). \tag{14.55}$$

Thus, using again the first identity in display (14.53), this term is curl free. In summary, after applying the curl operator, we have proved that the velocity field must satisfy the equation

$$(\psi \cdot \nabla)U = 0. \tag{14.56}$$

The vector ψ (as in Eq. (14.45)) is given by the transpose of the vector $(0, 0, \dot{\theta})$; thus, in component form, Eq. (14.56) states exactly the desired result [Eq. (14.52)]. This completes the proof.

Returning to our rotating cylindrical water tank, let us suppose that the hypotheses of the Proudman–Taylor theorem are valid. At the surface of the water, we expect that $U_3 = 0$. By the theorem, $U_3 = 0$ *everywhere* and the fluid velocity $U = (U_1, U_2, 0)$ does not depend on R_3; in other words, the flow is the same on every plane perpendicular to the axis of rotation. This is a remarkable claim. The mathematics is rigorous, but the result is based on several assumptions. Is the result true for real fluids? The answer is yes. G. I. Taylor performed several experiments to verify this result. In his first experiments (see [107]), small amounts of dye were injected into a small volume of a rotating cylinder filled with water. The dye was drawn out into two-dimensional sheets that were perpendicular to the bottom of the cylinder. In other words, the flow marked by the dye was the same in the vertical direction. In a second, more elaborate experiment, Taylor put a small solid cylindrical object on the bottom of a rotating tank and demonstrated by injecting dye into the flow that there is a stagnant cylindrical column that lay directly above the submerged cylinder. The two-dimensional flow on every horizontal plane knows that cylinder is there! A beautiful description of this experiment is in Taylor's original paper [108]:

> ...*the only possible two dimensional motion satisfying the required conditions is one in which a cylinder of fluid moves as if fixed to the body. The boundary of such a cylinder would act as a solid body, and the liquid outside would behave as though a solid cylindrical body were being moved through it. No fluid would cross this boundary, and the liquid inside it would, in general, be at rest relative [to] the solid body. This idea appears fantastic, but the experiments now to be*

described show that the true motion does, in fact, approximate to this curious type.

Exercise 14.14. Prove identity (14.41).

Exercise 14.15. Prove identities (14.53).

Exercise 14.16. Prove that fluids move from high-pressure regions toward low-pressure regions.

Exercise 14.17. Prove the vector identity $A \times (B \times C) = (A \cdot C)B - (A \cdot B)C$ and use it to determine the magnitude and direction of the centripetal force.

Exercise 14.18. Show that the centripetal force near the equator due to the rotation of the Earth is approximately $1/300$th of the gravitational acceleration.

Exercise 14.19. State the hypotheses required for identity (14.55) to be true and prove the identity.

CHAPTER 15

Water Waves

The mathematical formulation of water wave theory and a classical route to the Boussinesq approximate equations are discussed in this chapter (see [55, 60, 118]).

15.1 THE IDEAL WATER WAVE EQUATIONS

Imagine a two-dimensional slice of an ideal fluid with velocity u. The vorticity of the fluid is defined to be $\nabla \times u$, and a fluid is called irrotational, if its vorticity vanishes. A fluid is called incompressible if the divergence of its velocity field vanishes. Equivalently, the material derivative of its density vanishes. A fluid of constant density (by the continuity equation or conservation of mass) is incompressible, but the converse is not true; an incompressible fluid might not have constant density. A fluid—more precisely, an idealization of a fluid—is called inviscid if its viscosity vanishes.

Consider a two-dimensional region $\Omega := \{(x, y, t) : -\operatorname{wd} < y < \eta(x, t)\}$ and imagine it is filled with an irrotational, incompressible, and inviscid fluid, where the solid bottom boundary is represented by the line $\mathcal{B} := \{(x, y) : y = -\operatorname{wd}\}$. Here, the symbol wd abbreviates a positive water depth, and the free surface of the fluid is assumed to be given by $\mathcal{S} := \{(x, y) : y = \eta(x, t)\}$ for some smooth function $\eta : \mathbb{R} \times (0, \infty) \to (-\operatorname{wd}, \infty)$.

As in Section 13.2, the velocity field u of an irrotational flow is given by a potential ϕ; that is, $u = \nabla\phi$ for some function $\phi : \Omega \to \mathbb{R}$. In this case, the flow velocity u is determined by the scalar potential function that satisfies the system of partial differential equations (PDEs)

$$\operatorname{grad}\Big(\frac{\partial \phi}{\partial t} + \frac{1}{2}|\operatorname{grad}\phi|^2 + \frac{p}{\rho} - B\Big) = 0, \qquad \Delta\phi = 0,$$

first derived in display (13.5).

Four additional, physically realistic assumptions are made: the free surface $\mathcal{S}$ is invariant under the fluid flow, the flow does not penetrate the

An Invitation to Applied Mathematics: Differential Equations, Modeling, and Computation.
http://dx.doi.org/10.1016/B978-0-12-804153-6.50015-4, Copyright © 2017 Elsevier Inc. All rights reserved.

bottom, the pressure of the atmosphere above the surface of the water is constant (that is, the atmospheric pressure is not affected by the fluid flow), and the pressure above and below the surface of the fluid are in balance.

Using Bernoulli's law in the form of Eq. (13.6)

$$\frac{\partial \phi}{\partial t} + \frac{1}{2}|\operatorname{grad}\phi|^2 + \frac{p}{\rho} - B = C,$$

these assumptions translate to the basic set of water wave equations

$$\Delta\phi = 0 \quad \text{on } \Omega, \tag{15.1}$$

$$\eta_t + \phi_x\eta_x - \phi_y = 0 \quad \text{on } \mathcal{S}, \tag{15.2}$$

$$\phi_y = 0 \quad \text{on } \mathcal{B}, \tag{15.3}$$

$$\phi_t + \frac{1}{2}(\phi_x^2 + \phi_y^2) + g\eta = 0 \quad \text{on } \mathcal{S}, \tag{15.4}$$

where g is the acceleration due to gravity.

The existence theory for water waves is not trivial—it is a free boundary value problem. But, this problem has been solved for two- and three-dimensional waves in the case where the underlying space Ω has infinite extent in the horizontal directions (see [120, 121]) and infinite depth. There is a vast literature on this subject. Existence of water waves in many other circumstances has been proved, but this subject remains an area of active research.

To render the water wave equations dimensionless, introduce the scaling

$$\phi = a\tilde{\phi}, \quad \eta = b\tilde{\eta}, \quad x = \ell\tilde{x}, \quad y = \mathrm{wd}\,\tilde{y}, \quad t = \tau\tilde{t}, \tag{15.5}$$

where the dimensions of the scaling constants (with L the dimension length and T the dimension time) are given by

$$[a] = \frac{L^2}{T}, \quad [b] = L, \quad [\ell] = L, \quad [\tau] = T \tag{15.6}$$

and use the (traditional) dimensionless parameters

$$\alpha := \frac{b}{\mathrm{wd}}, \qquad \beta := \frac{\mathrm{wd}^2}{\ell^2}. \tag{15.7}$$

With these choices and the sets $\tilde{\mathcal{B}} := \{(\tilde{x}, \tilde{y}) : \tilde{y} = -1\}$, $\tilde{\mathcal{S}} := \{(\tilde{x}, \tilde{y}) : \tilde{y} = \alpha\tilde{\eta}(\tilde{x}, \tilde{t})\}$, and $\tilde{\Omega} := \{(\tilde{x}, \tilde{y}) := -1 < \tilde{y} < \alpha\tilde{\eta}(\tilde{x}, \tilde{t})\}$, the water wave equations take the dimensionless form

$$\beta\tilde{\phi}_{\tilde{x}\tilde{x}} + \tilde{\phi}_{\tilde{y}\tilde{y}} = 0 \qquad \text{on } \tilde{\Omega}, \tag{15.8}$$

$$\tilde{\eta}_{\tilde{t}} + \alpha\tilde{\phi}_{\tilde{x}}\tilde{\eta}_{\tilde{x}} - \frac{1}{\beta}\tilde{\phi}_{\tilde{y}} = 0 \qquad \text{on } \tilde{\mathcal{S}}, \tag{15.9}$$

$$\tilde{\phi}_{\tilde{y}} = 0 \qquad \text{on } \tilde{\mathcal{B}}, \tag{15.10}$$

$$\tilde{\phi}_{\tilde{t}} + \frac{1}{2}(\alpha\tilde{\phi}^2_{\tilde{x}} + \frac{\alpha}{\beta}\tilde{\phi}^2_{\tilde{y}}) + \tilde{\eta} = 0 \qquad \text{on } \tilde{\mathcal{S}}, \tag{15.11}$$

as long as

$$\frac{\tau a}{\mathrm{wd}\, b} = \frac{1}{\beta}, \quad \frac{\tau a}{\mathrm{wd}^2} = \frac{\alpha}{\beta}, \quad \frac{g\tau b}{a} = 1. \tag{15.12}$$

Here, g is a fixed parameter and wd is fixed to be the average water depth. The scaling parameter b (respectively, ℓ) is chosen to represent an expected characteristic wave amplitude (respectively, a characteristic wave length). Although these choices would seem to presuppose measurements of the phenomena we wish to determine, they need not be precise. In practice, they are taken to be guesses of the expected amplitude and wavelength. The dimensionless groups α and β are determined by the fixed parameters together with the guessed amplitude and wavelength. This leaves the scaling parameters a and τ to be determined from Eqs. (15.12), which imply that

$$\frac{a}{\tau} = bg, \qquad a\tau = \frac{b\,\mathrm{wd}}{\beta}.$$

The solution of this system of algebraic equations is

$$a = b\ell\sqrt{\frac{g}{\mathrm{wd}}}, \qquad \tau = \frac{\ell}{\sqrt{g\,\mathrm{wd}}}. \tag{15.13}$$

The ultimate choice of scaling would be made to ensure that α and β are small enough to justify the first-order approximation that will be made state variables expanded in Taylor series with respect to these parameters. Unfortunately, there is no known way to determine how small they should be. The strategy in applications is to make reasonable choices and hope that the simplified model thus obtained captures some of the observed behavior. For the moment, simply choose b to be a characteristic length related to the depth of the water and ℓ a characteristic length in the horizontal direction such that α and β are smaller than unity.

15.2 THE BOUSSINESQ EQUATIONS

Boussinesq's approximation is obtained under the assumption that $\alpha \ll 1$ and $\beta \ll 1$ (which is often interpreted as the regime of small amplitude waves over shallow water). The basic idea is to expand the (scaled) potential into a series of the form

$$\tilde{\phi}(\tilde{x},\tilde{y},\tilde{t}) = \sum_{n=0}^{\infty} (\tilde{y}+1)^k f_n(\tilde{x},\tilde{t}),$$

where the sequence of functions $\{f_n\}_{n=0}^{\infty}$ is to be determined. By imposing the boundary condition on $\tilde{\mathcal{B}}$ given by Eq. (15.10), it follows that f_1 must vanish. Because the potential must also satisfy the scaled Laplace equation (15.8), we have the recursion

$$\begin{aligned} f_0 &\quad \text{arbitrary}, \\ f_1 &= 0, \\ \beta(f_n)_{\tilde{x}\tilde{x}} + (n+2)(n+1)f_{n+2} &= 0 \quad \text{for } n \geq 2. \end{aligned}$$

By an easy induction argument and the notational replacement $f := f_0$, the velocity potential is seen to be formally (that is, without considering convergence) represented by the infinite series

$$\tilde{\phi}(\tilde{x},\tilde{y},\tilde{t}) = \sum_{n=0}^{\infty} \frac{(-1)^k}{(2n)!}(\tilde{y}+1)^{2n}\frac{\partial^{2n} f}{\partial \tilde{x}^{2n}}(\tilde{x},\tilde{t})\beta^k. \tag{15.14}$$

Having a representation of the solution $\tilde{\phi}$ of the (scaled) Laplace equation, we seek the first-order (in α and β) approximation to the free surface conditions (15.9) and (15.11).

Substitution and truncation in Eq. (15.9) results in the equation

$$\tilde{\eta}_{\tilde{t}} + \alpha f_{\tilde{x}}\tilde{\eta}_{\tilde{x}} + (\tilde{y}+1)f_{\tilde{x}\tilde{x}} - \frac{\beta}{6}(\tilde{y}+1)^3 f_{\tilde{x}\tilde{x}\tilde{x}\tilde{x}} = 0,$$

which must hold for $\tilde{y} = \alpha\tilde{\eta}$. Hence, we obtain the equation

$$\tilde{\eta}_{\tilde{t}} + \alpha f_{\tilde{x}}\tilde{\eta}_{\tilde{x}} + (1+\alpha\tilde{\eta})f_{\tilde{x}\tilde{x}} - \frac{\beta}{6} f_{\tilde{x}\tilde{x}\tilde{x}\tilde{x}} = 0. \tag{15.15}$$

In a similar manner, the approximation to Eq. (15.11) is

$$\tilde{\eta} + f_{\tilde{t}} - \frac{\beta}{2} f_{\tilde{x}\tilde{x}\tilde{t}} + \frac{\alpha}{2} f_{\tilde{x}}^2 = 0. \tag{15.16}$$

Tradition and good sense dictate defining a new function

$$w = f_{\tilde{x}}$$

(which is therefore the leading-order approximation of the fluid velocity) and replacing Eq. (15.16) by the equation obtained from it by differentiation with respect to $\tilde{x}$. This procedure results in the Boussinesq model

$$\tilde{\eta}_{\tilde{t}} + \{(1 + \alpha\tilde{\eta})w\}_{\tilde{x}} - \frac{\beta}{6} w_{\tilde{x}\tilde{x}\tilde{x}} = 0, \tag{15.17}$$

$$w_{\tilde{t}} + \tilde{\eta}_{\tilde{x}} + \alpha w w_{\tilde{x}} - \frac{\beta}{2} w_{\tilde{x}\tilde{x}\tilde{t}} = 0 \tag{15.18}$$

for the approximate scaled free surface (graph of η) and fluid velocity w.

15.3 KDV

In this subsection we derive the Korteweg–de Vries (KdV) equation (Diederik Korteweg and Gustav de Vries, 1895), a scalar PDE for an approximation to the (scaled) free surface $\tilde{\eta}$ of a shallow small amplitude water wave.

Set $\alpha = \beta = 0$ in the Boussinesq model [Eqs. (15.17) and (15.18)] to obtain the equations

$$\tilde{\eta}_{\tilde{t}} + w_{\tilde{x}} = 0, \quad w_{\tilde{t}} + \tilde{\eta}_{\tilde{x}} = 0.$$

They are the zero-order approximations to the Boussinesq equations with respect to the small parameters α and β. By differentiating the first equation with respect to $\tilde{t}$ and the second with respect to $\tilde{x}$, it is easy to see that (at this order of approximation) $\tilde{\eta}$ satisfies the wave equation

$$\tilde{\eta}_{\tilde{t}\tilde{t}} = \tilde{\eta}_{\tilde{x}\tilde{x}}.$$

It has traveling wave solutions of the form $\tilde{\eta}(\tilde{x}, \tilde{t}) = N(\tilde{x} - \tilde{t})$, where N is an arbitrary scalar function, corresponding to waves moving to the right along the real line.

Note that (for these traveling waves)

$$\tilde{\eta}_{\tilde{t}}(\tilde{x}, \tilde{t}) = -N'(\tilde{x} - \tilde{t}) = -\tilde{\eta}_{\tilde{x}}(\tilde{x}, \tilde{t}) \tag{15.19}$$

and

$$\tilde{\eta}_{\tilde{t}}(\tilde{x}, \tilde{t}) = -N'(\tilde{x} - \tilde{t}) = -w_{\tilde{x}}(\tilde{x}, \tilde{t}).$$

Using this last equation,

$$w(\tilde{x}, \tilde{t}) = N(\tilde{x} - \tilde{t}) = \tilde{\eta}(\tilde{x}, \tilde{t})$$

up to a constant that is taken here to be zero.

In view of Eq. (15.19), derivatives of $\tilde{\eta}$ with respect to $\tilde{t}$ can be replaced by derivatives with respect to $\tilde{x}$ plus terms of order one in α and β. Thus, the higher-order approximation to w in the Boussinesq model can be expressed in the form

$$w = \tilde{\eta} + \alpha A(\tilde{\eta}, \tilde{\eta}_{\tilde{x}}, \ldots) + \beta B(\tilde{\eta}, \tilde{\eta}_{\tilde{x}}, \ldots) + O(\alpha^2, \beta^2),$$

where A and B are expressions in $\tilde{\eta}$ and its derivative with respect to $\tilde{x}$. After substitution into Eqs. (15.17) and (15.18) and truncation at first-order in α and β, the following equations are obtained:

$$\tilde{\eta}_{\tilde{t}} + \tilde{\eta}_{\tilde{x}} + \alpha(A_{\tilde{x}} + 2\tilde{\eta}\tilde{\eta}_{\tilde{x}}) + \beta(B_{\tilde{x}} - \frac{1}{6}\tilde{\eta}_{\tilde{x}\tilde{x}\tilde{x}}) = 0,$$
$$\tilde{\eta}_{\tilde{t}} + \tilde{\eta}_{\tilde{x}} + \alpha(A_{\tilde{t}} + \tilde{\eta}\tilde{\eta}_{\tilde{x}}) + \beta(B_{\tilde{t}} - \frac{1}{2}\tilde{\eta}_{\tilde{x}\tilde{x}\tilde{t}}) = 0.$$

Again, all derivatives with respect to $\tilde{t}$ *in the first-order terms* can be replaced by derivatives with respect to $\tilde{x}$; in particular, $A_{\tilde{t}} = -A_{\tilde{x}} + O(\alpha, \beta)$ and $B_{\tilde{t}} = -B_{\tilde{x}} + O(\alpha, \beta)$. To be compatible (that is, for the first-order terms to agree), we must have $A = -\tilde{\eta}^2/4$ and $B = \tilde{\eta}_{\tilde{x}\tilde{x}}/3$. Using these values for A and B results in the KdV equation,

$$\tilde{\eta}_{\tilde{t}} + \tilde{\eta}_{\tilde{x}} + \frac{3}{2}\alpha\tilde{\eta}\tilde{\eta}_{\tilde{x}} + \frac{1}{6}\beta\tilde{\eta}_{\tilde{x}\tilde{x}\tilde{x}} = 0.$$

By seeking a traveling wave solution of the KdV equation of the form

$$\tilde{\eta}(x, t) = N(x - ct),$$

it is not difficult to show that there is a family of solutions given by

$$\tilde{\eta}(x, t) = \frac{2(c-1)}{\alpha}\operatorname{sech}^2\left(\left(\frac{3(c-1)}{2\beta}\right)^{1/2}(x - ct)\right). \tag{15.20}$$

Solutions of this type correspond to the solitary water waves first observed by John Scott Russell:

> *I was observing the motion of a boat which was rapidly drawn along a narrow channel by a pair of horses, when the boat suddenly stopped — not so the mass of water in the channel which it had put in motion; it accumulated round the prow of the vessel in a state of violent agitation, then suddenly leaving it behind, rolled forward with great velocity, assuming the form of a large solitary elevation, a rounded, smooth and well-defined heap of water, which continued its course along the channel apparently without change of form or diminution of speed. I followed it on horseback, and overtook it still rolling on at a rate of some eight or nine miles an hour, preserving its original figure some thirty feet long and a foot to a foot and a half in height. Its height gradually diminished, and after a chase of one or two miles I lost it in the windings of the channel. Such, in the month of August 1834, was my first chance interview with that singular and beautiful phenomenon which I have called the Wave of Translation.*[1]

Exercise 15.1. Derive the family of traveling wave solutions of the KdV equation given in Eq. (15.20).

Exercise 15.2. (This exercise was suggested by Samuel Walsh.) The KdV equation may be viewed as a combination of two other famous equations: The linear part corresponds to Airy's equation

$$\eta_t + \eta_x + \frac{\beta}{6}\eta_{xxx} = 0$$

and the nonlinear part to Burgers's equation

$$\eta_t + \frac{3\alpha}{2}\eta\eta_x = 0.$$

(a) Show that there are solutions of Airy's equation of the form $\cos(k(x - ct))$ for real numbers c and k provided that these parameters are related by a certain algebraic relation. This relation, which you are asked to find, is called the dispersion relation..
(b) Reconsider the dimensionless form of the water wave equations (15.8)–(15.11). Set $\alpha = 1$ and $\beta = 1$ and show that there is a solution of the form

$$\eta(x,t) = \cos(k(x - ct)), \qquad \phi(x,y,t) = \Phi(y)\sin(k(x - ct))$$

[1] J. Scott Russell, *Report on Waves*, Fourteenth meeting of the British Association for the Advancement of Science, 1844.

for some function Φ if and only if

$$c = \sqrt{\frac{\tanh k}{k}}.$$

Compare this result with the great wave of translation for KdV and interpret the solution of the water wave equations for k large and k small.
(c) Write a numerical code to approximate solutions of Burgers's equation for the case $\alpha = 2/3$ on the spatial interval $[0, 10]$ with zero Dirichlet boundary conditions. Choose initial data $(x \mapsto \eta(x, 0))$ that is zero near the end points of the interval and with $\eta_x(a, 0) < 0$ a some interior point $x = a$. Discuss the results of your numerical experiments. Note in particular the extent T of the time interval $0 \leq t \leq T$ for which your computation suggests that solutions exist. Theoretically, $\eta_x(a, t) \to -\infty$ in finite time; in particular, solutions of Burgers's equation can blow up in finite time. Is there numerical evidence for this fact in your numerical computations? How does one distinguish between blowup of a solution and blowup in a numerical experiment due to an unstable numerical method? There is much more on Burgers's equation in this book starting on page 431).
(d) Discuss the view that KdV is a combination of the equations of Airy and Burgers.

15.4 BOUSSINESQ STEADY STATE WATER WAVES

In steady state, Eqs. (15.17) and (15.18) reduce to the system

$$\{(1 + \alpha\tilde{\eta})w\}_{\tilde{x}} = \frac{\beta}{6} w_{\tilde{x}\tilde{x}\tilde{x}}, \tag{15.21}$$

$$\tilde{\eta}_{\tilde{x}} = -\alpha w w_{\tilde{x}}. \tag{15.22}$$

Replace Eq. (15.22) by its integrated form

$$\tilde{\eta} = -\frac{\alpha}{2} w^2 + k, \tag{15.23}$$

where k is a constant. The free surface must correspond to the water depth in case the flow velocity vanishes; that is, the free surface in this case is given by $\eta = 0$. Thus, $k = 0$ and

$$\tilde{\eta} = -\frac{\alpha}{2} w^2. \tag{15.24}$$

Eq. (15.21), after substitution of Eq. (15.24) and integration with respect to $\tilde{x}$, is a form of Duffing's equation given by

$$\frac{\beta}{6} w_{\tilde{x}\tilde{x}} + c = w - \frac{\alpha^2}{2} w^3, \tag{15.25}$$

where c is a constant.

Imagine that the fluid surface has a local maximum or minimum at $x = 0$, which may be viewed as a crest or trough of a surface wave. In this case, the second component of the fluid velocity vector $(\phi_x(0, \eta(0)), \phi_y(0, \eta(0)))$ vanishes when evaluated at this point; thus, it may be considered to have coordinates $(v_0, 0)$.

Let the magnitude of this velocity vector be the characteristic velocity for the flow and define the (dimensionless) Froude number

$$\mathrm{Fr} := \frac{v_0}{\sqrt{g\,\mathrm{wd}}}. \tag{15.26}$$

It is the important dimensionless quantity that appears in the study of surface waves on flows of finite depth.

There are at least three interesting flow regimes:

$$\begin{aligned} &\textbf{I.} \quad 0 < \mathrm{Fr}^2 < \tfrac{2}{3}; \\ &\textbf{II.} \quad \tfrac{2}{3} < \mathrm{Fr}^2 < \tfrac{8}{3}; \\ &\textbf{III.} \quad \tfrac{8}{3} < \mathrm{Fr}^2 . \end{aligned} \tag{15.27}$$

In the scaled variables (see Eqs. (15.5)–(15.13)), the amplitude of the characteristic velocity is given by

$$\frac{\ell v_0}{a} = \frac{v_0}{\alpha\sqrt{g\,\mathrm{wd}}} = \frac{\mathrm{Fr}}{\alpha},$$

and by Eq. (15.14),

$$\tilde{\phi}_{\tilde{x}}(0, \tilde{\eta}(0)) = w(0) - \frac{1}{2}(\tilde{\eta}(0) + 1)^2 w_{\tilde{x}\tilde{x}}(0)\beta + O(\beta^2) = \frac{\mathrm{Fr}}{\alpha}, \tag{15.28}$$

$$\tilde{\phi}_{\tilde{y}}(0, \tilde{\eta}(0)) = -(\tilde{\eta}(0) + 1)w_{\tilde{x}}(0) + O(\beta) = 0. \tag{15.29}$$

To simplify the discussion, assume that the velocity $(v_0, 0)$ at $(x, y) = (0, \eta(0))$ is independent of the water depth, or equivalently, this choice holds for all β. Under this assumption, Eqs. (15.28) and (15.29) imply that

$$w(0) = \frac{\mathrm{Fr}}{\alpha}, \quad w_{\tilde{x}}(0) = 0, \quad w_{\tilde{x}\tilde{x}}(0) = 0. \tag{15.30}$$

These values agree with the actual values up to $O(\beta)$ corrections.

The integration constant c in the differential equation (15.25) is given by

$$c = w(0) - \frac{\alpha^2}{2} w^3(0) = \frac{\gamma}{\alpha}, \tag{15.31}$$

where

$$\gamma = \frac{1}{2}\,\mathrm{Fr}(2 - \mathrm{Fr}^2).$$

Consider the first-order nonlinear system

$$w' = z, \qquad z' = \frac{6}{\beta}\Big(w - \frac{\alpha^2}{2} w^3 - \frac{\gamma}{\alpha}\Big), \tag{15.32}$$

which is equivalent to differential equation (15.25).

The rest points of the system of differential equations (15.32) correspond to the solutions of the cubic polynomial equation

$$\frac{\alpha^2}{2} w^3 - w + \frac{\gamma}{\alpha} = 0. \tag{15.33}$$

Because of the choice of c given in Eq. (15.31), $w(0)$ is a root of the polynomial; therefore, $w - w(0)$ is a factor. In fact,

$$\frac{\alpha^2}{2} w^3 - w + \frac{\gamma}{\alpha} = (w - w(0))\Big(\frac{\alpha^2}{2} w^2 + \frac{\alpha^2}{2} w(0) w + \frac{\alpha^2}{2} w^2(0) - 1\Big),$$

and its remaining roots

$$w_{\pm} = \frac{1}{2}\Big(-w(0) \pm \Big(\frac{8}{\alpha^2} - 3w^2(0)\Big)^{1/2}\Big) = \frac{1}{2\alpha}\big(-\mathrm{Fr} \pm \sqrt{8 - 3\,\mathrm{Fr}^2}\,\big) \tag{15.34}$$

are obtained from the quadratic formula.

These roots [Eq. (15.34)] are real in Regimes I and II that are discussed here; these roots are complex in Regime III (see Exercise 15.3).

Using the representation of $w(0)$ in Eq. (15.30) and the representation of w_+ in Eq. (15.34) together with a simple calculation, it follows that the three real roots in Regime I lie on the w-axis in the order of increasing size $[w_-, w(0), w_+]$ in Regime I and $[w_-, w_+, w(0)]$ in Regime II (see Fig. 15.1). At $\mathrm{Fr}^2 = 2/3$ the roots $w(0)$ and w_+ coincide.

In the (w, z) plane, system (15.32) has three rest points on the w-axis corresponding to the roots $w_\pm$ and $w(0)$. In Regime I, the rest points $(w_\pm, 0)$

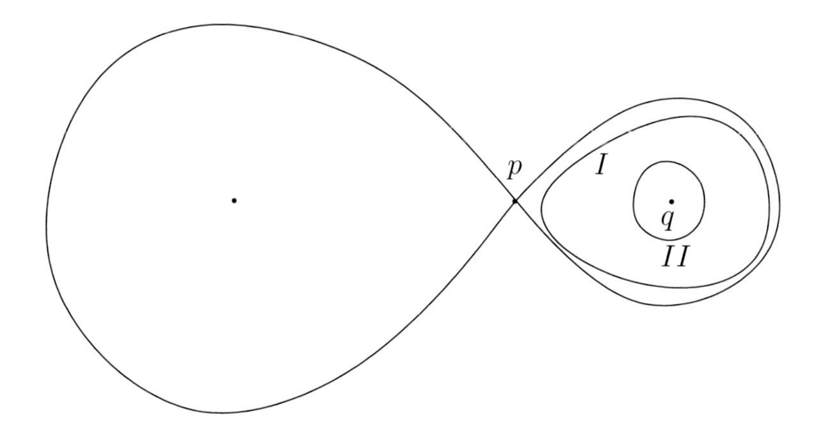

Fig. 15.1 A schematic phase portrait of system (15.32) *is depicted. The saddle point is marked* p*, one center is marked* q*. A typical orbit for Regime I is marked I and a typical orbit for Regime II is marked* II*. Orbits in Regime I pass close to the saddle point. Orbits in Regime II pass close to the center.*

are centers and (the middle rest point) $(w(0), 0)$ is a hyperbolic saddle; in Regime II, w_- and $w(0)$ are centers and w_+ is a hyperbolic saddle point. To prove these statements, note first that the linearization of the differential equation (15.32) at $(w, 0)$ is given by the Jacobian matrix

$$\begin{pmatrix} 0 & 1 \\ \frac{6}{\beta}(1 - \frac{3}{2}\alpha^2 w^2) & 0 \end{pmatrix}, \tag{15.35}$$

whose eigenvalues are the square roots of the component

$$\frac{6}{\beta}(1 - \frac{3}{2}\alpha^2 w^2). \tag{15.36}$$

At $w(0)$, this quantity is given by

$$\frac{6}{\beta}(1 - \frac{3}{2}\alpha^2 w^2) - \frac{6}{\beta}(1 - \frac{3}{2}\,\mathrm{Fr}^2). \tag{15.37}$$

Hence, in Regime I, the eigenvalues are real with one eigenvalue positive and the other negative. In Regime II the eigenvalues are pure imaginary.

By the Grobman–Hartman theorem (see Appendix A.7), $(w(0), 0)$ is a saddle point for the nonlinear system in Regime I. In Regime II, the rest point $(w(0), 0)$ is a center for the nonlinear system. The proof of this fact can be easily constructed with two main ingredients: the existence of a first integral (a function that is constant along solutions) for system (15.32) and the Morse lemma (see Appendix A.13). The total energy

$$E := \frac{\beta}{12}z^2 + \frac{\gamma}{\alpha}w - \frac{1}{2}w^2 + \frac{\alpha^2}{8}w^4 \tag{15.38}$$

of Hamiltonian system (15.32) is constant along solutions. In other words, solutions lie on the one-dimensional level sets of E.

To determine the nature of the level sets of the energy [Eq. (15.38)] near the rest point in Regime II for the Hamiltonian system, consider the slightly more general context of the Hamiltonian system

$$w' = z, \qquad z' = -f(w) \tag{15.39}$$

with Hamiltonian given by

$$\tilde{H}(w, z) = \frac{1}{2}z^2 + F(w)$$

where $F' = f$. Suppose that $f(w_0) = 0$ and define the new Hamiltonian

$$H(w, z) = \frac{1}{2}z^2 + F(w) - F(w_0),$$

which vanishes at the rest point $(w_0, 0)$ of the corresponding Hamiltonian system (15.39). Of course, H is also constant along the solutions of this system of differential equations.

The linearization of our Hamiltonian system at the rest point $(w_0, 0)$ is given by the matrix

$$\begin{pmatrix} 0 & 1 \\ -f'(w_0) & 0 \end{pmatrix}. \tag{15.40}$$

Hence, under the assumption that there is no eigenvalue with zero real part, we have that $f'(w_0) \neq 0$, with real eigenvalues in case $f'(w_0) < 0$ and pure imaginary eigenvalues in case $f'(w_0) > 0$. By the Morse lemma applied to the function $F(w) - F(w_0)$, which has the Taylor series expansion

$$\begin{aligned} F(w) - F(w_0) &= F'(w_0)(w - w_0) + \frac{1}{2}F''(w_0)(w - w_0)^2 + O((w - w_0)^3) \\ &= \frac{1}{2}f'(w_0)(w - w_0)^2 + O((w - w_0)^3), \end{aligned}$$

there is a change of coordinates that transforms the Hamiltonian to the form

$$H(\omega, z) = \frac{1}{2}z^2 + \frac{1}{2}f'(w_0)\omega^2.$$

In the case of the linear center (pure imaginary eigenvalues), the level sets in the new coordinates are ellipses surrounding the rest point that has been transformed to the origin. In other words, all the solutions lie on closed orbits of the original differential equation. This is almost the end of the proof. The last detail is to prove that there are no rest points in some neighborhood of the original rest point. But, this fact follows immediately because $f(w_0) = 0$ and $f'(w_0) \neq 0$. Thus, all solutions in some neighborhood of the rest point are closed orbits. A rest point of this type is called a center.

We have proved that $(w(0), 0)$ is a saddle point in Regime I and a center in Regime II.

In Regime I, the level set of the Hamiltonian [Eq. (15.38)] given by all (w, z) such that

$$\frac{\beta}{12}z^2 + \frac{\gamma}{\alpha}w - \frac{1}{2}w^2 + \frac{\alpha^2}{8}w^4 = \big(\frac{\gamma}{\alpha}w(0) - \frac{1}{2}w^2(0) + \frac{\alpha^2}{8}w^4(0)\big) \quad (15.41)$$

is a (horizontal) figure eight in the (w, z) plane with its double point at $(w(0), 0)$. The rest point $(w_-, 0)$ is surrounded by the left loop of the figure eight and the rest point $(w_+, 0)$ is surrounded by the right loop. Note that the right and left points where this figure eight crosses the w-axis can be determined by writing Eq. (15.41) in the form

$$\frac{\beta}{12}z^2 + \frac{\alpha^2}{8}(w-w(0))^2((w-w(0))^2+4w(0)(w-w(0))+(6w^2(0)-\frac{4}{\alpha^2})) = 0,$$

setting $z = 0$, dividing by $(w - w(0))^2$, and finding the roots of the remaining quadratic equation. In fact, these roots (corresponding to the right and left crossing points) are

$$w_L = -\omega(0) - \frac{2}{\alpha}\sqrt{1 - \frac{\alpha^2 w(0)^2}{2}},$$
$$w_R = -\omega(0) + \frac{2}{\alpha}\sqrt{1 - \frac{\alpha^2 w(0)^2}{2}}. \quad (15.42)$$

With the mathematical analysis completed, note that the surface waveform is given by the function η or, in scaled variables, $\tilde{\eta}$. We are interested in the solution of differential equation (15.32) whose initial value is $(w(0), 0)$ (more precisely, $(w(0) + O(\beta), 0)$) on the interval $0 \le \tilde{x} \le 1$. This solution determines the surface wave that corresponds to the graph of the function η given by

$$\eta(x) = -\frac{\mathrm{wd}\,\alpha^2}{2} w^2(x/\ell), \quad 0 \le x \le \ell. \tag{15.43}$$

In Regime I, the solution of system (15.32) starts near the saddle point $(w(0), 0)$. If β is sufficiently small, then the waveform would take a few different shapes depending on where the initial point is in relation to the saddle (either at the rest point or left, right, above, or below the rest point). For small enough β, the solution will not wander too far from the rest point on its finite domain of definition $0 \le \tilde{x} \le 1$. This statement is not precise. All that can be said, without a much more serious analysis, is that for given parameter values (contained in some compact set), given a neighborhood of the rest point, there is a smaller neighborhood such that our solution starting in the smaller neighborhood stays in the given neighborhood on the domain $0 \le \tilde{x} \le 1$. This follows from the continuity of solutions of ordinary differential equations with respect to initial conditions and parameters. Also, for sufficiently small β, surface waveforms will not have several crests; rather, the expected waveform will be a single dip, a single hump, a rise, or a fall.

For some fixed values of the parameters α and β, with β not too small, the waveform could arise from an excursion along a periodic orbit near the homoclinic loops formed by the stable and unstable manifolds of the saddle point (see Fig. 15.1). There is a slight surprise here: the left-hand loop corresponds to a surface wave whose amplitude exceeds the water depth. The right-hand loop corresponds to a wave whose amplitude is physical provided that the Froude number exceeds $2 - \sqrt{2} \approx 3/5$. To see this fact, consider the lowest-order approximations

$$w_L = -w(0) - \frac{2}{\alpha}, \qquad w_R = -w(0) + \frac{2}{\alpha}. \tag{15.44}$$

By Eq. (15.43), the water wave amplitude is less than the water depth whenever

$$|w| < \frac{\sqrt{2}}{\alpha}.$$

Note that $w = w_R$ satisfies this requirement whenever $\mathrm{Fr} > 2 - \sqrt{2}$. On the other hand, with $w = w_L$ the requirement reduces to $\mathrm{Fr} + 2 < \sqrt{2}$, which is never satisfied. Thus, under our assumptions, the physically realistic surface waves correspond to solutions of system (15.32) that start near and to the right of the saddle point $(w(0), 0)$.

In Regime II, the solution starts on or near a center. We might expect several waves provided that the period of the periodic solutions near the center are sufficiently small so that the solution traverses the closed orbit more than once on its domain. This period is obtained immediately from the square root of an eigenvalue of the Jacobian matrix (15.35) at $w(0)$ to be

$$\frac{2\pi\sqrt{\beta}}{\sqrt{6(3\,\mathrm{Fr}^2/2 - 1)}}, \tag{15.45}$$

which is $O(\sqrt{\beta})$ and therefore small. Thus, in Regime II, waveforms with multiple crests and troughs are predicted.

15.5 A FREE-SURFACE FLOW

Consider water flowing from a reservoir over a flat-bottomed horizontal open trough. An excellent example, which is discussed here, is the Tiger Fountain at the University of Missouri campus in Columbia, Missouri. The main feature of the flow in this fountain, before it cascades over the lip of its trough, is the existence of a steady state surface waveform. The waveform stretches from bank to bank across the trough in a roughly parabolic shape, with the parabola opening in the upstream direction. This shape seems to be explained by the Poiseuille flow in a pipe. At least, the parabolic shape of the velocity field of the Poiseuille flow is in agreement with the parabolic shape of the surface waveform observed in the Tiger Fountain flow. The most basic question about the observed waves is why are they there? A much more difficult question is the stability of these waves. After destruction of the wave pattern at the Tiger Fountain by disturbing the flow with hand movements in the water, the wave is observed to always return to its original steady state form after a very short time interval. This experiment suggests that the waveform is stable to small perturbations. A good applied mathematics problem is to prove that some realistic models of the Tiger Fountain flow exhibit stable steady state waveforms with shapes as observed in the physical flow. In this case, the purpose of the modeling and mathematical analysis is to help answer the question: Why does the waveform exist and why is it stable?

To gain some insight, consider the two-dimensional flow in a thin vertical slice parallel to the bulk Tiger Fountain flow velocity vector.

The basic measured quantities for the Tiger Fountain channel flow are

$$\begin{array}{ll} \text{water depth (11/16 inch)} & 0.0174625\,\mathrm{m} \\ \text{plate slice length (18.75 inch)} & 0.47625\,\mathrm{m} \\ \text{plate width (132 inch)} & 3.3528\,\mathrm{m} \\ \text{pump flow rate (200 gallon/minute)} & 0.012618\,\mathrm{m}^3\,/\,\mathrm{sec} \\ \text{wave amplitude (2/16 inch)} & 0.003175\,\mathrm{m} \\ \text{wave length (3 inch)} & 0.0762\,\mathrm{m} \\ \text{bulk speed (0.70707 foot/second)} & 0.215515\,\mathrm{m}\,/\,\mathrm{sec} \\ \text{surface speed (1 foot/second)} & 0.3048\,\mathrm{m}\,/\,\mathrm{sec} \end{array} \tag{15.46}$$

where the bulk speed is computed to be the flux through the outflow rectangle whose area is given by the product of plate width and water depth (that is, bulk speed equals pump flow rate divided by the area of this rectangle) and the surface speed is computed by timing floating objects that traverse the plate.

Using the plate slice length as the characteristic length, the bulk velocity as the characteristic velocity, and the kinematic viscosity of water to be approximately $1.005 \times 10^{-6}\,\mathrm{m}^2\,/\,\mathrm{sec}$, the Reynolds number is

$$\mathrm{Re} \approx 1.02128 \times 10^5.$$

Thus, this flow is in the upper range of laminar flows. In this regime, the flow remembers it is viscous near the flat plate over which it flows. Thus, a boundary layer is expected.

A standard approximation to the boundary layer thickness for flow over a flat plate is given by

$$5\sqrt{\frac{\nu x}{V}},$$

where x is the distance downstream from the leading edge of the plate, ν is the kinematic viscosity, and V is the characteristic velocity. This result is derived from an approximation of the Navier–Stokes equations called the boundary layer equations (see [93] and Section 17.3). At half the slice

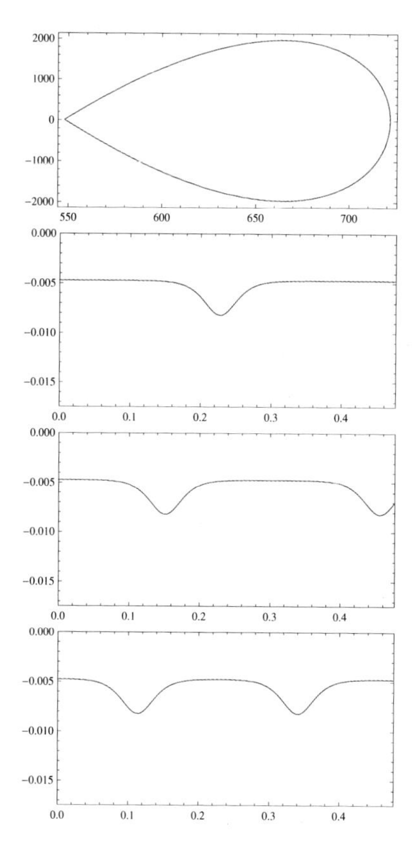

Fig. 15.2 The top panel is a graph of the homoclinic loop for system (15.32) *with* $\mathrm{Fr} = 0.736798$ *and* $\alpha = \beta = 0.001344$. *The bottom three panels are surface waveforms computed as graphs of* η *(with both axes calibrated in meters) using the Eq.* (15.43), *where* w *is the solution of system* (15.32) *with initial condition* $(w, z) = (\mathrm{Fr}/\alpha + n\beta, 0)$, *and from top to bottom* $n = 1, 100, 1000$.

length, the approximate layer thickness is a third of the water depth. Thus, the boundary layer is significant for this flow and is discussed more fully later.

What about the Boussinesq equations?

The ideal water wave equations in two-dimensions and Boussinesq's equations cannot be expected to be predictive of the precise quantitative features of a real flow. But, as usual in applied mathematics, let us assume that the steady state Boussinesq equations [Eqs. (15.21) and (15.22)] constitute the model equations for our flow. What does this model predict for the Tiger Fountain flow? In particular, does the predicted waveform agree with the observed waveform?

As mentioned previously, the correct numerical values for the characteristic lengths b and ℓ used to make the model dimensionless are not known a priori. The only obvious restriction is that b must be less than the characteristic water depth; that is, $b < \mathrm{wd}$ or $\alpha < 1$, and ℓ must be larger. It seems natural to take ℓ equal to the plate slice length. This is a natural horizontal length scale for our problem and β is determined by this choice. There does not seem to be a natural candidate for b. Simply to make α and β approximately the same size, suppose that b is chosen so that $\alpha = \beta$; that is, $b = \mathrm{wd}\,\beta$.

In view of the data in display (15.46), the choices just mentioned, and choosing the characteristic velocity to be the surface flow rate, we obtain the Froude number and the small parameters

$$\alpha = \beta, \quad \beta = 0.001344, \quad \mathrm{Fr}^2 = 0.542872. \tag{15.47}$$

As the Froude number is smaller than the critical value ($\mathrm{Fr}^2 = 2/3$), the flow is in Regime I (see display (15.27)).

The approximate wave amplitude, computed using Eqs. (15.42) and (15.43) , is

$$-\frac{\mathrm{wd}\,\alpha^2}{2}\Big(-w(0)+\frac{2}{\alpha}\sqrt{1-\frac{\alpha^2 w(0)^2}{2}}\Big)^2 \approx 0.00822067.$$

This value agrees (reasonably well) with the observed value 0.003175 in display (15.46).

The prediction of wavelength is not obvious because the initial condition is unknown; rather, the initial value for scaled differential equation system (15.32) should be an order β perturbation of the saddle point at $(w(0), z(0) = (\mathrm{Fr}/\alpha, 0)$. Fig. 15.2 shows the results of some numerical experiments based on this observation. Although the size of the $O(\beta)$ perturbation is not known, mathematical consideration of this size only makes sense in calculations that include limits as β goes to zero. In principle, every multiple of β is an order β perturbation. Using our parameter values, $1000\beta \approx 1.34$. Perhaps this is not such a large perturbation given the size of the homoclinic loop in scaled variables, which has a width of approximately 750 dimensionless units.

What have we learned? Boussinesq's approximate equations appear to be consistent with the observed flow phenomena. This fact provides some evidence that the equations of fluid dynamics (even the rather crude shallow water wave model) reflect physical reality. Of course the predictive value of the model is limited; it is at best a low-order approximation of the flow observed at the Tiger Fountain.

Exercise 15.3. Discuss steady state surface waves in case the Froude number exceeds $\sqrt{8/3}$.

Exercise 15.4. Discuss the sensitivity of the computed wavelength and wave amplitude to changes in the measured surface and bulk flow velocities. Show that a small change in the pump flow rate and hence the bulk velocity produces results much closer to observed values for the wave length and amplitude. Note: The pump flow rate was obtained from a manufacturers specification, not a measurement.

CHAPTER 16

Numerical Methods for Computational Fluid Dynamics

16.1 APPROXIMATIONS OF INCOMPRESSIBLE NAVIER–STOKES FLOWS

A main objective of computational fluid dynamics (CFD) is to approximate solutions of the Navier–Stokes equations [Eqs. (11.33)]

$$\begin{aligned} \rho(u_t + (u \cdot \nabla)u) &= -\nabla p + \mu \Delta u + \frac{\mu}{3}\nabla(\nabla \cdot u) + \rho b, \\ \rho_t + \nabla \cdot (\rho u) &= 0, \\ u &= 0 \text{ on } \partial\mathcal{R}, \end{aligned} \tag{16.1}$$

which arise in many important applications including aerodynamics, hydrodynamics, meteorology, geology, biomedical science, acoustics, industrial processes, and others.

The Navier–Stokes equations are supposed to model all the fluid motions that might be imagined. Thus, determining a good approximation to a particular motion is usually not easy. Indeed, there is no generally accepted best numerical method. Rather, CFD has developed into a vast subject that remains an active area of research. The best way to approach the subject is to write a code and apply it to problems that interest you. Most likely, this experience will lead to an appreciation of the challenges that must be met and overcome to produce useful results. This section is an introduction to some of the most basic ideas of the subject. The ingredients of some viable numerical algorithms are discussed and applied to a few simple applications. The purpose of the section is to teach you how to understand and write a simple CFD code. Although there are many excellent readily available CFD codes, the best practice is to write a code yourself. Some sage advice: Do not rely on a code unless it is well-tested and *you have a basic understanding of how it works*.

To focus on one important set of applications, the discussion here is limited to incompressible flow that arises in case the fluid has constant

An Invitation to Applied Mathematics: Differential Equations, Modeling, and Computation.
http://dx.doi.org/10.1016/B978-0-12-804153-6.50016-6, Copyright © 2017 Elsevier Inc. All rights reserved.

density ρ. In this case, the dimensionless Navier–Stokes equations are

$$\begin{aligned} u_t &= -u \cdot \nabla u + \frac{1}{\mathrm{Re}} \Delta u - \nabla p + b, \\ 0 &= \nabla \cdot u, \end{aligned} \tag{16.2}$$

where u is the scaled fluid velocity, p the scaled pressure, Re the Reynolds number, and b is the scaled body force (see Section 11.1). The objective is to approximate solutions of this system with appropriate boundary conditions ($u = 0$ at solid stationary boundaries, u tangent to free surfaces, u specified at inlets, and the normal derivative of u equal to zero at outlets) and initial conditions: a specified divergence-free velocity field that satisfies the boundary conditions.

Because the time derivative of pressure does not appear in system (16.2), the incompressible Navier–Stokes equations do not form a system of (infinite-dimensional) ordinary differential equations (ODEs) for the time evolution of velocity and pressure. This fact plays a central role in the development of numerical algorithms for approximating solutions. The main difficulty is approximation of the gradient of pressure, sometimes called the velocity-pressure coupling problem.

An equation of state is employed for compressible flow: pressure is a function of density. For an incompressible flow, the density is not changing in the direction of the flow and is usually assumed to be constant. Thus, the pressure must be determined from the equations of motion [Eqs. (16.2)] and the initial and boundary data. A compressible flow responds to changes in pressure; an incompressible fluid flows in response to pressure gradients.

Perhaps the most natural idea for solving the velocity-pressure coupling problem is to seek a set of governing equations for the fluid motion that are equivalent to system (16.2) and include an explicit equation for pressure. One way to do this is to compute the divergence of the momentum balance to obtain

$$(\nabla \cdot u)_t = \nabla \cdot (-u \cdot \nabla u + \frac{1}{\mathrm{Re}} \Delta u - \nabla p + b). \tag{16.3}$$

Using the continuity equation, the time derivative of the divergence vanishes. Solving for the pressure and assuming the velocity and pressure are sufficiently smooth, the pressure is given explicitly by

$$\Delta p = \nabla \cdot (-u \cdot \nabla u + \frac{1}{\mathrm{Re}} \Delta u + b). \tag{16.4}$$

Clearly, every (sufficiently smooth) solution of the Navier–Stokes equations satisfies the new system of equations

$$u_t = -u \cdot \nabla u + \frac{1}{\mathrm{Re}} \Delta u - \nabla p + b, \tag{16.5}$$

$$\Delta p = \nabla \cdot (-u \cdot \nabla u + \frac{1}{\mathrm{Re}} \Delta u + b), \tag{16.6}$$

$$(\nabla \cdot u)(x, 0) = 0. \tag{16.7}$$

Suppose that u and p satisfy the new system [Eqs. (16.5)–(16.7)]. The first equation in the original system (16.2) is satisfied. What about the second equation? The initial value of the divergence is specified to be zero by the third equation in the new system. By substituting Eq. (16.4) into Eq. (16.3), the time derivative $(\nabla \cdot u)_t$, for a velocity u satisfying the new system, vanishes as long as the solution exists. It follows that the divergence of u is always zero. Thus, the second equation in system (16.2) is also satisfied. In summary, system (16.5)–(16.7) is equivalent to the Navier–Stokes equations [Eqs. (16.2)] whenever their solutions are sufficiently smooth.

A numerical method—sometimes called the pressure equation method—for solving the Navier–Stokes equations arises from inspection of the equivalent system (16.5)–(16.7):

- Start with a divergence-free velocity field u that satisfies the boundary conditions.
- Substitute the velocity field into Eq. (16.6) and solve for the pressure field.
- Employ a time discretization on Eq. (16.5) (perhaps Euler's method), update the velocity field by computing forward one step in time using the computed pressure field and the boundary conditions, replace the starting velocity with the updated velocity field, and go to the first step.

This method is attractive for its conceptual simplicity; it can be used to obtain good approximate solutions to many fluid dynamics problems. The second step lies at the heart of the method: solving for the pressure field using Eq. (16.6), which has the form

$$\Delta p = f,$$

where the right-hand side f is known. Equations of this type are called Poisson (Siméon Denis Poisson, 1781–1840) equations, and for this reason the equation at hand is called the pressure Poisson equation. This well-studied partial differential equation (PDE) has unique solutions for Dirichlet boundary conditions and solutions unique up to an additive constant for compatible Neumann boundary conditions.

What are the correct boundary conditions?

The Navier–Stokes equations are well posed with the previously stated boundary conditions (for example, the no-slip condition at solid boundaries) imposed on the velocity field. But, no boundary conditions are imposed on the pressure field. The relation of this fact to numerical methods has been the subject of much discussion, most of which requires mathematics beyond the scope of this book. The best resolution of the problem seems to be very simple: as no boundary condition is imposed on the pressure in the Navier–Stokes model, arbitrary boundary conditions *that are consistent with the model* and imply the existence of a unique solution of the pressure Poisson equation up to an additive constant are appropriate.

A convenient way to determine consistent boundary conditions is to integrate the pressure Poisson equation over the domain Ω occupied by the fluid. Every term in the equation is the divergence of some quantity. Using the divergence theorem and the outer normal on the boundary of Ω, each integral is converted to a surface integral over $\partial\Omega$. The result is the equation

$$\int_{\partial\Omega} \nabla p \cdot \eta dS = \int_{\partial\Omega} (-u \cdot \nabla u + \frac{1}{\text{Re}} \Delta u + b) \cdot \eta \, dS.$$

Thus, a consistent boundary condition for pressure, which can be used in the second step of the pressure equation method, is

$$\nabla p \cdot \eta = (-u \cdot \nabla u + \frac{1}{\text{Re}} \Delta u + b) \cdot \eta \text{ on } \partial\Omega. \tag{16.8}$$

To implement the algorithm, the differential equations must be appropriately discretized on the computational domain corresponding to the region filled with fluid and a numerical method must be chosen from among many viable possibilities to approximate solutions of the pressure Poisson equation. Once these tasks are accomplished, the rest is writing code.

It seems that all problems have been solved. Why would anyone look for an alternative method to approximate solutions of the Navier–Stokes equations? The answer is probably best appreciated after writing and applying a pressure equation code. A major challenge is designing and implementing an accurate and efficient Poisson solver. Although efficient Poisson solvers exist, your code might still run for a very long time when it is used to try to approximate solutions of an interesting fluid dynamics problem. Another difficulty is ensuring sufficient accuracy to maintain a divergence-free velocity field at each time step. Although the algorithm guarantees this result for exact arithmetic and no discretization error, neither of these requirements will be maintained in practical numerical computations. After many time steps the computed velocity may no longer be divergence free. The numerical approximation might drift away from the solution of the fluid dynamics model: the Navier–Stokes equations with initial and boundary conditions.

An obvious remedy for loss of the divergence-free condition is to seek a correction after each time step of the pressure equation method that would replace the approximate velocity by a nearby divergence-free velocity field. Fortunately, there is a mathematical theorem that can be used to formulate a viable procedure.

An important result in vector analysis, first formulated by Hermann von Helmholtz and later generalized and proved by W. V. D. Hodge, is called the Helmholtz–Hodge decomposition theorem: *A smooth vector field X on a region Ω of space with smooth boundary $\partial\Omega$ and outer normal N can be decomposed uniquely as*

$$X = \nabla\phi + Y \tag{16.9}$$

where ϕ is a scalar valued function and Y is a vector field such that $\nabla \cdot Y = 0$ and $Y \cdot N = 0$ on $\partial\Omega$ (that is, Y is divergence free and parallel to the boundary of the region), Moreover,

$$\Delta\phi = \nabla \cdot X \text{ in } \Omega, \qquad \nabla\phi \cdot N = X \cdot N \text{ in } \partial\Omega,$$

and $Y = X - \nabla\phi$. In other words, every vector field can be written as a sum of two vector fields, one divergence free and the other curl free. The curl-free field is the gradient of a potential determined as the solution of a Poisson equation.

Using the notation of Eq. (16.9), define the Helmholtz–Leray (Jean Leray) projection $\mathcal{P}$ by $\mathcal{P}X = Y$. It projects a vector field onto the divergence-free vector fields. The operator is linear ($\mathcal{P}(X + Z) = \mathcal{P}X + \mathcal{P}Z$) and idempotent ($\mathcal{P}\mathcal{P}X = \mathcal{P}X$). Linear transformations of this type should be familiar from basic vector analysis. For example, every vector in two-dimensional space can be written as a linear combination of the usual basis vectors e_1 and e_2. More precisely, a vector v can be expressed as $v = v_1e_1 + v_2e_2$. The vector projection of v onto the linear space of vectors generated by e_1 is the vector v_1e_1. This projection is also given by $v \mapsto (v \cdot e_1)e_1$; it is linear and idempotent.

The pressure equation algorithm could be modified to include one additional step:

- Start with a divergence-free velocity field u that satisfies the boundary conditions.
- Substitute the velocity field into Eq. (16.6) and solve for the pressure field.
- Employ a time discretization on Eq. (16.5) and update the velocity field by computing forward one step in time using the computed pressure field and the boundary conditions.
- Project the updated velocity field (via Helmholtz–Leray projection) to a divergence-free field, replace the starting velocity with the updated velocity field, and go to the first step.

The divergence-free conditions will be enforced, but the projection requires the solution of a Poisson equation. Thus, two Poisson equations are solved to complete one time step. The additional effort will likely double the time required to complete the computation.

A better projection method, first introduced by Alexandre Chorin, requires just one Poisson solution per step. In its most elegant from, the method arises from a recasting of the Navier–Stokes equations. The idea is to apply the Helmholtz–Leray projection to the momentum equation of the Navier–Stokes system, where the velocity field is divergence free (by the continuity equation) and parallel to the boundary of the domain (because $u = 0$ on the boundary). By the Helmholtz–Hodge theorem, $\mathcal{P}u = u$, and similarly $\mathcal{P}u_t = u_t$. (Why?) Again, by the same theorem, $\mathcal{P}(\nabla p) = 0$.

Using the linearity of the projection, it follows that

$$u_t = \mathcal{P}((-u \cdot \nabla)u + \frac{1}{\mathrm{Re}}\Delta u + b). \tag{16.10}$$

Pressure does not appear explicitly, there is only one dynamical variable u, and solutions of this differential equation can be approximated using standard time discretization methods such as Euler's method. Of course, the pressure reappears in the computation of the projection, but writing the incompressible Navier–Stokes equations in this compact form leads to a beautiful algorithm for numerical approximations:

- Choose a time step Δt and start with a divergence-free velocity field u^k at time step k.
- Using u^k as initial data for the PDE

$$u_t = -(u \cdot \nabla)u + \frac{1}{\mathrm{Re}}\Delta u + b \tag{16.11}$$

  with the Navier–Stokes boundary conditions imposed, compute one forward time step, and call the computed velocity field $\tilde{u}$. (The PDE is the fluid momentum equation with the pressure gradient removed.)
- Compute the Helmholtz-Hodge decomposition

$$\tilde{u} = \nabla\phi + u^{k+1},$$

  where u^{k+1} is the name of the divergence-free part, and let

$$p^{k+1} := \frac{\phi}{\Delta t}.$$

- Go to the first step with u^k replaced by u^{k+1}.

Note: In Chorin's original algorithm the momentum equation is rewritten in the form

$$u_t + \nabla p = -(u \cdot \nabla)u + \frac{1}{\mathrm{Re}}\Delta u + b$$

and the right-hand side of this equation is decomposed via the Helmholtz-Hodge decomposition to obtain $u^* + \nabla\phi$. The velocity update is computed by time discretization of $u_t = u^*$ and the pressure is taken to be ϕ.

To justify the computation of pressure in the last step of the algorithm, consider using a forward Euler time step at the second step of the algorithm to obtain

$$\tilde{u} = u^k + \Delta t(-(u^k \cdot \nabla)u^k + \frac{1}{\text{Re}}\Delta u^k + b^k).$$

By substituting the decomposition of $\tilde{u}$ and multiplying and dividing by Δt, a new update equation is obtained:

$$u^{k+1} + \Delta t \nabla(\frac{\phi}{\Delta t}) = u^k + \Delta t(-(u^k \cdot \nabla)u^k + \frac{1}{\text{Re}}\Delta u^k + b^k).$$

After the obvious rearrangement and using the definition of the approximate updated pressure p^{k+1}, the algorithm produces the updated velocity field

$$u^{k+1} = u^k + \Delta t(-(u^k \cdot \nabla)u^k - \nabla p^{k+1} + \frac{1}{\text{Re}}\Delta u^k + b^k),$$

which is taken as an approximate divergence-free solution of the momentum equation.

The updated velocity u^{k+1} is divergence free, but it might not satisfy the correct boundary conditions because it is obtained from a Helmholtz-Hodge decomposition where a Neumann boundary condition is enforced. At a solid boundary, for example, u may not vanish (no slip) but it will have a zero normal component at this boundary (no penetration). As $\tilde{u}$ does satisfy the correct boundary condition, the tangential component is expected to be small for small time steps. The updated pressure p^{k+1} is associated with the field $\tilde{u}$, which is not expected to be divergence free. A better approximation of the pressure can be obtained by solving the (full) pressure Poisson equation

$$\Delta p = \nabla \cdot (-u_t - u \cdot \nabla u + \frac{1}{\text{Re}}\Delta u + b) \tag{16.12}$$

with u replaced by u^{k+1} without the assumption that u is divergence free. This extra computation might be necessary at specific time steps when accurate pressures are desired as output. But, of course, solving the additional pressure Poisson equation at each time step defeats the advantage of the algorithm. In practice, good approximations of the velocity field are obtained; the pressure field is not as accurate, especially at solid boundaries.

The issues concerning accurate computation of the pressure field, higher-order accurate velocity field computations, and more efficient algorithms have been addressed with many proposed modifications and refinements that will continue to appear into the future. To approach these developments in depth would require a more extensive discussion of CFD, which is beyond the scope of this book. The goals here are more modest: explore some simple implementations of the projection method and illustrate some of the successes and challenges of CFD that might inspire the reader to explore more fully this important and fascinating subject.

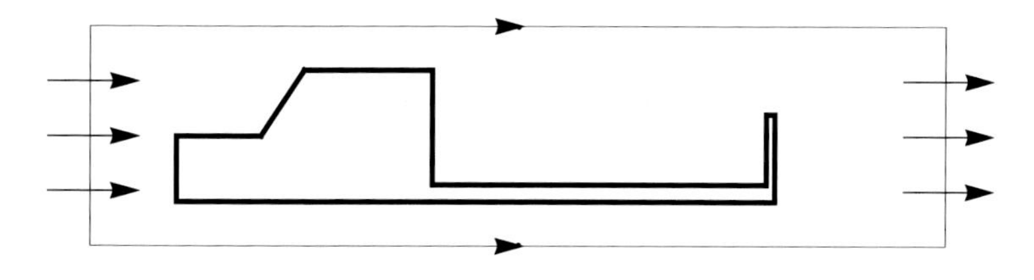

Fig. 16.1 The figure depicts a schematic diagram of pickup truck in constant wind field in the tailgate up configuration.

A basic implementation of the method using a finite difference approach starts with a grid covering the computational domain (where the fluid resides). The fluid velocity and the pressure are discretized by sampling these functions at the vertices of the grid. By finite difference approximations of derivatives, this finite set of unknowns (three velocity components and one pressure at each vertex) is determined by solving a set of linear equations with the same number of equations as the number of unknowns. Clearly, the number of such equations will be large. For a $10 \times 10 \times 10$ cell grid there are over 1000 vertices and thus over 4000 unknowns. What problem in CFD could be solved with such a small grid? Moreover, for time-dependent flow problems, every time step requires solving a large set of equations. The challenge of large-scale computation must be met for realistic flow problems. Even with very fast computation, numerical approximation for many flow problems in three-dimensional space are not feasible at a fine enough resolution (large number of grid cells) to reach reliable predictions.

Imagine, as an illustrative example, a constant velocity fluid flow that meets and passes over a body fixed with respect to the motion of the fluid. For definiteness, consider the following problem.

Problem 16.1. [Tailgate Problem] A pickup truck is parked on a road where a 70 mile per hour wind is blowing over the truck from front to back with the truck facing directly into the wind. What is the velocity and pressure of the fluid flow near the truck? Suppose further that the truck has a tailgate and consider the drag on the truck due to the fluid motion. Which configuration has smaller drag: tailgate up or tailgate down?

Although the physical tailgate problem is certainly three-dimensional and there might be fluid motions near the truck that are time dependent, there is an apparent symmetry in the problem: the flows in the regions bounded by an imaginary plane cutting the truck in half lengthwise should be identical, and the flow over the truck should be in steady state. Thus, some insight should be gained by considering the two-dimensional steady state flow in this plane. This is a typical problem, where reduction to two-dimensional steady state flow is natural.

The components of the dimensionless two-dimensional Navier–Stokes equations for incompressible flow, *with the scaled gravitational body force*, are

$$\begin{aligned} u_t &= -uu_x - vu_y + \frac{1}{\mathrm{Re}}(u_{xx} + u_{yy}) - p_x, \\ v_t &= -uv_x - vv_y + \frac{1}{\mathrm{Re}}(v_{xx} + v_{yy}) - p_y - g, \\ 0 &= u_x + v_y, \end{aligned} \tag{16.13}$$

where u and v denote the components of the fluid velocity in the horizontal and vertical directions, respectively. The current objective is to approximate steady state solutions of this system for specified boundary conditions.

The geometry of the problem must be specified. The usual approach is to consider first the simplest possible geometry that maintains the essential feature(s) of the physical problem. For the truck with tailgate, the essential feature is the configuration of the bed of the truck bounded by the cab and the tailgate. A simple geometry is depicted in Fig. 16.1. Although the exact boundaries are not specified in the problem statement, the computational domain might be the complement of the schematic truck in the depicted rectangular region where the upper and lower boundaries of the rectangle are streamlines of the constant velocity wind that enters and leaves the rectangle at the left and right boundaries. The wind velocity is assumed to be constant at some distance away from the truck body. Thus boundary conditions on the rectangle are set. For Navier–Stokes, no-slip boundary conditions are required on the surface of the truck body.

A strategy for numerical experiments that might provide insight into the physics and be useful for making predictions about real trucks traveling on a highway requires many steps, but in broad outline, we seek approximations to the steady state two-dimensional incompressible Navier–Stokes equations on the computational domain Ω where the fluid velocity is specified on its outer boundary and vanishes on its inner boundary. A rectangular grid is to be defined on Ω, the differential equations to be solved via the projection method are to be discretized using this grid, and the projection method is to be employed to determine an approximate solution of the Navier–Stokes equations.

The projection method is designed to solve the time-dependent incompressible Navier–Stokes equations. To find a steady state solution requires a leap of faith: The steady state is asymptotically stable. Thus, the plan is to reach the steady state by marching along an (approximate) solution until it reaches steady state. In this scenario, the accuracy of the time integration is not important as long as the procedure reaches a verifiable steady state.

To make the implementation simple, consider a rectangular computational region, with width W and height H, that will eventually include the truck body and consider a grid constructed in the usual manner with increments $\Delta x = W/m$ and $\Delta y = H/n$ for positive integers m and n. There are m cells in the horizontal direction and n cells in the vertical direction with nodes in the horizontal direction numbered $i = 1, 2, 3, \ldots, m + 1$ and in the vertical direction $j = 1, 2, 3, \ldots, n + 1$. Anticipating the necessity to treat boundary data, this grid is augmented with a border that is one cell thick. The two additional nodes in the horizontal direction are numbered $i = 0$ and $i = m + 2$; likewise the new nodes in the vertical direction are $j = 0$ and $j = n + 2$. With this numbering scheme, which is one of many choices, there are mn cells in the computational domain (which by definition does not include the ghost cells in the border). In the computer, the storage requirement for each variable is an $(m + 2) \times (n + 2)$ array. No variables are stored with address $i = 0$ or $j = 0$.

An influential paper of Francis Harlow and J. Eddie Welch in 1965 [48] introduced the idea of a staggered grid where the pressure is discretized at cell centers and the velocity components on the cell sides as in Fig. 16.3. Although there are many other grid configurations, each with advantages and disadvantages, the staggered grid is a useful choice for the first draft of

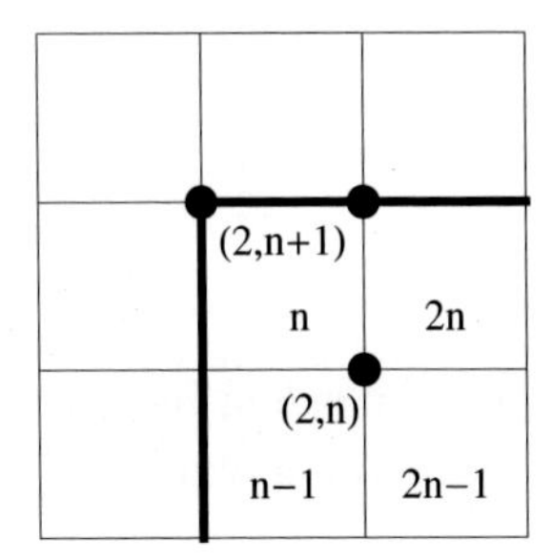

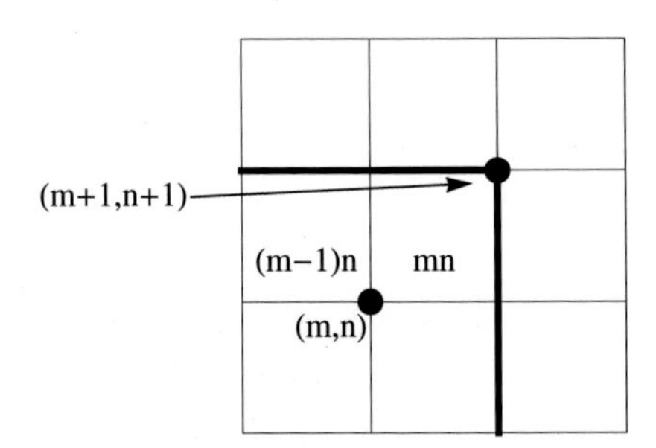

Note cell order numbers and node ordered pair addresses.

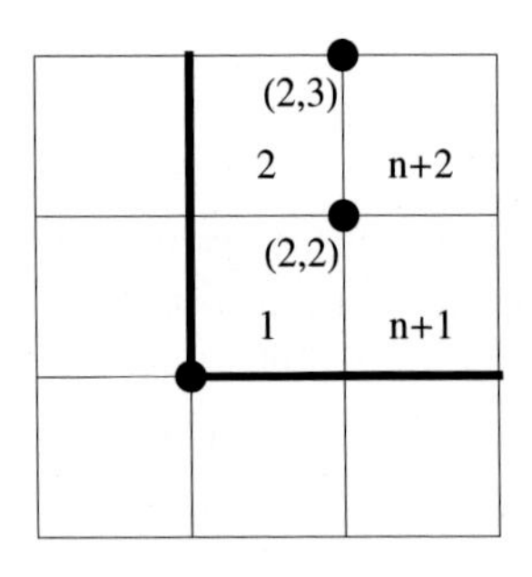

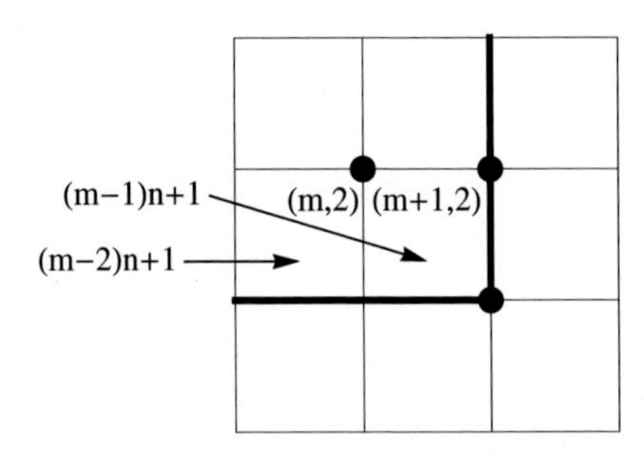

Fig. 16.2 Selected staggered grid cells near corners are depicted.

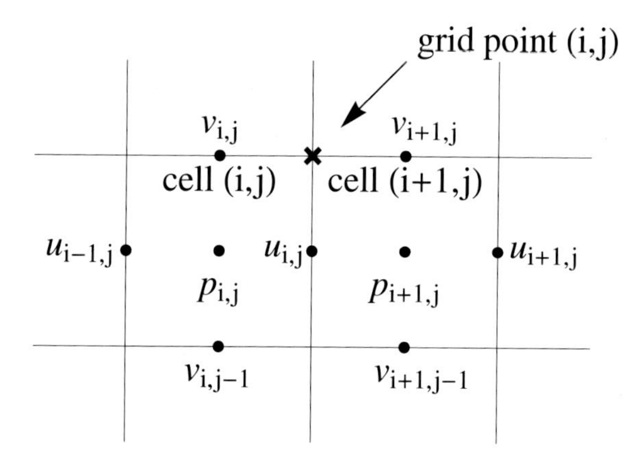

Fig. 16.3 A generic part of a staggered grid is depicted. Pressure is computed at the centers of cells, the first component of the velocity field at the midpoints of the left and right cell boundaries, and the second component of the velocity field at the midpoints of the top and bottom boundaries.

a viable code. At least one good reason to use a staggered grid will become apparent as the discretization process unfolds.

The first step in the projection method is to solve the PDE formed from the Navier–Stokes equations (16.13) while ignoring the pressure. By using the continuity equation $u_x + v_y = 0$ (which should be enforced whenever possible in the approximation procedure), the system can immediately be recast in the conservation form

$$\begin{aligned} u_t &= -(uu)_x - (uv)_y + \frac{1}{\text{Re}}(u_{xx} + u_{yy}), \\ v_t &= -(uv)_x - (vv)_y + \frac{1}{\text{Re}}(v_{xx} + v_{yy}) - g, \end{aligned} \tag{16.14}$$

that turns out to be convenient for discretization over staggered grids using centered differences.

On the staggered grid, the discretized state variables are defined by

$$\begin{aligned} u_{ij}^k &= u(i\Delta x, (j - 0.5)\Delta y, k\Delta t), \\ v_{ij}^k &= v((i - 0.5)\Delta x, j\Delta y, k\Delta t), \end{aligned}$$

$$p_{ij}^k = p((i - 0.5)\Delta x, j\Delta y, k\Delta t).$$

Be sure to understand the notation before proceeding with the discretization.

Time stepping toward the steady state solution, as previously mentioned, need not be accurate to second order in the time step size. The natural choice is Euler's method, which is first order. As usual, superscripts $k = 1, 2, 3, \ldots, k_{\text{max}}$ denote time steps corresponding to the discrete times $k\Delta t$. Thus, the time derivatives are discretized by

$$u_t(i\Delta x, (j - 0.5)\Delta y, k\Delta t) \approx \frac{u_{ij}^{k+1} - u_{ij}^k}{\Delta t},$$
$$v_t((i - 0.5)\Delta x, j\Delta y, k\Delta t) \approx \frac{v_{ij}^{k+1} - v_{ij}^k}{\Delta t}.$$

At least second-order accurate spatial discretization is desired. Thus, central differences are used for first- and second-order partial derivatives. The Laplacian of u at $(i\Delta x, (j - 0.5)\Delta y, k\Delta t)$ is approximated by

$$(u_{xx} + u_{yy}) \approx \frac{u_{i+1,j}^k - 2u_{ij}^k + u_{i-1,j}^k}{\Delta x^2} + \frac{u_{i,j+1}^k - 2u_{ij}^k + u_{i,j-1}^k}{\Delta y^2}, \tag{16.15}$$

and the Laplacian of v at $((i - 0.5)\Delta x, j\Delta y, k\Delta t)$ is approximated by

$$(v_{xx} + v_{yy}) \approx \frac{v_{i+1,j}^k - 2v_{ij}^k + v_{i-1,j}^k}{\Delta x^2} + \frac{v_{i,j+1}^k - 2v_{ij}^k + v_{i,j-1}^k}{\Delta y^2}. \tag{16.16}$$

Approximations of the first-order partial derivatives in system (16.14) is more complicated. The partial derivative $(uu)_x$ at $(i\Delta x, (j - 0.5)\Delta y, k\Delta t)$ is to be centered at the position of u_{ij} on the staggered grid. One possibility is to use the discretization

$$(uu)_x \approx \frac{(u_{i+1,j}^k)^2 - (u_{i-1,j}^k)^2}{2\Delta x}. \tag{16.17}$$

An alternative is to imagine evaluations of the quantity u^2 at the cell centers, where the pressure is discretized. The advantage would be in taking a central difference over a discretization interval of length Δx rather than an interval of length $2\Delta x$ as in the former discretization. The velocity variables are not stored at the cell centers, but a good approximation of these values could be obtained by averaging the velocities on the two cell walls. In other words, an alternative to approximation (16.17) is

$$(uu)_x \approx \frac{1}{\Delta x}\left(\left(\frac{u^k_{i+1,j}+u^k_{ij}}{2}\right)^2 - \left(\frac{u^k_{i-1,j}+u^k_{ij}}{2}\right)^2\right). \tag{16.18}$$

A third possibility is to use the power rule to write

$$(uu)_x = 2uu_x \approx 2u^k_{ij}\left(\frac{u^k_{i+1,j}-u^k_{i-1,j}}{2\Delta x}\right). \tag{16.19}$$

The last choice is the most accurate of these discretizations (see Exercise 16.2). It is recommended for use in writing a computer code. There is an analogous approximation of $(vv)_y$.

For $(uv)_x$, there are also several possible approximations. This derivative appears in the momentum equation for the v component of velocity where discretization in time at the (i,j) grid cell is at v_{ij}, which resides at the middle of the top of the cell. The most natural discretization for the derivative would be a centered difference of uv values at the top right and top left corners of this cell. The required u and v values at these points are approximated by averages. For example, the u component at the top right corner is approximated by $(u_{i,j+1}+u_{ij})/2$, which is the average of the discretized u values on cell faces just above and below the top right corner of the (i,j) grid cell. Using this idea, the desired approximation is

$$(uv)_x \approx \frac{1}{\Delta x}\Bigg(\frac{(u^k_{i,j+1}+u^k_{ij})(v^k_{ij}+v^k_{i+1,j})}{4} - \frac{(u^k_{i-1,j}+u^k_{i-1.j+1})(v^k_{i-1,j}+v^k_{ij})}{4}\Bigg). \tag{16.20}$$

The approximation for $(uv)_y$ is similar:

$$(uv)_y \approx \frac{1}{\Delta y}\Bigg(\frac{(u^k_{i,j+1}+u^k_{ij})(v^k_{i+1,j}+v^k_{ij})}{4} - \frac{(u^k_{ij}+u^k_{i.j-1})(v^k_{i,j-1}+v^k_{i+1,j-1})}{4}\Bigg). \tag{16.21}$$

Staggered grids are designed in part to solve for pressure. Indeed, to obtain the Helmholtz–Hodge decomposition of the velocity vector $\tilde{u}$ computed in the second step of the projection algorithm requires solving the Poisson equation $\Delta\phi = \nabla\cdot\tilde{u}$ in the computational domain for the potential ϕ with Neumann boundary condition $\nabla\phi\cdot N = \tilde{u}\cdot N$. This potential ϕ is

Δt times the desired pressure, which is discretized with values specified at the cell centers. Thus, it is natural to approximate the divergence of the velocity field $\tilde{u}$ at the cell centers with centered differences that span exactly the length of one cell; the discretized values of the components of the velocity vector, $\tilde{u}$ and $\tilde{v}$, are conveniently located on the cell walls and the finite difference from cell center to adjacent cell center is a *second-order* approximation to the velocity component's value at the node on the cell wall when the ordinary difference is viewed as a centered difference with step size set at half the width of a cells. In this sense, a staggered grid leads to second-order approximations of the velocity field at no extra expense. These considerations lead to the discretized Poisson equation

$$\frac{\phi_{i-1,j} - 2\phi_{ij} + \phi_{i+1,j}}{\Delta x^2} + \frac{\phi_{i,j-1} - 2\phi_{ij} + \phi_{i,j+1}}{\Delta y^2} = \frac{\tilde{u}_{ij} - \tilde{u}_{i-1,j}}{\Delta x} + \frac{\tilde{v}_{ij} - \tilde{v}_{i,j-1}}{\Delta y}. \tag{16.22}$$

All the ingredients are in place to write a code that can be used to approximate two-dimensional flow problems.

A natural test problem is lid-driven cavity flow, which is the classic test problem for CFD computer codes. Imagine a rectangular domain, usually taken to be the unit square, filled with a fluid and the top boundary of the rectangle is a solid wall that moves horizontally (to the right) at a constant speed. What is the motion of the fluid in the rectangle? This is an idealized example of the tailgate problem where the airflow over the truck bed is assumed to remain at constant velocity; that is, the flow over the truck bed is not affected by the motion of the air in the bed. The interaction is unidirectional: the airflow drives the flow in the bed; the flow in the bed does not change the airflow over the bed.

Coding the projection method is not trivial, even for the simple geometry of the lid-driven cavity flow over a unit square cavity. A recommended strategy is to break up the code into (at least) four parts: a main program (`main`) and three subroutines (`advdiff`, `project`, `linsolve`).

The main program serves several purposes. It is used to set the parameters (gravity, Reynolds's number, grid size, time step, and so on) and the arrays that will contain the state variables (the velocity components u and v and the pressure p); it is also used to initialize the arrays and to read and write data to the screen or to files. The output data files will later be used to visualize the flow. The program `main` calls `advdiff` and `project`.

The subroutine `advdiff` computes a single time step of the advection-diffusion evolution equation [Eq. (16.11)] with appropriate Navier–Stokes boundary conditions and stores the computed velocity field components in the arrays u and v. For the cavity flow problem, no-slip boundary conditions are imposed on the three walls of the cavity and constant horizontal velocity at the top boundary. This later boundary condition is also a no-slip condition: the flow sticks to the moving lid.

The subroutine `project` takes as input the updated state variable arrays and returns these same arrays updated further with the divergence-free part of the velocity field and the pressure obtained from the Helmholtz-Hodge decomposition of the input velocity field. `Project` is also used to determine the matrix equation $Ax = b$ obtained by discretization of the Poisson equation $\Delta\phi = \nabla \cdot \tilde{u}$, where b is used here to denote the discretized divergence of $\tilde{u}$, the vector field output of `advdiff`.

For most applications (including the cavity flow problem), the matrix A does not change with the time step; it can be computed from the geometry of the problem and stored for use during time stepping. The vector b does depend on the velocity at the current time step and the geometry of the problem when the boundary contains velocities not tangent to the boundary; for example, when fluid enters through the boundary. This vector must be updated at each time step.

`Project` calls `linsolve` to approximate the solution x of the system of linear equations. The values stored in the array x are the approximations to the potential ϕ evaluated at cell centers. These are used to obtain the divergence-free part of $\tilde{u}$ and the corresponding pressure p. `Project` stores the updates in the arrays u, v, and p and returns them to `main`.

This completes the approximation of the velocity field and the pressure over one time step with the state variable arrays u, v, and p set to start the computation over another time step or for output to a file.

The staggered grid with ghost cells surrounding the computational domain and the experience gained by solving reaction-diffusion problems in Chapter 5 should make the task of writing the `advdiff` subroutine easy. The required methods are indeed exactly the same as for reaction-diffusion problems.

The subroutine `project` is perhaps slightly more challenging to write due to the Neumann boundary condition for the Poisson equation. There is one pitfall that must be avoided: the Poisson equation with Neumann boundary condition does not have a unique solution. In abstract form, the PDE is

$$\Delta \phi = f \text{ in } \Omega, \qquad \nabla \phi \cdot \eta = g \text{ in } \partial\Omega,$$

where η is the outer unit normal on $\partial\Omega$. Suppose that ϕ is a solution. Because the derivative of a constant is zero, $\phi + c$ for every real number c is also a solution.

This nonuniqueness will be approximated in the discretization that results in a linear system $Ax = b$. The system matrix A will be singular or very close to a singular matrix. For this reason the linear solver encoded in `linsolve` will likely fail to approximate a solution of the ill-conditioned linear system.

One possible cure is to specify the value of ϕ at some (grid) point. A good choice is to set $\phi = 0$ at the center of some cell adjacent to the upper boundary of the computational rectangle. This choice will affect the construction of the matrix A: the second-order centered differences used to approximate the partial derivatives of ϕ in cells adjacent to the chosen cell with zero potential require the value of ϕ at the center of the chosen cell that is set to zero. The computation of the rows of A, one for each cell, has to take the special value of ϕ in the chosen cell into account. Of course, the construction of A requires some ordering of the cells in the grid. The rows of A might naturally correspond to the chosen grid cell ordering.

A simple way to order the cells is to determine and employ a function that maps each pair of integers (corresponding to nodes in the grid) to the integer corresponding to the placement of the corresponding cell in the ordering. With the staggered grid as illustrated in Fig. 16.2, the first cell in the ordering has address $(2, 2)$ and the last cell in the ordering has address $(m+1, n+1)$. The function

$$(i, j) \mapsto (i - 2)n + j - 1$$

determines the place in the ordering of the cell with address (i, j). There are many other possibilities. One possible choice for the special cell with fixed potential value zero is the last cell in the ordering. The matrix A has dimensions $(mn - 1) \times (mn - 1)$.

With the suggested ordering, A is a banded matrix with bandwidth $2n + 1$. A natural question arises: What is the smallest possible bandwidth that can be achieved by reordering the cells? Although it is possible to achieve a smaller bandwidth with perhaps a more complicated ordering function, it is not possible to achieve bandwidth equal to 5. Why? Common sense dictates that the banded matrix A not be stored in its entirety as there is no good reason to store elements in A that are known to be zero. Although only five diagonals have nonzero elements, a compromise that maintains simplicity is to store the band in an $(mn - 1) \times (2n + 1)$ array where each diagonal in the banded part of A is stored in a column of the new matrix by respecting the original row address of each component. The main diagonal is stored in column $(n+1)$. The first superdiagonal of A is stored in column $(n + 2)$, the first subdiagonal in column n, and the diagonals that bound the band in A are stored in column 1 and $(2n + 1)$, respectively, of the storage matrix. Elements in the storage matrix that are not assigned nonzero elements of A are assigned the value zero. This arrangement is simple to program and understand. Clearly, there are more efficient storage schemes that could be employed if a more efficient code is important or in case large arrays overwhelm the available storage capacity of available computers.

The subroutine `linsolve` should take as input the $(mn - 1) \times (2n + 1)$ storage array for the banded matrix A and the $(mm - 1)$ vector b and return an approximate solution of the matrix system $Ax = b$. As `linsolve` need not take into account the PDE or the computational grid, it may be written as a subroutine using one of many algorithms available for solving general matrix equations. Modern numerical methods, called multigrid methods, *do* take into account the computational grid; in fact, they incorporate several computational grids with different mesh sizes. But, for simplicity and to reuse the iterative methods introduced in Chapter I, successive overrelaxation (SOR) is recommended as a viable method for basic fluid dynamics codes. As a bonus, this method plays a central role in multigrid methods that might be considered for implementation to improve the speed of convergence of a basic code.

The Courant–Friedrichs–Lewy (CFL) stability condition, viewed as a restriction on the time step size to avoid numerical instability, may be obtained approximately by imposing the condition that nothing important should happen during a time step unless the happening is confined to a computational cell. In the case of fluid motion, one interpretation of this restriction is that no fluid particle should travel across more than one cell during one time step. Thus the maximum absolute value of the u-component of the fluid velocity multiplied by the time step size should not exceed the width of a cell and the product of the maximum absolute value of the v-component and the time step size should not exceed a cell height. These conditions do not guarantee numerical stability, but they offer a rough guide to choosing a suitable step size.

It is time to write and test a code.

Using exactly the methodology explained so far, a code was written, debugged, and used in an attempt to simulate behavior, especially steady state behavior, of lid-driven cavity flow. For testing, the cavity was taken to be the unit square discretized with a 128×128 square cell grid. The SOR stopping procedure—the norm of the difference of two successive iterates less than a preassigned tolerance—was implemented with tolerance 1.0×10^{-3} and a maximum number of 500 iterations. The overrelaxation parameter for SOR was set to 1.5 and the time step to 1.0×10^{-3}. The characteristic velocity, in this case the velocity of the lid, was taken to be one unit. Thus the CFL condition is a time step no larger than $1/128$. The time step 1.0×10^{-3} satisfies this inequality and respects the approximate nature of the CFL estimate. Recall that in simpler cases (for example, Euler's method applied to the heat equation) the exact CFL condition can be obtained by computing the eigenvalues of a matrix. For the nonlinear Navier–Stokes model, the stability analysis is more complicated and not presented here. The code was written so that the Reynolds number is requested at run time.

Fig. 16.4 depicts a numerical experiment for the lid-driven cavity flow problem at Reynolds number 1000. As might be expected from physical intuition, at steady state the dominant flow is a clockwise rotating vortex. Counter rotating vortices appear in the corners of the cavity. This figure was make by postprocessing the output from the Chorin projection method code,[1] which in part produces an approximate flow velocity field (u, v) defined at a discrete set of points on a staggered grid. The depicted streamlines are solutions of the ODE system

$$\dot{x} = u(x, y), \qquad \dot{y} = v(x, y).$$

[1]The figure was made using the StreamFunction command in *Mathematica 10*.

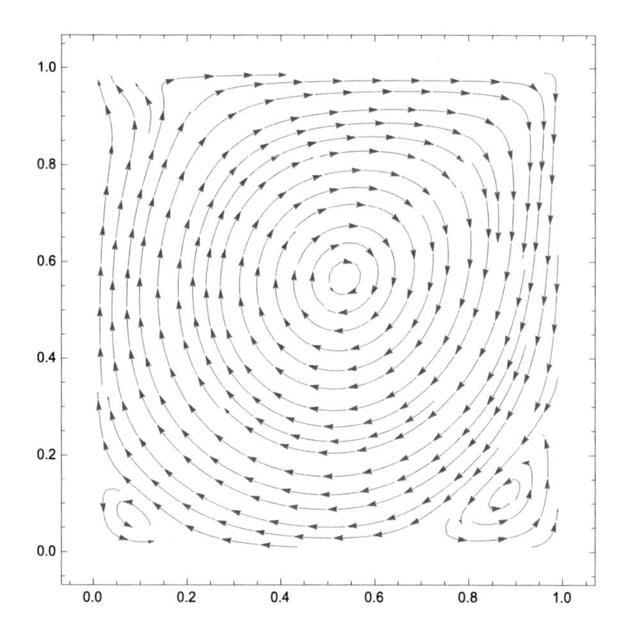

Fig. 16.4 Streamlines are depicted for steady state lid-driven cavity flow (after forward integration for approximately 20 dimensionless time units) with unit lid speed, grid size 128×128*, and Reynolds number 1000.*

Its solutions may be approximated by a numerical method, Euler's method for example, using interpolation of the values of u and v at the grid points. Spatial evaluation of the components u and v should take into account the staggered grid. The values of the vector field $(x, y) \mapsto (x, y, u(x, y), v(x, y))$ are approximated on the grid points as follows:

$$\Big((i-1)dx, (j-1)dy, \frac{u(i,j)+u(i,j+1)}{2.0}, \frac{v(i,j)+v(i+1,j)}{2.0}\Big).$$

This discretized vector field is used for the interpolation required to determine streamlines.

Postprocessing and visualization are important topics in computational fluid dynamics and for analyzing data obtained from many other types of experiments. Further treatment of this subject is beyond the scope of this book.

The cavity might be considered as a very simple model for a truck bed where the lid is not a solid but instead a model of the air flowing over the bed. What is the Reynolds number for such a flow? For air flowing over a 7 foot long truck bed moving at 70 miles per hour and taking the air kinematic viscosity to be 1.5×10^5, the Reynolds number obtained (using the bed length and air speed as the characteristic length and velocity) is on the order of 10^6. This is a typical Reynolds number for realistic flows. But, it seemed too large for an initial numerical test. The Reynolds number 1000 was used instead. The lid speed was set at the dimensionless value 1.0 for no particular reason except to keep values within what seemed to be a reasonable range. With this choice, $1/\mathrm{Re}$ may be viewed as a measure of the viscosity of the fluid. After a careful debugging using coarse grids to test assignments of boundary conditions, checking for coding errors in the discretizations, and testing the `linsolve` subroutine on matrix systems with known solutions, the code was compiled and executed; it ran successfully.

The next step was to test with more realistic Reynolds numbers produced by larger lid speeds. Because the initial velocity is set to zero, the code should be able to (successfully) deal with an abrupt change in velocity from zero to the lid speed for at least the first few time steps. To avoid this problem, a better choice is to write the code to allow a gradual increase in lid speed up to the desired speed, which is more physically realistic, or to gradually increase the Reynolds number with a fixed lid speed. The latter scenario was followed with some success for Reynolds numbers approaching 10^5. To reach Reynolds numbers of this magnitude, the tolerance in the SOR solver was reduced to 10^{-2}, the flow was integrated forward for several dimensionless seconds until the velocity field did not seem to be changing, and after a (somewhat) steady flow was reached, the tolerance was slowly increased toward 10^{-3}. The main issue was the speed of convergence of the SOR solver. The number of iterations required for convergence increased until it was impractical to try larger Reynolds numbers. The cause of the slow convergence was traced to abrupt increases in the magnitude of the computed velocity field, a type of numerical instability.

Reaching a limitation can be frustrating, but *this experience is typical when applying a numerical code to a physically realistic model.* What is causing the numerical problems?

Before continuing, some remarks are in order. The numerical experiments just described are rather limited for several reasons, in particular the coarse mesh size of the grid. Although it is satisfying that the code produces qualitatively appealing results, *how do we know that a numerical experiment produces a good approximation to the solution of the Navier–*

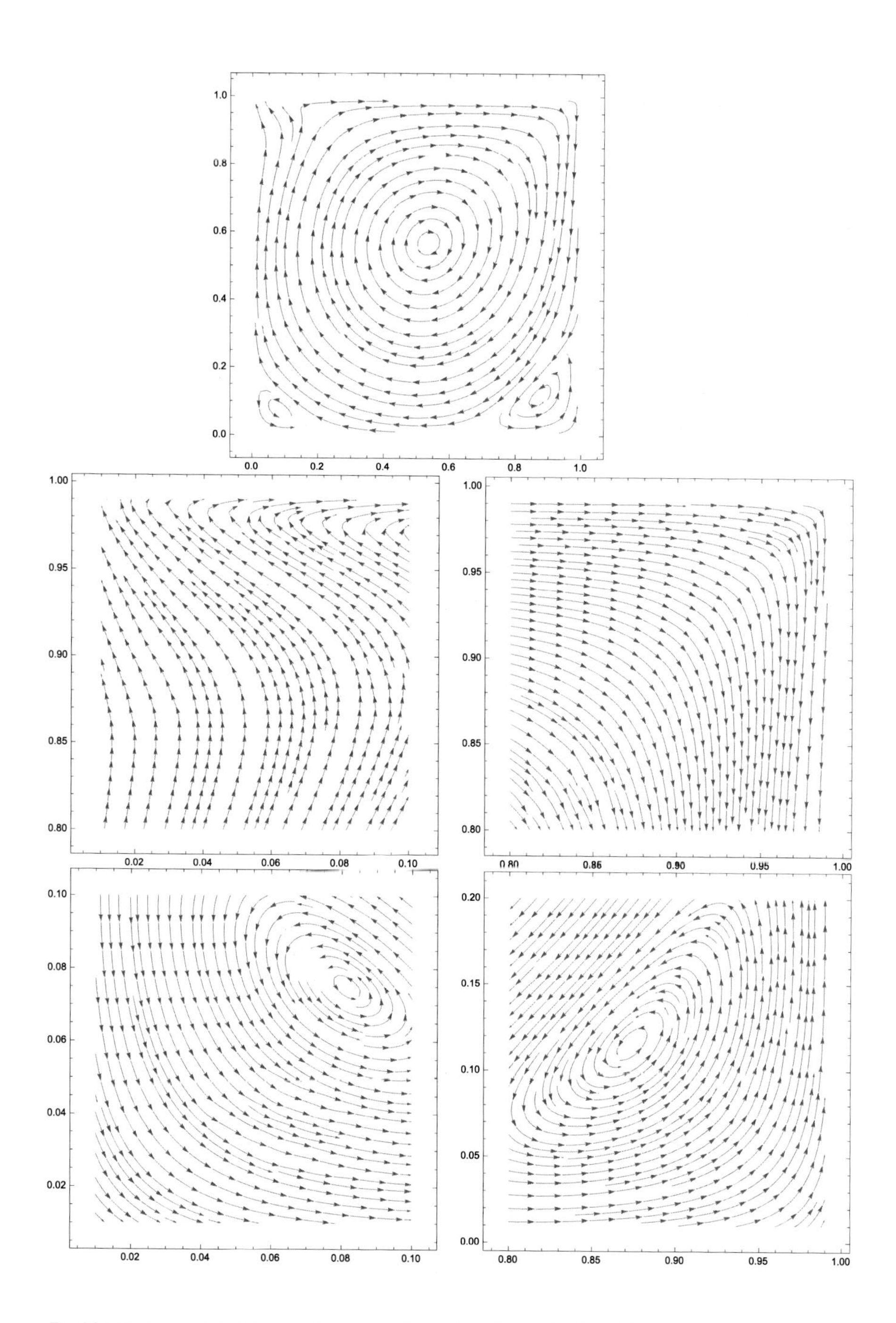

Fig. 16.5 The top panel depicts streamlines for steady state lid-driven cavity flow with unit lid speed, grid size 256×256*, and Reynolds number 1000. The four bottom panels depict blowups of the flow near the top left corner, the top right corner, the bottom left corner, and the bottom right corner of the cavity.*

Stokes model? The simple answer is that we do not know. At present the best we can do is to gather evidence that our numerical experiments produce good approximations. One way to do this is to test the code against known exact solutions; another is to refine the mesh with the intention of showing the answer is essentially the same as the mesh size decreases. Of course, the same model can be discretized in several different ways and different numerical methods can be employed. Confidence in a result increases as more numerical experiments suggest the same conclusion.

Fig. 16.5 depicts the approximate streamlines for the lid-driven cavity at $\text{Re} = 1000$ with a grid of 256×256, which is a refinement of the grid used to produce Fig. 16.4. The result is essentially the same. Thus, there is some reason to believe the numerical approximation is viable. Lid-driven cavity flow is widely used as a test case for numerical fluid dynamics codes (see, for example, [15]). Up to Reynolds numbers of a few thousand, numerical codes all produce similar results. For high Reynolds numbers numerical results are not all the same. One reason is that the flow seems to be in steady state after runs of dimensionless time units of about 20, as in Figures 16.4 and 16.5. For high Reynolds numbers the flow may not settle to a steady state. At this time there does not seem to be theoretical results that would answer the following question: For which Reynolds numbers does the lid-driven cavity flow settle to a steady state? The simple code discussed here is competitive up to a grid size of 256×256 (which was used to obtain the figure) and with a SOR solver used in the algorithm for approximating the solution of Poisson's equation. For larger matrices that appear in this code SOR does not converge or speed of convergence is too slow to be practical. Finer grids can be considered when more sophisticated linear solvers (multigrid methods in particular) are employed.

One obvious issue with the performance of numerical experiments with large Reynolds numbers is starting with the lid speed not zero and initially zero velocity in the cavity; a situation that is not physically realistic: the air near the lid would not be moving with zero velocity at the initialization of the experiment. Either the lid speed must be slowly built up to the desired running velocity and this scenario modeled and coded, the tolerance for the SOR stopping procedure can be relaxed until the air velocity near the lid is nearly the lid velocity after which the tolerance can be incrementally increased to the desired value, or the Reynolds number can be initialized at a much lower value and slowly increased to a desired value. Also, it is wise to store the velocity and pressure from a successful computation and read the velocity and pressure into the code as an initialization from which the Reynolds number might be incrementally changed.

What is causing the numerical approximations to blow up (be unstable) as the Reynolds number increases beyond some critical value?

Perhaps there is a bug in the code? Perhaps SOR would converge if more iterations were allowed? Maybe the time step is too large relative the spatial discretization size as measured by a CFL number? Maybe the spatial grid is too coarse to capture the fluid motion? Perhaps the first-order accuracy of Euler's method used to march forward in time is too low? All of the above might be partly responsible for the numerical instability, or some other problem that is not yet addressed might be the true culprit.

The observed instability is manifested by the computation of large values for velocity components. This phenomenon would seem to be unrelated to the projection part of the algorithm. If so, it must be caused by applying Euler's method with central discretization of spatial derivatives to the advection-diffusion equation. Is there a way to test this hypothesis?

Perhaps the best approach is to analyze simple examples where explicit solutions can be computed.

A useful example is provided by the linear PDE

$$u_t + \alpha u_x = \delta u_{xx}. \tag{16.23}$$

The term αu_x is a simple model for advection and δu_{xx} is a diffusion term. Using Euler's method and central differences for the spatial derivatives and ignoring boundary conditions, the discretized equation is

$$u_i^{k+1} = u_i^k + \Delta t(-\alpha \frac{u_{i+1}^k - u_{i-1}^k}{2\Delta x} + \delta \frac{u_{i+1}^k - 2u_i^k + u_{i-1}^k}{\Delta x^2}).$$

In a more compact form and with the obvious notation, this equation is given by

$$u_i^{k+1} = (d-a)u_{i+1}^k + (1-2d)u_i^k + (a+d)u_{i-1}^k.$$

In matrix form, time marching is given by an iteration scheme

$$U^{k+1} = TU^k + b,$$

where b is a vector that might be needed when boundary conditions are imposed and T is the tridiagonal system matrix whose first subdiagonal has every component equal to $d+a$, main diagonal components equal to $1-2d$, and first superdiagonal components $d-a$. For the case of pure diffusion ($a=0$), recall that the Gerschgorin theorem (see Appendix A.5) can be applied to show that if $d<1/2$, then every eigenvalue of T lies in the unit disk in the complex plane and the iteration scheme converges to $U=T^{-1}b$. The inequality $d<1/2$ is exactly the condition $\delta\Delta t/\Delta x^2<1/2$; that is, the CFL stability condition for the relation between time step size and the spatial step size. The addition of advection changes the radius of the Gerschgorin circle centered at $1-2d$ to $|a-d|+|a+d|$, which for $a>d$—when advection dominates—has value $2a$. In this case, the circle includes the number 1 on the real axis and there is the possibility that T has an eigenvalue outside the unit disk. Indeed the radius of the Gerschgorin circle grows with a, which increases the likelihood that T has an eigenvalue with absolute value larger than 1 as a is increased. The exact stability condition can be determined by using the results of Appendix A.19. In fact, the eigenvalues of T are given by

$$1-2d+2\sqrt{(d+a)(d-a)}\cos\frac{k\pi}{N+1}, \qquad k=1,2,3,\ldots,N$$

where N is the dimension of T. For $a>d$ and for N larger than 2, there is an eigenvalue such that the square of its absolute value is larger than

$$(1-2d)^2+2(a^2-d^2).$$

Thus, as a grows, there is an eigenvalue outside the unit disk in the complex plane. In this case, the iteration scheme is unstable. It will produce arbitrarily large components of the vector U^k for sufficiently large n. Numerical experiments can be used to verify this claim (see Exercise 16.3). The lesson here is that centered differences are associated with numerical instability in advection-diffusion problems. This is unfortunate because centered differences give second-order accuracy.

Does second-order accuracy have to be abandoned to maintain stability? Are there problems with forward or backward spatial differences?

To gain some insight, consider the simplest pure advection equation $u_t+\alpha u_x=0$, also called the one-way wave equation. As an illustrative example, consider this PDE on the entire real line with initial condition $u(x,0)=f(x)$ and seek a solution for $t>0$. The geometry of the equation is simple:

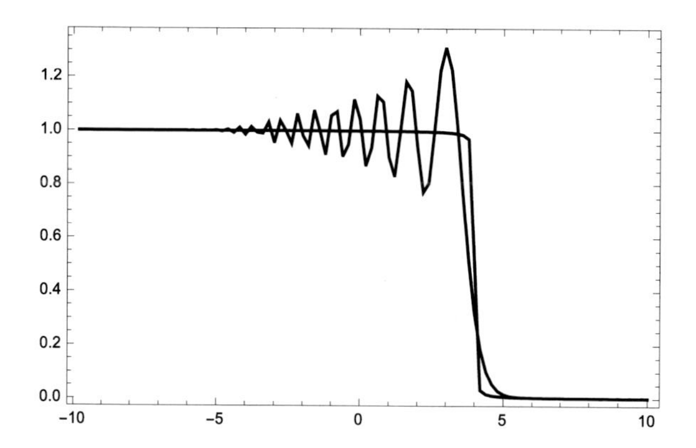

Fig. 16.6 The graph of $x \mapsto u(x, \tau)$ for the exact solution [Eq. (16.24)] of the one-way wave equation $u_t + 2u_x = 0$ with initial data f given in Eq. (16.25) is overlaid on the graph of a numerically approximated solution using the central difference scheme [Eq. (16.26)] with $\Delta x = 0.2$ and $\Delta t = 0.005$. The computational domain is $(-10, 10)$ with Dirichlet boundary conditions $u(-10, t) = f(-10)$ and $u(10, t) = f(10)$ and $\tau = 2$. The numerical scheme is unstable.

viewed as a function of two variables the inner product of the gradient of the function u (which has components u_x and u_t) with the vector $(\alpha, 1)$ is always zero. Thus, the level sets of u are lines parallel to the latter vector. All such lines are parameterized by curves of the form $s \mapsto (\xi + \alpha s, s)$, where $(\xi, 0)$ is some point on the x-axis. Given an arbitrary point (x, t) in space-time, the curve $s \mapsto (\xi + \alpha s, s)$ with $\xi = x - \alpha t$ passes through this point. Because u has the same value at every point of this curve,

$$u(x, t) = u(x - \alpha t, 0) = f(x - \alpha t), \tag{16.24}$$

where f is the specified initial condition. The solution is a traveling wave with profile f that moves right for $\alpha > 0$ and left for $\alpha < 0$.

For a specific example, consider initial data that is a smooth approximation to a step function:

$$u(x, 0) = f(x) := \frac{1}{\pi}(\frac{\pi}{2} - \arctan(\sigma(x - \tau))), \tag{16.25}$$

where $\sigma > 0$ determines the steepness of the continuous step from zero to one and τ is the center of the step. Fig. 16.6 shows the results of a numerical experiment where the computed solution is depicted together with the exact solution [Eq. (16.24)]. The numerical method is an implementation of the central difference scheme

$$u_i^{k+1} = u_i^k - \alpha \frac{\Delta t}{2\Delta x}(u_{i+1}^k - u_{i-1}^k). \tag{16.26}$$

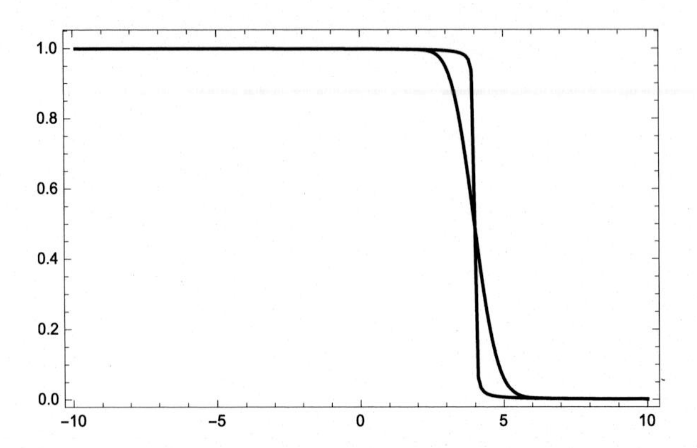

Fig. 16.7 The graph of $x \mapsto u(x, \tau)$ for the exact solution [Eq. (16.24)] of the one-way wave equation $u_t + 2u_x = 0$ with initial data f given in Eq. (16.25) is overlaid on the graph of a numerically approximated solution using the backward difference scheme [Eq. (16.28)] with $\Delta x = 0.2$ and $\Delta t = 0.005$. The computational domain is $(-10, 10)$ with Dirichlet boundary conditions $u(-10, t) = f(-10)$ and $u(10, t) = f(10)$ and $\tau = 2$. The numerical method produces a traveling wave that compares favorably with the known solution.

Boundary conditions are imposed at the ends of the computational domain to allow the numerics to proceed; they are not part of the formulation of the initial value problem for the PDE imposed on the whole real line. The idea is that as long as experiments stop before the boundary conditions play a significant role, the results should be reliable. Of course, this belief should be tested by computing over several nested intervals with increasing length to approximate the solution on the infinite line. The proved numerical instability of the method is manifested in the figure.

Central differencing is preferred because it uses a second-order approximation of the (spatial) derivative. What about first-order approximations?

By checking eigenvalues for the (triangular) system matrix of the iteration scheme, forward differencing,

$$u_i^{k+1} = u_i^k - \alpha \frac{\Delta t}{\Delta x}(u_{i+1}^k - u_i^k), \tag{16.27}$$

is also unstable when applied to the one-way wave equation with $\alpha > 0$.

The correct choice, backward differencing,

$$u_i^{k+1} = u_i^k - \alpha \frac{\Delta t}{\Delta x}(u_i^k - u_{i-1}^k), \tag{16.28}$$

is stable when applied to the one-way wave equation with $\alpha > 0$: all eigenvalues of the system matrix of this iteration scheme lie in the open unit disk in the complex plane (see Exercise 16.4). Results of a numerical experiment that agrees with this fact are depicted in Fig. 16.7.

The backward difference scheme, for the case where $\alpha > 0$ and the wave propagates to the right, is called an upwind scheme because it takes into account nodes in the direction from which the wind would be blowing a wind-driven wave. The forward difference is called downwind. In case $\alpha < 0$ and the wave propagates to the left, the roles are reversed: the forward difference is a stable upwind scheme and the backward difference is an unstable downwind scheme.

The numerical experiments and analysis of the one-way wave equation suggest that upwind schemes perform better for advection problems. The method is stable for the one-way wave equation. Perhaps an upwind scheme for the advection part of the fluid code would make it more robust with respect to increases in the Reynolds number. But it is not yet clear how to write an upwind scheme for a nonlinear PDE. How would the code know which way the wind is blowing?

An excellent one-dimensional test example is provided by the viscous Burgers's equation (Johannes M. Burgers, 1948)

$$u_t + uu_x = \delta u_{xx}. \tag{16.29}$$

It has a nonlinear advection term (the one-dimensional $u \cdot \nabla u$ that appears in fluid models) and a linear diffusion term (akin to Δu).

The nonviscous Burgers's equation ($\delta = 0$) is considered in detail later in this chapter (see page 544); it is an example of a nonlinear conservation law, which in one spatial dimension is a PDE of the form

$$u_t + (f(u))_x = 0 \tag{16.30}$$

for some differentiable scalar valued function f, called the flux function. Of course, the name for PDEs of this type comes from interpreting them as the differential forms of conservation laws. More precisely, let $u(x, t)$ denote the amount of some substance at position x at time t. The total amount of the substance in some interval $\Omega = [a, b]$ is

$$\frac{d}{dt} \int_{\Omega} u \, dx = - \int_{\partial\Omega} X \cdot \eta \, dp,$$

where X is the vector field that determines the flow carrying the substance, dp is the point measure on the boundary of the interval, and η is the outer unit normal at the ends of the interval. In case X is $f(u)$, the equation reads

$$\frac{d}{dt}\int_{\Omega} u\,dx = -(f(u(b,t)) - f(u(a,t)), \tag{16.31}$$

or by the fundamental theorem of calculus,

$$\int_{\Omega} u_t\,dx + \int_{\Omega}(f(u(x,t)))_x\,dx = 0.$$

PDE (16.30) holds because Ω is arbitrary. For Burgers's equation, the flux function is $f(u) = \frac{1}{2}u^2$; for the one-way wave equation, $f(u) = \alpha u$.

A new idea that leads naturally to upwind schemes and other numerical algorithms is called the finite volume method. In the context of conservation laws, the basic idea is to approximate

$$u_i^k := u(i\Delta x, k\Delta t)$$

by the average of u over some finite volume, usually a grid cell. For example, define

$$\bar{u}_i(t) = \frac{1}{\Delta x}\int_{(i-1/2)\Delta x}^{(i+1/2)\Delta x} u(x,t)\,dx$$

and

$$\bar{u}_i^k = \bar{u}_i(k\Delta t).$$

The spatial average of u over one cell width centered at the grid node $(i\Delta x, j\Delta t)$, namely $\bar{u}_i^k$, may be taken as an approximation of u_i^k—which is the value of the function u at this grid node.

To make a numerical method for time stepping to approximate the solution of a conservation law, approximations of the unknown u at different discrete times must be related. For the usual case where time step k is related to time step $k+1$ so that known values at the current time $k\Delta t$ can be propagated to time $(k+1)\Delta t$, a good idea is to integrate the conservation law over a space-time cell one spatial increment Δx wide and one time increment Δt tall with the spatial cell centered at $i\Delta x$. More precisely, the space-time cell is

$$C_i^k = [(i-1/2)\Delta x, (i+1/2)\Delta x] \times [k\Delta t, (k+1)\Delta t]$$

and the mentioned integration gives equalities

$$\begin{aligned}0 &= \int\int_{C_i^k} u_t + f(u)_x \, dxdt \\ &= \int_{(i-1/2)\Delta x}^{(i+1/2)\Delta x} u(x,(k+1)\Delta t) - u(x,k\Delta t)\, dx \\ &+ \int_{k\Delta t}^{(k+1)\Delta t} f(u((i+\frac{1}{2})\Delta x, t)) - f(u((i-\frac{1}{2})\Delta x, t))\, dt.\end{aligned}$$

Using the $\bar{u}$ notation, the result of the integrations is summarized in the fundamental formula

$$\Delta x(\bar{u}_i^{k+1} - \bar{u}_i^k) = -\int_{k\Delta t}^{(k+1)\Delta t} f(u((i+\frac{1}{2})\Delta x, t)) - f(u((i-\frac{1}{2})\Delta x, t))\, dt. \tag{16.32}$$

The left side of Eq. (16.32) is in discretized form and the right side is given by integrations along the sides of the space-time cell, where one side is the line connecting the nodes $((i-\frac{1}{2})\Delta x, k\Delta t)$ and $((i-\frac{1}{2})\Delta x, (k+1)\Delta t)$, and the other side connects $((i+\frac{1}{2})\Delta x, k\Delta t)$ and $((i+\frac{1}{2})\Delta x, (k+1)\Delta t)$. The finite volume method leaves open the possibility of many different approximations of these line integrals. Of course, desired approximations of these integrals are functions, called numerical flux functions, of the space-averaged approximations of u given by the $\bar{u}_i^k$.

One possibility is to approximate the line integral along the left side by a Riemann rectangle with base length Δt and height $f(\bar{u}_{i-1/2}^k)$ and the line integral of the right side using the height $f(\bar{u}_{i+1/2}^k)$. In turn, the unknown averages $\bar{u}_{i-1/2}^k$ and $\bar{u}_{i+1/2}^k$ at the cell-side midpoints may be approximated by $\bar{u}_{i-1}^k$ and $\bar{u}_{i+1}^k$. This leads back to the (unstable) central difference method

$$\Delta x(\bar{u}_i^{k+1} - \bar{u}_i^k) = -\Delta t(f(\bar{u}_{i+1}^k) - f(\bar{u}_{i-1}^k)).$$

For the one-way wave equation, the direction of the wave can be taken into account by approximating the line integral over the left-hand face of the cell by

$$\Delta t(\frac{\alpha}{2}(\bar{u}_i^k + \bar{u}_{i-1}^k) - \frac{|\alpha|}{2}(\bar{u}_i^k - \bar{u}_{i-1}^k)).$$

The integral is approximated by $\Delta t \bar{u}_{i-1}^k$ when $\alpha > 0$ and the wave is coming from the left and by $\Delta t \bar{u}_i^k$ when $\alpha < 0$ and the wave is coming from the right. Using the same idea to approximate the line integral over the right-hand face by

$$\Delta t(\frac{\alpha}{2}(\bar{u}_{i+1}^k + \bar{u}_i^k) - \frac{|\alpha|}{2}(\bar{u}_{i+1}^k - \bar{u}_i^k)),$$

a general upwind scheme is produced that works for $\alpha < 0$ and $\alpha > 0$:

$$\Delta x(\bar{u}_i^{k+1} - \bar{u}_i^k) = -\Delta t(\frac{\alpha}{2}(\bar{u}_{i+1}^k - \bar{u}_{i-1}^k) - \frac{|\alpha|}{2}(\bar{u}_{i+1}^k - 2\bar{u}_i^k + \bar{u}_{i-1}^k)).$$

Indeed, the right-hand side reduces to $-\Delta t\alpha(\bar{u}_i^k - \bar{u}_{i-1}^k)$ when $\alpha > 0$ and $-\Delta t\alpha(\bar{u}_{i+1}^k - \bar{u}_i^k)$ when $\alpha < 0$. This numerical scheme may be interpreted as approximating the first-order spatial derivative by the average of a centered difference for the spatial derivative and a centered difference approximation for an artificial viscosity term $|\alpha|u_{xx}/2$. The additional diffusion tends to stabilize the numerical method for the same reason that central difference schemes that satisfy the CFL condition are stable for the diffusion equation.

Although the one-way wave equation is a good example, it is too special in the sense that the sign of the coefficient α in the flux function $f(u) = \alpha u$ determines the upwind direction. For nonlinear conservation laws (and the advection part of the Navier–Stokes equations) the situation is not as explicit. But, taking into account the success of the upwind scheme for the one-way wave equation suggests consideration of the general scalar conservation law written in the form

$$u_t + f'(u)u_x = 0,$$

which would be a one-way wave equation if $f'(u)$ were constant. For this reason the sign of $f'(u)$ in the nonlinear case is naturally associated with the sign of α, which is the sign of the first derivative of the one-way wave equation's flux function $f(u) = \alpha u$. This observation suggests the upwind scheme that uses a backward difference to approximate the spatial derivative when $f'(u) > 0$ and a forward difference when $f'(u) < 0$.

In the finite volume framework, the flux along the left side of the space-time cell would be approximated by $\Delta t f(\bar{u}_{i-1}^k)$ in case $f'(\bar{u}_i^k) > 0$ and by $\Delta t f(\bar{u}_i^k)$ in case $f'(\bar{u}_i^k) < 0$. Likewise, the line integral along the right-

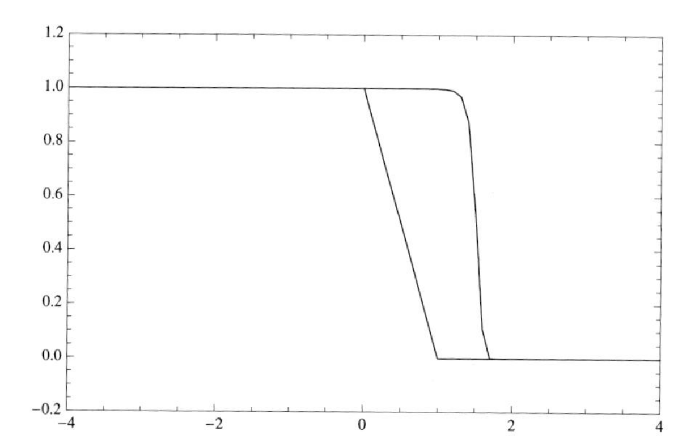

Fig. 16.8 Graphs are depicted of the initial data ($u(x,0) = 1$ for $x < 0$, $u(x,0) = 1 - x$, for $0 \le x \le 1$, and $u(x,0) = 0$ for $x > 1$) and the computed solution at time $t = 2$ of the inviscid Burgers's equation using upwind discretization.

hand boundary of the space-time cell is approximated by $\Delta t f(\bar{u}_i^k)$ in case $f'(\bar{u}_i^k) > 0$ and by $\Delta t f(\bar{u}_{i+1}^k)$ in case $f'(\bar{u}_i^k) < 0$.

To implement the proposed upwind scheme for Burgers's equation, a logical test is used at each time step to determine the sign of the derivative. The same result is obtained (approximately) in finite volume language by defining the numerical flux function

$$F_{i-1/2}^k = \begin{cases} f(u_{i-1}^k), & \frac{f(u_i^k)-f(u_{i-1}^k)}{u_i^k-u_{i-1}^k} \geq 0; \\ f(u_i^k), & \frac{f(u_i^k)-f(u_{i-1}^k)}{u_i^k-u_{i-1}^k} < 0 \end{cases} \tag{16.33}$$

and using the numerical scheme

$$\Delta x(u_i^{k+1} - u_i^k) = -\Delta t(F_{i+1/2} - F_{i-1/2}).$$

This approximation might be used in case the derivative of f is not available or is computationally inefficient to evaluate.

Numerical flux functions depend on the indices—in this case under discussion i and k—and the elements of the finite sequence of states (the u_i^k). The definition of the numerical flux function [Eq. (16.33)] includes the quotient of two differences that approximates $(f(u))_x/u_x = f'(u)$. In view of this fact, it is easy to see that this numerical flux function can be used to replace (up to a close approximation) the direct logical test for the sign of this derivative.

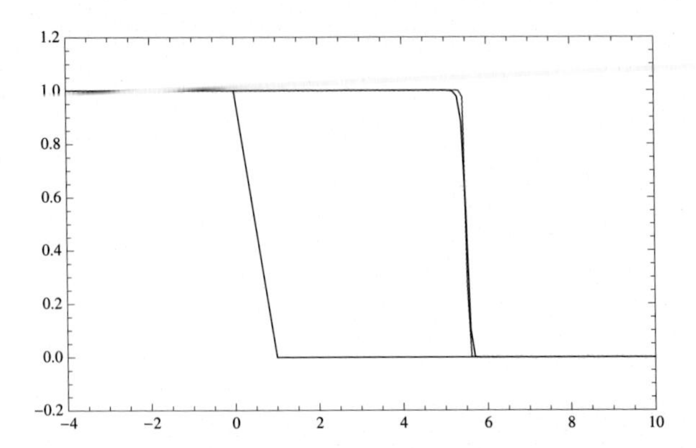

Fig. 16.9 Graphs are depicted of the initial data ($u(x,0) = 1$ for $x < 0$, $u(x,0) = 1 - x$, for $0 \leq x \leq 1$, and $u(x,0) = 0$ for $x > 1$), the computed solution at time $t = 10$ of the viscous Burgers's equation using upwind discretization for the advection term and the usual second-order centered difference for the diffusion term, and the solution computed from the exact solution [Eq. (16.34)*]. The diffusion coefficient is $\delta = 0.01$, $\Delta t = 0.01$, and $\Delta x = 0.1$ for the computation done over the spatial interval* $[-10, 10]$.

The choice of a good test problem for the inviscid Burgers's equation requires some mathematical analysis of conservation laws, which is done in a different context later in this chapter and in Section 17.6. One example is provided by the exact solution depicted in Fig. 17.5, where the initial data is $u(x,0) = 1$ for $x < 0$, $u(x,0) = 1 - x$, for $0 \leq x \leq 1$, and $u(x,0) = 0$ for $x > 1$. The solution for $t > 1$ is a discontinuous step: $u(x,t) = 1$ for $x < (t+1)/2$ and $u(x,t) = 0$ for $x > (t+1)/2$. Fig. 16.8 shows the result of a numerical experiment where the integration is carried to $t = 2$. The discontinuity of the exact solution is at $x = 3/2$; this jump is matched with reasonable accuracy by the numerical approximation (see Exercise 16.5).

Upwinding gives good results for the test cases of conservation laws already presented. The viscous Burgers's equation (16.29) is an advection-diffusion equation. Thus, it serves as a test case exhibiting both features. By a gem of mathematical analysis, there is an explicit formula for the solution of the initial value problem for the viscous Burgers's equation valid for $t > 0$:

$$u(x,t) = \frac{\int_{-\infty}^{\infty} \frac{x-y}{t} \exp\left(\frac{-(x-y)^2}{4\delta t} - \frac{1}{2\delta}\int_0^y u(\eta,0)\,d\eta\right) dy}{\int_{-\infty}^{\infty} \exp\left(\frac{-(x-y)^2}{4\delta t} - \frac{1}{2\delta}\int_0^y u(\eta,0)\,d\eta\right) dy}. \tag{16.34}$$

(see Exercise 16.6). The graphs in Fig. 16.9 computed from this formula and from upwind discretization for the advection term (with the logical test coded into the computer program) and the usual second-order centered

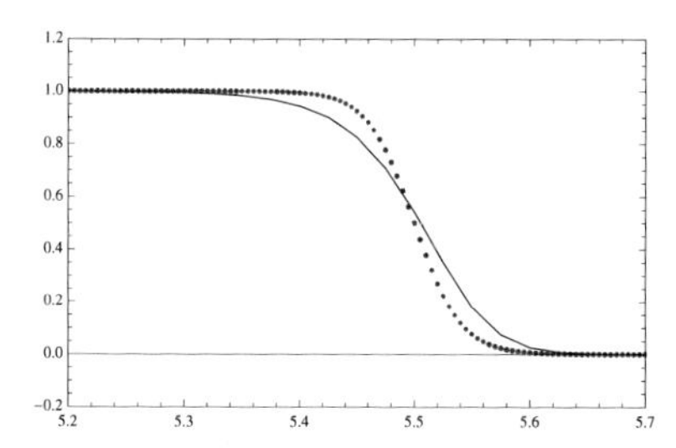

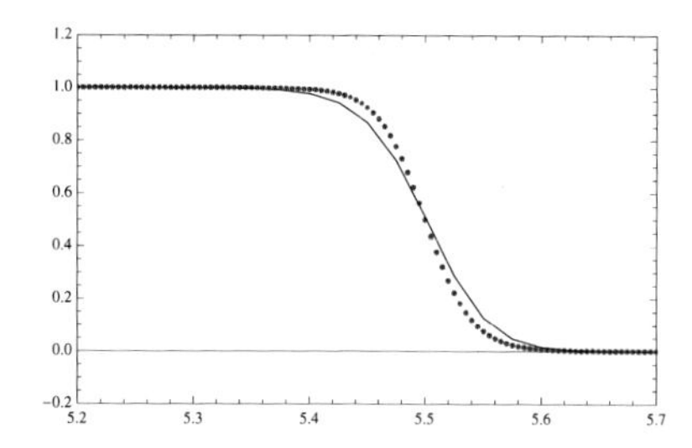

Fig. 16.10 The left panel shows the graphs of the approximate solution of the viscous Burgers's equation with initial data ($u(x,0) = 1$ *for* $x < 0$, $u(x,0) = 1 - x$, *for* $0 \leq x \leq 1$, *and* $u(x,0) = 0$ *for* $x > 1$) *computed at time* $t = 10$ *using upwind discretization for the advection term and second-order centered difference for the diffusion term (solid line), and the solution computed from the exact solution [Eq.* (16.34)*]. The right panel shows the solution computed from the exact solution and the approximation computed using the average of the first-order upwind scheme and the centered difference for the advection term, and the same approximation for the diffusion term. The diffusion coefficient is* $\delta = 0.01$, $\Delta t = 0.001$, *and* $\Delta x = 0.025$ *for the computation done over the spatial interval* $[-10, 10]$.

difference for the diffusion term in Burgers's equation are almost indistinguishable. This test again evidences the utility of upwinding.

There is a downside to the upwind schemes employed here: they are only first-order accurate with respect to the spatial derivative. A useful idea that might improve accuracy is to use a weighted average of the first-order upwind scheme and the second-order central difference approximation for the advection term. The results of tests using the initial data for the viscous Burgers's equation as in Fig. 16.9 are depicted in Fig. 16.10. Averaging (with equal weights) the upwind and second-order difference approximations for the advection term produces a more accurate result. Averaging the two methods is of course still first-order accurate, but at least for the discretization used for the tests, averaging gives a better result than pure upwind (see Exercise 16.9).

One way to incorporate upwinding in the numerical approximation of solutions of the advection-diffusion part of the Navier–Stokes model is to use the suggested averaging of centered difference approximations and upwinding for the discretization of the partial derivatives $(u^2)_x$, $(uv)_y$, $(uv)_x$, and $(v^2)_y$ that appear in the advection terms of PDEs (16.14).

The term $(uv)_y$ appears in the partial differential equation for the time derivative of the first velocity component u. Isolating these two terms, we may consider the conservation law $u_t + (vu)_y = 0$. In view of the discussion of the one-way wave equation and more general conservation laws, the flux function in this case is $f(u) = vu$ where v is considered to be a given function of space and time. The sign of the derivative of f is determined by the sign of v. Thus, an upwind scheme for approximating solutions of this conservation law will involve switching between forward and backward difference approximation of the spatial derivative depending on the sign of v. The time discretization of u_t at the spatial location p where u has value u_{ij} should be balanced by a spatial discretization of $(vu)_y$ at this same point. But, due to the staggered grid, v and u are not specified at the same spatial locations. Refer to Fig. 16.3 and consider the vertical axis through u_{ij}. The adjacent u-values along this axis are $u_{i,j-1}$ and $u_{i,j+1}$ each values of u at distance Δx from p. The natural adjacent v-values along this axis are approximated at the spatial locations with nodal addresses (i,j) and $(i,j-1)$ by

$$v_{ij}^{\text{above}} = \frac{v_{ij} + v_{i+1,j}}{2}, \qquad v_{ij}^{\text{below}} = \frac{v_{i,j-1} + v_{i+1,j-1}}{2},$$

which are $\Delta x/2$ units from p. The value of v at p is not given. Although it could be approximated by another average, the preferred upwind scheme is to consider the signs of v_{ij}^{above} and v_{ij}^{below} to determine the upwind direction. This choice leads to the definitions

$$u_{ij}^{\text{above}} = \begin{cases} u_{ij}, & v_{ij}^{\text{above}} \geq 0; \\ u_{i,j+1}, & v_{ij}^{\text{above}} < 0, \end{cases} \qquad u_{ij}^{\text{below}} = \begin{cases} u_{i,j-1}, & v_{ij}^{\text{below}} \geq 0; \\ u_{ij}, & v_{ij}^{\text{below}} < 0, \end{cases}$$

and the upwind discretization

$$((uv)_y)_{ij} = \frac{v_{ij}^{\text{above}} u_{ij}^{\text{above}} - v_{ij}^{\text{below}} u_{ij}^{\text{below}}}{\Delta y}. \tag{16.35}$$

By combining the upwind [Eq. (16.35)] and central difference [Eq. (16.21)] approximations with a weighted sum, we obtain the family of useful discretizations

$$\begin{aligned} ((uv)_y)_{ij} = (1-\gamma)\frac{1}{\Delta y}\Big(&\frac{(u_{i,j+1}^k + u_{ij}^k)(v_{i+1,j}^k + v_{ij}^k)}{4} \\ &- \frac{(u_{ij}^k + u_{i.j-1}^k)(v_{i,j-1}^k + v_{i+1,j-1}^k)}{4}\Big) \\ &+ \gamma\frac{1}{\Delta y}\Big(v_{ij}^{\text{above}} u_{ij}^{\text{above}} - v_{ij}^{\text{below}} u_{ij}^{\text{below}}\Big), \end{aligned} \tag{16.36}$$

for $0 \leq \gamma \leq 1$. Of course, $\gamma = 0$ corresponds to central difference and $\gamma = 1$ to upwind. This formula may be written in a more compact form by employing the absolute value instead of a sign test:

$$((uv)_y)_{ij} = \frac{1}{\Delta y}\Big(\frac{(u^k_{i,j+1} + u^k_{ij})(v^k_{i+1,j} + v^k_{ij})}{4} - \frac{(u^k_{ij} + u^k_{i.j-1})(v^k_{i,j-1} + v^k_{i+1,j-1})}{4}\Big) + \frac{\gamma}{\Delta y}\Big(\frac{|v_{ij} + v_{i+1,j}|(u_{ij} - u_{i,j+1})}{4} - \frac{|v_{i,j-1} + v_{i+1,j-1}|(u_{i,j-1} - u_{ij})}{4}\Big) \tag{16.37}$$

(see [46] for an extended, excellent treatment of this scheme and much more; and see Exercise 16.10).

Following the same prescription, $(u^2)_x$ may be discretized by

$$((u^2)_x)_{ij} = \frac{1}{\Delta x}\Big(\frac{(u^k_{ij} + u^k_{i+1,j})^2}{4} - \frac{(u^k_{i-1,j} + u^k_{ij})^2}{4}\Big) + \frac{\gamma}{\Delta x}\Big(\frac{|u^k_{ij} + u^k_{i+1,j}|(u^k_{ij} - u^k_{i+1,j})}{4} - \frac{|u^k_{i-1,j} + u^k_{ij}|(u^k_{i-1,j} - u^k_{ij})}{4}\Big). \tag{16.38}$$

Similar considerations are used to discretize $(v^2)_y$ and $(uv)_x$. For example, the upwind scheme for $(uv)_x$ is best approached by considering the conservation law $v_t + (uv)_x = 0$. The formulas in compact form are

$$((v^2)_y)_{ij} = \frac{1}{\Delta y}\Big(\frac{(v^k_{ij} + v^k_{i,j+1})^2}{4} - \frac{(v^k_{i,j-1} + v^k_{i.j})^2}{4}\Big) + \frac{\gamma}{\Delta y}\Big(\frac{|v^k_{ij} + v^k_{i,j+1}|(v^k_{ij} - v^k_{i,j+1})}{4} - \frac{|v^k_{i,j-1} + v^k_{ij}|(v^k_{i,j-1} - v^k_{ij})}{4}\Big) \tag{16.39}$$

and

$$((uv)_x)_{ij} = \frac{1}{\Delta x}\Big(\frac{(u^k_{ij} + u^k_{i,j+1})(v^k_{ij} + v^k_{i+1,j})}{4} - \frac{(u^k_{i-1,j} + u^k_{i-1,j+1})(v^k_{i-1,j} + v^k_{ij})}{4}\Big)$$

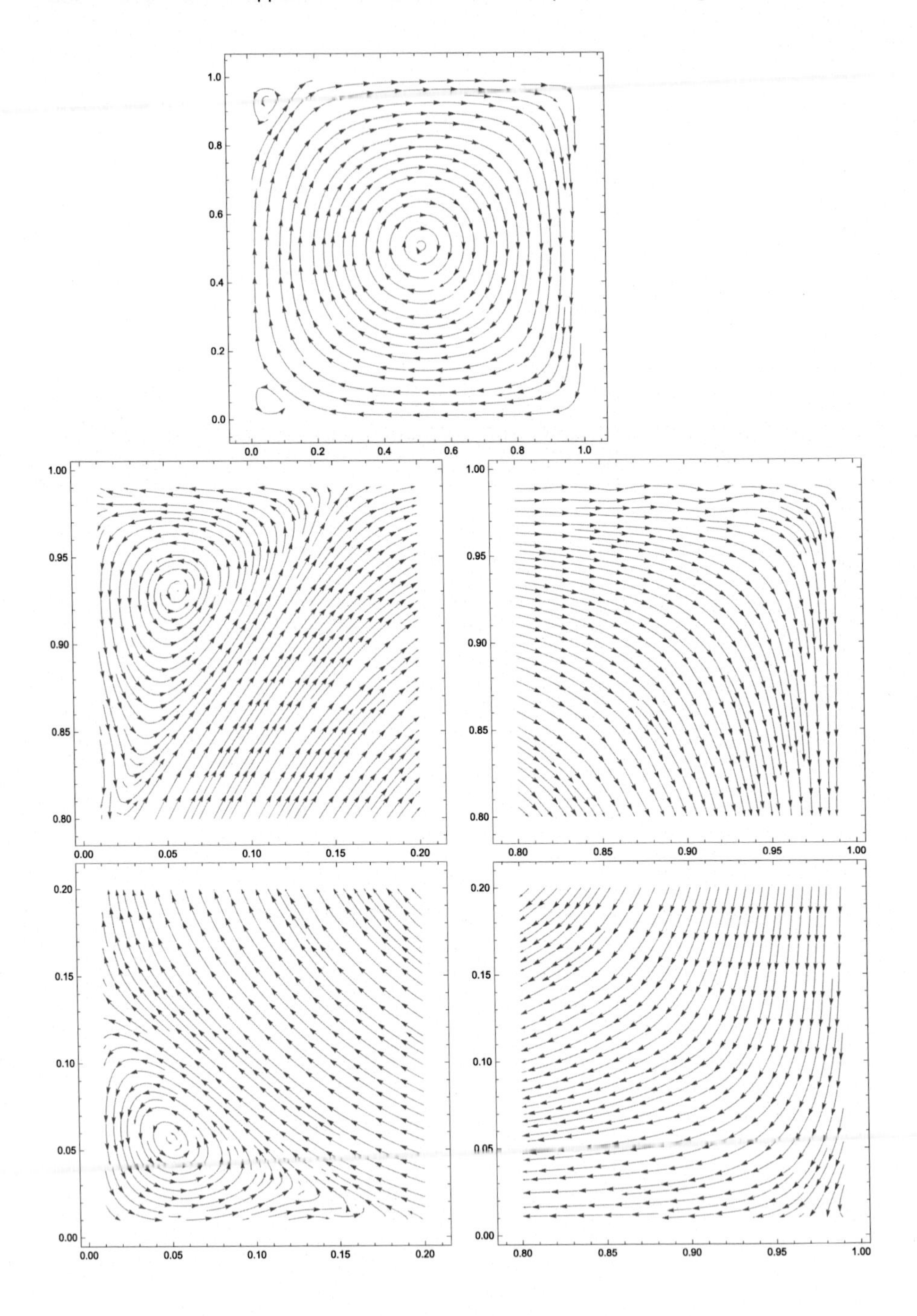

Fig. 16.11 The top panel depicts streamlines for steady state lid-driven cavity flow with unit lid speed, grid size 128×128*, and Reynolds number 128000. The four bottom panels depict blowups of the flow near the top left corner, the top right corner, the bottom left corner, and the bottom right corner of the cavity.*

$$+\frac{\gamma}{\Delta x}\Big(\frac{|u_{ij}^k+u_{i,j+1}^k|(v_{ij}^k-v_{i+1,j}^k)}{4} - \frac{|u_{i-1,j}^k+u_{i-1,j+1}^k|(v_{i-1,j}^k-v_{ij}^k)}{4}\Big). \tag{16.40}$$

Using these weighted averages of central difference and upwind discretizations to compute the advection-diffusion step in the projection algorithm, for example with $\gamma = 0.9$ so that upwinding is dominant, lid-driven cavity flow simulations at Reynolds numbers in the hundreds of thousands are stable (as in Fig. 16.11). This improvement of our original implementation of the Chorin projection method, which is unstable for Reynolds numbers less than about 10^5, is dramatic.

To implement the suggested method, previous considerations should be taken into account: start with a lower Reynolds number and ramp up to a desired higher value incrementally by increasing the Reynolds number slowly with (dimensionless) time integration with, perhaps, an increase of 5% over a dimensionless time interval of 0.5. In addition, the choice of SOR parameter should be tuned to minimize the number of SOR iterations. One way to do this is to implement a time step adjustment scheme that takes into account the number of SOR iterations. For example, a starting time step of 1.0×10^{-5} might be modified as follows: As the integration proceeds, if the number of SOR iterations does not exceed four, then the time step size is increased by 20%; if the number of SOR iterations is at least six, then the step size is decreased by 70%. Set a SOR parameter and run your code over some number of time steps, perhaps 10,000 or 100,000 while monitoring the total elapsed dimensionless time. The SOR parameter that produces the largest elapsed time performs the best. Such a test was performed for the computations that produced Fig. 16.11. A SOR parameter of 1.6 produced a dramatic improvement over other values in the usual range $[0, 2]$. Also, the number of SOR iterations remained in the range of one to eight. Other values of the SOR parameter performed well over some regions and required a large number of SOR iterations over other regions to converge to a test tolerance of 5×10^{-3}. Of course, this test is not definitive. Perhaps you can do better (see Exercise 16.12).

As might be expected, implementation of a weighted first-order upwind and central difference scheme is not nearly the end of the story for simulating fluid flow. For higher Reynolds numbers and more realistic physical problems, the code will not produce physically realistic results or it will

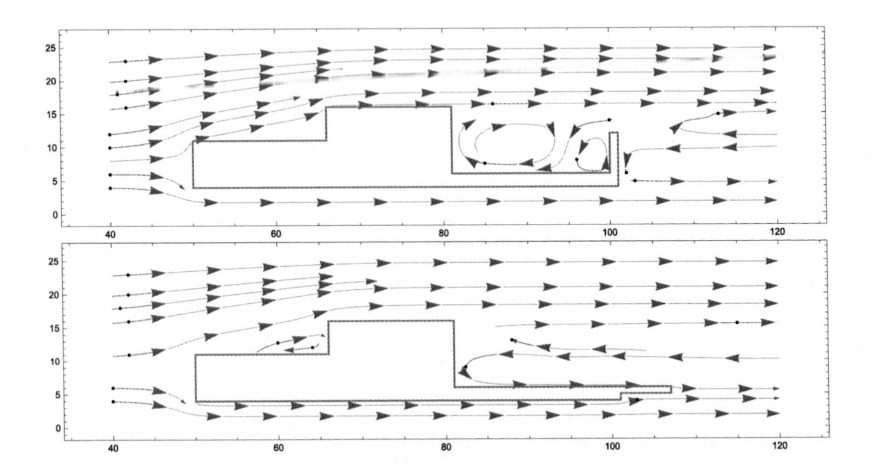

Fig. 16.12 The two panels depict streamlines for flow over the boxy truck moving left at 70 miles per hour. The top panel is for the tailgate up configuration and the bottom panel for tailgate down. The horizontal and vertical scales are relative to the 250×25 *computational grid.*

be unstable. Depending on the desired application, further improvements will be necessary: second-order-in-time numerical schemes, more robust methods for approximating solutions of large and space matrices, other approaches to discretizing the Navier–Stokes equations, and so on. Thus, a journey into the world of CFD begins.

As mentioned several times, in the world of CFD and scientific computing in general, there is a fundamental question: How do we know the simulated result (for example, the streamlines in Fig. 16.11) is correct? This question does not have a simple answer. The ultimate verification is the formulation and proof of a theorem. Short of that, evidence of a correct prediction is gathered by seeking to obtain the same result from computations using different methodologies, different discretizations, different codes, and so on (see Exercise 16.11).

Returning to Problem 16.1, the projection method code for lid-driven cavity flow can be easily modified to approximate flow around a truck moving on a highway. Streamlines for velocity fields computed from numerical experiments are depicted in Fig. 16.12. A boxy truck model is chosen for convenience in coding and to avoid the important, more advanced topic of mesh generation in CFD for curved boundaries. The qualitative features of the flow seem to be correct even for the coarse mesh used in the experiments. In particular, a large vortex forms in the truck bed when the tailgate is up. This is the key difference between the tailgate up and tailgate down configurations. Your first challenge is to repeat these numerical experiments to obtain comparable velocity fields. Use the following data to compare results.

The computational rectangular grid is 250×25 and the tailgate-up boxy truck is defined by connecting the grid points

$$(50,4),(101,4),(101,12),(100,12),(100,6),(81,6),\\(81,16),(66,16),(66,11),(50,11),(50,4)$$

in order with line segments. Corresponding grid points for the tailgate down configuration are

$$(50,4),(101,4),(101,5),(107,5),(107,6),(100,6),\\(81,6),(81,16),(66,16),(66,11),(50,11),(50,4).$$

The dimensionless discretization size $dx = dy = 1/50$ is used in the compuation.

Ghost cells surround the computational domain in which the velocity and pressure fields are changed. The free-stream velocity is left-to-right and specified at the upstream entrance by fixing the velocity in left boundary ghost cells. Similarly, free-stream velocity is set in the ghost cells above and below the computational rectangle. Outflow is modeled by a do-nothing boundary condition: the velocities in the ghost cells along the right-hand boundary match the velocities in adjacent computational cells. No-slip boundary conditions are imposed on the truck boundary by setting interior ghost cell velocities equal to zero. As usual, the discretized Poisson equation for pressure leads to a system of linear equations that do not have a unique solution because pressure is defined up to an additive constant. To specify a solution, pressure was set to zero in the upper-right computational cell (which is the natural choice when cell ordering is bottom-to-top and left-to-right).

The code was initialized with zero initial velocity and small free-stream velocity. The free-stream velocity is slowly ramped up to the desired free-stream velocity. Also the SOR parameter ω was tuned by numerical experiment to optimize the speed of convergence. The (code-dependent) value of $\omega = 1.9$ was used in the final computations together with the time step size 0.5×10^{-4} and upwind parameter 0.8.

Your second (grand) challenge is to compare pressure drag for the tailgate-up and tailgate-down configurations to determine which has smaller magnitude (see Exercise 19.5). Spatial resolution with a 250×25 grid is too coarse to obtain reliable approximations of pressure on the truck body. In fact, at the time of this writing, the computation of drag from CFD approximations of the Navier–Stokes equations is problematic. As computational power increases and algorithms are improved, CFD approximations will certainly produce more accurate results. Can you meet the grand challenge? Perhaps working on the tailgate drag problem will be your first step on a path to learning more advanced CFD where you will make a useful contribution.

Exercise 16.1. Show that the gradient of a function is curl free; that is, the curl of the gradient of a function is zero.

Exercise 16.2. Use Taylor expansions about $\Delta x = 0$ to show that the accuracy of discretizations (16.17)–(16.19) increases in the order they are presented.

Exercise 16.3. (a) Perform numerical experiments to verify that Euler's method for the advection-diffusion equation (16.23) with central differencing for the spatial derivatives is unstable when advection dominates the diffusion. (b) Prove that for sufficiently large advection with fixed diffusion, the numerical method as in part (a) is unstable.

Exercise 16.4. Show by determining eigenvalues of an appropriate matrix that upwind differencing is indeed stable (for appropriate choices of the time step size) when applied to the one-way wave equation with $\alpha > 0$ (and also for $\alpha < 0$). Discuss the range of viable time step sizes for given spatial discretizations.

Exercise 16.5. (1) Recreate Fig. 16.8 using upwind differencing for the spatial derivative and a logical test of the sign of the derivative of the flux function. Hint: Be careful concerning the case $f'(u_i^k) = 0$. It is sometimes desirable to write code tailored to a specific problem. But, to write a general code, perhaps a different choice of direction can be made at each time step. Discuss this issue. (2) Repeat part (1) using the numerical flux function defined in Eq. (16.33).

Exercise 16.6. The exact solution (16.34) of the viscous Burgers's equation is obtained in two basic steps: a change of variables to the heat equation and use of the known exact solution of the heat equation. (1) Step 1 is called the Hopf–Cole (Eberhard Hopf and Julian Cole) transformation. Let u be a solution of Burgers's equation. Show that there is a function ψ of the same two variables such that

$$\psi_x = u, \qquad \psi_t = \delta \psi_{xx} - \frac{1}{2}\psi_x^2.$$

The new function

$$\phi := e^{-\psi/(2\delta)}$$

is a solution of the PDE

$$\phi_t = \delta\phi_{xx}.$$

Also, up to an additive constant,

$$\phi(x,0) = \exp(-\frac{1}{2\delta}\int_0^x u(y,0)\,dy).$$

(2) Step 2 is the exact solution of the initial value problem for $\phi_t = \delta\phi_{xx}$:

$$\phi(x,t) = \frac{1}{\sqrt{4\pi\delta t}}\int_{-\infty}^{\infty} e^{\frac{-(x-y)^2}{4\delta t}}\phi(\xi,0)\,dx.$$

This formula should be familiar to students who have studied PDE. It can be checked by substitution into the heat equation and derived using invariance properties of the heat equation. Show that it is correct by direct substitution.
(3) Complete the derivation of the exact solution using the results of (1) and (2).
(4) Recreate Fig. 16.9. Discuss the computational overhead incurred by using the exact solution and the finite difference approximation.

Exercise 16.7. The initial value problem for the invicid Burgers's equation $u_t + uu_x = 0$ has an exact solution, which can be derived from the result of Exercise 16.6:

$$u(x,t) = \frac{\partial}{\partial x}(\min_y(\int_0^y u(\eta,0)\,d\eta + \frac{(x-y)^2}{2t})). \tag{16.41}$$

(a) Use this result to check the accuracy of the upwind method for some of the initial data used to create Fig. 16.9. (b) Derive exact solution (16.41) from the exact solution of the viscous Burgers's equation. Warning: This exercise seems to be mathematically challenging.

Exercise 16.8. Consider the one-way wave equation $u_t + \alpha u_x = 0$ with $\alpha > 0$ and for $x \geq 0$. (a) Find the exact solution with the boundary condition $u(0,t) = f(t)$. Hint: Reverse the roles of x and t in the derivation of the solution with initial data $u(x,0) = f(x)$ given in this section. (b) Reformulate and redo part (a) in case $\alpha < 0$. (c) Verify with numerical experiments that upwinding produces a good approximation of the exact solution derived in part (a).

Exercise 16.9. [Upwinding Project] (a) Reproduce Fig. 16.10. (b) A weighed average of two quantities Q_1 and Q_2 is an expression of the form $\lambda Q_1 + (1-\lambda)Q_2$. When $\lambda = 1/2$, the usual average is obtained. Write a numerical scheme where advection is treated as a weighted average of upwinding and central differences. Apply your code to Burgers's equation as in Fig. 16.10. Which choice of weight λ produces the best results? Hint: This question is open to some interpretation: the discretization sizes for time and space play a role. Set up an experiment to test some range of these values and for each choice of these parameters optimize over the weight. Report on the results of this experiment. Can you draw a general conclusion? (c) Alternate second-order finite difference approximations of the first derivative (which are not central differences) are obtained by incorporating two nodes to the left or right of the node where the derivative is desired; for example, the three-point formula

$$f'(x) = \frac{f(x - 2\Delta x) - 4f(x - \Delta x) + 3f(x)}{2\Delta x}$$

is a second-order approximation to the derivative of f at x that uses two points to the left of x. Prove this. Write an upwind scheme based on this approximation together with its counterpart that uses values to the right of x and apply it to Burgers's equation. Discuss the efficacy of this approach compared with the weighed averages of parts (a) and (b).

Exercise 16.10. Show that the discretizations in Eqs. (16.36) and (16.37) for $(uv)_y$ give exactly the same result.

Exercise 16.11. [CFD for Lid-Driven Cavity Flow I] *This exercise requires nontrivial coding.* Repeat the numerical experiment reported in Fig. 16.4 and make a figure for comparison. Does your result agree with the result reported in the figure? Provide evidence that *your* result is correct.

Exercise 16.12. [CFD for Lid-Driven Cavity Flow II] *This exercise requires nontrivial coding.* (a) Reproduce Fig. 16.11, which was made from output obtained after forward integration of more than 200 dimensionless time steps. At least make a comparable figure by using *your* code to produce the velocity flow for lid-driven cavity flow at Reynolds number 128000 and integrating forward for at least 100 dimensionless time units. Gather evidence that your result is correct. (b) Run tests to tune the SOR parameter used in your code as suggested on page 441. Note: The test suggested in the narrative suggests a place to start, but you should try to formulate and perform other tests.

16.2 A NUMERICAL METHOD FOR WATER WAVES

The ideal water wave equations [Eqs. (15.1)–(15.4)]

$$\begin{aligned} \Delta\phi &= 0 \quad \text{on } \Omega, \\ \eta_t + \phi_x\eta_x - \phi_y &= 0 \quad \text{on } \mathcal{S}, \\ \phi_y &= 0 \quad \text{on } \mathcal{B}, \\ \phi_t + \frac{1}{2}(\phi_x^2 + \phi_y^2) + g\eta &= 0 \quad \text{on } \mathcal{S}, \end{aligned} \tag{16.42}$$

may be approximated by a variety of numerical methods, some of which are simpler to implement than general purpose algorithms such as the finite difference methods described in Section 16.1. The main difficulty is dealing with the free surface (see Exercise 19.8 for a more general alternative than the simplistic approach presented here).

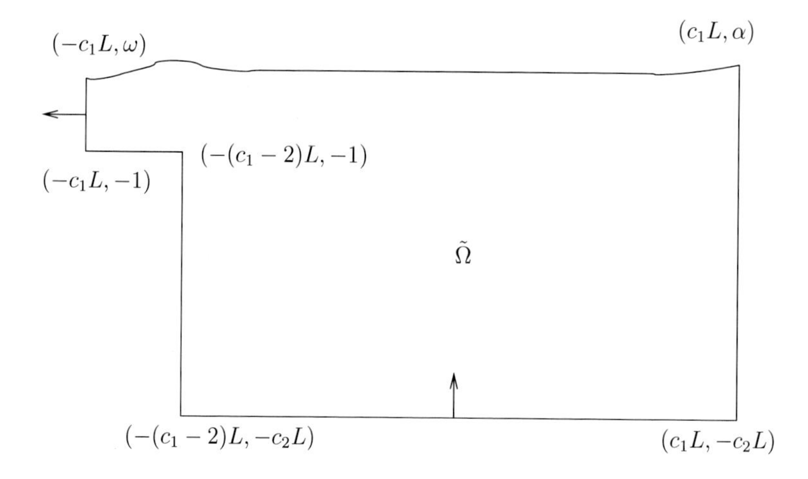

Fig. 16.13 A schematic boundary of the domain $\tilde{\Omega}$ is depicted for Problem 16.3. The arrows correspond to the direction of the fluid velocity at the inflow at the reservoir bottom to the outflow at the left end of the plate. The coordinates of the corners are relative to L as in Problem 16.2.

Numerical experiments will be described for the dimensionless water wave equations obtained using the scaling in display (15.5) with $b = \mathrm{wd}$ and $\ell = \mathrm{wd}$; that is,

$$\phi = g^{1/2}\,\mathrm{wd}^{3/2}\,\tilde{\phi}, \quad \eta = \mathrm{wd}\,\tilde{\eta}, \quad x = \mathrm{wd}\,\tilde{x}, \quad y = \mathrm{wd}\,\tilde{y}, \quad t = \Big(\frac{\mathrm{wd}}{g}\Big)^{1/2}\tilde{t}. \tag{16.43}$$

This choice gives $\alpha = 1$ and $\beta = 1$ in the dimensionless equations (15.8)–(15.11). Using the sets

$$\begin{aligned}
\tilde{\mathcal{B}} &:= \{(\tilde{x}, \tilde{y}) : \tilde{y} = -1\},\\
\tilde{\mathcal{S}} &:= \{(\tilde{x}, \tilde{y}) : \tilde{y} = \tilde{\eta}(\tilde{x}, \tilde{t})\},\\
\tilde{\Omega} &:= \{(\tilde{x}, \tilde{y}) : -1 < \tilde{y} < \tilde{\eta}(\tilde{x}, \tilde{t})\},
\end{aligned}$$

these equations have the form

$$\begin{aligned}
\tilde{\phi}_{\tilde{x}\tilde{x}} + \tilde{\phi}_{\tilde{y}\tilde{y}} &= 0 \quad \text{on } \tilde{\Omega},\\
\tilde{\eta}_{\tilde{t}} + \tilde{\phi}_{\tilde{x}}\tilde{\eta}_{\tilde{x}} - \tilde{\phi}_{\tilde{y}} &= 0 \quad \text{on } \tilde{\mathcal{S}},\\
\tilde{\phi}_{\tilde{y}} &= 0 \quad \text{on } \tilde{\mathcal{B}},\\
\tilde{\phi}_{\tilde{t}} + \frac{1}{2}(\tilde{\phi}_{\tilde{x}}^2 + \tilde{\phi}_{\tilde{y}}^2) + \tilde{\eta} &= 0 \quad \text{on } \tilde{\mathcal{S}}.
\end{aligned} \tag{16.44}$$

To explore, for example, approximations of the Tiger Fountain flow, consider the following problems.

Problem 16.2. [Flow Over A Plate] For a given dimensionless length $L > 0$ (for example, physical length divided by the depth wd in the same units), approximate the function $\tilde{\eta}$ whose graph is the steady state free surface for the solution of system (16.44) over the (redefined, finite) domain $\tilde{\Omega}$ with bottom $\tilde{\mathcal{B}}$ and surface $\tilde{\mathcal{S}}$ given by

$$\begin{aligned}\tilde{\mathcal{B}} &:= \{(\tilde{x},\tilde{y}) : \tilde{y} = -1,\ -L < \tilde{x} < L\},\\ \tilde{\mathcal{S}} &:= \{(\tilde{x},\tilde{y}) : \tilde{y} = \tilde{\eta}(\tilde{x}),\ -L < \tilde{x} < L\},\\ \tilde{\Omega} &:= \{(\tilde{x},\tilde{y}) : -1 < \tilde{y} < \tilde{\eta}(\tilde{x}),\ -L < \tilde{x} < L\},\end{aligned}$$

with the gradient $(\partial\tilde{\phi}/\partial\tilde{x}, \partial\tilde{\phi}/\partial\tilde{y})$ prescribed on the left and right vertical boundaries of $\tilde{\Omega}$, which are the sets

$$\begin{aligned}\tilde{\mathcal{L}} &:= \{(\tilde{x},\tilde{y}) : \tilde{x} = -L, -1 \le \tilde{y} \le \omega\},\\ \tilde{\mathcal{R}} &:= \{(\tilde{x},\tilde{y}) : \tilde{x} = L, -1 \le \tilde{y} \le \alpha\},\end{aligned}$$

where α and ω are given numbers in the interval $(-1, 0)$, $\tilde{\eta}$ is such that $\tilde{\eta}(L) = \alpha$ and $\tilde{\eta}(-L) = \omega$, and the sum of the fluxes of the flow through the left and right vertical boundaries is zero.

Problem 16.3. [Flow Over A Plate With Reservoir] For a given dimensionless length $L > 0$ (as in Problem 16.2 where it is half the length of the plate), approximate the function $\tilde{\eta}$ whose graph is the steady state free surface for the solution of system (16.44) over the two-dimensional domain $\tilde{\Omega}$ bounded by line segments connecting in order the corners

$$(-c_1L,\omega),\ (-c_1L,-1),\ (-(c_1-2)L,-1),\ (-(c_1-2)L,-c_2L),\ (c_1L,-c_2L),\ (c_1L,\alpha)$$

(for $c_1 > 2$ and $c_2 > 1$) and the curve

$$\tilde{\mathcal{S}} := \{(\tilde{x},\tilde{y}) : \tilde{y} = \tilde{\eta}(\tilde{x}),\ -c_1L < \tilde{x} < c_1L\},$$

where α and ω are given numbers in the interval $(-1, 0)$, $\tilde{\eta}$ is such that $\tilde{\eta}(c_1L) = \alpha$ and $\tilde{\eta}(-c_1L) = \omega$, and where the gradient $(\partial\tilde{\phi}/\partial\tilde{x}, \partial\tilde{\phi}/\partial\tilde{y})$ is specified on the left vertical boundary (connecting the first two corners) of $\tilde{\Omega}$ and the horizontal bottom of the reservoir (connecting the fourth and fifth corners) such that the sum of its fluxes, into the bottom of the reservoir and out of the left boundary, is zero.

For the Tiger Fountain, the physical length is half the plate slice length

$$0.238125\,\mathrm{m},$$

the water depth at the upstream end of the plate is

$$\mathrm{wd} = 0.0174625\,\mathrm{m},$$

and the water depth at the downstream end is

$$\text{dswd} = 0.0142874\,\text{m}\,.$$

Using wd for scaling, the dimensionless length L is

$$L = 13.6364,$$

the coefficients for the reservoir case are

$$c_1 = 4.84, \quad c_2 = 6.4$$

(corresponding to a width of 6 feet and a depth of 4 feet), and the velocity conversion factor γ from dimensionless velocity to physical velocity (measured in meters/sec) is

$$\gamma = \sqrt{g\,\text{wd}} = 0.413682.$$

There are side conditions for the Tiger Fountain Problem 16.2 that arise from physical considerations. The *dimensionless* quantity corresponding to water depth is approximately unity along the flat plate. The dimensionless quantity $s_{\mathcal{R}}$ corresponding to surface speed at the right boundary—which is the quantity $((\tilde{\phi}_{\tilde{x}})^2 + (\tilde{\phi}_{\tilde{y}})^2)^{1/2}$ evaluated at the intersection of $\mathcal{R}$ and the (dimensionless) surface—is less than the corresponding quantity $s_{\mathcal{L}}$ at the left boundary. Also, $s_{\mathcal{R}}$ (on average) is approximately 0.736798 (surface speed/γ). The dimensionless boundary condition corresponding to the velocity profile on $\mathcal{R}$ is of the form $(-a_{\mathcal{R}}, b_{\mathcal{R}})$, where $a_{\mathcal{R}} : [-1, \tilde{\eta}(L)] \to [0, \infty)$ and $b_{\mathcal{R}} : [-1, \tilde{\eta}(L)] \to (-\infty, \infty)$ with $a_{\mathcal{R}}(\tilde{\eta}(L))^2 + b_{\mathcal{R}}(\tilde{\eta}(L))^2 = s_{\mathcal{R}}^2$. Likewise, the velocity profile on $\mathcal{L}$ is of the form $(a_{\mathcal{L}}, b_{\mathcal{L}})$ with $a_{\mathcal{L}} : [-1, \tilde{\eta}(-L)] \to \mathbb{R}$ and $b_{\mathcal{L}} : [-1, \tilde{\eta}(-L)] \to (-\infty, \infty)$ with $a_{\mathcal{L}}(\tilde{\eta}(-L))^2 + b_{\mathcal{L}}(\tilde{\eta}(-L))^2 = s_{\mathcal{R}}^2$. Also, the total flux through $\tilde{\Omega}$ must vanish; that is,

$$\int_{-1}^{\tilde{\eta}(-L)} a_{\mathcal{L}}(\tilde{y})\, d\tilde{y} = \int_{-1}^{\tilde{\eta}(L)} a_{\mathcal{R}}(\tilde{y})\, d\tilde{y}.$$

Similar considerations apply in Problem 16.3.

The choice of the side conditions is important if we wish to reproduce observations. The main reason to conduct numerical experiments with our model is to determine how the side conditions affect the steady state surface.

We should not expect Euler flow to be physical at the plate boundary where the physical flow velocity must be zero. Euler flow might produce a nonzero velocity parallel to the plate. In the Euler approximation, a (supposedly thin) boundary layer that respects the viscosity is ignored and velocity fields that have nonzero horizontal components at the bottom are allowed. For Problem 16.2, we might imagine a profile that has a positive vertical component near the boundary on the right boundary to mimic the flow moving up over the plate and a negative vertical component near the surface corresponding to water falling from the reservoir onto the plate due to the gravitational force. The velocity profile at the upstream end of the plate is determined automatically in Problem 16.3 .

16.3 THE BOUNDARY ELEMENT METHOD (BEM)

We will describe a numerical method that is well-suited to solving the Laplace equation in a bounded domain. The idea is to take advantage of a basic principle from analysis: A harmonic function on a bounded domain is determined by its behavior on the boundary. Using this fact, we can reduce the dimension of the computation on a three-dimensional domain to two dimensions and on a two-dimensional domain to one dimension.

16.4 BOUNDARY INTEGRAL REPRESENTATION

Let Ω be a bounded open domain in two- or three-dimensional space whose boundary $\partial\Omega$ is locally the graph of a class C^1 function (that is, a function all of whose first-order partial derivatives are defined and continuous), and suppose that u is a harmonic function on Ω; that is, $\Delta u = 0$, where Δ is the Laplace operator. In three dimensions with *rectangular coordinates* (x, y, z), the Laplacian Δ of the function u is given by $u_{xx} + u_{yy} + u_{zz}$. We will also use the divergence and gradient operators. The divergence div acts on vector fields. In rectangular coordinates it is given on the vector field X with components (X^1, X^2, X^3) by $\operatorname{div} X = X^1_x + X^2_y + X^3_z$. The gradient grad acts on functions. In rectangular coordinates it is given on the function u by $\operatorname{grad} u = (u_x, u_y, u_z)$. Recall also the alternative notation $\nabla \cdot X$ for the divergence and ∇u for the gradient.

We will need an important result about harmonic functions: A harmonic function has continuous partial derivatives of all orders. The simplest proof uses a basic fact from complex analysis: A harmonic function is the real part of a holomorphic function. This result was used previously to construct the stream function.

Our goal is to construct a representation of harmonic functions on Ω by their behavior on $\partial\Omega$.

The first step of the construction is a basic identity in vector calculus called Green's second identity. Let us start with two functions v and w, which are class C^2 on Ω (all partial derivatives up to and including second-order partials exist and are continuous), and note the following version of Leibniz's rule:

$$\operatorname{div}(w \operatorname{grad} v) = \operatorname{grad} w \cdot \operatorname{grad} v + w\Delta v. \tag{16.45}$$

Of course, we also have

$$\operatorname{div}(v \operatorname{grad} w) = \operatorname{grad} v \cdot \operatorname{grad} w + v\Delta w$$

and

$$\operatorname{grad} w \cdot \operatorname{grad} v = \operatorname{grad} v \cdot \operatorname{grad} w.$$

Thus, it follows that

$$v\Delta w - w\Delta v = \operatorname{div}(v \operatorname{grad} w) - \operatorname{div}(w \operatorname{grad} v)$$

and by integration over Ω,

$$\int_\Omega v\Delta w - w\Delta v \, dV = \int_\Omega \operatorname{div}(v \operatorname{grad} w) - \operatorname{div}(w \operatorname{grad} v) \, dV.$$

By an application of the divergence theorem to the right-hand side of the last equation (where N is the outer unit normal on $\partial\Omega$), we obtain Green's second identity

$$\int_\Omega v\Delta w - w\Delta v \, dV = \int_{\partial\Omega} v \operatorname{grad} w \cdot N - w \operatorname{grad} v \cdot N \, dS, \tag{16.46}$$

which is valid in two- and three-dimensional spaces as long as the boundary integral is computed using the *positive* orientation on $\partial\Omega$. For this reason, we will assume that all boundary integrals are computed using positive orientation.

The second step in our construction involves the fundamental solution of Laplace's equation on the space in which Ω resides. In two dimensions, view $q \in \mathbb{R}^2$ as a parameter and define $u^* : \mathbb{R}^2 \to \mathbb{R}$ by

$$u^*(p,q) = \begin{cases} -\frac{1}{2\pi} \ln |p-q|, & p \neq q, \\ 0, & p = q, \end{cases} \tag{16.47}$$

where $|p-q|$ is the length of the vector $p-q$. A direct calculation shows that, for each q, the gradient of the function $p \mapsto u^*(p,q)$ is

$$\operatorname{grad} u^*(p,q) = -\frac{1}{2\pi|p-q|^2}(p-q).$$

In three dimensions, with $q \in \mathbb{R}^3$ viewed as a parameter, define $u^* : \mathbb{R}^3 \to \mathbb{R}$ by

$$u^*(p,q) = \begin{cases} \frac{1}{4\pi|p-q|}, & p \neq q, \\ 0, & p = q, \end{cases} \tag{16.48}$$

and note that

$$\operatorname{grad} u^*(p,q) = -\frac{1}{4\pi|p-q|^3}(p-q).$$

In two- and three-dimensions the function $p \mapsto u^*(p,q)$ is harmonic for all $p \neq q$. This fact is proved by simply computing the divergence of $\operatorname{grad} u^*$ (see Exercise 16.16).

Proposition 16.4. For two or three dimensions, suppose that B_ϵ is the ball of radius $\epsilon > 0$ centered at a point q and N is the outer unit normal on ∂B_ϵ. If ϕ is a class C^1 function defined on an open set containing the closure of B_ϵ, then

$$\lim_{\epsilon \to 0} \int_{\partial B_\epsilon} u^*(p,q) \operatorname{grad} \phi(p) \cdot N(p) - \phi(p) \operatorname{grad} u^*(p,q) \cdot N(p)\, dS(p) = \phi(q).$$

Proof. We will prove the result for two-dimensional space; the idea of the proof is the same in three-dimensional space (see Exercise 16.15).

For $p \in \partial B_\epsilon$, we have

$$N(p) = \frac{1}{|p-q|}(p-q) = \frac{1}{\epsilon}(p-q).$$

Using the definition of u^*, the formula for N, the Cauchy–Schwarz inequality, and the length $2\pi\epsilon$ of ∂B_ϵ, note that

$$\begin{aligned}|\int_{\partial B_\epsilon} u^*(p,q)\,\mathrm{grad}\,\phi(p)\cdot N(p)\,dS(p)| &\leq \frac{1}{2\pi}\sup_{p\in B_\epsilon}|\mathrm{grad}\,\phi(p)|\int_{\partial B_\epsilon}|\ln\epsilon|\,dS(p)\\ &\leq \sup_{p\in B_\epsilon}|\mathrm{grad}\,\phi(p)|\epsilon|\ln\epsilon|.\end{aligned}$$

Thus,

$$\lim_{\epsilon\to 0}\int_{\partial B_\epsilon} u^*(p,q)\,\mathrm{grad}\,\phi(p)\cdot N(p)\,dS(p) = 0.$$

Using the formula for the grad $u^*(p)$, we have that

$$\int_{\partial B_\epsilon} -\phi(p)\,\mathrm{grad}\,u^*(p,q)\cdot N(p)\,dS(p) = \frac{1}{2\pi\epsilon}\int_{\partial B_\epsilon}\phi(p)\,dS(p)$$

and

$$\begin{aligned}\frac{1}{2\pi\epsilon}\int_{\partial B_\epsilon}\phi(p)\,dS(p) &= \frac{1}{2\pi\epsilon}\Big(\int_{\partial B_\epsilon}\phi(p)-\phi(q)\,dS(p)+\int_{\partial B_\epsilon}\phi(q)\,dS(p)\Big)\\ &= \frac{1}{2\pi\epsilon}\Big(\int_{\partial B_\epsilon}\phi(p)-\phi(q)\,dS(p)+\phi(q)\int_{\partial B_\epsilon}dS(p)\Big)\\ &= \frac{1}{2\pi\epsilon}\int_{\partial B_\epsilon}\phi(p)-\phi(q)\,dS(p)+\phi(q).\end{aligned}$$

Also,

$$|\frac{1}{2\pi\epsilon}\int_{\partial B_\epsilon}\phi(p)-\phi(q)\,dS(p)| \leq \sup_{p\in B_\epsilon}|\phi(p)-\phi(q)|.$$

Using the continuity of ϕ and passing to the limit as ϵ goes to zero, it follows that

$$\lim_{\epsilon\to 0}\int_{\partial B_\epsilon} -\phi(p)\,\mathrm{grad}\,u^*(p,q)\cdot N(p)\,dS(p) = \phi(q).$$

□

The following result is the desired representation theorem for harmonic functions.

Theorem 16.5. Suppose that Ω is a bounded open domain in two- or three-dimensional space whose boundary $\partial\Omega$ is locally the graph of a C^1 function. If ϕ is harmonic in Ω and C^1 on some open set containing the closure of Ω and $q \in \Omega$, then

$$\phi(q) = \int_{\partial\Omega} u^*(p,q)\,\text{grad}\,\phi(p)\cdot N(p) - \phi(p)\,\text{grad}\,u^*(p,q)\cdot N(p)\,dS(p), \tag{16.49}$$

where u^* is given by Eq. (16.47) in two-dimensional space and by Eq. (16.48) in three dimensions.

Proof. Let $\epsilon > 0$ be sufficiently small so that the closure of the ball B_ϵ centered at q with radius ϵ is contained in Ω. With B_ϵ excised from Ω and N the outer unit-normal on the boundary of the resulting set, Green's second identity applied to the harmonic functions ϕ and $p \mapsto u^*(p,q)$—under the additional assumption that ϕ is harmonic on an open set containing the closure of Ω—yields the identity

$$\begin{aligned} 0 = &\int_{\partial\Omega} u^*(p,q)\,\text{grad}\,\phi(p)\cdot N(p) - \phi(p)\,\text{grad}\,u^*(p,q)\cdot N(p)\,dS(p) \\ &+ \lim_{\epsilon\to 0}\int_{\partial B_\epsilon} u^*(p,q)\,\text{grad}\,\phi(p)\cdot N(p) - \phi(p)\,\text{grad}\,u^*(p,q)\cdot N(p)\,dS(p). \end{aligned} \tag{16.50}$$

The limit in the last formula is $-\phi(q)$ by Proposition 16.4 because here the outer normal on the set $\Omega \setminus B_\epsilon$ is the inner normal on B_ϵ. By substituting this value for the indicated limit and rearranging, we obtain the desired representation. The assumption that ϕ is harmonic on an open set containing the closure of Ω is eliminated by applying the result just proved to a family of domains $\tilde{\Omega}$ (with C^1 boundaries) whose closures are contained in Ω and such that the family converges to Ω in a suitable sense by passing to the limit in the resulting representations as $\tilde{\Omega}$ converges to Ω. □

Remark 1. The function $p \mapsto u^*(p,q)$ is not a solution of Laplace's equation on $\mathbb{R}^2$ in the classical sense because p is discontinuous at $p = q$. Our redefinition of $u^*(q,q) = 0$ is simply a notational convenience. A more accurate statement, which requires the theory of distributions to make precise, is

$$\Delta u^*(p,q) = -\delta(p-q)$$

where δ is the Dirac-delta function. This theory can also be used to obtain our representation.

Exercise 16.13. Suppose that ϕ is harmonic in a domain containing the ball of radius r centered at a. Show that the value of ϕ at a is the average of ϕ over the boundary of the ball. This is called the mean value property of harmonic functions.

16.5 BOUNDARY INTEGRAL EQUATION

Representation formula (16.49) tells us how to compute the value of a harmonic function at an interior point of a domain when its values and normal derivative—that is, the dot product of the gradient of the function and the normal—are known on the boundary of the domain. The BEM is based on an additional limit process applied to our representation formula that determines a new representation in the limit as interior points approach a point on the boundary. This process works equally well in two and three dimensions. We will carry out the procedure for the two-dimensional case.

To determine the behavior of the representation as interior points approach a boundary point p, we enlarge the domain Ω by adding an open disk centered at p with radius ϵ. Our representation formula [Eq. (16.49)] is valid on the new domain Ω_ϵ with its new boundary (whose generic point is again called p) and the original boundary point (which is an interior point of Ω_ϵ) relabeled q. The idea is to pass to the limit as ϵ goes to zero.

Let Γ_ϵ denote the portion of the original boundary of Ω remaining with the disk B_ϵ removed and γ_ϵ the closure of the portion of the boundary of B_ϵ not contained in the closure of Ω. The new boundary is $\Gamma_\epsilon \cup \gamma_\epsilon$ and

$$\begin{aligned}\phi(q) &= \int_{\Gamma_\epsilon \cup \gamma_\epsilon} u^*(p,q) \operatorname{grad} \phi(p) \cdot N(p) - \phi(p) \operatorname{grad} u^*(p,q) \cdot N(p)\, dS(p) \\ &= \int_{\Gamma_\epsilon} u^*(p,q) \operatorname{grad} \phi(p) \cdot N(p) - \phi(p) \operatorname{grad} u^*(p,q) \cdot N(p)\, dS(p) \\ &\quad + \int_{\gamma_\epsilon} u^*(p,q) \operatorname{grad} \phi(p) \cdot N(p) - \phi(p) \operatorname{grad} u^*(p,q) \cdot N(p)\, dS(p).\end{aligned}$$

To finish the construction, we use exactly the same ideas as in the proof of Proposition 16.4. Indeed, the computation of the second limit is nearly identical except that the length of the perimeter of the disk is replaced by the length of γ_ϵ. In case q is at a point where the original boundary is C^1, which we have been assuming so far, the length is asymptotically $\pi\epsilon$ because there is a well-defined tangent line that will be a diameter of the disk as ϵ decreases to zero. In case q is a corner point of a piecewise C^1 boundary, which will work equally well for what we have proved so far with some technical additions to our proofs, the length of γ_ϵ is asymptotically $a\epsilon$, where a is the outer angle at the corner. Using these observations and steps in the proof of Proposition 16.4, it follows that

$$\lim_{\epsilon\to 0}\int_{\gamma_\epsilon} u^*(p,q)\,\text{grad}\,\phi(p)\cdot N(p) - \phi(p)\,\text{grad}\,u^*(p,q)\cdot N(p)\,dS(p) = \frac{a}{2\pi}\phi(q).$$

Let us examine separately the limits

$$\lim_{\epsilon\to 0}\int_{\Gamma_\epsilon} u^*(p,q)\,\text{grad}\,\phi(p)\cdot N(p)\,dS(p),$$
$$\lim_{\epsilon\to 0}\int_{\Gamma_\epsilon} -\phi(p)\,\text{grad}\,u^*(p,q)\cdot N(p)\,dS(p). \tag{16.51}$$

We must take into account the singularity of the integrands at $p = q$. The first integral has a logarithmic singularity. After parameterizing the boundary appropriately, we will be faced with computing two limits

$$\lim_{\epsilon\to 0}\int_{-a}^{-\epsilon} \ln|t|h(t)\,dt, \qquad \lim_{\epsilon\to 0}\int_{\epsilon}^{a} \ln|t|k(t)\,dt$$

where h and k are continuous functions with $h(0) = k(0)$, and where we can assume $0 < a < 1$. Both limits exist. For the second integral, the continuous function $|k|$ is bounded by some number $M > 0$; therefore, we have

$$\begin{aligned}|\int_{\epsilon}^{a} \ln|t|k(t)\,dt| &\le M\int_{\epsilon}^{a}(-\ln|t|)\,dt\\ &\le -M(a\ln a - a - \epsilon\ln\epsilon + \epsilon).\end{aligned}$$

Using the comparison test and a basic fact from analysis (absolute convergence implies convergence), it follows that the integral converges. The same is true for the first integral. Thus, we have that

$$\lim_{\epsilon\to 0}\int_{\Gamma_\epsilon} u^*(p,q)\,\text{grad}\,\phi(p)\cdot N(p)\,dS(p) = \int_{\partial\Omega} u^*(p,q)\,\text{grad}\,\phi(p)\cdot N(p)\,dS(p),$$

where the right-hand side is an improper convergent integral.

For the second integral in display (16.51) we again face two limits

$$\lim_{\epsilon \to 0} \int_{-a}^{-\epsilon} \frac{1}{|t|} h(t)\, dt, \qquad \lim_{\epsilon \to 0} \int_{\epsilon}^{a} \frac{1}{|t|} k(t)\, dt$$

where h and k are again continuous with $h(0) = k(0)$. In general, such limits do not exist. For example, suppose h is a nonzero constant and note that the integral

$$\int_0^1 \frac{1}{t}\, dt$$

is divergent. To proceed, we must obtain more properties of the functions h and k.

Let q be a point on $\partial\Omega$ and choose coordinates so that $\partial\Omega$ near q is the graph of a continuous function $f : (-a, a) \to \mathbb{R}$ with $f(0) = 0$ that is continuously differentiable except possibly at the origin and such that its left-hand and right-hand derivatives exist at the origin.

The original integral (the second integral in display (16.51)) over the curve Γ_ϵ can be rewritten as a sum of three integrals: one integral over the portion of the boundary not parametrized by f (which is not a singular integral) and a sum of two integrals over the portion parametrized by f. This latter sum (up to a change of sign depending on the *direction* of the outer normal, which is either $(f'(t), -1)$ or $(-f'(t), 1)$ for both integrals) is

$$I_1 + I_2 := \int_{-a}^{-\epsilon} \phi(t, f(t)) \frac{tf'(t) - f(t)}{t^2 + (f(t))^2}\, dt + \int_{\epsilon}^{a} \phi(t, f(t)) \frac{tf'(t) - f(t)}{t^2 + (f(t))^2}\, dt.$$

Because f is continuously differentiable and $f(0) = 0$, we have (for $t \neq 0$) the formula

$$f(t) = f(0) + t \int_0^1 f'(ts)\, ds = t \int_0^1 f'(ts)\, ds.$$

By substitution of this formula into I_2 and some simplification, this integral is rewritten as

$$I_2 = \int_{\epsilon}^{a} \phi(t, f(t)) \frac{\int_0^1 (f'(t) - f'(ts))\, ds}{t(1 + (\int_0^1 f'(ts)\, ds)^2)}\, dt.$$

For a continuously differentiable function f, we do not remove the singularity $1/t$ and it is possible that $\lim_{\epsilon\to 0} I_2$ does not exist. On the other hand, if f is just slightly smoother, then the improper integral does converge.

What should we mean by the phrase "slightly smoother?" Answer: The derivative of f is Hölder. A function g is called Hölder of order α if $0 < \alpha \leq 1$ and there is a positive constant M such that

$$|g(x) - g(y)| \leq M|x - y|^{\alpha}$$

for all x and y in the domain under consideration. In the special case $\alpha = 1$, the function g is called Lipschitz. We will say that a function (or the graph of a function) is class $C^{1,\alpha}$ if the function is continuously differentiable and its derivative is Hölder of order α.

If f is class $C^{1,\alpha}$, then

$$\Big(\int_0^1 (f'(t) - f'(ts))\, ds\Big) \leq M|t|^{\alpha} \int_0^1 |1 - s|\, ds$$

and our singularity in I_2 is $1/t^{1-\alpha}$. Because for $0 \leq \alpha \leq 1$ the improper integral $\int_0^1 t^{1-\alpha}$ converges, it follows that the improper integral I_2 converges. The same will be true for I_1.

By combining our results for the limit process we arrive at an important result.

Theorem 16.6 (Boundary Integral Formula for Laplace Equation). If Ω is a bounded open domain in $\mathbb{R}^2$ whose boundary $\partial\Omega$ is piecewise $C^{1,\alpha}$ with outer normal N and ϕ is a harmonic function on Ω that is continuous on $\partial\Omega$ and has a continuous normal derivative on $\partial\Omega$, then ϕ is given on $\partial\Omega$ by

$$\begin{aligned}(1 - \frac{a}{2\pi})\phi(q) &= \frac{1}{2\pi}\int_{\partial\Omega} \phi(p)\frac{1}{|p-q|^2}(p-q)\cdot N(p)\, dS(p)\\ &\quad - \frac{1}{2\pi}\int_{\partial\Omega} \ln|p-q| \operatorname{grad}\phi(p)\cdot N(p)\, dS(p), \qquad (16.52)\end{aligned}$$

where $a = \pi$ in case q is a noncorner point of the boundary and a is the angular measure of the exterior angle at q in case q is a corner point. The boundary integrals are singular (at q), but they are both convergent as improper integrals.

Remark 2. The boundary integral formula for Laplace's equation in three dimensions is

$$\begin{aligned}\phi(q) &= \frac{1}{2\pi}\int_{\partial\Omega}\phi(p)\frac{1}{|p-q|^3}(p-q)\cdot N(p)\,dS(p)\\ &\quad -\frac{1}{2\pi}\int_{\partial\Omega}\frac{1}{|p-q|}\operatorname{grad}\phi(p)\cdot N(p)\,dS(p). \qquad (16.53)\end{aligned}$$

Eqs. (16.52) and (16.53) are the theoretical basis for the boundary element numerical method. The idea is to approximate the boundary values of ϕ and its normal derivatives at the boundary points of the domain. Once these are determined, the interior values of the harmonic function ϕ are recovered (in the two-dimensional case) using the result of Theorem 16.5; that is, for $q \in \Omega$, we have

$$\phi(q) = \frac{1}{2\pi}\int_{\partial\Omega}\phi(p)\frac{1}{|p-q|^2}(p-q)\cdot N(p) - \ln|p-q|\operatorname{grad}\phi(p)\cdot N(p)\,dS(p), \qquad (16.54)$$

which is of course essentially the same as Eq. (16.52) except that the boundary integral is not singular.

16.6 DISCRETIZATION FOR BEM

For the practical implementation of a numerical method we must discretize Eq. (16.52) (or in the three-dimensional case Eq. (16.53)). There are many ways to accomplish this step for the two-dimensional case. We will consider only one of the simplest: We will approximate the boundary of Ω using line segments and approximate ϕ and $\operatorname{grad}\phi \cdot N$ by constants on each segment. This leads to a system of linear equations for the unknown values of ϕ on the boundary, which can be solved by Gaussian elimination.

Water Wave BEM Reformulation

In anticipation of applying the BEM to the water wave Problem 16.2, let us consider $L > 0$, a twice continuously differentiable function $\eta : [-L, L] \to (-1, 0)$, and the domain

$$\Omega := \{(x, y) : -1 < y < \eta(x), \quad -L < x < L\}.$$

Note that the left, right, and bottom boundaries of Ω are line segments; the top boundary is the graph of η. We will also specify the normal derivatives of the unknown harmonic function ϕ everywhere on the boundary. In other words, we will pose the Neumann problem for the Laplace equation on Ω. We could consider the Dirichlet problem (where the values of ϕ are specified on the boundary), or some combination of such boundary conditions. The essential feature is that one or the other of the two conditions is specified on the boundary. Also, in the special case of the Neumann problem, we must have a compatibility condition; it is the content of the next proposition.

Proposition 16.7. If ϕ is a harmonic function in Ω with unit outer normal N and normal derivative $\operatorname{grad}\phi \cdot N$ defined everywhere on $\partial\Omega$, then

$$\int_{\partial\Omega} \operatorname{grad}\phi(p) \cdot N(p)\, dS(p) = 0.$$

Proof. Apply the divergence theorem as follows

$$0=\int_{\Omega} \Delta\phi(p)\, dV(p)=\int_{\Omega} \operatorname{div}(\operatorname{grad}\phi)(p)\, dV(p)=\int_{\partial\Omega} \operatorname{grad}\phi(p)\cdot N(p)\, dS(p).$$

□

As in Problem 16.2, the top and bottom boundary conditions are the zero Neumann condition $\operatorname{grad}\phi(p)\cdot N(p) = 0$. In fact, on the bottom, $\operatorname{grad}\phi(p)\cdot N(p) = \phi_y(p)$, and on the top,

$$(\operatorname{grad}\phi\cdot N)(x,\eta(x)) = -\frac{1}{\sqrt{1+(\eta'(x))^2}}(\eta'(x)\phi_x(x,\eta(x))-\phi_y(x,\eta(x))).$$

Both of these normal derivatives are zero according to the steady state water wave equations.

We have some freedom in specifying the inlet and outlet conditions; the only restriction is the compatibility condition of Proposition 16.7 that requires the influx and outflux to be equal. The normal derivative at the inlet (the right-hand boundary $\mathcal{R}$) is given by a scalar function $\rho : [-1,\eta(L)] \to (-\infty,0)$ and at the outlet (left-hand boundary $\mathcal{L}$) by $\lambda : [-1,\eta(-L)] \to (0,\infty)$. In other words, the dot product of the gradient of ϕ and N is ρ at the right-hand boundary and λ at the left-hand boundary. The compatibility condition is

$$\int_{\mathcal{L}} \operatorname{grad}\phi(p)\cdot N(p)\, dS(p) + \int_{\mathcal{R}} \operatorname{grad}\phi(p)\cdot N(p)\, dS(p) = 0,$$

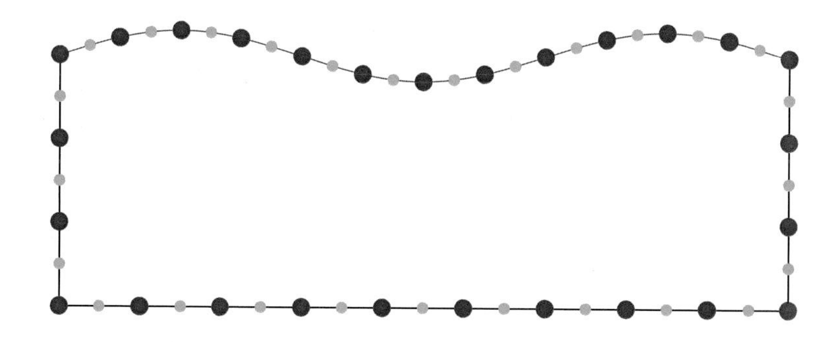

Fig. 16.14

where the orientations of $\mathcal{L}$ and $\mathcal{R}$ are compatible with the positive orientation of $\partial\Omega$. To compute the second line integral, we can parameterize $\mathcal{R}$ by the function $\gamma : [0, 1] \to \mathbb{R}^2$ given by

$$\gamma(t) = (L, t\eta(L) - (1 - t))$$

so that

$$\int_{\mathcal{R}} \operatorname{grad} \phi(p) \cdot N(p)\, dS(p) = \int_0^1 \rho(t\eta(L) - (1 - t))|1 + \eta(L)|\, dt.$$

With $y = t\eta(L) - (1 - t)$, the integral is

$$\int_1^{\eta(L)} \rho(y)\, dy.$$

The integral over $\mathcal{L}$ (with $\gamma(t) = (-L, -t + (1 - t)\eta(-L))$ is

$$\int_{-1}^{\eta(-L)} \lambda(y)\, dy.$$

Thus, the compatibility condition is

$$\int_{-1}^{\eta(L)} \rho(y)\, dy + \int_{-1}^{\eta(-L)} \lambda(y)\, dy = 0. \tag{16.55}$$

Boundary Elements and Nodes

The next step is to choose meshes on the four parts of the boundary, where the integration is counterclockwise around the boundary starting and ending at the top-left point. We will divide $\mathcal{L}$ into $m_{\mathcal{L}}$ equal length line segments (called boundary elements), $\mathcal{B}$ into $m_{\mathcal{B}}$ segments, and $\mathcal{R}$ into $m_{\mathcal{R}}$ segments. This makes perfect sense because each of these three parts of the boundary are themselves line segments. The surface $\mathcal{S}$ might be curved, so we must specify what we mean by equal length segments. Perhaps marking points on $\mathcal{S}$ at equal arc-lengths along this curve is a good way to discretize, but it is simpler to divide the parameterization interval into $m_{\mathcal{S}}$ equal line segments. The natural parameterization interval (for nonbreaking waves) is $[-L, L]$ corresponding to the parameterization $x \mapsto (x, \eta(x))$ of $\mathcal{S}$. Each of the $m_{\mathcal{S}}$ segments on $[-L, L]$ corresponds to a segment of $\mathcal{S}$. For simplicity, we will replace this segment by a line interval connecting the end points of the segment. Let us also choose the midpoint of each boundary element as in Fig. 16.14 and call it the corresponding node. There are exactly

$$M := m_{\mathcal{L}} + m_{\mathcal{B}} + m_{\mathcal{R}} + m_{\mathcal{S}}$$

nodes.

The value of a in the boundary integral formula [Eq. (16.52)] is $a = \pi$ because each boundary node lies on a line segment. Also, from the Neumann boundary condition, we know the value of the normal derivative $\operatorname{grad}\phi \cdot N$ at each node. We wish to determine the unknown values of ϕ at each node. To do this, we will make one final approximation: The value of ϕ and its normal derivative $\operatorname{grad}\phi \cdot N$ are constant on each boundary element according to their values at the corresponding node.

BEM Linear System

Suppose that $\{\mathcal{J}_i\}_{i=1}^{M}$ is the set of boundary elements with corresponding nodes $\{q_i\}_{i=1}^{M}$. Also, for notational convenience, let $\phi_i := \phi(q_i)$ and $d\phi_i := (\operatorname{grad}\phi_i \cdot N)(q_i)$. Under our assumptions, the boundary integral formula [Eq. (16.52)] yields

$$\phi_i = \frac{1}{\pi}\sum_{j-1}^{M}\Big(\phi_j \int_{\mathcal{J}_i} \frac{(p - q_i)\cdot N(p)}{|p - q_i|^2}\, dS(p) - d\phi_j \int_{\mathcal{J}_j} \ln|p - q_i|\, dS(p)\Big). \tag{16.56}$$

Note that there are exactly M equations for the M unknowns $\{\phi_i\}_{i=1}^{M}$. We would expect that this system of *linear* equations has a unique solution, which is our desired approximation to ϕ on $\partial\Omega$. Unfortunately, there is a complication: Our Neumann boundary value problem (BVP) does not have a unique solution; rather, it has an infinite number of solutions because the addition of a nonzero constant to a given solution produces a different solution. For this reason, the system of equations we have just defined—whose solutions are supposed to approximate solutions of the Neumann problem—will be singular or nearly singular; thus, our linear system will either not have a solution or be highly ill-conditioned. A remedy—which we will adopt—is to assign a value to ϕ at one of the nodes and replace the known value of the normal derivative $d\phi$ at this node by a new unknown.

The linear system is made into a numerical linear system by evaluating the integrals in Eq. (16.56):

$$\mathcal{P} := \int_{\mathcal{J}} \frac{(p-q)\cdot N(p)}{|p-q|^2}\, dS(p),$$
$$\mathcal{Q} := \int_{\mathcal{J}} \ln|p-q|\, dS(p).$$

For this computation, suppose that (x_1, y_1) and (x_2, y_2) are the endpoints of the element $\mathcal{J}$ with the orientation direction from the first to the second point, and parameterize $\mathcal{J}$ with the function $\gamma : [0,1] \to \mathbb{R}^2$ given by

$$\gamma(t) = (\gamma_1(t), \gamma_2(t)) = ((1-t)x_1 + tx_2, (1-t)y_1 + ty_2).$$

The outer normal along $\mathcal{J}$ is obtained by rotating the direction vector of the line segment $\mathcal{J}$ clockwise by 90° to obtain the normal direction $(y_2 - y_1, -(x_2 - x_1))$. The normal N along $\mathcal{J}$ is the unit vector in this direction.

With some simplification, and the notation

$$r_1 := x_1 - q_1, \quad r_2 := y_1 - q_2, \quad z_1 := x_2 - x_1, \quad z_2 := y_2 - y_1,$$

we have

$$\mathcal{P} = (r_1 z_2 - r_2 z_1)\int_0^1 \frac{1}{(r_1 + z_1 t)^2 + (r_2 + z_2 t)^2}\, dt,$$
$$\mathcal{Q} = \frac{\sqrt{z_1^2 + z_2^2}}{2}\int_0^1 \ln((r_1 + z_1 t)^2 + (r_2 + z_2 t)^2)\, dt. \tag{16.57}$$

The integrals in display (16.57) are regular if q does not belong to the element and singular if q does belong to the element. In all cases, there are efficient numerical methods for evaluating the integrals. As you might expect, numerical integration is a vast and well-developed subject that deserves careful attention because it is a core issue encountered in scientific computing. Indeed, the implementation of the BEM is a prime example. We will avoid a discussion of numerical integration here due to a pleasant surprise: for the special case we are considering, all integrals can be evaluated exactly. We will list the results and leave the verifications as exercises. Warning: Be careful not to integrate across a singularity without taking the singularity into account.

If q belongs to the line containing the element, which includes the case where the integral is singular, then

$$\mathcal{P} = 0.$$

Indeed, if q is on the line containing the element, then $\gamma(t) - q$ is a vector parallel to the linear element; therefore, $(\gamma(t) - q) \cdot N(\gamma(t)) = 0$. On the other hand, if q is not on the line containing the element, then

$$\mathcal{P} = \arctan\left(\frac{z_1^2 + z_2^2 + r_1 z_1 + r_2 z_2}{r_1 z_2 - r_2 z_1}\right) - \arctan\left(\frac{r_1 z_1 + r_2 z_2}{r_1 z_2 - r_2 z_1}\right) \tag{16.58}$$

If q is not on the line containing the element, then

$$\begin{aligned}\mathcal{Q} = \frac{\sqrt{z_1^2 + z_2^2}}{2}\Big(\Big(\frac{r_1 z_1 + r_2 z_2}{z_1^2 + z_2^2} + 1\Big) \ln((r_1 + z_1)^2 + (r_2 + z_2)^2)\\ - 2 - \frac{r_1 z_1 + r_2 z_2}{z_1^2 + z_2^2} \ln(r_1^2 + r_2^2)\\ + 2\mathcal{P}\frac{(z_1^2 + z_2^2)(r_1^2 + r_2^2) - (r_1 z_1 + r_2 z_2)^2}{(z_1^2 + z_2^2)(r_1 z_2 - r_2 z_1)}\Big).\end{aligned} \tag{16.59}$$

If q is on the line containing the element but not on the element, then

$$\begin{aligned}\mathcal{Q} = \frac{\sqrt{z_1^2 + z_2^2}}{2}\Big(\Big(\frac{r_1 z_1 + r_2 z_2}{z_1^2 + z_2^2} + 1\Big) \ln((r_1 + z_1)^2 + (r_2 + z_2)^2)\\ - 2 - \frac{r_1 z_1 + r_2 z_2}{z_1^2 + z_2^2} \ln(r_1^2 + r_2^2)\\ + 2\frac{(z_1^2 + z_2^2)(r_1^2 + r_2^2) - (r_1 z_1 + r_2 z_2)^2}{(z_1^2 + z_2^2 + r_1 z_1 + r_2 z_2)(r_1 z_1 + r_2 z_2)}\Big).\end{aligned} \tag{16.60}$$

And if q is on the element, then

$$\mathcal{Q} = \frac{\sqrt{z_1^2 + z_2^2}}{2}(\ln(r_1^2 + r_2^2) - 2). \tag{16.61}$$

We have all the ingredients necessary to implement the BEM for our special case. Although there are many possible ways to write a computer code to solve our problem, they will all share some common features in the following list.

Specification of the mesh and nodes: The mesh consists of the end points of elements and the nodes are the midpoints of the elements.
Function evaluations: Natural functions for the code have input consisting of the two end points of an element and the corresponding node. The output must incorporate logic to use the explicit values of $\mathcal{P}$ and $\mathcal{Q}$ according to the relation between the node and the element. For example, if the node is on the line containing the element, then $\mathcal{P}$ is zero, otherwise it is computed using the formula in Eq. (16.58). The logic for $\mathcal{Q}$ is slightly more complicated: $\mathcal{Q}$ is computed using Eq. (16.59) in case the node is not on the line containing the element; $\mathcal{Q}$ is computed using Eq. (16.60) in case the node is on the line but not on the element; and $\mathcal{Q}$ is computed using Eq, (16.61) in case the node is on the element.
Preliminary matrix assembly: The linear equations obtained from the discretization using Eq. (16.56) must be assembled. They may be expressed in the matrix form $(I - \frac{1}{\pi}B)\Phi = \frac{1}{\pi}K$, where

$$b_{ij} = \int_{\mathcal{J}_j} \frac{(p - q_i) \cdot N(p)}{|p - q_i|^2}\, dS(p),$$

$$c_{ij} = \int_{\mathcal{J}_j} \ln|p - q_i|\, dS(p),$$

and $K := Cd\Phi$. For our Neumann boundary conditions, the matrix $I - \frac{1}{\pi}B$ will be nearly singular. This fact may be used as an internal check while debugging code.
Neumann boundary data: The bottom and surface have zero Neumann data; that is, $d\phi_j = 0$ at all corresponding nodes. The vertical sides have given Neumann data.
System matrix assembly: The system of linear equations in component form is given by

$$\phi_i - \frac{1}{\pi}\sum_{j=1}^{m-1} b_{ij}\phi_j + \frac{1}{\pi}c_{im}d\phi_m = -\frac{1}{\pi}\Big(- b_{im}\phi_m + \sum_{j=1}^{m-1} c_{ij}d\phi_j\Big),$$

$$-\frac{1}{\pi}\sum_{j=1}^{m-1} b_{mj}\phi_j + \frac{1}{\pi}c_{mm}d\phi_m = -\frac{1}{\pi}\Big(- (b_{mm} - \pi)\phi_m + \sum_{j=1}^{m-1} c_{mj}d\phi_j\Big),$$

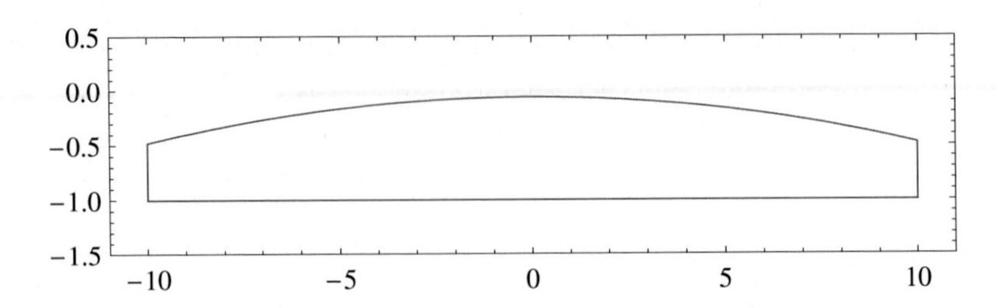

Fig. 16.15 The boundary of the domain for the test example corresponding to data (16.64) *is depicted.*

where the first equation holds for $i = 1, 2, \ldots, m-1$ and the second for $i = m$. In other words, the system matrix A is obtained from the matrix $I - \frac{1}{\pi}B$ by replacing its last column by $\{c_{im}/\pi\}_{i=1}^{m}$.
System right-hand side: The vector b in our system $Ax = b$ is obtained by first replacing the last column of the matrix C with the transpose of the vector $(-b_{1m}, -b_{2m}, \ldots, -b_{m-1,m}, -(b_{mm} - \pi))$ and then muliplying $-1/\pi$ times the product of the resulting matrix and the vector $d\Phi$ with its last element replaced by ϕ_m.
Solution of linear equations: The linear system $Ax = b$ can be solved by Gaussian elimination. The numerical approximation of solutions of linear systems is a fundamental issue in numerical analysis (and in abstract mathematics). The fundamental algorithm is Gaussian elimination. It is the method of choice for BEM because the linear systems that arise are full (that is most elements of the system matrix are nonzero) and the matrices have no special structure (for example, they are not symmetric or banded). Special methods of solution are advantageous for special types of matrices.

Test Examples

A BEM code may be tested with many possible examples as exact solutions of the Laplace equation abound.

A code should be debugged with an example where all calculations can be easily checked by hand. For instance, a simple choice is the potential $\phi(x, y) = -x$ in the rectangular domain with corners $(-10, -1/2)$, $(-10, -1)$, $(10, -1)$ and $(10, -1/2)$. In this case the harmonic conjugate (the stream function) is $\psi(x, y) = -y$. Thus, the top and bottom of the domain are streamlines. In particular, the function η is given by $\eta(x) = -1/2$. The functions λ and ρ are given by $\lambda(y) = 1$ and $\rho(y) = -1$.

A more realistic and interesting example, which is important in fluid dynamics, is potential flow with one source and one sink placed on a horizontal line. The suggested complex potential is a modification of the complex function $z \mapsto 1/2 \log((z-1)/(z+1))$, which has singularities—the source and sink—at $z = 1$ and $z = -1$, respectively. To fit more naturally into the context of our water wave problem, we may use instead the function

$$z \mapsto \frac{1}{2} \log \Big(\frac{z+i-h}{z+i+h}\Big),$$

which has singularities at $z = -h - i$ and $z = h - i$. The analysis requires some knowledge of complex variables, but the result used for testing code does not.

The real part of the complex potential (the potential for our flow) is

$$\phi(x,y) = \frac{1}{2} \ln \Big(\frac{(h-x)^2 + (1+y)^2}{(h+x)^2 + (1+y)^2}\Big) \tag{16.62}$$

and the imaginary part (the stream function) is

$$\phi(x,y) = \arctan \Big(\frac{2(1+y)h}{x^2 - h^2 + (1+y)^2}\Big). \tag{16.63}$$

The level sets of ϕ are the streamlines. Choose a positive value of h, say $h = 15$, and construct a domain (see Fig. 16.15) with vertical sides symmetric with respect to the y-axis that would meet the x-axis at points with x-coordinates having absolute value less than h, say -10 and 10, a horizontal bottom along the line $y = -1$, and a top consisting of part of a streamline that lies below the x-axis, say the level set $\{(x,y) : \phi(x,y) = -1/8\}$. Let us also note that the bottom of our domain is part of the streamline $\{(x,y) : \phi(x,y) = 0\}$. The relevant data for these choices is

$$\begin{aligned}
\phi(x,y) &= \frac{1}{2} \ln \Big(\frac{(15-x)^2 + (1+y)^2}{(15+x)^2 + (1+y)^2}\Big),\\
\eta(x) &= -\cot(\frac{1}{8})\big(15 + \tan(\frac{1}{8}) - (225(1+\tan^2(\frac{1}{8})) - \tan^2(\frac{1}{8})x^2)^{1/2}\big),\\
\rho(y) &= -\frac{30(125 + (1+y)^2)}{50626 + (100 + (1+y)^2)^2 + 225(2(1+y)^2 - 200)},\\
\lambda(y) &= -\rho(y).
\end{aligned} \tag{16.64}$$

Nodes Per Side	Absolute Error	Relative Error
$n = 2$	2.28167	0.664819
$n = 4$	2.40385	0.483266
$n = 8$	2.18281	0.305717
$n = 16$	1.60163	0.157182
$n = 32$	0.913901	0.0630917
$n = 64$	0.445372	0.0216804
$n = 128$	0.211173	0.00725832
$n = 256$	0.102029	0.00247789

Table 16.1 Errors are listed for test case (16.64) on the domain depicted in Fig. 16.15. The absolute error is the Euclidean distance between the exact and computed values of the potential ϕ on the boundary at all but the last node. The relative error is this distance divided by the length of the vector of exact values at these nodes.

Errors are listed in Table 16.1 for a numerical experiment with data (16.64) on the domain depicted in Fig. 16.15. The relative error is less than 0.5% with 32 nodes per side. On the other hand, the method seems to be approximately first order (that is, the error is proportional to the mesh size). Indeed, doubling the number of nodes reduces the error by approximately 1/2.

An Algorithm to Approximate Steady State Water Waves

The BEM is a tool that can be used to approximate solutions of the Laplace equation with Neumann (Dirichlet or mixed) boundary conditions on a bounded domain. For Problem 16.2, we can approximate solutions of a version of BVP (16.44): Given $\tilde{\eta}$, we can approximate the function $\tilde{\phi}$ that satisfies the first three conditions and given Neumann data on $\tilde{\mathcal{R}}$ and $\tilde{\mathcal{L}}$. But, how do we find η such that the last condition $\tilde{\eta} + 1/2|\operatorname{grad}\tilde{\phi}|^2 = 0$ is satisfied? Although there are many possibilities, we will explore a natural idea in this section: For a suitable space of functions $\mathcal{N}$ and a corresponding norm, minimize the functional $F : \mathcal{N} \to \mathbb{R}$ given by

$$F(\tilde{\eta}) = |\tilde{\eta} + \frac{1}{2}|\operatorname{grad}\tilde{\phi}|^2|^2, \tag{16.65}$$

where ϕ is the solution of our Newmann BVP in the domain $\tilde{\Omega}$ such that $\tilde{S}$ is given by the graph of $\tilde{\eta}$. The solution of our problem (if it exists) is a zero of F.

To implement a minimization procedure, we will need a method to compute $|\operatorname{grad}\tilde{\phi}|^2$ along the surface $\tilde{S}$, given as the graph of the function $\tilde{\eta}$, using the output of the BEM. Fortunately, we do not need to compute the partial derivatives of $\tilde{\phi}$ separately. As usual let N denote the outer unit normal on $\tilde{S}$ and T the unit tangent. Because N and T are a basis for $\mathbb{R}^2$ at each point on $\tilde{S}$, we have that

$$\operatorname{grad}\tilde{\phi} = (\operatorname{grad}\tilde{\phi}\cdot N)N + (\operatorname{grad}\tilde{\phi}\cdot T)T.$$

The Neumann boundary condition on $\tilde{S}$ implies the simplification

$$\operatorname{grad}\tilde{\phi} = (\operatorname{grad}\tilde{\phi}\cdot T)T;$$

therefore,

$$|\operatorname{grad}\tilde{\phi}|^2 = (\operatorname{grad}\tilde{\phi}\cdot T)^2.$$

Using the chain rule, note that

$$\begin{aligned}\frac{d}{dx}\tilde{\phi}(x,\tilde{\eta}(x)) &= \operatorname{grad}\tilde{\phi}\cdot\begin{pmatrix}1\\ \tilde{\eta}'(x)\end{pmatrix}\\ &= \operatorname{grad}\tilde{\phi}\cdot T\sqrt{1+(\tilde{\eta}'(x))^2}.\end{aligned}$$

Thus,

$$\operatorname{grad}\tilde{\phi}\cdot T = \frac{1}{\sqrt{1+(\tilde{\eta}'(x))^2}}\frac{d}{dx}\tilde{\phi}(x,\tilde{\eta}(x))$$

and

$$|\operatorname{grad}\tilde{\phi}(x,\tilde{\eta}(x))|^2 = \frac{1}{1+(\tilde{\eta}'(x))^2}\Big(\frac{d}{dx}\tilde{\phi}(x,\tilde{\eta}(x))\Big)^2. \tag{16.66}$$

Eq. (16.66) is useful to compute the value of our functional F because both derivatives can be approximated by numerical differentiation of the scalar functions $\tilde{\eta}$ and $x \mapsto \tilde{\phi}(x,\tilde{\eta}(x))$, which requires only the values of these functions at the nodes on $\tilde{S}$.

We have described a method to evaluate the functional F for a choice of $\tilde{\eta}$ given by its values at the nodes on $\tilde{S}$.

To turn the idea of minimizing over a function space (which should contain the desired solution $\tilde{\eta}$) into a numerical method, we must discretize the space of functions. Our candidate solutions are (smooth) functions on the interval $[-L, L]$ with values in $(-1, 0)$ and with end point values α at $x = L$ and ω at $x = -L$, as in Problem 16.2.

Using the BEM discretization, we may consider discretized candidate functions as vectors in $\mathbb{R}^{m_{\tilde{S}}}$; that is, one function value for each node on $\tilde{S}$. In this case, the discretization of the functional F is a function from the product neighborhood $\Pi_{i=1}^{m_{\tilde{S}}}(-1,0)$ in $\mathbb{R}^{m_{\tilde{S}}}$ to the real numbers. This discretization is straightforward, but the minimization problem is over a space of (perhaps large) dimension $m_{\tilde{S}}$.

Another natural way to represent functions on a bounded interval is by Fourier series. Recall one of the basic facts of the subject: If f is a twice continuously differentiable (real valued) function on $[-L, L]$ such that $f(-L) = f(L)$ and two sequences of (real) numbers $\{a_n\}_{n=0}^{\infty}$ and $\{b_n\}_{n=1}^{\infty}$—called the Fourier coefficients—are defined by

$$a_n = \frac{1}{L}\int_{-L}^{L} f(x)\cos\left(\frac{n\pi}{L}x\right)dx, \qquad b_n = \frac{1}{L}\int_{-L}^{L} f(x)\sin\left(\frac{n\pi}{L}x\right)dx,$$

then

$$f(x) = \frac{1}{2}a_0 + \sum_{n=1}^{\infty} a_n\cos\left(\frac{n\pi}{L}x\right) + \sum_{n=1}^{\infty} b_n\sin\left(\frac{n\pi}{L}x\right) \tag{16.67}$$

and the series converges uniformly on $[-L, L]$.

In our situation, we do not have the required condition $f(-L) = f(L)$, but it is easy to modify the problem so that this condition is satisfied. Simply choose the candidate functions to be of the form

$$\tilde{\eta}(x) = \omega + \frac{\alpha - \omega}{2L}(x + L) + f(x)$$

where $f(-L) = 0$ and $f(L) = 0$. Each such f is represented by a uniformly convergent Fourier series as in Eq. (16.67) with

$$\sum_{n=1}^{\infty}(-1)^k a_n = 0$$

to ensure the end-point conditions.

An advantage of the Fourier series representation is the expectation (but not the guarantee) that the first few terms (Fourier modes) of the series representation will give a good approximation to the desired minimum. On the other hand, the constraint that the image of $\tilde{\eta}$ be in $(-1, 0)$ is more difficult to impose.

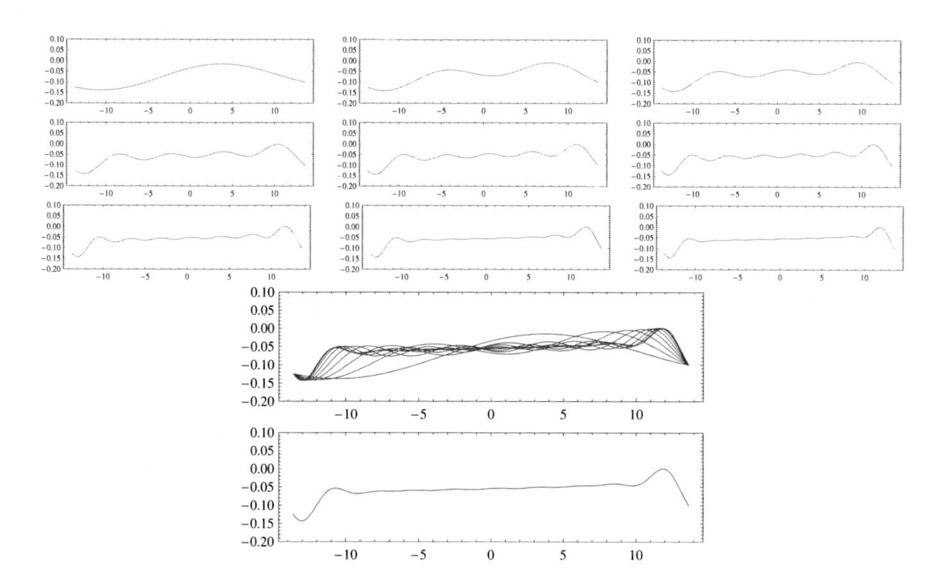

Fig. 16.16 The top panels show the results of pattern-search minimizations to approximate the steady state free surface for the Tiger Fountain flow for Problem 16.2 using inflow given by $\rho(\tilde{y}) = -v_{\tilde{\mathcal{R}}}(\tilde{y}+1)/(1+\alpha)$ *and outflow* $\lambda(\tilde{y}) = v_{\tilde{\mathcal{L}}}(\tilde{y}+1)/(1+\omega)$*, where* $v_{\tilde{\mathcal{R}}} = 2v(1+\omega)/(\alpha+\omega+2)$, $v_{\tilde{\mathcal{L}}} = 2v(1+\alpha)/(\alpha+\omega+2)$*, and* v *is the dimensionless quantity corresponding to surface velocity (*≈ 0.7367984698*). The panels from left to right and top to bottom are for 1–9 Fourier modes (corresponding to 2–18 coefficients of cosines and sines). The middle panel is an overlay to show the convergence, and the bottom panel is an enlargement of the surface profile for nine modes.*

One more important ingredient is needed: A numerical minimization algorithm (which can be successfully applied to the functional (16.65)). The premier numerical minimization algorithm is based on Newton's method (see A.14); but its application requires knowledge of the derivative of the function that is to be minimized.

A viable alternative is to use a pattern search (see, for example, [82]). The idea could not be simpler: Suppose we wish to minimize a function $G : \mathbb{R}^k \to \mathbb{R}$. Choose a finite set of vectors in $\mathbb{R}^k$ such that every vector in $\mathbb{R}^k$ can be written as a linear combination of the elements in this set with nonnegative coefficients. Such a set—which must have at least $n + 1$ elements—is called a positive spanning set. A useful example of a positive spanning set (with $2n$ elements) is given by the union of a basis and the set obtained by multiplying every vector in the basis by -1. To implement the algorithm, choose a guess $p \in \mathbb{R}^k$ for the minimizer of G and a step size $\lambda > 0$. Compute $G(p + \lambda v)$ for every element v in the spanning set. Select the smallest new value obtained for G, replace p by the corresponding vector, and repeat the process until $G(p)$ is at least as small as all the new values or a preassigned number of iterations is reached. The algorithm may be refined in several ways, for example, by decreasing the step size after the algorithm halts with the previous step size, taking steps only in directions near the previously successful direction, or changing the positive spanning set. Of course, it is possible for the algorithm to halt at a local minimum of G that is not the global minimum. There is no general method known that would always avoid this possibility. On the other hand, confidence in solutions may be increased by applying the algorithm to several different initial guesses. Problem 16.3 reflects a more physically realistic situation than does Problem 16.2, but it is computationally more expensive. On the other hand, no new numerical algorithms are needed. Only the geometry of the boundary used in the BEM needs to be changed to modify a code used to approximate the solution of Problem 16.2. The important change is the treatment of inflow and outflow velocity profiles.

A reasonable choice for the inflow velocity profile for Problem 16.3 is a parabolic velocity field along the bottom of the reservoir whose flux corresponds to the known pump flow rate. Using the geometry of the boundary (depicted in Fig. 16.13), the bottom of the reservoir is a horizontal line segment with end coordinates $-(c_1 - 2)L$ and $c_1 L$. We may choose the

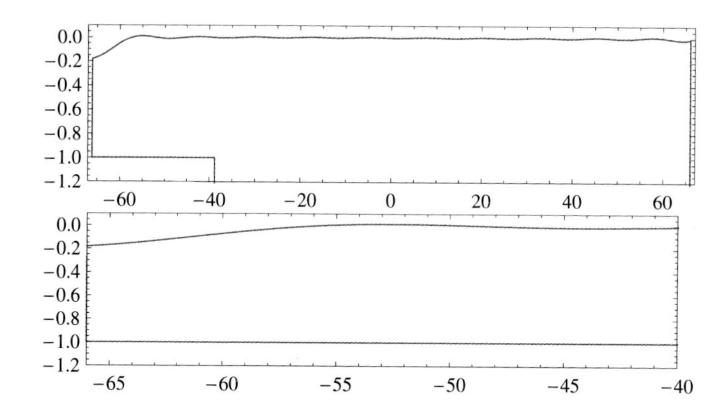

Fig. 16.17 The panels show the results of a pattern-search minimization (using 10 Fourier modes) to approximate the steady state free surface for the Tiger Fountain flow for Problem 16.3 with inflow and outflow velocity profiles as in displays (16.68) *and* (16.69)*. The top panel depicts the free surface together with a portion of the plate and reservoir boundaries. The bottom panel shows the same free-surface flow over the plate.*

inflow velocity field to be given by

$$x \mapsto -h\Big(\frac{(x-L)^2}{((c_1-1)L)^2} - 1\Big)e_2, \tag{16.68}$$

so that the velocity field is normal to the boundary and vanishes at the two vertical walls of the reservoir with h a positive constant to be determined by the flow rate. By choosing the outflow field magnitude to be the bulk speed bs along the entire outflow (that is, the outflow velocity field is $-\,\mathrm{bs}\,e_1$), we may compute the value of h so that the total flux is zero. In this scenario,

$$h \approx 0.00542675. \tag{16.69}$$

The results of a corresponding numerical experiment using BEM and a pattern search for the coefficients of 10 Fourier modes (20 parameters) are presented in Fig. 16.17. A standing wave over the plate is clearly visible. Thus, this experiment suggests that our model captures some of the observed phenomena. The Tiger Fountain flow exhibits a wave train with three or four maxima with diminishing amplitude in the downstream direction. These additional peaks do not seem to be predicted by our numerical experiment.

The inclusion of the reservoir into the computational domain gives a more accurate result than the flow over a plate when compared with the physical flow. On the other hand, for the experiment reported in Fig. 16.16, the inflow velocity profile is not derived from physical measurement. What would happen if the inflow profile from the (numerical) reservoir flow were inserted into the plate flow (see Exercise 16.27).

Exercise 16.14. Approximate the steady state temperature at a focus of an elliptical plate situated in an environment with constant temperature under the assumptions that the heat in the plate obeys Fourier's law, the heat flow at the boundary obeys Newton's law of cooling, and the temperature at the other focus is held fixed. For a specific example, take the boundary of the plate to be given by $x^2/4 + y^2 = 1$, the ambient temperature T_a to be zero, the heat transfer coefficient $\lambda = 1$, and the temperature at the focus $(-\sqrt{3}, 0)$ to be $u = 1$. Hint: The steady state temperature u must statisfy the boundary value problem $\Delta u = 0$ in the region bounded by the ellipse and $\nabla u \cdot N = \lambda(u - T_a)$ on the ellipse, where N is the outer unit normal.

Exercise 16.15. Prove Proposition 16.4 for the three-dimensional case.

Exercise 16.16. For u^* defined as in Eq. (16.47) or (16.48), prove that the function $p \mapsto u^*(p, q)$ is harmonic for all $p \neq q$.

Exercise 16.17. Verify Eq. (16.58).

Exercise 16.18. Verify Eq. (16.59).

Exercise 16.19. Verify Eq. (16.60).

Exercise 16.20. Verify Eq. (16.61). Hint: In our case, q is the midpoint of the element; therefore, the integral is singular at $t = 1/2$. Split the integral into two improper integrals on the intervals $[0, 1/2]$ and $[1/2, 1]$. The integral of the logarithm is best handled using integration by parts; that is, $\int u\, dv = uv - \int v\, du$ where u is the logarithm. Take $v = t - 1/2$.

Exercise 16.21. Discuss the claim: The most efficient method of evaluating the integrals in display (16.57) is numerical integration.

Exercise 16.22. Verify Eqs. (16.62) and (16.63).

Exercise 16.23. Verify the formulas in display (16.64).

Exercise 16.24. [BEM Coding I] Write a BEM code and reproduce Table 16.1.

Exercise 16.25. (a) Find the (exact) minimum of the function $g : \mathbb{R}^2 \to \mathbb{R}$ given by

$$g(x, y) = \frac{1}{2}y^2 - \frac{1}{2}x^2 + \frac{1}{3}x^3 + \frac{1}{4}x^4.$$

Approximate the minimum using Newton's method and a pattern search. Discuss the relative efficiencies of these numerical methods.
(b) Approximate the function that minimizes the functional

$$F[\gamma](t) := \int_0^{2\pi} \dot{\gamma}(t)^2 - \gamma(t)^2\, dt$$

over all piecewise C^1 functions $\gamma : [0, 2\pi] \to \mathbb{R}$ such that $\gamma(0) = 1$ and $\gamma(2\pi) = 1$ using a pattern search together with a discretization of the functions in the domain of the functional. (Note: The function $\gamma(t) = \cos t$ is the minimizer.)

Exercise 16.26. Verify the approximate value of h given in display (16.69).

Exercise 16.27. [BEM Coding II] (a) Repeat a version of the numerical experiment reported in Fig. 16.17. (b) Use the BEM integral representations to compute the velocity profile along a vertical line at the upstream end of the plate. (c) Use the result of (b) to solve Problem 16.2 and compare your result with your computation for part (a).

16.7 SMOOTHED PARTICLE HYDRODYNAMICS

The equations of fluid motion, in particular the water wave equations, seem to be too complicated to admit exact general solutions. Although some qualitative predictions can be determined from these systems, we must often resort to numerical computations to obtain detailed predictions. The numerical treatment of the equations of fluid motion is a vast subject, which often goes by the name CFD (computational fluid dynamics). One approach is by finite difference methods, which were introduced in Chapter I. A simpler, but less developed approach, is called smoothed particle hydrodynamics (SPH) (see [51, 66, 77]). This method is conceptually easy and it produces good results; its main disadvantage is a lack of a corresponding rigorous numerical analysis. Wc will discuss it here as a plausible approach to the approximate simulation of free-surface problems and as an introduction to Lagrangian methods in fluid mechanics.

16.7.1 Euler Versus Lagrange

Recall Euler's equations of fluid motion

$$\begin{aligned} \rho_t + \nabla \cdot (\rho u) &= 0, \\ \rho(u_t + (u \cdot \nabla)u) &= -\nabla p. \end{aligned} \tag{16.70}$$

In this Eulerian formulation, the state variables (density) ρ, (pressure) p, and (velocity) u are viewed as fields on space-time; they are functions of position x and time t. Euler's equations are a system of PDEs for these fields. The motion of fluid particles is ignored. The Eulerian observer resides at a point (x, t) in space-time and asks, for example, what is the velocity of the fluid *at my* position x and current time t?

In the Lagrangian formulation, each fluid particle or perhaps a set of fluid parcels is labeled—some authors say marked—at a reference time taken here to be $t = 0$. Usually each particle is labeled by its starting position in space, but a set of fluid parcels might be marked by some other attribute (see, for example, Exercise 16.31).

Once a particle is marked by a Lagrangian coordinate ξ, its future position is determined by solving the initial value problem

$$\dot{x} = u(x,t), \qquad x(0) = \xi,$$

where u is the Eulerian fluid velocity field. The hallmark of the Lagrangian formulation is the Lagrange flow map defined to be the future position of ξ at each time t. More precisely, the Lagrangian flow map X is the function of ξ and t such that $t \mapsto X(\xi,t)$ is the solution of the ODE $\dot{x} = u(x,t)$ with initial condition $X(\xi,0) = \xi$. Warning: Because the vector field u is time dependent, X does not satisfy the semigroup property in time as does the flow of an autonomous vector field (see Exercise 16.30). The Lagrangian observer is moving with a marked fluid particle and asks what *is my* current position and velocity?

In more prosaic language, an Eulerian observer sits on a riverbank to view its flow; a Lagrangian observer drifts on the river in a boat.

To begin the transformation of the equations of motion from Eulerian to Lagrangian coordinates, consider the density ρ, which is a scalar state variable and let $R(\xi,t)$ be its Lagrangian representation. In other words,

$$R(\xi,t) = \rho(X(\xi,t),t). \tag{16.71}$$

Let D denote differentiation with respect to the spatial variable (either x or ξ according to the context). The Lagrangian representation U of the velocity u, a vector state variable, is given by

$$U(\xi,t) = DX^{-1}(X(\xi,t),t)u(X(\xi,t),t). \tag{16.72}$$

To understand this formula, consider a point (ξ,t), map it to the spatial coordinates by $X(\xi,t)$, evaluate the velocity field at this point $u(X(\xi,t),t)$, and transform this vector back to the Lagrangian markers via the derivative of the inverse of $x \mapsto X(x,t)$. Of course, $DX^{-1}(X(\xi,t)) = [DX(\xi,t)]^{-1}$;

thus, we also have that

$$U(\xi, t) = [DX(\xi, t)]^{-1} u(X(\xi, t), t). \tag{16.73}$$

This is the way a vector field changes coordinates.

A fundamental and useful formula relates time derivatives in the Lagrangian and Eulerian formulations. Recall, from the definition of the flow map X, that

$$X_t(\xi, t) = u(X(\xi, t), t), \qquad X(\xi, 0) = \xi.$$

For an Eulerian scalar quantity f, the partial derivative of its Lagrangian representation $F(\xi, t) = f(X(\xi, t), t)$ with respect to time is

$$\begin{aligned} F_t(\xi, t) &= Df(X(\xi, t), t) X_t(\xi, t) + f_t(X(\xi, t), t) \\ &= f_t + \nabla f \cdot u \\ &= f_t + (u \cdot \nabla) f. \end{aligned} \tag{16.74}$$

In other words, the partial time derivative of the Lagrangian scalar state is the material derivative of the Eulerian state.

In the interpretation of the Lagrangian formulation used here for SPH, the equations of motion may be viewed as ODEs for the states *along the path of a fluid particle in space-time*. Although the Lagrangian equations of motion can be derived directly from physical considerations, they are discussed here as alternative formulations of the Eulerian field equations.

Each state variable for the fluid is treated as a Lagrangian variable. For simplicity of notation the Lagrangian marker is suppressed and the density, pressure, and velocity *along a fluid particle trajectory* are denoted by $R(t) = \rho(X(t), t)$, $P(t) = p(X(t), t)$, and $V(t) = u(X(t), t)$. The quantities ∇p, $\nabla \cdot u$, and Δu are also named to suggest the functions they define along the path: $\mathrm{gradP}(t) = (\nabla p)(X(t), t)$, $\mathrm{divV}(t) = (\nabla \cdot u)(X(t), t)$, and $\mathrm{LapV}(t) = \Delta u(X(t), t)$. There is a possible confusion concerning the definition of V, which gives the value of the fluid velocity along the path of a fluid particle. This vector function is not the Lagrangian velocity field V properly defined in Eq. (16.73); in fact, the two fields are related by

$$V(\xi, t) = DX(\xi, t) U(\xi, t).$$

To obtain the Lagrangian formulation of fluid mechanics, consider (for instance) the Eulerian equation of continuity

$$\rho_t + \nabla \cdot (\rho u) = 0,$$

expand the second term, and rewrite this equation in the form

$$\rho_t + (u \cdot \nabla)\rho = -\rho \nabla \cdot u,$$

or alternatively, as

$$\rho_t + (u \cdot \nabla)\rho = (u \cdot \nabla)\rho - \nabla \cdot (\rho u).$$

Using Eq. (16.74), the left-hand sides of these equations are the time derivative dR/dt in Lagrangian notation. Thus, the corresponding Lagrangian equations may be expressed as

$$\frac{dR}{dt} = -R\,\mathrm{divV} \tag{16.75}$$

and

$$\frac{dR}{dt} = V \cdot \mathrm{gradR} - \mathrm{divRV}. \tag{16.76}$$

Similarly, one of the Lagrangian forms of the full Eulerian equations of fluid motion is the system of ODEs

$$\frac{dx}{dt} = V, \tag{16.77}$$

$$\frac{dR}{dt} = -R\,\mathrm{divV}, \tag{16.78}$$

$$R\frac{dV}{dt} = -\mathrm{gradP} + Rg, \tag{16.79}$$

where we have also included the gravitational force.

The Lagrangian system [Eqs. (16.77)–(16.79)] is underdetermined: there are 11 variables (three components of x, V, and gradP plus the scalars R and divV), but only 7 equations. This should be expected. It is not possible to transform the equations of fluid motion to a well posed system of ODEs. If we could do so, existence and uniqueness of solutions in fluid dynamics would follow from the existence theory for ODEs. But at a fundamental level, because there are two equations (conservation of mass and momentum balance) for three variables (density, velocity, and pressure), we need one more equation to close the system. The missing equation is the mathematical statement of the conservation of energy. But because we will consider only regimes where heat dissipation (for example) is not important, we will close the system in a different manner. We will employ an equation of state; that is, a constitutive relation between pressure and density together with a method (called SPH) for expressing divV and gradP using the state variables x, V, and ρ but not their spatial derivatives. Using these ideas, we will *approximate* the Lagrangian equations of motion with a well posed system of ODEs. In fact, our formulation is more general. We will carry out the smoothed particle approximation theory for the Lagrangian formulation of the Navier–Stokes model for three spatial dimensions. The corresponding formulas reduce immediately to the case of two spatial dimensions, the case we will consider in our numerical experiments.

In principle, the quantities, for example divV, can be more explicitly determined in Lagrangian formulations. To see how this is accomplished, recall the space-time coordinates are (x, t) and these are related to the Lagrangian coordinates (ξ, t) as before by the Lagrangian flow map X. Also, recall the fundamental identities

$$\frac{\partial X}{\partial t}(\xi, t) = u(X(\xi, t), t), \qquad X(\xi, 0) = \xi.$$

By differentiation with respect to ξ and using D to denote differentiation with respect to the spatial variable, the variational equation for the spatial derivativc of the Lagrangian flow map is

$$\frac{\partial}{\partial t} DX(\xi, t) = Du(X(\xi, t), t) DX(\xi, t), \qquad DX(\xi, 0) = I. \tag{16.80}$$

Consider the Eulerian continuity equation in the form

$$\rho_t + (u \cdot \nabla)\rho = -\rho \nabla \cdot u$$

and transform to Lagrangian coordinates to obtain

$$\frac{\partial R}{\partial t} = -R \mathrm{divV}.$$

There is one remaining question: What is the explicit expression for RdivV in Lagrangian coordinates?

By reinterpreting the notation or recomputing,

$$\frac{\partial R}{\partial t} = -R\nabla \cdot u.$$

The divergence is easily seen to be the trace of the spatial derivative. Here,

$$\nabla \cdot u = \operatorname{tr} Du(x,t) = \operatorname{tr} Du(X(\xi,t),t).$$

Applying Liouville's Theorem A.10 to the variational equation [Eq. (16.80)],

$$\nabla \cdot u = \frac{1}{\det DX(\xi,t)} \frac{\partial}{\partial t} \det DX(\xi,t),$$

and using j for the Jacobian determinant $\det DX(\xi,t)$, the Lagrangian form of the continuity equation is

$$\frac{\partial R}{\partial t}(\xi,t) = -R(\xi,t)\frac{1}{j(\xi,t)}\frac{\partial}{\partial t} j(\xi,t),$$

or more compactly,

$$\frac{\partial}{\partial t}(jR) = 0. \tag{16.81}$$

This means

$$R = \frac{\rho_0}{j},$$

which is a concise and useful way to describe the evolution of the initial density field ρ_0.

Exercise 16.28. Show that J in PDE (16.81) is constant if and only if $\nabla \cdot u = 0$.

Exercise 16.29. (a) Consider the partial differential equation

$$\nabla \cdot (u_t + (u.\nabla)u) = h(\rho, u, t),$$

where h is a continuous function. For X the Lagrangian flow map corresponding to the velocity u, define $J(\xi,t) = DX(p,t)$ and $j(\xi,t) = \det DX(\xi,t)$. Convert the partial differential equation to Lagrangian variables. Answer:

$$\frac{\partial^2}{\partial t^2} \ln j + \operatorname{tr}((\frac{\partial J}{\partial t} J^{-1})^2) = h(\rho(X(\xi,t), u(X(\xi,t),t),t).$$

(b) Part (a) is used in cosmology, where a simple model for the matter (as a cloud of dust) in the universe is given by

$$\rho_t + \nabla \cdot (\rho u) = 0,$$

$$u_t + (u \cdot \nabla)u = -\nabla\Phi,$$
$$\Delta\Phi = 4\pi G\rho.$$

The function Φ is the Newtonian gravitational potential and G is the universal gravitational constant. One way to eliminate Φ is to compute the divergence of both sides of the second equation and substitute using the third equation. A useful first step in the analysis of this model is to find a family of exact solutions. Show that

$$\rho = \frac{\rho_0}{a^3(t)}, \qquad u = \frac{\dot{a}}{a}\xi, \qquad \Phi = -\frac{1}{2}\frac{\ddot{a}}{a}|\xi|$$

solves the system exactly provided that the function a satisfies a certain second-order ODE. What is this ODE? Here a is called the expansion constant of the universe and the family of solutions is called the Hubble flow (Edwin Hubble, 1923). Yakov Zel'dovich found a more interesting family of solutions whose Lagrangian flow maps are of the form

$$X(\xi, t) = a(t)(\xi - f(\xi_1, t)e_1),$$

where f if an element of a certain class of scalar functions and e_1 is the usual first basis vector. Challenge: Find a nontrivial f that produces a Lagrangian flow map (see [94] or [85] for more about cosmology).

Exercise 16.30. (a) Let $t \mapsto X(\xi, t)$ denote the flow of the vector field u given by $u(x) = x$. Solve the initial value problem

$$\dot{x} = x, \qquad x(0) = \xi$$

and show that $X(X(\xi, s), t) = X(\xi, s + t)$. This is called the flow property. (b) Let $t \mapsto X(\xi, t)$ denote the flow of the vector field u given by $u(x) = tx$. Solve the initial value problem

$$\dot{x} = tx, \qquad x(0) = \xi$$

and show that X does not satisfy the flow property. (c) Solve the initial value problem

$$\dot{x} = tx, \qquad x(s) = \xi,$$

where s is a parameter, and let $t \mapsto X(\xi, t, s)$ denote its solution. The initial condition is given by $X(\xi, s, s) = \xi$. Show that

$$X(X(\xi, t, \tau), \tau, s) = X(\xi, t + s).$$

(d) Show that the properties of X mentioned in (a) and (c) are true in general for autonomous and time-dependent vector fields.

Exercise 16.31. Reconsider the Eulerian continuity equation

$$\rho_t + (\rho u)_x = 0$$

for one spatial dimension with coordinate x, where ρ is the density and u is the fluid velocity, and define the transformation ϕ from the Eulerian coordinates (x,t) to the coordinate system with coordinates (ξ, t), where ξ is akin to a Lagrangian marker, by the rule

$$\xi = \phi(x) = \int_0^x \rho(z,0)\,dz.$$

The new coordinate ξ is the total density of the parcel of fluid corresponding to the interval $[0,x]$.

Let ψ denote the inverse of ϕ and, as usual, let $X(\psi(\xi),t)$ be the spatial position of the particle with Lagrangian marker ξ at time t. Also, recall that the Lagrangian density R and *scalar* velocity V along the path of a fluid particle are given by

$$R(\xi,t) = \rho(X(\psi(\xi),t),t), \qquad V(\xi,t) = u(X(\psi(\xi),t),t).$$

(1) Show that

$$R_t(\xi,t) = -R(\xi,t)u_x(X(\psi(\xi),t),t).$$

(2) Show that

$$V_\xi(\xi,t) = u_x(X(\psi(\xi),t),t)X_x(\psi(\xi),t)\psi_\xi(\xi).$$

(3) Show that

$$\psi_\xi(\xi) = \frac{1}{R(\xi,0)}.$$

(4) Show that the functions $t \mapsto X_x(\psi(\xi),t)\psi_\xi(\xi)$ and $t \mapsto 1/\rho(X(\psi(\xi),t),t)$ are equal. Hint: Show that both functions solve the ODE $\dot{w} = u_x(X(\psi(\xi),t),t)w$ with the same initial condition $w(0) = 1/R(\xi,0)$.
(5) Use the previous parts of this exercise to show that

$$R_t(\xi,t) = -R^2(\xi,t)V_\xi(\xi,t).$$

(6) Define $r = 1/R$ and show that

$$r_t = V_\xi.$$

This is a way to express the conservation of mass as a PDE using a variant of the usual Lagrangian coordinates.
(7) Is there a more efficient way to obtain the same result?

16.7.2 Localization

Let a be a positive number. We call a sequence $\{W_n\}_{n=1}^\infty$ of smooth functions $W_n : \mathbb{R}^m \to \mathbb{R}$ (each at least class C^1 with m a positive integer that may be taken to be two or three) localizing if for some positive number a and each positive integer n, the function W_n has support in the closed ball with center at the origin and radius a,

$$\int_{\mathbb{R}^m} W_n(x)\,dx = 1, \tag{16.82}$$

$$\nabla W_n(0) = 0, \tag{16.83}$$

$$\nabla W_n \text{ is uniformly bounded,} \tag{16.84}$$

and (for every continuous function $g : \mathbb{R}^3 \to \mathbb{R}$)

$$\lim_{n\to\infty} \int_{\mathbb{R}^m} g(x) W_n(x - x_0)\,dx = g(x_0) \tag{16.85}$$

(see Exercise 16.32). For the remainder of this chapter, we will assume that we have determined a sequence of functions $w_n : \mathbb{R} \to \mathbb{R}$ such that the sequence given by $W_n(x) = w_n(|x|)$ is localizing.

To determine a field f on space-time, we must specify its value at each point in space-time. Usually this is not possible when the desired field is given as a solution of a differential equation. Instead, we may seek a sequence of approximations $\{f_n\}_{n=1}^{\infty}$ that converges pointwise to the unknown field f. The basic approximation in SPH produces such a sequence:

$$f_n(x, t) := \int_{\mathbb{R}^m} f(y, t) w_n(|y - x|)\,dy. \tag{16.86}$$

Exercise 16.32. Prove that there is a sequence of C^∞ functions $\{W_n\}_{n=1}^{\infty}$ with the properties stated in Eqs. (16.82), (16.83), (16.84), and (16.85).

16.7.3 Discretization

Consider N parcels of fluid in space and, as an approximation to reality, let us suppose that each has a fixed position, velocity, volume, and mass. In particular, the (spatial) position of the jth particle is x_j. The value of a field

f defined on space-time is approximated at the point (x,t) by

$$\int_{\mathbb{R}^3} f(y,t)w(|y-x|)\,dy \approx \sum_{j=1}^{k} f(x_j,t)w(|x-x_j|)\mathrm{vol}(x_j), \tag{16.87}$$

where w denotes some element of the sequence of localizing functions and $\mathrm{vol}(x_j)$ denotes the volume of the jth parcel.

In fluid mechanics it is inconvenient to carry the volume with each fluid parcel; instead, we let m_j denote the mass and ρ_j the density of the jth parcel. Using this notation, an approximation of our field is given by

$$\sum_{j=1}^{k} \frac{m_j}{\rho_j} f(x_j,t)w(|x-x_j|). \tag{16.88}$$

This is the smoothed particle approximation (which perhaps should be called the smoothed parcel approximation) of the field. The accuracy of the approximation depends on the choice of the localizing function and the number of parcels. Also, our approximation carries the same units as the field f, as it should, because the localization function has units of inverse volume by property (16.82).

For our applications to fluid dynamics, we must also approximate derivatives of functions and fields. We cannot simply differentiate both sides of Eq. (16.85), at least we cannot do so without checking that the convergence is (locally) uniform. On the other hand, if we wish to approximate a (continuous) derivative of a function, we may apply the limit process of Eq. (16.85) with g replaced by the desired derivative.

Let us approximate ∇g where g is a function or a vector field. Recall that in case g is a vector field, we apply ∇ to each component of g. According to Eq. (16.85) and using the assumption that ∇g is continuous, we have that

$$\nabla g(x) = \lim_{n\to\infty} \int_{\mathbb{R}^m} \nabla g(y)W_n(y-x)\,dy$$

Using the definition of W_n, there is a closed ball $B(x)$ centered at each x (which is the translation to x of a closed ball at the origin that contains the support of W_n in its interior) such that

$$\int_{\mathbb{R}^m} \nabla g(y)W_n(y-x)\,dy = \int_{B(x)} \nabla g(y)W_n(y-x)\,dy.$$

For every smooth function $\omega : \mathbb{R}^m \to \mathbb{R}$, Leibniz's rule implies

$$\nabla(\omega g) = \omega \nabla g + (\nabla \omega) g. \tag{16.89}$$

With x arbitrary but fixed and $\omega(y) := W_n(y - x)$, the last integral may be recast as

$$\int_{B(x)} \nabla g(y) W_n(y - x)\, dy = \int_{B(x)} \nabla(\omega g)(y)\, dy - \int_{B(x)} \nabla \omega(y) g(y)\, dy. \tag{16.90}$$

Recall the divergence theorem: If f is a smooth vector field on a bounded region $B \subset \mathbb{R}^m$ with a smooth boundary and η is the outer unit normal field on the boundary of B , then

$$\int_B \nabla \cdot f\, dy = \int_{\partial B} f \cdot \eta\, dy. \tag{16.91}$$

An easy corollary of this result states that if f is a smooth function on B and η is the outer unit normal field on the boundary of B, then

$$\int_B \nabla f\, dy = \int_{\partial B} f \eta\, dy \tag{16.92}$$

(see Exercise 16.33). Because $\omega(y) = W_n(y - x)$ vanishes on the boundary of $B(x)$, the first integral on the right-hand side of Eq. (16.90) vanishes and we have the approximation

$$\nabla g(x) \approx \int_{\mathbb{R}^m} \nabla g(y) W_n(y - x)\, dy = - \int_{B(x)} g(y) \nabla_y W_n(y - x)\, dy, \tag{16.93}$$

where ∇_y denotes differentiation with respect to y.

To use the approximation (16.93), we need the derivative of the function h given by $h(y) = W_n(y - x)$. To compute it, recall the assumption that W_n is an element of a localizing sequence that consists of the function of the form $W(y - x) := w(|y - x|)$ for some scalar function w.

Claim: If $h : \mathbb{R}^m \to \mathbb{R}$ is given by $h(y) = W(y - x)$, then

$$Dh(y) = \frac{w'(|y - x|)}{|y - x|}(y - x)^T, \tag{16.94}$$

where the superscript T denotes the transpose. The claim follows from the chain rule and the computation of the derivative of the function $k : \mathbb{R}^m \to \mathbb{R}$

given by $k(y) = |y - x|$, which is obtained by using the formula

$$k^2(y) = \langle y - x, y - x \rangle.$$

The directional derivative of k^2 in the direction v is

$$2k(y)Dk(y)v = 2\langle y - x, v \rangle.$$

Hence, the derivative of k may be represented as the vector

$$Dk(y) = \frac{1}{|y - x|}(y - x)^T.$$

The derivative of k is not defined at $y = x$. On the other hand, it is easy to prove that

$$\lim_{x \to y} \frac{w'(|y - x|)}{|y - x|}(y - x) = 0.$$

It follows that the derivative of h vanishes at y as it must because W is a localizing function.

We have obtained the approximation

$$\nabla g(x) \approx -\int_{\mathbb{R}^m} g(y)\frac{w'(|y - x|)}{|y - x|}(y - x)\, dy = \int_{\mathbb{R}^m} g(y)\frac{w'(|x - y|)}{|x - y|}(x - y)\, dy. \tag{16.95}$$

The smoothed particle approximation for ∇g is

$$\nabla g(x) \approx \sum_{j=1}^{k} \frac{m_j}{\rho_j} g(x_j) \frac{w'(|x - x_j|)}{|x - x_j|}(x - x_j). \tag{16.96}$$

The approximation for the divergence of a field g is obtained in the same way we approximated the gradient of a function:

$$\begin{aligned} \nabla \cdot g(x) &= \lim_{n \to \infty} \int_{\mathbb{R}^k} \nabla \cdot g(y) w_n(|x - y|)\, dy \\ &= \lim_{n \to \infty} \int_{B(x)} \nabla \cdot g(y) w_n(|x - y|)\, dy, \end{aligned}$$

where $B(x)$ is the closed ball centered at x that is the translate of the ball B in $\mathbb{R}^m$ that contains the supports of the localizing functions. Let us view x as arbitrary but fixed, ignore the subscript n, and define $h(y) := w(|x - y|)$.

Using the identity

$$\nabla \cdot (hg) = \nabla h \cdot g + h\nabla \cdot g,$$

we have that

$$\int_{B(x)} \nabla \cdot g(y)h(y)\,dy = \int_{B(x)} \nabla \cdot (hg)\,dy - \int_{B(x)} \nabla h \cdot g\,dy.$$

By an application of the divergence theorem,

$$\int_{B(x)} \nabla \cdot g(y)h(y)\,dy = \int_{\partial B(x)} hg \cdot \eta\,dy - \int_{B(x)} \nabla h \cdot g\,dy,$$

where η is the outer unit normal vector on the boundary of $B(x)$. Because the localizing function vanishes on the boundary of B, the function h vanishes on the boundary of $B(x)$ and the integral over $\partial B(x)$ vanishes. Thus, for every x, we have

$$\nabla \cdot g(x) = -\lim_{n\to\infty} \int_{\mathbb{R}^m} \nabla h(y) \cdot g(y)\,dy. \tag{16.97}$$

As before, let us approximate $\nabla \cdot g(x)$ using the fluid parcels. It is important to note that, as a consequence of Eq. (16.94),

$$Dh(y) = \frac{w'(|x-y|)}{|x-y|}(y-x)^T.$$

In particular, pay attention to the sign of the transposed vector to write

$$\nabla \cdot g(x) \approx \int_{\mathbb{R}^m} g(y) \cdot \frac{w'(|x-y|)}{|x-y|}(x-y)\,dy.$$

The smoothed particle approximation is

$$\nabla \cdot g(x) \approx \sum_{j=1}^{k} \frac{m_j}{\rho_j} \frac{w'(|x-x_j|)}{|x-x_j|}(x-x_j) \cdot g(x_j). \tag{16.98}$$

In the application to fluid dynamics, the fields are all time dependent, and we will be interested only in the approximations evaluated at the locations of the fluid parcels. As a convenient reference, the appropriate approximations are

$$g(x_i,t) \approx \sum_{j=1}^{k} \frac{m_j}{\rho_j} g(x_j,t) w(|x_i - x_j|), \tag{16.99}$$

$$\nabla g(x_i,t) \approx \sum_{\substack{j=1 \\ j\neq i}}^{k} \frac{m_j}{\rho_j} g(x_j,t) \frac{w'(|x_i - x_j|)}{|x_i - x_j|} (x_i - x_j), \tag{16.100}$$

$$\nabla \cdot g(x_i,t) \approx \sum_{\substack{j=1 \\ j\neq i}}^{k} \frac{m_j}{\rho_j} \frac{w'(|x_i - x_j|)}{|x_i - x_j|} (x_i - x_j) \cdot g(x_j,t). \tag{16.101}$$

16.7.4 Conservation of Mass

A discretization of the Lagrangian conservation of mass (as expressed in Eq. (16.78)) is obtained directly from Eq. (16.100):

$$\dot{\rho}_i = -\rho_i \sum_{\substack{j=1 \\ j\neq i}}^{k} \frac{m_j}{\rho_j} \frac{w'(|x_i - x_j|)}{|x_i - x_j|} (x_i - x_j) \cdot u_j. \tag{16.102}$$

Similarly, a (different) discretization is obtained from the alternative expression given in Eq. (16.76):

$$\dot{\rho}_i = \sum_{\substack{j=1 \\ j\neq i}}^{k} m_j \frac{w'(|x_i - x_j|)}{|x_i - x_j|} (x_i - x_j) \cdot (u_i - u_j). \tag{16.103}$$

Density also has the direct smoothed particle approximation

$$\rho_i = \sum_{j=1}^{k} m_j w(|x_i - x_j|), \tag{16.104}$$

which does not require the solution of a differential equation.

The approximation (16.103) is often preferred. It is symmetrical in x and u (which is advantageous for programming the evaluation of this formula), the density does not appear on the right-hand side, and the density can be updated in the momentum balance in a computational efficient manner. Also, it is satisfying that the time derivative of the density approximation (16.104) agrees with the approximation (16.103) (see Exercise 16.34). On the other hand, the optimal choice of approximation method for a given problem is the subject of current research.

The reader should consider why we need to formulate a dynamical equation for density if we intend to apply our discretization to a fluid like water, which could be taken (in a reasonable approximation) to have constant density. There are several reasons. Real fluids are compressible. Perhaps this is reason enough, but this answer does not relate to our discretization method. The main reason for the introduction of (artificial) compressibility is to obtain an efficient method to compute pressure. For example, recall that for an incompressible fluid, the Euler equations are

$$\rho\frac{Du}{Dt} = -\nabla p, \qquad \nabla \cdot u = 0.$$

To solve these equations (and the more complicated equations of motion that include viscosity and gravity) with a numerical method, we must compute the pressure p during the numerical procedure. We cannot simply discretize with finite differences and solve a system of ODEs because the time derivative of pressure does not appear. Instead, we must use both equations to solve for the unknown velocity and pressure. The determination of pressure is a problem that must be solved whatever the choice of numerical method. In the SPH approach used here, this problem will be overcome by evolving the density and computing the pressure via an equation of state that gives the pressure as a function of density.

Exercise 16.33. Let B be a bounded region with a smooth boundary in $\mathbb{R}^3$ and η the outer unit normal field on its boundary. Assume the divergence theorem [Eq. 16.91].
(i) Prove: If f is a smooth vector field on $B \subset \mathbb{R}^3$, y_i is one of the three coordinate functions on $\mathbb{R}^3$ and η_i is the corresponding component of the normal field, then

$$\int_B \frac{\partial f}{\partial y_i}\, dy = \int_{\partial B} f\eta_i \, dy.$$

(ii) Prove the result stated in Eq. (16.92).

Exercise 16.34. Show that the time derivative of density approximation (16.104) agrees with the approximation (16.103).

16.7.5 Momentum Balance

The general (Lagrangian) form of the momentum balance is given by

$$\frac{dV}{dt} = \frac{1}{R}\mathrm{divS} + g, \qquad (16.105)$$

where divS is the divergence of the stress tensor σ along a fluid particle path in space-time (see Eq. (11.11)). It is not important at this stage of the analysis to know the form of the stress tensor; it suffices to work formally. As in the alternate form of the continuity equation [Eq. (16.76)], a symmetrical form of the smoothed particle approximation of the momentum balance is obtained by using the identity

$$\nabla \cdot (\frac{1}{\rho}\sigma) = \frac{1}{\rho}\nabla \cdot \sigma - \frac{1}{\rho^2}\sigma \cdot \nabla \rho \tag{16.106}$$

to recast the momentum balance in the form

$$\frac{du}{dt} = \frac{1}{\rho}\nabla \cdot \sigma - \frac{1}{\rho^2}\sigma \cdot \nabla \rho + g. \tag{16.107}$$

After some algebra, the corresponding smoothed particle approximation is

$$\dot{u}_i = \sum_{\substack{j=1 \\ j\neq i}}^{k} m_j \frac{w'(|x_i - x_j|)}{|x_i - x_j|}\Big(\frac{1}{\rho_j^2}\sigma_j + \frac{1}{\rho_i^2}\sigma_i\Big)(x_i - x_j) + g. \tag{16.108}$$

Stress Tensor

The stress tensor for a fluid is usually taken to have the form

$$\sigma^{\alpha\beta} = -p\,\delta^{\alpha\beta} + \tau^{\alpha\beta}, \tag{16.109}$$

where p is the pressure, $\delta^{\alpha\beta} = 1$ if $\alpha = \beta$ and $\delta^{\alpha\beta} = 0$ otherwise, τ is the viscous stress tensor, and the superscripts range over the indices $\{1, 2, 3\}$. The exact form of the viscous stress depends on properties of the fluid. A Newtonian fluid has

$$\tau^{\alpha\beta} = 2\mu\varepsilon^{\alpha\beta}, \tag{16.110}$$

where μ is the viscosity (which we will assume is a constant) and

$$\varepsilon^{\alpha\beta} = \frac{1}{2}(\frac{\partial u^\beta}{\partial x^\alpha} + \frac{\partial u^\alpha}{\partial x^\beta}) - \frac{1}{3}(\nabla \cdot u)\delta^{\alpha\beta} \tag{16.111}$$

is the strain tensor. The upper subscript denotes the corresponding component of the vectors u and x.

Eq. (16.108) is a vector equation. Using the tensors defined for a Newtonian fluid and the summation convention (for the multiplication of the stress tensor matrix and the component representations of vectors in space), the smoothed particle momentum balance equation is

$$\dot{u}_i^\alpha = g^\alpha - \sum_{\substack{j=1 \\ j\neq i}}^{k} m_j \frac{w'(|x_i - x_j|)}{|x_i - x_j|}\left(\frac{p_i}{\rho_i^2} + \frac{p_j}{\rho_j^2}\right)(x_i - x_j)^\alpha$$

$$+ \mu \sum_{\substack{j=1 \\ j\neq i}}^{k} m_j \frac{w'(|x_i - x_j|)}{|x_i - x_j|}\left(\frac{\varepsilon_i^{\alpha\beta}}{\rho_i^2} + \frac{\varepsilon_j^{\alpha\beta}}{\rho_j^2}\right)(x_i - x_j)^\beta. \tag{16.112}$$

For the ith parcel,

$$\varepsilon_i^{\alpha\beta} = \left(\frac{\partial u_i^\beta}{\partial x_i^\alpha}\right) + \left(\frac{\partial u_i^\alpha}{\partial x_i^\beta}\right) - \frac{2}{3}(\nabla \cdot u_i)\delta^{\alpha\beta}. \tag{16.113}$$

Also, for an arbitrary function G,

$$\left(\frac{\partial G}{\partial x_i^\alpha}\right) = \sum_{\substack{j=1 \\ j\neq i}}^{k} \frac{m_j}{\rho_j} G_j \frac{w'(|x_i - x_j|)}{|x_i - x_j|}(x_i - x_j)^\alpha.$$

Thus, we have that

$$\varepsilon_i^{\alpha\beta} = \sum_{\substack{j=1 \\ j\neq i}}^{k} \frac{m_j}{\rho_j}\left(u_j^\beta (x_i - x_j)^\alpha + u_j^\alpha (x_i - x_j)^\beta - \frac{2}{3}\delta^{\alpha\beta} u_j \cdot (x_i - x_j)\right)\frac{w'(|x_i - x_j|)}{|x_i - x_j|}. \tag{16.114}$$

Eq. (16.114) can be symmetrized in u by making some further reasonable approximations. To accomplish the symmetrization, note that

$$0 = \nabla 1 \approx \sum_{\substack{j=1 \\ j\neq i}}^{k} \frac{m_j}{\rho_j} \frac{w'(|x_i - x_j|)}{|x_i - x_j|}(x_i - x_j).$$

For a field g, it follows immediately that

$$\sum_{\substack{j=1 \\ j\neq i}}^{k} \frac{m_j}{\rho_j} g_i \frac{w'(|x_i - x_j|)}{|x_i - x_j|}(x_i - x_j) \approx 0.$$

Thus, for instance,

$$\sum_{\substack{j=1\\ j\neq i}}^{k} \frac{m_j}{\rho_j} u_i^\beta \frac{w'(|x_i - x_j|)}{|x_i - x_j|}(x_i - x_j)^\alpha \approx 0.$$

By subtracting three similar expressions from Eq. (16.114), we obtain the new symmetrized approximation

$$\begin{aligned} \varepsilon_i^{\alpha\beta} = &-\sum_{\substack{j=1\\ j\neq i}}^{k} \frac{m_j}{\rho_j}\Big((u_i - u_j)^\beta (x_i - x_j)^\alpha + (u_i - u_j)^\alpha (x_i - x_j)^\beta \\ &- \frac{2}{3}\delta^{\alpha\beta}(u_i - u_j)\cdot(x_i - x_j)\Big)\frac{w'(|x_i - x_j|)}{|x_i - x_j|}. \end{aligned} \tag{16.115}$$

Smoothed Particle Equations of Motion

The complete set of discretized equations of motion for a Newtonian fluid are Eqs. (16.103) and (16.112)–(16.115), the parcel motion equation

$$\dot{x}_i^\alpha = u_i^\alpha - \mathrm{mc}\sum_{\substack{j=1\\ j\neq i}}^{k} \frac{2m_j}{\rho_i + \rho_j}(u_i - u_j)^\alpha w(|x_i - x_j|), \tag{16.116}$$

and the equation of state

$$P = B\Big(\Big(\frac{R}{\rho_b}\Big)^\gamma - 1\Big). \tag{16.117}$$

Here, the second term on the right-hand side of Eq. (16.116) is called the Monaghan correction [75] with $0 \le \mathrm{mc} < 1$. It is put in to smooth out the computation using an average velocity that takes into account neighboring parcels. For the equation of state, P is the pressure and R is the density along a fluid path, ρ_b is a bulk (or reference) density (which is usually taken to be the density at the initial fluid surface so that the pressure vanishes on this set), γ is a material dependent exponent ($\gamma = 7$ is traditional for water), and B is a constant that has units of pressure ($\mathrm{kg}/(\mathrm{m}\,\mathrm{sec}^2)$). As we will show in the next section, a viable choice for B is $B = \rho_b c^2/\gamma$ (see Eq. (16.121)).

16.7.6 Artificial Viscosity

The equations of motion for SPH derived so far suffer from as least one implementation problem: The momentum balance involves a sum within a sum due to the required computation of the tensor ε when the viscosity μ is not zero. This adds a layer of complexity in computer code and an additional computational expense. In addition, as should be clear, all the physics concerning viscosity (what makes a fluid a fluid) is modeled in the stress tensor. It is not clear how to obtain the best results. Should the SPH approximation be as faithful as possible to the continuous Navier–Stokes model, or should some new model for viscosity be implemented that takes into account the discretization into parcels that lies at the heart of SPH? This issue is not yet settled and remains an area of active research.

J. J. Monaghan [75] introduced an artificial viscosity into SPH—perhaps influenced by J. von Neumann and R. D. Richmeyer who introduced[2] in 1950 a form of artificial viscosity to study shock waves—which has been modified subsequently in several directions. The commonly used artificial viscosity Π replaces the stress tensor divided by the density in the momentum balance [Eq. (16.112)], which may then be rearranged into the form

$$\dot{u}_i^\alpha = g^\alpha - \sum_{\substack{j=1\\ j\neq i}}^{k} m_j \frac{w'(|x_i - x_j|)}{|x_i - x_j|}\left(\frac{p_i}{\rho_i^2} + \frac{p_j}{\rho_j^2} + \Pi_{ij}\right)(x_i - x_j)^\alpha. \quad (16.118)$$

The definition of this artificial viscosity Π requires some consideration of pressure and sound speed. We note without further explanation that the pressure in the equations of motion for a fluid is the thermodynamic pressure; it can be scaled so that it vanishes at the fluid surface (which is supposed to be exposed to the atmosphere). In this case, the scaled pressure is called the gauge pressure (simply pressure minus atmospheric pressure). The static pressure of the fluid (that is, the pressure in case the fluid is not moving) changes with depth

$$p = \rho g \times \text{depth} \quad (16.119)$$

and the atmospheric pressure is taken to be zero.

We will determine the pressure using the equation of state [Eq. (16.117)]. The derivative of pressure with respect to density evaluated at the reference

[2] *Journal of Applied Physics*, **21**, 232–247.

density is given by

$$\left.\frac{dp}{d\rho}\right|_{\rho=\rho_b} = \frac{B\gamma}{\rho_b}\left(\frac{\rho}{\rho_b}\right)^{\gamma-1}\Bigg|_{\rho=\rho_b} = \frac{B\gamma}{\rho_b}. \tag{16.120}$$

It has the units of the square of velocity. The square root of this quantity is defined to be the sound speed c; that is,

$$c^2 := \frac{B\gamma}{\rho_b}. \tag{16.121}$$

By following this description for water, we have that $B \approx 2 \times 10^8$. This large value of B reflects the near incompressibility of water. If water were incompressible, the speed of sound would be infinite.

In keeping with the SPH methodology, we may also consider the sound speed to vary over the parcels of fluid due to their separations in space. The fluid parcels carry density; therefore, we may consider the sound speed at the ith parcel to be

$$c_i := \left(\frac{B\gamma}{\rho_b}\right)^{1/2}\left(\frac{\rho_i}{\rho_b}\right)^{(\gamma-1)/2}. \tag{16.122}$$

We are now ready to define the (standard or Monaghan) artificial viscosity as

$$\Pi_{ij} = \begin{cases} \dfrac{-ac_{ij}\mu_{ij} + b\mu_{ij}^2}{\rho_{ij}}, & (u_i - u_j)\cdot(x_i - x_j) \le 0; \\ 0, & (u_i - u_j)\cdot(x_i - x_j) > 0, \end{cases} \tag{16.123}$$

where a and b are artificial viscosity coefficients,

$$\begin{aligned} c_{ij} &= (c_i + c_j)/2, \\ \rho_{ij} &= (\rho_i + \rho_j)/2, \\ \mu_{ij} &= \frac{2\,\mathrm{sr}(u_i - u_j)\cdot(x_i - x_j)}{|x_i - x_j|^2 + (0.2\,\mathrm{sr})^2}, \end{aligned}$$

and sr is the support radius of the localization function. The factor 0.2 is somewhat arbitrary; it is there to prevent the tensor μ from blowing up when the distance between two parcels is small. We note that in much of the literature on SPH the support *diameter* is taken as fundamental; here we have used the support radius. One motivation for this choice of viscosity tensor is that approaching particles are repelled by the artificial force. This choice also approximates the Navier–Stokes viscosity (see [76]).

16.7.7 Localization of Navier–Stokes Viscosity

A last approach to viscosity is to localize the viscosity terms in the momentum Navier–Stokes equation

$$\rho(u_t + (u \cdot \nabla)u) = -\nabla p + \mu \Delta u + \frac{\mu}{3}\nabla(\nabla \cdot u) + \rho b, \tag{16.124}$$

which in keeping with Eq. (16.105), we will divide by density. The relevant terms are

$$\frac{\mu}{\rho}\Delta u \quad \text{and} \quad \frac{\mu}{3\rho}\nabla(\nabla \cdot u). \tag{16.125}$$

Theoretically, the sum of these terms is a representation of the divergence of the stress tensor (divided by the density), but starting with these terms leads to a new SPH approximation.

The SPH approximation of Δu is obtained as before using the divergence theorem and its corollaries. The basic SPH approximation is

$$\Delta u(x,t) \approx \int_{\mathbb{R}^m} \Delta u(y,t) W_n(y-x)\,dy = \int_{\mathbb{R}^m} \nabla \cdot \nabla u(y,t) W_n(y-x)\,dy.$$

Using the divergence theorem as before followed by an application of Green's first identity (see Exercise 16.35), it is not difficult to see that

$$\int_{\mathbb{R}^m} \nabla \cdot \nabla u(y,t) W_n(y-x)\,dy = \int_{\mathbb{R}^m} u(y,t)\Delta_y W_n(y-x)\,dy,$$

where the subscript y means that the Laplacian applies to the function $y \mapsto W_n(y-x)$ with x fixed. With our usual localization $W_n(x) = w(|x|)$ and some calculation, we find that

$$\Delta_y W_n(y-x) = \frac{w'(|x-y|)}{|x-y|} + w''(|x-y|).$$

Therefore, we have the approximation

$$\Delta u(x,t) \approx \sum_{j=1}^{k} \frac{m_j}{\rho_j}\left(\frac{w'(|x-x_j|)}{|x-x_j|} + w''(|x-x_j|)\right)u(x_j,t). \tag{16.126}$$

Of course, this approximation can be symmetrized using the same approximation for $\Delta v(x)$ for each of the constant unit basis vectors v (in place of u). It follows that

$$0 \approx \sum_{j=1}^{k} \frac{m_j}{\rho_j}(\frac{w'(|x-x_j|)}{|x-x_j|} + w''(|x-x_j|))u^{\alpha}(x,t)v_{\alpha};$$

therefore, we may subtract

$$\sum_{j=1}^{k} \frac{m_j}{\rho_j}(\frac{w'(|x-x_j|)}{|x-x_j|} + w''(|x-x_j|))u(x,t)$$

from the original approximation (16.126) to obtain the new symmetrized approximation

$$\Delta u(x,t) \approx -\sum_{j=1}^{k} \frac{m_j}{\rho_j}(\frac{w'(|x-x_j|)}{|x-x_j|} + w''(|x-x_j|))(u(x,t) - u(x_j,t)). \tag{16.127}$$

The SPH approximation of $\nabla(\nabla \cdot u)$ can be determined by a slight modification of the same procedure. We will first simplify the notation, as we did previously, by considering a field g and using $h(y)$ in place of $W_n(y-x)$. We start with the approximation

$$\nabla(\nabla \cdot g)(x) \approx \int_{\mathbb{R}^m} \nabla(\nabla \cdot g)(y)h(y)\,dy.$$

Using the product rule, we obtain the equality

$$\int_{\mathbb{R}^m} \nabla(\nabla \cdot g)(y)h(y)\,dy = -\int_{\mathbb{R}^m} \nabla \cdot g(y)\nabla h(y)\,dy.$$

Perhaps the simplest way to proceed is to compute in components; that is, to write

$$-\int_{\mathbb{R}^m} \nabla \cdot g(y)\nabla h(y)\,dy = -\int_{\mathbb{R}^m} (g^1_{,1} + \cdots + g^m_{,m}) \begin{pmatrix} h_{,1} \\ \vdots \\ h_{,m} \end{pmatrix} dy,$$

where the superscripts denote the components of g and the notation $,j$ in subscripts denotes partial differentiation with respect to the jth component of the space variable y. Integration by parts—yet another version of the divergence theorem—in the component form

$$\int_B u_{,i} v\,dx = \int_{\partial B} uv\,dS - \int_B uv_{,i}\,dx$$

may now be applied to each term in each component of the field being integrated with the observation that all boundary terms vanish when B is an appropriate ball containing the support of the localization function. After some algebra the result is the equality

$$-\int_{\mathbb{R}^m} \nabla \cdot g(y)\nabla h(y)\,dy = \int_{\mathbb{R}^m} \operatorname{Hess} h(y) g(y)\,dy,$$

where $\operatorname{Hess} h$ is the Hessian matrix $h_{,ij}$ of second partial derivatives. Thus we have outlined a derivation of the approximation

$$\nabla(\nabla \cdot u)(x,t) \approx \int_{\mathbb{R}^m} \operatorname{Hess}_y w(|x-y|) u(y,t)\,dy.$$

Here, the Hessian matrix of the localization function has components

$$\operatorname{Hess}_y^{\alpha\beta} w(|x-y|) = \frac{w'(|x-y|)}{|x-y|}\delta^{\alpha\beta} + \Big(\frac{w''(|x-y|)}{|x-y|^2} - \frac{w'(|x-y|)}{|x-y|^3}\Big)(x-y)^\alpha (x-y)^\beta$$

After symmetrization, the SPH approximation is

$$\nabla(\nabla \cdot u)(x,t) \approx -\sum_{j=1}^{k} \frac{m_j}{\rho_j} \operatorname{Hess}_y^{\alpha\beta} w(|x-x_j|)(u(x,t) - u(x_j,t)). \tag{16.128}$$

Exercise 16.35. Prove Green's first identity

$$\int_B \nabla v \cdot \nabla u\,dx = -\int_B u\Delta v\,dx + \int_{\partial B} u\nabla v \cdot \eta\,dS,$$

where η is the outer unit normal on ∂B and both u and v are smooth functions (compare Eq. (16.46)).

Localization Functions

There are an infinite number of choices for the localizing function w. A natural and early choice is called the Lucy function (see [68]). After a support radius $\mathrm{sr} > 0$ is specified, the Lucy function $w : \mathbb{R} \to \mathbb{R}$ is given by

$$w(s) = \begin{cases} \mathrm{lnc}\,(1 + 3\frac{|s|}{\mathrm{sr}})(1 - \frac{|s|}{\mathrm{sr}})^3, & 0 \le |s| < \mathrm{sr}; \\ 0, & |s| \ge \mathrm{sr}, \end{cases} \tag{16.129}$$

where the Lucy normalizing constant lnc (chosen so that condition (16.82) holds) is $5/(4\,\mathrm{sr})$ if the function is to be used for one-dimensional flow, $5/(\pi\,\mathrm{sr}^2)$ for two dimensions, and $105/(16\pi\,\mathrm{sr}^3)$ for three dimensions. The corresponding family of functions $W_{\mathrm{sr}}(s) := w(|x|)$ is localizing as $\mathrm{sr} \to 0$ (see Exercise 16.36).

The Gaussian localization

$$w(s) = \mathrm{gnc}\, e^{-(\lambda s/\,\mathrm{sr})^2} \tag{16.130}$$

for a suitable $\lambda > 0$ with gnc equal to $\lambda^2(\pi\,\mathrm{sr}^2)^{-1}$ for two dimensions is often used even though this function is not compactly supported (see Exercise 16.38 for three dimensions). Of course, in numerical computations, the Gaussian is compactly supported because the exponential converges rapidly to zero as $s \to \infty$.

A function suggested by Wendland [115] is given by

$$w(s) = \begin{cases} \mathrm{wnc}\,(1 - \frac{|s|}{\mathrm{sr}})^4(4\frac{|s|}{\mathrm{sr}} + 1), & 0 \le |s| < \mathrm{sr}; \\ 0, & |s| \ge \mathrm{sr}, \end{cases} \tag{16.131}$$

where wnc is equal to $7/(\pi\,\mathrm{sr}^2)$ in two dimensions (see Exercise 16.38 for three dimensions).

The localizing function

$$w(s) = \begin{cases} \mathrm{cc}\,(1 - \frac{|s|}{\mathrm{sr}})^3(1 + 3\frac{|s|}{\mathrm{sr}} + 6(\frac{|s|}{\mathrm{sr}})^2), & 0 \le |s| < \mathrm{sr}; \\ 0, & |s| \ge \mathrm{sr}, \end{cases} \tag{16.132}$$

where cc is equal to $7/(2\pi\,\mathrm{sr}^2)$ in two dimensions is useful for implementation with localization of the Navier–Stokes viscosity where second derivatives of the localizing function appear. The second derivative of the function vanishes at $s = 0$, which avoids blowup for particles that are close together.

Because the integral over space is normalized to unity, *the localizing function carries units*: inverse length, inverse area, or inverse volume depending on the dimension.

Exercise 16.36. Prove that Lucy's family of functions is localizing.

Exercise 16.37. Verify the normalizing constants in the Lucy and Gaussian localizing functions.

Exercise 16.38. Find the normalization constant for the Gaussian and Wendland localization functions for the case of three dimensions.

16.7.8 SPH Numerical Implementation

We have in place all the ingredients for SPH discretization of fluid motion. In essence, after a choice of smoothing radius h for the localization function, the constants in the equation of state, and the other parameters in the system (mass, viscosity, and gravity) we have a (large) system of ODEs for the positions x_i, velocities u_i, and densities ρ_i of the fluid parcels, which we can approximate by a numerical method. In this section, we will discuss issues related to the practical implementation of SPH into a computer code.

Water

Let us agree to work in the standard units kg (kilograms), m (meters), and sec (seconds). The physical properties of water at room temperature together with other required physical constants are given approximately as follows:

$$\begin{array}{ll} \text{density} & 997\,\mathrm{kg}\,/\,\mathrm{m}^3, \\ \text{viscosity} & 1.002\times10^{-3}\,\mathrm{kg}\,/(\mathrm{m}\,\mathrm{sec}), \\ \text{kinematic viscosity} & 1.005\times10^{-6}\,\mathrm{m}^2\,/\,\mathrm{sec}, \\ \text{sound speed} & 1.5\times10^{3}\,\mathrm{m}\,/\,\mathrm{sec}, \\ \text{atmospheric pressure} & 10^{5}\,\mathrm{kg}\,/(\mathrm{m}\,\mathrm{sec}^2), \\ \text{gravity} & 9.8\,\mathrm{m}\,/\,\mathrm{sec}^2\,. \end{array} \tag{16.133}$$

The symbol g has been used so far in this chapter to denote the gravitational field. In rectangular coordinates (x, y, z), where the positive z-axis points in the direction of the outer normal of the surface of the Earth, the gravitational field near the surface of the Earth is taken to be

$$g\begin{pmatrix} 0 \\ 0 \\ -1 \end{pmatrix}, \tag{16.134}$$

where the symbol g is redefined to be the gravitational constant near the surface of the Earth; that is, $g \approx 9.8\,\mathrm{m}\,/\,\mathrm{sec}^2$.

Mach Number and Density Fluctuation

Starting with the momentum balance (in its simplest Eulerian form)

$$\rho \frac{Du}{Dt} = -\nabla p,$$

we have (using Eqs. (16.120) and (16.121) together with the chain rule differentiation $\frac{dp}{dx} = \frac{dp}{d\rho}\frac{d\rho}{dx}$) the approximation

$$\left.\frac{Du}{Dt}\right|_{\rho=\rho_b} \approx -\frac{1}{\rho_b} c^2 \frac{\rho - \rho_b}{\Delta x} \tag{16.135}$$

(where Δx denotes the change in x and ρ_b is the bulk density). The material derivative in units is length/time^2, which we may view as bulk (or characteristic) velocity u_b per time; that is,

$$\left.\frac{Du}{Dt}\right|_{\rho=\rho_b} \approx \frac{u_b}{\Delta t}.$$

Employing the (dimensionless) Mach number

$$M := \frac{u_b}{c},$$

using the approximation $u_b = \Delta x/\Delta t$, and rearranging Eq. (16.135), we have that

$$\frac{|\rho - \rho_b|}{\rho_b} \approx M^2. \tag{16.136}$$

Because the speed of sound in water is large, in many models the sound speed is taken to be infinite. This corresponds to the relative fluctuation in density being zero; that is, the fluid is assumed to be incompressible. In SPH modeling, the fluid is assumed to be slightly compressible as it should be because it is discretized into parcels that we imagine as discrete particles in the computational domain. Thus, for example, as suggested by Monaghan [75], if we take the relative density fluctuations to be about about 1%, then M is about 0.1.

In view of the underlying ideas for SPH, we will compute with parcels of fluid instead of a continuous fluid. The finite number of parcels have locations; therefore, they are separated. Thus, the density *of the parcels* is smaller than the fluid density. Because the second derivative of the equation of state with respect to density is positive, the sound speed decreases with a decrease in density. For this reason we should expect that the sound speed for our SPH approximation is slower than the sound speed of a real fluid.

For the SPH simulation of water, we may take

$$M = 0.1, \qquad \gamma = 7, \qquad \rho_b = 10^3,$$

and consequently the pressure coefficient in the equation of state is approximately

$$B = \frac{c^2 \rho_b}{\gamma} = \frac{\rho_b v_b^2}{M^2 \gamma} = 10^5 \frac{v_b^2}{7} \approx 10^4 v_b^2.$$

The value of B is therefore determined by our choice for the bulk velocity.

In practice, the choice of bulk velocity for numerical simulation is problem dependent and not obvious. Our construction relates the sound speed and the bulk velocity $v_b = Mc$. By fixing the Mach number, we are also relating the bulk velocity to the compressibility of the fluid. The more compressible the fluid, the lower the sound speed. Thus, for example, the bulk velocity should not be taken too small for SPH simulation of water, where the fluid is in reality only slightly compressible. Some experimentation is required to find a suitable choice of bulk velocity for SPH simulation.

Numerical Integration

Our discretized equation of motion is a system of ODEs; its solutions may be approximated by using a numerical method, some of which are described in Chapter I. In practice, the implementation of the smoothed particle method requires using a large number of fluid parcels (at least several hundred and perhaps several thousand parcels). The scale of such a problem will soon overwhelm available computational power and storage space, especially if several function evaluations are required per step. To achieve at least second-order accuracy, the explicit improved Euler method is viable, which requires two function evaluations per step. A second-order multistep method, such as the Adams–Bashforth two-step method requires only one function evaluation per step. Indeed, for the initial value problem

$$\dot{y} = f(y, t), \qquad y(t_0) = y_0$$

with time step Δt, we may use another method, such as the explicit improved Euler, to compute y_1 with subsequent approximations given by

$$y_{n+1} = y_n + \frac{\Delta t}{2}(3f(y_n, t_n) - f(y_{n-1}, t_{n-1})).$$

Here, of course, $t_{n+1} = t_n + \Delta t$.

Although good results can be obtained by employing a general ODE solver, there are special methods available for the second-order differential equations encountered in mechanics; that is, initial value problems of the form

$$\ddot{y} = f(y,t), \quad y(t_0) = y_0, \quad \dot{y}(t_0) = z_0.$$

One such algorithm is second order and requires only one function evaluation per step of the position variable y. To describe the algorithm, let us first write the differential equation as the equivalent first-order system

$$\dot{y} = z, \qquad \dot{z} = f(y,t).$$

As for the Adams–Bashforth method, preliminary computations are required to start the algorithm. Euler approximations may be used for the initial velocity advanced by half of a time step and the initial position a full time step

$$z_{1/2} = z_0 + \frac{\Delta t}{2} f(y_0, t_0), \qquad y_1 = y_0 + \frac{\Delta t}{2} z_{1/2}$$

to start the process. The quantities

$$z_{1/2}, \qquad \text{fv} := f(y_0, t_0), \qquad y_1$$

are used in the next step. For $n \geq 1$, the time stepping is

$$\begin{aligned} z_n &= z_{n-1/2} + \frac{\Delta t}{2} \text{fv}, \\ \text{fv} &= f(y_n, t_n), \\ z_{n+1/2} &= z_{n-1/2} + \Delta t \, \text{fv}, \\ y_{n+1} &= y_n + \Delta t z_{n+1/2}. \end{aligned} \tag{16.137}$$

A variant of the numerical method for SPH, when the Monaghan correction [Eq. (16.116)] is employed, is the following predictor-corrector algorithm (see [75]). For the initial value problem

$$\dot{y} = f(y,t), \qquad y(t_0) = y_0,$$

choose a time step Δt and use a second-order startup routine (for instance, the explicit improved Euler method) to approximate and store two quantities: the initial state advanced by a full time step y_1, and the function evaluation $f(y_{1/2}, t_{1/2})$ where $y_{1/2}$ is the state advanced by half of a time step (that is, the approximate state at time $t_{1/2} = t_0 + \Delta t/2$). Subsequent states are approximated using the predictor-corrector scheme (Adam's–Bashforth)

$$\begin{aligned} y_{n+1/2} &= y_n + \frac{\Delta t}{2} f(y_{n-1/2}, t_{n-1/2}), \\ y_{n+1} &= y_n + \Delta t f(y_{n+1/2}, t_{n+1/2}), \end{aligned} \tag{16.138}$$

which requires storage of the new state y_{n+1} *and* the function evaluation $f(y_{n+1/2}, t_{n+1/2})$ at each time step. Only one additional function evaluation is required to compute each new time step. This numerical method is second order (see Exercise 16.39), which is perhaps a surprise given that the natural predictor step is $y_{n+1/2} = y_n + \frac{\Delta t}{2} f(y_n, t_n)$ because of the half time step. Of course, the natural predictor with the same corrector step is also second order, but this alternate scheme requires two function evaluations per step.

Because we are approximating the solution of a PDE that depends on space and time, we should expect that the step size for stable numerical integration must respect the CFL condition (as in inequality (5.57))

$$\Delta t \leq \frac{1}{c} \Delta x.$$

In practice some experimentation is necessary to determine an appropriate step size.

Exercise 16.39. (i) Prove that numerical method (16.137) is second order. (ii) Write a code to implement the numerical integration and verify that it is second order for the Duffing equation $\ddot{y} = y - y^3$. The following information might be useful for debugging: For the initial data $y(0) = 0.2$ and $\dot{y}(0) = 0.1$, $y(10) \approx 1.4036365$.

Exercise 16.40. (i) Prove that the predictor-corrector method [Eq. (16.138)] is second order. (ii) Write a code to implement this method and verify that it is second order for Duffing's equation as in Exercise 16.39.

Exercise 16.41. (i) Prove that the Adams–Bashforth two-step method [Eq. (16.138)] is second order. (ii) Write a code to implement this method and verify that it is second order for Duffing's equation as in Exercise 16.39.

Boundary Conditions

The treatment of boundary conditions is a major problem for SPH because there seems to be no natural way to incorporate the fluid boundary conditions into the system of ODEs produced by the discretization. On the other hand, the methods discussed in this section produce excellent results.

In fluid dynamics the appropriate boundary conditions depend on the treatment of viscosity. Experiments with fluids reveal that the velocity field vanishes at the boundary. Also, a mathematical argument can be used to show that conservation of mass requires the velocity field to be tangent to solid boundaries. This later boundary condition is sufficient for the inviscid approximation (Euler's equations) to have unique solutions. On the other hand, for the viscous case, the extra physical condition is required: the velocity field must vanish at the boundary. Much of the advanced application of fluid dynamics is predicated on the notion that viscosity must be considered in a thin layer adjacent to the boundary (the boundary layer), but away from the boundary, the fluid may be treated as if it were invicid. The interaction between these two regimes is the source of complicated (but interesting) behavior; for example, boundary layers are the source of vorticity (rotation in the fluid) and many other effects (see Section 17.3).

An essential problem with SPH is that fluid parcels might penetrate boundaries as the integration proceeds.

A useful method to mitigate against (but not ensure) that a boundary is not crossed is to incorporate stationary fictitious parcels of fluid on the boundary and fictitious particles (placed outside the physical region near the boundary) that move in the opposite direction from the fluid flow on the other side of the boundary.

A second method (which is also often used in addition to fictitious particles) is to incorporate into the equations of motion a (short range) repulsive force at the boundary to mimic the repulsive force between molecules. A standard way to do this is to use a Lennard–Jones type force that is cut off so that it is purely repulsive. This force can be implemented as follows: In case x_i is a fluid parcel and x_j is a fictitious parcel on the boundary, d_{ij} is the distance between them, d_0 is a characteristic parcel spacing, and m and n are positive integers with $m > n$, the parcel x_i is subjected to the force

$$\begin{cases} K\left[\left(\frac{d_0}{d_{ij}}\right)^m - \left(\frac{d_0}{d_{ij}}\right)^k\right]\frac{1}{d_{ij}^2}(x_i - x_j), & \frac{d_0}{d_{ij}} \geq 1; \\ 0, & \frac{d_0}{d_{ij}} < 1, \end{cases} \tag{16.139}$$

where K is a constant proportional to the mass times the square of a characteristic velocity (for instance $\rho\,\mathrm{sr}^3\,v_b^2$ with v_b the bulk velocity). Typical choices for the exponents are $m = 12$ and $n = 6$ and $m = 4$ and $n = 2$. Note: We have written the fluid acceleration on the left-hand side of the SPH equation of motion [Eq. (16.118)]; therefore, boundary or other forces that might act on fluid parcels must be divided by the mass of the parcel before they are included on the right-hand side of this equation.

The cutoff Lennard–Jones force suffers from not being smooth at $d_0 = d_{ij}$. A simple remedy is to use instead

$$\begin{cases} K\Big[m\big(\frac{d_0}{d_{ij}} - 1\big)^{m-1}\frac{d_0}{d_{ij}}\Big]\frac{1}{d_{ij}^2}(x_i - x_j), & \frac{d_0}{d_{ij}} \geq 1; \\ 0, & \frac{d_0}{d_{ij}} < 1, \end{cases} \tag{16.140}$$

where $m \geq 2$.

Another viable boundary force is derived from the following considerations: The initial placement of parcels should determine the future distance of parcels from the boundary. An ideal initial parcel adjacent to the boundary should remain at its initial distance. The boundary force should blow up at the boundary, smoothly go to zero as the distance from the boundary approaches some multiple $\delta > 0$ of the initial parcel spacing, and remain zero for all larger distances from the boundary. Such a force is given by

$$\begin{cases} K\Big[\big(\frac{d_{ij}}{d_0}\big)^{-1/2}\big(\delta - \frac{d_{ij}}{d_0}\big)^2\frac{1}{d_0 d_{ij}}(x_i - x_j), & \frac{d_{ij}}{d_0} < 1; \\ 0, & \frac{d_{ij}}{d_0} > 1. \end{cases} \tag{16.141}$$

The boundary treatments suggested so far are in keeping with SPH methodology: the boundary is discretized by replacing it with boundary parcels. This treatment is the correct approach if there are interactions between the fluid and the boundary in both directions; that is, the fluid moves the boundary. In case the boundary motion is not influenced by the fluid, the boundary force on a fluid parcel may be considered to be some function of the state variables of a fluid parcel and its distance to the boundary. The boundary forces mentioned above may be employed from this point of view, In effect, d_{ij} would be replaced by the distance of the ith parcel to the boundary. The coefficient K might also be a function of the state of this fluid parcel. Of course, a hybrid method might be appropriate if parts of the boundary are moved by the fluid and the rest of the boundary is fixed.

An ODE model should be defined so that the physical region of interest is an invariant set as the system evolves in forward time. The SPH approximate fluid model generally does not automatically preserve the region that contains the fluid. Perhaps there is a way to redefine the discretization so that the physical region is invariant?

Exercise 16.42. (i) Suppose that f is a scalar function of a scalar variable and define the potential U by the rule

$$U(p) = f\Big(\frac{\sigma}{|p - q|}\Big),$$

where q is some fixed point and σ is a fixed scalar. Determine the force determined by this potential (that is, the negative gradient of the potential).
(ii) Find the potential whose gradient is the force given in display (16.140).

Programming

The appearance of the localization function in the discretized equations of motion makes it unnecessary to sum over all parcels, only those that are within the support radius of the parcel under consideration. On the other hand, the parcels are moving, so some check must be made to determine which parcels are nearby.

An effective method for tracking neighboring parcels is to construct a stationary grid of boxes (with edge length the support radius) over the physical region and its boundaries. After each time step, the parcels are assigned to boxes according to their spatial positions. In the summations only the parcels in boxes with at least one corner in common with the box containing the parcel under consideration are summed. In two dimensions, each summation is over nine boxes.

The SPH method can be easily parallelized: different processors can simultaneously compute time step updates of different parcels.

16.8 SIMULATION OF A FREE-SURFACE FLOW

By tradition, the first application to consider is the dam break problem: Imagine still water behind a vertical dam on a river with a horizontal bottom extending downstream from the dam. Determine the profile of the escaping water and the position of its leading edge after the dam instantaneously disappears.

The dam break problem is an excellent test bed to debug code. For the experiments reported here, an SPH code was written that employs the Monaghan correction [Eq. (16.116)] to the velocity with mc $= 0.5$, Wendland localization [Eq. (16.131)], artificial viscosity with $a = 0.01$ and $b = 0.0$, the repulsive boundary force (16.141) with $K = 1000.0$ that acts normal to the boundary, and the predictor-corrector numerical integration scheme [Eq. (16.138)].

An initial 25 meter by 25 meter reservoir is modeled using a 54 by 54 evenly distributed rectangular grid of fluid parcels. Boundary parcels with the same vertical spacing as the fluid parcels are placed along the left-hand reservoir vertical wall (taken to be 40 meters high), the river bottom (taken to be 100 meters long), and a 40 meters high vertical wall at the 100 meter mark downstream, which might model an obstruction. The initial velocities of the fluid parcels are set to zero while their densities are computed using equation of state (16.117) (with $\gamma = 7$). In fact, the density ρ is computed from the equation substituted for the pressure given as a function of depth via Bernoulli's law:

$$B\Big(\big(\frac{\rho}{\rho_0}\big)^\gamma - 1\Big) = \rho_0 g(\mathrm{wd} - y).$$

The mass of the fluid parcels is equally distributed: mass is reference density (10^3 kg / m^3) times the area of the initial region divided by the number of parcels, and the mass does not change with time. With this choice, units are not consistent. If desired, the inconsistent units can be repaired by replacing the area of the region by the volume of a slab whose face area is the area of the two-dimensional region and whose thickness is one meter.

The smoothing length (radius of support of the smoothing function) was taken to be 0.4 and the Mach number is 0.1.

The system was integrated 4.5 seconds forward in time using 9000 steps (equivalently the time step size is 0.0005 seconds). Some results of the numerical experiment are shown in Fig. 16.18. The leading edge of the flow travels at the average velocity of approximately $18.75\ \mathrm{m}/\mathrm{sec}$ (or about 42 miles per hour). If the reservoir behind the dam were infinitely long, it is possible to show that the leading edge of the flow moves at approximately $2\sqrt{gH}$, where H is the water depth in the reservoir; that is, about $32\ \mathrm{m}/\mathrm{sec}$ (see [101]). Our reservoir is finite and thus the speed is expected to be lower.

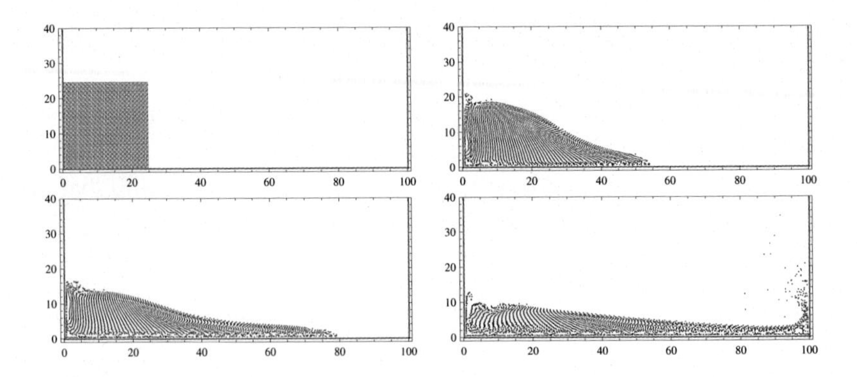

Fig. 16.18 The top left panel shows the initial configuration of water parcels and the boundary parcels for the dam break problem whose description begins on page 506, the top right panel shows the flow after 2 seconds, the bottom left panel shows the flow after 3 seconds, and the bottom right panel shows the flow after 4.5 seconds.

Exercise 16.43. [Dam Break Problem] Write an SPH code and use it to approximate the dam break problem described in the text. Currently there are good open source codes available for SPH simulations, but there is no substitute—if you wish to learn the subject—for writing your own code. To make life simple, you may start by writing a code that runs through all pairs of parcels (including boundary parcels) to compute interactions between parcels whose separation is less than the support radius of the smoothing function.

Exercise 16.44. [Still Water] Apply your SPH code to the 25 meter by 25 meter reservoir with no dam break. The initial placement of fluid parcels is an approximation of zero velocity water. There is no reason to believe that your configuration is a steady state of the SPH equations. Describe the flow that is obtained in the reservoir by numerical simulation with your SPH code. Does the flow approach a steady state? Discuss your results.

Let us return to the Tiger Fountain free-surface flow discussed in Section 15.5.

Simulation of the full three-dimensional flow problem requires high-performance computing, that may be unavailable to students. Thus, we will discuss a two-dimensional flow problem that still seems to capture the essential features of the real flow. We will simulate a thin slice of the flow parallel to the bulk flow velocity vector as in Section 15.5.

For the implementation of SPH, it is convenient to choose rectangular Cartesian coordinates (x, y) with the x-axis parallel to the flow and the y-axis the vertical direction of the slice. As in Table 15.46 but with names for some of the variables, recall that

$$\begin{array}{lll} \text{water depth} & \text{wd} & = 0.01746\,\text{m} \\ \text{plate slice length} & \ell & = 0.47625\,\text{m} \\ \text{bulk speed} & v_b & = 0.296333\,\text{m}\,/\,\text{sec} \\ \text{bulk density} & \rho_b & = 997\,\text{kg}\,/\,\text{m}^3 \\ \text{viscosity} & \mu & = 1.002 \times 10^{-3}\,\text{kg}\,/(\text{m}\,\text{sec}) \\ \text{kinematic viscosity} & \nu & = 1.005 \times 10^{-6}\,\text{m}^2\,/\,\text{sec} \\ \text{sound speed} & c & = 1.5 \times 10^3\,\text{m}\,/\,\text{sec} \\ \text{gravity} & g & = 9.8\,\text{m}\,/\,\text{sec}^2 \end{array} \tag{16.142}$$

The flat bottom of the rectangular region containing fluid is $\{(x, y) : y = 0,\ 0 \leq x \leq \ell\}$ and the fluid moves from right to left.

Initial Data

We must prescribe initial data for the state variables: position, velocity, density, and pressure. Because the pressure and the density are related via the equation of state [Eq. (16.117)], it suffices to determine only one of these variables.

Of the many possible choices for the initial data, let us imagine an initial steady flow that is parallel to the bottom. In Eulerian variables, the equations of motion are

$$\begin{aligned} &\begin{pmatrix} u_1 u_{1x} \\ 0 \end{pmatrix} = -\frac{1}{\rho}\begin{pmatrix} p_x \\ p_y \end{pmatrix} + \begin{pmatrix} 0 \\ -g \end{pmatrix} + \nu \begin{pmatrix} u_{1xx} + u_{1yy} \\ 0 \end{pmatrix}, \\ &(\rho u_1)_x = 0. \end{aligned} \tag{16.143}$$

The equation of continuity implies that the product ρu_1 is a function of y alone. Let us assume that the velocity and the density are functions of y alone. Using this assumption and the equation of state, it follows that p is a function of y alone; and, the first component of the momentum balance reduces to

$$u_{1yy} = 0.$$

Thus $u_1(y) = Ay + B$ for some constants A and B. The no-slip boundary condition at the bottom implies that $B = 0$. Under the assumption that the surface is moving with bulk velocity, we have that

$$u_1(y) = \frac{v_b}{\text{wd}} y. \tag{16.144}$$

The second component of the momentum balance together with the equation of state can be recast as a differential equation for the density:

$$\frac{B\gamma}{\rho_b \rho}\Big(\frac{\rho}{\rho_b}\Big)^{\gamma-1}\rho_y + g = 0, \tag{16.145}$$

which has the general solution

$$\Big(\frac{\rho}{\rho_b}\Big)^{\gamma-1} = \frac{\rho_b(\gamma-1)}{B\gamma}(A - gy), \tag{16.146}$$

where A is a constant. A useful rearrangement of this equation is given by

$$B\Big(\Big(\frac{\rho}{\rho_b}\Big)^{\gamma} - 1\Big) = B\Big(\frac{(\gamma-1)\rho}{B\gamma}(c - gy) - 1\Big). \tag{16.147}$$

Both sides of Eq. (16.147) must give the internal pressure, which is defined up to a constant. Let us take advantage of this fact by specifying the pressure to be zero at the surface of the fluid. In this case (with $y = \mathrm{wd}$ in the right-hand side of the equation), it follows that

$$A = \frac{B\gamma}{\rho(\gamma-1)} + g\,\mathrm{wd};$$

therefore, for the equation of state to be compatible with our choice of steady flow, we must have

$$B\Big(\Big(\frac{\rho}{\rho_b}\Big)^{\gamma} - 1\Big) = \frac{(\gamma-1)}{\gamma}\rho g(\mathrm{wd} - y). \tag{16.148}$$

In particular, density is given implicitly as a function of depth. The value of the density for a given depth can be efficiently approximated by Newton's method.

CHAPTER 17

Channel Flow

Imagine a three-dimensional flow whose most important features are observed in a fixed direction. The prototypical case is channel flow. But, flow in a pipe, in an artery (perhaps with flexible walls), or in a river all share this feature. The main purpose of this chapter is to show how to obtain partial differential equation (PDE) models for such flows where the model equations depend on time and exactly one spatial dimension. Because such models are obtained by making approximations, many different models may be derived.

The main motivation for reducing the fluid equations to simpler models, such as those to be obtained for channel flows, is to gain a foothold on understanding the underlying physical principles that produce observed phenomena; for example, hydraulic jumps (to be discussed), roll waves, solitary waves, or tidal bores. Historically, simplifications were made to derive approximate formulas useful for engineering calculations. For instance, in the case of channel flows, formulas (by Chézy and Manning mentioned below) were obtained to compute the velocity of the flow from channel configurations. Given the advances in numerical methods for computational fluid mechanics, modern numerical approximations for engineering applications should be made using the Navier–Stokes equations when possible; that is, when the Reynolds number is not too large. But, there is certainly a need for numerical computations using simplified equations for the flows observed in nature: most of them have large Reynolds numbers (see [40] and Exercise 17.2). One such simplified model for channel flow is discussed in this section: open channel flow. Although

A basic assumption restricts the geometry of the channel:

Assumption 17.1. The flow is confined to a channel with a straight axis. Each cross section of the fluid filling the channel is bounded by the channel wall that is convex and oriented to hold water. The upper boundary of the flow is the fluid surface, which is exposed to the atmosphere.

An Invitation to Applied Mathematics: Differential Equations, Modeling, and Computation.
http://dx.doi.org/10.1016/B978-0-12-804153-6.50017-8, Copyright © 2017 Elsevier Inc. All rights reserved.

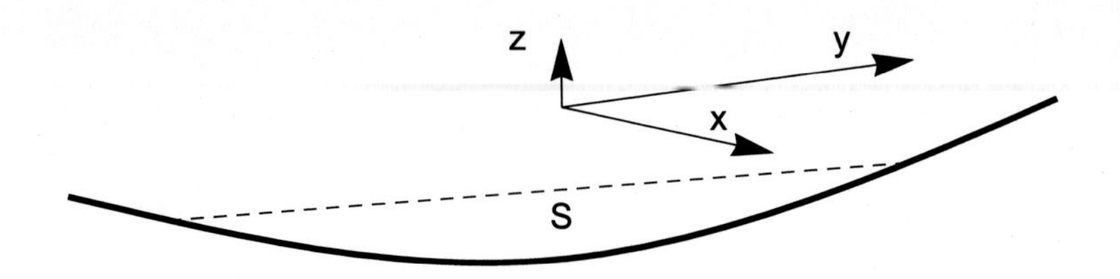

Fig. 17.1 Schematic diagram of channel cross section with channel coordinate system, channel bottom (thick curve), fluid surface (dashed), and wet part of the section labeled S.

Cartesian coordinates x, y, and z with direction vectors e_1, e_2, and e_3 forming a right-hand coordinate system (depicted in Fig. 17.1) are fixed so that e_1 points in the downstream direction, e_2 points in a horizontal direction, and e_3 has a positive inner product with the upward vertical direction defined by the direction of the gravitational field. Time is denoted by the variable t. The channel axis may be inclined with respect to the horizontal defined by the gravitational field. Cross sections of the channel (which play a prominent role in the analysis) are taken perpendicular to the channel axis, which has direction e_1.

The wet part of the channel cross section at (x, t) is denoted $S(x, t)$ and its area by $A(x, t)$. The unit normal to $S(x, t)$ in the downstream (positive x) direction is the Cartesian unit vector e_1, which is in the direction of the channel axis.

A control volume is defined to be a portion of fluid in the channel bounded by two (channel) cross sections.

Assumption 17.2. The fluid density ρ is constant.

Assumption 17.3. The flow velocity

$$u = (u_1, u_2, u_3)$$

(with components taken with respect to the channel coordinate system) is constant with respect to the second (horizontal) coordinate y on each cross section. The surface of the fluid, which is the top boundary of the wet cross section, is assumed to be a line segment with constant x and z coordinates. The channel bottom need not be flat, but its profile is assumed to be concave up. At the lowest point of the bottom, the tangent line is of course perpendicular to the channel axis. The heights of the fluid surface and the tangent line at the lowest point are given by $z = \zeta(x, t)$ and $z = B(x, t)$, respectively.

Although, under this assumption, the no-slip boundary condition is not satisfied at the horizontal boundaries of the flow unless the flow velocity is zero, it can be enforced at the bottom of the channel. For most of the modeling in this section the height of the bottom of the channel may change with time. But, for the main applications to follow, the bottom of the channel depends only on the position of the cross section as measured by the axial coordinate.

To complete notation for the dimensions of the section, let $W(x, z, t)$ denote the width of the section at (x, t) at height z. For prismatic channels (those with fixed solid boundaries that have the same cross-sectional profile at all points along the channel axis), the width is independent of x, and for prismatic rectangular channels, which will later be the simplest application, the width is constant.

Assumption 17.4. The inner product of the channel flow velocity field u and the unit vector e_1 (in the direction of the channel axis) is everywhere positive; that is, the flow velocity has no upstream component.

Under Assumption 17.3, the flow velocity field is taken to be constant with respect to the y coordinate; that is, $u = u(x, z, t)$.

The flux (measured in units of volume/time) of the fluid velocity field through a cross section with respect to the downstream unit normal is called the discharge of the channel at the cross section; it is a function of position and time given by

$$Q(x,t) := \int_{S(x,t)} u \cdot e_1 \, dS = \int_{S(x,t)} u_1 \, dS. \tag{17.1}$$

The discharge Q and *wet* cross-sectional area A are the state variables for the channel flow model that will be constructed.

17.1 CONSERVATION OF MASS

Suppose that Ω is a control volume. To set notation, let the horizontal coordinate of the upstream (respectively, downstream) cross-sectional part of the boundary of Ω (denoted as usual by $\partial\Omega$) be $x = x_U$ (respectively, $x = x_D$); that is, $S(x_U, t)$ (respectively, $S(x_D, t)$) is the upstream (respectively, downstream) cross-sectional boundary of Ω. The cross sections are at fixed locations and do not vary with time; the wet parts of these cross sections might vary with time. In particular, Ω might vary with time; for example, the position of the fluid surface might change with time while the positions in space of the upstream and downstream boundaries of Ω remain fixed. The lateral boundary of Ω is

$$\Sigma := \partial\Omega \setminus (S(x_U, t) \cup S(x_D, t)).$$

By conservation of mass, the time rate of change of the total mass of the fluid in Ω is the negative fluid flux through $\partial\Omega$ with respect to the outer unit normal N on this boundary. No mass is created or destroyed in a control volume. More precisely, conservation of mass is encoded in the equation

$$\frac{d}{dt}\int_{\Omega} \rho \, d\mathcal{V} = -\int_{\partial\Omega} \rho u \cdot N \, d\mathcal{S}. \tag{17.2}$$

For sufficiently smooth flows, there is an equivalent statement of this conservation law that is expressed as a PDE relating the state variables Q and A. Its derivation is the subject of the remainder of this section.

The time derivative of the total mass can be expressed as an iterated integral:

$$\begin{aligned} \frac{d}{dt}\int_{\Omega} \rho \, d\mathcal{V} &= \rho \frac{d}{dt}\int_{x_U}^{x_D}\int_{S(x,t)} d\mathcal{S}\, dx \\ &= \rho \int_{x_U}^{x_D} A_t(x,t)\, dx, \end{aligned}$$

where, as previously defined, A is the wet section area.

The flux term is separated into integrals over the upstream and downstream wet cross sections and the lateral boundary. Using the definition of discharge,

$$\begin{aligned} \int_{\partial\Omega} \rho u \cdot N \, d\mathcal{S} &= \rho\Big(\int_{S(x_D,t)} u_1 \, d\mathcal{S} - \int_{S(x_U,t)} u_1 \, d\mathcal{S}\Big) + \rho \int_{\Sigma} u \cdot N \, d\mathcal{S} \\ &= \rho(Q(x_D,t) - Q(x_U,t)) + \rho \int_{x_U}^{x_D}\int_{\partial S(x,t)} u \cdot N \, d\mathcal{L}\, dx \\ &= \rho \int_{x_U}^{x_D} Q_x(x,t)\, dx + \rho \int_{x_U}^{x_D}\int_{\partial S(x,t)} u \cdot N \, d\mathcal{L}\, dx. \end{aligned}$$

For notational convenience define the negative (volumetric) flux through the boundary of a wet cross section to be

$$q(x,t) = -\int_{\partial S(x,t)} u \cdot N \, d\mathcal{L}. \tag{17.3}$$

It has units of area per time with its sign is chosen so that fluid enters the channel when $q > 0$ and exits when $q < 0$. Auxiliary flow due to rain, evaporation, seepage through the channel walls, flooding, oil spills, influx via pipes with openings in the channel, or other causes, can be modeled by specifying q. An alternative possibility is to model (for example, evaporation) by some constitutive relation so that q is a function of A and Q.

Because the conservation of mass holds for all choices of x_U and x_D, its differential form is

$$A_t + Q_x = q, \tag{17.4}$$

where each term has units of area per time.

17.2 MOMENTUM BALANCE

Newton's second law for a fluid parcel Ω (as given by Eq. (11.5) but with new notation where the parcel of fluid A is now Ω, the flow is γ, Cauchy stress tensor σ, and body force per mass b) is the equation

$$\frac{d}{dt}\int_{\gamma(\Omega,t)} \rho u \, d\mathcal{V} = \int_{\partial\gamma(\Omega,t)} \sigma N \, d\mathcal{S} + \int_{\gamma(\Omega,t)} \rho b \, d\mathcal{V}, \tag{17.5}$$

which (by applying the transport theorem (A.2) to each vector component of the integral on the left-hand side of Eq. (17.5)) can be put into the more convenient equivalent form

$$\int_{\gamma(\Omega,t)} (\rho u)_t \, d\mathcal{V} + \int_{\partial\gamma(\Omega,t)} \rho u \cdot N u \, d\mathcal{S} = \int_{\partial\gamma(\Omega,t)} \sigma N \, d\mathcal{S} + \int_{\gamma(\Omega,t)} \rho b \, d\mathcal{V}.$$

Because this equation holds for every fluid parcel, so does the equation

$$\int_{\Omega} (\rho u)_t \, d\mathcal{V} + \int_{\partial\Omega} \rho u \cdot N u \, d\mathcal{S} = \int_{\partial\Omega} \sigma N \, d\mathcal{S} + \int_{\Omega} \rho b \, d\mathcal{V}. \tag{17.6}$$

Thus, we have derived an integral form for momentum balance where integration is over fluid parcels (which in this context are called control volumes) fixed in time and space.

The goal of channel flow *theory* is to replace the vector equation for momentum balance [Eq. (17.6)] in three spatial dimensions with a scalar equation in one spatial dimension that preserves, as much as possible, the main features of observed channel flows. As might be expected, there is no natural way to achieve this goal; there are several possible scalar PDEs that can be derived using different approximations. An instructive approach is discussed in this section.

Define the average velocity over a cross section to be

$$U(x,t) := \frac{1}{A(x,t)} \int_{S(x,t)} u_1 \, dS. \tag{17.7}$$

In other words, this definition states that

$$Q = UA.$$

To obtain the desired one-dimensional channel flow approximation of momentum balance, the fluid parcel in this conservation law is taken to be a control volume and the momentum balance is replaced by its dot product with the downstream unit normal. Using the notation of the previous section and the constant density assumption, the replacement for the first integral on the left-hand side of Eq. (17.6) is

$$\begin{aligned}\int_\Omega (\rho u)_t \cdot e_1 \, d\mathcal{V} &= \rho \int_{x_U}^{x_D} \int_{S(x)} u_t \cdot e_1 \, dS \, dx \\ &= \rho \int_{x_U}^{x_D} \frac{d}{dt} \int_{S(x)} u_1 \, dS \, dx \\ &= \int_{x_U}^{x_D} \rho Q_t \, dx.\end{aligned} \tag{17.8}$$

Because the channel flow is assumed not to penetrate the walls or the top fluid surface of the control volume, $u \cdot N$ vanishes on these portions of its boundary. On the upstream cross-sectional boundary $S(x_U, t)$, the dot product of the fluid velocity with the normal is $u \cdot N = -u \cdot e_1$ and on $S(x_D, t)$ the dot product is $u \cdot N = u \cdot e_1$. Therefore, the replacement for the second integral on the left-hand side of Eq. (17.6) is

$$\int_{\partial\Omega} \rho u_1 u \cdot N \, d\mathcal{S} = \int_{S(x_D,t)} \rho u_1^2 \, d\mathcal{S} - \int_{S(x_U,t)} \rho u_1^2 \, d\mathcal{S}$$
$$= \int_{x_U}^{x_D} \rho \frac{d}{dx}\Big(\int_{S(x,t)} u_1^2 \, d\mathcal{S} \Big) \, dx. \tag{17.9}$$

Assumption 17.5. When integrating a function of the dot product of the velocity and the downstream normal over a wet cross section, this quantity may be replaced by its (constant) average value over the section. More precisely, the approximation is

$$u(x,y,z,t) \cdot e_1 = u_1(x,y,z,t) \approx U(x,t) = \frac{1}{A(x,t)} \int_{S(x,t)} u_1 \, d\mathcal{S}.$$

Using Eq. (17.9) and Assumption 17.5, we have the approximation

$$\int_{\partial\Omega} \rho u_1 u \cdot N \, d\mathcal{S} \approx \int_{x_U}^{x_D} \rho \frac{d}{dx}\Big(\int_{S(x,t)} U(x,t)^2 \, d\mathcal{S} \Big) \, dx$$
$$= \int_{x_U}^{x_D} \rho \Big(\frac{Q^2}{A} \Big)_x \, dx. \tag{17.10}$$

Thus, for channel flow,

$$\int_{\Omega} (\rho u)_t \, d\mathcal{V} + \int_{\partial\Omega} \rho u \cdot N u \, d\mathcal{S} \approx \int_{x_U}^{x_D} \rho (Q_t + \Big(\frac{Q^2}{A} \Big)_x) \, dx. \tag{17.11}$$

This result is the standard approximation of the left-hand side of the momentum balance [Eq. (17.6)]. Obtaining physically meaningful approximations of the right-hand side of this equation is a more challenging task where much variation is possible.

Gravity acts vertically downward; it drives the flow in an inclined channel. Let θ denote the angle between the tangent line to the channel bottom in the axial direction and channel axis. For a horizontal bottom $\theta = 0$. For a bottom that tilts toward the center of Earth in the downstream direction, this angle is positive. Treating the bottom as an inclined plane, and resolving the gravitational force into components with respect to the unit vector in the downstream axial direction (which is tangent to the bottom) and the normal vector to the channel bottom pointing into the fluid, the magnitude of the gravitational force per mass in the direction of the channel bottom is $b \cdot e_1 = g \sin\theta$. Using this model of the gravitational (body) force on the channel flow, approximation

$$\begin{aligned}\int_\Omega \rho b\cdot e_1\, d\mathcal{V} &= \int_\Omega \rho g \sin\theta\, d\mathcal{V}\\ &= \int_{x_U}^{x_D}\int_{S(x,t)} \rho g\sin\theta\, dS dx\\ &= \int_{x_U}^{x_D} \rho g A\sin\theta\, dx. \end{aligned}\tag{17.12}$$

The next derivation of a one-dimensional model for the normal stress integrated over the boundary of the control volume Ω requires new approximations.

Using the Navier–Stokes stress (for example, Eq. (11.31)), the total stress induced by the divergence-free velocity field u is

$$\int_{\partial\Omega} \sigma N\, dS = \int_{\partial\Omega} -pN + \mu(\nabla u + (\nabla u)^T)N\, dS,$$

where p is the pressure, μ is the viscosity, and the deformation tensor

$$\frac{1}{2}(\nabla u + (\nabla u)^T)$$

(with the superscript T denoting the transpose) is the symmetric part of ∇u.

Assumption 17.6. The pressure p for the moving fluid is the steady state Eulerian pressure (given by Bernoulli's law [Eq. (13.7)])

$$p := C - g\rho(z-\zeta) - \frac{\rho}{2}u\cdot u,$$

where $\zeta = \zeta(x,t)$ is the surface height of the flow and C is a constant.

The pressure assumption [Eq. (17.6)] should be realistic for steady state channel flows. For nonsteady state flows it is problematic. Some successful models simply assume that the pressure is hydrostatic; that is, $p := C - g\rho(z-\zeta)$. At least, disregarding the translation, pressure would be zero at the surface, negative above the surface, and positive below the surface; moreover, the pressure is maximum at the bottom.

Desired approximations are most easily carried out by treating the integral of the velocity term in the pressure function as a surface integral and transforming the viscosity term in the stress to a volume integral. Using the divergence theorem, *the divergence-free assumption*, and treating, for example, the components of pN in the form $pe_i \cdot N$, the total stress is given by

$$\int_{\partial\Omega} \sigma N \, d\mathcal{S} = \int_{\Omega} \nabla(g\rho(z-\zeta)) \, d\mathcal{V} + \int_{\partial\Omega} \frac{\rho}{2} u \cdot uN \, d\mathcal{S} + \int_{\Omega} \mu \Delta u \, d\mathcal{V}, \tag{17.13}$$

where as usual Δu is shorthand notation for the Laplace operator applied to each component of u. There are two pressure terms and one viscosity term.

After taking the dot product of the right-hand side of Eq. (17.13) with the downstream normal e_1, each integral on the right-hand side will be approximated.

Proceeding in order, note that

$$\begin{aligned} \int_{\Omega} \nabla(g\rho(z-\zeta)) \cdot e_1 \, d\mathcal{V} &= -g\rho \int_{x_U}^{x_D} \int_{S(x)} \zeta_x \, d\mathcal{S} \, dx \\ &= -g\rho \int_{x_U}^{x_D} A\zeta_x \, dx. \end{aligned} \tag{17.14}$$

The integrand of the dot product of the second pressure term in Eq. (17.13) with e_1 vanishes except on the inflow and outflow cross-section boundaries of Ω. Thus,

$$\begin{aligned} \int_{\partial\Omega} \rho \frac{u \cdot u}{2} N \cdot e_1 \, d\mathcal{S} &= \rho\Big(\int_{S(x_D,t)} \frac{u \cdot u}{2} \, d\mathcal{S} - \int_{S(x_U,t)} \frac{u \cdot u}{2} \, d\mathcal{S} \Big) \\ &= \rho \int_{x_U}^{x_D} \frac{d}{dx} \Big(\int_{S(x,t)} \frac{u \cdot u}{2} \, d\mathcal{S} \Big) \, dx. \end{aligned} \tag{17.15}$$

By the channel flow assumptions $u \cdot u \approx (u_1)^2$; therefore,

$$\int_{\partial\Omega} \rho \frac{u \cdot u}{2} N \cdot e_1 \, d\mathcal{S} \approx \rho \int_{x_U}^{x_D} \frac{1}{2} \Big(\frac{Q^2}{A} \Big)_x \, dx. \tag{17.16}$$

Channel flow approximation of the dot product of the viscosity part of the stress with the downstream normal starts by applying the divergence theorem:

$$\int_{\Omega} \Delta u \cdot e_1 \, d\mathcal{V} = \int_{\Omega} \Delta u_1 \, d\mathcal{V}$$

$$= \int_{\partial\Omega} \nabla u_1 \cdot N \, d\mathcal{S},$$

where N is the outer unit normal on the boundary of the control volume Ω. The latter integration may be written as a sum of integrals over the lateral boundary LB and the union of the upstream and downstream cross-sectional boundaries; that is,

$$\int_{\partial\Omega} \nabla u_1 \cdot N \, d\mathcal{S} = \int_{\text{LB}} \nabla u_1 \cdot N \, d\mathcal{S} + \int_{S(x_U,t)\cup S(x_D,t)} \nabla u_1 \cdot N \, d\mathcal{S}. \tag{17.17}$$

At this point, the shape of the channel seems to be important. The main example is a rectangular channel with vertical sides. But, to include the possibility of different channel profiles, consider the lateral boundary of the control volume to be split into two parts: the union TB of the top boundary at the free surface and the (flat part) of the channel bottom (which could consist of a single point) and the sides SS of the channel. Also, for future reference, let RB denote the part of the boundary of a wet cross section that is on the channel boundary; that is, the portion of the boundary that does not include the fluid surface.

The orientation of the control volume Ω is taken to be positive with respect to the usual positive orientation with respect to previously chosen Cartesian coordinates.

Integration over the lateral boundary of Ω with respect to the orientations induced by the outer unit normal N is the sum of the integrals of $\nabla u_1 \cdot N$ over TB and SS. The function $\nabla \cdot N = (u_1)_z$ is integrated over the top boundary and $\nabla \cdot N = -(u_1)_z$ over the bottom boundary with respect to the usual orientation in xy coordinates; that is, the functions $\pm(u_1)_z(x, y, z, t)$ are integrated over a region in the xy plane with its usual orientation. The integration over SS is simply indicated because it will vanish for prismatic rectangular channels. Thus,

$$\begin{aligned}
\int_{\text{LB}} \mu\Delta u \cdot e_1 \, d\mathcal{V} &= \int_{x_U}^{x_D} \mu \int_{\partial S(x,t)} \nabla u_1 \cdot N \, d\mathcal{L} \, dx \\
&= \int_{x_U}^{x_D} \mu\Big(\int_{\text{TB}} \mu\nabla u_1 \cdot N \, d\mathcal{L} + \int_{\text{SS}} \nabla u_1 \cdot N \, d\mathcal{L}\Big) \, dx \\
&= \int_{x_U}^{x_D} -\mu(u_1)_z(x, B(x,t), t) W(x, B(x,t), t) \, dx
\end{aligned}$$

$$+\int_{x_U}^{x_D} \mu(u_1)_z(x,\zeta(x,t),t)W(x,\zeta(x,t),t)\,dx$$
$$+\int_{x_U}^{x_D} \mu \int_{\mathrm{SS(x,t)}} \nabla u_1 \cdot N\, d\mathcal{L}\, dx. \tag{17.18}$$

Integration over the wet cross sections that form the upstream and downstream boundaries of Ω is unified by an application of the fundamental theorem of calculus:

$$\begin{aligned}\int_{S(x_U,t)\cup S(x_D,t)} \nabla u_1 \cdot N\, d\mathcal{S} &= \int_{S(x_D,t)} \nabla u_1 \cdot e_1\, d\mathcal{S} - \int_{S(x_U,t)} \nabla u_1 \cdot e_1\, d\mathcal{S}\\ &= \int_{S(x_D,t)} (u_1)_x\, d\mathcal{S} - \int_{S(x_U,t)} (u_1)_x\, d\mathcal{S}\\ &= \int_{x_U}^{x_D} \frac{\partial}{\partial x}\int_{S(x,t)} (u_1)_x\, d\mathcal{S}\, dx. \end{aligned} \tag{17.19}$$

Proposition 17.7. Suppose the boundary of each wet cross section S consists of two parts: the portion at the fluid surface and the curve RB along the solid channel. Let Γ denote the function whose graph (given by $z = \Gamma(x,y,t)$) is the channel boundary and assume that this graph is everywhere below the free surface. The first partial derivatives of the discharge Q are given by

$$\begin{aligned} Q_t(x,t) &= \int_{S(x,t)} (u_1)_t\, d\mathcal{S} + \zeta_t(x,t)W(x,\zeta(x,t),t)u_1(x,\zeta(x,t),t),\\ Q_x(x,t) &= \int_{S(x,t)} (u_1)_x\, d\mathcal{S} + \zeta_x(x,t)W(x,\zeta(x,t),t)u_1(x,\zeta(x,t),t)\\ &\quad + \int_{\mathrm{RB(x,t)}} u_1(\xi,\zeta(\xi,t),t)\Gamma_x(\xi,\eta,t)\, d\mathcal{L}. \end{aligned}$$

Proof. To determine $Q_t(x,t)$, we may fix x and compute

$$\frac{d}{ds}\int_{S(x,t+s)} u_1\, d\mathcal{S}\Big|_{s=0}.$$

According to the transport theorem A.2,

$$\frac{d}{ds}\int_{S(x,t+s)} u_1\, d\mathcal{S}\Big|_{s=0} = \int_{S(x,t)} (u_1)_t\, d\mathcal{S} + \int_{\partial S(x,t)} u_1(Y\cdot N)\, d\mathcal{L},$$

where N is the normal to the boundary of $S(x,t)$ and Y is the vector field tangent at $s = 0$ to the one-parameter family of diffeomorphisms induced by the flow from $S(x,t)$ to $S(x,t+s)$. The lateral (solid) channel boundary is mapped to itself by each such diffeomorphism. Thus, the vector field Y is tangent to the lateral boundary and therefore orthogonal to N. In view of this fact, the only possible nonzero contribution to the second integral on the right-hand side is along the fluid surface in the cross section. By Assumption 17.3, the fluid velocity field is constant along this boundary, which is a line segment of length $W(x,\zeta(x,t),t)$. The top boundary is simply moved vertically by the one-parameter family of diffeomorphisms, which by assumption acts on sections where the velocity field is constant across the channel. The quantity $Y \cdot N$ is the normal derivative in the vertical direction at x. In other words

$$Y \cdot N = \zeta_t(x,t).$$

Thus, we have that

$$\begin{aligned}\frac{d}{ds}\int_{S(x,t+s)} u_1\,dS\Big|_{s=0} &= \int_{S(x,t)} (u_1)_t\,dS \\ &\quad + \zeta_t(x,t)W(x,\zeta(x,t),t)u_1(x,\zeta(x,t),t).\end{aligned}$$

Because

$$Q_t(x,t) = \frac{d}{ds}\int_{S(x,t+s)} u_1\,dS\Big|_{s=0},$$

the first equality of the proposition follows after a rearrangement.

The proof of the second equation in the statement of the theorem is similar but technically more difficult because the fluid flow at some fixed time may not carry an entire wet section to a wet section.

Note that, because the first coordinate x is fixed on a wet cross section,

$$Q(x+s,t) = \int_{S(x+s,t)} u_1(x+s,y,z,t)\,dS.$$

Thus, we have that

$$Q_x(x,t) = \frac{d}{ds}\int_{S(x+s,t)} u_1(x+s,y,z,t)\,dS\Big|_{s=0}$$

The transport theorem can be applied to represent the indicated derivative whenever a flow ϕ_s is chosen such that $\phi_s(S(x,t)) = S(x+s,t)$. A time t is fixed for the remainder of the proof.

Using the assumption that the channel bottom is everywhere below the free surface ($\Gamma(x,y,t) < \zeta(x,t)$) and fixing t, the autonomous system of differential equations

$$\begin{aligned}\frac{dx}{ds} &= 1,\\ \frac{dy}{ds} &= 0,\\ \frac{dz}{ds} &= \frac{z-\Gamma(x,y,t)}{\zeta(x,t)-\Gamma(x,y,t)}\zeta_x(x,t) + \frac{z-\zeta(x,t)}{\zeta(x,t)-\Gamma(x,y,t)}\Gamma_x(x,y,t),\end{aligned}$$

has the boundary of the fluid as an invariant set because the vector field Y defining the differential equation is everywhere tangent to the boundary including the free surface. For example, a tangent vector in the x direction on the free surface given by $z = \zeta(x,t)$ is the transpose of the vector $(1, 0, \zeta_x(x,t))$. This is exactly the value of Y on the free surface at x, which is the set

$$\{(x,y,z) : z = \zeta(x,t) \text{ and } |y| \le \tfrac{1}{2}W(x,\zeta(x,t),t)\}$$

where W is the width of the wet section. By this property and because $dx/ds = 1$, the flow ϕ_s of this differential equation is such that $\phi_s(S(x,t)) = S(x+s,t)$, as desired.

By an application of the transport theorem using the flow ϕ_s,

$$\begin{aligned}Q_x(x,t) &= \frac{d}{ds}\int_{S(x+s,t)} u_1(x+s,y,z,t)\,dS\Big|_{s=0}\\ &= \int_{S(x,t)} (u_1)_x\,dS + \int_{\partial S(x,t)} u_1 Y\cdot N\,d\mathcal{L}.\end{aligned}$$

On the free surface the scalar $Y \cdot N$ is $\zeta_x(x,t)$ and on the remainder of the boundary of the wet cross section RB it is $\Gamma_x(x,y,t)$. Thus,

$$Q_x(x,t) = \int_{S(x,t)} (u_1)_x\,dS + u_1(x,\zeta(x,t),t)W(x,\zeta(x,t),t)\zeta_x(x,t)$$

$$+\int_{\mathrm{RB(x,t)}} u_1(\xi,\zeta(\xi,t),t)\Gamma_x(\xi,\eta,t)\,d\mathcal{L}.$$

□

By the proposition and Eq. (17.19) together with the abbreviation $\mathbb{S} := S(x_U,t)\cup S(x_D,t)$,

$$\begin{aligned}\int_{\mathbb{S}} \nabla u_1\cdot N\,d\mathcal{S} &= \int_{x_U}^{x_D} \frac{\partial}{\partial x}\big[Q_x(x,t)\\ &\quad - \zeta_x(x,t)W(x,\zeta(x,t),t)u_1(x,\zeta(x,t),t)\\ &\quad + \int_{\mathrm{RB(x,t)}} u_1(\xi,\zeta(\xi,t),t)\Gamma_x(\xi,\eta,t)\,d\mathcal{L}\big]\,dx\\ &= \int_{x_U}^{x_D} \big[Q_{xx} - (\zeta_x(x,t)W(x,\zeta(x,t),t)u_1(x,\zeta(x,t),t))_x\\ &\quad + \frac{\partial}{\partial x}\int_{\mathrm{RB(x,t)}} u_1(\xi,\zeta(\xi,t),t)\Gamma_x(\xi,\eta,t)\,d\mathcal{L}\big]\,dx. \end{aligned} \tag{17.20}$$

Adding the integrations over LB and the union of the wet cross sections, integration of the viscosity term over the entire control volume may be represented by the formula

$$\begin{aligned}\int_{\Omega} \mu\Delta u_1\,d\mathcal{V} &= \int_{x_U}^{x_D} -\mu(u_1)_z(x,B(x,t),t)W(x,B(x,t),t)\,dx\\ &\quad + \int_{x_U}^{x_D} \mu(u_1)_z(x,\zeta(x,t),t)W(x,\zeta(x,t),t)\,dx\\ &\quad + \int_{x_U}^{x_D} \mu\int_{\mathrm{SS(x,t)}} \nabla u_1\cdot N\,d\mathcal{L}\,dx\\ &\quad + \int_{x_U}^{x_D} \mu\big(Q_{xx} - (\zeta_x(x,t)W(x,\zeta(x,t),t)u_1(x,\zeta(x,t),t))_x\big)\,dx\\ &\quad + \int_{x_U}^{x_D} \mu\frac{\partial}{\partial x}\int_{\mathrm{RB(x,t)}} u_1(\xi,\zeta(\xi,t),t)\Gamma_x(\xi,\eta,t)\,d\mathcal{L}\,dx. \end{aligned} \tag{17.21}$$

Taking into account Eqs. (17.6), (17.11), (17.12), (17.14), (17.16), and (17.21) to obtain the differential form of the momentum balance, and the differential equation for conservation of mass; the equations of motion in the channel flow approximation with pressure as in Bernoulli's law are

$$\begin{aligned}
A_t + Q_x &= q, \\
\rho(Q_t + \Big(\frac{Q^2}{A}\Big)_x) &= -g\rho A\zeta_x + \frac{\rho}{2}\Big(\frac{Q^2}{A}\Big)_x \\
&\quad - \mu(u_1)_z(x, B(x,t), t)W(x, B(x,t), t) \\
&\quad + \mu(u_1)_z(x, \zeta(x,t), t)W(x, \zeta(x,t), t) \\
&\quad + \mu \int_{\mathrm{SS(x,t)}} \nabla u_1 \cdot N \, d\mathcal{L} \\
&\quad + \mu(Q_{xx} - (\zeta_x(x,t)W(x, \zeta(x,t), t)u_1(x, \zeta(x,t), t))_x) \\
&\quad + \mu \frac{\partial}{\partial x} \int_{\mathrm{RB(x,t)}} u_1(\xi, \zeta(\xi,t), t)\Gamma_x(\xi, \eta, t) \, d\mathcal{L} \\
&\quad + \rho g A \sin(\theta(x,t)). \qquad (17.22)
\end{aligned}$$

These model equations are variants of systems that appear in several different forms in the literature on channel flow (see, for example, [19]).

The right-hand side of the second equation of system (17.22) is too complicated to yield a tractable simplification of the Navier–Stokes model. Also, as might be expected, the system is not closed. Because the flow velocity and the free-surface height are not eliminated from the system and the channel shape is not specified, there are too many state variables. Fortunately, the new model can and will be simplified for some special cases of channel flow.

The important viscosity term $\mu(u_1)_z(x, B(x,t), t)W(x, B(x,t), t)$ is related to the shear stress at the channel bottom caused by the no-slip boundary condition. A useful approximation

$$\mu(u_1)_z(x, B(x,t), t)W(x, B(x,t), t) \approx c\rho\omega\Big(\frac{Q}{A}\Big)^2, \qquad (17.23)$$

where c is a dimensionless parameter and $\omega = \omega(x,t) := W(x, B(x,t), t)$, is a by-product of the discussion of boundary layers presented in Section 17.3. The shear stress coefficient c can be adjusted to model (to some extent) the shear stress at the sides of the channel when the integral over the wet side boundary SS is not taken into account or when the sides of the channel are vertical and this integral vanishes. The approximation is best for prismatic channels with flat bottoms. The latter case is discussed in Section 17.4.

Name	Variable Description	Name	Variable Description
A	Wet Section Area	Q	Discharge Flux Through Wet Section
U	Q/A	ρ	Density
μ	Viscosity	W	Wet Section Width
B	Wet Section Bottom Height	ζ	Wet Section Top Height
ω	Wet Section Bottom Width	q	Negative Boundary Flux
Fr	Froude Number	SS	Wet Channel Side Boundary
RB	Wet Channel Boundary \ SS	θ	Channel Inclination Angle
c	Roughness Constant (17.23)	Re	Reynolds's Number
u_1	Axial Fluid Velocity	h	Dimensionless Fluid Depth
Γ	Wet Channel Boundary	N	Outer Unit Normal on Section Boundary
λ	Dimensionless Constant (17.41)	ι	Dimensionless Constant
d	A Characteristic Velocity	δ	Dimensionless Constant (17.42) ≥ 1
g	Gravitational Acceleration	b	Body Force
β	Scaled Bottom Height	H	Fluid Depth
κ	Dimensionless Parameter	γ	Dimensionless Parameter
σ	Dimensionless Parameter		

Table 17.1 Table of Variable Names For Channel Flow.

Using hydrostatic pressure (no $u \cdot u$ term), ignoring the viscosity except at the channel walls, using the shear stress approximation, and with no additional fluid sources or sinks, the model reduces to the simpler form

$$\begin{aligned} A_t + Q_x &= 0, \\ Q_t + \Big(\frac{Q^2}{A}\Big)_x &= -gA\zeta_x + gA\sin\theta - c\omega\Big(\frac{Q}{A}\Big)^2. \end{aligned} \tag{17.24}$$

This system of equations (where $c\omega$ may be replaced by an appropriately dimensioned parameter) is called the Saint-Venant model (introduced by Adhémar Jean Claude Barré de Saint-Venant, *circa* 1845).

Channel flow model (17.24) is not closed because there are three state variables A, Q, and ζ and only two equations. But, this issue is easily resolved when the shape of the channel is known. For example, if the width W of the channel and the bottom B are known and every wet cross section is assumed to be rectangular, then

$$W(\zeta - B) = A$$

and

$$\zeta = \frac{A}{W} + B.$$

By substitution in the second Saint-Venant equation, the model in case of rectangular cross sections is closed and given by

$$
\begin{aligned}
A_t + Q_x &= 0, \\
Q_t + \Big(\frac{Q^2}{A}\Big)_x &= -gA\Big(\frac{A}{W}\Big)_x - gAB_x + gA\sin\theta - c\omega\Big(\frac{Q}{A}\Big)^2. \qquad (17.25)
\end{aligned}
$$

The scalar ω is the width at the bottom of the wet section.

Several alternatives to the shear stress approximation (last term in the second equation in display (17.25)) for channel friction have been proposed. For example, the denominator A^2 (which is also called Chézy friction) is sometimes replaced by $A^{7/3}$ (Manning friction) together with an appropriate change of the coefficient c.

For the case of a rectangular channel ($W = \omega$) tilted with angle θ where the bottom profile is a line with slope $-\tan\theta$, the Saint-Venant model simplifies to

$$
\begin{aligned}
A_t + Q_x &= 0, \\
Q_t + \Big(\frac{Q^2}{A}\Big)_x &= -\frac{g}{\omega}AA_x + gA(\sin\theta + \tan\theta) - c\frac{\omega}{\sqrt{\mathrm{Re}}}\Big(\frac{Q}{A}\Big)^2. \qquad (17.26)
\end{aligned}
$$

This is the natural model simplification for hydraulics.

Exercise 17.1. Assume the channel is prismatic (all channel cross sections are the same). Is

$$
\begin{aligned}
Q_{xx}(x,t) = \int_{S(x,t)} &(u_1)_{xx}\, dS + \zeta_x(x,t)\omega(x,t)(u_1)_x(x,\zeta(x,t),t) \\
&+ (\zeta_x(x,t)\omega(x,t)u_1(x,\zeta(x,t),t))_x?
\end{aligned}
$$

Exercise 17.2. (a) Determine the Reynolds number for a river with mean depth 1.5 meters flowing on average at one meter per second. Discuss using depth as the length scale for river flow. Is this always the best choice? (b) Discuss the Reynolds numbers for some examples (of rivers) from nature including examples whose Reynolds numbers are less than and greater than 2500.

17.3 BOUNDARY LAYER THEORY

A physical flow moving rapidly along a fixed solid boundary has zero velocity at the boundary. Thus, there must be an abrupt transition between the main flow and the flow near the boundary. Modeling this transition is the subject of boundary layer theory.

Recall that the term $\mu W(u_1)_z$ in the channel flow model [Eq. (17.22)] provides a measure of the interaction of the flow with the channel wall. A full explanation of the meaning of this term and a method to approximate it are results of one of the most important developments in fluid mechanics: the Prandtl boundary layer theory. It is the subject of this section.

The original context for the boundary layer theory discussed here is aerodynamics. In the late 19th century basic fluid mechanics was well understood. Using the two-dimensional Euler equations to model airflow over a wing profile, pressure near the wing's surface was computed to sufficiently high accuracy to be useful in design. But, as can be proved, Euler flow predicts zero drag on a body (for example a cross section of an airplane wing) in a moving fluid. This fact is called d'Alembert's paradox and in more modern form the Kutta–Zhukovski theorem (see Section 13.4, [21], or [60]). Prandtl recognized that although the Euler equations give a good approximation of the flow away from the wing's surface, the no-penetration boundary condition (fluid velocity parallel to the boundary) in the Euler model is not correct near the surface. Indeed, the no-slip boundary condition (zero fluid velocity) is valid at the surface of the wing. Thus, there is a thin layer near the wing's surface—the boundary layer—that must be taken into account in the computation of the drag force. When this is done, the drag force can be predicted in principle with a high degree of precision from the mathematical model. Although Prandtl's fundamental observation seems simple from a modern perspective, its impact was revolutionary. Prandtl did more: he determined a simplification of the Navier–Stokes equations that give a good approximation of the flow in the boundary layer. His students (H. Blasius, in particular) extended the theory and found exact solutions of the boundary layer equations in some special cases that can be used to obtain practical estimates of the drag force. Some of these results are discussed in this section.

Prandtl's theory—in its original form—is for incompressible two-dimensional Navier–Stokes flow. Assigning, as in the case of channel flow, a Cartesian coordinate system where the first coordinate x is in the direction of the main flow, imagine that the flow boundary is the xy coordinate plane and the third coordinate z is perpendicular to this plane. In addition, assume that the part of the xz coordinate plane corresponding to $z > 0$ is an invariant slice of fluid: a particle of fluid in this plane stays in the plane as the fluid moves. In the invariant plane, the Navier–Stokes equations for the fluid velocity field (u, w) are

$$\begin{aligned} u_t + uu_x + wu_z &= -p_x + \frac{1}{\mathrm{Re}}(u_{xx} + u_{zz}), \\ w_t + uw_x + ww_z &= -p_z + \frac{1}{\mathrm{Re}}(w_{xx} + w_{zz}), \\ u_x + w_z &= 0. \end{aligned} \tag{17.27}$$

The basic problem of boundary layer approximation theory is to approximate the Navier–Stokes equations for flow near its boundary (in this case the points with coordinate $z = 0$) where the effects of viscosity are manifested by the transition from zero velocity at the boundary to the nonzero velocity of the main flow.

At the boundary $z = 0$ both velocity components u and w vanish. Away from the boundary in the main flow, u is big and w is small. The boundary layer where the transition takes place has thickness proportional to some small parameter $\epsilon > 0$. To magnify (or stretch) the flow profile near the boundary so its properties are easier to detect, define a new coordinate $\zeta = z/\epsilon$ so that ζ is big relative to z if ϵ is small. Likewise, magnify the vertical component w of the flow by scaling $W = w/\epsilon$. (The symbol W is reused here and should not be confused with the width of the channel in the channel flow model.) The remaining variables are left unaltered: $\xi = x$, $U = u$, $P = p$, and $t = t$. System 17.27 in the new variables (after a simple rearrangement) is

$$\begin{aligned} \mathrm{Re}\,\epsilon^2(U_t + UU_\xi + WU_\zeta + P_\xi) &= \epsilon^2 U_{\xi\xi} + U_{\zeta\zeta}, \\ \mathrm{Re}\,\epsilon^2(W_t + UW_\xi + WW_\zeta) &= -\mathrm{Re}\,P_\zeta + \epsilon^2 W_{\xi\xi} + W_{\zeta\zeta}, \\ U_\xi + W_\zeta &= 0. \end{aligned} \tag{17.28}$$

The idea is to only retain the dominate terms in case Re is large and ϵ is small. A relation between these two quantities is assumed. Suppose, to illustrate the process of finding the correct relationship, that $\mathrm{Re} = 1/\epsilon^3$. In this case, Re is large relative to $1/\epsilon^2$. When the corresponding small terms are discarded from the system, the model reduces to

$$\begin{aligned} U_t + UU_\xi + WU_\zeta + P_\xi &= 0, \\ 0 &= P_\zeta, \end{aligned}$$

$$U_\xi + W_\zeta = 0,$$

which has no viscosity terms. This simplification certainly does not capture the motion of viscous flow near the boundary! On the other hand, if $\text{Re} = 1/\epsilon$ so that Re is small compared with $1/\epsilon^2$, the model reduces to

$$\begin{aligned}
0 &= U_{\zeta\zeta}, \\
0 &= P_\zeta, \\
U_\xi + W_\zeta &= 0,
\end{aligned}$$

a system where there is no interaction between the viscosity terms and the advection terms. Thus, the natural choice is to take Re and $1/\epsilon^2$ to have the same order. The simplest way to accomplish this is to simply set $\text{Re} = 1/\epsilon^2$. By keeping the dominant terms and then returning to the original dimensionless variables, we obtain the Prandtl boundary layer equations

$$\begin{aligned}
u_t + uu_x + wu_z &= -p_x + \frac{1}{\text{Re}} u_{zz}, \\
0 &= -p_z, \\
u_x + w_z &= 0,
\end{aligned} \tag{17.29}$$

where u and w also vanish when $z = 0$. Away from the boundary, where viscous effects are less important, the flow may be modeled by the Euler equations. For this reason, the desired solution of the boundary layer equations is required to match (at least approximately) the Euler flow at the interface between the boundary layer and the main flow.

Note that the pressure changes linearly with z through the boundary layer. Thus, for a thin boundary layer, the pressure may be determined by the Euler flow that is supposed to model the fluid away from the boundary surface, a fact that was known before Prandtl's work. The scaling $\text{Re} \approx 1/\epsilon^2$, or perhaps more properly $\epsilon \approx 1/\sqrt{\text{Re}}$, implies that the boundary layer thickness is proportional to $1/\sqrt{\text{Re}}$. In particular, as the Reynolds number increases, the boundary layer thickness decreases.

The boundary layer equations are a simplified version of the Navier–Stokes equations, but no general solution is known. An important special case was analyzed by Blasius [9]. He considered flow over a flat plate (modeled by the plane $z = 0$), where the ambient Euler flow is constant and parallel to the plate.

Following Blasius, suppose that the ambient fluid-velocity field (in the same dimensionless variables used to define the Reynolds number) is given in components by $(u^\infty, 0)$. The pressure difference across the boundary layer, as previously mentioned, is small. Thus, the dependence of pressure on z may be ignored. Using Euler's equation, this velocity field is related to pressure via the dimensionless equation

$$u_t^\infty + u^\infty u_x^\infty = -p_x.$$

Thus, the boundary layer equations are reduced to

$$\begin{aligned} u_t + uu_x + wu_z &= u_t^\infty + u^\infty u_x^\infty + \frac{1}{\text{Re}} u_{zz}, \\ u_x + w_z &= 0. \end{aligned} \tag{17.30}$$

For a constant u^∞, the steady state boundary layer equations and boundary conditions are

$$\begin{aligned} uu_x + wu_z &= \frac{1}{\text{Re}} u_{zz}, \\ u_x + w_z &= 0, \\ u(x, 0) &= w(x, 0) = 0, \\ u(x, \infty) &= u^\infty. \end{aligned} \tag{17.31}$$

The Blasius problem is to find a solution of this boundary value problem (BVP). Of course, the last equation is the idealization of the Euler match; it is a shorthand for $\lim_{z \to \infty} u(x, z) = u^\infty$.

Blasius's solution of system (17.31) is a famous example of a similarity solution of a nonlinear PDE. His first observation is standard: The two-dimensional flow is incompressible; therefore, the flow is given by a potential ϕ that has an associated stream function ψ such that

$$u = \psi_z(x, z), \qquad w = -\psi_x(x, z).$$

Of course, if a solution of this form is found, it automatically satisfies the second PDE of the system; that is, the flow is divergence free.

Blasius looks for a solution of the first PDE that is invariant under scaling of the independent variables and the potential. A variant of this method, which is worth trying when seeking special solutions of a PDE, is to look for a solution invariant under what is called a dilatation scaling. This is the approach suggested here.

Suppose that there is a solution given by the stream function ψ and note that in this case

$$\psi_z \psi_{xz} - \psi_x \psi_{zz} = \frac{1}{\text{Re}} \psi_{zzz}. \tag{17.32}$$

Let λ, a, and b be real parameters, and seek the most general solution $\tilde{\psi}$ of this partial differential equation that can be expressed in the form

$$\tilde{\psi}(x,z) := \tilde{\psi}(x,z,\lambda,a,b) := \lambda^a \psi(\lambda x, \lambda^b z).$$

Substituting $\tilde{\psi}(x,z)$ for ψ in PDE (17.32) to obtain (up to a scalar multiple) the new equation for ψ given by

$$\lambda^{1+a}(\psi_z \psi_{xz} - \psi_x \psi_{zz}) = \lambda^b \frac{1}{\text{Re}} \psi_{zzz}.$$

Clearly, $\tilde{\psi}$ is a solution whenever $1 + a = b$ and ψ solves PDE (17.32).

The second step is to consider the possibility that there is a solution *invariant* under the most general admissible dilatation; that is, a solution ψ such that

$$\psi(x,z) = \lambda^a \psi(\lambda x, \lambda^{1+a} z)$$

for all λ and a. One way to find such a function ψ is to set $\lambda = 1/x$. This produces the equation

$$\psi(x,z) = \frac{1}{x^a} \psi(1, \frac{z}{x^{1+a}}).$$

Apparently, the invariant choices for ψ are

$$\psi(x,z) = \frac{1}{x^a} g(\frac{z}{x^{1+a}}) \tag{17.33}$$

for some unknown function g and parameter a. It is easy to check that ψ given in this form is indeed invariant under the dilation.

The third step is to substitute this form for the unknown ψ into the PDE to determine the differential equation that g must satisfy so that ψ will be a solution of the PDE. After some algebra, this substitution yields the *ordinary differential equation* (ODE)

$$\frac{1}{\text{Re}\, x^{3+4a}}(-\text{Re}(1+2a)(g')^2 + a\,\text{Re}\, g g'' - g''') = 0. \tag{17.34}$$

This ODE must be supplemented with the boundary conditions for the Blasius problem. The Euler match requires that

$$\psi_z(x,\infty) = \lim_{z\to\infty} \frac{1}{x^{2a+1}} g'(\frac{z}{x^{1+a}}) = u^\infty.$$

Our choice, which places the plate parallel to a constant ambient flow, requires the limit to be independent of the horizontal position denoted by x. There is a natural way to achieve this boundary condition: set $a = -1/2$ and require $g'(\infty) = u^\infty$. In this case, Eq. (17.34) reduces to the *ordinary differential equation*

$$2g''' + \mathrm{Re}\, gg'' = 0. \tag{17.35}$$

To make the BVP for ODE (17.35) mathematically elegant, the condition $g'(\infty) = 1$ is preferable. This normalization and the removal of the Reynolds number from the ODE can be achieved by a simple modification of the stream function. For example, insert new parameters p and q into the proposed stream function to obtain

$$\psi(x,z) = p\sqrt{x} f(q\frac{z}{\sqrt{x}})$$

and note that the choices $p = \sqrt{u^\infty}/\sqrt{\mathrm{Re}}$ and $q = \sqrt{u^\infty\,\mathrm{Re}}$ yield the stream function

$$\psi(x,z) = \sqrt{\frac{u^\infty x}{\mathrm{Re}}}\, f(\sqrt{u^\infty\,\mathrm{Re}}\frac{z}{\sqrt{x}}). \tag{17.36}$$

It solves the Blasius problem provided that f solves the BVP

$$\begin{aligned} 2f''' + ff'' &= 0, \\ f(0) = f'(0) &= 0, \\ f'(\infty) &= 1. \end{aligned} \tag{17.37}$$

Blasius chose to remove the factor 2 by yet another change of variables; his ODE (called the Blasius equation) is $f''' + ff'' = 0$ (see Exercise 17.3).

For use in the channel flow model, we will need the derivative of the velocity field with respect to the variable for the vertical direction. The analysis in this section is performed using dimensionless variables. To match previous work, the derivative must be expressed in dimensioned variables. Recall from Section 11.1 that the dimensioned velocity is obtained from the dimensionless velocity u used in this section by the expression

$$Vu(\frac{\tilde{x}}{L},\frac{\tilde{z}}{L}),$$

where V is the choice of characteristic velocity, L is a characteristic length, and $\tilde{x}$ and $\tilde{z}$ now denote the original dimensioned variables. The partial derivative of the dimensioned stream function Ψ with respect to $\tilde{z}$ is given by

$$\Psi_{\tilde{z}}(\tilde{x},\tilde{z}) = Vu(\frac{\tilde{x}}{L},\frac{\tilde{z}}{L}) = Vu^\infty f'(\sqrt{\frac{\mathrm{Re}\, u^\infty}{L}}\frac{\zeta}{\sqrt{\tilde{x}}}).$$

Thus, for the dimensioned velocity component $\tilde{u}$, dimensioned $\tilde{x}$, and dimensioned Euler velocity U^∞, the product of the viscosity μ and the partial derivative of $\tilde{u}$ with respect to $\tilde{z}$ evaluated at the solid boundary for the Blasius solution (the shear stress at the boundary surface) is

$$\begin{aligned}\mu\tilde{u}_{\tilde{z}}(\tilde{x},0) &= \mu\sqrt{\frac{\mathrm{Re}}{LV\tilde{x}}}\, f''(0)U^\infty\sqrt{U^\infty}\\ &= \rho\sqrt{\frac{LV}{\mathrm{Re}\,\tilde{x}}}\, f''(0)U^\infty\sqrt{U^\infty}.\end{aligned} \tag{17.38}$$

The usual choice for the characteristic velocity is $V = U^\infty$; it results in a shear stress that depends on the square of this characteristic velocity.

A good estimate for $f''(0)$ is

$$f''(0) \approx 0.332$$

(see Exercise 17.4).

The boundary layer theory is successful in predicting skin friction drag. This is indeed one of the triumphs of theoretical fluid dynamics. Another phenomenon called boundary layer separation, whose analysis is beyond the scope of this book, is the next natural subject in boundary layer theory (see, for example, [60] or [72]). As fluid passes over a body at high Reynolds number, a boundary layer develops. One boundary of the boundary layer would seem to be the solid boundary that contains the fluid. This is the case near the leading edge of a body immersed in a flow. Further downstream, the boundary layer (viewed as the portion of the flow where there is a rapid transition from zero velocity to the free-stream velocity of the ambient flow) may separate from the body and a slower flow, perhaps one that has some velocity vectors pointing upstream, might exist downstream of the separation. When this occurs, there is usually a sharp increase in drag and a loss of lift force on the body. Because of the obvious application to airplane flight, separation phenomena have been widely studied. Although much progress has been made, the subject is still not completely understood. For example, the determination of the position of the first separation point remains an open problem.

The Blasius solution was generalized by Victor Falkner and Sylvia Skan (1930) to the case where the ambient flow is in the horizontal direction but not constant with respect to the horizontal coordinate; that is, the velocity component u_∞ defined in the above analysis is allowed to be a function of x (see Exercise 17.6).

Exercise 17.3. Show that by a modification of stream function (17.36), the corresponding ODE for BVP (17.37) is $f''' + ff'' = 0$.

Exercise 17.4. (a) Use a numerical method to verify the estimate $f''(0) \approx 0.332$ for BVP (17.37). Hint: Solve the ODE with initial data $f(0) = f'(0) = 0$ and $f'(0) = \lambda$ where λ is some real number. Adjust the parameter λ until the desired additional boundary condition $f'(\infty) = 1$ is closely approximated. This is called the shooting method. Can you devise a different numerical method that obtains the result more efficiently? (b) Determine the value of $f''(0)$ correct to five decimal places and defend your result.

Exercise 17.5. (a) Use numerical approximations to draw a portrait of the streamlines for the Blasius solution. Also, plot a portrait of its velocity field. Describe in words the behavior of the fluid in the boundary layer. (b) How thick is the boundary layer in this solution?

Exercise 17.6. (a) Suppose that the ambient flow horizontal component u^∞ is given by $u^\infty(x) = \gamma x^\alpha$ for constants γ and α. Repeat the derivation of the Blasius solution but with this more general replacement of u^∞. The case $\alpha = 0$ corresponds to the Blasius solution. Show that there is a solution of the boundary layer equations of the form in Eq. (17.33) provided that there is a solution of the Falkner–Skan BVP

$$\begin{aligned} f''' + ff'' + \beta(1 - (f')^2) &= 0, \\ f(0) = f'(0) &= 0, \\ f'(\infty) &= 1, \end{aligned}$$

for an appropriate choice of β. Hint: Start with Eqs. (17.30) and note that now $u^\infty u_x^\infty$ does not vanish. (b) For which choices of β does the Falkner–Skan BVP have a solution? Is the solution unique when at least one solution exists? Hint: This is not an easy problem. Begin with a numerical exploration. To pursue the analysis further, consider the paper [24].

17.4 FLOW IN PRISMATIC CHANNELS WITH RECTANGULAR CROSS SECTIONS OF CONSTANT WIDTH

To consider applications of the channel flow model [Eq. (17.22)] where closed systems of equations can be obtained, suppose the flow with viscosity μ is confined by a prismatic channel with rectangular cross sections, bottom height $z = B(x,t)$, surface height $z = \zeta(x,t)$, and constant width ω. The area A of a (wet) channel cross section S is $A = (\zeta - B)\omega$. The model system reduces to

$$
\begin{aligned}
A_t + Q_x &= q, \\
\rho(Q_t + \Big(\frac{Q^2}{A}\Big)_x) &= -\frac{g\rho}{\omega} A A_x - g\rho A B_x + \frac{\rho}{2}\Big(\frac{Q^2}{A}\Big)_x \\
&\quad - \mu\omega(u_1)_z(x, B(x,t), t) \\
&\quad + \mu\omega(u_1)_z(x, \zeta(x,t), t) \\
&\quad + \mu(Q_{xx} - (\zeta_x(x,t)\omega u_1(x, \zeta(x,t), t))_x) \\
&\quad + \rho g A \sin(\theta(x,t)),
\end{aligned}
\tag{17.39}
$$

where the first two terms on the right-hand side of the second model equation are derived from the Bernoulli pressure, the third term models the wall friction at the sides and bottom of the channel, the terms on the next line model viscosity effects in the moving fluid, and the next to last term models the effect due to gravity when the incline θ is not zero.

Using the shear stress at the channel wall [Eq. (17.38)] derived from the Blasius solution, the most natural characteristic velocity V appearing there is the average velocity

$$
U(x,t) := \frac{1}{A(x,t)} \int_{S(x,t)} u_1 \, dS = \frac{Q(x,t)}{A(x,t)}.
$$

The appropriate choice of length scale is more problematic. For the Blasius solution, the length scale L is usually taken to be the downstream width of the plate considered in the Blasius problem. An adaptation to channel flow is to simply take $L = x$, where x is the distance in the Blasius solution to the leading edge of the plate. The basic idea here is that the boundary layer thickness should not grow indefinitely with x in the channel flow approximation. Of course, there is no reason why L should be exactly x; thus, to allow for this error and in view of the boundedness of $f''(0)$, we may take the shear stress approximation

$$\mu\omega(u_1)_z(x, B(x,t), t) \approx c\omega\rho U^2 = c\omega\rho\Big(\frac{Q}{A}\Big)^2, \tag{17.40}$$

where c is a positive dimensionless coefficient that decreases with an increase in Reynolds number. This dimensionless constant would have to be measured by experiment in a channel flow application.

The viscosity term $\mu\omega(u_1)_z(x, \zeta(x,t), t)$ would seem to be related to shear stress at the fluid surface, which is usually taken to be negligible. But, literally this term would vanish only if the fluid velocity were not changing with elevation. Physical experiments with channel flows indicate two facts: For low Reynolds number flows (laminar flows) the maximum speed is at the free surface, but for high Reynolds number flows (turbulent flows), the maximum speed is below the surface. Thus, the change in u_1 in the vertical z direction at the surface should be positive for low-speed flows and negative for high-speed flows. This observation suggests including this term using the phenomenological model

$$\mu\omega(u_1)_z(x, \zeta(x,t), t) \approx \frac{\lambda\mu}{d}(dU - U^2) = \frac{\lambda\mu}{d}\Big(d\frac{Q}{A} - \Big(\frac{Q}{A}\Big)^2\Big), \tag{17.41}$$

where d has the dimensions of length per time and λ is a dimensionless parameter. At $U = d$ the sign changes from positive to negative in accordance with the experimental evidence for the sign of $(u_1)_z$. The parameter λ measures the strength of this viscosity term.

Likewise, the viscosity term $\mu(\zeta_x(x,t)\omega u_1(x, \zeta(x,t), t))_x$ does not seem to have a natural channel flow approximation. Perhaps the most obvious treatment is to replace u_1 at the surface by some multiple of the average velocity U (see Exercise 17.7). Using this approximation,

$$\mu(\zeta_x(x,t)\omega u_1(x, \zeta(x,t), t))_x \approx \mu\delta\Big(A_x\frac{Q}{A}\Big)_x, \tag{17.42}$$

where δ is a dimensionless constant. For most applications, the speed of the flow at the surface is greater than the average speed; thus, for such applications, $\delta \geq 1$.

Taking into account the approximations discussed in this section, the model system reduces to

$$A_t + Q_x = q,$$

$$\begin{aligned}\rho\Big(Q_t + \Big(\frac{Q^2}{A}\Big)_x\Big) = & -\frac{g\rho}{\omega}AA_x - \rho g AB_x + \frac{\lambda\mu}{d}(d\frac{Q}{A} - \Big(\frac{Q}{A}\Big)^2) \\ & - c\rho\omega\Big(\frac{Q}{A}\Big)^2 + \frac{\rho}{2}\Big(\frac{Q^2}{A}\Big)_x \\ & + \mu(Q_{xx} - \delta(A_x\frac{Q}{A})_x) \\ & + \rho g A\sin(\theta(x,t)). \end{aligned} \tag{17.43}$$

The fifth term on the right-hand side of the second equation in system (17.43) should perhaps be moved to the left-hand side to simplify the equation. Its presence is meant to aid the reader in tracking the derivations of the various terms.

Recall that channel flow theory is the study of one-space dimensional models of three-dimensional flows confined to a channel. Clearly it is impossible to do this without making approximations. Hence, channel flow models are not exact derivations from fundamental physical laws. They are useful in gaining some insight into the mechanisms underlying observed channel flows.

What phenomena are predicted by model system (17.43)?

Exercise 17.7. Suppose the derivative of a function f is to be approximated in case an approximation g of f is known. (a) Show that the approximation f' by g' can be arbitrarily bad no matter how close g is to f. (b) Express part (a) in precise mathematical language and discuss the precise mathematical conditions on the approximation of f by g that ensures g' is close to f'. Hint: The statement of the problem is intentionally vague as an aid to understanding one reason for defining different norms to measure the difference between functions. Review and consider the uniform (or C^0) norm and the C^1 norm.

Exercise 17.8. [Falling Fluid Films] Derivation of the channel flow equations is excellent background for similar modeling of thin films of fluid moving down an incline or a vertical wall. There are many industrial applications of these flows; for example, in the design of evaporators and some chemical plants. At a more fundamental level, the existence of a wide variety of observed surface waves has produced a lot of experimental, numerical, and theoretical work. The project is to read and understand some of the literature on this subject, repeat the derivation of a model and at least one numerical experiment found in the literature, and discuss the state of current knowledge (see, for example, the review [17], the papers [97] and [98], and the book [18]). One important theme that motivates inclusion of this project here is the desire to produce reductions of the Navier–Stokes equations to simpler models that might be amenable to analysis. Simple models are less reliable for making predictions but they might be analyzed completely; fundamental models are reliable but might be too complicated for analysis and perhaps too expensive for numerical computation. There is no easy choice on how to proceed as long as making predictions from the fundamental model remains out of reach. By now, viable computer experiments on falling films using the full Navier–Stokes equations are possible. But, at this time, no one knows how to prove the existence of such waves or reliably determine their behavior directly from the Navier–Stokes equations. Thus, there are certainly many good research problems related to this topic.

17.5 HYDRAULIC JUMP

A fundamental phenomenon in channel flow, observable in many natural and man-made channels, is a change in flow depth not obviously accounted for by a change in the bottom profile of the channel. For example, the outflow from a dam, weir, or sluice gate often causes a marked increase in depth in the outflow channel; rainwater drains on a sidewalk and standing waves appear; or water flows from a tap into a kitchen sink, spreads in all directions, and creates a circular standing wave. A generic term for such behavior is "hydraulic jump." There are many variations on this theme. The standing wave at the Tiger Fountain (see page 397) is also an example of this phenomenon.

The most basic result for hydraulic jumps is an approximate formula for the downstream fluid depth for steady state flow in a horizontal and rectangular channel as a function of the upstream Froude number and upstream depth. Recall that the Froude number is given by the characteristic flow velocity V divided by the square root of the product of a characteristic length scale L and the gravitational constant g; that is, $\mathrm{Fr} := V/\sqrt{Lg}$. Choose characteristic dimensions at the upstream end of the channel flow. For this, let the subscripts U and D reference upstream and downstream positions. If the characteristic velocity and length are respectively $V = Q_U/A_U$ and $L = \mathrm{wd}_U$ (the depth), then the classic formula often used to relate upstream and downstream fluid depths is

$$\mathrm{wd}_D = \frac{\mathrm{wd}_U}{2}\left(\sqrt{1 + 8\,\mathrm{Fr}_U^2} - 1\right). \tag{17.44}$$

To derive Eq. (17.44), consider a steady state flow with hydrostatic pressure and ignore all viscous effects. In steady state, the discharge is constant (from conservation of mass) and the equation that expresses conservation of momentum [Eq. (17.43)] reduces to the single ODE

$$\left(\frac{Q^2}{A}\right)_x = -\frac{g}{\omega}AA_x, \tag{17.45}$$

which has the general family of solutions

$$\frac{Q^2}{A} + \frac{g}{2\omega}A^2 = C \tag{17.46}$$

parameterized by the constant of integration C. Because the expression on the left-side of the equation is constant, its upstream and downstream values are equal; that is,

$$\frac{Q^2}{A_U} + \frac{g}{2\omega}A_U^2 = \frac{Q^2}{A_D} + \frac{g}{2\omega}A_D^2. \tag{17.47}$$

Rearrange this equality to obtain

$$Q^2\frac{A_D - A_U}{A_U A_D} = \frac{g}{2\omega}(A_D^2 - A_U^2).$$

One solution for the unknown wet areas is $A_D = A_U$. In this case and for a channel of fixed width, $\zeta_D = \zeta_U$; that is, the up and downstream depths are equal and no jump occurs. In case $A_D \neq A_U$, there is another possibility:

$$\frac{Q^2}{A_U A_D} = \frac{g}{2\omega}(A_D + A_U).$$

The solution of this latter equation for A_D as a function of A_U is obtained using the quadratic formula and choosing the physically realistic positive root. The desired result [Eq. (17.44)] is derived from this function after substituting $A_D = \zeta_D\omega$, $A_U = \zeta_U\omega$, dividing by ω, and some rearrangement to make the square of the Froude number appear in the formula.

Under the assumptions mentioned in its derivation, the simplified flow model [Eq. (17.45)] seems to predict at least two possible configurations of steady state flow: (1) constant depth, and (2) a depth change from upstream to downstream given by Eq. (17.44). In the latter case, which is taken as bedrock hydraulics, $\mathrm{Fr}_U = 1$ is a critical value. For $\mathrm{Fr}_U < 1$ (called subcritical flow), the depth decreases downstream; for $\mathrm{Fr}_U > 1$ (called supercritical flow), the depth increases and there is a hydraulic jump. The derivation does not specify which alternative occurs.

Although observations in nature and experiment confirm that hydraulic jumps occur and Eq. (17.44) gives a reasonable approximation to reality, the usual derivation of the formula just presented contains a serious flaw. The discharge Q is assumed to be constant (as it should be by conservation of mass) and the left-hand side of Eq. (17.46) is also assumed to be constant so that it has the same upstream and downstream values. Under these assumptions for the approximate momentum conservation model [Eq. (17.45)], the wet area A must also be constant (see Exercise 17.9). Thus, the correct prediction from the simplified model is that $A_D = A_U$. No hydraulic jump occurs. This result is perfectly reasonable simply because in the model nothing happens between the upstream and downstream observations to cause a change in the flow.

An alternative viewpoint is that something happens to the flow conditions between the upstream and downstream observation points that is not modeled (perhaps the bottom profile changes), but at the observation points the momenta happen to be the same so that Eq. (17.47) is satisfied. By design, the model allows A to change with the downstream coordinate. In this scenario, hydraulic jumps are possible and the approximate formula (17.44) is obtained when they occur. This suggests that hydraulic jumps are not determined by upstream flow conditions; rather, they are the result of some change in flow conditions between the observation points.

The channel flow equations allow for modeling the channel bed and other features between the upstream and downstream observation points. More accurate results can be obtained by applying the full model equations, but to do so requires a more complicated analysis, speculative assumptions, or the use of numerical approximations (compare Exercise 17.10).

Although a large variation in depth through a hydraulic jump is predicted by much used formula (17.44), it does not suggest the flow profile during this change in depth. The remarkable fact is that the change in depth can occur over a short distance along the channel between the upstream and downstream stations—thus the name hydraulic jump. More complete channel flow models are required to approach the true flow profile of a hydraulic jump.

Exercise 17.9. Prove that if Q is constant in ODE (17.45), then A is constant.

Exercise 17.10. Consider the steady state channel flow momentum balance model obtained by including the bottom profile of the channel:

$$\left(\frac{Q^2}{A}\right)_x = -\frac{g}{\omega}AA_x - gAB_x, \tag{17.48}$$

(a) Under the assumption of constant discharge (steady state conservation of mass), find the general solution of the differential equation. Determine physically realistic conditions such that the predicted hydraulic jump agrees with model equation (17.47) (and therefore formula (17.44)). (b) The general solution states that a certain function of A is constant. Thus, the function of A is conserved. By its derivation the expectation is that momentum is conserved. But, for the simplified model a more direct link is to Bernoulli's law, which may be viewed as energy conservation. Show that under the assumption that the pressure is hydrostatic, the momentum balance model [Eq. (17.48)] is a restatement of Bernoulli's law. Hint: Consider the general solution and rewrite it using the channel flow velocity $U := Q/A$. (c) Show that to obtain the general formula for A_D as a function of A_U, according to model (17.48), requires solution of a cubic polynomial. Choose reasonable (sets of) values for the system parameters, the discharge, the channel width, and the upstream wet area. Use your numbers to compare the predictions for the downstream wet area using the present model and formula (17.44). Verify that the more precise model sometimes gives the same result as the simpler model. Make a statement based on your calculations and argue that your observation would be useful to a hydraulic engineer.

17.6 SAINT-VENANT MODEL AND SYSTEMS OF CONSERVATION LAWS

By simplifying to the Saint-Venant equations (17.25), some of the mathematics of conservation laws is developed in a context where it can be used to make useful predictions. Under the restriction to horizontal channel flow, the simplification discussed here is given by system (17.43) with $\mu = 0$ (no internal viscosity), the term $(\rho/2)(Q^2/A)_x$ removed (hydrostatic pressure only), a bottom profile not changing along the channel, and (for the moment) no source term. After the common factor ρ is removed from the second equation of this system and the resulting equation is rearranged, the model reduces to

$$\begin{aligned} A_t + Q_x &= 0, \\ Q_t + \left(\frac{Q^2}{A} + \frac{g}{2\omega}A^2\right)_x &= -c\omega\left(\frac{Q}{A}\right)^2, \end{aligned} \tag{17.49}$$

where viscosity at the boundary is taken into account by the presence of the last term of the second equation.

Recall that the channel flow models (and the more general fluid flow models) are derived from conservation of mass and Newton's second law. A review of the derivation of these models reveals that they all have the same general form. There is some quantity w (perhaps the vector quantity $w = (A, Q)$) measured in a volume Ω, and the rate of change of the amount of w in Ω is determined by the flux of the vector field X that determines the motion of w through $\partial\Omega$ with outer normal η; that is,

$$\frac{d}{dt}\int_{\Omega} w\, d\mathcal{V} = -\int_{\partial\Omega} X(w)\cdot\eta\, dS = -\int_{\Omega} \operatorname{div} X(w)\, d\mathcal{V}.$$

Often the dependence of X on w (for example, when w is a density and X generates the flow carrying the underlying substance) involves the negative gradient of w and the divergence term becomes the Laplacian. When the dependence is given directly as a function of w and there is exactly one space dimension x (so that the divergence is simply the partial derivative with respect to the space variable), the conservation law becomes

$$\int_{\Omega} w_t\, d\mathcal{V} = -\int_{\Omega} (X(w))_x\, d\mathcal{V},$$

or in differential form,

$$w_t + (X(w))_x = 0.$$

Note that the left-hand side of system (17.49) has this form. In case there is a source term given by a vector function Y, the system of conservation laws takes the form

$$w_t + (X(w))_x = Y(w, x, t),$$

which is precisely the form of system (17.49) where Y is a function that does not depend explicitly on x or t.

Conservation laws are fundamental, but the analysis of general model systems of conservation laws is not completely understood. Fortunately, much is known for systems of one or two conservation laws in case the conserved quantities depend on exactly one spatial dimension. The channel flow model [Eq. (17.49)] is an important example where these assumptions are satisfied.

There is a surprise at the most basic level of the theory of conservation laws that motivates further mathematical investigation: bounded and continuous initial data may evolve in finite time to a bounded state that cannot be

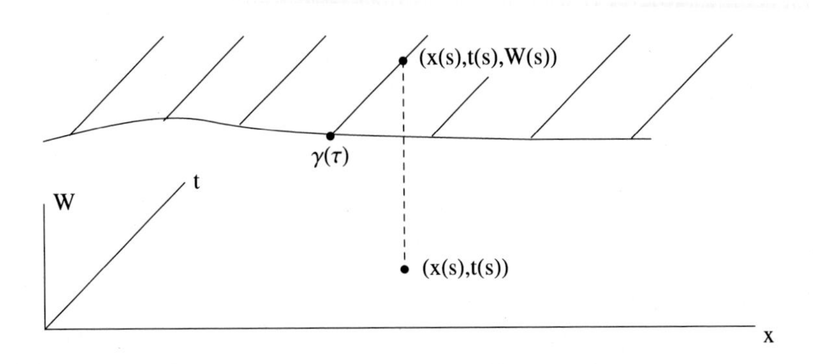

Fig. 17.2 Schematic diagram of a surface consisting of a union of characteristic trajectories starting on the image of a noncharacteristic curve γ.

continuously extended in time. A classic example is provided by Burgers's equation

$$w_t + ww_x = 0, \tag{17.50}$$

where $w = w(x, t)$ is a scalar function of position x in (one-dimensional) space and time t. Note that this equation may be alternately expressed as the conservation law

$$w_t + (\frac{1}{2}w^2)_x = 0. \tag{17.51}$$

A natural setting for study of this first-order nonlinear PDE is to seek solutions defined for the spatial variable x on the whole real line and the temporal variable $t > 0$. More precisely, the physically relevant mathematical formulation is Cauchy's problem: Determine the evolution of an initial function $w_0 : \mathbb{R} \to \mathbb{R}$ (that is, $w(x, 0) = w_0(x)$) as it evolves forward in time under the conservation law [Eq. (17.51)]. The surprising result is that there is a solution of Cauchy's problem for the initial function

$$w_0(x) = \begin{cases} 1 & \text{whenever} \quad x < 0, \\ 1 - x & \text{whenever} \quad 0 \le x \le 1, \\ 0 & \text{whenever} \quad x > 1. \end{cases} \tag{17.52}$$

that cannot be evolved continuously for $t > 1$.

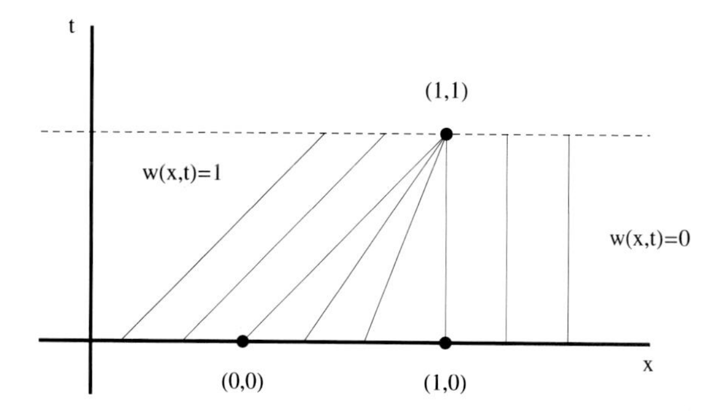

Fig. 17.3 The figure depicts a portion of the (x,t) plane, a sample of projected characteristic curves for Burgers's equation, and the coordinates of three points. In the triangle with vertices at these points, the solution of Burgers's equation with initial data given by the function f defined in Eq. (17.52) *is constant along the projected characteristics, the solution has distinct values along these curves, and all of them pass through the point with coordinates* $(1,1)$.

Smooth solutions of first-order PDEs of the form

$$a(x,t,w)w_t + b(x,t,w)w_x = c(x,t,w), \tag{17.53}$$

which includes Burgers's equation, are directly related to invariant manifolds of the associated first-order system of three scalar ODEs

$$\frac{dx}{ds} = b(x,t,W), \qquad \frac{dt}{ds} = a(x,t,W), \qquad \frac{dW}{ds} = c(x,t,W) \tag{17.54}$$

whose trajectories are called characteristic curves (or simply characteristics) of the PDE. In fact, if there is a function $W = w(x,t)$ defined on an open subset of space-time whose graph (in the three-dimensional space with coordinates (x,t,W)) is contained in an invariant manifold (a union of solutions) for system (17.54), then w is a solution of the PDE. To prove this statement, note that a normal to the surface given by the graph of w is

$$(x,t) \mapsto (x,t,w(x,t),w_x(x,t),w_t(x,t),-1),$$

where this vector field is given by the vector $(w_x(x,t),w_t(x,t),-1)$ at the point $(x,t,w(x,t))$ on the surface. As the invariant surface is a union of solutions, the velocity vector along a solution (namely, the vector whose components are the right-hand sides of the ODEs (17.54)) must be orthogonal to the surface normal; that is,

$$w_x(x,t)b(x,t,w(x,t)) + w_t(x,t)a(x,t,w(x,t)) - c(x,t,w(x,t)) = 0,$$

as required.

To construct an invariant surface for system (17.54), the idea is to start with a curve in the three dimensional space, consider this curve as a set of initial conditions for the ODEs whose solutions are characteristics, and trace out a surface by moving the curve of initial data via the flow determined by the ODEs (see Fig. 17.2). More precisely, given a curve

$$\tau \mapsto \gamma(\tau) = (\gamma_1(\tau), \gamma_2(\tau), \gamma_3(\tau)),$$

the desired surface consists of the union of the trajectories of system (17.54) that correspond to the solutions of the ODEs with initial data

$$x(0) = \gamma_1(\tau), \qquad t(0) = \gamma_2(\tau), \qquad W(0) = \gamma_3(\tau)$$

for each choice of τ in the domain of γ. To ensure that the surface obtained in this way is a graph over the space-time plane, at least in some (perhaps small) neighborhood of the image of γ, the initial data must be a noncharacteristic curve; that is, at each point in the image of γ, the projections of the tangent vector to γ and the velocity vector of the system of ODEs at this point into the space-time plane must not be parallel. In symbols, the curve γ is noncharacteristic if

$$\dot{\gamma}_1(\tau)a(\gamma_1(\tau), \gamma_2(\tau), \gamma_3(\tau)) - \dot{\gamma}_2(\tau)b(\gamma_1(\tau), \gamma_2(\tau), \gamma_3(\tau)) \neq 0. \quad (17.55)$$

Using the implicit function theorem and the smoothness of solutions of ODEs with respect to initial data, it is possible to prove the existence of a solution of PDE (17.53) defined on an open subset of the space-time plane containing the projection of a noncharacteristic curve (see, for example, [20]) by constructing a function w whose graph is contained in an invariant surface. The values of the solution w of the PDE along the characteristic $s \to (x(s), t(s), W(s))$ are given by $W(s) = w(x(s), t(s))$. Thus, solutions of the PDE are determined by the characteristics.

Surfaces constructed from characteristics starting on a noncharacteristic curve of initial data may develop folds in the three-dimensional space (as the initial noncharacteristic curve is evolved in time by the system of ODEs) so that several points on the folded surface project to the same point in the space-time plane. When this happens, additional conditions must be imposed to choose the correct characteristic to define the solution w of the PDE.

For Cauchy's problem, a noncharacteristic curve is chosen so that the solution of the PDE given by the graph constructed from characteristics satisfies the initial conditions. Under the assumption that the initial data is bounded and given by a noncharacteristic curve, Cauchy's problem has a continuous bounded solution that exists at least for some finite time interval. The possible development of folds in the characteristic surface, which is the usual case, may obstruct the extension of the solution beyond this finite time interval.

Some methods used to construct solutions for IVPs for first-order PDEs will be illustrated by several examples.

For Burgers's equation (17.50), the characteristics are solutions of the ODE system

$$\frac{dx}{ds} = W, \qquad \frac{dt}{ds} = 1, \qquad \frac{dW}{ds} = 0.$$

The correct choice for the initial curve is

$$\gamma(\xi) = (\xi, 0, w_0(\xi));$$

it corresponds to the initial data (17.52) along the spatial axis. Note that γ is noncharacteristic; in fact, in this case, $a(x, t, W) = 1$, $b(x, t, W) = W$, and

$$\dot{\gamma}_1(\xi)a(\gamma_1(\xi), \gamma_2(\xi), \gamma_3(\xi)) - \dot{\gamma}_2(\xi)b(\gamma_1(\xi), \gamma_2(\xi), \gamma_3(\xi)) = 1 \neq 0.$$

The solutions of the ODEs starting on γ are

$$x(s) = w_0(\xi)s + \xi, \qquad t(s) = s, \qquad W(s) = w_0(\xi). \tag{17.56}$$

The solution w of the IVP is determined along characteristics via $W(s) = w(x(s), t(s))$. Using Eqs. (17.56), the desired solution w (defined in the region where the characteristic surface is a graph over space-time) is constant along each projected characteristic in the space-time plane; in other words,

$$w_0(\xi) = w(w_0(\xi)s + \xi, s).$$

The projected characteristic is the line $x = w_0(\xi)t + \xi$ that crosses the horizontal axis at $(\xi, 0)$; therefore, the value $w(x, t)$ is obtained at every point (x, t) that can be connected back in time to the horizontal axis by a unique projected characteristic. This is how the initial data profile is propagated forward by the PDE.

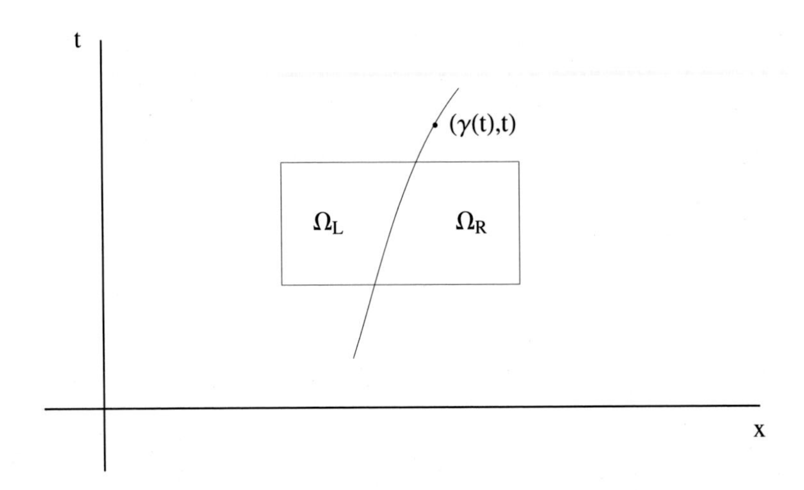

Fig. 17.4 A rectangle in space-time is separated by a curve $t \mapsto (\gamma(t), t)$.

According to initial data (17.52), $w_0(\xi) = 1$ whenever $\xi < 0$. The corresponding characteristic is $s \mapsto (s + \xi, s, 1)$. The curves $s \mapsto (s + \xi, s)$ are straight lines in (x, t) space with unit slope (see Fig. 17.3). A point (x, t) (with $t \geq 0$) is on one of these curves (that is, $(s + \xi, s) = (x, t)$) when $\xi = x - t$ is negative. In this case, $w(x, t) = 1$. In other words, the value of w at (x, t) is determined by following the projection of a characteristic backward (with respect to the positive direction of the coordinate s) until it meets the x-axis and assigning the value of w_0 at the x coordinate of this point. Simple enough.

The initial data at $(\xi, 0)$ with $\xi > 1$ is $w_0(\xi) = 0$, the corresponding characteristic is $s \mapsto (\xi, s, 0)$ (which is a vertical line in the (x, t) plane); hence, $w(x, t) = 0$ along the projected characteristic.

What happens for initial data with $0 \leq \xi \leq 1$? The corresponding projections of characteristics have the form $s \mapsto ((1 - \xi)s + \xi, s)$. If (x, t) is on this curve, then $x = (1 - \xi)t + \xi$ and $w(x, t)$ should be assigned the value $(1 - \xi)$; that is, $w(x, t) = (1 - x)/(1 - t)$ in the interior of the triangle in space-time with vertices $(0, 0)$, $(1, 0)$, and $(1, 1)$ (see, Fig. 17.3). But, this result leads to a problem: The value of w at the point $(x, t) = (1, 1)$ is not well-defined. Indeed, for each ξ in the unit interval, the curve $s \mapsto ((1 - \xi)s + \xi, s)$ passes through the point $(1, 1)$. There is no consistent way to define $w(1, 1)$. Thus, the evolution of the initial profile $t \mapsto w(\cdot, t)$ cannot be extended continuously beyond $t = 1$.

Burgers's equation illustrates a ubiquitous phenomenon: the evolution of a continuous initial profile by a conservation law remains bounded but is not continuously extendible for all positive time. Thus there is a basic problem: Although conservation laws are natural models of physical phenomena derived from fundamental physics (conservation of mass, momentum balance, and conservation of energy), continuous solutions of these models usually do not exist on sufficiently long timescales to be physically relevant. Fortunately, there is a way around this problem: *Give up the requirement that solutions must be continuous*.

There are many discontinuous processes in nature. Consider, for instance, a piece of chalk used to write on a blackboard. Hold the chalk by its ends and apply a force intended to bend the piece of chalk. The likely outcome of this experiment is clear: the chalk breaks into two pieces. The amount of chalk is conserved during this process, but the evolution of the initial profile of the chalk is not continuous through the break. A model of chalk breaking must allow discontinuous solutions.

Perhaps a physical process does not have discontinuities, but it exhibits abrupt changes that are adequately modeled by discontinuous functions up to the accuracy of available measuring devices. The quintessential example is a shock wave or shock front emanating from the tip of a supersonic projectile or the leading edge of an airplane wing. Conservation law models are idealizations that do not account for all aspects of these complex physical phenomena; thus, discontinuous solutions of such models may be viewed as approximations of the true physical processes.

Acceptance of discontinuous solutions of continuous PDE models is a fruitful approach in many applications of mathematics.

Consider the possibility of discontinuous solutions of Burgers's equation. How should we extend the solution we have already constructed beyond $t = 1$? The answer requires a few more observations and the introduction of some new ideas.

Our discussion mentions discontinuous solutions of a PDE. This is an oxymoron: A solution of a PDE must have partial derivatives to satisfy the PDE. Thus, the usage "discontinuous solutions of a PDE" does not have its literal meaning. A precise definition requires reconsideration of the integral equations used to derive conservation laws.

Conservation laws, defined as PDEs, are derived from mathematical formulations of physical conservation laws by *integration* over volumes. Discontinuous functions can be integrated. Indeed, a bounded piecewise continuous function defined on a finite interval has a finite integral. To employ this observation to solve PDEs, consider a function ϕ defined on space-time that has continuous partial derivatives of all orders and whose support (the set $\{(x,t) : \phi(x,t) \neq 0\}$) is contained in a compact (that is, closed and bounded) subset of the plane. Such functions exist (in abundance); they are called test functions. In other words, test functions are infinitely smooth functions with compact support.

A far-reaching idea is to multiply both sides of a PDE by a test function ϕ, integrate over all of space, and use integration by parts to move the partial derivatives from the desired solution of the PDE to ϕ.

Consider the PDE

$$w_t + g(w)w_x = 0 \tag{17.57}$$

with initial condition $w(x,0) = h(x)$, where g and h are (piecewise) continuous functions and let G denote an antiderivative of g. The same PDE in conservation form is given by

$$w_t + (G(w))_x = 0. \tag{17.58}$$

Suppose that w is a classical solution of the IVP (that is, a solution that is continuously differentiable). Multiply Eq. (17.57) by a test function ϕ and integrate to obtain the equality

$$\int_0^\infty \int_{-\infty}^\infty w_t(x,t)\phi(x,t) + g(w(x,t))w_x(x,t)\phi(x,t)\,dxdt = 0.$$

Using integration by parts, we have that

$$\begin{aligned} 0 &= -\int_0^\infty \int_{-\infty}^\infty w\phi_t\,dxdt + \int_{-\infty}^\infty w\phi\Big|_{t=0}^{t=\infty}\,dx - \int_0^\infty \int_{-\infty}^\infty G(w)\phi_x\,dxdt \\ &= -\int_0^\infty \int_{-\infty}^\infty w\phi_t\,dxdt - \int_{-\infty}^\infty h\phi(x,0)\,dx - \int_0^\infty \int_{-\infty}^\infty G(w)\phi_x\,dxdt, \end{aligned}$$

or equivalently,

$$\int_0^\infty \int_{-\infty}^\infty w\phi_t + G(w)\phi_x\,dxdt + \int_{-\infty}^\infty h\phi(x,0)\,dx = 0. \tag{17.59}$$

Thus, a classical solution satisfies the integral equation (17.59) for every test function ϕ.

The important new observation is that integral equation (17.59) makes sense for bounded (measurable) functions $w(x, t)$ that may be discontinuous. This leads to an important definition: a bounded (measurable) function w that satisfies the integral equation *for every test function* is called a weak solution of the conservation law. Clearly, every classical solution is a weak solution. Thus, by relaxing the smoothness requirement, the door is opened to seek discontinuous functions that are weak solutions of the PDE.

What is an example of a weak solution that is not a classical solution? Do weak solutions always exist? Are they unique? If they exist, how are they constructed? Are weak solutions physically realistic? Can weak solutions be approximated by numerical methods? These important questions can all be answered with further mathematical analysis.

The construction of weak solutions that are piecewise smooth—the most important class of weak solutions for applications—begins with a crucial observation about the nature of such a solution near a curve of jump discontinuities. There is a surprise: the definition of weak solutions restricts the nature of their discontinuities.

Consider a curve in space-time of the from $(x, t) = (\gamma(t), t)$, where γ is a smooth function and $t \geq 0$. Note that the curve is a graph with respect to the time axis, which is taken as usual in this subject to be the vertical axis. The curve separates the space-time plane into two components called the left-hand and right-hand sides of the curve. For the analysis to follow, it is not important that the curve separates the entire space; rather, it suffices for there to be an open set Ω (which can be taken to be the interior of a rectangle with sides parallel to the axes) situated such that the curve separates Ω into left (Ω_L) and right (Ω_R) open components as in Fig. 17.4.

Let w be a weak solution of conservation law (17.58) (that is, a bounded solution of integral equation (17.59)); and assume in addition that w is continuously differentiable in each of the sets Ω_L and Ω_R and w has definite left- and right-hand limits (w_L and w_R) at each point of the curve $s \mapsto (\gamma(t), t)$. In other words, w has (at worst) jump discontinuities on this curve. The left- and right-hand limits must satisfy the relation

$$G(w_L(t)) - G(w_R(t)) = (w_L(t) - w_R(t))\gamma'(t), \tag{17.60}$$

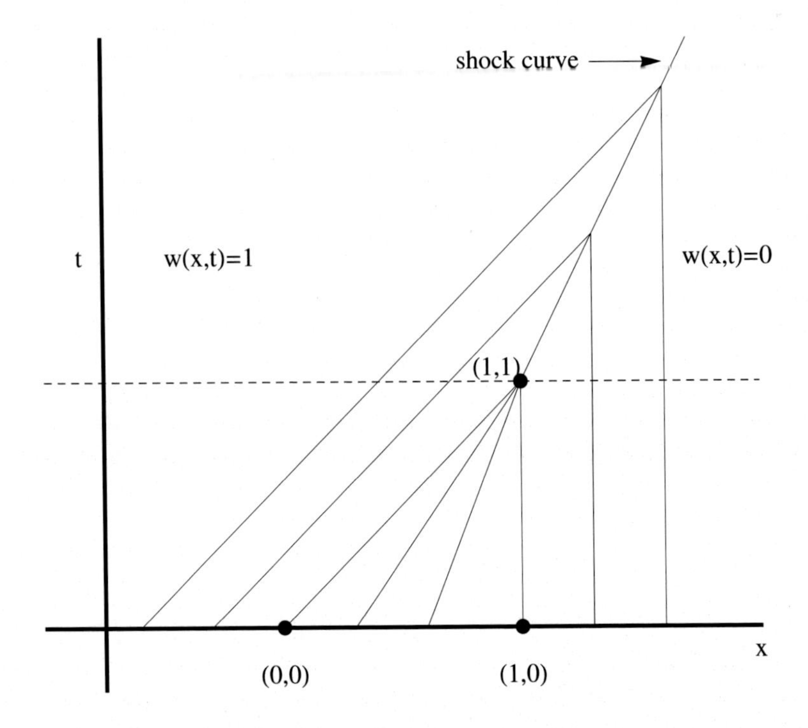

Fig. 17.5 The figure depicts a weak solution of Burgers's equation (17.51) *with initial data* (17.52). *After time* $t = 0$ *the solution is unity on the left of the shock emanating upward from the space-time point* $(x, t) = (1, 1)$ *and zero on the right.*

which is called the Rankine–Hugoniot (jump) condition after William John Macquorn Rankine and Pierre Henri Hugoniot first developed this relation in the period 1870–1890.

To see why the Rankine–Hugoniot condition must hold for a weak solution w, let ϕ be a test function with support in Ω_L. By definition, ϕ is smooth and vanishes outside of Ω_L. Using integral equation (17.59), the choice of ϕ, and the observation that Ω_L does not contain a point in space-time with time coordinate $t = 0$, we have that

$$\int_{\Omega_L} w\phi_t + G(w)\phi_x \, d\mathcal{A} = 0,$$

where, of course, $d\mathcal{A}$ is the usual Euclidean area element. In Ω_L, integration by parts is valid because all functions involved are smooth. Moreover, the boundary terms vanish because ϕ vanishes on the boundary. Thus, it follows that

$$\int_{\Omega_L} (w_t + (G(w))_x)\phi \, d\mathcal{A} = 0$$

for every test function with support in Ω_L; hence,

$$w_t + (G(w))_x = 0$$

when the left-hand side of the equation is evaluated in Ω_L. A similar argument proves that the equation holds for every space-time point in Ω_R.

Next, consider a test function ϕ with support in Ω. Maybe its support includes a portion of the curve $t \mapsto (\gamma(t), t)$. Because w satisfies integral equation (17.59) and the support of ϕ is in Ω,

$$\int_{\Omega} w\phi_t + G(w)\phi_x \, d\mathcal{A} = 0.$$

The integral can be written as a sum over the two components of Ω and integration by parts (as in Appendix A.4) is valid on each component provided we redefine the values of w on the separating curve to be the left- or right-hand limit corresponding to the components. Note that the outer unit normal (in components) on the portion C of $\partial\Omega_L$ given by the separating curve is

$$\eta_L = (\eta_L^1, \eta_L^2) := \frac{1}{\sqrt{1 + (\gamma'(t))^2}}(1, -\gamma'(t))$$

and the outer unit normal on the corresponding part of $\partial\Omega_R$ is

$$\eta_R = (\eta_R^1, \eta_R^2) := \frac{1}{\sqrt{1 + (\gamma'(t))^2}}(-1, \gamma'(t)).$$

In symbols, these facts are expressed by the equalities

$$\begin{aligned} 0 &= \int_{\Omega_L} w\phi_t + G(w)\phi_x \, d\mathcal{A} + \int_{\Omega_R} w\phi_t + G(w)\phi_x \, d\mathcal{A} \\ &= -\int_{\Omega_L} w_t\phi + (G(w))_x\phi \, d\mathcal{A} + \int_C w_L\phi\eta_L^2 + G(w_L)\phi\eta_L^1 \, d\ell \\ &\quad - \int_{\Omega_R} w_t\phi + (G(w))_x\phi \, d\mathcal{A} + \int_C w_R\phi\eta_R^2 + G(w_R)\phi\eta_R^1 \, d\ell. \end{aligned}$$

By the first part of the argument, the integrals over Ω_L and Ω_R both vanish. Moreover, because the last equation holds for every test function with support in Ω,

$$\begin{aligned} 0 &= w_L\eta_L^2 + G(w_L)\eta_L^1 + w_R\eta_R^2 + G(w_R)\eta_R^1 \\ &= -w_L\gamma'(t) + w_R\gamma'(t) + G(w_L) - G(w_R). \end{aligned} \tag{17.61}$$

The Rankine–Hugoniot condition [Eq. (17.60)] is obtained by rearranging Eq. (17.61).

Return to Burgers's equation (17.51) with initial data (17.52) and its solution defined on $0 \le t < 1$ as depicted in Fig. 17.3, and imagine a separating curve $s \mapsto (\gamma(t), t)$ emanating at $t = 1$ from the point $(1, 1)$ where the projected characteristic curves intersect. Left-hand limits for a solution w should all be unity, and right-hand limits should all be zero as these are the values along the corresponding projected characteristics in this case. According to the Rankine–Hugoniot condition, the function γ must then be determined from the differential equation

$$\gamma'(t) = \frac{(w_L(t))^2 - (w_R(t))^2}{2(w_L(t) - w_R(t))} = \frac{1}{2}$$

with initial condition $\gamma(1) = 1$. Hence, $\gamma(t) = t/2 + 1/2$ and the separating curve $t \mapsto (\gamma(t), t)$, for $t \ge 1$ is a ray on the straight line in the space-time plane with equation $t = 2(x - 1/2)$. We may define w arbitrarily along this line, let it be unity on the left side of the line and zero on the right side to extend the classical solution (which exists for $0 \le t < 1$) as a weak solution for all $t > 0$, as depicted in Fig. 17.5.

A weak solution of a conservation law with jump discontinuities must satisfy the Rankine–Hugoniot condition. As the last example illustrates, this condition can also be used to help construct weak solutions.

Unfortunately, weak solutions with only jump discontinuities may not be unique (see Exercise 17.12). Much of the further development of the theory of conservation laws is concerned with restricting the class of weak solutions so that there is a unique weak solution of the conservation law within the restricted class for each choice of initial data. The correct restriction is determined by physical considerations based on the second law of thermodynamics, which asserts that entropy must increase in thermodynamic processes. The historical development of the appropriate way to define the entropy condition for weak solutions of conservation laws is inextricably related to gas dynamics and is best described after a thorough treatment of classical thermodynamics. Alas, the usual explanation of the connection of the mathematical notion to physical entropy is tenuous because the physical theory is based on equilibrium thermodynamics. Perhaps a better understanding of the role of entropy will be obtained after a theory of nonequilibrium thermodynamics is accepted. By a fair assessment, the mathematical notion of entropy used in the theory of conservation laws *is* related to entropy in gas dynamics, but its purpose is to make rigorous definitions of properties that can be used to restrict the class of weak solutions so that IVPs have unique solutions in the restricted class. These restrictions are believed to determine the physically relevant weak solutions. A further discussion of the physical basis for the entropy conditions discussed here is beyond the scope of this book.

In the class of piecewise smooth weak solutions with only jump discontinuities, the entropy condition requires the values of weak solutions on each side of a curve of discontinuities to be obtained by a connection to the initial data without passing through additional discontinuities. In cases where the solution is constant on one side of the curve of discontinuities, this constant value must be the value of the solution on a part of its domain bounded by the horizontal axis. In case the solution is not constant, its limiting values on the curve must be the value obtained along a unique projected characteristic that can be traced backward in time to the initial data without passing through a discontinuity. A curve of discontinuities of a weak solution that separates the space-time plane in such a way that this entropy condition is satisfied on both sides is called a shock, a shock curve, or a shock wave. Of course, because a shock is a curve of jump discontinuities of a weak solution, the Rankine–Hugoniot condition is satisfied at each point of a shock. The derivative $\gamma'(t)$ in Eq. (17.60) is called the shock speed at time t. For instance, the separating curve of discontinuities $t \mapsto (t/2 + 1/2, t)$ found for Burgers's equation (17.51) with initial data (17.52) is a shock with constant shock speed $1/2$.

Shocks are ubiquitous in weak solutions of conservation laws. Weak solutions can be constructed in simple examples (for instance, Burgers's equation with an appropriate choice of initial data) that do not satisfy the entropy condition, but as previously mentioned, solutions that satisfy the entropy condition are deemed to be more physically realistic. The proof of uniqueness of entropy solutions, which is essential for applications, requires a careful mathematical formulation of the entropy condition and some sophisticated mathematics (see, for example, [64]).

There are several ways to express the restrictions on weak solutions that are imposed by enforcing the entropy condition. One of the simplest is obtained for the conservation law

$$w_t + g(w)w_x = 0 \tag{17.62}$$

with initial data $w(x,0) = h(x)$. Recall that its characteristics are solutions of the ODEs

$$\frac{dx}{ds} = g(W), \qquad \frac{dt}{ds} = 1, \qquad \frac{dW}{ds} = 0.$$

For $x(0) = \xi$ and $W(0) = h(\xi)$, the characteristic curve (starting at $t = 0$) is given by

$$x(s) = g(h(\xi))s + \xi, \qquad t(s) = s, \qquad W(s) = h(\xi).$$

Suppose that two such projected characteristics, one starting at $(\xi_L, 0)$ and the other at $(\xi_R, 0)$, meet at a point (x_m, t_m) in space-time (which must then be on a curve of discontinuities that satisfies the entropy condition at this point). Also, denote the left- and right-hand limits of the solution w at the curve of discontinuities by w_L and w_R, which are functions of position along this curve. For $s = t_m$ two equations hold:

$$x_m = g(h(\xi_L))t_m + \xi_L, \qquad x_m = g(h(\xi_R))t_m + \xi_R.$$

By simple algebra and the observation that $\xi_L < \xi_R$, it follows that

$$(g(h(\xi_L)) - g(h(\xi_R)))t_m = \xi_R - \xi_L > 0;$$

therefore,

$$g(h(\xi_L)) > g(h(\xi_R)).$$

Equivalently,

$$g(w_R) < g(w_L) \text{ or } G'(w_R) < G'(w_L). \tag{17.63}$$

In the special case where G is convex ($G''(w) > 0$), the only possibility is that $w_R < w_L$ at a shock. Also, by the Rankine–Hugoniot condition written in the form

$$\frac{G(w_L) - G(w_R)}{w_L - w_R} = \gamma'(t)$$

and the mean value theorem, $\gamma'(t)$ is a value of G' at some w such that $w_R < w < w_L$. By the convexity of G,

$$G'(w_R) < \gamma'(s) < G'(w_L).$$

This condition (called the Lax entropy condition after Peter Lax) is equivalent to the entropy condition for convex G.

For not necessarily convex G, the Oleinik entropy condition (after Olga Oleinik) is

$$\frac{G(w) - G(w_L)}{w - w_L} > \gamma'(t) > \frac{G(w) - G(w_R)}{w - w_R} \tag{17.64}$$

for all w between w_L and w_R.

Although the entropy condition does not easily generalize to systems of conservation laws, it is sufficient to ensure unique weak solutions of IVPs in the case of one conservation law in one spatial dimension.

The Rankine–Hugoniot condition and the uniqueness of weak solutions that satisfy the entropy condition remain unchanged for scalar conservation laws with a source, which are PDEs of the form

$$w_t + (G(w))_x = f(w, x, t)$$

(see Exercise 17.14). This fact is used to analyze some simplified channel flow models.

Our channel flow model is a system of conservation laws for the wet area A of channel cross sections and the discharge Q through these sections. Suppose we believed that Q is determined (or closely approximated) by a function of A, x, and t, say $Q = F(A, x, t)$. In view of the model for conservation of mass with a source given in differential form by the equation $A_t + Q_x + q = 0$, there would be a single scalar conservation law for A given by

$$A_t + (F(A, x, t))_x + q = 0;$$

or, with $f := F_A$,

$$A_t + f(A, x, t)A_x = -q - F_x(A, x, t). \tag{17.65}$$

Indeed, several choices for F have been proposed in the long history of channel flow modeling. A classic example is essentially Chézy's formula

(1775), which states that the channel flow velocity U is a constant times the square root of the hydraulic radius (ratio of wet area and wet perimeter) times the slope of the bottom of the channel. In simpler language, $U = K\sqrt{A}$, for some constant K. Using $U = Q/A$, it follows that $Q = A^{3/2}$. Chézy's formula is an approximation based on observations. An alternative derived from a formula due to Manning states that, for a rectangular channel, $Q = A^{5/3}$.

To apply the theory of hyperbolic conservation laws developed so far to a simple channel flow model that respects the contributions of Chézy and Manning, let $p > 0$ be given and consider the discharge function

$$F(A) = \frac{c}{p+1} A^{p+1}$$

and the corresponding conservation law

$$A_t + cA^p A_x = -q. \tag{17.66}$$

Suppose that the flow in a rectangular channel is augmented by a lateral constant inflow along a finite portion of the channel, the flow has constant initial depth downstream from the inflow region, and a constant depth in and upstream of the inflow area. What depth is predicted in and downstream from the lateral inflow region at later times?

A mathematical model corresponding to the problem description is differential equation (17.66), together with the inflow flux function q given by $q(x) = -b$ for $0 \le x \le x_{\text{in}}$, where $b > 0$ is the constant inflow flux and $x_{\text{in}} > 0$ is the length of the inflow region $[0, x_{\text{in}}]$, and $q(x) = 0$ for x not in the inflow region. The initial wet cross-sectional area function is assumed to be piecewise constant: $A(x, 0) = A_{\text{up}} > 0$ for $x \le x_{\text{in}}$ and $A(x, 0) = A_{\text{down}} > 0$ for $x > x_{\text{in}}$. As the fluid pours into the channel, the initial water depths upstream and downstream of the influx region could be taken arbitrarily. But because the process may not start exactly at the beginning of the influx, let us take $A_{\text{down}} < A_{\text{up}}$. As we will see, the model has a physically meaningful solution for this case.

By scaling the wet area A, the time t, and the spatial coordinate x, the model can be made dimensionless. A convenient scaling is

$$A = \Big(\frac{bx_{\text{in}}}{c}\Big)^{1/(p+1)} w, \qquad x = x_{\text{in}}\xi, \qquad t = \frac{1}{b}\Big(\frac{bx_{\text{in}}}{c}\Big)^{1/(p+1)} \tau.$$

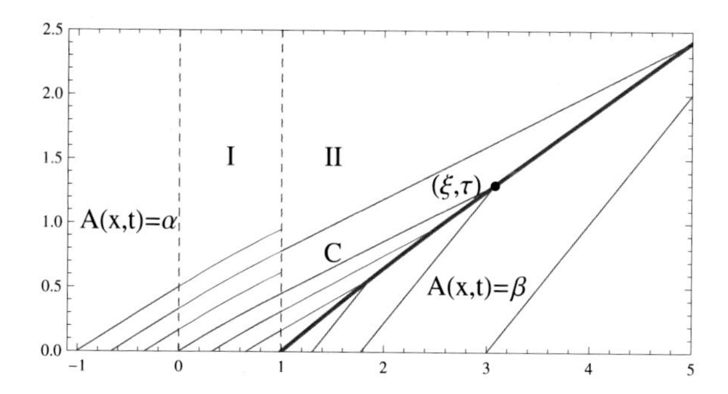

Fig. 17.6 The figure depicts characteristics of model (17.67) *projected into the space-time plane for the case where* $\alpha = 2$, $\beta = 1$, *and* $p = 1$. *The regions where* $A(x,t) = \alpha$ *and where* $A(x,t) = \beta$ *are marked. The thick curve passing through the points* $(1,0)$ *and* $(\xi,\tau) \approx (3.079, 1.298)$ *is a shock. The symbol* C *marks the projected characteristic emanating from the origin in space-time. Region* I *is filled with the extensions of projected characteristics ending along the vertical axis* $\{x = 0\}$, *and in this region, the wet cross-sectional area is in the steady state* $A(x,t) = ((p+1)x+\alpha^{p+1})^{1/(p+1)}$. *In Region* II *the wet area is in the constant steady state* $A(x,t) = (p+1+\beta^{p+1})^{1/(p+1)}$.

Using it produces the model

$$\begin{aligned} w_\tau + w^p w_\xi &= r(\xi), \\ w(\xi,0) &= w_0(\xi), \end{aligned}$$

where $r(x) = 1$ on the interval $0 \le \xi \le 1$ and $r(x) = 0$ otherwise, and

$$w_0(x) := \begin{cases} \alpha := A_{\text{up}}\left(\frac{c}{bx_{\text{in}}}\right)^{1/(p+1)}, & \xi \le 1; \\ \beta := A_{\text{down}}\left(\frac{c}{bx_{\text{in}}}\right)^{1/(p+1)}, & \xi > 1. \end{cases}$$

Following usual practice, let us revert to the original variable names A, x, and t and analyze the model

$$\begin{aligned} A_t + A^p A_x &= r(x), \\ A(x,0) &= A_0(x), \end{aligned} \tag{17.67}$$

where r is now unit inflow on the unit interval $0 \le x \le 1$ and A_0 is the wet area (essentially the water depth for our rectangular channel) at time $t = 0$.

The characteristic system of ODEs for model (17.67) are

$$x'(s) = a^p(s), \qquad t'(s) = 1, \qquad a'(s) = r(x).$$

Recall that on the region of space-time where the characteristic surface is a graph, we have the relation $a(s) = A(x(s), t(s))$.

For $x = \xi$ (where ξ is now a generic name for the x coordinate of a point on the x-axis) and $0 \leq \xi \leq 1$ as in region I of Fig. 17.6, the inflow is $r = 1$, and the corresponding characteristic starting at $t = 0$ is given by

$$x(s) = \frac{1}{p+1}\big((s+\alpha)^{p+1} - \alpha^{p+1}\big) + \xi, \quad t(s) = s, \quad a(s) = s + \alpha \tag{17.68}$$

as long as $x(s) \leq 1$. This latter condition is satisfied for s in the interval

$$0 \leq s \leq s_*(\xi) := \big((p+1)(1-\xi) + \alpha^{p+1}\big)^{1/(p+1)} - \alpha.$$

The projected characteristic bends toward the right in the space-time plane until it reaches $x = 1$. Thereafter (that is, for $s > s_*(\xi)$), the characteristic starting at $(x, t, a) = (1, s_*(\xi), s_*(\xi) + \alpha)$ is in the region where $x > 1$ and $r = 0$; it is given by

$$x(s) = (s_*(\xi) + \alpha)^p (s - s_*(\xi)) + 1, \quad t(s) = s, \quad a(s) = s_*(\xi) + \alpha. \tag{17.69}$$

The projected characteristic curve C corresponding to $\xi = 0$ is depicted in Fig. 17.6.

For $x = \xi$ and $\xi < 0$, the characteristic starting at $x = \xi, t = 0, a = \alpha$ is

$$x(s) = \alpha^p s + \xi, \qquad t(s) = s, \qquad a(s) = \alpha$$

as long as $x(s) \leq 0$; that is, for $0 \leq s \leq s^* := -\xi/\alpha^p$. In particular, A has the constant value α on the projected characteristic, which is a line with slope $1/\alpha^p$ in the space-time plane as in Fig. 17.6. This characteristic extends into the region $x > 0$—marked I in Fig. 17.6—where $r(x) = 1$. In fact, the extension

$$x(s) = \frac{1}{p+1}\big((s - s^* + \alpha)^{p+1} - \alpha^{p+1}\big), \quad t(s) = s, \quad a(s) = (s - s^*) + \alpha,$$

is defined until $x(s) = 1$; that is, on the interval

$$s^* \leq s \leq s_* := (p + 1 + \alpha^{p+1})^{1/(p+1)} + s^* - \alpha.$$

By inspection of the formulas for the projected characteristic in this region, it is easy to see that $x(s)$ may be viewed as a function of $a(s)$. After solving for $a(s)$, note that the wet area is in the steady state

$$A(x, t) = \big((p+1)x + \alpha^{p+1}\big)^{1/(p+1)}$$

at points above the curve C in this region. The extension of the characteristic into region II in Fig. 17.6 where $x > 1$ and $r(x) = 0$ is

$$\begin{aligned} x(s) &= (p+1+\alpha^{p+1})^{p/(p+1)}(s-s_*)+1, \\ t(s) &= s, \\ a(s) &= (p+1+\alpha^{p+1})^{1/(p+1)}. \end{aligned}$$

In the portion of this region filled by such projected characteristics, the solution A is in the (constant) steady state

$$A(x,t) = \left((p+1)+\alpha^{p+1}\right)^{1/(p+1)}.$$

For $x = \xi > 1$, the characteristics starting at time $t = 0$ are given by

$$x(s) = \beta^p s + \xi, \qquad t(s) = s, \qquad a(s) = \beta.$$

The projected characteristics are all lines with the same slope $1/\beta^p$ and $A(x,t) \equiv \beta$ in the region filled with the corresponding projected characteristics, which is the lower portion of region II in Fig. 17.6.

The projected characteristics starting with $0 \le \xi \le 1$ and $\xi > 1$ cross in the region of space-time where $x > 1$. To determine a well-defined state A requires the identification of a shock; that is, a curve of discontinuities $t \mapsto (\gamma(t), t)$ where the Rankine–Hugoniot conditions and the entropy condition are satisfied. Near the line $x = 1$ we have the projected extensions of characteristics starting at $(x,t,a) = (\xi, 0, \alpha)$ with ξ in the unit interval and those starting at $(x,t,a) = (\xi, 0, \beta)$ with $\xi > 1$. A shock should exist in the space-time plane starting at $(x,t) = (1,0)$ and extending into the region where $x > 1$. Using the Rankine–Hugoniot condition, γ must be a solution of the IVP

$$\gamma'(t) = \frac{A_L^{p+1} - A_R^{p+1}}{(p+1)(A_L - A_R)}, \qquad \gamma(0) = 1,$$

where the subscripts L and R refer to the left-hand and right-hand limits of A at the jump discontinuity at the shock. The right-hand limit must be β as this is the value of A on the entire region covered by the projected characteristics starting with $\xi > 1$. Although it is not possible to find an explicit elementary formula for A_L except in special cases (for example, $p = 1$), the projected characteristic [Eq. (17.69)] is a line that must meet the shock. That is, we must have the equality

$$\gamma(t) = (s_*(\xi) + \alpha)^p (t - s_*(\xi)) + 1$$

and by rearranging its right-hand side,

$$\gamma(t) = (s_*(\xi) + \alpha)^p (t + \alpha - (s_*(\xi) + \alpha)) + 1.$$

Along this curve, $a(t) = s_*(\xi) + \alpha$; therefore, A_L is given implicitly by the equation

$$\gamma(t) = A_L^p (t + \alpha - A_L) + 1.$$

At $t = 0$, the implicit root is either $A_L = 0$ or $A_L = \alpha$. The desired choice is $A_L = \alpha$, as this agrees with the value for the family of projected characteristics on the left of the desired shock.

To determine γ seems to require solving the differential algebraic equation (DAE)

$$\begin{aligned} \gamma'(t) &= \frac{A_L^{p+1}(t) - \beta^{p+1}}{(p+1)(A_L(t) - \beta)}, \\ \gamma(t) &= A_L^p(t)(t + \alpha - A_L(t)) + 1, \\ \gamma(0) &= 1, \\ A_L(0) &= \alpha. \end{aligned} \tag{17.70}$$

Under the assumption that $\alpha > \beta$, it is possible to show that the DAE IVP has a solution that produces the desired shock (see Exercise 6.8 for a numerical method that can be used to approximate solutions of this DAE). The solution γ is valid on an interval $0 \leq t \leq \tau$ up to the time $t = \tau > 0$ when the curve C meets the shock. The meeting point (ξ, τ) for the case $p = 1$ is shown in Fig. 17.6. This shock can be extended for $t > \tau$ by taking into account the values of A already obtained. In fact, by taking into account the jump condition and the new values of A_L and A_R, the differential equation for γ becomes

$$\begin{aligned} \gamma'(t) &= \frac{A_L^{p+1} - A_R^{p+1}}{(p+1)(A_L - A_R)} = \frac{(p + 1 + \alpha^{p+1}) - \beta^{p+1}}{(p+1)((p + 1 + \alpha^{p+1})^{1/(p+1)} - \beta)}, \\ \gamma(\tau) &= \xi. \end{aligned}$$

The solution is

$$\gamma(t) = \frac{(p + 1 + \alpha^{p+1}) - \beta^{p+1}}{(p + 1)((p + 1 + \alpha^{p+1})^{1/(p+1)} - \beta)}(t - \tau) + \xi$$

and the shock is a line as depicted for $p = 1$ in Fig. 17.6.

The water depth, at a point in the section of the rectangular channel downstream from the inflow region, maintains its constant initial value for a finite time until the shock front reaches this point. After this time, the depth in the upper part of the channel (upstream from the position with $x = \xi$ determined by the meeting point of the shock and curve C) jumps to a greater depth and then continuously increases in depth until the steady state depth is reached. In the lower part of the channel, the depth jumps to the steady state value as the shock passes. In the inflow region, the depth increases continuously until the steady state (which depends on the spatial position along the channel) is reached. The height H of the shock wave in the lower part of the channel is

$$H = p + 1 + \alpha^{p+1} - \beta.$$

Its time of arrival can also be approximately computed.

Exercise 17.11. Consider Burgers's equation with initial data h such that $h(x) = 0$ for $x < 0$ and $h(x) = 1$ for $x > 0$, which is a version of Riemann's problem. Show that there are at least two possible weak solutions of this IVP and check which of your solutions satisfy the Lax entropy condition.

Exercise 17.12. Consider Burgers's equation with the initial data $w(x, 0) = w_0(x)$ given by $w_0(x) = 0$ for $x < 0$ and $w_0(x) = 1$ for $x > 0$. (a) Show that the value of w in the wedge bounded by the vertical axis and the line $t = x$ is not uniquely determined by integrating along a characteristic. (b) Show that w can be defined in the wedge in at least two different ways so that w is piecewise smooth and w is a weak solution. In particular, Burgers's equation with the given initial data does not have a unique weak solution.

Exercise 17.13. Suppose that G is convex ($G''(w) > 0$). Show that the Rankine–Hugoniot equation implies that $\gamma'(s)$ lies between $G'(w_L)$ and $G'(w_R)$; that is, the Lax entropy condition holds automatically. Hint: There is a wedge in space-time that is not covered by projected characteristics. Fill in the wedge with line segments where w is constant in such a manner that the Rankine–Hugoniot conditions are satisfied on the boundaries of the wedge and on any curves of discontinuity that you introduce.

Exercise 17.14. Show that the Rankine–Hugoniot jump condition holds for scalar conservation laws with a source.

Exercise 17.15. (a) Show that DAE (17.70) has a solution in case $\alpha > \beta$. (b) Give a mathematical reason for the inequality $\alpha > \beta$. (c) Is there a physical reason require $\alpha > \beta$?

Exercise 17.16. Assume the validity of model (17.67). Suppose a river, whose bed is well approximated by a rectangular channel, is 30 m wide, 4 m deep, and currently has a flow depth of 1 m. Storm water flows into the river in the region between $8,000$–$10,000\, m$ upstream of a bridge whose road bed is 4.5 m above the river bottom. Assume rain covers a $2,000$ m by $2,000$ m area at the rate of r cm per cm^2 per hour and the entire area drains immediately into the river (that is, as soon as a raindrop falls it is assumed to have drained into the river). (a) What is the (approximate) maximum rate r such that the roadbed remains dry? At this maximum rate, what is the height of the flood wave at the bridge?

Exercise 17.17. Make a figure similar to Fig. 17.6 for the case $\alpha = 2$, $\beta = 1$, and $p = 1/2$. Discuss the numerical methods you use.

Exercise 17.18. [Mississippi River Flow Rate] Reread Exercise 13.3 on the velocity of river flow. Write a detailed derivation of the formula

$$u = \sqrt{\frac{4gRsin\theta}{f}},$$

where u is the river velocity, g is the acceleration of gravity near the surface of Earth, R is the hydraulic radius of the river cross section, θ is the tilt angle of the river bottom as in Exercise 13.3, and f is an empirical constant (which for large rivers is taken to be $f = 0.35$) that appears in the shear stress of the river flow $\tau = \frac{1}{8} f \rho u^2$, where ρ is the density of the river water. Follow the outline suggested by Sean Sweany: Start with the $1 - d$ Navier–Stokes equation:

$$\rho(u_t + uu_x) = -p_x + \mu u_{xx} + b,$$

where x is the coordinate in the direction of river flow. Assume steady state flow, hydrostatic pressure, negligible viscosity, and constant river depth. Show that the Navier–Stokes equation reduces to

$$\rho u u_x = b.$$

Show that the body force due to gravity is $\rho g \sin\theta$ and a reasonable choice of frictional stress is $-PL\tau$, where P is the wetted perimeter and L is some suitable choice of length along the river. To obtain a body force with the correct units, divide by the hydraulic radius. Show that these considerations lead to the equation

$$\rho u u_x = \rho g \sin\theta - \frac{f}{4R} \rho u^2.$$

Solve this equation for u and obtain the desired formula for the river velocity by approximating the solution under the assumption that the x coordinate is large (to take into account the length of the river). Using this result as well as data given here and in Exercise 13.3, estimate the speed of the flow in the Mississippi River. Check your result against observed flow rates.

17.7 SURFACE WAVES

Imagine the wave profiles that might appear on the surface of a flow in a prismatic channel with rectangular cross section. There are many examples in nature; for example, traveling waves have been discussed. Another type of wave called a roll wave occurs over shallow water moving on an inclined surface; for example, rainwater draining over a sidewalk down a shallow incline. Similar waves appear over horizontal channels with a steady water supply, perhaps fed by a pump. Some aspects of roll waves are discussed in this section (see [81] and [54] and the references therein for a more sophisticated treatment of this subject and especially for existence and stability theory).

The channel flow model [Eq. (17.43)] for the wet area A, discharge Q, bottom profile B, and external flow through the boundary q (area per time)—repeated here for convenience—is

$$
\begin{aligned}
A_t + Q_x &= q, \\
\rho\Big(Q_t + \Big(\frac{Q^2}{A}\Big)_x\Big) &= -\frac{g\rho}{\omega}AA_x - \rho g A B_x \\
&\quad - c\rho\omega\Big(\frac{Q}{A}\Big)^2 + \frac{\iota\rho}{2}\Big(\frac{Q^2}{A}\Big)_x \\
&\quad + \frac{\lambda\mu}{d}\Big(d\frac{Q}{A} + \Big(\frac{Q}{A}\Big)^2\Big) \\
&\quad + \mu(Q_{xx} - \delta(A_x\frac{Q}{A})_x) \\
&\quad + \rho g A \sin(\theta(x,t))
\end{aligned}
\tag{17.71}
$$

with one modification: a new parameter ι is inserted so that $\iota = 0$ corresponds to using only hydrostatic pressure in the derivation and $\iota = 1$ corresponds to using the fluid velocity term in Bernoulli's law. Table 17.1 contains descriptions of the remaining system parameters.

A dimensionless version of model system (17.71) is desired for simplification and numerical computation. Although there are many possible scalings that render the channel flow equations dimensionless, the traditional choice for scaling in fluids is to choose length (ℓ) and velocity (V) scales, define the timescale $\tau = \ell/V$, and rewrite dimensionless groups using the Reynolds and Froude numbers ($\mathrm{Re} = V\ell/(\mu/\rho)$ and $\mathrm{Fr} = V/\sqrt{g\ell}$). With the change of variables

$$A(x,t) = \ell^2 \mathcal{A}(\frac{x}{\ell}, \frac{t}{\tau}), \quad Q(x,t) = \ell^2 V \mathcal{Q}(\frac{x}{\ell}, \frac{t}{\tau}), \quad q(x,t) = \ell V p(\frac{x}{\ell}, \frac{t}{\tau}),$$
$$B(x,t) = \frac{V^2}{g}\mathcal{B}(\frac{x}{\ell}, \frac{t}{\tau}), \quad \theta(x,t) = \vartheta(\frac{x}{\ell}, \frac{t}{\tau}), \tag{17.72}$$

and the assignments

$$d = \frac{\ell}{\tau}\bar{d}, \qquad \omega = \ell\bar{\omega}, \tag{17.73}$$

the channel flow model is transformed to the dimensionless form

$$\begin{aligned} \mathcal{A}_s &= -\mathcal{Q}_\xi + p, \\ \mathcal{Q}_s &= -(1-\frac{\iota}{2})(\frac{\mathcal{Q}^2}{\mathcal{A}})_\xi - \frac{1}{\bar{\omega}\,\mathrm{Fr}^2}(\frac{\mathcal{A}^2}{2})_\xi - \mathcal{A}\mathcal{B}_\xi - c\bar{\omega}\frac{\mathcal{Q}^2}{\mathcal{A}^2} \\ &\quad + \frac{\lambda}{\bar{d}\,\mathrm{Re}}(\bar{d}\frac{\mathcal{Q}}{\mathcal{A}} - \frac{\mathcal{Q}^2}{\mathcal{A}^2}) + \frac{1}{\mathrm{Re}}(\mathcal{Q}_{\xi\xi} - \delta(\frac{\mathcal{Q}\mathcal{A}_\xi}{\mathcal{A}})_\xi) + \mathcal{A}\sin\vartheta. \end{aligned} \tag{17.74}$$

Analysis and applications of the channel flow model is a rich subject that is not completely understood, but many important features of the model have been explored. Some simple analysis is discussed in the remainder of this section, but before this presentation is made, a desire to apply the model serves to motivate a few comments about realistic models that arise in applied mathematics.

Multiparameter models require some method of fixing values of parameters that correspond to the intended application. In an ideal world this is a straightforward task. Measure all the parameters and fix them in the model at these values. But, in the real world of applications there are always parameters whose values are not known, or parameters whose values are known only within a crude approximation. A good example, in the channel flow model, is the value of c that determines the strength of the bottom shear force. The model could be used to try to measure this quantity by fitting parameters to experimental data. Indeed, this is a subject of great interest. At a basic level, the model equations have solutions that depend (continuously) on all the parameters in the system. By solving the equations (numerically) and minimizing the squares of the differences between measured and modeled state variables, the parameters should in principle be determined. Noise in measurements is a serious problem. Also, there is no guarantee that the parameters are uniquely determined by a given data set.

In most nonlinear systems, such as the channel flow model, there are multiple timescales that are not simply determined. For this discussion, the existence of multiple timescales means that in some regimes one of the state variables (in this case $\mathcal{A}$ or $\mathcal{Q}$) is changing rapidly with respect to the other. This ubiquitous phenomena in nonlinear dynamics causes inaccuracies in numerical computations (stiff differential equations) and difficulty in fitting the model to observed phenomena.

Closely related to problems with multiple timescales is sensitivity with respect to the choice of parameters: a small change in a parameter can produce a large change in the solution of the equations. This is in addition to bifurcations due to changes in parameters that produce changes in the qualitative nature of solutions. This subject—sensitivity analysis—is essential in making quantitative predictions from models. Although basic sensitivity can be ascertained from the linear system obtained by differentiation with respect to a parameter, the resulting linearized system might be as difficult to study as the original system. For this reason (and others), a unified and useful treatment of sensitivity remains a difficult unsolved problem for multiparameter nonlinear systems.

All of the difficulties mentioned here occur in the analysis of the channel flow model.

To illustrate some of the challenges that must be met in dealing with realistic nonlinear systems, consider again the popular and reasonable Saint-Venant simplification of the channel flow model with no lateral inflow, zero surface shear stress, and with viscous dissipation:

$$\begin{aligned} \mathcal{A}_s &= -\mathcal{Q}_\xi, \\ \mathcal{Q}_s &= -\Big(\frac{\mathcal{Q}^2}{\mathcal{A}}\Big)_\xi - \frac{1}{\bar{\omega}\,\mathrm{Fr}^2}\Big(\frac{\mathcal{A}^2}{2}\Big)_\xi - \mathcal{A}\mathcal{B}_\xi + \mathcal{A}\sin\vartheta - c\bar{\omega}\frac{\mathcal{Q}^2}{\mathcal{A}^2} \\ &\quad + \frac{1}{\mathrm{Re}}\Big(\mathcal{Q}_{\xi\xi} - \delta\Big(\frac{\mathcal{Q}\mathcal{A}_\xi}{\mathcal{A}}\Big)_\xi\Big). \end{aligned} \tag{17.75}$$

For the case of a rectangular trough tilted downward in the downstream direction so that

$$\mathcal{A}\mathcal{B}_\xi + \mathcal{A}\sin\vartheta = \mathcal{A}(\tan\vartheta + \sin\vartheta)$$

and with $\delta = 1$, the system becomes

$$\begin{aligned}\mathcal{A}_s &= -\mathcal{Q}_\xi,\\ \mathcal{Q}_s &= -\Big(\frac{\mathcal{Q}^2}{\mathcal{A}}\Big)_\xi - \frac{1}{\bar{\omega}\,\mathrm{Fr}^2}\Big(\frac{\mathcal{A}^2}{2}\Big)_\xi - c\bar{\omega}\frac{\mathcal{Q}^2}{\mathcal{A}^2}\\ &\quad + \frac{1}{\mathrm{Re}}\Big(\mathcal{A}\Big(\frac{\mathcal{Q}}{\mathcal{A}}\Big)_\xi\Big)_\xi + \mathcal{A}(\tan\vartheta + \sin\vartheta).\end{aligned} \tag{17.76}$$

The flow is driven by gravity because the channel is inclined to the horizontal.

Observed roll waves, for example in the channel flow caused by rain runoff on a hard sidewalk, generally travel downstream with a periodic profile. A challenging problem is to show that the model equations (17.76) have such a solution. Choosing model parameters to match roll wave behavior for a measured real flow is a grand challenge.

To look for traveling waves, replace the state variables by wave forms traveling at speed C with profiles F and G:

$$\mathcal{A}(\xi, s) = F(\xi - Cs), \qquad \mathcal{Q}(\xi, s) = G(\xi - Cs),$$

and note that the first equation in system (17.76) becomes

$$-CF' = G'.$$

Thus,

$$G = -CF - b, \tag{17.77}$$

where b is the constant of integration and the minus sign is simply taken for convenience. Inserting this latter equation into the second differential equation produces a second-order ODE for the unknown profile F. If G is desired, it may be recovered from Eq. (17.77). The challenge is to find a periodic solution of this second-order differential equation; it would produce the desired periodic traveling wave profile.

One way to meet the challenge is to use phase plane analysis. The first step, as always, is to treat the second-order profile equation as a first-order system of ODEs. With $x := F$, $y := x'$, and

$$\alpha_1 := \bar{\omega}\,\mathrm{Fr}^2\,\mathrm{Re}, \quad \alpha_2 := \bar{\omega}\,\mathrm{Fr}^2, \quad \alpha_3 := \alpha_1(\tan\vartheta + \sin\vartheta),$$

the first-order system for the unknown profile function x and its derivative is

$$\dot{x} = y,$$
$$\dot{y} = \frac{-b^2 c\alpha_1 + 2bCc\alpha_1 x - C^2 c\alpha_1 x^2 + \alpha_3 x^3 + b^2\alpha_1 y - \operatorname{Re} x^3 y - b\alpha_2 y^2}{\alpha_2 x(-b - x + Cx)}. \tag{17.78}$$

What is an effective method for finding a periodic solution of a nonlinear system of ODEs? The short answer: look for a Hopf bifurcation. Recall that Hopf bifurcations occur when a rest point changes stability type due to a continuous change in a single system parameter. More precisely, consider a rest point of the nonlinear system and the eigenvalues of the system matrix of the linearization of the system at this rest point. As some system parameter changes, the position of the rest point and the eigenvalues of the corresponding system matrix of its linearization change. Tracking these eigenvalues in the complex plane, a Hopf bifurcation occurs when a pair of complex conjugate eigenvalues crosses the imaginary axis. The real parts of the eigenvalues determine the stability of the rest point. So, at the crossing, the rest point changes its stability type from a spiral sink to a spiral source or from a spiral source to a spiral sink. Under generic conditions, a limit cycle is born from the rest point or dies at the rest point as the parameter changes. Exactly which scenario occurs can be determined by local computations at the rest point (see, for example, [20]). The case of interest here is an unstable limit cycle surrounding a spiral sink that disappears into the rest point at the parameter value for which the rest point changes stability from a sink to a source. In fancy language this is called a subcritical Hopf bifurcation. Knowing that this scenario exists is a key to finding a limit cycle and thus a traveling wave with a periodic profile. A solution of the differential equation corresponding to the unstable limit cycle gives the desired wave profile.

Nonlinear system (17.78) has some standard features that should be taken into consideration: from the first equation ($\dot{x} = y$) we know that solutions drift to the right in the upper half-plane and to the left in the lower half-plane; the vertical lines given by the equations $x = 0$ and $x = b/(C-1)$ correspond to singularities of the system where solutions do not exist, and rest points (if any) are on the horizontal coordinate axis ($y = 0$). The variable x represents the scaled area of a wet cross section of the flow; therefore, it must be positive. Physically relevant solutions all lie in the right half-plane ($x > 0$).

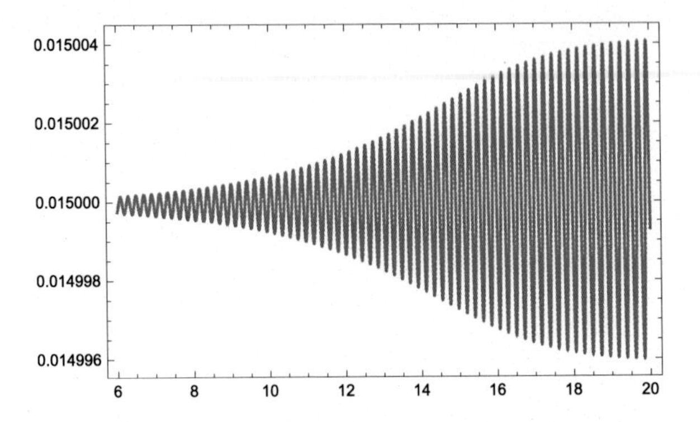

Fig. 17.7 Amplitude (meters) versus time (seconds) for the area component A of a traveling wave that at first has a nearly constant profile, undergoes a rapid increase in amplitude, and then remains oscillatory for all future time as it passes an observation point.

The horizontal coordinates of rest points are solutions of the cubic polynomial obtained by substituting $y = 0$ in the right-hand side of the second equation in system (17.78). Because $-b^2 c\alpha_1 > 0$ (except in the case $b = 0$ that can be handled separately) and $\alpha_3 > 0$ for a moderate channel incline, there is always at least one positive real root. At least the system has rest points to analyze and the possibility of Hopf bifurcation is not excluded.

From an applied perspective, the desired periodic profile must occur in a physically realistic regime that corresponds to some measurements. Imagine, for example, water flowing down an inclined hard surface after a rainfall. The flow velocity is measured to be $0.2\,\mathrm{m}\,/\,\mathrm{sec}$ and the water depth is $0.015\,\mathrm{m}$. A natural choice for the characteristic velocity is of course the measured flow velocity, but there is no natural choice for the characteristic length. Water depth is a reasonable choice, but so is some expected wave length for waves, an arbitrary unit of measurement (for example one meter), or some other length. A conventional characteristic length is called the hydraulic diameter, which is defined to be four times the area divided by the wet perimeter of a cross section. For the present case, assume the channel is one meter wide. This gives the hydraulic diameter $0.0295567\,\mathrm{m}$ taken here to be the characteristic length. With this choice the Reynolds number is 5911.33 and the Froude number is 1.17514.

Absent a measurement, the bottom friction parameter c remains unknown.

With the system parameters (except for c) set, the wave speed C and integration constant b remain free. These parameters may be adjusted to find a Hopf bifurcation by first seeking a rest point whose linearization has a pair of pure imaginary eigenvalues. Once such a rest point is identified, one of the free parameters may be adjusted to see if a pair of complex conjugate eigenvalues crosses the imaginary axis at the parameter values corresponding to the rest point. Under generic conditions this scenario will occur.

A viable strategy is to solve a system of equations to determine the unknown values of c, C, b, and x. The first equation to solve is the right-hand side of the second equation in system (17.78) set to zero after substituting $y = 0$; it is a cubic equation for the unknown position of the rest point. Next, linearize at the unknown rest point by computing the Jacobian matrix T of the right-hand side of system (17.78). Then set $y = 0$, and recall that the eigenvalues of A are roots of the quadratic polynomial

$$z^2 - \operatorname{tr} T z + \det T.$$

Thus, these eigenvalues are pure imaginary when $\operatorname{tr} T = 0$ and $\det T > 0$. Also, the square root of the determinant determines the (circular) frequency of the linearized oscillations. This number may be set to yet another parameter $k > 0$ so that the third equation is $\det T = k$.

After solving the three equations for C, b, and x, each of these is given as a function of k and c. The value of x (which determines the position of the rest point) should be near the value of the scaled wet area, which is the quantity that x represents. Making x be exactly this value produces an equation whose solution specifies k as a function of c. Because a quadratic equation must be solved to achieve this goal, there are two solutions. The strategy is to choose parameters that correspond to a Hopf bifurcation. Although both solutions have this property, only one of the choices produces low frequency (k small) oscillations. For the parameters chosen here, the value $c = 1.75$ produces $k \approx 3$. Using this value of c seems reasonable. But, there is freedom to choose c so that k is as small as desired. In fact, $k = 0$ for $c \approx 1.75265$. With these values of k and c, the free parameters are fixed at

$$b \approx 14.7723, \qquad C \approx 2.576, \qquad x \approx 17.1704.$$

A few numerical computations show that the corresponding rest point changes stability from a spiral sink to a spiral source as the parameter C passes through the given value in the positive direction. The phase portrait of system (17.78) undergoes other nearby bifurcations. In particular, the number of rest points changes (in a saddle-node bifurcation) from one to three near the given parameter values.

For

$$C = 2.576 + 10^{-8},$$

there are three rest points on the horizontal coordinate axis at

$$x \approx 4.3, \qquad x \approx 17.1703, \qquad x \approx 17.2374.$$

The first value corresponds to a source, the second to a spiral sink, and the third to a saddle.

Recall that the traveling wave solution, for the scaled wet area, is given by $\mathcal{A}(\xi, s) = F(\xi - Cs)$. Thus, as the temporal variable in system (17.78) increases in the positive direction, the temporal parameter s for the PDE decreases; that is, $\xi - Cs$ increases (for $C > 0$, of course) as s decreases.

Numerical computations indicate that there is a subcritical Hopf bifurcation resulting in an unstable limit cycle surrounding the sink. This periodic orbit corresponds to a periodic traveling wave solution, as desired. By similar analysis there are infinitely many other choices for periodic solutions. The periodic profile traveling wave solution corresponding to the numerical values given in this discussion would be obtained by choosing an initial value for a solution of the system of ODEs that lies on the limit cycle. In practice, finding a point on a limit cycle is usually not possible using numerical experiments: an error (no matter how small) will cause the solution to leave the vicinity of the limit cycle in either forward or backward time. Instead, choosing an initial condition near the spiral sink ensures a traveling wave profile that in positive time acquires the periodic profile of the limit cycle and in negative time the constant profile of the sink. A numerical approximation showing this behavior for $A(x, t)$ versus t for fixed $x = 1$ is depicted in Fig. 17.7. An observer at $x = 1$ would see a flat surface at their location that eventually changes quickly to an oscillating surface wave that continues to oscillate for all future time.

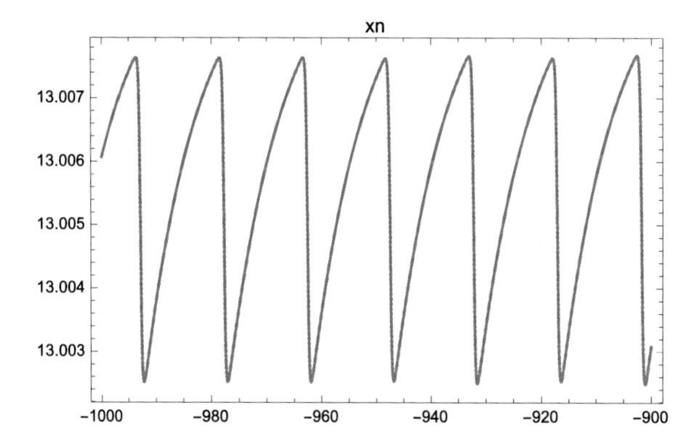

Fig. 17.8 The figure shows a plot of x versus the (unnamed) temporal variable for system (17.78) *with parameter values* wd $= 0.01\,\mathrm{m}$, *characteristic velocity* $1.0\,\mathrm{m}$, *channel width* $0.5\,\mathrm{m}$, $\tan\vartheta + \sin\vartheta = 0.3$, $c = 1.7187664596060788$, $b = 9.79938796267978$, *and* $C = 2.2601418989288105$. *Note the fast-slow (relaxation) oscillation.*

The periodic traveling wave profile just described is mathematically interesting, but from a physical point of view it is not satisfying because the amplitude of the wave is probably too small to be observed. In the regime discussed here, which was determined by the choice of Reynolds and Froude numbers and the bottom shear stress, the model predicts at best the existence of a tiny surface ripple but not the roll waves that are easily observed on flows over inclined channels where the amplitude of the waves is a few millimeters (see, for example, [6] where a laboratory experiment is discussed). Are there realistic choices of the system parameters that produce close approximations of waves observed in nature?

Part of Exercise/Project 17.19 explores pulse-type traveling wave profiles corresponding to saddle (homoclinic) loops that occur at some saddle points for some choices of the system parameters for the first-order system (17.78). Solutions of this system corresponding to periodic traveling wave profiles were identified near Hopf bifurcations. As the wave speed parameter changes in the correct direction, these limit cycles disappear in a subcritical Hopf bifurcation. Changing the parameter in the opposite direction causes the limit cycles to grow in amplitude until they disappear in a saddle-loop (homoclinic-loop) bifurcation. Limit cycles near a saddle-loop bifurcation have the correct profile for roll waves: during each period there is an abrupt change in amplitude as the limit cycle moves away and then back to the vicinity of the saddle point followed by a slow change as the periodic orbit passes the vicinity of the saddle point. A second fast-slow mechanism is due to the compression and expansion near the saddle point. For the tested example, the expansion rate was much larger than the compression. This configuration has the effect that in backward time and near the saddle, solutions rapidly approach the stable manifold of the saddle. Such periodic solutions also have larger amplitudes than the limit cycles near Hopf bifurcations. Fig. 17.8 shows an example of this type of unstable limit cycle, which seems to require some delicate numerics to produce. The fast-slow dynamics is mostly due to the compression and expansion rates near the saddle. The existence of such traveling wave profiles evidences the validity of the channel flow model. But, as the discussion here is meant to show, simply using system parameters and initial data at some measured values in the channel flow model is unlikely to produce a solution that agrees with the observed flow. At best, the observed flow would be modeled by some nearby set of parameters and some nearby initial data. And, solutions obtained from the model will likely be sensitive to changes in the data. On the positive side, the model does predict wave forms that agree with observation. Thus, there is strong evidence that the mechanism producing the waves (shear stress at the channel bottom, viscosity, and the gravitational driving force) is correctly understood.

Exercise 17.19. [Channel Flow Modeling] (a) Repeat the search for a subcritical Hopf bifurcation and the existence of an unstable limit cycle wave profile (as in Fig. 17.7) but with the basic scales wd $= 0.015\,\mathrm{m}$, characteristic velocity $1\,\mathrm{m}/\,\mathrm{sec}$, channel width $0.5\,\mathrm{m}$, and bottom shear-stress parameter $c = 1.15$. Give evidence, based on numerical computations, that there is an unstable limit cycle surrounding a spiral sink for the model equations in the regime close to these parameter values. Choosing a traveling wave profile by starting with initial data on this limit cycle produces a periodic traveling wave profile. Starting with initial data inside the region bounded by the limit cycle produces a traveling wave profile that is eventually constant in negative time and periodic in forward time. Explain.
(b) Choosing initial data outside (but near) the region bounded by the limit cycle usually does not produce a traveling wave profile. Why not? But the special case, where initial data is chosen on the stable manifold of the saddle rest point (which exists in the traveling wave profile ODEs), does produce a traveling wave profile that is eventually constant in negative time and periodic in forward time. Explain. Make some hand drawn figures to illustrate the phase portrait.
(c) The expected behavior of the limit cycle under discussion as a single parameter (in this case the wave speed C) changes is governed by the Hopf bifurcation for C near the parameter value corresponding to pure imaginary eigenvalues of the system matrix. As C changes in one direction from this value, no limit cycle exists, but as this parameter changes in the opposite direction, the limit cycle grows in size. Of course, this growth is usually not unlimited. The expected behavior, given the presence of the saddle point, is growth until the limit cycle disappears in a saddle-loop bifurcation. Give evidence that a

saddle-loop bifurcation occurs. To do this, use numerical computations to approximate the positions of the portions of the stable and unstable manifolds of the saddle point that lie to the left of the saddle point. For parameter values of C corresponding to existence of the unstable limit cycle, the unstable manifold of the saddle crosses the horizontal coordinate axis to the left of the (infinitely many) crossings of the stable manifold, which in backward time is asymptotic to the limit cycle. After the limit cycle disappears, the positions of the crossings are interchanged. Evidence for this interchange of crossing positions implies the existence of a choice for the wave speed parameter such that the stable and unstable manifolds cross the coordinate axis with the same horizontal coordinate so that these manifolds form a saddle-loop. Choosing the traveling wave profile by taking initial data on this saddle-loop produces a traveling wave of pulse type; it has the same constant values as time goes to plus or minus infinity.
(d) Discuss the role of ι in the channel flow model equations. This parameter was defined to turn on or off a term in the channel flow model by taking the values zero or one, but it could reasonably take on values between these two numbers. Do periodic traveling wave profiles exist for $\iota = 1$? Give evidence for your answer.
(e) Discuss the design of a physical experiment that could be performed to decide between $\iota = 0$ and $\iota = 1$ in the channel flow model equations. Simulate the experiment and report some data that an experimenter could compare with physical measurements.
(f) Reconsider the term in display (17.41) meant to model the change in downstream velocity near the fluid surface. Should this term be ignored as in the text? Discuss this in view of possible physical scenarios. Can you imagine a physical problem where this term would be important? If so, make simulations of steady configurations and discuss the predicted effects of the suggested approximation. Is there a better approximation for use in channel flow models?
(g) Discuss the role of λ (surface shear strain) in the channel flow model equations. Do periodic traveling wave profiles exist for $\lambda > 0$? Give evidence for your answer.
(h) Discuss the design of a physical experiment that could be performed to decide between $\lambda = 0$ and $\lambda > 0$ in the channel flow model equations. Simulate the experiment and report some data that an experimenter could compare with physical measurements.
(i) A research problem: Suppose the channel flow model is used to make a prediction; for example, the existences of a periodic traveling wave solution in a certain flow regime. Would the Navier–Stokes equations imply the same prediction?

CHAPTER 18

Elasticity: Basic Theory and Equations of Motion

Imagine a closed and bounded region Ω in three-dimensional space filled with matter—which is called a body—such that Ω has a smooth boundary surface $\partial\Omega$. A deformation of the body is a function $F : \Omega \times J \to \mathbb{R}^3$, where J is a time interval of real numbers. This function is assumed to be infinitely differentiable. And, for each fixed time, the corresponding function $x \to F(x, t)$ is assumed to be invertible and orientation preserving. The direct problem of elasticity theory is to determine the deformation F from the stresses, body forces (gravity or electromagnetism), and forces that act only on the surface of Ω.

Deformations of the body are functions of position and time. For simplicity and in keeping with the classical theory, we will assume that a Cartesian coordinate system is given in $\mathbb{R}^3$. Be warned that this is an important assumption; the equations of the theory take different forms in different coordinate systems.

Physical deformation is measured by relative changes in lengths called strains. For this reason, the theory is usually formulated for the displacement, given by $u(p,t) := F(p,t) - p$, rather than the deformation. (Note the change in notation—meant to comply with some of the literature on elasticity—from the chapter on fluid motion: here *u denotes displacement, not velocity*.) The relation between stress and strain lies at the heart of elasticity theory. Their functional relationship is too complicated to be reduced to the fundamental laws of nature; constitutive laws are used instead. The most important of these is Hooke's law. It states that (for small displacements) stress is proportional to strain. This relation leads to the classical (linear) theory of elasticity.

Consider a point p in Ω and a curve γ in this body with parameter s passing through p at $s = 0$. The deformation F, with its time dependence suppressed, deforms γ into a new curve $F \circ \gamma$. To measure the distortion, denote the tangent vector to γ at p by v and note that the vector tangent to

An Invitation to Applied Mathematics: Differential Equations, Modeling, and Computation.
http://dx.doi.org/10.1016/B978-0-12-804153-6.50018-X, Copyright © 2017 Elsevier Inc. All rights reserved.

the distorted curve at $F(p)$ is

$$\frac{d}{ds}F(\gamma(s))\big|_{s=0} = DF(p)v,$$

the length along γ starting at p is

$$\ell(s) := \int_0^s |\dot{\gamma}(t)|\, dt,$$

and the length along the distorted curve is

$$L(s) := \int_0^s |DF(\gamma(t))\dot{\gamma}(t)|\, dt.$$

The difference of the squares of these lengths is easily computed using the Taylor approximation to be

$$\begin{aligned} L^2(s) - \ell^2(s) &= (|DF(p)v|^2 - |v|^2)s^2 + O(s^3) \\ &= (\langle DF(p)v, DF(p)v\rangle - \langle v, v\rangle)s^2 + O(s^3) \\ &= \langle (DF(p)^T DF(p) - I)v, v\rangle s^2 + O(s^3), \end{aligned}$$

where the superscript T denotes the matrix transpose.

To take into account the size of the displacement at the lowest order of approximation, the linear transformation $DF(p)^T DF(p) - I$ is recast in the form

$$\begin{aligned} DF(p)^T DF(p) - I &= (Du(p) + I)^T (Du(p) + I) - I \\ &= Du(p)^T Du(p) + Du(p)^T + Du(p). \end{aligned}$$

The Lagrange–Green (strain) tensor E is defined by

$$\begin{aligned} E(p,t)(v,w) &= \frac{1}{2}\langle (DF(p,t)^T DF(p,t) - I)v, w\rangle \\ &= \frac{1}{2}\langle (Du(p)^T Du(p) + Du(p)^T + Du(p))v, w\rangle, \end{aligned}$$

where the factor $\frac{1}{2}$ is inserted to agree with the usual definition of this tensor. Informally or in a physical application, a displacement (or deformation) is called small if the product $Du(p)^T Du(p)$ (which involves quadratic nonlinearities) may be safely neglected. From a mathematical perspective, the same result is achieved by simply declaring the strain tensor ε of linear elasticity theory to be the linear approximation of the Lagrange–Green tensor; that is, the linear strain tensor is given by

$$\varepsilon(p,t)(v,w) = \frac{1}{2}\langle (Du(p)^T + Du(p))v, w\rangle.$$

In the continuum mechanics literature, the expression $(Du(p)^T + Du(p))/2$ is often called the strain tensor. This tradition might lead the reader to believe that the word "tensor" is simply a way of referring to a family of linear transformations (parameterized by the base point p) of matrices in some special circumstances. This is not the case. Although the strain tensor is determined by the linear transformation $(Du(p)^T + Du(p))/2$, this transformation and the strain tensor are not the same mathematical objects. To see this distinction more clearly, consider a constant family of linear transformations A on $\mathcal{R}^k$ and the tensor S defined by $S(v,w) = \langle Av, w\rangle$. Suppose that A is represented by a matrix with respect to the basis $\mathcal{B}_1 := \{e_1, e_2, \dots, e_n\}$, and let B denote the linear transformation taking the new basis $\mathcal{B}_2 := \{f_1, f_2, \dots, f_n\}$ to the original basis $\mathcal{B}_1$. A vector v represented by the coordinates $v_1, v_2, \dots v_n$ with respect to the basis $\mathcal{B}_2$ is given by $v = \sum v_i f_i$. This same vector is represented in the basis $\mathcal{B}_1$ by the coordinates $\nu_1, \nu_2, \dots, \nu_n$ where $\nu = Bv$. To transform A to a matrix with respect to the basis $\mathcal{B}_2$ (that is, to express the same linear transformation on vectors expressed in coordinates with respect to the basis $\mathcal{B}_2$) we may consider an arbitrary vector v expressed in coordinates with respect to the basis $\mathcal{B}_2$, multiply by B to express the vector in coordinates with respect to $\mathcal{B}_1$, multiply by A to apply the linear transformation, and multiply by B^{-1} to express the transformed vector in coordinates with respect to the basis $\mathcal{B}_2$. In symbols, the linear transformation expressed in the basis $\mathcal{B}_2$ is $B^{-1}AB$. In classical language, we have explained how a matrix changes coordinates and how a vector changes coordinates: by transformations of the form $B^{-1}AB$ and $B^{-1}ABv$. How does a tensor change coordinates? For A expressed in the basis $\mathcal{B}_1$, the tensor T acts on two vectors v and w expressed in coordinates with respect to this basis according to the formula $S(v,w) = \langle Av, w\rangle$. Suppose α and β are vectors expressed in coordinates with respect to the basis $\mathcal{B}_2$. To obtain the same tensor in the new basis simply express the vectors in coordinates with respect to the basis $\mathcal{B}_1$ via $v = B\alpha$ and $w = B\beta$. The value of the tensor S on the pair of vectors α and β is $\langle AB\alpha, B\beta\rangle$. Using the properties of the usual inner product, we have that

$$\langle AB\alpha, B\beta\rangle = \langle B^T AB\alpha, \beta\rangle.$$

Thus, the matrix representing the tensor S expressed in coordinates with respect to the basis $\mathcal{B}_2$ is $B^T AB$, not $B^{-1}AB$. Our tensor changes coordinates via its corresponding matrix A being transformed by $B^T AB$. This is an example of the classical approach most physicists or engineers take in defining and using tensors. It is perfectly legitimate. The modern view is to define tensors as multilinear maps (linear in each argument) from a cross product of vector spaces to the real numbers. The tensor S is a bilinear map from $\mathbb{R}^k \times \mathbb{R}^k$ to $\mathbb{R}$ called a rank-two tensor. To reiterate the point of this discussion: A rank-two tensor is determined by an associated linear transformation A, and the matrix representing a rank-two tensor changes coordinates according to the rule $B^T AB$. In particular, a rank-two tensor and its associated linear transformation are not the same mathematical objects. This distinction is often suppressed in applied elasticity theory, until the moment after it becomes important. The reader should also note that the tensors considered in continuum mechanics are actually tensor *fields*; that is, they are families of tensors parameterized by space and time. For instance, at each point p and time t the strain tensor field is the multilinear map that takes vectors v and w defined at p at time t to the real number $\frac{1}{2}\langle (Du(p,t)^T + Du(p,t))v, w\rangle$.

For a second approach to the strain tensor, let p and q be points in the body Ω and note that these points move under the deformation to $P := F(p,t)$ and $Q := F(q,t)$. The position of q relative to p is the vector $q - p$ and after the deformation it is $Q - P$. Using the definitions of F and u, the change in these vectors is

$$(Q - P) - (q - p) = u(q,t) - u(p,t).$$

For q close to p, the expansion via Taylor series of the function $q \mapsto u(q,t)$ at p may be used to recast the last equation into the form

$$(Q - P) - (q - p) = \nabla u(p,t)(q - p) + O(|q - p|^2).$$

Linear elasticity theory is developed by ignoring the second-order term. Thus, the difference of the deformed and undeformed position vectors is approximated via the derivative of the function $q \mapsto \nabla u(q,t)$ at p. In this context, ∇u is called the displacement gradient.

The displacement gradient (or an arbitrary matrix) may be decomposed into the sum of a symmetric and a skew-symmetric matrix. Indeed,

$$\nabla u(p,t) = \varepsilon(p,t) + \omega(p,t)$$

where

$$\varepsilon := \frac{1}{2}(\nabla u + (\nabla u)^T), \qquad \omega := \frac{1}{2}(\nabla u - (\nabla u)^T), \tag{18.1}$$

where, as always, the superscript T denotes the transpose, ε is symmetric, and ω is skew-symmetric. The matrix ε corresponds to the linear approximation of local distortion and, as we have seen, determines the strain tensor; the matrix ω corresponds to the local infinitesimal rigid rotation imparted by the deformation.

What is meant by infinitesimal rotation? The word "infinitesimal" in this context refers to the linear approximation given by the first derivative. Thus, "infinitesimal rotation" means differentiation of rotation. More precisely, recall that a rotation is given by an orthogonal matrix that may be defined as a matrix whose transpose is its inverse, or more generally, by the geometric property that the matrix preserves length; that is, $\mathcal{O}$ is an orthogonal matrix if $|\mathcal{O}w| = |w|$ for every vector w.

The polarization identity states that the inner product is related to the norm by

$$\langle w, z\rangle = \frac{1}{4}(|w+z|^2 - |w-z|^2) \tag{18.2}$$

for every pair of vectors w and z, and by an easy exercise [Exercise 18.4],

$$\langle \mathcal{O}w, \mathcal{O}z\rangle = \langle w, z\rangle \tag{18.3}$$

for every w and z. Let $t \mapsto \mathcal{O}(t)$ be a curve of orthogonal matrices passing through the identity matrix at $t = 0$, compute using the product rule

$$0 = \frac{d}{dt}\langle \mathcal{O}(t)w, \mathcal{O}(t)z\rangle = \langle \dot{\mathcal{O}}(0)w, z\rangle + \langle w, \dot{\mathcal{O}}(0)z\rangle,$$

and rearrange the identity to obtain the equation

$$\langle \dot{\mathcal{O}}(0)w, z\rangle = -\langle w, \dot{\mathcal{O}}(0)z\rangle. \tag{18.4}$$

By definition, a matrix (in this case $\dot{\mathcal{O}}(0)$) that satisfies the last identity is skew symmetric. In other words, skew-symmetric matrices (such as ω in display (18.1)) are infinitesimal rotations (see Exercise 18.5).

Points in the deformed body, called material points, move as the body is deformed. Let p be a material point. The partial derivative $\frac{\partial}{\partial t}u(p,t)$ is the velocity of the point labeled p as it moves with the body; it is called the material velocity of p. This corresponds to a time-dependent vector field $\mathcal{V}$ defined by $\mathcal{V}(p,t) = \frac{\partial}{\partial t}u(p,t)$. A different point of view is to consider the corresponding spatial point P corresponding to p at some time t_0. The curve of points $u(p,s+t)$ passes through P at time $s = 0$. Its velocity, given by

$$\frac{d}{ds}u(p,s+t)\big|_{s=0},$$

is called the spatial velocity at P at time t and is denoted $v(P,t)$. Clearly, $\mathcal{V}(p,t) = v(u(p,t),t)$. These velocities reflect two views of the motion: The Eulerian point of view, where the velocity field is v and the Lagrangian point of view, where it is $\mathcal{V}$. In the Eulerian view, we consider the motion through P; in the Lagrangian view, we assign some reference configuration, label its (material) points, and follow them as they move in space. The essential difference in the Eulerian and Lagrangian viewpoints is apparent from the expressions for the acceleration of a point particle:

$$\frac{\partial \mathcal{V}}{\partial t}(p,t) = \frac{\partial v}{\partial t}(u(p,t),t) + (v(u(p,t),t)\cdot\nabla)v(u(p,t),t) = v_t + v\cdot\nabla v = \frac{Dv}{Dt}.$$

where Dv/Dt is called the material derivative.

The Eulerian point of view was used in the derivation of the equations of fluid motion and the Lagrangian point of view was applied in the discussion of smoothed particle hydrodynamics. For elasticity, the Eulerian description will be employed to take advantage of Cauchy's partial differential equation (PDE) [Eq. (11.11)]

$$\rho\frac{Dv}{Dt} = \nabla\cdot\sigma + \rho b \tag{18.5}$$

for the velocity v of a body, with density ρ, subjected to stress modeled by the stress tensor σ and the body force per mass b. This equation was previously derived in the context of a moving fluid from the momentum balance; the same equation holds for the motion of a (deformable) body. As in fluid motion, a theory of elasticity for solid bodies is determined by specifying the stress tensor and its relation to the surface forces in and on the body.

To approach the subject of elasticity, we begin by defining traction τ to be the force per area at points on two-dimensional surfaces within the deformed body. Recall that pressure is a scalar field defined to be the normal component of force per area at a point on the surface. Thus, pressure is the normal component of the traction. Clearly, traction is a function of position in space $P \in \mathbb{R}^3$, time, and the normal η to the surface under consideration at P. Using momentum balance, it is possible to prove that traction is a linear function of the normal η (see [69]) along the surface. We will assume this fact. Thus, at a point p in Ω, the traction τ corresponding to a surface with normal η at P is the vector obtained by multiplying the normal vector η by the matrix defining the Cauchy stress tensor σ, whose components depend only on the point p and time; that is,

$$\tau(p,t,\eta) = \sigma(p,t)\eta(p,t). \tag{18.6}$$

Of course, as explained previously, the Cauchy stress tensor is a multilinear map; in fact, it is the bilinear map taking vectors v and w at p at time t to $\langle \sigma(P,t)v, w\rangle$. We may call σ the stress tensor, but we must remember that more precisely σ is the linear transformation that determines the stress tensor, which is a family of bilinear maps parameterized by points in the body and time.

The conservation of angular momentum states that the rate of change of the angular momentum of a body is equal to the sum of the moments on the body. The moments are simply cross products with the position vector from the origin of the coordinate system to the spatial points in the body, which we identify with the Cartesian point P. The balance for conservation of angular momentum is

$$\frac{d}{dt}\int_{\Omega(t)} P \times \rho v\, dV = \int_{\partial\Omega(t)} P \times \tau\, dS + \int_{\Omega(t)} P \times \rho b\, dV, \tag{18.7}$$

where the vector quantities are functions of position $P = u(p,t)$ and time.

We will show that the Cauchy stress tensor is symmetric as a consequence of the balance of angular momentum. The simplest proof is a computation in coordinates. The (index) notation usually employed to make the computation is worth learning because it is used extensively in the literature on elasticity.

A point in space is given in the Cartesian coordinates by $P = (x_1, x_2, x_3)$. We will denote the point by x_i, with the index understood to range over the integers 1, 2, and 3. Likewise, a vector v is specified by its components v_i. A matrix A is denoted by its array of components a_{ij}, where both indices range over 1, 2, and 3. Matrix multiplication Av is denoted $a_{ij}v_j$. Here the index i is free over the usual range, but the Einstein summation convention is used for the *repeated* index j; that is, the sum $\sum_{j=1}^{3} a_{ij}v_j$ in this single term is assumed. In other words, $a_{ij}v_j$ may be viewed as the ith component of the vector Av. Differentiation with respect to x_i is denoted by the subscript $_{,i}$. For example, $u_{i,j}$ denotes the Jacobian matrix of deformation u. The Kronecker delta δ_{ij} in this notation is the 3×3 identity matrix and the permutation symbol ϵ_{ijk} is defined to be zero if two indices are equal, plus one if ijk is an even permutation of 1, 2, 3, and minus one if it is an odd permutation; for example, $\epsilon_{123} = \epsilon_{231} = \epsilon_{321} = 1$. The basic vector operations in this notation are

$$w \cdot z = w_i z_i, \qquad w \times z = \epsilon_{ijk} w_k z_j, \quad \nabla f = f_{,i},$$

$$\nabla \cdot v = v_{i,i}, \qquad \nabla \times v = \epsilon_{ijk} v_{k,j}, \qquad \nabla \cdot \nabla f = f_{,ii}.$$

The time derivative in index notation must be handled with some care. We will write $\dot{v}$ to denote the material derivative of the vector v; that is,

$$\dot{v} = \frac{Dv}{dt} = \frac{d}{dt} v(u(p,t),t) = v_t + (v \cdot \nabla)v$$

Usually the arguments of functions are suppressed.

The momentum balance in index notation if given by

$$\rho \dot{v}_i = \sigma_{ij,j} + \rho b_i. \tag{18.8}$$

Here, the stress term perhaps should be written $\sigma_{ji,j}$ because the divergence is taken with respect to the columns of the matrix. As the matrix is symmetric, the two forms of the stress term are equal.

Conservation of angular momentum [Eq. (18.9)] may be expressed in the form

$$\frac{d}{dt} \int_{\Omega(t)} \rho \epsilon_{ijk} v_k x_j \, dV = \int_{\partial\Omega(t)} \epsilon_{ijk} \tau_k x_j \, dS + \int_{\Omega(t)} \rho \epsilon_{ijk} b_k x_j \, dV. \tag{18.9}$$

By an argument similar to the derivation of the momentum balance, which uses the transport theorem, the time derivative can be moved inside the first integral to obtain the equivalent formula

$$\int_{\Omega(t)} \rho \epsilon_{ijk} \dot{v}_k x_j \, dV = \int_{\partial\Omega(t)} \epsilon_{ijk} \tau_k x_j \, dS + \int_{\Omega(t)} \rho \epsilon_{ijk} b_k x_j \, dV. \tag{18.10}$$

The divergence theorem applied to the boundary term implies that

$$\int_{\partial\Omega(t)} \epsilon_{ijk}\tau_k x_j \, dS = \int_{\partial\Omega(t)} \epsilon_{ijk} x_j \sigma_{k\ell}\eta_\ell \, dS = \int_{\Omega(t)} (\epsilon_{ijk} x_j \sigma_{k\ell})_{,\ell} \, dV. \tag{18.11}$$

The derivative in the last integrand may be expanded by the product rule to derive the desired identity

$$\int_{\partial\Omega(t)} \epsilon_{ijk}\tau_k x_j \, dS = \int_{\Omega(t)} \epsilon_{ijk}\sigma_{kj} + \epsilon_{ijk} x_j \sigma_{k\ell,\ell} \, dV. \tag{18.12}$$

By substituting Eq. (18.12) into Eq. (18.10) and replacing $\rho\dot{v}_k$ with $\sigma_{k\ell,\ell} + \rho b_k$ from the momentum balance [Eq. (18.8)], it follows that

$$\int_{\Omega(t)} \epsilon_{ijk}\sigma_{kj} \, dV = 0.$$

Because this equation is true for all choices of $\Omega(t)$,

$$\epsilon_{ijk}\sigma_{kj} = 0.$$

Using the definition of the permutation symbol, we have the desired result: if $k \neq j$, then $\sigma_{kj} = \sigma_{jk}$; that is, σ is symmetric.

Cauchy's equation of motion [Eq. (18.5)] is the mathematical formulation of momentum balance for continuum mechanics. It is of course the same for fluid and elastic motions. For a fluid, the stress tensor σ is related to the strain rates defined from the velocity of the fluid (see Eq. (11.31)). Also, take note of the change in notation: here u is a displacement; for fluids, u denotes the fluid velocity. For elasticity, the stress is related to the infinitesimal strain ε derived from the deformation gradient u (see Eq. (18.1)). Exactly how this is done is the subject of elasticity theory, which is extensive, deep, and rich. In the simplest case of a homogeneous and isotropic material—the same at each point with the same physical properties in every direction—the basic form of the stress-strain relation is given by

$$\sigma = \lambda \operatorname{tr}(\varepsilon) I + 2\mu\varepsilon \tag{18.13}$$

or

$$\sigma_{ij} = \lambda\varepsilon_{kk}\delta_{ij} + 2\mu\varepsilon_{ij}, \tag{18.14}$$

where λ and μ are real numbers called the Lamé constants. This linear relation is a version of Hooke's law, which states that stress is linearly related to strain. In this notation, the strain tensor defined in display (18.1) is

$$\varepsilon_{ij} = \frac{1}{2}(u_{i,j} + u_{j,i}). \tag{18.15}$$

The final step is to substitute the stress-strain relation into the equation of motion (18.5) to obtain the fundamental equation of motion for homogeneous and isotropic materials:

$$\rho \dot{v}_i = (\lambda \varepsilon_{kk} \delta_{ij} + 2\mu \varepsilon_{ij})_{,j} + \rho b_i. \tag{18.16}$$

By carrying out the differentiation (see Exercise 18.1), the equations of motion for the displacement and its velocity can be recast in component form

$$\begin{aligned} \dot{u}_i &= v_i, \\ \rho \dot{v}_i &= (\lambda + \mu) u_{j,ji} + \mu u_{i,jj} + \rho b_i \end{aligned} \tag{18.17}$$

or in vector form,

$$\begin{aligned} \frac{\partial u}{\partial t} &= v, \\ \rho \frac{Dv}{Dt} &= (\lambda + \mu) \nabla (\nabla \cdot u) + \mu \Delta u + \rho b. \end{aligned} \tag{18.18}$$

Small deformations are assumed in the choice of the strain tensor, which is the linear approximation to the deformation. The material derivative contains nonlinear terms in the velocities and hence in the deformation. Thus, in keeping with the assumption of small deformations, there is some reason to retain only linear terms in the material derivative. With this approximation, the dynamic equation of linear elasticity for a homogeneous and isotropic body is

$$\rho \ddot{u}_i = (\lambda + \mu) u_{j,ji} + \mu u_{i,jj} + \rho b_i, \tag{18.19}$$

or in vector form,

$$\rho \frac{\partial^2 u}{\partial^2 t} = (\lambda + \mu) \nabla (\nabla \cdot u) + \mu \Delta u + \rho b. \tag{18.20}$$

The displacement is to be determined from the material properties (which determine λ and μ), given body forces (gravity or electromagnetism), forces applied to the surface of the body (which are modeled by boundary conditions), and a specification of the initial displacement and (displacement) velocity of the body. Thus, a basic problem is to determine the displacements, stresses, and strains when body forces, surface forces, or surface displacements are given, and the initial displacement and velocity are specified.

The Euler and Navier–Stokes equations of fluid mechanics are first-order in time and depend only on the fluid velocity; the equation of motion of linear elasticity is second-order in time and depends on the deformation. This is an important difference that reflects the contrast between solids and liquids. As we have seen, the equation of heat transfer $u_t = c^2 \Delta u$ is akin to the fluid equations. The wave equation $u_{tt} = c^2 \Delta u$, which is the basic equation for small amplitude wave propagation, is a special case of the dynamic equation of elasticity.

In the development of the theory so far, the density ρ need not be constant. To continue modeling for solid materials (that is, not liquids or gases), let us assume that the density is constant, a reasonable assumption for many physical models.

Recall the Helmholtz–Hodge decomposition theorem: *A smooth vector field X on a region Ω of space with smooth boundary $\partial\Omega$ and outer normal N can be decomposed uniquely as*

$$X = \nabla \Phi + Y,$$

where ϕ is a scalar valued function and Y is a vector field such that $\nabla \cdot Y = 0$ and $Y \cdot N = 0$ on $\partial\Omega$ (that is, Y is divergence free and parallel to the boundary of the region).

Suppose that the Helmholtz decompositions of the displacement u and force b that appear in the linear elastodynamic equation (18.20) have the form

$$u = \nabla \phi + w, \qquad b = \nabla \beta + a.$$

Substitution into Eq. (18.20), taking into account that $\nabla \cdot w = 0$, yields the formula

$$\rho \nabla \ddot{\phi} + \rho \ddot{w} = (\lambda + \mu) \nabla (\Delta \phi) + \mu \Delta \nabla \phi + \rho \nabla \beta + \mu \Delta w + \rho a. \quad (18.21)$$

Under the assumption that all fields are sufficiently smooth and by an easy calculation, it follows that

$$\Delta\nabla\phi = \nabla\Delta\phi.$$

Using this fact and some simple algebra, Eq. (18.21) can be rearranged to read

$$\nabla(\rho\ddot{\phi} - (\lambda + 2\mu)\Delta\phi - \rho\beta) + (\rho\ddot{w} - \mu\Delta w - \rho a) = 0.$$

The first set of parentheses of the last equation enclose a scalar field. By Exercise 18.3, the second set of parentheses enclose a divergence-free vector field. Thus, by the uniqueness of the Helmholtz decomposition, and the observation that the zero vector field may be decomposed as $\nabla 0 + 0$ with the second summand divergence free, the decomposition produces an equivalent form of the linear equations of elasticity *for the case of constant density*:

$$\begin{aligned} \rho\ddot{\phi} &= (\lambda + 2\mu)\Delta\phi + \rho\beta, \\ \rho\ddot{w} &= \mu\Delta w + \rho a \end{aligned} \tag{18.22}$$

where the displacement is $u = \nabla\phi + w$ and the sum of the forces b acting on the body is decomposed as $b = \nabla\beta + a$.

As a preview of the analysis to follow, pay attention to the forms of the PDEs (18.22). The important and basic PDE

$$U_{tt} = k^2\Delta U,$$

where k is a nonzero scalar parameter, plays a fundamental role. It is called the (linear) wave equation because its nontrivial solutions are all waves that travel with fixed profiles at speed k. Note that the decomposition of the displacement [Eqs. (18.22)] implies that there are at least two types of elastic waves that travel at different speeds $((\lambda + 2\mu)/\rho)^{1/2}$ and $(\mu/\rho)^{1/2}$, respectively, depending on the material properties measured by the Lamé constants λ and μ. This is a beautiful example of applying mathematics to derive a prediction (which has been verified by physical experiments). The constant μ (also called the shear modulus) must be positive. In principle, λ could be negative, but for most materials it is positive. Thus, waves corresponding to the divergence-free field w travel slower than waves that involve volumetric changes.

Exercise 18.1. Show that Eqs. (18.17) and (18.18) follow from Eq. (18.16).

Exercise 18.2. The most general form of Hooke's law is expressed by $\sigma_{ij} = C_{k\ell ij}\varepsilon_{kl}$ where C, called the elasticity tensor, is symmetric; that is, the components remain unchanged for every permutation of the indices. Suppose, for example, that the strain tensor were modified in the presence of a symmetric matrix B as follows: the strain resulting from two vectors u and v is given by

$$\langle \sigma Bu, Bv \rangle.$$

What are the components of C in this case?

Exercise 18.3. Show that for a divergence-free vector field, the divergence of the Laplacian of the vector field vanishes.

18.1 THE TAUT WIRE: SEPARATION OF VARIABLES AND FOURIER SERIES FOR THE WAVE EQUATION

Consider a taut wire in space. It is a deformable body. In reality, a wire is a three-dimensional object, perhaps nearly a round cylinder, which we may assume is homogenous and isotropic. As a first approximation, idealize the wire and imagine a one-dimensional body at rest along the first coordinate axis, which is assumed to be horizontal. Suppose that a small load is applied. The wire undergoes a small deformation that should be correctly modeled by linear elasticity theory. For simplicity, suppose the displacement is confined to just one direction, say in the direction of the second coordinate axis, which is also assumed to be horizontal. Thus, the displacement function u is given by $u(x_1, t) = (0, u_2(x_1, t), 0)$, and the equation of motion (18.19) reduces to

$$\rho \ddot{u}_2 = (\lambda + 2\mu) u_{2,11}. \tag{18.23}$$

Taut wire dynamics—a subject that is certainly intrinsically interesting—has many practical applications (see, for example, Exercise 18.20 on vibrating wire sensors).

Our simplifying assumptions—the wire is one-dimensional, taut, and restricted to move in the horizontal plane in a direction normal to the wire—eliminate for the moment consideration of the gravitational force that acts in the downward vertical direction. The equation of motion, recast in its more usual abstract form, is the one-dimensional wave equation

$$\frac{\partial^2 u}{\partial t^2} = c^2 \frac{\partial^2 u}{\partial x^2}, \tag{18.24}$$

where u_2 is replaced by u (which is reinterpreted to measure displacement in the direction of the second coordinate), x_2 replaced by x, and $c^2 := (\lambda + 2\mu)/\rho$. Here, the density ρ in this idealized model is the mass of the wire per length, which is assumed to be a positive constant. Although Lamé's constant λ could be negative, it and the second constant μ (called the shear modulus) are assumed to combine so that c^2 is a positive constant.

We should always consider the dimensions in a new model. For Eq. (18.24), u has the dimension of length; hence, c has the dimension of length per time. It is a speed. Thus, something should be moving at this speed. In fact, as its name suggests, the wave equation has solutions that are waves traveling at speed c.

Consider a (profile) function $f : \mathbb{R} \to \mathbb{R}$ that is twice continuously differentiable, and use it to construct a new function $u : \mathbb{R}^2 \to \mathbb{R}$ given by

$$u(x,t) = f(x - ct).$$

This function u is a (traveling wave) solution of the wave equation; in fact,

$$\frac{\partial^2 u}{\partial t^2}(x,t) = c^2 f(x - ct) \quad \text{and} \quad \frac{\partial^2 u}{\partial x^2}(x,t) = f(x - ct).$$

Likewise, $u(x,t) = f(x + ct)$ is a solution. Because the wave equation is linear, it is easy to check that the superposition (sum) of two solutions is a solution. Hence, if f and g are twice continuously differentiable functions, then

$$u(x,t) = f(x - ct) + g(x + ct)$$

is a solution of the wave equation.

To specify the particular solution of the (second-order in time) wave equation that corresponds to some specific physical problem, the initial position and velocity of the wire must be specified for the same reason this data is required to uniquely solve Newton's equation $F = ma$; that is, for the wave equation $x \mapsto u(x,0)$ and $x \mapsto u_t(x,0)$ must be given functions, say α and β, of the position x along the wire. To satisfy the initial data, the wave profile functions f and g must be such that

$$\alpha(x) = f(x) + g(x), \qquad \beta(x) = -cf'(x) + cg'(x).$$

These requirements uniquely determine the solution u. Indeed, by differentiating both sides of the first relation and subsequently solving for f' and g',

$$f = -\frac{1}{2}(\frac{\beta}{c} - \alpha'), \qquad g = \frac{1}{2}(\frac{\beta}{c} + \alpha')$$

and by an integration,

$$f(x) = f(0) - \frac{1}{2c}\int_0^x \beta(s)\,ds + \frac{1}{2}(\alpha(x) - \alpha(0)),$$
$$g(x) = g(0) + \frac{1}{2c}\int_0^x \beta(s)\,ds + \frac{1}{2}(\alpha(x) - \alpha(0)).$$

Using the relation $\alpha(0) = f(0) + g(0)$ and some simplification, we obtain d'Alembert's solution

$$u(x,t) = f(x-ct)+g(x+ct) = \frac{1}{2c}\int_{x-ct}^{x+ct} \beta(s)\,ds+\frac{1}{2}(\alpha(x-ct)+\alpha(x+ct)). \tag{18.25}$$

The physical interpretation is that u is the superposition of two traveling waves, $f(x - ct)$ traveling to the right and $g(x + ct)$ traveling to the left with the wave profiles f and g determined via d'Alembert from the initial data.

The wire model and its solution are more interesting and physically relevant when the finite length of the wire is taken into account. For definiteness, suppose that one end of the wire is at $x = 0$, the other at $x = L > 0$, and these ends are clamped so that they cannot be displaced. To take these conditions into account, the mathematical model is augmented by requiring that the solutions of the wave equation satisfy the Dirichlet boundary conditions

$$u(0,t) = 0, \qquad u(L,t) = 0. \tag{18.26}$$

Although it is possible (but not easy) to determine a solution of this boundary value problem (BVP) using d'Alembert's solution of the wave equation, we will discuss another important solution method.

A powerful idea for solving linear PDEs on rectangular domains is to seek a solution by separation of variables. To this end, suppose u has the form

$$u(x,t) = X(x)T(t),$$

where X and T are functions to be determined. If there is such a solution, these functions must satisfy the relation

$$X(x)T''(t) = c^2X''(x)T(t)$$

for all $x \in (0, L)$ and all $t > 0$. Without regard to division by zero for the moment, the functions may be separated as follows:

$$\frac{T''(t)}{c^2T(t)} = \frac{X''(x)}{X(x)}.$$

Clearly, if this identity holds, then both sides of the equation must be equal to the same constant value denoted here by $-\Lambda$, where Λ is some arbitrary real number and the minus sign is chosen for mathematical convenience. For the case $\Lambda > 0$, define $\ell = \sqrt{\Lambda}$ and note that if X and T are solutions of the respective ordinary differential equations (ODEs)

$$X''(x) + \ell^2X(x) = 0, \qquad T''(t) + c^2\ell^2T(t) = 0,$$

then $u(x,t) = X(x)T(t)$ is a solution of the wave equation. Thus, we have separated the variables x and t.

The ODEs for X and T are harmonic oscillators. Their general solutions are

$$X(x) = a\cos\ell x + b\sin\ell x, \qquad T(t) = d\cos c\ell t + e\sin c\ell t$$

for arbitrary real numbers a, b, d, and e. The function

$$u(x,t) = (a\cos\ell x + b\sin\ell x)(d\cos c\ell t + e\sin c\ell t)$$

is a solution of the wave equation for every choice of these coefficients.

Imposition of the boundary conditions (18.26) requires

$$a(d\cos c\ell t + e\sin c\ell t) = 0, \quad (a\cos\ell L + b\sin\ell L)(d\cos c\ell t + e\sin c\ell t) = 0$$

for all $t > 0$. Using the result of Exercise 18.6,

$$ad = 0, \quad ae = 0, \quad d(a\cos\ell L + b\sin\ell L) = 0, \quad e(a\cos\ell L + b\sin\ell L) = 0.$$

If $a \neq 0$, then $d = 0$ and $e = 0$. This corresponds to the zero solution of the wave equation. Thus, to obtain a nonzero solution, we must have $a = 0$. In this case, if $b = 0$, we again obtain the zero solution of the wave equation.

For $b \neq 0$, we must have $\sin \ell L = 0$, or equivalently,

$$\ell = \frac{n\pi}{L}$$

for some integer $n > 0$. In summary, a family of nonzero solutions is given by

$$u(x,t) = (d \cos c\ell t + e \sin c\ell t) \sin \ell x$$

for arbitrary real numbers c and d, and $\ell = n\pi/L$ for an arbitrary integer $n > 0$. By a slight change in notation, this infinite set of solutions (which all satisfy the boundary conditions) may be written in the form

$$u_n(x,t) = \big(a_n \cos \frac{n\pi c}{L} t + b_n \sin \frac{n\pi c}{L} t\big) \sin \frac{n\pi}{L} x$$

for $n > 0$.

The sum of solutions of the (linear) wave equation is again a solution. In general, the sum of two solutions that satisfy some boundary condition may not be a solution of the wave equation. But, in the case of the zero Dirichlet boundary conditions, the superposition of two solutions is a solution that satisfies the same boundary conditions (see Exercise 18.7). Using this result, every function u given by a finite sum of the form

$$u(x,t) = \sum_{n=1}^{k} \big(a_n \cos \frac{n\pi c}{L} t + b_n \sin \frac{n\pi c}{L} t\big) \sin \frac{n\pi}{L} x$$

is a solution of the wave equation that satisfies the Dirichlet boundary conditions.

What about the initial conditions: $u(x,0) = \alpha(x)$ and $u_t(x,0) = \beta(x)$? In case the initial conditions happen to be of the form

$$\alpha(x) = \sum_{n=1}^{k} a_n \sin \frac{n\pi}{L} x, \qquad \beta(x) = \sum_{n=1}^{k} b_n \frac{n\pi c}{L} \sin \frac{n\pi}{L} x$$

for some N and constants $\{a_n\}_{n=1}^{k}$ and $\{b_n\}_{n=1}^{k}$, the given u is the desired explicit solution of the initial value problem (IVP).

The last result suggests the question: Which functions can be expressed as finite sums of sines? Or more precisely, how large is the class of functions defined on the interval $(0, L)$ that are given by

$$f(x) = \sum_{n=1}^{k} a_n \sin \frac{n\pi}{L} x \tag{18.27}$$

for some integer $N > 0$ and numbers $\{a_n\}_{n=1}^{k}$. All these functions are periodic with period $2L$. So, for example, the function $f(x) = x$ is certainly not of the given form. Although an infinite class of solutions of the wave equation has been constructed whose linear combinations all satisfy the zero Dirichlet boundary conditions, perhaps only a special set of initial data can be modeled. One way to try to obtain additional solutions would be to investigate negative or zero values of the parameter Λ as part of the separation of variables technique. In fact, these values of Λ do not lead to new solutions. The corresponding solutions of the wave equation do not satisfy the boundary conditions (see Exercise 18.8). In fact, no additional solutions are needed.

Joseph Fourier (sometime before 1807) realized that almost all functions can be represented by *infinite sums* of sines (or cosines). His revolutionary discovery implies that almost all initial data for the wave equation can be represented in this way. Thus, solutions of the wave equation that satisfy the boundary and initial data can be represented by infinite series of sines and cosines. The remainder of this section is a discussion of the approximation of functions by Fourier series.

An infinite sum of the form

$$\sum_{n=1}^{\infty} a_n \sin \frac{n\pi}{L} x$$

is called a Fourier sine series. There are also Fourier cosine series

$$\sum_{n=1}^{\infty} a_n \cos \frac{n\pi}{L} x$$

and (full) Fourier series

$$a_0 + \sum_{n=1}^{\infty} (a_n \cos \frac{n\pi}{L} x + b_n \sin \frac{n\pi}{L} x).$$

These series are important in many areas of applied mathematics and mathematics. The branch of mathematics devoted to their study is called

Fourier analysis or harmonic analysis. The basic idea is that most functions have a Fourier series representation; that is, for a function f defined on $(0, L)$ there are (for example) corresponding Fourier coefficients $\{a_n\}_{n=1}^{\infty}$ such that

$$f(x) = \sum_{n=1}^{\infty} a_n \sin \frac{n\pi}{L} x.$$

Consider, for example, the initial data for the wave equation given by positively displacing the material point at $x = L/4$ to a distance $L/100$ and releasing the wire from rest at this displacement. What is the subsequent motion of the wire?

The obvious approximation of the initial displacement function

$$\alpha(x) := \begin{cases} \frac{x}{25}, & 0 \le x \le \frac{L}{4}, \\ \frac{L-x}{75}, & \frac{L}{4} < x \le L \end{cases} \tag{18.28}$$

is not a finite sum of sines. But, it may be approximated as closely as desired with a finite Fourier sum. How should the closeness of the approximation be measured?

We already know how to measure the closeness of two points P and Q in $\mathbb{R}^k$: we simply compute the (Euclidean) distance between them $|Q - P|$. The appropriate measurement of distance between two functions, for instance the distance between the function f defined by the sum of sines in Eq. (18.27) and α defined by Eq. (18.28), is not obvious. Many different measurements are possible; the correct choice depends on the context. A natural possibility is to measure the distance between the real numbers $f(x)$ and $\alpha(x)$ for each x and define the distance between the functions to be the maximum such distance. In symbols, we write

$$\|\alpha - f\|_0 = \sup_{x \in (0,L)} |\alpha(x) - f(x)|. \tag{18.29}$$

There is a delicate matter that requires writing $\sup$ instead of $\max$: perhaps the supremum exists when the maximum is not attained (see Exercise 18.10). Another, perhaps less obvious, choice of distance is defined using integration:

$$\|\alpha - f\|_2 = \left(\int_0^L |\alpha(x) - f(x)|^2 \, dx \right)^{1/2}. \tag{18.30}$$

In both cases the distance between functions is defined using a new concept callcd a norm. More precisely, the supremum norm of an arbitrary function f defined on $(0, L)$ is defined to be

$$\|f\|_0 = \sup_{x \in (0,L)} |f(x)| \tag{18.31}$$

and the L^2 norm (pronounced "L-two norm") is

$$\|f\|_2 = \left(\int_0^L |f(x)|^2 \, dx \right)^{1/2}. \tag{18.32}$$

Clearly the notion of a norm is a generalization of the Euclidean length of vectors in $\mathbb{R}^k$.

To determine how well finite sums of sines approximate the function α defined in Eq. (18.28), consider first the problem for the supremum norm: Determine the coefficients $\{a_n\}_{n=1}^k$ such that

$$\sup_{0<x<L} \begin{cases} |\frac{x}{25} - \sum_{n=1}^k a_n \sin \frac{n\pi}{L} x|, & 0 \le x \le \frac{L}{4}, \\ |\frac{L-x}{75} - \sum_{n=1}^k a_n \sin \frac{n\pi}{L} x|, & \frac{L}{4} < x \le L \end{cases}$$

is minimized. The complexity of this optimization problem is apparent for $N = 1$ and the performance of many numerical algorithms that might be used to solve this problem degrades as N increases (see Exercise 18.11). The supremum norm is not the natural choice for the approximation problem.

Fourier recognized that the trigonometric functions sine and cosine have special properties that can be exploited to solve the problem of approximating arbitrary functions with sums of sines (or cosines or both). Their most important property should be familiar from calculus: If $m \neq n$ are integers, then

$$\int_0^L \sin \frac{n\pi x}{L} \sin \frac{m\pi x}{L} \, dx = 0, \quad \int_0^L \cos \frac{n\pi x}{L} \cos \frac{m\pi x}{L} \, dx = 0, \tag{18.33}$$

$$\int_0^L \cos \frac{n\pi x}{L} \sin \frac{m\pi x}{L} \, dx = 0.$$

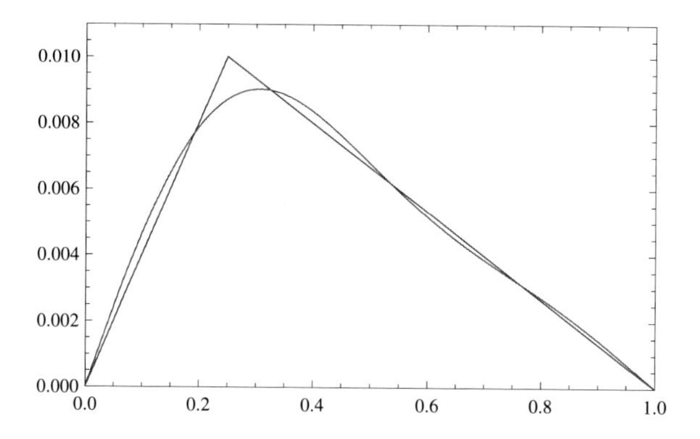

Fig. 18.1 The figure depicts a plot of the function α defined in Eq. (18.28) *and the corresponding third-order Fourier sum* (18.35) *for the case $L = 1$.*

Suppose that a function $f : (0, L) \to \mathbb{R}$ is given by a sum of sines

$$f(x) = \sum_{n=1}^{k} a_n \sin \frac{n\pi x}{L}.$$

The trigonometric integral formulas can be used to determine the coefficients $\{a_n\}_{n=1}^{k}$ that determine f. To find the jth coefficient, simply integrate the product of f and $\sin \frac{j\pi x}{L}$ as follows

$$\begin{aligned}
\int_0^L f(x) \sin \frac{j\pi x}{L}\, dx &= \int_0^L \Big(\sum_{n=1}^{k} a_n \sin \frac{n\pi x}{L}\Big) \sin \frac{j\pi x}{L}\, dx \\
&= \sum_{n=1}^{k} a_n \int_0^L \sin \frac{n\pi x}{L} \sin \frac{j\pi x}{L}\, dx \\
&= a_j \int_0^L \sin^2 \frac{j\pi x}{L}\, dx \\
&= \frac{L}{2} a_j
\end{aligned}$$

to obtain the formula

$$a_j = \frac{2}{L} \int_0^L f(x) \sin \frac{j\pi x}{L}\, dx. \tag{18.34}$$

This number, for obvious reasons, is called the jth Fourier (sine) coefficient of the function f. A similar formula exists for sums of cosines.

For an arbitrary (integrable) function, its Fourier coefficients may be computed up to some order N to form the corresponding trigonometric sum. For example, the first three Fourier sine coefficients of the function α defined in Eq. (18.28) are

$$\frac{2}{75L\pi^2}\left(4\sin\frac{L\pi}{4} - \sin L\pi\right), \quad \frac{1}{150L\pi^2}\left(4\sin\frac{L\pi}{2} - \sin 2L\pi\right),$$

$$\frac{2}{675L\pi^2}\left(4\sin\frac{3L\pi}{4} - \sin 3L\pi\right)$$

and the corresponding Fourier sum is

$$\begin{aligned} f(x) = {} & \frac{2}{75L\pi^2}\left(4\sin\frac{L\pi}{4} - \sin L\pi\right)\sin\frac{\pi x}{L} \\ & + \frac{1}{150L\pi^2}\left(4\sin\frac{L\pi}{2} - \sin 2L\pi\right)\sin\frac{2\pi x}{L} \\ & + \frac{2}{675L\pi^2}\left(4\sin\frac{3L\pi}{4} - \sin 3L\pi\right)\sin\frac{3\pi x}{L}. \end{aligned} \tag{18.35}$$

The graphs of α and f for the case $L = 1$ are depicted in Fig. 18.1 where the close approximation is apparent. Higher-order Fourier sums are better approximations. Why should Fourier sums be such good approximations?

Although a complete answer to the last question is beyond the scope of this book, the most important fact is easy to understand: The best approximation of an integrable function by a sum of sines, measured by the L^2 norm, is the Fourier sum.

The underlying idea that leads to the desired result should be familiar from the geometry of m-dimensional Euclidean space. Suppose that $\mathcal{M}$ is a plane that happens to contain the point Q. The shortest distance from a point P in space to $\mathcal{M}$ is the magnitude of the orthogonal projection of the vector $P - Q$ onto one of the unit normals of $\mathcal{M}$ at Q (see Exercise 18.12). This fact can be used to write a formula for the minimum distance.

Note that among all points $Q \in \mathcal{M}$, the distance from P to Q is least when the vector $P - Q$ is orthogonal to all vectors tangent to $\mathcal{M}$. There are several methods that might be used to determine this minimizer Q.

For simplicity and relevance to the minimization problem with sine series, assume that $\mathcal{M}$ passes through the origin. There are $k < m$ vectors $e^1, e^2, e^3, \ldots, e^k$ in the Euclidean space that are pairwise orthogonal and such that every point in $\mathcal{M}$ is given by $\sum_{n=1}^{k} a_n e^n$ for some choice of coefficients $\{a_n\}_{n=1}^{k}$. In other words, these vectors form an orthogonal basis for $\mathcal{M}$, which is a k-dimensional subspace of the m-dimensional Euclidian space. Vectors tangent to the plane all have the same form $\sum_{n=1}^{k} a_n e^n$. Thus, the orthogonality condition is $(P - Q) \cdot \sum_{n=1}^{k} a_n e^n = 0$ for all $\{a_n\}_{n=1}^{k}$. Clearly, for the orthogonality condition to hold, it suffices to have $(P - Q) \cdot e^j = 0$ for each $j = 1, 2, 3, \ldots, k$.

Suppose that $Q = \sum_{n=1}^{k} q_n e^n$. Then, the sufficient condition is $(P - \sum_{n=1}^{k} q_n e^n) \cdot e^j = 0$ for each $j = 1, 2, 3, \ldots, k$. It implies that

$$q_j = P \cdot \frac{e^j}{e^j \cdot e^j}.$$

Thus, the desired point in the plane that minimizes the distance to P is

$$Q = \sum_{n=1}^{k} \frac{P \cdot e^n}{e^n \cdot e^n} e^n.$$

In case the basis elements $\{e^n\}_{n=1}^{k}$ are orthonormal (that is, mutually orthogonal and all elements of unit length), the last sum simplifies to $Q = \sum_{n=1}^{k} (P \cdot e^n) e^n$. In both cases, $|P - Q|^2 = |P|^2 - |Q|^2$ by Pythagorus's theorem.

Perhaps it is also instructive to derive the same result and express the final formula more directly using a computation with inner products:

$$\begin{aligned}
|P - Q|^2 &= (P - Q) \cdot (P - Q) \\
&= |P|^2 - 2P \cdot Q + |Q|^2 \\
&= |P|^2 - 2P \cdot \sum_{n=1}^{k} \frac{P \cdot e^n}{|e^n|^2} e^n + \sum_{n=1}^{k} \frac{P \cdot e^n}{|e^n|^2} e^n \cdot \sum_{n-1}^{k} \frac{P \cdot e^n}{|e^n|^2} e^n \\
&= |P|^2 - 2 \sum_{n=1}^{k} \frac{(P \cdot e^n)^2}{|e^n|^2} + \sum_{n=1}^{k} \frac{(P \cdot e^n)^2}{|e^n|^2} \\
&= |P|^2 - |Q|^2,
\end{aligned}$$

or

$$|P - Q|^2 = |P|^2 - \sum_{n=1}^{k} \frac{(P \cdot e^n)^2}{|e^n|^2}.$$

Instead of Euclidean space, consider the set $\mathcal{L}$ of all square integrable functions on the interval $[0, L]$; that is, all functions $g : [0, L] \to \mathbb{R}$ such that

$$\|g\|_2^2 := \int_0^L |g(x)|^2 \, dx$$

has a finite value. For this discussion, there is no essential difference between functions defined on the open interval $(0, L)$ and the closed interval $[0, L]$; indeed, changing the value of a function at two points does not change its integral. The closed interval is used here simply to make this point.

The set $\mathcal{L}$ has much (perhaps unexpected) structure. We will assume two important inequalities concerning pairs of functions g and h in $\mathcal{L}$: Schwarz's inequality,

$$\int_0^L g(x)h(x)\,dx \le \Big(\int_0^L (g(x))^2\,dx\Big)^{1/2}\Big(\int_0^L (h(x))^2\,dx\Big)^{1/2} \tag{18.36}$$

and Minkowski's inequality,

$$\Big(\int_0^L g(x)+h(x))^2\,dx\Big)^{1/2} \le \Big(\int_0^L (g(x))^2\,dx\Big)^{1/2} + \Big(\int_0^L (h(x))^2\,dx\Big)^{1/2}. \tag{18.37}$$

Minkowski's inequality implies that the sum of two functions in $\mathcal{L}$ is a function in $\mathcal{L}$. Thus, $\mathcal{L}$ is similar to $\mathbb{R}^k$ in some respects. Consider functions in $\mathcal{L}$ to be abstract vectors. Note first that all linear combinations of these functions (that is, expressions of the form $ag + bh$ for real numbers a and b and functions g and h in $\mathcal{L}$) remain in the set $\mathcal{L}$. In addition, the distance between functions in $\mathcal{L}$ is derived from an inner product just as the Euclidean distance between vectors P and Q in $\mathbb{R}^k$ can be defined using the usual dot product. The norm of a vector P is defined using the dot product to be $|P| = \sqrt{P \cdot P}$. The distance between two vectors P and Q is defined to be the norm of their difference $|Q - P|$. Likewise, the inner product of two functions g and h in $\mathcal{L}$ is defined to be

$$g \cdot h = \int_0^L g(x)h(x)\,dx.$$

The norm—called the L^2 norm or the root mean square (RMS)—of g is $\|g\|_2 := \sqrt{g \cdot g}$, and the distance between functions g and h is $\|h - g\|_2$. By tradition, the inner product is rarely written $g \cdot h$. Most often it is denoted by $\langle g, h\rangle$ or (g, h). We will use the latter notation here. Schwartz's inequality implies that the inner product is indeed a number whenever g and h are in $\mathcal{L}$. The inner product defined this way satisfies all the other properties of the inner product in $\mathbb{R}^k$:

$$\begin{aligned}
&(g,g) \geq 0 && \text{for every } g \in \mathcal{L};\\
&(g,g) = 0 && \text{if and only if } g = 0 \text{ except}\\
& && \text{perhaps on a set of zero measure;}\\
&(g,h) = (h,g) && \text{for all } g \text{ and } h \text{ in } \mathcal{L};\\
&(g+k,h) = (g,h) + (k,h) && \text{for all } g,\ h, \text{ and } k \text{ in} \mathcal{L};\\
&(ag,h) = a(g,h) && \text{for all } g \text{ and } h \text{ in } \mathcal{L} \text{ and } a \text{ in } \mathbb{R}
\end{aligned} \tag{18.38}$$

(see Exercise 18.13).

In analogy with Euclidean space, the inner product on $\mathcal{L}$ is used to define the notion of orthogonality: Two functions in $\mathcal{L}$ are orthogonal if their inner product is zero. The main point of the discussion so far is that $x \mapsto \sin \frac{n\pi x}{L}$ (and $x \mapsto \cos \frac{n\pi x}{L}$) are functions in $\mathcal{L}$, and every pair of such functions is orthogonal whenever they have different frequencies ($m \neq n$) as in display (18.33).

Consider the subset $\mathcal{M}$ of $\mathcal{L}$ consisting of all the functions in $\mathcal{L}$ that can be obtained by taking finite sums of multiples of the functions

$$e^1 := \sin \frac{\pi x}{L}, e^2 := \sin \frac{2\pi x}{L}, e^3 := \sin \frac{3\pi x}{L}, \ldots, e^k := \sin \frac{k\pi x}{L}.$$

Of course, $\mathcal{M}$ is exactly the set of all functions of the form

$$\sum_{n=1}^{k} a_n e^n,$$

where as before $\{a_n\}_{n=1}^k$ is a set of real numbers.

Problem 18.1. Suppose g is a function in $\mathcal{L}$. Determine the minimum of $\|g - h\|_2$ over all $h \in \mathcal{M}$ and the Fourier coefficients $\{a_n\}_{n=1}^k$ of the minimizer.

Problem 18.1 is akin to the problem of finding the minimum distance from a point to a plane. In fact, the answer is exactly the same. The minimizer is

$$h = \sum_{n=1}^{k} \frac{2}{L}(g, e_n)e_n$$

and the minimum distance is

$$\inf_{h\in\mathcal{M}} \|g-h\|_2^2 = \|g\|^2 - \sum_{n=1}^{k} \frac{2}{L}(g, e_n)^2, \tag{18.39}$$

where

$$e^n := \sin\frac{n\pi x}{L}.$$

A proof of the stated solution of Problem 18.1 is similar to, but not the same as, the proof used for determining the distance from a point to a plane in Euclidean space. The difficulty is that the new finite-dimensional set of functions $\mathcal{M}$ is contained in the infinite-dimensional space $\mathcal{L}$. Thus, the normal directions form an infinite-dimensional set. What does this mean? Recall that a vector space has dimension m if there is a set of elements $\{v^i\}_{i=1}^m$ such that every element of the space is a linear combination of these elements (that is, every g in the space is given by $g = a_i v^i$, for real numbers $\{a_i\}_{i=1}^m$) and there is no smaller set of elements with this same property. A set of such elements is called a basis of the space. A space is called infinite-dimensional if no finite basis exists. The existence of the orthogonal set $\{e^n\}_{n=1}^\infty$, where e^n is defined to be the function $x \mapsto \sin n\pi x/L$ can be used to show that $\mathcal{L}$ is infinite-dimensional (see Exercise 18.15).

In fact, for Problem 18.1, a unique minimizer $h \in \mathcal{M}$ exists and $g - h$ is orthogonal to every function in $\mathcal{M}$. But this result requires completeness of $\mathcal{L}$; that is, every Cauchy sequence of functions in the space (using the L^2 norm to measure distance) converges to a function in the space. Recall that completeness is also a fundamental property of the real numbers with respect to the norm given by absolute value. Thus, it should not be a surprise that this property is needed to endow $\mathcal{L}$ with properties that are analogous to Euclidean space.

Strictly speaking, the set $\mathcal{L}$ is not complete, but the space L^2 (obtained from $\mathcal{L}$ by equating functions that agree on $[0, L]$ except for a set of measure zero) is complete. The desired existence, uniqueness, and orthogonality result for the minimization problem is true in L^2 (see, for example, [91]). For simplicity, assume this result. The formula for the minimizer is then found exactly as in the problem for Euclidean space.

Suppose that $h = \sum_{n=1}^{k} h_n e^n$. Due to the orthogonality, $(g - h, e^n) = 0$ for each $n = 1, 2, 3, \ldots, k$ because each function e^n is in $\mathcal{M}$. Thus,

$$(g, e^n) = (\sum_{j=1}^{k} h_j e^j, e^n) = \frac{L}{2} h_n$$

and the desired formula for the minimizer is

$$h = \sum_{n=1}^{k} \frac{2}{L}(g, e^n)e^n,$$

where

$$e^n := \sin \frac{n\pi x}{L}.$$

Moreover, the minimum distance is

$$(g - k, g - k) = (g - k, g) = \|g\|^2 - \sum_{n=1}^{k} \frac{2}{L}(g, e_n)^2.$$

The solution of Problem 18.1 tells us that the best approximation with respect to the L^2 norm of a function g by the finite set of functions $\{e^n := \sin \frac{n\pi x}{L}\}_{n=1}^{k}$ is given by its Fourier sum

$$\sum_{n=1}^{k} \frac{2}{L}(g, e^n)e^n.$$

Also, the exact error measured by the L^2 norm is available using Eq. (18.39) (see, Exercise 18.16).

As mentioned perviously, more is true: *If g is square integrable on a finite open interval and $\epsilon > 0$ is given, then there is some positive integer K such that*

$$\|g - \sum_{n=1}^{k} \frac{2}{L}(g, e^n)e^n\|_2 < \epsilon$$

for every $k \geq K$. Or, in other words, the Fourier sine series of g converges to g in the L^2 norm.

Much more is known about convergence of Fourier series. Two classical results are the following.

If g is twice continuously differentiable, $g(0) = 0$, and $g(1) = 0$, then its Fourier sine partial sums converge to g in the supremum norm (which is called uniform convergence).

If g is piecewise twice continuously differentiable and x_0 is a point where g is continuous, then the Fourier partial sums evaluated at $x = x_0$ converge to $g(x_0)$ (which is called pointwise convergence). *If x_0 is at a jump discontinuity of g, then the partial sums evaluated at x_0 converge to*

$$\frac{1}{2}\left(\lim_{x \to x_0^+} g(x) + \lim_{x \to x_0^-} g(x)\right).$$

Similar statements are true for the cosine and full Fourier series.

Fourier sine series are the natural choice when zero Dirichlet boundary conditions are required at the end points of the interval $(0, L)$, cosine series are the natural choice for zero Neumann boundary conditions, and full Fourier series are used to approximate $2L$-periodic functions on the interval $(-L, L)$ (see [5, 11, 103]).

Exercise 18.4. Prove the polarization identity [Eq. (18.2)] and Eq. (18.3).

Exercise 18.5. Prove that every skew-symmetric matrix is an infinitesimal rotation.

Exercise 18.6. Show that if $A \sin x + B \cos x = 0$ for all x in some open interval of real numbers, then $A =$ and $B = 0$.

Exercise 18.7. (a) Show that the sum of two (arbitrary) solutions of the wave equation that each satisfy zero Dirichlet boundary conditions is again a solution that satisfies the same boundary conditions. (b) Show, for example, that the superposition of solutions that satisfy the boundary conditions $u(0, t) = 0$ and $u(L, t) = 1$ do not satisfy the same boundary conditions.

Exercise 18.8. Continue the separation of variables method for the wave equation with zero Dirichlet boundary conditions using zero and negative values for the separation constant λ. Show that nonzero solutions produced by the method do not satisfy the boundary conditions.

Exercise 18.9. The temperature u of a heated bar of length L positioned along the x-axis with insulated ends and initial temperature α is approximated by the solution of the boundary, IVP

$$\frac{\partial u}{\partial t} = c^2 \frac{\partial^2 u}{\partial x^2}, \qquad t \geq 0, \quad 0 < x < L,$$

$$\frac{\partial u}{\partial x}(0,t) = 0, \qquad \frac{\partial u}{\partial x}(L,t) = 0, \qquad t \geq 0,$$
$$u(x,0) = \alpha(x), \qquad 0 < x < L.$$

Use Fourier series to determine a formula for the temperature u.

Exercise 18.10. Construct a function on the interval $(0,1)$ such that its supremum exists but not its maximum.

Exercise 18.11. (a) Determine the minimum number a and the minimum value of $\|\alpha - f\|$, where α is defined by Eq. (18.28) and $f(x) = a\sin \pi x$. (b) Write a numerical code to approximate the minimizer $\{a_n\}_{n=1}^k$ for the corresponding k-dimensional case where $f(x) = \sum_{n=1}^k a_n \sin n\pi x$. Use your code to find the minimizers with an accuracy less than 10^{-5} for $k = 1, 2, 3, \dots, 20$.

Exercise 18.12. Use the Pythagorean theorem to prove that the shortest distance from a point P to a plane containing the point Q is the magnitude of the orthogonal projection of the vector $P - Q$ onto one of the unit normals, say η, of the plane. Also show that for the hyperplane $\{x_i \in \mathbf{R}^k : a_i x_i = B\}$ (where the summation convention is used),

$$(P - Q)\cdot \eta = \frac{p_i a_i - B}{\sqrt{a_i a_i}}.$$

In particular, the distance does not depend on the choice of Q.

Exercise 18.13. (a) Show that the inner product on $\mathcal{L}$ satisfies the properties in display (18.38). (b) Give an alternate proof that the inner product of two functions in $\mathcal{L}$ is finite based on the observation that $(a - b)^2 \geq 0$ for every pair of real numbers a and b.

Exercise 18.14. (a) Find an orthonormal basis that generates the hyperplane in $\mathbb{R}^3$ given by the equation $2x + y + z = 0$. (b) Show that the points in $\mathbb{R}^4$ that satisfy the equations $2x + y + z + w = 0$ and $x + 2y + z + 2w = 0$ form a two-dimensional plane. Determine an orthonormal basis that generates this plane. (c) Show that the points in $\mathbb{R}^4$ that satisfy the equation $x + 2y + z + 2w = 0$ form a three-dimensional plane. Determine an orthonormal basis that generates this plane.

Exercise 18.15. Show that the set $\mathcal{L}$ of all square integrable functions on the interval $[0, L]$ is infinite-dimensional.

Exercise 18.16. (a) Let g be the function on the interval $(0,1)$ that has value zero on the subinterval $(0, 1/2)$ and value one on the interval $[1/2, 0)$. Determine its Fourier sine series. (b) Make (computer-generated) graphs showing the function g and its best approximations by Fourier sums up to the sum of at least 40 sines. According to the theory, these sums must converge to g in the L^2 norm. Also, the series converges pointwise to g except at $x = 1/2$, where it converges to $1/2$. Describe what you see near the points $x = 1/2$ and $x = 1$. The behavior has a name: the Gibbs phenomenon. This behavior is expected near points of discontinuity. (c) Show by numerical experiments that the Fourier series converges to zero at $x = 2/16$, one-half at $x = 1/2$, and to unity at the point $x = 9/16$. (d) Conclude that the rate of convergence of the Fourier series to a point near $x = 1/2$ is slow.

Exercise 18.17. (a) Least squares linear regression in its simplest form is the problem of determining the line $y = mx + b$ that best fits a finite set of data points $\{(x_n, y_n)\}_{n=1}^k$ by minimizing the function $F(m, b) := \sum_{n=1}^k (mx_n + b - y_n)^2$. Solve this problem by finding explicit formulas for m and b that depend on the given data. Hint: Use calculus. Appendix A.12 has a general discussion of this problem. (b) Find the best fitting line to the data points $\{(x_n, y_n)\}_{n=1}^k$ where $x_n = n/50$ and $y_n = e^{-0.15n/50} \sin(100\pi x_n)$ for $n = 1, 2, 3, \ldots, 50$.

Exercise 18.18. Let $\mathcal{M}$ denote the two-dimensional plane generated by the vectors $P = (1, 1, 1, 1, 1)$ and $Q = (-1, 1, -1, 1, -1)$ in five-dimensional Euclidean space. Find the point on this plane closest to the point $R = (8, 0, 8, 0, 8)$.

Exercise 18.19. [Approximation and Orthogonal Polynomials] Suppose that f is a continuous function defined on the interval $[-1, 1]$. Which polynomial of degree k is closest to f? This question is not well posed because closeness is not defined. To make the problem precise, suppose that distance is measured in the sense of least squares (the L^2 norm). (a) The question then is to minimize the quantity

$$\int_{-1}^{1} |f(x) - \sum_{n=0}^{k} a_n x^n|^2 \, dx$$

over sets of real coefficients $\{a_n\}_{n=0}^k$. Use calculus to write out the system of linear equations (normal equations) that must be solved to obtain the minimum. Also, determine (numerically) the polynomials of degrees $k = 1, 2, 3, \ldots, 10$ that best fit the function $x \mapsto \cos \pi x$. Draw graphs that illustrate the fit. Use an algebraic processor to determine the exact minimizing polynomials (at least for the first few values of k) and check the accuracy of your numerical approximations against the exact solutions. Discuss your algebraic and numerical experiments. Hint: Numerical approximations are notoriously difficult (ill-conditioned) for this problem because the system matrices become nearly singular as the degree of the polynomial increases.

(b) Could there be more than one polynomial minimizer? Explain.

(c) Consider the polynomials

$$L_0 = 1, \quad L_1 = x, \quad L_2 = \frac{1}{2}(-1 + 3x^2), \quad L_3 = \frac{1}{2}(-3x + 5x^3),$$
$$L_4 = \frac{1}{8}(3 - 30x^2 + 35x^4).$$

Show that every polynomial of degree four is a linear combination of these polynomials. Can you make a more general statement? What if the polynomials were instead your favorite polynomials of degree zero, degree one, degree two, and so on up to degree four? Is there anything special about the number four? What is the most general statement you can make?

(d) There is something special about the polynomials in part (c): they are orthogonal with respect to the inner product

$$(f, g) = \int_{-1}^{1} f(x) g(x) \, dx.$$

In other words $(L_i, L_j) = 0$ whenever $i \neq j$.
(e) A great idea: Instead of looking for the polynomial minimizer as in part (a) directly, find the minimizer of

$$\int_{-1}^{1} |f(x) - \sum_{n=0}^{k} a_n L_n(x)|^2 \, dx$$

over all sets of real coefficients $\{a_n\}_{n=0}^{k}$ for $k = 0, 1, 2, 3, 4$. Show that the corresponding minimizing polynomial and the one sought in part (a) are the same. Moreover, show that the minimizer is given exactly by

$$a_n = \frac{\int_{-1}^{1} L_n(x) f(x) \, dx}{\int_{-1}^{1} L_n(x) L_n(x) \, dx}.$$

(f) Use the result of part (e) to solve the problem in part (a) concerning the polynomial approximation of the cosine with a polynomial of degree four. Does the new theory produce a more efficient and accurate answer?
(g) Of course, this entire exercise is meant to introduce you to something much more general: the world of orthogonal polynomials. As you might guess, the L polynomials are simply the first five polynomials in an infinite sequence of polynomials. These are called Legendre polynomials. There are many other classes of orthogonal polynomials. Although all such classes can be used to find polynomial approximations of functions, orthogonal polynomials have many other uses in applied mathematics and numerical analysis. Note also that the Legendre polynomials, for example, are solutions of a one-parameter family of second-order ODEs: $(1 - x^2)y'' - 2xy' + n(n + 1)y = 0$. The parameter n corresponds to the degree of the respective Legendre polynomial. Check this fact and use the ODE to determine L_5 up to a constant multiple.

Exercise 18.20. [Wave Modeling and Numerics] Tie a rope to a tree, hold the rope taut with one hand, and move your hand quickly up and down once. A wave will propagate down the rope toward the tree. How does the wave behave before and after it reaches the tree? Rope and flexible cord are complex materials. Also, the described experiment will likely produce a wave whose displacement is far from equilibrium. Thus, a realistic model would likely require taking into account nonlinear effects. But perhaps as a crude model, you might consider the linear wave equation to model the displacement u of the rope. Let x denote distance (say $0 \leq x \leq L$) along the taut rope from its position in hand to the tree, and let t denote time. Most likely, the Lamé constants for the rope are not known, so the usual modeling procedure would be to assume the equation of motion is the one-dimensional wave equation $u_{tt} = c^2 u_{xx}$, where c is the unknown wave speed to be determined by experiment. At time $t = 0$, the beginning of the experiment, the taut rope may be modeled by $u(x, 0) = 0$ (no displacement) and $u_t(x, 0) = 0$ (zero initial velocity). There is no displacement at the tree, so $u(L, t) = 0$. At the hand holding the rope, $u(0, t) = f(t)$, where the function f of time models the displacement imparted by the hand motion to the rope.
(a) Show that for $u(x, t) = LU(x/L, ct/L)$, the model takes the dimensionless form

$$U_{\tau\tau} = U_{\xi\xi}, \quad U(\xi, 0) = 0, \quad U_\tau(\xi, 0) = 0,$$

$$U(0,\tau) = g(\tau), \quad U(1,0) = 0,$$

where $g(\tau) := f(L\tau/c)/L$.
(b) Solve the dimensionless form of the model equations. Hint: Separation of variables does not work, at least not directly. Perhaps the best (and most beautiful) solution method, called the expansion method in [103], uses Fourier series to expand all functions with respect to the spatial variable. Pay attention to one important fact: unlike convergent Taylor series, convergent Fourier series cannot (in general) be differentiated term-by-term.
(c) Make a model of the hand motion used to produce a wave; that is, specify a function f (or better yet a family of functions) that might model this displacement. Also, specify the corresponding function g.
(d) Use a finite-difference numerical approximation scheme to produce graphs of the wave profile produced by your hand-motion model for part (c) at several carefully chosen times so that your graphs show the wave profile before and after the time at which the wave reaches the position of the tree. Hint: An excellent finite-difference method for the wave equation is described and analyzed in [61].
(e) Use truncations of the exact solution (which is likely an infinite series) for your choice of f to compare and verify the numerical results obtained in part (d). Hint: Perhaps in your work on this part of the problem, you will realize at least two important facts: (1) Making the required graphs will include numerical approximations of a truncation of the exact solution. (2) Perhaps from a different point of view, the exact solution method can be turned into a numerical method. This is evident when the suggested expansion method is employed. You will be led to solve a number of second-order ODEs (one for each Fourier mode in the truncated expansion), which are harmonic oscillators with forcing given by certain multiples of g. The solutions of these ODEs can be approximated numerically with your favorite ODE solver, for example Euler's method, the trapezoidal method, or the Störmer–Verlet method (as in Exercise 10.9). Approximating the solution of the PDE in this manner is an example of a spectral method: the desired approximate solution is sought as a truncated Fourier series with unknown coefficients and these coefficients are approximated by solving a system of linear equations or, in this time-dependent case, by solving a system of ODEs. By simply changing the point of view, some beautiful classical mathematics developed before the advent of high-speed computers, can be turned into a numerical method.
(f) What would happen if the end of the rope at the tree were free to move in the direction of the displacement? Perhaps the rope is tied to a ring that is free to slide on a vertical wire.
(g) Suppose that a wave is initiated by hand motion as part (c), the hand is held still to allow the wave to reach the vicinity of the midpoint of the taut rope, and then another wave is initiated by a second up-and-down hand motion. Would there be an interaction between the two waves at some later time? What does the model predict? Use graphs generated from numerical experiments or the exact solution to illustrate a discussion of this interaction.
(h) Repeat the project for the case of a plucked string; that is, change the model so that the left-hand boundary condition is $u(0,t) = 0$. Use the function f (or in dimensionless form g) to set an initial profile for the string and suppose that the string is let go from this

profile at time $t = 0$. Describe the wave profile as time increases. Compare the period of the oscillation of the wave profile for the plucked string and the rope tied to a tree under the assumption that the wave speeds are the same. Be sure to illustrate your answers with graphs made using the exact solution or numerical approximations.

Exercise 18.21. [Vibrating Wire Sensors] Suppose you want to build a sensor. The fundamental starting point is to use a mechanism that interacts with the phenomenon that is to be measured. Recall the model PDE (18.23) for the deformation u of a taut wire with mass per length ρ (a change of notation) and tension (or traction) τ:

$$\rho u_{tt} = \tau u_{xx}.$$

In this context, τ has units of force. With more physics incorporated, it becomes

$$\rho u_{tt} + a u_t = \tau u_{xx} + \rho f(u, x, t), \tag{18.40}$$

where a is a coefficient of damping and f is a body force per mass. Suppose the wire is rigidly attached at both ends and has length L. From the analysis of the undamped and unforced equation, the lowest frequency deformation is given by

$$u(x,t) = \sin(\frac{\pi\sqrt{\tau}}{L}t + \alpha)\sin\frac{\pi}{L}x,$$

where α is a phase angle. This fundamental mode is sinusoidal with spatial profile $x \mapsto \sin\frac{\pi}{L}x$. It has (natural) frequency

$$\omega_0 = \frac{1}{2L}\sqrt{\frac{\tau}{\rho}}.$$

Plucking the wire will usually produce this fundamental vibration as the dominant (highest amplitude) mode. There are likely to be higher frequency terms in the theoretical solution; but, these will have small amplitude. (a) Write the theoretical solution of a model for the plucked wire. You choose the initial data. Also, use your solution to predict the dominant mode.
(b) One type of vibrating wire sensor (for which there is a thriving commercial industry) is configured as a strain gauge: the length of the wire is changed by the phenomenon that is to be measured. The wire might be strung over a crack in some material that is changing in size over time, or perhaps one end of the wire is attached to a diaphragm that moves when there is a change in gas pressure. In these configurations, the length of the wire changes when the crack space or gas pressure changes. How sensitive is the natural frequency to a change in wire length? One way to approach this problem is to recall the basic stress-strain relation of linear elasticity: stress is proportional to strain. In the context of this problem, the relation is $\sigma = E\epsilon$, where E is the modulus of elasticity (Young's modulus). Note that E has units of pressure. Tension has units of force. Let A denote the cross-sectional area of the wire. Tension per area is the same as stress. Using this notation and defining τ_0 to be the base operating wire tension, the changed wire

tension should be (up to linear approximation)

$$\tau = \tau_0 + AE\frac{\Delta L}{L},$$

where of course $\Delta L/L$ is the relative change in length (which in other words is the strain).

Using this notation, the wire frequency as a function of the change in length is

$$\omega(\Delta L) = \omega_0 + \frac{1}{2(L_0 + \Delta L)}\sqrt{\frac{\tau_0 + AE\Delta L/L_0}{\rho}}.$$

The relative change in frequency (which is measured in the sensor to determine the stain) is

$$\frac{\omega - \omega_0}{\omega_0} \approx \frac{L_0}{\omega_0}\omega'(0)\frac{\Delta L}{L_0}.$$

Thus the amplification factor (gauge factor) is the quantity

$$\text{gf} := \frac{L_0}{\omega_0}\omega'(0).$$

Some typical values of the system parameters for a metal wire with diameter d might be

$$E = 150\,\text{GPa}, \qquad \tau = 980\,\text{gm}\,\text{m}\,/\,\text{sec}^2, \qquad d = 0.1\,\text{mm}, \qquad L_0 = 0.1\,\text{m}\,.$$

Compute the amplification factor for the given values. Is this value favorable for vibrating wire sensors? Explain. Is there a simpler formula for gf?

(c) Vibrating wire sensors usually have electronic mechanisms to keep the wire vibrating and to record changes in frequency. To fully understand how they work requires some understanding of electromagnetism, which might be acquired by reading further in this book. In essence, the device design sets up an oscillating electromagnetic field that affects the motion of a metal wire (perhaps one that supports an electric current) and in turn changes in the vibrations of the wire affect the electromagnetic field. These later changes can be used to produce an electric current that can be analyzed by a postprocessor to determine the frequency of the vibrating wire. Advanced engineering is required to make a precision working device, but the operating principles could be easily demonstrated in a physics or engineering lab. Although electromagnetic effects are used in vibrating wire strain gauge applications, vibrating wire sensors can also be configured to detect electromagnetic fields. These are used in the nuclear industry. An early example of such a sensor configuration and its operation, with an accessible and mostly self-contained mathematical analysis, is described in paper [110]. Read this article, fill in the details, and discuss its contents. The model described therein is in the form of Eq. (18.40) with an electromagnetic force and the force due to gravity taken into account. Demonstrate the resonance response described in the paper via numerical experiments.

(d) The frequency of a vibrating wire sensor should be sensitive to changes in temperature. How might a VWS thermometer be configured? Make a model, discuss the mathematical aspects of its intended operation, and demonstrate your findings with numerical experiments.

18.2 LONGITUDINAL WAVES IN A ROD WITH VARYING CROSS SECTION

In addition to wires, elasticity theory has been specialized to rods, plates, and shells. The underlying idea is to reduce the spatial dimension of the equation of motion by taking advantage of symmetry or the relative (small) size of one or more spatial dimensions. Perhaps the simplest model of this type is for the longitudinal motion of a rod with varying cross-sectional area in case the only forces on the surface of the rod are at its ends and these forces act parallel to the axis of the rod.

Consider a three-dimensional rod whose central axis is on a line and whose bounded cross sections are taken perpendicular to the central axis. Choose coordinates so that the central axis is the x-axis of a rectangular coordinate system such that each cross section lies in a plane parallel to the plane defined by $x = 0$ and the left end of the rod is at $x = 0$. The cross section of the rod at x is denoted $S(x)$ and its area by $A(x)$.

Recall the basic equation of motion

$$\rho \ddot{u}_i = \sigma_{ij,j} + \rho b_i \tag{18.41}$$

for linear elasticity, where σ is the stress tensor, ρ is the constant density of the rod, u the displacement vector, b_i the body force per mass, and summation on repeated indices is enforced (see page 318). The objective of this section is to reduce this basic model to a more tractable form, which might still capture the essential features of longitudinal waves, by making several simplifying assumptions.

Let x be the first coordinate of a point in the rod and h a positive real number such that $x + h$ is also the first coordinate of such a point. Consider the result of integrating the first component of the equation of motion—the component in the axial direction—over the portion of the rod Ω bounded by the cross sections $S(x)$ and $S(x + h)$:

$$\rho \int_\Omega \ddot{u}_1 \, d\mathcal{V} = \int_\Omega \sigma_{1j,j} \, d\mathcal{V} + \rho \int_\Omega b_1 \, d\mathcal{V}.$$

The first integral on the right-hand side of this equation is the integral of the divergence of a vector field over the solid Ω. By the divergence theorem, it is the integral over the boundary of Ω of the inner product of this vector field with the outer unit normal η; therefore,

$$\rho \int_{\Omega} \ddot{u}_1 \, d\mathcal{V} = \int_{\partial\Omega} \sigma_{1j}\eta_j \, d\mathcal{V} + \rho \int_{\Omega} b_1 \, d\mathcal{V}.$$

Notice that the integral over the boundary of Ω can be split into a sum of three parts: The integral over $S(x)$, the integral over $S(x+h)$, and the integral over the lateral boundary cut off by these cross sections. Physical intuition dictates that the dominant displacement is in the axial direction, or in other words, the component of the stress field in the normal direction over the lateral boundary should be small.

Assumption: The integral of the stress field over the lateral boundary is negligible.

As soon as the assumption is implemented, the equation of motion is no longer equivalent to the derived theory of linear elasticity; rather, it is a new simplified model. Because the (outer) normal η on the cross sections is parallel to the x-axis and points right on $S(x+h)$ and left on $S(x)$, this new model takes the preliminary form

$$\rho \int_{\Omega} \ddot{u}_1 \, d\mathcal{V} = \int_{S(x+h)} \sigma_{11} \, dS - \int_{S(x)} \sigma_{11} \, dS + \rho \int_{\Omega} b_1 \, d\mathcal{V}.$$

Each volume integral can be written as an iterated integral by integrating over each cross section and then along the axis. Also, the mean value theorem applies to the function

$$x \mapsto \int_{S(x)} \sigma_{11} \, dS$$

on the interval $[x, x+h]$. Thus, there is a point $x(h)$ in the open interval $(x, x+h)$ such that the preliminary form of the new model can be recast as

$$\rho \int_{x}^{x+h} \int_{S(\xi)} \ddot{u}_1 \, dS \, d\xi = h \frac{\partial}{\partial x} \int_{S(x)} \sigma_{11} \, dS \Big|_{x=x(h)} + \rho \int_{x}^{x+h} \int_{S(\xi)} b_1 \, dS \, d\xi.$$

Using the mean value theorem for integrals applied to the iterated integrals, division by h, and passage to the limit as h approaches zero from the right, the new equation of motion is

$$\rho \int_{S(x)} \ddot{u}_1 \, d\mathcal{S} = \frac{\partial}{\partial x} \int_{S(x)} \sigma_{11} \, d\mathcal{S} + \rho \int_{S(x)} b_1 \, d\mathcal{S},$$

or with

$$U_1(x,t) := \int_{S(x)} \ddot{u}_1 \, d\mathcal{S},$$

$$\rho \ddot{U}_1 = \frac{d}{dx} \int_{S(x)} \sigma_{11} \, d\mathcal{S} + \rho \int_{S(x)} b_1 \, d\mathcal{S}. \tag{18.42}$$

For a rod made of a homogeneous and isotropic material,

$$\begin{aligned} \sigma_{11} &= \lambda(u_{1,1} + u_{2,2} + u_{3,3}) + 2\mu u_{1,1} \\ &= (\lambda + 2\mu) u_{1,1} + \lambda(u_{2,2} + u_{3,3}). \end{aligned}$$

With this model of the stress field, the equation of motion is

$$\rho \ddot{U}_1 = (\lambda+2\mu) \frac{\partial}{\partial x} \int_{S(x)} u_{1,1} \, d\mathcal{S} + \lambda \frac{\partial}{\partial x} \int_{S(x)} u_{2,2} + u_{3,3} \, d\mathcal{S} + \rho \int_{S(x)} b_1 \, d\mathcal{S}. \tag{18.43}$$

The sum $u_{2,2} + u_{3,3}$ is the divergence with respect to the second and third variables of the vector field with components (u_2, u_3). By the divergence theorem, the corresponding integral is the flux of this field over the boundary of $S(x)$.

Assumption: The flux of the field (u_2, u_3) over the boundary of each of the rod's cross sections is negligible.

Assumption: There is a planar vector field X defined on the plane $\{(x, y, z) : x = 0\}$ such that its flow ϕ parametrized by x takes the left-end cross section of the rod (which is $S(0)$) to the projection in this plane of each $S(x)$ via $\phi_x(S(0)) = S(x)$.

By taking these assumptions into account and by the Reynolds transport theorem applied to the first integral on the right-hand side of Eq. (18.43), a new equation of motion is

$$\rho \ddot{U}_1 = (\lambda + 2\mu)\frac{\partial}{\partial x}\Big(\frac{d}{dx}\int_{S(x)} u_1\, d\mathcal{S} - \int_{S(x)} \operatorname{div}(u_1 X)\, d\mathcal{S}\Big) + \rho \int_{S(x)} b_1\, d\mathcal{S},$$

where the divergence operator is with respect to the spatial coordinates y and z. Leibniz's product rule (with the gradient operator taken with respect to the spatial coordinates y and z) implies that

$$\rho \ddot{U}_1 = (\lambda + 2\mu)\frac{\partial}{\partial x}\Big(\frac{d}{dx}\int_{S(x)} u_1\, d\mathcal{S} - \int_{S(x)} \operatorname{grad}(u_1)\cdot X + u_1 \operatorname{div} X\, d\mathcal{S}\Big)$$
$$+ \rho \int_{S(x)} b_1\, d\mathcal{S}, \tag{18.44}$$

Assumption: The gradient of the first component of the displacement field u_1 with respect to the spatial coordinates y and z is negligible.

Using this assumption and the mean value theorem for integrals applied to the second integral on the right-hand side of Eq. (18.44), there is a point $q_1 = q_1(x)$ in the section $S(x)$ such that

$$\rho \ddot{U}_1 - (\lambda + 2\mu)\frac{\partial}{\partial x}\Big(\frac{d}{dx}\int_{S(x)} u_1\, d\mathcal{S} - \operatorname{div} X(q_1(x))\int_{S(x)} u_1\, d\mathcal{S}\Big)$$
$$+ \rho \int_{S(x)} b_1\, d\mathcal{S}$$
$$= (\lambda + 2\mu)\frac{\partial}{\partial x}\Big(U_1' - \frac{d}{dx}(\operatorname{div} X(q_1(x))U_1)\Big) + \rho \int_{S(x)} b_1\, d\mathcal{S}. \tag{18.45}$$

Note that

$$A(x) := \int_{S(x)} d\mathcal{S},$$

and by the transport theorem, there is some $q_2 = q_2(x)$ so that

$$A'(x) := \int_{S(x)} \operatorname{div} X\, d\mathcal{S} = \operatorname{div} X(q_2(x)) \int_{S(x)} d\mathcal{S} = \operatorname{div} X(q_2(x))A(x).$$

Assumption: The divergence of the vector field X with respect to the variables y and z is nearly constant.

Under this last assumption the model equation takes the form

$$\begin{aligned}\rho\ddot{U}_1 &= \frac{\partial}{\partial x}(U_1' - \frac{A'(x)}{A(x)}U_1)) + \rho\int_{S(x)} b_1\, d\mathcal{S} \\ &= (\lambda + 2\mu)\frac{\partial}{\partial x}(A(x)\frac{\partial}{\partial x}\frac{U_1}{A(x)}) + \rho\int_{S(x)} b_1\, d\mathcal{S}.\end{aligned} \tag{18.46}$$

By defining the average displacement and average body force per mass

$$U(x,t) := \frac{1}{A(x)}\int_{S(x)} u_1\, d\mathcal{S}, \qquad B(x,t) := \frac{1}{A(x)}\int_{S(x)} b_1\, d\mathcal{S}$$

and using the product rule, the final form of the new model equation for longitudinal motion is

$$\rho\ddot{U} = (\lambda + 2\mu)(\frac{\partial^2 U}{\partial x^2} + \frac{A'(x)}{A(x)}\frac{\partial U}{\partial x}) + \rho B(x,t). \tag{18.47}$$

Although there are many other ways to derive this model and some of them are more efficient than the method described here (compare Exercise 18.22), the reader might agree that the simplifying assumptions made to reduce the general linear theory to the new model are clearly specified. This one-dimensional model is not equivalent to the more basic three-dimensional model of linear elasticity [Eq. (18.41)], but it is much simpler to analyze. An application is made in the next section. Of course, the model must be supplemented with initial and boundary data.

Exercise 18.22. [Calculus of Variations: Hamilton's Principle] (a) The kinetic energy of a small slice of a tapered rod (the solid part of the rod between two cross sections that are close together) is approximately $\rho A(x)\dot{u}^2/2$, where A is the cross-sectional area at a point in the slice and u is the displacement in the horizontal direction. Explain. (b) The potential energy of a small slice is approximately $EA(x)u_x^2/2$, where E (Young's modulus) is a constant that gives the elastic properties of the rod. Explain. (c) The Lagrangian for the motion of the rod (which is assumed to have its central axis the x-axis between the origin and L) is

$$\frac{1}{2}\int_0^L \rho A(x)\dot{u}^2 - EA(x)u_x^2\, dx.$$

The action is the integral of the Lagrangian with respect to time:

$$\int_0^T \frac{1}{2}\Big(\int_0^L \rho A(x)\dot{u}^2 - EA(x)u_x^2\,dx\Big)\,dt.$$

According to the principle of least action (Hamilton's principle), the action corresponding to the physical path is stationary with respect to variations of the path. That is, if $u(x,t)$ is the physical path, the derivative with respect to δ at $\delta = 0$ of the action along the path $u(x,t) + \delta w(x,t)$ vanishes, for every choice of w with $w(x,0) = 0$ and $(w(x,T) = 0$ for all x and $w(0,t) = w(L,t) = 0$ for all t. Put the variation into the action integral and differentiate with respect to δ as described. Show that the action principle says

$$\int_0^T \Big(\int_0^L \rho A(x)\dot{u}\dot{w} - EA(x)u_x w_x\Big)\,dt = 0$$

for all such w. (d) Continuing with part (c), show by changing the order of integration as necessary and integration by parts, that

$$\int_0^T \Big(\int_0^L -\rho A(x)\ddot{u}w + (EA(x)u_x)_x w\Big)\,dt = 0$$

for all w. Conclude that

$$-\rho A(x)\ddot{u} + (EA(x)u_x)_x = 0$$

and compare with model (18.47).

18.3 ULTRASONICS

Imagine a solid horn (a rod that is tapered continuously with decreasing cross-sectional area) attached at its left end to a transducer that imparts a sinusoidal oscillation in the axial direction. To achieve ultrasonic vibration, the method of choice is via piezoelectric transduction (which is actuated by a material that expands and contracts in the presence of an appropriately generated electromagnetic field). High-frequency oscillations are easily achieved with this technology, but they have small amplitude. A hypothetical *rigid* horn, which might be exposed to a chemical bath, a sample of material to be tested for fractures, or a test tube containing bacteria, would vibrate with the same amplitude as the transducer and not do much work. Amplification is desirable and achieved with an *elastic* horn (also called a sonotrode). Note that the mass of the horn is greatest near its large end, which is vibrating at the amplitude of the transducer. The small end has less mass but the same force is applied. It should move farther. This physical insight is predicted by model (18.47). The actual amplification of amplitude by elastic horns is a fundamental component of ultrasonic technology.

Amplitude amplification is illustrated by a horn designed with exponentially decreasing cross-sectional area; that is, the cross-sectional area A is given by

$$A(x) = A_0 e^{-\gamma x}$$

for some positive constants A_0 and γ. This profile is very special:

$$A'(x)/A(x) = -\gamma,$$

a constant; it reduces the mathematical model to a linear differential equation with constant coefficients. Let us suppose in addition that the transducer imparts a sinusoidal oscillation $\alpha \sin(\omega t)$ at the left end of the horn, and the right end of the horn is traction free.

The change of variables

$$U = \frac{\alpha}{\gamma(\lambda + 2\mu)} W, \quad t = \Big(\frac{\rho}{\gamma^2(\lambda + 2\mu)}\Big)^{1/2} s, \quad x = \frac{\xi}{\gamma},$$
$$\Omega := \omega \Big(\frac{\rho}{\gamma^2(\lambda + 2\mu)}\Big)^{1/2}$$

renders the elasticity model [Eq. (18.47)], with exponentially decreasing cross-sectional area, dimensionless:

$$\begin{aligned} W_{ss} &= W_{\xi\xi} - W_{\xi}, \\ -W_{\xi}(0, s) &= \sin \Omega s, \\ W_{\xi}(L, s) &= 0, \end{aligned} \tag{18.48}$$

where L is the scaled length of the horn. The Neumann boundary conditions arise from tractions at the ends of the rod. Recall Eq. (18.6) (which reads $\sigma_{ij}\eta_i = \tau_j$ in components) that was used to construct the fundamental equations of elasticity by considering the tractions on the surface of a small cube. The stress in the normal direction to a face is the traction on that face. Simply stated, the Neumann boundary condition arises from the same equation applied to a surface that happens to be on the boundary of the material. Linear stress tensor (18.14) applied to the normals at the boundaries for the case at hand (which is one-dimensional in space) reduces to $\pm(\lambda + 2\mu)U_x$, where the sign is positive at the right end of the rod and negative at the left end. No traction is applied at the right end of the rod, and a periodic traction (assumed to be a sinusoid) is acting on its left end. If instead, the displacement of the left end of the bar were specified, the left end boundary condition would be $W(0, s) = \sin \Omega s$ (see Exercise 18.24).

To make model (18.48) well posed, the initial (scaled) displacement $W(\xi, 0)$ and velocity $W_s(\xi, 0)$ must be specified. In fact, these quantities are not known for the physical application. When the transducer is constructed and the system is turned on, this initial data is not easily defined or measured. In this case, physical intuition suggests that the solution, after transients have died away, should be the real or imaginary part of a complex function of the form

$$W(\xi, s) = w(\xi)e^{i\Omega s};$$

that is, the displacement of the rod at each coordinate ξ is oscillating about some fixed displacement $w(\xi)$ at the frequency of the sinusoidal driving force. The boundary conditions are reduced to

$$w'(0) = -1, \qquad w'(L) = 0. \tag{18.49}$$

A measure of the efficiency of the horn is the amplification factor AF := $|w(L)/w(0)|$; it measures the ratio of the maximum (positive) displacements at the right and left ends of the rod.

To compute the amplification factor, insert the complex solution into the model equation to determine the linear equation with constant coefficients for w:

$$w'' - w' + \Omega^2 w = 0. \tag{18.50}$$

The solution of this equation with the given boundary conditions [Eqs. (18.49)] is easily computed using the usual methods for solving linear second-order ODEs. Under the assumption that $\Omega > 1/2$, with some algebraic simplification, and with

$$\omega := \sqrt{4\Omega^2 - 1},$$

the amplification factor is

$$\text{AF} = \frac{e^{L/2}\omega}{\left|\omega \cos(\frac{L\omega}{2}) + \sin(\frac{L\omega}{2})\right|}. \tag{18.51}$$

The amplification factor can be infinite. Thus, there is no bound to the amplification predicted by our model using linear elasticity. Of course, this prediction is not realizable. The situation is analogous to the usual model of a mass on a spring driven by a sinusoidal force. In scaled variables, the equation of motion for the mass displacement is

$$\ddot{u} + u = \sin \Omega t.$$

For $\Omega = 1$, all solutions oscillate with increasing amplitudes that grow without bound. We say the driving frequency is resonant with the natural frequency of the spring, which is given by its material properties. For this reason, the frequencies Ω (which include in their definition the elastic properties of the rod) and lengths L (which include in their definition the taper of the rod) such that AF is infinite are in resonance; in essence, the material properties of the rod are in resonance with the driving frequency. These models do not take into account dissipation of energy (damping), which is always present (see Exercises 18.27 and 19.8). A reasonable physical prediction is that the actual maximum amplification occurs near a resonance. This prediction may be used in the design process. For example, the (shortest) resonant horn length can be calculated using the formula

$$L = \frac{2}{\sqrt{4\Omega^2 - 1}}(\pi - \arctan \sqrt{4\Omega^2 - 1}) \tag{18.52}$$

if the operating frequency of the piezoelectric activator is known.

Exercise 18.23. (a) Verify Eq. (18.52). (b) Find the length of a stainless steel resonant rod (tapered exponentially with area $A_0 e^{-x/2}$ and $A_0 = 25\,\mathrm{cm}^2$ at $x = 0$) for an ultrasonics application where the actuator frequency is 50 kHz. Use the Lamé constants $\lambda = 1.2 \times 10^{11}\,\mathrm{kg}/(\mathrm{m\,sec}^2)$, $\mu = 7.7 \times 10^{10}\,\mathrm{kg}/(\mathrm{m\,sec}^2)$ and density $\rho = 7.6 \times 10^3\,\mathrm{kg}/\mathrm{m}^3$. Answer: $\approx 2.98\,\mathrm{cm}$. (c) Find the area of the tip of the rod. Answer: $\approx 5.64\,\mathrm{cm}^2$.

Exercise 18.24. Analyze the amplification factor for a bar with exponentially decreasing cross section where the tapered end is free and the large end is displaced sinusoidally.

Exercise 18.25. Show that the BVP (18.48) can be converted to a BVP for the Klein–Gordon equation

$$v_{ss} = v_{\xi\xi} + kv,$$

where k is a constant, via a change of variables of the form $w = e^{\ell\xi} v(\xi, s)$ for some choice of ℓ. Specify k and ℓ.

Exercise 18.26. Analyze the amplification factor for a bar with constant cross-sectional area.

Exercise 18.27. [Resonance Horn Amplification] The amplification factor for an exponentially tapered horn is given by the Eq. (18.51). As noted, this number can be infinite at resonance. A more realistic model would include damping, perhaps via the dissipation of heat due to internal friction. Taking inspiration from the usual model for viscous damping of a harmonic oscillator, consider the phenomenological model

$$\rho\ddot{U} + k\dot{U} = (\lambda + 2\mu)\Big(\frac{\partial^2 U}{\partial x^2} + \frac{A'(x)}{A(x)}\frac{\partial U}{\partial x}\Big), \tag{18.53}$$

where k is a positive constant, which might be called the damping factor. Assume the traction at the right end of the rod is zero and a sinusoid at the left end. Moreover, assume the rod is at rest before the forcing is applied. (a) Show that the model can be made dimensionless and put in the form

$$W_{ss} + W_s = W_{\xi\xi} + \frac{\mathcal{A}'(\xi)}{\mathcal{A}(\xi)}W_\xi$$

with boundary conditions $W_\xi(0,t) = \sin t$ and $W_\xi(1,0) = 0$ and initial data $W(\xi,0) = W_s(\xi,0) = 0$. (b) Find the steady state solution for the case where $\mathcal{A}$ has the exponential profile considered in this section. (c) Suppose the (scaled) profile function is a cubic with $\mathcal{A}(0) = 1$, $\mathcal{A}'(0) = 0$, $\mathcal{A}(1) = 1/100$, and $\mathcal{A}'(1) = 0$. What is the approximate value of the amplification factor? (d) Fix the boundary and initial data and consider the steady state amplification factor as a function of the cross-sectional area profile $\mathcal{A}$. Which (continuously differentiable) profile defined on $[0,1]$ corresponds to the largest amplification factor? This question may not have a simple answer. What can you say? At least compare the exponential profile with the members of a family of cubic profiles where a range of heights at the end points is considered. (e) Suppose the damping is caused by a traction at the right end of the rod that opposes the motion. Perhaps the right end is oscillating in a viscous fluid or in some other elastic material. Make a model for this situation, solve it for the case of an exponential horn profile, check that the motion is damped, and repeat the problems concerning the amplification factor.

18.4 A THREE-DIMENSIONAL ELASTOSTATICS PROBLEM: A COPPER BLOCK BOLTED TO A STEEL PLATE

Imagine a large steel plate with a round hole. A rectangular copper block is welded to a round steel bolt (which fits the hole in the steel plate) so that the bolt is centered at and perpendicular to a face of the copper block. The bolt is threaded through the hole in the plate and pulled (perhaps by tightening a nut) so that the copper block is forced against the steel plate. Problem: Determine the equilibrium deformation field of the copper block as a function of the force on the bolt and the dimensions of the block relative to the diameter of the bolt. Specifically, determine the distortion field in the copper block in the plane through the central axis of the bolt and perpendicular to the longest edges of the block.

The deformation of the steel plate is expected to be small compared with the deformation of the copper block. Thus, a reasonable assumption is zero deformation of the steel plate. In other words, the steel plate may be modeled as a rigid body in a fixed position. To simplify the mathematical model for the deformation, Cartesian coordinates are chosen so that the face of the plate adjacent to the copper block is in the plane $\{(x, y, z) : z = 0\}$, the hole in the plate is centered at the origin of the coordinate system, and the copper block resides in the half-space $\{(x, y, z) : z \geq 0\}$.

The steady state deformation field in the interior of the copper block must satisfy the differential equation

$$(\lambda \varepsilon_{kk} \delta_{ij} + 2\mu \varepsilon_{ij})_{,j} + \rho b_i = 0, \tag{18.54}$$

which is the steady state field equation obtained from the fundamental model [Eq. (18.16)], and the boundary conditions imposed by the geometry and the force induced by tightening the bolt.

Boundary data must be specified at each (noncorner or edge) point on the surface of the copper block. Each point on the surface of the block lies in a three-dimensional space, which is generated by two independent surface directions (tangents to the surface) and one (outer) normal direction. The external surface forces tangent to the surface are called surface tractions; the force in the other direction is called the normal force or normal traction.

The boundary forces are most easily modeled using the stresses, simply because stress is force per area. Recall that the basic relation between stress σ, traction τ, and the surface outer normal η is

$$\tau_i = \sigma_{ij} \eta_j. \tag{18.55}$$

(see Eq. (18.6)). The corresponding boundary conditions are obtained from this relation using the material relations between stress, strain, and displacement given in our model by the relations in Eqs. (18.14) and (18.15):

$$\sigma_{ij} = \lambda \varepsilon_{kk} \delta_{ij} + 2\mu \varepsilon_{ij}, \qquad \varepsilon_{ij} = \frac{1}{2}(u_{i,j} + u_{j,i}).$$

Thus, external traction on the body surface is related to boundary displacements via

$$\tau_i = \lambda u_{k,k}\eta_i + \mu(u_{i,j} + u_{j,i})\eta_j. \tag{18.56}$$

The force on a surface due to its contact with an external surface (such as the surface of our steel plate) is theoretically some force equal and opposite to the force exerted by the deformable body on the external surface. There is such a force, but rather than attempting to model this force for the case where the external surface is assumed to be rigid and not moving, the correct model is to assume that the displacement is zero in the normal direction of surface contact. In case the external surface is moving, the corresponding displacement may be imposed at the boundary of the elastic surface. Contact with an elastic media might also be modeled with a traction τ that depends on the displacement. Perhaps deformation of the body causes an external pressure field to change.

A boundary condition, which is a component of a traction force or the displacement, must be specified in each of the three coordinate directions at each point on the boundary. Careful treatment is required to ensure that predictions made from PDE models reflect physical reality, mathematical rigor is maintained, and algorithms used to produce approximate solutions are viable.

The undeformed rectangular copper block can be situated and dimensioned by specifying the block corners proximal to the (horizontal) steel plate

$$(-\frac{\mathcal{L}}{2}, -\frac{\mathcal{W}}{2}, 0), \quad (\frac{\mathcal{L}}{2}, -\frac{\mathcal{W}}{2}, 0), \quad (\frac{\mathcal{L}}{2}, \frac{\mathcal{W}}{2}, 0), \quad (-\frac{\mathcal{L}}{2}, \frac{\mathcal{W}}{2}, 0),$$

and those distal from the plate

$$(-\frac{\mathcal{L}}{2}, -\frac{\mathcal{W}}{2}, \mathcal{H}), \quad (\frac{\mathcal{L}}{2}, -\frac{\mathcal{W}}{2}, \mathcal{H}), \quad (\frac{\mathcal{L}}{2}, \frac{\mathcal{W}}{2}, \mathcal{H}), \quad (-\frac{\mathcal{L}}{2}, \frac{\mathcal{W}}{2}, \mathcal{H}). \tag{18.57}$$

The bolt is attached to the block at a disk of radius a centered at the origin of the three-dimensional coordinate system and contained in the proximal block face.

The material properties of copper are given by Young's modulus

$$E \approx 120 \times 10^9 \,\text{kg} / \text{m}\,\text{sec}^2$$

(at approximately 120 GPa =GigaPascal) and Poisson's ratio

$$\nu \approx 0.34.$$

These translate to Lamé constants via the identities

$$\lambda = \frac{E\nu}{(1-2\nu)(1+\nu)}, \qquad \mu = \frac{E}{2(1+\nu)}$$

to

$$\lambda \approx 95 \times 10^9 \,\mathrm{kg}\,/\,\mathrm{m\,sec}^2, \qquad \mu \approx 45 \times 10^9 \,\mathrm{kg}\,/\,\mathrm{m\,sec}^2. \tag{18.58}$$

The density of copper is approximately

$$\rho \approx 8.94 \,\mathrm{g}\,/\,\mathrm{cm}^3. \tag{18.59}$$

Of course, the material properties are obtained by physical experiments.

A simple steady state model for our copper block tightened with a bolt against the steel plate is given by assuming the displacement vector u satisfies the linear elasticity equation (18.54); the traction vanishes everywhere on the block's surface except on its proximal face; and, on the proximal face, the tractions tangent to the surface vanish everywhere, the normal displacement vanishes outside the bolt attachment disk, and the normal traction is constant in the downward vertical direction (toward the steel plate) everywhere in the attachment disk. Some of the simplifying assumptions are the zero frictional force between the copper and steel surfaces the shaft of the bolt and torque on the bolt are ignored, and the deformation is assumed to remain elastic; that is, the distortion in the copper block due to tightening the bolt is small and reversible.

The copper block will distort against the steel plate and the deformation field will reach a steady state. Is this state uniquely determined by the force applied on the bolt? It should be, but this is not obvious. It is also not obvious (but true) that the mathematical model is well posed. In fact, there is a unique solution of the mathematical model that depends continuously on the boundary conditions. Thus, the model predicts a unique solution. The problem set here is to find a useful approximation of this solution.

There are many possible approximation methods for elasticity models. In the following sections we will introduce the finite element method (FEM) and apply it to our model problem. Boundary element methods, finite difference methods, as well as mesh-free methods (akin to smoothed particle hydrodynamics) are also viable.

18.5 A ONE-DIMENSIONAL ELASTICITY MODEL

A basic understanding of the FEM is foundational in applied continuum mechanics; it is widely used and well studied. The best approach to the subject is by applying it to simple model problems. We will return to the three-dimensional model for the distortion of a copper block bolted to a steel plate from Section 18.4 after introduction of concepts and some experience with model problems in one space-dimension.

The BVP

$$u_{xx} + f(x) = 0, \qquad u(0) = 0, \quad u_x(1) = a, \tag{18.60}$$

where a is some real parameter, will be used to illustrate most of the central ideas of the FEM. It may be considered the dimensionless form of a one-dimensional linear elasticity model. For example,

$$(\lambda + 2\mu)\tilde{u}_{\tilde{x}\tilde{x}} + \rho b(\tilde{x}) = 0, \qquad \tilde{u}(0) = 0, \quad (\lambda + 2\mu)\tilde{u}_{\tilde{x}}(L) = \tilde{a},$$

where λ and μ are the Lamé constants, ρ is the density, the spatial coordinate is $\tilde{x}$, the length of the elastic material is L, the body force per mass is b, and the (normal) traction (at $x = L$) is $\tilde{a}$. With the rescaling $u = \tilde{u}/\kappa$ and $x = \tilde{x}/L$, where κ has dimensions of length, we have the dimensionless BVP (18.60) with

$$f(x) := \alpha b(x)$$

and

$$\alpha := \frac{\rho L^2}{\kappa(\lambda + 2\mu)}, \qquad a := \frac{L\tilde{a}}{\kappa(\lambda + 2\mu)}.$$

BVP (18.60) is explicitly solvable by integrating twice with respect to the space variable. For example, in case the elastic material is hanging vertically with zero traction force and the body force per mass is gravity ($b(x) = g$), the scaling constant κ may be chosen to make $f(x) \equiv 1$ and the displacement field is

$$u(x) = x - \frac{1}{2}x^2, \tag{18.61}$$

where x is taken to be positive in the downward vertical direction. The model predicts quadratic displacement of the material points, which is given by u as a function of distance from the place where the material is clamped.

The largest displacement occurs at $x = 1$, where the displacement is $1/2$. Does this agree with your physical intuition? It would seem that just the opposite is true: the effect of the mass distribution of the hanging elastic body—all mass below the point where it is clamped—should stretch the body most near the point where it is attached. This is indeed the case. But what is the definition of displacement? Remember that displacement means distorted position minus original position. Imagine the elastic body hanging but not influenced by the gravitational field. This is the undeformed state. Positions of the material points are measured on the interval $[0, 1]$ with respect to a fixed coordinate system. When the gravitational field is turned on, the body deforms. The bottom material point is displaced the most relative to its initial position as measured by the fixed coordinate system because its distorted position is an accumulation of all the stretching along the body. Distortion relative to nearby points *in the body* is measured by the strain. Or, via Hooke's law, it is proportional to stress. For the one-dimensional body, the stress tensor has exactly one component:

$$\sigma_{11} = \frac{\kappa}{L}(\lambda + 2\mu)u_x.$$

Thus, the stress field of the deformed body is

$$\sigma_{11}(x) = \frac{\kappa}{L}(\lambda + 2\mu)(1 - x)$$

The maximum stress occurs (as it should) where the elastic body is clamped at $x = 0$.

Imagine the elastic material clamped as before but stretched horizontally due to a normal traction at the unclamped end. In this case, gravity does not play a role; the body force may be taken to vanish in the model. The traction force enters the model as a Neumann boundary condition at $x = 1$. For example, a scaled model might be

$$u_{xx} = 0, \qquad u(0) = 0, \quad u_x(1) = 1.$$

In this case the displacement is predicted to be

$$u(x) = x$$

and the stress field is constant along the elastic body, in agreement with physical intuition.

As mentioned previously, boundary conditions cannot be specified arbitrarily. For example, the BVP

$$u_{xx} = -1, \qquad u_x(0) = 0, \quad u_x(1) = 2$$

has no solution. A physical realization of this model would be a vertical piece of elastic material influenced by gravity with a nonzero traction force at its downward end. But, the body is *not in equilibrium* when the forces are applied; indeed, the total (scaled) force on the body does not vanish:

$$\int_0^1 (-1)\,dx - u_x\Big|_0^1 = -1 + u_x(1) - u_x(0) = 1.$$

Note that for BVP (18.60), the forces on the body are in balance when

$$\int_0^1 -f(x)\,dx + u_x(1) - u_x(0) = 0.$$

Given $u_x(1) = a$, the unspecified boundary traction $u_x(0)$ is determined from the solution of the BVP, but it is not required in the statement of the BVP because of the Dirichlet condition.

Compatibility conditions are necessary for the BVPs of linear elasticity to be well posed. They are, however, not sufficient. The BVP

$$u_{xx} = 0, \qquad u_x(0) = 0, \quad u_x(1) = 0$$

meets the force balance compatibility condition, but every constant function u is a solution. Thus, the problem is ill posed; it has more than one solution. The physical interpretation is a one-dimensional piece of elastic material situated horizontally with zero traction at its boundary. The formulation of the model allows every rigid displacement of the position of the body to be a deformation. This result happens to give an acceptable physical prediction from an ill-posed model, a circumstance that should always be viewed with caution.

18.6 WEAK FORMULATION OF ONE-DIMENSIONAL BOUNDARY VALUE PROBLEMS

The FEM for approximating solutions of (well posed) boundary value problems consists of two essential ingredients: a reformulation of the PDE to a system of integral equations and a method for discretization of these integral equations. This section introduces the reformulation.

Although BVP (18.60) has an explicit solution, it will be used to illustrate several new ideas. The goal is to introduce new concepts in a simple context where they are easily understood. Later, they can be generalized to aid in approximating solutions of BVPs that do not have explicit solutions. The ultimate goal here is the methodology for approximating solutions of BVPs for a PDE of the form

$$u_{xx} + u_{yy} + u_{zz} + f = 0,$$

where f is a given function defined on some domain Ω in three-dimensional space, together with Dirichlet or Neumann boundary conditions imposed on the boundary of Ω. For example, the unknown function u might be required to vanish on the boundary of Ω. As an intermediate step toward this goal, the reader should be able to use the methods discussed here in the context of BVP (18.60) to approximate solutions of BVPs formulated for second-order linear ODEs of the form

$$u_{xx} + g(x)u + h(x) = 0$$

where g and h are given function and explicit solutions might not exist (see Exercise 18.32).

A classical solution of a differential equation such as in BVP (18.60) is a twice continuously differentiable function. Although the ultimate goal is to approximate classical solutions for BVPs, perhaps some BVPs are simpler to analyze by first setting aside this requirement. Also, recall that the fundamental conservation laws that lead to such problems are statements about equality of certain integrals; their differential forms are derived under the assumption that solutions are sufficiently smooth. But, the integral forms of these equations might be satisfied by functions that are not differentiable. Integration does not require smoothness of the integrand. Thus, as a general mathematical idea, perhaps the solution of a BVP would be easier to find or approximate using some reformulation to an integral form. This is a far-reaching idea whose implementation requires some new mathematical concepts.

Suppose β is a vector in $\mathbb{R}^k$, A is an $k \times k$ matrix, and we wish to solve the matrix equation $Av = \beta$ for an unknown vector $v \in \mathbb{R}^k$. Recall the notation $\langle v, w\rangle$ for the usual inner product of the vectors $v, w \in \mathbb{R}^k$, and note that if v is a solution of the matrix equation, then

$$\langle Av, w\rangle = \langle \beta, w\rangle$$

for every w in $\mathbb{R}^k$. The converse statement is also true: If there is some v such that

$$\langle Av, w\rangle = \langle \beta, w\rangle$$

for every w in $\mathbb{R}^k$, then $Av = \beta$. As a corollary, if for some basis $\mathcal{B}$ of $\mathbb{R}^k$ (for example, the usual unit basis vectors) there is a vector v such that

$$\langle Av, e\rangle = \langle \beta, e\rangle$$

for every e in $\mathcal{B}$, then $Av = \beta$. There is no reason in most cases to seek solutions of linear systems of equations as an application of this corollary because there are more efficient methods of solving linear equations, but the theoretical basis of the corollary is sound. Thus, the same idea can be applied in other similar situations where better solution methods may not be known.

The differential equation $u_{xx} = -f$ shares some of the features of the matrix equation $Av = \beta$ at a formal level. To see this, simply write it as

$$\Delta u = -f,$$

where Δ is the one-dimensional Laplacian, and treat this differential operator as if it were a matrix operating on the vector u. The main difference from the matrix case is that the desired vector solution u is a function of a real variable instead of a vector in $\mathbb{R}^k$.

To continue the analogy with the finite-dimensional vector equation, the idea is to create a vector space of functions and an inner product on this vector space. Exactly which vector space of functions and inner product to choose is part of the mathematical analysis that will be discussed.

Let us start with perhaps a familiar example: the set of all continuous functions on the interval $[0, 1]$. It is a vector space because linear combinations of continuous functions with real coefficients are again continuous functions. There is also an inner product obtained by generalizing the usual inner product to a continuous variable: for two continuous functions ϕ and ψ, simply multiply their values at each point of the unit interval and sum over all these products; that is, define the inner product of these functions to be

$$\langle \phi, \psi \rangle = \int_0^1 \phi(x)\psi(x)\,dx. \tag{18.62}$$

At least this definition produces a pairing of elements of the vector space of continuous functions that has all the usual properties of an inner product: the paring is bilinear and produces a scalar, the pairing of a function with itself is nonnegative with value zero if and only if the zero function is paired with itself.

A natural question now arises from the analogy with matrix equations: Suppose that there is a u such that

$$\int_0^1 u_{xx}(x)\phi(x)\,dx = -\int_0^1 f(x)\phi(x)\,dx \tag{18.63}$$

for every continuous function ϕ. Is u a solution of the differential equation? The answer, ignoring the boundary conditions, is yes. But, it turns out that this preliminary reformulation of the problem is too crude to help with the goal of finding such a function u. One problem is immediate: a continuous function is not twice continuously differentiable, so the second derivative would not make sense if u is to be sought as an element of the vector space of continuous functions. A second difficulty is that the boundary conditions must somehow be incorporated.

As mentioned, the space of continuous functions is not the best place to seek a solution because its elements are not differentiable. To incorporate more smoothness, the obvious next choice is the set of all twice continuously differentiable functions on the unit interval. This is again a vector space with inner product (18.62). At a formal level, this choice has the advantage of at least containing viable candidates for u that might satisfy integral equation (18.63) for every twice continuously differentiable ϕ. Unfortunately, this vector space is also not the correct choice. The problem is not obvious and will require more mathematics to explain. A hint of the difficulty is the observation that inner product (18.62) does not take into account differentiation.

To begin the required analysis that leads to the correct function space, suppose that u and ϕ are smooth enough so that the integral on the left-hand side of Eq. (18.63) can be integrated by parts. This leads to the identity

$$u_x(x)\phi(x)\big|_0^1 - \int_0^1 u_x(x)\phi_x(x)\,dx = -\int_0^1 f(x)\phi(x)\,dx. \qquad (18.64)$$

There would be an advantage if this equation was to be satisfied instead of Eq. (18.63): only first derivatives appear and, at least, the desired function space need not consist of twice continuously differentiable functions. A good idea is to abandon the original smoothness requirement in favor of seeking a function u that satisfies Eq. (18.64) for every continuously differentiable function ϕ. In fact, the correct equation has been obtained; but, the smoothness requirement is still too strong.

Setting smoothness aside for the moment, consider the boundary term $u_x(x)\phi(x)|_0^1$ in Eq. (18.64). The spatial derivative of u with respect to x (the boundary traction) appears, but not the pure displacement u. Thus, boundary values for spatial derivatives (Neumann boundary conditions) appear naturally in the integration-by-parts formula for this type of BVP. For this reason, they are called natural boundary conditions. Boundary values for the displacement u (Dirichlet boundary conditions) must be built into the space of functions where the solution is to be found. They are called essential boundary conditions.

Zero Dirichlet boundary conditions (like $u(0) = 0$) are easily incorporated into a function space without destroying its vector space structure. Indeed, taking linear combinations of functions that satisfy these conditions results in a new function that satisfies the same boundary condition. More precisely, if ϕ_1 and ϕ_2 are functions that both vanish at the same point and c_1 and c_2 are scalars, then $c_1\phi_1 + c_2\phi_2$ vanishes at the same point. This fact is obviously false for nonzero Dirichlet conditions. For the case of nonzero Dirichlet boundary conditions, which will not be considered in detail here, the original BVP may be reformulated by looking for a solution $u = v + g$, where the function g is chosen to satisfy the boundary conditions so that v satisfies zero Dirichlet conditions. The function v is found in the manner that will be described and the true displacement is then recovered by adding g.

To incorporate the Dirichlet boundary condition $u(0) = 0$, every function in the desired function space must vanish at $x = 0$. The desired solution u has this property.

With the zero Dirichlet boundary condition at $x = 0$ in force so that the element ϕ of the yet to be defined function space vanishes at $x = 0$, Eq. (18.64) reduces to

$$\begin{aligned} -\int_0^1 f(x)\phi(x)\,dx &= u_x(1)\phi(1) - \int_0^1 u_x(x)\phi_x(x)\,dx \\ &= a\phi(1) - \int_0^1 u_x(x)\phi_x(x)\,dx. \end{aligned} \tag{18.65}$$

In the special case where the (scaled) traction vanishes ($a = 0$), the boundary term vanishes and the equation reduces to

$$\int_0^1 f(x)\phi(x)\,dx = \int_0^1 u_x(x)\phi_x(x)\,dx. \tag{18.66}$$

Integral equation (18.66) suggests the correct choice for the desired function space, which is denoted here by $H_D^1(0,1)$. Each function in the space must satisfy the (zero) Dirichlet boundary conditions where they are imposed; and, in addition, it must be square integrable on $[0,1]$ and have a square integrable derivative. The reason is that for two such functions, u and ϕ, both sides of Eq. (18.66) are finite. The relevant result, called the Schwarz inequality (Hermann A. Schwarz, circa 1888), states that

$$\int_0^1 \phi\psi\,dx \le \Big(\int_0^1 \phi^2\,dx\Big)^{1/2}\Big(\int_0^1 \psi^2\,dx\Big)^{1/2};$$

thus, if each integral on the right is finite (that is, both functions are square integrable), then the inner product on the left is also finite.

In view of the square integrability requirement, the natural norm (called the H^1 norm) for measuring the sizes of functions is

$$\|\phi\|_1 = \Big(\int_0^1 \phi^2(x)\,dx + \int_0^1 \phi_x^2(x)\,dx\Big)^{1/2}. \tag{18.67}$$

Remember that the distance between two functions ϕ and ψ, measured with respect to this norm, is simply $\|\phi - \psi\|_1$. By definition, ϕ approximates ψ with absolute error δ with respect to the H^1 norm if $\|\phi - \psi\|_1 < \delta$. Among the many possible ways to measurement distance between functions, the H^1

norm is perfectly suited to BVPs of the type discussed here. In mathematical analysis of BVPs, the correct choice of a norm produces a corresponding function space—consisting of functions that have finite norms—where a solution can be proved to exist.

Continuously differentiable functions on $[0, 1]$ are square integrable and have square integrable derivatives. Why not, as previously suggested, define the desired function space $H^1_D(0, 1)$ to be all such functions that satisfy the zero Dirichlet boundary conditions? The problem (which has been in the background until now) is that this set of functions is not complete. What does this mean?

Recall that the set of real numbers is complete. The definition of this property uses the notion of a Cauchy sequence: A sequence of real numbers $\{x_i\}_{i=1}^{\infty}$ is called a Cauchy sequence if for every $\epsilon > 0$ there is some integer J such that $|x_i - x_j| < \epsilon$ for every pair of integers i and j each of which is larger than J. The real numbers are complete in the sense that every Cauchy sequence converges to a real number. This property is essential to prove the basic theorems of calculus (for example, the intermediate value theorem). The definition of a Cauchy sequence of functions in some function space is the same as for real numbers except that the absolute value, which measures distance between real numbers, is replaced by the norm associated with the function space. The function space is called complete if every Cauchy sequence of its elements converges to some function in the same function space. Completeness is the key property needed to prove theorems using limit processes (which is the mathematically precise way to make arbitrarily close approximations), and limit processes are needed to prove existence theorems.

Cauchy sequences of continuously differentiable functions do not always converge to a continuously differentiable function when distances between functions are measured with H^1 norm (18.67). In other words, if the function space $H^1_D(0, 1)$ consisted entirely of continuously differentiable functions it would not be complete.

A remedy for the lack of completeness is to define a notion of differentiation for square integrable functions (which agrees with the usual definition when functions happen to be differentiable in the usual sense) in such a way that the inclusion of all such functions that have square integrable derivatives makes $H^1_D(0, 1)$ complete. This new notion, called the weak derivative, will be introduced in the next section where some nondifferentiable (in the usual sense) square integrable functions are defined that must be included in $H_D^1(0,1)$ to construct the discretization that gives the FEM.

To make the precise definition of the complete function space $H_D^1(0,1)$, another subtle issue must be resolved: the correct notion of integration must be specified. Riemann integration from calculus is inadequate for this purpose; it must be replaced by a more general concept called Lebesgue integration. Also, instead of functions, the elements of $H_D^1(0,1)$ must be equivalence classes of Lebesgue integrable functions that differ at most on a set of Lebesgue measure zero. Although some sophisticated mathematical analysis with new concepts is unavoidable in the rigorous construction of the weak formulation of BVPs, the full mathematical analysis is not required to understand the rules that must be followed to correctly apply the resulting methodology when seeking to specify or approximate weak solutions. In analogy, the true understanding of calculus as a mathematical theory of limit processes does not have to be understood to use this tool to solve applied problems. In fact, the name of this subject, the calculus, is meant to convey the meaning that it is a method useful for calculation. In both cases, the required mathematical facts can simply be stated and accepted without proof while the rules for applying the theory are learned. As the mathematical sophistication of the practitioner increases, these rules can be appreciated at a deeper level as their proofs are studied and understood. In applications, the elements of $H_D^1(0,1)$ may be considered to be functions and Riemann integration may be used to evaluate integrals.

Once the proper definitions of differentiation and integration are made and equivalence classes are properly defined, the function space $H_D^1(0,1)$ is defined to be the set of all (equivalence classes of) integrable functions with finite H^1 norm that satisfy the Dirichlet boundary conditions. *The important mathematical theorem states that $H_D^1(0,1)$ is complete with respect to the H^1 norm.* Actually, the space $H_D^1(0,1)$ is too simple compared with the analogous spaces defined when the unit interval is replaced by a region Ω in a higher-dimensional Euclidean space (usually $\mathbb{R}^3$): the elements of $H_D^1(0,1)$ are (represented by) continuous functions on the unit interval, but this result is not true for $H_D^1(\Omega)$. Because the purpose of the discussion is to introduce the concepts used for the general definition of such spaces, the special properties of $H_D^1(0,1)$ were ignored.

An important fact is that the H^1 norm can be derived from the H^1 inner product

$$\langle \phi, \psi \rangle_1 = \int_0^1 \phi(x)\psi(x)\,dx + \int_0^1 \phi_x(x)\psi_x(x)\,dx;$$

indeed,

$$\|\phi\|_1 = (\langle \phi, \psi \rangle_1)^{1/2}.$$

This fact, plays a fundamental role in the FEM where three different products are required: $\langle \phi, \psi \rangle_1$, the L^2 inner product

$$\langle \phi, \psi \rangle := \int_0^1 \phi(x)\psi(x)\,dx,$$

and the pairing

$$(\phi, \psi) := \int_0^1 \phi_x(x)\psi_x(x)\,dx.$$

The most important result of the discussion so far is the weak reformulation of the model BVP problem

$$u_{xx} + f = 0, \qquad u(0) = 0, \qquad u_x(1) = 0.$$

Find a function $u \in H_D^1(0,1)$ *such that*

$$(u, \phi) = \langle f, \phi \rangle + a\phi(1) \tag{18.68}$$

for every $\phi \in H_D^1(0,1)$. Such a function u is called a weak solution. BVPs for PDEs with several space variables have similar weak formulations.

A classical solution is obviously a weak solution. Also, if a weak solution is found that happens to be twice continuously differentiable, it is a classical solution of the original BVP. A deeper and amazing theorem states that a weak solution (when appropriately redefined on a set of measure zero) is automatically the unique classical solution of the original BVP. *This mathematical result is essential; it justifies seeking a weak solution of the BVP.* To fully appreciate it requires some graduate-level analysis (called elliptic regularity theory) explained in courses on PDEs (see [20] for more detail in the style of the present discussion, [102] for in context discussion of the FEM, or [35] for a general treatment).

The weak formulation of a BVP (such as in Eq. (18.68)) may be discretized in several different ways for the purpose of making finite-dimensional approximations of the unknown function u. This is an important remark. In particular, the weak formulation of a BVP is not the FEM; rather, the FEM is one among many possibilities of taking advantage of the weak formulation to make approximations (see Exercise 18.36).

18.7 ONE-DIMENSIONAL FINITE ELEMENT METHOD DISCRETIZATION

The discretization that produces the FEM is obtained by seeking a solution that belongs to a finite-dimensional subspace of the (infinite-dimensional) function space in which the problem is posed; for instance, BVP (18.68) is posed in the function space is $H_D^1(0,1)$.

For BVPs posed in $H_D^1(0,1)$, a possible choice for an appropriate subspace might be the d-dimensional subspace of all polynomials of maximum degree d that have no constant term (to ensure the Dirichlet boundary condition at $x = 0$). This set is obviously a subspace of $H_D^1(0,1)$. Why? Using the weak formulation of the BVP, the idea would be to find a degree-d polynomial u such that Eq. (18.68) is satisfied for every polynomial ϕ of degree d. This requirement reduces to solving a linear system of $d+1$ equations. This can be a viable approximation method (see Exercise 18.36). Another choice is to split the spatial domain $[0,1]$ into a finite set of closed subintervals so that this set has a finite number of elements and consider the continuous functions that are piecewise polynomials (all of some fixed degree) with respect to the chosen intervals. Each of these functions (once the meaning of differentiation for such functions is properly defined) has a finite H^1 norm, and by restricting to those that vanish at $x = 0$, the resulting set of functions forms a finite-dimensional subspace of $H_D^1(0,1)$. By the same procedure (using Eq. (18.68)), an element of this class u can be found by solving a finite-dimensional linear system of equations. This is the FEM for approximating solutions of the BVPs. The approximation can be improved by using larger and larger numbers of subintervals.

For one spatial dimension, the usual way to form finite elements is via a partition of the spatial domain (in this case $[0,1]$)

$$0 = x_0 < x_1 < x_2 < \cdots < x_{n-1} < x_n = 1. \tag{18.69}$$

The corresponding finite elements are the closed subintervals $[x_{i-1}, x_i]$ for $i = 1, 2, \ldots, n$. The simplest viable set of piecewise polynomial functions is the finite-dimensional vector space $\mathcal{P}_n$ consisting of all continuous *piecewise linear polynomials* with respect to this partition that vanish at $x = 0$; more precisely, each member of the set $\mathcal{P}_n$ is a continuous function that vanishes at $x = 0$ and is also a polynomial of degree one on each finite element. When the correct notion of differentiation is defined, each element of $\mathcal{P}_n$ has finite H^1 norm and satisfies the zero Dirichlet boundary condition. Thus, this finite-dimensional vector space, $\mathcal{P}_n$, is a subspace of $H^1_D(0,1)$.

A function $\phi \in \mathcal{P}_n$ is by definition continuous on the closed unit interval; thus, (by the completeness property of the real numbers) such a function must be bounded. (This is a theorem from advanced calculus). These two properties (continuous and bounded on $[0,1]$) imply that such a function is square integrable. Actually, the advanced calculus theorem on boundedness is not needed in the simple case of a piecewise linear function on $[0,1]$ with respect to a finite collection of closed subintervals. Such a function is obviously bounded. Why? In any case, if $\phi \in \mathcal{P}_n$, then

$$\int_0^1 \phi^2\, dx < \infty;$$

that is, the function is square integrable.

The derivative of a function $\phi \in \mathcal{P}_n$ exists in the usual sense at each point in the interior of each finite element; in fact, the derivative is constant over the interior of each such interval. But at an end point of an element (called a node) the slope of ϕ may not be defined. A new notion of differentiation is required to make sense of the integral involving derivatives in the definition of the H^1 norm. The new notion, which has far-reaching utility, is not designed specifically for piecewise linear functions. They merely serve as an example to set a context where the new definition is useful.

The motivation for the new definition is integration by parts, which has already played a central role in the analysis leading to the definition of $H^1_D(0,1)$. Let ϕ be a continuously differentiable function on $[0,1]$ and suppose that ψ is a function that can be differentiated as many times as desired and has the additional property that it vanishes at the end points of the interval. Functions with these latter properties are called test functions.

Using integration by parts,

$$\int_0^1 \phi'\psi\,dx = \phi\psi|_0^1 - \int_0^1 \phi\psi'\,dx = -\int_0^1 \phi\psi'\,dx.$$

The left-hand side involves a derivative of ϕ; the right-hand side does not. Moreover, when paired with a test function, an integrable function ϕ always produces a finite value. Thus, the paring defines a function from the set of test functions to the real numbers:

$$\psi \mapsto \int_0^1 \phi\psi\,dx. \tag{18.70}$$

A function whose domain is a set of functions and whose range is a set of scalars is often called a functional; for instance, function (18.70) is a functional on the space of test functions. Using this prescription, an integrable function ϕ produces a functional on the space of test functions. The same function ϕ can also be used to define another functional,

$$\psi \mapsto -\int_0^1 \phi\psi'\,dx,$$

called its the distributional derivative. Using this definition, every functional defined by an integrable function via rule (18.70) has a distributional derivative. In case ϕ is differentiable (in the usual sense), integration by parts implies that the distributional derivative of the functional it produces is also given by

$$\psi \mapsto \int_0^1 \phi'\psi\,dx.$$

Thus, the distributional derivative of ϕ in this case is what it should be: the functional on test functions produced by its derivative. The concept of a distributional derivative turns out to be far-reaching.

Suppose that ϕ is square integrable. Repeating what has already been said, ϕ produces a functional on test functions that has a distributional derivative. It might happen that there is a square integrable function γ defined on [0,1] such that

$$\int_0^1 \gamma\psi\,dx = -\int_0^1 \phi\psi'\,dx. \tag{18.71}$$

In other words, the functionals

$$\psi \mapsto \int_0^1 \gamma\psi\, dx, \qquad \psi \mapsto -\int_0^1 \phi\psi'\, dx$$

are the same. In this case, we say that γ is a square integrable weak derivative of ϕ or that ϕ has a square integrable derivative in the sense of distribution. As always, if ϕ is differentiable in the usual sense, then its derivative will be a square integrable weak derivative. In fact, a square integrable weak derivative is unique (up to changes on a set of measure zero). So, the usual derivative (if it exists) and the weak derivative coincide.

The space of functions $H_D^1(0,1)$ is the set (of equivalence classes) of all square integrable functions with domain $[0,1]$ (which differ at most on a set of measure zero) that have a square integrable weak derivative and vanish at $x = 0$. Replacing ϕ_x in the definition of the H^1 norm [Eq. (18.67)] by its weak square integrable derivative removes the ambiguity in case ϕ is not differentiable in the usual sense and allows a wider class of functions to have finite H^1 norms. *This wider class is exactly the space $H_D^1(0,1)$, which is complete with respect to the H^1 norm.*

Every continuous piecewise polynomial function on the unit interval has a square integrable weak derivative (see Exercise 18.28). (Warning: This fact is not true without the continuity assumption.) Thus, every continuous piecewise polynomial function that vanishes where a zero Dirichlet boundary condition is imposed belongs to the function space $H_D^1(0,1)$. In particular, $\mathcal{P}_n$ may be considered as a candidate for the finite-dimensional subspace used to make the desired finite element discretization. For u to be a weak solution of the BVP, Eq. (18.68) must hold for every $\psi \in H_D^1(0,1)$. To obtain an approximate of u, simply require the equation to hold for every $\psi \in \mathcal{P}_n$. By increasing n toward infinity, the approximation should (and will) approach the desired weak solution. By the elliptic regularity theory, this solution is actually the desired classical solution of the BVP. In its simplest form, this is the underlying theory for the FEM.

The most basic implementation of the FEM is an efficient way to obtain the approximation to u over the subspace $\mathcal{P}_n$. It is not necessary to check Eq. (18.68) on every element of the finite-dimensional subspace. As pointed out previously, it suffices to check the equation on a basis of this finite-dimensional vector space. A careful choice of basis, a key element in the implementation of the FEM, lessens the computational overhead required to produce a FEM approximation.

Recall partition (18.69), assume that $n > 2$, and consider the interval $[x_0, x_2]$ (which includes three nodes). The first basis function in $\mathcal{P}_n$ is

$$\phi_1(x) = \begin{cases} \frac{x-x_0}{x_1-x_0}, & x \in [x_0, x_1]; \\ \frac{x-x_2}{x_1-x_2}, & x \in (x_1, x_2]; \\ 0, & x \in (x_2, x_n]. \end{cases} \tag{18.72}$$

Note that $\phi_1(0) = 0$, $\phi(x_1) = 1$ and $\phi(x_2) = 0$. The graph looks like a tent supported by a tent pole of unit height at the node $x = x_1$. For $i = 2, \ldots n-1$, take similar tent functions defined by

$$\phi_i(x) = \begin{cases} 0, & x \in [x_0, x_{i-1}); \\ \frac{x-x_{i-1}}{x_i-x_{i-1}}, & x \in [x_{i-1}, x_i]; \\ \frac{x-x_{i+1}}{x_i-x_{i+1}}, & x \in (x_i, x_{i+1}]; \\ 0, & x \in (x_{i+1}, x_n]. \end{cases}$$

And, for $i = n$, use the half-tent

$$\phi_i(x) = \begin{cases} 0, & x \in [x_0, x_{n-1}); \\ \frac{x-x_{n-1}}{x_n-x_{n-1}}, & x \in [x_{n-1}, x_n]. \end{cases}$$

The set $\mathcal{B}_n := \{\phi_1, \phi_2, \ldots, \phi_n\}$ is a basis for the vector space $\mathcal{P}_n$ (see Exercise 18.31). As a consequence, this vector space is n-dimensional.

The primary reason for choosing the basis $\mathcal{B}_n$ is localization: the basis functions are zero except on sets that are three nodes wide. As a result, many of the pairings in the weak formulation [Eq. (18.68)] of the BVP vanish. Indeed, if $|i - j| \geq 2$, then $(\phi_i, \phi_j) = 0$ and $\langle \phi_i, \phi_j \rangle = 0$.

The final step in setting up the FEM is to seek a weak solution u as a linear combination of the basis functions; that is,

$$u = \sum_{i=1}^{k} u_i \phi_i,$$

or equivalently, $u = u_i \phi_i$ using the sum rule, where u_i are the unknown components of an n-dimensional vector of real numbers. There are n equations in n unknowns:

$$(u_i \phi_i, \phi_j) = \langle f, \phi_j \rangle + a\phi_j(1),$$

or equivalently,

$$(\phi_i, \phi_j)u_i = \langle f, \phi_j \rangle + a\phi_j(1).$$

In matrix form, the problem is to determine the vector $U := (u_1, u_2, \ldots, u_n)$ such that

$$AU = \beta \tag{18.73}$$

where $a_{ij} := (\phi_j, \phi_i)$ and $\beta_j := \langle f, \phi_j \rangle + a\phi_j(1)$. Note that the correct definition of A is given. It is the transpose of the matrix with components $c_{ij} := (\phi_i, \phi_j)$. But because the pairing (ϕ_i, ϕ_j) is symmetric (that is, $(\phi_i, \phi_j) = (\phi_j, \phi_i)$), the ordering does not matter.

The final matrix equation $AU = \beta$ need not be solved by taking inner products with basis functions; instead, it may be solved using the most efficient available method. More accuracy is obtained by refining the mesh. In fact, if the original BVP is well posed and the sequence of approximation vectors U^k corresponding to $(n+1)$ node partitions is computed, the corresponding functions $u^k(x) := u_i^k(x)\phi_i(x)$ will converge to a (classical) solution of the original BVP.

Exercise 18.28. (a) Show that the function ϕ defined by $\phi(x) = 2x$ for $0 \le x < 1/2$ and $\phi(x) = 2-2x$ for $1/2 \le x \le 0$ is square integrable and has a weak square integrable derivative. Also, determine the weak derivative explicitly and compute the H^1 norm of ϕ. (b) Generalize part (a) to arbitrary continuous piecewise linear functions on $[0, 1]$. (c) Is the function ϕ given by $\phi(x) = 0$ for $0 \le x < 1/2$ and $\phi(x) = 1$ for $1/2 \le x \le 0$ square integrable? Does it have a weak square integrable derivative?

Exercise 18.29. Instead of using tent functions at each node, many other choices are possible for the FEM. One important choice is localized cubic splines. Note that the general cubic polynomial has four coefficients. Thus for instance, there are exactly enough coefficients to specify the height and slope of a cubic polynomial at two points on the line. (a) Suppose some data is given over a finite set of nodes on some interval. Construct a *twice* continuously differentiable piecewise cubic interpolating function for the data. Hint: Construct a cubic through each pair of nodes such that the left- and right-hand first and second derivatives match at each interior node. This exercise is an invitation to learn more about spline interpolation, an important subject in applied mathematics and numerical analysis. (b) Construct a finite-element basis consisting of cubic splines of three types: (1) smooth functions that vanish outside of the union of two adjacent linear elements (intervals) and have unit value at the interior common node, (2) smooth functions that vanish outside two adjacent linear elements and have unit slope at the interior common node and (3) smooth functions that vanish outside one of the end linear elements and also satisfy the specified essential boundary condition at the

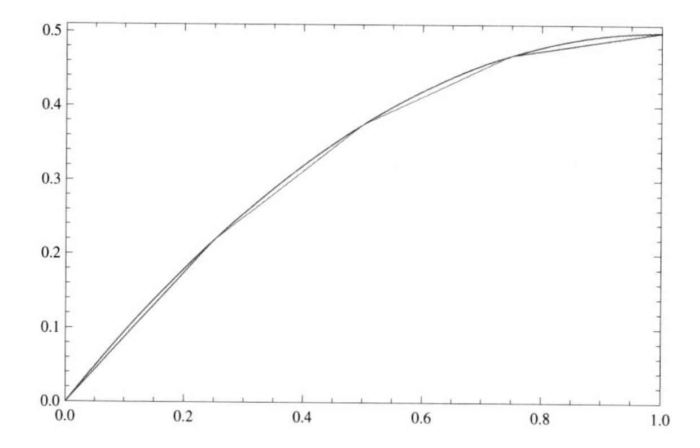

Fig. 18.2 The figure depicts the graph of the exact displacement function $u(x) = x - x^2/2$ predicted by model (18.60) for a one-dimensional elastic material (clamped at one end and free at the other) subjected to a constant body force and the graph of its finite element approximation using the four-dimensional tent function basis with four (closed interval) elements.

end point of that element. Can these basis function be made to have continuous second derivatives? (c) Compare the accuracy and efficiency of using cubic splines versus tent functions for FEM approximations of one-dimensional steady state BVPs with known solutions. Hint: There is a vast literature on this subject and many textbooks are available. Read more about it before completing your project. (d) Write a computer code using the ideas in the next section to illustrate your findings. Note: Spline interpolation can be generalized to higher dimensional settings.

18.8 CODING FOR THE ONE-DIMENSIONAL FINITE ELEMENT METHOD

Coding the FEM for elasticity problem (18.60) via its weak formulation [Eq. (18.68)] reduces to computing the matrix A and the vector β in the finite-dimensional approximation as in Eq. (18.73) and then solving the matrix equation $AU = \beta$ for U.

Recall that for the choice of basis (tent) functions $\mathcal{B}_n := \{\phi_1, \phi_2, \ldots, \phi_n\}$, the condition $|i - j| \geq 2$ ensures that $(\phi_i, \phi_j) = 0$ and $\langle \phi_i, \phi_j \rangle = 0$. Thus, the matrix A is tridiagonal; that is, all nonzero entries lie on the main diagonal, the first superdiagonal, and the first subdiagonal. This fact is one of the main reasons for the efficiency of the FEM.

Basic results in the one-dimensional case may be obtained with rudimentary codes. The reader is invited to consider model (18.60) for $f(x) \equiv 1$ and $a = 0$ (which may be interpreted as a constant body force [for example gravity] and zero traction) and compute the approximate solution using the tent function basis with five nodes $0, 1/4, 2/4, 3/4, 1$ and the corresponding four tent basis functions. In this case, the 4×4 matrix A, the 4-vector β, and the 4-dimensional vector solution U of the matrix equation $AU = \beta$ are

$$A = \begin{pmatrix} 8 & -4 & 0 & 0 \\ -4 & 8 & -4 & 0 \\ 0 & -4 & 8 & -4 \\ 0 & 0 & -4 & 4 \end{pmatrix}, \qquad \beta = \begin{pmatrix} 1/4 \\ 1/4 \\ 1/4 \\ 1/8 \end{pmatrix}, \qquad U = \begin{pmatrix} 7/32 \\ 3/8 \\ 15/32 \\ 1/2 \end{pmatrix}.$$

The graphs of the exact and approximate displacement functions are depicted in Fig. 18.2.

A systematic approach to coding for the FEM begins with the choice of elements, their nodes, and the basis functions. The objective of the code is to construct the matrix A and the vector β. Once this task is completed, the linear system $AU = \beta$ is ported to a linear equation solver, which returns an approximation of the vector U. The desired approximate displacement is the sum $u_i\phi_i$, where u_i is the ith component of U and ϕ_i is the ith basis function. The process (called meshing) of determining the elements and nodes may be done partly by hand or it may be fully automated. Perhaps a course mesh is designed by hand and a computer program is used to refine the mesh. Once the mesh is constructed, the key part of the code (called assembly) produces approximations of the integrals required to form the matrix A and the vector β. The solution of the assembled matrix system is approximated using an appropriate linear solver to make the FEM approximation of the solution of the BVP. In most applications, this approximate displacement is sent to a postprocessor that might produce a graph or otherwise utilize the result.

Assembling A requires the integration of products of basis functions or products of derivatives of these functions. This task is usually accomplished by integrating over each element separately followed by addition of the computed values.

Suppose that there are n basis functions ϕ_j, for $j = 1, 2, 3, \ldots, n$ and m elements I_j, for $j = 1, 2, 3, \ldots, m$. In the one-dimensional case $m = n$, but generally there are more basis functions than elements. A typical computation of a component of the matrix A requires a paring of basis functions; for example,

$$\langle \phi_i, \phi_j \rangle = \int_0^1 \phi\psi \, dx = \sum_{k=1}^{m} \int_{I_k} \phi_i \phi_j \, dx. \tag{18.74}$$

Most of the summands vanish due to the local nature of the basis functions. The remaining integrals are often organized and evaluated using one new concept: shape functions.

Recall that the basis (tent functions) are defined with respect to the nodes. Indeed, there is one basis function defined for each node where no zero Dirichlet boundary condition is imposed. Each basis function has value one at its corresponding node, is continuous and piecewise linear, and has value zero on every element not containing this node as a vertex. The restriction of one of these tent functions to an element I is called a shape function associated with I. For the case of one-dimensional elements, each of them has two possible shape functions corresponding to its two vertices p and q at the end points of the element. Shape functions are linear. Moreover, for the one-dimensional case, the shape function on I associated with vertex p has value one at p and value zero at q; the other shape function has value one at q and value zero at p.

A typical integral

$$\int_{I_k} \phi_i \phi_j \, dx$$

that appears as a summand in Eq. (18.74) is computed after checking the nodes associated with the element I_k. The ith and jth nodes correspond to basis functions. If I_k does not contain both nodes as vertices, then the integral vanishes. If both nodes are vertices, then

$$\int_{I_k} S_i^k S_j^k \, dx, \tag{18.75}$$

where S_i^k is the shape function associated with the ith node and kth element and S_j^k is the shape function associated with the jth node and kth element. Thus, the method used to evaluate the integrals required to compute the elements of A may be described using shape functions instead of basis elements. The advantage of this point of view is that shape functions are associated directly with elements and their nodes. Of course, the reason for using shape functions is to more efficiently organize required computations to avoid the wasted expense of computing an integral that is known to be zero.

The usual procedure starts by defining a standard reference element. For the one-dimensional case, this might be the interval $[0, 1]$ with vertices 0 and 1. The usual orientation is assumed; that is, from 0 to 1. The shape function corresponding to 0 is $s_0(x) = 1 - x$ and the shape function corresponding to 1 is $s_1(x) = x$. An element I_k in the discretization of the body Ω is an interval (x_{k-1}, x_k); the orientation is built into the notation: the node x_{k-1} corresponds to the first node on the element and is referenced to 0; the second node x_k is referenced to 1. The orientation preserving linear transformation T given by

$$T(x) = \frac{x - a}{b - a}$$

maps the element I_k to the standard element, where a and b are the nodes (in the correct order) on the element. The shape function S^k_{k-1} (node number $k - 1$ and element number k) is defined by

$$S^k_{k-1}(x) = s_0(T(x)) = s_0((x - x_{k-1})/(x_k - x_{k-1})) = 1 - \frac{x - x_{k-1}}{x_k - x_{k-1}}$$

and

$$S^k_k(x) = \frac{x - x_{k-1}}{x_k - x_{k-1}}.$$

Exactly the same procedure is used in the higher-dimensional cases where this added structure is more advantageous.

To summarize: The collection of nodes and elements are stored in the computer along with the orientations of the elements and the correspondence between the nodes and elements. A standard element is defined along with shape functions for each of its nodes. The values of a basis function on an element are determined via transformation to the standard element, the node associated with the basis function, and its corresponding shape function on the standard element. Efficient implementation of this method of calculating appropriate integrals of products of basis functions is in general an interesting and challenging problem in computer programming, but in one space-dimension a simple implementation is straightforward.

Exercise 18.30. (a) Show that the space $\mathcal{P}_n$ of continuous piecewise linear polynomials with respect to the partition

$$0 = x_0 < x_1 < x_2 < \cdots < x_{n-1} < x_n = 1$$

of the interval $[0, 1]$ that vanish at the origin is an n-dimensional subspace of $H_D^1(0, 1)$.
(b) Does it matter that the polynomials are continuous?

Exercise 18.31. Show that the set $\mathcal{B}_n := \{\phi_1, \phi_2, \ldots, \phi_n\}$, defined in this section, is a basis for the vector space $\mathcal{P}_n$ defined in Exercise 18.30.

Exercise 18.32. [One Dimensional Finite Element Coding] (a) Write a finite element code to approximate solutions of one-dimensional BVPs for PDEs of the form

$$u_{xx} + g(x)u + h(x) = 0$$

defined on an interval $[a, b]$ where Dirichlet or Neumann boundary conditions are imposed at the end points of the interval.
(b) Debug your code by repeating the numerical experiment reported in Fig. 18.2.
(c) Solve the BVP $u_{xx} = 1$ with zero Dirichlet boundary conditions $u(0) = 0$ and $u(1) = 0$ and use your code to approximate the solution with an accuracy of less than 10^{-5} in the supremum norm (that is, the absolute difference between the approximate and exact solutions should be less than 10^{-5} at every point). What is the least number of equal length elements required to achieve this accuracy?
(d) Could better accuracy be achieved with fewer elements if the elements are not restricted to be all the same length? (e) Repeat parts (c) and (d) for the boundary conditions $u(0) = 0$ and $u(1) = 2$. Be careful: What are you going to do about the nonzero Dirichlet boundary condition at $x = 1$?
(e) Debug your code a second time by creating a BVP with a known solution by choosing a function u defined on the interval $[0, 1]$ that satisfies (for example) zero Dirichlet boundary conditions at the end points of the interval, choosing a nonzero function g defined on the same interval, and defining $h = -(u_{xx} + gu)$. Use your code to approximate the known function u as a solution of the BVP with PDE $u_{xx} + g(x)u + h(x) = 0$ together with the boundary conditions satisfied by the choice of u. Discuss the results of your numerical experiments.
(f) Compare your approximations for part (e) with an approximation made using a finite-difference method.
(g) Compare your approximations for part (e) with an approximation made using the shooting method. That is, set up a method to approximate the solution of the ODE as an IVP with $u(0) = 0$ and $u_x(0) = \lambda$, and adjust λ with the goal of making $u(1, \lambda) = 0$.
(h) Approximate the solution of the BVP $u_{xx} + \sin(x)u = 1$ with zero Dirichlet boundary conditions at $x = 0$ and $x = 2\pi$ using your finite element code.
(i) Approximate the solution of the BVP $u_{xx} + \sin(x)u = 1$ with zero Dirichlet boundary condition at $x = 0$ and zero Neumann boundary condition at $x = 2\pi$.

Exercise 18.33. (a) Create a mathematical model (using linear elasticity) for the distortion in a one-dimensional elastic material that is hung vertically by attaching one of its ends to a rigid support and its other end to a mass. (b) Determine the distortion in equilibrium. (c) Determine the dynamic motion if the mass is pulled downward and then released from rest. Hint: Perhaps your model has an exact solution. If not, solutions of the dynamic BVP can be approximated by spatial discretization using the FEM and stepping forward in time using, for example, Euler's method or the Störmer–Verlet method (as in Exercise 10.9).

Exercise 18.34. Imagine a one-dimensional elastic material clamped at one end and immersed in some other elastic material at the other end. Determine a boundary condition at the latter end that might be used as a model. Solve the model equation and determine the displacement and stain field. Hint: The traction force at the end depends on the displacement.

Exercise 18.35. A two kilogram mass is attached to a hanging copper wire that is one meter long (and 10 milimeters in diameter). Determine the displacement field and the stress field along the wire in the vertical direction.

Exercise 18.36. [Alternative Galëkin Method]. The weak formulation of BVPs is itself often called Galërkin's method. As mentioned in the text, the FEM is just one possibility for making approximations of the weak solution. A key ingredient is choosing a finite-dimensional subspace of the test functions, represent the approximate solution as the components of a linear combination of basis functions, and turn the weak equations into a linear system ($AU = \beta$) for the unknown vector U of components. Of course, the finite-dimensional subspace should be chosen as a member of a sequence of increasing larger subspaces (the closure of) whose union is the entire space. For example, instead of taking a space consisting of linear combinations of tent functions associated with finite elements, we might consider all polynomial test functions of some fixed degree that also satisfy the Dirichlet boundary conditions. (a) Consider the hanging elastic material pinned at a point and with no traction. We already know that the exact solution is a quadratic polynomial. Choose the set of all quadratic polynomials that vanish at $x = 0$, and look for a solution u of the form $u(x) = U_1 x + U_2 x^2$. Put this into the weak formulation and test it against the basis functions $\phi_1(x) = x$ and $\phi_2(x) = x^2$ to obtain a system of two equations in the two unknowns U_1 and U_2. Solve the system and recover the known solution.
(b) Note that the coefficients of the system matrix for your system of two equations is full; that is, all the coefficients are not zero. If you were to ignore the known solution and look for u as a 10th-degree polynomial that vanishes at the origin, you will end up with a 10×10 system of linear equations. The system matrix will be full. Thus, the solution method for solving such a matrix system cannot take advantage of the sparse (tridiagonal) matrix that appears in the FEM. This is one reason why the FEM is used: it is computationally efficient. Now for the problem: Is it possible to choose the basis functions for spaces of polynomial test functions (all polynomials of some fixed degree that vanish corresponding to the Dirichlet boundary conditions) in a manner that would make the system matrix be sparse? Hint: There is a mathematical subject called "orthogonal polynomials."
(c) The additional computation required to solve a matrix system with a full system matrix may be manageable in some applications; for example, extra computation time may be acceptable in case a system needs to be solved only once. Full or not, solving matrix systems in applied problems may present additional difficulties. Note that the

system matrix for polynomial trial functions in the usual polynomial basis $\{x^i\}_{i=1}^k$ (for polynomials that vanish at $x = 0$) has components

$$a_{ij} = \int_0^1 x^i x^j \, dx = \frac{1}{1+i+j}.$$

For the n-dimensional vector b with components $b_i = 1$, $i = 1, 2, 3, \ldots, n$, use an algebraic processor to solve the system $Ax = b$ for $n = 2, 3, 4, \ldots, 20$. Use a numerical solver (that is floating point arithmetic up to a fixed finite number of decimal places) to solve the same system. Compare, for example, the exact and numerical sequence of values of x_n, the nth component of the solution. Report on your findings. You should see a rapid deterioration in accuracy as n increases. The coefficient matrix (called the Hilbert matrix) is highly ill-conditioned. A small roundoff error, for example, will be magnified in the solution by the structure of the matrix. What is the underlying reason for the ill-conditioning? How can ill-conditioned system matrices be detected? Hint: Consult a book on numerical linear algebra.

18.9 WEAK FORMULATION AND FINITE ELEMENT METHOD FOR LINEAR ELASTICITY

Background for this section includes an understanding of the general equations of linear elasticity, the one-dimensional weak reformulation of one-dimensional BVPs, and the one-dimensional FEM. By way of motivation, the reader should know that the FEM was invented to approximate solutions of linear elasticity models.

The fundamental problem of steady state elasticity is to determine the displacement u of a three-dimensional elastic body Ω with a (piecewise) smooth boundary $\partial\Omega$ taking into account body forces per mass b, constrains on the motion of the boundary of the body, and normal and surface tractions on the boundary of the body. In case the deformations are not too large, this physical problem is usually modeled using linear elasticity theory; that is, the mathematical model is formulated as a BVP for the PDE

$$\sigma_{ij,j} + \rho b_i = (\lambda \varepsilon_{kk} \delta_{ij} + 2\mu\varepsilon_{ij})_{,j} + \rho b_i = 0, \tag{18.76}$$

where

$$\varepsilon_{ij} = \frac{1}{2}(u_{i,j} + u_{j,i}),$$

the Lamé constants λ and μ reflect the elastic properties of the body, summation on repeated indices is assumed, and this PDE is required to hold everywhere in the interior of Ω. The displacement field u is a vector function of a vector variable with respect to a fixed three-dimensional rectangular coordinate system. Appropriate boundary conditions must also be imposed.

Each smooth point on the piecewise smooth boundary $\partial\Omega$ has an outer unit normal η and a choice of two orthogonal unit vectors s_1 and s_2 tangent to the surface at this point.

Two types of boundary conditions may be imposed at each boundary point. (1) Dirichlet conditions are given by

$$u \cdot s_1 = d_1, \qquad u \cdot s_2 = d_2, \qquad u \cdot \eta = d_3, \tag{18.77}$$

where d_1, d_2, and d_3 are constants and the dot denotes the usual inner product in three-dimensional space. Each formula states that the displacement is specified in one direction. Nonzero choices of d_1, d_2, or d_3 are allowed, but only zero Dirichlet boundary conditions (that is, $d = 0$) will be discussed here. (2) Neumann boundary conditions are given by

$$\begin{aligned}
a_1 &= \tau \cdot s_1 = \sigma_{ij}\eta_j(s_1)_i = [\lambda(\nabla \cdot u)\eta + \mu((\nabla u)^T + \nabla u)\eta] \cdot s_1, \\
a_2 &= \tau \cdot s_2 = \sigma_{ij}\eta_j(s_2)_i = [\lambda(\nabla \cdot u)\eta + \mu((\nabla u)^T + \nabla u)\eta] \cdot s_2, \\
a_3 &= \tau \cdot \eta = \sigma_{ij}\eta_j\eta_i = [\lambda(\nabla \cdot u)\eta + \mu((\nabla u)^T + \nabla u)\eta] \cdot \eta,
\end{aligned} \tag{18.78}$$

where a_1, a_2, and a_3 are specified tractions in the corresponding directions. Six possible boundary conditions are available at each boundary point, and each boundary condition is associated with a direction.

The fundamental BVP is to find a displacement u that solves PDE (18.76) *when exactly one of the six available boundary conditions [Eqs.* (18.77) *and* (18.78)*] is specified for each of the three associated directions at each point on the smooth part of the piecewise smooth boundary.* In particular, there are three (scalar) boundary conditions at each such point.

Careful consideration of boundary conditions is essential to the modeling process and the application of the mathematics used to solve BVPs. An appropriate choice of rectangular coordinate system might simplify some assignments of boundary conditions and the mathematical analysis. Of course, circular, cylindrical, or spherical symmetry is often best approached by using polar, cylindrical, or spherical coordinates together with corresponding transformations of the differential equations and boundary conditions.

A weak formulation of the fundamental BVP is obtained (as in the one-dimensional case) via integration by parts. As we will see, the natural (Neumann) boundary conditions appear automatically in the resulting formulas. As in the one-dimensional case, the correct function space is $H^1_D(\Omega)$. Here, its elements are triples $\phi = (\phi_1, \phi_2, \phi_3)$ of functions where each component function is square integrable with square integrable first partial derivatives (over Ω with respect to the space variables) and satisfies the zero essential (Dirichlet) boundary conditions that are specified in the particular BVP under consideration.

The weak BVP is derived by taking products with test functions and integrating by parts. For an arbitrary test function ϕ (which will later be replaced by an element in $H^1_D(\Omega)$), take the usual inner product of its value at each point with the value at the same point of the left- and right-hand sides of the PDE and integrate over the body Ω to obtain in components

$$\int_\Omega \sigma_{ij,j}\phi_i + \rho b_i \phi_i \, d\mathcal{V} = 0, \tag{18.79}$$

where of course summation on repeated indices is implied. Using the product rule, we have the identity

$$\int_\Omega (\sigma_{ij}\phi_i)_{,j} \, d\mathcal{V} = \int_\Omega \sigma_{ij,j}\phi_i \, d\mathcal{V} + \int_\Omega \sigma_{ij}\phi_{i,j} \, d\mathcal{V}. \tag{18.80}$$

The first integrand is the divergence of the vector field $\sigma_{ij}\phi_i$, given in components. By the divergence theorem

$$\int_\Omega (\sigma_{ij}\phi_i)_{,j} \, d\mathcal{V} = \int_{\partial\Omega} \sigma_{ij}\phi_i \eta_j \, d\mathcal{S}, \tag{18.81}$$

where η is the outer unit normal. Using these results, Eq. (18.79) is equivalent to

$$\int_\Omega \sigma_{ij}\phi_{i,j} \, d\mathcal{V} = \int_{\partial\Omega} \sigma_{ij}\eta_j \phi_i \, d\mathcal{S} + \int_\Omega \rho b_i \phi_i \, d\mathcal{V} \tag{18.82}$$

provided that all of the functions involved are smooth enough to reverse the integration by parts. This equation is correct, but the left-hand side is not expressed symmetrically. To symmetrize, note that

$$\int_\Omega \sigma_{ij}\frac{1}{2}(\phi_{i,j} + \phi_{j,i}) \, d\mathcal{V} = \frac{1}{2}\int_\Omega \sigma_{ij}\phi_{i,j} \, d\mathcal{V} + \frac{1}{2}\int_\Omega \sigma_{ij}\phi_{j,i} \, d\mathcal{V}$$

$$= \frac{1}{2} \int_\Omega \sigma_{ij} \phi_{i,j} \, d\mathcal{V} + \frac{1}{2} \int_\Omega \sigma_{ji} \phi_{i,j} \, d\mathcal{V}$$

and the stress tensor σ is symmetric ($\sigma_{ij} = \sigma_{ji}$); thus,

$$\int_\Omega \sigma_{ij} \frac{1}{2} (\phi_{i,j} + \phi_{j,i}) \, d\mathcal{V} = \int_\Omega \sigma_{ij} \phi_{i,j} \, d\mathcal{V}.$$

Thus, the symmetric form of Eq. (18.82) is

$$\int_\Omega \sigma_{ij} \frac{1}{2} (\phi_{i,j} + \phi_{j,i}) \, d\mathcal{V} = \int_{\partial\Omega} \sigma_{ij} \eta_j \phi_i \, d\mathcal{S} + \int_\Omega \rho b_i \phi_i \, d\mathcal{V}, \tag{18.83}$$

or with the definitions,

$$\begin{aligned}
(u, \phi) &:= \int_\Omega \lambda u_{k,k} \phi_{\ell,\ell} + \frac{\mu}{2} (u_{i,j} + u_{j,i})(\phi_{i,j} + \phi_{j,i}) \, d\mathcal{V}, \\
[u, \phi] &:= \int_{\partial\Omega} (\lambda u_{k,k} \delta_{ij} + \mu(u_{i,j} + u_{j,i})) \eta_j \phi_i \, d\mathcal{S}, \\
\langle \phi, \psi \rangle &:= \int_\Omega \phi_i \psi_i \, d\mathcal{V},
\end{aligned} \tag{18.84}$$

it is the equivalent equation

$$(u, \phi) = \langle \rho b, \phi \rangle + [u, \phi] \tag{18.85}$$

that plays a fundamental role in the analysis to follow.

The vector form of Eq. (18.83) requires a new definition: The Frobenius inner product of two matrices A and B with components a_{ij} and b_{ij}, respectively, is the scalar

$$A : B := a_{ij} b_{ij}.$$

Using this definition, the vector forms of the round and square bracket parings are

$$\begin{aligned}
(u, \phi) &= \int_\Omega \lambda \nabla \cdot u \nabla \cdot \phi + \frac{\mu}{2} (\nabla u + (\nabla u)^T) : (\nabla \phi + (\nabla \phi)^T) \, d\mathcal{V}, \\
[u, \phi] &= \int_{\partial\Omega} (\lambda (\nabla \cdot u) \eta + \mu ((\nabla u)^T + \nabla u) \eta) \cdot \phi \, d\mathcal{S}.
\end{aligned} \tag{18.86}$$

The weak BVP is to find a (vector) function u in the space $H^1_D(\Omega)$ of (vector) square integrable functions on Ω that satisfies Eq. (18.85) for all (vector) functions ϕ in $H^1_D(\Omega)$, where zero Dirichlet boundary *conditions are incorporated into the definition of $H^1_D(\Omega)$ and surface tractions (Neumann boundary conditions) are specified so that $[u, \phi]$ can be computed from this data (before solving for u) for every ϕ at every node in every direction where no Dirichlet boundary condition is imposed.*

Dimensionless variables are usually best for numerical computation. Following the usual procedure, let T denote a characteristic time, L a characteristic length, δ a characteristic displacement, and define new dimensionless variables by

$$w := \frac{u}{\delta}, \qquad (\xi, \eta, \zeta) := \frac{1}{L}(x, y, z), \qquad s = \frac{t}{T}. \tag{18.87}$$

With

$$\kappa := \frac{\mu}{\lambda}, \qquad T := L\sqrt{\frac{\rho}{\lambda}}, \qquad \iota := \frac{\rho L^2}{\lambda\delta}\mathbf{g}, \tag{18.88}$$

and the assumption that the only body force is gravity, the dynamic equation of linear elasticity is recast in the form

$$\frac{d^2 w_i}{ds^2} = (w_{k,k} + \kappa(w_{i,j} + w_{j,i}))_{,j} + |\iota|\delta_{3i}. \tag{18.89}$$

In these dimensionless variables, the brackets can be redefined as

$$\begin{aligned}
(w, \phi) &:= \int_\Omega w_{k,k}\phi_{\ell,\ell} + \frac{\kappa}{2}(w_{i,j} + w_{j,i})(\phi_{i,j} + \phi_{j,i})\, d\mathcal{V},\\
[w, \phi] &:= \int_{\partial\Omega} (w_{k,k}\delta_{ij} + \kappa(w_{i,j} + w_{j,i}))\eta_j\phi_i\, d\mathcal{S},\\
\langle\phi, \psi\rangle &:= \int_\Omega \phi_i\psi_i\, d\mathcal{V},
\end{aligned} \tag{18.90}$$

and the fundamental equation for the steady states is essentially unchanged:

$$(w, \phi) = \langle\iota, \phi\rangle + [w, \phi]. \tag{18.91}$$

In this formulation, the body Ω and its boundary are measured using the scaled variables.

By inspection of the boundary term (the square brackets in display (18.90)), the three components of the traction vector

$$\sigma_{ij}\eta_j$$

must be specified at each point on $\partial\Omega$ except in the presence of a Dirichlet boundary condition. More precisely, a component may be left unspecified provided that the corresponding component of the test function ϕ vanishes; that is, the zero Dirichlet boundary condition is imposed on that component. The mathematics of the weak formulation dictates the correct boundary conditions, which might also have been determined from physical considerations in the original derivation of the BVP for linear elasticity.

A procedure to approximate solutions of the weak BVP requires an appropriate discretization and a method to reduce the displacement with respect to this discretization to the solution of a system of linear algebraic equations. For finite element analysis, an appropriate set of elements must be defined along with corresponding basis functions for $H^1_D(\Omega)$. The unknown function u is approximated by a linear combination of these basis functions with undetermined coefficients. This expression for u is substituted into Eq. (18.91) and a system of linear equations for the unknown coefficients is obtained by replacing ϕ in turn by each basis function. An industry has been built on this basic idea. Every aspect of what has just been mentioned, from designing elements to solving large systems of linear equations, has been refined to produce excellent results with general codes.

The reader should be aware that the approach to finite elements taken here is from the point of view of continuum mechanics. The equations of linear elasticity, derived from continuum mechanics, come first; the finite element analysis is merely a method to obtain approximations of the distortion field. The history and practice of the FEM in engineering is closely tied to the analysis of hinged structures; for example, some bridges, the interior constructions of tall buildings, and the structures of some machines. Perhaps each structural element of the physical structure is a beam, and these elements are held together by hinges. In this application, the continuum model for elasticity is ignored in favor of a model constructed directly from idealizations of the structural elements (perhaps intervals, rectangles, or rectangular solids). This approach to the physical problem is sometimes called the direct stiffness method (DSM). Continuum mechanics engineers soon realized that a similar method works well for modeling the elastic properties of materials: they could and did bypass the equations of continuum mechanics by making finite element models directly with elastic materials materials considered as conglomerations of large numbers of structural elements allowed to flex at their boundaries. Of course, the two approaches to making the models can be made equivalent for continuum problems, and where both approaches apply, similar results are obtained. Although the continuum mechanics approach taken here is satisfying because it leads to equations of motion derived from basic physical laws (the incorporation of constitutive laws like Hooke's law not withstanding), the utility of direct modeling with elements should be appreciated when discussing elasticity problems with engineers.

Exercise 18.37. (a) Show that $A : B = \mathrm{tr}(AB^T) = \mathrm{tr}(A^T B)$ and this pairing defines an inner product on the vector space of $n \times n$ matrices. (b) Show that $\|A\| := \sqrt{A : A}$ is a norm on the vector space of $n \times n$ matrices.

Exercise 18.38. Why is it correct to write the integrals in display (18.90) without mentioning the Jacobian of the change to scaled variables? Hint: Which equation was used to determine the effect of the scaling on the model?

18.10 A THREE-DIMENSIONAL FINITE ELEMENT APPLICATION

A simple finite element approach to approximating solutions of the model problem posed in Section 18.4—a copper block bolted to a steel plate—is explored in this section.

The construction of elements and the associated basis functions for finite-dimensional subspaces of $H^1_D(\Omega)$ is a vast topic that remains an area of continued research. It should be clear that the choice of elements for a particular problem is open to experimentation and judgment. In principle, elements should be chosen so that an infinite sequence of refinements (more elements of the same type but with decreasing sizes) would lead to an exact solution of the BVP. In general, an infinite sequence of approximations of the solutions of a model equation with boundary and initial data in an applied problem should be known to converge to the true solution. Otherwise, the value of the predictions made from this model are problematic. In physical applications, predictions based on approximations may be tested by experiments to produce evidence for the utility of the model. Experience and judgment in element design complement theoretical considerations.

Recall the construction of one-dimensional elements and tent function bases (see the discussion concerning Eq. (18.72)). The same underlying ideas can be generalized to construct two- or three-dimensional finite elements.

In one-dimension, the body Ω is modeled by an interval that is partitioned by nodes into (closed) subintervals, which play the role of the finite elements. The corresponding basis functions, one associated with each node, are constructed using tent functions that span two such elements. The reader will check (see Exercise 18.42) that every linear combination of these tent functions is a *continuous* function. In other words, each function in the subspace of $H^1_D(\Omega)$ generated by linear combinations of tent functions is continuous on Ω and linear when its domain is restricted to one of the finite elements. In addition, each such function vanishes at the end points of the interval where essential (Dirichlet) boundary conditions are imposed. Continuity is required by the mathematical setting: Although the function space $H^1_D(\Omega)$ (in case Ω is an interval on the real line) is defined with reference to square integrability, the requirement that the first derivative of each function in this space is square integrable implies that every such function (perhaps after redefinition on a set of zero measure) is continuous. Integrations of products of basis functions, which are required to form the linear system of equations whose solution is the approximate displacement, can be carried out using a standard element (perhaps the unit interval) and its associated shape functions, which are simply the restrictions to this element of the two tent functions that would be defined if this standard element were situated between two adjacent standard elements. This construction of shape functions will be explained with more detail for two- and three-dimensional elements.

As mentioned previously, trial functions that are piecewise linear with respect to a choice of finite elements may be replaced by other subspaces of $H^1_D(\Omega)$: piecewise quadratic and piecewise cubic polynomials are often used.

For a two-dimensional body Ω, a useful generalization of the one-dimensional case is achieved using triangular elements that form a mesh covering Ω with the condition that a vertex of an element in the mesh intersects another element only at a vertex. The triangle with vertices $(0, 0)$, $(1, 0)$, and $(0, 1)$ is taken to be the standard element with counterclockwise orientation. Three shape functions are defined with respect to this element and these vertices according to the formulas

$$\begin{aligned} s_{(0,0)}(x, y) &= 1 - x - y, \\ s_{(1,0)}(x, y) &= x, \\ s_{(0,1)}(x, y) &= y. \end{aligned} \tag{18.92}$$

These functions have unit value at the specified vertex and vanish on the edge containing the other two vertices similar to the restriction of a tent function to a one-dimensional element, which has unit value at one end point and vanishes at the other.

The linear transformation $\mathcal{T}$ from points in an arbitrary triangular element, with vertices (a_1, a_2), (b_1, b_2), and (c_1, c_2) listed in the order of the counterclockwise orientation, to the standard element is given by

$$\mathcal{T}\begin{pmatrix} x \\ y \end{pmatrix} = L\begin{pmatrix} x \\ y \end{pmatrix} + \begin{pmatrix} a_1 \\ a_2 \end{pmatrix}, \tag{18.93}$$

where

$$L := \begin{pmatrix} b_1 - a_1 & c_1 - a_1 \\ b_2 - a_2 & c_2 - a_2 \end{pmatrix}.$$

Each element in the mesh is uniquely determined by its vertices, and three shape functions are defined on each element by composing the standard shape functions with the inverse of the associated transformation $\mathcal{T}$.

Trial functions (that is, elements of the space $H^1_D(\Omega)$) are vector functions of a vector variable. In the two-dimensional case, each trial function is a map from Ω into $\mathbb{R}^2$; thus, each of these functions has two components (which each map Ω into $\mathbb{R}$). They may be constructed as direct analogues of the tent functions used in the one-dimensional case. The component function associated with a node is the continuous, piecewise linear function that has value one at this node and is zero on all elements not having this node as a vertex. Its values in an element that contains the node as a vertex are given by the associated shape function associated with this vertex and this element. The set of vector trial functions is constructed from these two-dimensional tent functions after taking into account the boundary conditions. In case no zero Dirichlet boundary condition is to be enforced at a boundary node, it is assigned two basis functions $(\phi, 0)$ and $(0, \phi)$, where ϕ is the component tent function associated with this node. When one zero Dirichlet boundary condition is enforced at the node, say $\phi \cdot v = 0$ with $v = (v_1, v_2)$, the basis function with first component $v_2\phi$ and second component $-v_1\phi$ is added to the basis. No basis function is added in case there are two (independent) zero Dirichlet boundary conditions at the node.

A three-dimensional tent function basis for $H^1_D(\Omega)$ is constructed using the tetrahedron with vertices $(0,0,0)$, $(1,0,0)$, $(0,1,0)$, and $(0,0,1)$ as the standard element (see Exercise 18.43).

Rectangular elements are an alternative choice for meshing. These will be used to approximate the displacement field for the copper block model problem. In two dimensions, the standard rectangular element is the rectangle with vertices $(0,0)$, $(1,0)$, $(1,1)$, and $(0,1)$ and counterclockwise orientation. Linear shape functions of the form

$$s(x,y) = a + bx + cy.$$

are not viable. There are three coefficients but four nodes. For this reason, it is not possible to define the coefficients a, b, and c so that s has unit value at one node and vanishes at all other nodes. A remedy is to add one quadratic term and use shape functions of the form

$$s(x,y) = a + bx + cy + dxy.$$

In fact, this choice leads to the four shape functions

$$\begin{aligned}
s_{(0,0)}(x,y) &= (x-1)(y-1),\\
s_{(1,0)}(x,y) &= x(1-y),\\
s_{(1,1)}(x,y) &= xy,\\
s_{(0,1)}(x,y) &= y(1-x).
\end{aligned}$$

The transformation from the standard rectangular element to an arbitrary rectangle (or parallelogram) with vertices (in counterclockwise order) (a_1,a_2), (b_1,b_2), (c_1,c_2), and (d_1,d_2) is given by

$$\mathcal{T}\begin{pmatrix} x \\ y \end{pmatrix} = L\begin{pmatrix} x \\ y \end{pmatrix} + \begin{pmatrix} a_1 \\ a_2 \end{pmatrix}, \qquad (18.94)$$

where

$$\mathcal{L} = \begin{pmatrix} b_1 - a_1 & d_1 - a_1 \\ b_2 - a_2 & d_2 - a_2 \end{pmatrix}.$$

For the three-dimensional case, the standard element is the unit cube with vertices

$$(0,0,0),\ (1,0,0),\ (1,1,0),\ (0,1,0),\ (0,0,1),\ (1,0,1),\ (1,1,1),\ (0,1,1) \tag{18.95}$$

taken in this order to correspond to the positive (right-hand rule) orientation of space. By appropriately determining the coefficients of the functional form

$$s(x,y,z) = a + bx + cy + dxy + exz + fyz + gxyz,$$

the corresponding shape functions are

$$\begin{aligned}
s_{(0,0,0)}(x,y,z) &= -(x-1)(y-1)(z-1),\\
s_{(1,0,0)}(x,y,z) &= x(y-1)(z-1),\\
s_{(1,1,0)}(x,y,z) &= -xy(z-1),\\
s_{(0,1,0)}(x,y,z) &= (x-1)y(z-1),\\
s_{(0,0,1)}(x,y,z) &= (x-1)(y-1)z,\\
s_{(1,0,1)}(x,y,z) &= -x(y-1)z,\\
s_{(1,1,1)}(x,y,z) &= xyz,\\
s_{(0,1,1)}(x,y,z) &= -(x-1)yz.
\end{aligned}$$

The transformation from the standard element to an arbitrary rectangle (or parallelopiped) with vertices (a_1,a_2,a_3), (b_1,b_2,b_3), (c_1,c_2,c_3), (d_1,d_2,d_3), (e_1,e_2,e_3), (f_1,f_2,f_3), (g_1,g_2,g_3), and (h_1,h_2,h_3), where $b-a$, $d-a$, and $e-a$ is the ordered basis at the vertex a with positive orientation (right-hand rule), is given by

$$\mathcal{T}\begin{pmatrix} x\\ y\\ z\end{pmatrix} = L\begin{pmatrix} x\\ y\\ z\end{pmatrix} + \begin{pmatrix} a_1\\ a_2\\ a_3\end{pmatrix}, \tag{18.96}$$

where

$$\mathcal{L} = \begin{pmatrix} b_1-a_1 & d_1-a_1 & e_1-a_1\\ b_2-a_2 & d_2-a_2 & e_2-a_2\\ b_3-a_3 & d_3-a_3 & e_3-a_3\end{pmatrix}.$$

Corresponding shape functions on an element in the mesh are defined by composing the standard shape functions with the inverse of the transformation $\mathcal{T}$.

As prescribed by the FEM, the weak solution of the BVP is approximated by a function w in the finite-dimensional subspace $H^1_D(\Omega)$ spanned by the set of basis functions denoted by $\{\phi_1, \phi_2, \phi_3, \ldots, \phi_n\}$, where n is determined by the spatial dimension and the type of standard element. In mathematical form,

$$w = w_i \phi_i, \tag{18.97}$$

where the summation rule is in effect and the scalar coefficients w_i are to be determined. (There is a possible confusion arising from the notation for basis functions: ϕ_i is the name of the ith basis function in this context; in other contexts it is the ith component of the basis function called ϕ. The appropriate meaning should always be clear.)

An unavoidable and essential task required to implement the FEM—to approximate steady state solutions of the weak formulation of the basic BVP—is the assembly of the linear system whose solution is the set of coefficients w_i in the representation of w [Eq. (18.97)]. This linear system (in scaled variables) is the set of n equations

$$w_i(\phi_i, \phi_j) = \langle \iota, \phi_j \rangle + [w, \phi_j], \tag{18.98}$$

where the scaled surface tractions $[w, \phi_i]$ are all specified. In matrix form, this system is $Aw = \beta$ where

$$A_{ij} = (\phi_j, \phi_i), \qquad \beta_j = \langle \iota, \phi_j \rangle + [w, \phi_j].$$

The $n \times n$ matrix A and n vector β must be assembled from the choice of elements and the basis for the chosen subspace of $H^1(\Omega)$.

One viable approach to the assembly problem is to reduce it to local calculations on each element in the mesh. The underlying idea is to take advantage of the linearity of the operation of integration. Indeed, the integrals corresponding to the three brackets used in Eq. (18.98) may be written as sums over the elements, or boundaries of elements, whose union is the domain Ω. The quantities being summed are the integrations, in the definitions of the various brackets, of functions defined on a single element. The problem is to determine the appropriate components A_{ij} and β_j to which to add the result of each local calculation. One way to do this, requires the design of a bookkeeping protocol (a data structure) that takes as input three quantities: the node number, the dimension number (which, for this section, is defined to be the number one for the x dimension, two for the y dimension, and three for the z dimension), and the element number; and, as output, assigns the corresponding basis number (that is, the position in the list of basis elements $\{\phi_1, \phi_1, \phi_3, \ldots, \phi_n\}$. Using this protocol, it is possible to assign each bracket summand (an integral over the standard element) to the appropriate component A_{ij} or β_j.

The first task is to assign numbers to each node. This step is accomplished simultaneously with the construction of the mesh.

For the rectangular copper block situated as described on page 622, choose the point $(-\frac{\mathcal{L}}{2}, -\frac{\mathcal{W}}{2}, 0)$ as the lower left corner of the block and call the rectangle with corners

$$(-\frac{\mathcal{L}}{2}, -\frac{\mathcal{W}}{2}, 0), \qquad (\frac{\mathcal{L}}{2}, -\frac{\mathcal{W}}{2}, 0), \qquad (\frac{\mathcal{L}}{2}, -\frac{\mathcal{W}}{2}, \mathcal{H}), \qquad (-\frac{\mathcal{L}}{2}, -\frac{\mathcal{W}}{2}, \mathcal{H})$$

the front face. The top face, bottom face, right face, left face, and back face are defined relative to the right-hand orientation of the coordinate axes. A choice of orientation is essential for the bookkeeping. Clearly, the FEM does not require a uniform mesh; in fact, this is a major strength of the method. But, for simplicity, a uniform mesh of rectangular solid elements is considered here.

Choose the number of subdivisions xn, yn, and zn in each direction and compute the corresponding increments δx, δy, and δz so that xn $\times$ δx is the length of the block, yn $\times$ δy is its width, and zn $\times$ δz its height. The coordinates of the nodes can then be determined as

$$(-\frac{\mathcal{L}}{2} + i\delta x, -\frac{\mathcal{W}}{2} + j\delta y, k\delta z)$$

in a triple loop nested over the indices k, j, and i with $k = 0, 1, 2, \ldots,$ zn, $j = 0, 1, 2, \ldots,$ yn, and $i = 0, 1, 2, \ldots,$ xn. The nodes are numbered by the natural order of the list of points produced by the triple loop and stored in a list or an array called `Nodes`.

Each interior node corresponds to three dimensions (which may also be interpreted as directions); thus, each interior node corresponds to three basis functions. The boundary nodes must be treated separately. Each of these also corresponds to three dimensions. Each of these dimensions that corresponds to a natural boundary condition, which will be called a natural direction, corresponds to a basis function. The remaining dimensions corresponding to essential boundary conditions, called essential directions, are assigned basis number zero (which is a placeholder that does not correspond to an actual basis function). To implement a numbering of the basis functions, simply loop through `Nodes` and incorporate logic to assign basis function numbers in the numerical order given by the list of nodes and each associated direction in order, except that `0` is assigned to the essential directions at boundary nodes. The information is stored in a list or array called `Direction`. For example, `Direction` might be a list whose ith member is a list of the three numbers corresponding to the basis functions assigned to the three directions at the ith node.

For the copper block model, all nodes except those on its bottom face (which lies on the steel plate) spawn three basis functions. The nodes on the bottom face, which are not in the disk where the bolt is attached, spawn two basis functions corresponding to the tangential directions where (as a first approximation) no frictional force is imposed. Those nodes inside this disk spawn one basis function corresponding to the normal direction of the traction force due to tightening the bolt. Essential boundary conditions—zero deformation due to the rigid weld that is supposed to be used to attach the steel bolt—are enforced in the tangential directions. The total number of nodes is

$$\texttt{NumNodes} = (\mathrm{xn}+1)(\mathrm{yn}+1)(\mathrm{zn}+1).$$

The number of basis elements depends on the size of the mesh and the radius of the disk; this number is computed after a mesh is specified. For example, in case there is no bolt, the number of basis elements is

$$\texttt{NumEqs} = 3\,\texttt{NumNodes} - (\mathrm{xn}+1)(\mathrm{yn}+1).$$

The next part of the data structure is a list or array `ElementNodes` that gives the eight node numbers (called the global node numbers) assigned to the corners of each element in correspondence to the orientation of the eight nodes on the standard element, called the local node numbers $1, 2, 3, \ldots, 8$. This part of the structure may be viewed as a function from the pair (local node number, element number) to the global node number.

For our rectangular domain and rectangular elements, the number of elements is $\mathrm{nx} \times \mathrm{ny} \times \mathrm{nz}$. Each node not on the top face, the right face, or the back face corresponds to an element that is the rectangular solid that has the given node as its bottom left corner; that is, the other nodes on the element are obtained by adding positive increments in the three positive coordinate directions.

To compute `ElementNodes`, loop through `Nodes` in order. A logical test excludes the current node if it lies on the right, top, or back face. The first node that is not excluded defines the first element whose remaining nodes are specified by adding appropriate positive increments in the appropriate order, an operation that can be incorporated in a subroutine. The points thus obtained are nodes in the list or array `Nodes`. Their global numbers are recorded in order according to the standard order of the eight nodes on the standard element. Perhaps `ElementNodes` is stored as a list of rows with the row numbers in the list corresponding to the element numbers. A typical row is a list of eight numbers, which are the global node numbers of the eight corners of the element stored in the order specified by the standard element, for instance the order given in display (18.95).

The final data structure `Location` (which can be constructed from `Nodes`, `Direction`, and `ElementNodes`) is an array or list whose order corresponds to the element numbers. For each element number, `Location` specifies the basis function number assigned to each of the eight nodes and their corresponding three directions. A typical row (for example the tenth row, which corresponds to element number 10) might contain the information, written here in the form of a list,

$$\{\{1,2,0\},\{33,34,0\},\{41,42,0\},\{9,10,0\},\\ \{3,4,5\},\{35,36,37\},\{43,44,45\},\{11,12,13\}\},$$

where its sublist $\{41,42,0\}$ specifies that the basis function number of the second direction at the third node of element 10 is 42. The third direction at the fourth node of element 10 corresponds to an essential boundary condition and is assigned the fictitious basis function number zero.

The structure `Location` sets up the bookkeeping protocol that is the main tool used to assemble the matrix A and vector β. Indeed, for each element, the basis number associated with its nodes and their directions are specified. The component A_{ij} of the matrix A corresponds to the basis function numbers i and j. Because A is symmetric, the ordering of the indices i and j does not matter; indeed, $A_{ij} = A_{ji}$.

The idea for the rest of the assembly is to consider each interaction between nodes on a given element; for example, for constructing the round bracket, the interaction of the shape function corresponding to node 6 and direction 3 with the shape function corresponding to node 37 and direction 1. The data structure `Location` is used to find perhaps that node 6 and direction 3 has basis function number 121 while node 37 and direction 1 has basis function number 73. The result of the interaction (an integration over the element corresponding to the round brackets) is added to the component $A_{73,121}$. Only the main diagonal and superdiagonals of the symmetric matrix A need to be computed. The final sum $A_{73,121}$ can be assigned to the element $A_{121,73}$.

With the location data structure in hand, the next step is to consider the integrations—interactions between nodes—that contribute to the brackets in the weak formulation restricted to a particular element. Recall that there is an affine mapping $\mathcal{T}$ (translation plus linear transformation) from the standard element—the unit cube denoted by $\mathbb{U}$—to the given element for our rectangular example. Elements in the mesh may have curved boundaries. In this case, the user would define a smooth, invertible, and orientation-preserving transformation $\mathcal{T}$ with smooth inverse (an orientation-preserving diffeomorphism) from the standard element to the given element in the mesh.

Integration over the mesh element $\mathcal{T}(\mathbb{U})$ can be written symbolically in the form

$$\int_{\mathcal{T}(\mathbb{U})} f\, d\mathcal{V},$$

where f denotes a function (perhaps a product of shape functions) defined on the element. The change of variables formula states that this integral is equal to an integral over the standard element $\mathbb{U}$; in fact,

$$\int_{\mathcal{T}(\mathbb{U})} f\, d\mathcal{V} = \int_{\mathbb{U}} f \circ \mathcal{T} \det(D\mathcal{T})\, d\mathcal{V},$$

where $D\mathcal{T}$ is the derivative of the transformation $\mathcal{T}$. The determinant of this derivative, which is positive when $\mathcal{T}$ is orientation preserving, is the Jacobian of the transformation. All integrations are thus referred to the standard element.

For a nonuniform mesh, the Jacobian will change with the choice of element. For a uniform rectangular mesh and an affine $\mathcal{T}$, the Jacobian is a constant. In fact, with reference to the transformation defined in Eq. (18.96) and the uniform rectangular mesh over the (idealization of the) copper block, the columns of the Jacobian matrix of $\mathcal{T}$ are the usual basis vectors multiplied by δx, δy, and δz, respectively. Thus, the Jacobian is simply the product $\delta x \delta y \delta z$ for every volume integral independent of the choice of element. Moreover, the shape functions are defined on the elements in the mesh by composing the standard shape functions with the inverse of $\mathcal{T}$. Thus, the required integrations can be carried out for the shape functions defined on the standard cube. For the case of constant Jacobian, the result of each integration can be simply multiplied by the constant value of the Jacobian and summed at the address given by `Location`. In general, these integrations may still be carried out over the standard element after the appropriate transformation to the given element and its Jacobian are taken into account. This key observation allows all calculations to be made locally (that is, on the reference cube) using the standard shape functions and the transformation to the given element.

For the eight-node standard cube with three directions at each node, 300 integrations are sufficient to determine the round brackets. Indeed, the interactions between nodes are represented by the pairs (i, j) with the indices ranging over eight choices. This corresponds to an 8×8 matrix. By symmetry, it suffices to compute 36 parings: (i, j) with $i \geq j$. There are three directions at each node; thus, nine pairings of directions for each of the node parings. The parings corresponding to diagonal elements (i, i) are symmetric; thus, six integrations suffice for these pairings. In total, $28 \times 9 + 8 \times 6 = 300$.

Suppose that υ and ν are the shape functions corresponding to two nodes of the standard element $\mathbb{U}$. These functions have unit value at their corresponding nodes and vanish on the opposite faces of the element. Together they also correspond to six basis functions, three for each node in the three directions at that node. The basis functions are

$$\begin{pmatrix} \upsilon \\ 0 \\ 0 \end{pmatrix}, \begin{pmatrix} 0 \\ \upsilon \\ 0 \end{pmatrix}, \begin{pmatrix} 0 \\ 0 \\ \upsilon \end{pmatrix}, \begin{pmatrix} \nu \\ 0 \\ 0 \end{pmatrix}, \begin{pmatrix} 0 \\ \nu \\ 0 \end{pmatrix}, \begin{pmatrix} 0 \\ 0 \\ \nu \end{pmatrix}. \tag{18.99}$$

To maintain consistent notation (where for example υ_i denotes the ith component of the vector υ), note that the ith component of the first three basis vectors is $\upsilon\delta_{\alpha i}$ for the index α taking values 1, 2, or 3 corresponding to the first, second, and third basis functions. Likewise, the second three vectors are given by $\nu\delta_{\gamma i}$ for γ is 1, 2, or 3.

For a choice of two basis functions

$$\phi = \upsilon\delta_{\alpha i}, \qquad \psi = \nu\delta_{\gamma i},$$

the round brackets of the corresponding pair of basis functions restricted to the standard element reduces to

$$\begin{aligned}
(\phi, \psi) &= \int_{\mathbb{U}} \nabla \cdot \phi \nabla \cdot \psi + \frac{\kappa}{2}(\nabla\phi + (\nabla\phi)^T) : (\nabla\psi + (\nabla\psi)^T)\, d\mathcal{V} \\
&= \int_{\mathbb{U}} \upsilon_{,\alpha}\nu_{,\gamma} + \frac{\kappa}{2}(\upsilon_{,j}\nu_{,j}\delta_{\alpha\gamma} + \upsilon_{,j}\nu_{,i}\delta_{\alpha i}\delta_{\gamma j} + \upsilon_{,i}\nu_{,j}\delta_{\alpha j}\delta_{\gamma i} + \upsilon_{,i}\nu_{,i}\delta_{\alpha\gamma})\, d\mathcal{V} \\
&= \int_{\mathbb{U}} \upsilon_{,\alpha}\nu_{,\gamma} + \kappa(\upsilon_{,i}\nu_{,i}\delta_{\alpha\gamma} + \upsilon_{,\gamma}\nu_{,\alpha})\, d\mathcal{V},
\end{aligned} \tag{18.100}$$

where of course summation on repeated indices is implied and the subscripted comma denotes partial differentiation.

Recall that the scaled body force ι is a three-dimensional vector. For the angle brackets restricted to the standard element and the special case considered here where the body force is gravity and the block is situated so that gravity acts in the negative direction of the third coordinate, the reduced integral is

$$\langle \iota, \psi \rangle = -|\iota| \int_{\mathbb{U}} \nu\delta_{3\gamma}\, d\mathcal{V}.$$

The boundary conditions on the surface of the block are all zero Neumann (natural) boundary conditions except on the bottom face. Recall that in the disk where the bolt emanates from this face, the Neumann boundary condition is specified in the (negative) z direction corresponding by the force on the block induced by tightening the bolt, and zero Dirichlet conditions are enforced in the tangential directions due to the rigid weld of the bolt to the block. On the complement of this disk in the bottom face, the essential (zero Dirichlet) boundary condition is enforced in the normal direction corresponding to the presence of the rigid steel plate, and zero Neumann conditions (no frictional force) are specified in the tangential directions. Of course, these boundary conditions are imposed here for specificity; they may be adjusted as part of the modeling process.

The local square brackets paring of the unknown local scaled displacement $w \circ \mathcal{T}$ and a local basis function ψ is computed from the product of the given scaled surface traction (composed with $\mathcal{T}$) and ψ by integration over $\mathbb{U}$. There is one difference from the computation of volume integrals that must be taken into account: The boundary integrals are area integrals over faces of the standard element; hence, the transformation $\mathcal{T}$ acts by restriction to a face. The Jacobian of the transformation must be computed accordingly. In the model problem, the scaled traction ω has constant magnitude and acts in the downward vertical direction on the block, which is the direction of the outer unit normal on the bottom face of the block. Also, the traction is nonzero only on the disk corresponding to the attachment of the bolt. On an element, whose bottom face is part of the bottom face of the block, the traction acts only on the part $\mathcal{T}(\mathcal{D})$ of the element face that lies inside this disk. The corresponding local square brackets paring $[w, \psi]$ on the standard element vanishes except when ψ is a basis function of the form $\psi = \nu\delta_{3,i}$ corresponding to a node on the bottom face, and in this case, the paring has local value

$$[w, \psi] = -\omega \int_{\mathcal{D}} \nu \, d\mathcal{S}.$$

The corresponding Jacobian for the model problem is constant and equal to $\delta x \delta y$. Each local square bracket paring must be multiplied by this constant and added into the appropriate address given by `Location`.

For definiteness, let us consider the block dimensions to be

$$d = 12\,\text{mm}, \qquad \mathcal{L} = 10d, \qquad \mathcal{W} = 5d, \qquad \mathcal{H} = 3d, \tag{18.101}$$

where d is the bolt diameter and $a := d/2$ is the bolt radius.

The calculation of clamping force of a treaded bolt as a function of torque due to tightening is a difficult modeling problem that is not considered here. The theory predicts that clamping forces in the range 2–6 kN are physically realistic for the given bolt radius. Using the metric units centimeter-gram-second for measurements, the Lamé constants (18.58), the density of copper (18.59), and the characteristic length $L = 10\,\text{cm}$ (chosen arbitrarily), the timescale is dictated by the definition of T in Eq. (18.88) to be

$$T \approx 3.068 \times 10^{-5}\,\text{sec}$$

and the dimensionless constant

$$\kappa = \frac{\mu}{\lambda} \approx \frac{9}{19} \approx 0.473684.$$

For the characteristic displacement

$$\delta = 10^{-2}\,\text{cm},$$

the scaled gravitational (body) force (from Eq. (18.88)) is

$$|\iota| \approx 9.222 \times 10^{-5}.$$

The traction on the disk that represents the region where the bolt is attached to the block is in units of force per area. The area is $\pi a^2 \approx 1.13097\,\text{cm}^2$. Thus, the maximum traction is in the approximate range 1.76839×10^8–5.30516×10^8 in units of $\text{g}/(\text{cm}\,\text{sec}^2)$. This quantity must be made dimensionless using the formula

$$\frac{\lambda\delta}{L}(w_{k,k}\delta_{ij} + \kappa(w_{i,j} + w_{j,i})\eta_j) = \tau.$$

Thus, the maximum scaled traction magnitude is

$$\omega := \frac{L}{\lambda\delta}\tau.$$

It lies in the range 0.186–0.558, which for the purpose of illustration is replaced by a maximum scaled traction of 0.5.

The assembly process for the system matrix A, using the approach in this section, is outlined by the pseudocode

For i from 1 to the number of elements
    For j from 1 to 8
        For k from 1 to 3
            If the global basis number `gbn1` corresponding to (i, j, k) in `Location` exceeds zero
                For ℓ from 1 to 8
                    For m from 1 to 3
                        If the global basis number `gbn2` corresponding to (i, ℓ, m) in `Location` satisfies `gbn2` $\geq$ `gbn1`
                            Compute local round bracket taking into account the transformation $\mathcal{T}$ corresponding to

nodes on the standard cube with addresses (j, k) and (ℓ, m)
Add the result to the system matrix component $A_{\texttt{gbn1gbn2}}$
End If
End If

The logical test $\texttt{gbn2} \geq \texttt{gbn1}$ is used to take advantage of the symmetry of the A. The output of the procedure is the upper triangular part of this symmetric matrix.

Assembly of β, the vector on the right-hand side of the linear system $Aw = \beta$ consists of two summands: the pointed brackets, which take into account the body force, and the square brackets, which take into account the surface tractions given by the boundary conditions.

The pointed brackets are assigned to components of β in the obvious manner: $\langle \iota, \phi_j \rangle$ is the first summand of β_j. The local pointed brackets are assembled in a manner similar to the assembly of A. The only difference is that instead of using the integral of the product of two local shape functions, the integral of the product of a local shape function with ι is used.

In general, the square brackets contributions to the assembly of β can be computed by considering each finite element in turn and reducing to local integrations using the standard block as usual, but for the square brackets, surface integrals are required instead of the volume integrals used to compute the round and pointed brackets. Note first that the square brackets $[w, \phi]$ vanish for every basis function ϕ associated with an interior node. Thus, in the loop over the finite elements, the surface integrals are computed only for those finite elements with at least one face on the boundary of Ω.

For our example, the copper block with a steel bolt, the surface tractions are all zero except for those acting on the portion of the boundary of the block in the bolt attachment disk. Thus, for the computation of square brackets, we may ignore all finite elements except those admissible elements with two properties: their bottom face, called the admissible face, is on the bottom boundary of the block and at least one node on this face, called an admissible node, is in the bolt attachment disk. The normal direction at each admissible node corresponds to a basis function. The corresponding

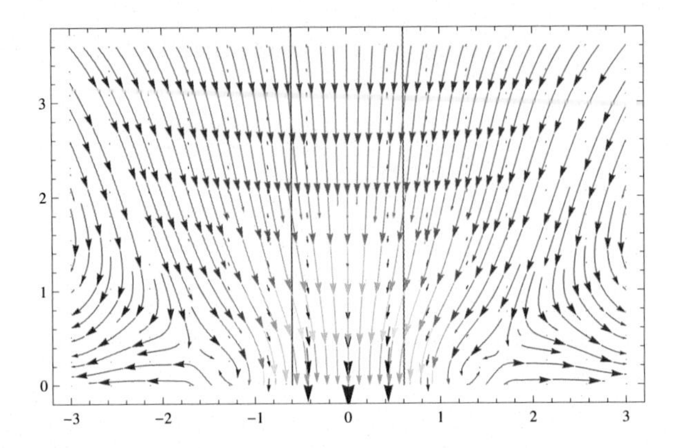

Fig. 18.3 The two-dimensional cross section described in this section with the integrated displacement field for a copper block bolted to steel plate is depicted. The horizontal axis corresponds to the y direction and the vertical axis to z. The block has dimensions $10 \times 5 \times 3$ in scaled bolt diameters. The block elements laid 20 wide, by 10 deep, by 6 high are used for a total of 1200 elements. The force of gravity is included and the dimensionless traction force on the bolt attachment disk is $\omega = 0.2$. The solid vertical lines correspond to the cylinder above the bolt, which is attached at the bottom center.

square brackets (of w and this basis function) is the (surface) integral of this basis function times the given constant normal traction over the bolt attachment disk. The contribution to this integral from the admissible element is computed by integrating the shape function times the constant traction times the Jacobian of the restriction of the transformation $\mathcal{T}$ to the admissible face over the set on this admissible face that is mapped by $\mathcal{T}$ inside the bolt attachment disk. One way to implement the calculation, for admissible faces that intersect the attachment disk but are not contained in this disk, is to simply multiply the shape function on the standard element by the characteristic function of the portion of the standard block bottom face that maps inside the attachment disk. In symbols, the local contribution is

$$\int_{\mathcal{U}} s\omega\chi \det(D\mathcal{T})\, dS,$$

where s is the shape function, ω is the given (scaled) traction, and χ is defined by $\chi(x, y) = 1$ if $\mathcal{T}(x, y)$ is in the attachment disk and $\chi(x, y) = 0$ otherwise. It is summed into the component of β corresponding to the global basis function number corresponding to the admissible node.

This completes the assembly of the matrix system $Aw = \beta$. It can be solved by iteration, or for the purposes of this section by an available solver built into a commercial software package. The system matrix A is symmetric and sparse (that is, most of its elements are zero). It is possible to take advantage of this property by avoiding multiplication by zero. This subject is covered in every treatment of numerical linear algebra.

The coefficients w_i of the approximate displacement $w = w_i\phi_i$ due to the traction force and the presence of the steel plate are given by the solution w of the linear system $Aw = \beta$. Note that at the ith node the approximate distortion is $w = w_i$. Indeed, ϕ_i has unit value at this node where all other basis functions vanish. Thus, the discretized distortion field is given by the w_i.

For computations that result in large data sets (in this case, the approximation of the displacement field at each node in the finite element mesh), the computed data should be rendered in a form that is useful for the intended application. As an example, consider Fig. 18.3 where some results of a finite element approximation of the displacement field for our copper block bolted to a steel plate are depicted. The approximate, dimensionless displacement field is shown for the plane that is perpendicular to the longest edges of the block and contains the bolt axis.

Inspection of Fig. 18.3 suggests that the displacement vanishes on a one-dimensional ring (surrounding the imaginary cylinder above the bolt) and nowhere else in the interior of the block. Perhaps the existence of this ring is important in some applications. If so, there is a natural question: Does the model predict that the zero set of the distortion field in the interior of the block is a simple closed curve that surrounds the imaginary cylinder above the position of the bolt? The genesis of this question illustrates a wise point of view: Computation is not only about producing numbers; it is a useful tool for gaining insight into physical phenomena. Answering the question requires making a deduction from the mathematical model. Gathering additional evidence for an answer can be accomplished by performing more numerical experiments (see Exercise 18.39). These are tasks for the applied mathematician. But, of course, the reality of a prediction remains unknown until it is confirmed by physical experiments.

All the ingredients are in place for the reader to write a basic finite element code. Although certainly not a simple task, writing and using one's own code is an invaluable learning experience.

Further study of elasticity theory and the FEM is certainly justified. Both are of great value in many practical applications.

Exercise 18.39. [FEM Three-Dimensional Coding] (a) Repeat the finite element computation that is used to make Fig. 18.3. (b) Where in the block is the strain maximal? (c) Compute the displacement field for several values of the traction due to tightening the bolt and consider the displacement at the centroid of the block. Is the magnitude of this displacement a linear function of the traction? Approximate this function. How does this function change with a change in dimensions of the block? Specifically, suppose the length and width of the block are constant and the thickness (z direction) is increased. How does the function depend on this change? What about a change in length? (d) Does the model predict that the zero set of the distortion field in the interior of the block is a simple closed curve that surrounds the imaginary cylinder above the position of the bolt? Hint: This might be a difficult question to answer. Discuss the numerical evidence for the existence of the curve provided by Fig. 18.3. To answer the question, at least provide new evidence in favor or against the existence of such a curve.

Exercise 18.40. A full three-dimensional finite element computation was used to produce the two-dimensional Fig. 18.3. Use a two-dimensional finite element computation to make a similar picture. Do the computations give the same answer?

Exercise 18.41. [Elastic Plate with Elliptical Hole] Describe the stress field for a thin square elastic plate with an elliptical hole centered at the center of the plate ignoring body forces but loading the plate on two of its opposite sides with equal and opposite constant normal tractions. In particular, determine the point(s) of maximum normal and tangential stress on the boundary of the hole. Note: This exercise has historical significance (see [23]).

Exercise 18.42. Show that every element in the span of the usual one-dimensional tent functions for a partition of an interval is a continuous function.

Exercise 18.43. Determine the shape functions for the standard three-dimensional tetrahedral element and determine the linear transformation from an arbitrary tetrahedron to this element.

Exercise 18.44. Consider the copper block model but with the top face of the block welded to a second steel plate. Determine the displacement field and compare it to the displacement field for the original model.

Exercise 18.45. Consider the copper block model but with the back face of the block welded to a second steel plate. Determine the displacement field and compare it to the displacement field for the original model.

CHAPTER 19

Problems and Projects: Rods, Plates, Panel Flutter, Beams, Convection-Diffusion in Tunnels, Gravitational Potential of a Galaxy, Taylor Dispersion, Cavity Flow, Drag, Low and High Reynolds Number Flows, Free-Surface Flow, Channel Flow

19.1 PROBLEMS: FOUNTAINS, TAPERED RODS, ELASTICITY,THERMOELASTICITY, CONVECTION-DIFFUSION, AND NUMERICAL STABILITY

Exercise 19.1. [Intermittent Fountain] Fig. 19.1 is a schematic representation of a water fountain. An appropriate constant inflow will cause the fountain to run intermittently. (a) Make a mathematical model of the fountain (based on flow rates and volumes) and show by simulation that the fountain runs intermittently. (b) The inflow rate and the elevations of the bends in the outflow are control parameters. How do changes in these parameters affect the fountain? (c) How would the fountain run in the presence of a periodic inflow?

Exercise 19.2. [Linearly Tapered Rod] Repeat Exercise 18.23 on the design of resonant tapered rods for the case where the rod is linearly tapered; that is, the cross-sectional area function is given by $A(x) = A_0 - A_1 x$. Hint: A review of Bessel functions might be helpful.

Exercise 19.3. [Elasticity versus Rod Equation] Compare a two- or three-dimensional linear elasticity analysis (perhaps using finite elements) for the longitudinal waves in a tapered bar with an analysis based on the one-dimensional model equation (18.47).

Exercise 19.4. [Barbell-Shaped Rod] Model and analyze the resonance lengths of barbell-shaped tapered bars (the area function decreases and then increases). What if there are several local maxima and minima of the area function? It would seem that a bar with sufficiently thin necks between regions where its cross-sectional area function has local maxima would behave elastically like masses connected by springs. Does this interpretation lead to a viable model for longitudinal waves for rods with moderate neck diameters?

Exercise 19.5. [Body versus Traction Force] A tapered rod is subjected to a sinusoidal body force acting in the axial direction of the rod. Discuss the motion of the rod tip. Compare your results with the motion of the same tapered bar driven by a sinusoidal traction force.

An Invitation to Applied Mathematics: Differential Equations, Modeling, and Computation.
http://dx.doi.org/10.1016/B978-0-12-804153-6.50019-1, Copyright © 2017 Elsevier Inc. All rights reserved.

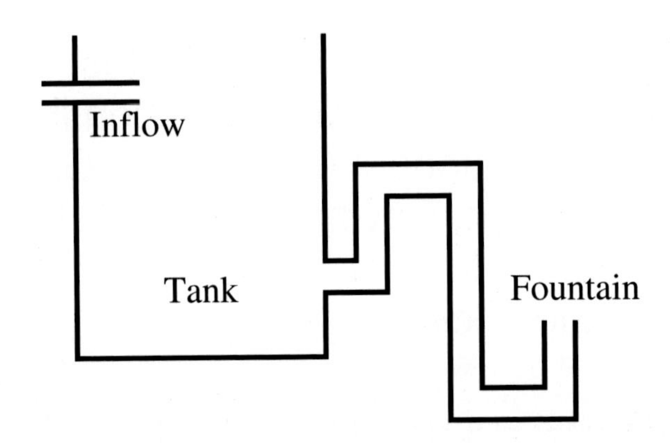

Fig. 19.1 Schematic diagram of intermittent water fountain.

Exercise 19.6. [Clamped Plate] Imagine a thin rectangular sheet of metal clamped on its two (short right and left) edges to an immovable frame. Determine the shape of the rectangular plate when a constant traction force is applied to a portion of its upper surface. (a) Determine the profile of the sheet after a load is applied in a small disk at the center of the upper face of the plate. (b) Determine the profile of the sheet after a load is applied in a strip from front to back on the upper face of the plate. (d) Determine the profile of the sheet after a periodic traction (of the form $A\sin(\Omega x)$) is applied across the upper face of the plate.

Exercise 19.7. [Dynamic Behavior of Elastic Solids-Panel Flutter] All the tools are in place to consider the dynamics of elastic materials. Start with the dimensionless Eq. (18.89). Use the same basis functions as for steady states, but consider the unknown dimensionless deformation to be $w = w_i(s)\phi_i$, where the coefficients depend on the dimensionless temporal parameter. Follow the weak formulation by taking inner product with respect to the basis functions and use the defined bracket parings to arrive at the fundamental weak form of the dynamic equations of linear elasticity:

$$\ddot{w}_i(t)\langle\phi_i, \phi_j\rangle = w_i(t)(\phi_i, \phi_j) - \langle\iota, \phi_j\rangle - w_i(t)[\phi_i, \phi_j]. \tag{19.1}$$

For the dynamic model, the assembly process creates a system of second-order ordinary differential equations (ODEs) instead of a system of linear equations that is created for the steady state. In case the boundary conditions are independent of time, they are imposed exactly as for steady state computations. This second-order system of ODEs can be solved by a variety of methods. But, the reader is warned that numerical integration of large-scale systems of ODEs requires some thought before writing code.

(a) Consider the copper block with a bolt through a steel plate but now suppose the traction force on the attachment disk is sinusoidal $A\sin(\Omega t)$. Determine (numerically) the displacement response in the block. Is the response sinusoidal? What is the frequency of the response? (b) Consider the setup for Exercise 19.6. Determine the response for a sinusoidal traction on the upper surface $A\sin(kx - \Omega t)$. Part (b) approaches perhaps the most important problem in elasticity: panel flutter. Air is moving over the top of a plate—think airplane wing. Air pressure fluctuations interact with the elastic plate. Perhaps for low flow velocities the steady state configuration of the plate is stable. In some cases, the steady state configuration of the plate becomes unstable as the airflow velocity increases, and the plate may begin to oscillate. Such oscillations may be catastrophic: an airplane wing might oscillate with growing amplitude due to energy pumped into the oscillation from the airflow until the wing structure fails. Less dramatic examples are Venetian blinds oscillating in an open window and rattling ducts in an air conditioning system. (c) Eq. (19.1) appears, with appropriate interpretation, as Eq. (1) in the article [73] summarizing work on disc brake squeal suppression. Read the article and explain the relation between these equations. (d) Imagine a structure in the form of a triangle where the legs of the triangle are rigid bars and the vertices are springs constructed and situated so that their restoring forces tend to make the interior angles smaller. One leg is clamped to a device that imparts a sinusoidal motion to this leg and hence to the structure. Determine the motion of the vertex opposite this leg. Treat the legs of the triangle as finite elements and include viscous damping. Generalize to a regular polygonal structure, which at rest is inscribed in a circle, with the ultimate goal of determining the motion of an elastic hoop driven by a sinusoidal motion imparted to one point on the hoop. For simplicity, consider the case where the direction of the driving motion is tangent to the idealized hoop modeled by a one-dimensional curve.

Exercise 19.8. [Thermoelastic Damping] (a) Consider a tapered rod with exponentially decreasing cross-sectional area subjected to a sinusoidal traction force at its large end. As the rod is excited, its compression and expansion increases its temperature. Thus, energy is extracted from the motion of the rod and dissipated in the form of heat. The subject concerned with coupling thermodynamics and elasticity is called thermoelasticity. In the case of the rod, heat loss is a form of thermoelastic damping. This damping effect is incorporated in the one-dimensional model for displacement

$$\rho\ddot{U} = (\lambda + 2\mu)(\frac{\partial^2 U}{\partial x^2} + \frac{A'(x)}{A(x)}\frac{\partial U}{\partial x}) - k\rho\dot{u}_{xx} + \rho B(x,t), \tag{19.2}$$

where k is the mechanical to heat rate constant. Discuss the motion of the rod tip and the amplification factor. In what sense does the heat dissipation term cause damping? (b) Research the physics of thermoelasticity and discuss the validity of the model used in part (a). (c) Are there additional sources of damping? How should they be modeled? Which source is dominant if, for example, the tapered end of the rod is inserted into a beaker full of water? Hint: This is a research problem. The complete answer is probably not simple.

Exercise 19.9. [Beam Theory] (a) The equations of linear elasticity are second-order with respect to spatial derivatives. Discussion with engineers on applied elastic problems might turn to equations of motion that are fourth-order with respect to spatial derivatives. Why the discrepancy? Most likely the engineers are using beam theory. Imagine a beam, which is a body where two of its dimensions (width and depth) are approximately the same and both are much smaller than the third dimension (length). As in channel flow (discussed in this book), the idea is to reduce the equations of linear elasticity from three space-dimensions to one, the dimension of length along the beam. To do this, approximations must be made. One way to approach the theory is to integrate the equations of linear elasticity over cross sections transverse to the beam axis. Read paper [27], fill in details, and explain how equations of motion arise that are fourth-order in the deformations along the length of the beam. Also, after some research in the literature, compare Timoshenko beam theory with Euler–Bernoulli beam theory. Will beam theory remain relevant as computational speed makes predictions from full three-dimensional models feasible? Gather evidence and discuss your findings. (b) Suppose one end of a beam is clamped to a rigid wall and its other end is free. A downward force is applied at the free end and held until the beam is in equilibrium. What is the equilibrium shape of the beam? Hint: The ultimate answer would compare predictions of three-dimensional elasticity theory and beam theory. (c) Suppose the applied force of part (b) is released from rest. What is the subsequent motion of the beam? Hint: Same as for part (b).

Exercise 19.10. [Convection-Diffusion in Tunnel] Recall the tunnel gas-diffusion model (5.74). As mentioned in the text, convection usually operates at a much faster timescale than diffusion. Construct a model for the motion of the air caused by the filter at the end of the tunnel, the increase in pressure at the tunnel entrance, or the motion that might be caused by the leak itself, and use your model to compare the sensor readings with and without air motion taken into account. What are the most important effects?

Exercise 19.11. [Stability of Numerical ODE Solvers] Review numerical algorithm (16.138) and consider the related method

$$\begin{aligned} y_{n+1/2} &= y_n + \frac{\Delta t}{2} f(y_n, t_n), \\ y_{n+1} &= y_n + \Delta t f(y_{n+1/2}, t_{n+1/2}). \end{aligned} \tag{19.3}$$

Read the paper [95] by Lawrence F. Shampine, which is a study of the stability properties of the midpoint method

$$y_{n+1} = y_{n-1} + 2\Delta t f(y_n, t_n),$$

and discuss the relation between his analysis and method (19.3). Modify Shampine's analysis so it applies to the predictor-corrector algorithm [Eq. (16.138)]. Discuss the strengths and weaknesses of these numerical methods. Create and run numerical experiments to test and verify your findings.

19.2 GRAVITATIONAL POTENTIAL OF A GALAXY

Let ρ be the density function for matter in a region of space and G the universal gravitational constant. The gravitational potential ϕ is the physically relevant solution of Poisson equation $\Delta\phi = 4\pi G\rho$:

$$\phi(x) = -\int_{\mathbb{R}^3} \frac{G}{|x-y|}\rho(y)\,dy.$$

The gravitational field produced by the mass with density ρ is $g = -\nabla\phi$ (where g is not necessarily the g used for the gravitational acceleration near the surface of Earth), and the motion of a test particle with mass m in this field is given by Newton's law $md^2x/dt^2 = F$, where the force is $F = mg$. A basic problem in physics is to determine the gravitational field of a body and the motion of test particles in the field. Here, a test particle might be an extended body that is assumed to move in the gravitational field according to how its center of mass moves under Newton's law. The project is to explore the motion of a star viewed as a test particle in the gravitational field of a galaxy. The problems stated here are merely starting points for further research.
(a) Consider an approximate (spiral) galactic shape in the form of a cylinder of height h and radius a where $a > h$. Suppose the density of the mass in the cylinder is a function of the distance from the axis of the cylinder. For definiteness, consider a Cartesian coordinate system such that the cylinder is the Cartesian product of the disk of radius a centered at the origin of the horizontal plane and the interval $[-h/2, h/2]$ on the vertical axis through the origin. Suppose the density function is a Gaussian

$$\rho(x,y,z) = be^{-c(x^2+y^2)}$$

for some positive constants b and c such that ρ on the lateral boundary of the cylinder $x^2 + y^2 = a^2$ is small to model the physical reality that almost all the mass of a galaxy is in the galactic bulge at its center. This density function may be taken to be zero outside the galaxy. For definiteness and simplicity, start with the values $h = 0.1$, $a = 6.0$, $b = 0.1$, and $c = 0.25$ and assume the density drops to zero outside the cylinder. Determine the approximate gravitational potential and gravitational field near the galaxy.

Notes: Which is better for a numerical treatment of the problem, solving the Poisson equation or performing the integration to obtain the gravitational potential? For the Poisson equation, consider the symmetry in the problem: the potential should be the same on every plane passing through the vertical axis. On one of these planes, a rectangular grid should suffice and it should be chosen so that the sides of the rectangular cross section corresponding to the galaxy lie on grid lines. The computational domain can be taken to be bounded by a large rectangle that contains the galactic cross section. Boundary conditions on the sides of the rectangle are not obvious. By physical considerations the gravitational potential must drop to zero at infinity. So, zero Dirichlet boundary conditions are reasonable. This choice might be tested by computing on a sequence of rectangles of increasing size until the computed potential does not change with increasing size up to some specified amount.

Discretization in the usual manner at the grid points leads to the solution of a linear system $Ax = b$, where A is a large, sparse (banded) *symmetric* matrix. The solution can be approximated by successive overrelaxation (SOR), a method already discussed in this book. Because A is symmetric, there are other (perhaps more efficient) methods of approximation. One of the best is the conjugate gradient method outlined in Section A.20. Which method is most efficient in the present application?
(b) Determine the gravitational field for the cylinder galaxy in part (a). Also, consider a star with some mass (small compared to the total galactic mass) and determine the motion of the star in case its initial velocity is perpendicular to the previously considered galactic cross sections. Do you expect periodic stellar orbits? If so, how do their periods depend on their initial positions from the galactic center?
(c) A cylindrical shape is a crude model for a spiral galaxy. Specify a more realistic galactic shape inside the large rectangular computational domain(s) or parts (a) and (b), and determine the gravitational field and star motion for the new shape.

Notes: A curved shape may be best. But, to avoid dealing with curved boundaries, consider a stair-stepped shape where the boundary of the shape is confined to grid lines. In principle, curved boundaries can be approximated as closely as desired by refining the grid and using stair-stepped shapes. Is this a viable approach to the problem? How should the finite-difference approximations be modified to take into account curved boundaries?

(d) Research Project: Spiral galaxies are usually surrounded by globular clusters. Find out what this means and conduct numerical experiments to determine the motions of the globular clusters and test particles in the gravitational field of the entire system.

19.3 TAYLOR DISPERSION

In 1953 Geoffrey Ingram Taylor (1886–1975) published an important paper [109] where he provided strong evidence that shear flows increase diffusion. In addition, he gave an estimate for the effective diffusion coefficient. His ideas, refined by other authors and very well studied, are applied in the design and function of devices used to measure diffusivity of liquids, in understanding biological processes including movement of tracers in blood flow, and in Earth science including the movement of environmental pollutants in rivers and in porous media. Although the true range of applicability and more rigorous treatment of his approximations are not completely understood, exploration of this topic is a tour into the mind of an exemplary applied mathematician.

Imagine the concentration c of some substance being carried by a moving fluid with velocity u where k^2 is the diffusivity of the substance in the flow. Using conservation of mass and constant diffusivity, the basic model for the concentration is the convection-diffusion equation

$$c_t + u \cdot \nabla c = k^2 \Delta c \tag{19.4}$$

with appropriate boundary and initial data. The usual (reasonable) assumption is that the substance under study does not affect the fluid flow. When it does (for example when the substance is particulate matter or another fluid), the convection-diffusion equation must be coupled to a fluid model (Navier–Stokes). This case is much more complicated and not discussed further here. In Taylor's theory the flow is taken to be specified and unidirectional. For definiteness, consider the flow to be all in the direction of the first coordinate of a Cartesian coordinate system. In this case $u = (v, 0, 0)$, where v is a function of space and time given by the coordinates (x, y, z, t). Also, suppose that measurements are made in bounded regions in plane sections perpendicular to the flow. The classic example (the one studied by Taylor) is Poiseuille flow in a round pipe whose axis is in the direction of u where the sections are disks perpendicular to the central axis of the pipe. Taylor considered the changes in average concentration over such cross sections. The section at the point $(x, 0, 0)$ is called Σ_x and its area $|\Sigma_x|$. By definition, the average concentration over this section is

$$\bar{c}(x,t) = \frac{1}{|\Sigma_x|}\int_{\Sigma_x} c(x,y,z,t)\,dydz \tag{19.5}$$

and the similarly defined average velocity is denoted $\bar{u}$.

The fluctuating part of c is defined to be $\tilde{c} = c - \bar{c}$. Note that the average of $\tilde{c}$ vanishes and, of course, the average of an average is the average.

Using the substitution $c = \bar{c} + \tilde{c}$, the model equation is transformed to

$$\bar{c}_t + \tilde{c}_t + v(\bar{c}_x + \tilde{c}_x) = k^2(\bar{c}_{xx} + \tilde{c}_{xx} + \tilde{c}_{yy} + \tilde{c}_{zz}). \tag{19.6}$$

By averaging each term over the section Σ_x, the latter equation is reduced to

$$\bar{c}_t + \bar{v}\bar{c}_x + \overline{v\tilde{c}_x} = k^2\bar{c}_{xx}, \tag{19.7}$$

where the third term on the left-hand side requires further explanation. With this term removed, the resulting partial differential equation (PDE) would simply be a convection-diffusion equation for the average concentration with diffusivity constant k^2 and Tayor's paper would not be so important. All of the interesting behavior comes from that third term.

Subtract Eq. (19.7) from Eq. (19.6) and rearrange the result in the form

$$\tilde{c}_t + (v - \bar{v})\bar{c}_x + v\tilde{c}_x - \overline{v\tilde{c}_x} = k^2(\tilde{c}_{xx} + \tilde{c}_{yy} + \tilde{c}_{zz}). \tag{19.8}$$

Taylor's idea (translated to the context of Eq. (19.8)) arises from physical reasoning: After a sufficiently long time has passed and in case the axial length over which the fluctuations are considered is large compared with the size of the cross-sectional regions over which the averages are computed, the quantities

$$\tilde{c}_t, \qquad v\tilde{c}_x - \overline{v\tilde{c}_x}, \qquad k^2\tilde{c}_{xx} \tag{19.9}$$

are all small relative to the other quantities in the differential equation. By neglecting the small terms, the model equation is reduced to

$$(v - \bar{v})\bar{c}_x = k^2(\tilde{c}_{yy} + \tilde{c}_{zz}). \tag{19.10}$$

This equation can be solved for $\tilde{c}$ and the result may then be substituted to tame the problematic third term in Eq. (19.7).

Returning to Taylor's original configuration, which he used to conduct physical experiments, recall that the velocity of a Poiseuille flow in a round pipe of radius a is given by

$$v = c(a^2 - r^2)$$

for some constant c. A simple calculation shows that in fact

$$v = 2\bar{v}(1 - \frac{r^2}{a^2}). \tag{19.11}$$

Check this!

Put v from Eq. (19.11) in Eq. (19.10) and change the Laplacian to polar coordinates. Notice that $\bar{c}_x$ does not depend on the radial polar variable r. For an axially symmetric initial concentration, the evolution of the concentration does not depend on the angular polar coordinate. Thus, the resulting equation for $\tilde{c}$ depends only on r with x playing the role of a parameter. This equation can be easily integrated twice with respect to r at the price of introducing two constants of integration. Check that the constant introduced by the first integration is the coefficient of $\ln r$ after the second integration. It must be set to zero to ensure a physically relevant solution that does not blow up at $r = 0$. The second constant of integration can be determined because the average value of $\tilde{c}$ vanishes over each cross-sectional disk. In fact, after these computations, the fluctuation is seen to be

$$\tilde{c} = \frac{\bar{v}\bar{c}_x}{k^2}(\frac{1}{4}r^2 - \frac{r^4}{8a^2} - \frac{1}{12}a^2). \tag{19.12}$$

In checking the computations be careful to use the polar area element $rdrd\theta$, and remember to divide by the area of the disk when computing the averages.

The problematic term in Eq. (19.7) is ready to be tamed. Insert the formulas for v and $\tilde{c}$ and compute. The result is

$$\overline{v\tilde{c}_x} = -\frac{a^2\bar{v}^2}{48k^2}\bar{c}_{xx}.$$

Substitute this into Eq. (19.7) and rearrange to obtain

$$\bar{c}_t + \bar{v}\bar{c}_x = (k^2 + \frac{a^2\bar{v}^2}{48k^2})\bar{c}_{xx}. \tag{19.13}$$

This remarkable result states that (up to the specified approximations) the average concentration satisfies a convection-diffusion equation with the effective diffusivity

$$k^2 + \frac{a^2 \bar{v}^2}{48k^2},$$

a quantity that is likely to be much larger than k^2. In practice, k^2 is small relative to a and $\bar{v}$.

There is a useful conclusion: The shear caused by the no-slip boundary condition at the pipe wall *increases* the effective diffusivity. Also, the effective model equation (19.13) can be solved exactly (see Eq. (9.29)).

Using carefully designed numerical experiments, discuss the evolution of an axially symmetric initial concentration of substance introduced into a round pipe at time $t = 0$. In particular, approximate the average concentration of the substance as it is carried downstream by the advection and diffusion processes governed by model equation (19.4), where u is a given Poiseuille flow. The PDE in this situation is two-dimensional; it depends on the radial coordinate in the transverse direction to the flow, the axial coordinate, and time. Discuss the sizes of the terms (19.9) in comparison to the terms in differential equation (19.10) to evidence the validity of Taylor's approximations.

In at least one experiment, consider a cylindrical plug of the substance introduced into the pipe that has the same radius as the pipe and discuss the geometric shape of the evolution of the plug as it is carried downstream by the flow.

Discuss the relevance of Taylor's result versus the use of numerical integration to approximate the full convection-diffusion model. As part of your report, be sure to include details confirming the computations outlined in this section.

As mentioned previously, Taylor's ideas are used to make measurements of diffusion coefficients. Consider the following quote (with internal citations removed) from a typical paper in the literature[1]

[1]Mohsen Ghanavati, Hassan Hassanzadeh, and Jalal Abedi (2014), Application of Taylor dispersion technique to measure mutual diffusion coefficient in hexane+bitumen system, *AIChE Journal,* **60**(7), 2670–2682, July 2014.

In a Taylor dispersion experiment, a minute amount of solute, called a pulse, is injected into a laminar carrier stream of a slightly different composition of solvent flowing in a long capillary tube. As the pulse travels through the tube, it spreads out into a nearly Gaussian profile under the combined actions of molecular diffusion and convection. The shape of the dispersed peak measured by an appropriate detector, commonly at the end of the tube, is used to determine the molecular diffusion coefficient D from the dispersion coefficient K

$$D + \frac{R^2\bar{u}^2}{48D},$$

where R *is the tube internal radius and inline image is the average velocity of the laminar flow in the tube. It is noteworthy that experimental conditions are usually designed such that the first term in comparison to the second term in [the equation] can be safely ignored.*

Describe in detail exactly what the authors are describing. What is meant by "The shape of the dispersed peak?"

Taylor performed physical experiments that verified to some extent his approximation. Many other physical experiments have been performed by other researchers, and as in the last paragraph, Taylor's ideas are widely used to make practical measurements. More sophisticated mathematical analysis has been carried out by several authors to justify Taylor's approximation in some operating regimes. Although Taylor's approximation is widely used, it does not always agree with reality. Nonetheless, it is a beautiful example of applied mathematics. The basic insight that shears in flows enhance diffusion is obviously important.

19.4 LID-DRIVEN CAVITY FLOW

Reconsider the numerical experiments for lid-driven cavity flow (starting on page 418). For low Reynolds numbers, the flow settles to a steady state. One way to quantify this fact, which is certainly not definitive, is to choose a spatial position (for example the center of the cavity) and plot the magnitude of the velocity field at this point as a function of time. In steady state, this quantity will remain constant. There seems to be a critical value of the Reynolds number such that for Reynolds numbers above this value there is no stable time-independent steady state indicated by the plot of the magnitude of the velocity field: the time trace oscillates with time. What is this critical value? The project is to use *your* code to detect the critical value and to give evidence that the value you determine is the correct prediction from your code. There are many papers and experiment reports on lid-driven cavity flow (see, for example, [15, 36, 59]). The oscillation might be periodic. If so, what is its period. The oscillation might be chaotic. Is there good evidence for this behavior?

19.5 AERODYNAMIC DRAG

Reconsider the tailgate-up versus tailgate-down Problem 16.1.

Recall that bodies placed in a flow field are generally of two types with respect to the drag force: streamlined and bluff. A streamlined body is surrounded by a flow that is approximately Eulerian; thus, the pressure drag is nearly zero and the drag is dominated by skin friction in a thin boundary layer (at least for moderate Reynolds numbers for which the boundary layer does not separate from the body). A bluff body has boundary layer separations even for moderate Reynolds numbers, and the drag is dominated by pressure drag. Indeed, the pressure on the downstream side of the body (where backwash, vortex shedding, or turbulent flow is prevalent) is lower than at the upstream side (where a more organized boundary layer is usually found). This pressure difference is responsible for the pressure drag. At least this is the basic scenario that is often recited. Such statements are not simple to justify using the Navier–Stokes model. Because good predictions of drag are so important in the design of airplanes and boats, this important problem has been studied extensively using theory and experimental measurements.

For this project, start by writing a computational fluid dynamics (CFD) code (based on the projection method) to approximate velocity and pressure fields for the two boxy truck configurations given in the text. Render the velocity field in a figure and approximate the pressure drag produced by a free-stream velocity of 70 miles per hour. At least treat this as a goal. Perhaps lower velocities are required to keep the Reynolds number sufficiently small for your code to produce reliable results. Compute an approximation of drag, perhaps by using the pressure Poisson equation [Eq. (16.12)] at the last step of a projection method computation, by integration of pressure times the outer normal $-p\eta$ over the truck surface followed by projection onto the free-stream direction. This is called a near-field method for approximating pressure drag.

By surrounding the truck inside the region of space bounded by a closed surface, consider the annular region bounded by the truck body and the surface; in three dimensions, this region is topologically a ball with an interior ball removed. Conservation of momentum inside the annular region leads to an expression for computing pressure drag by integration over the outer surface only, a far-field approximation. Develop your own formula or learn how this is done from the literature. Use the far-field theory to approximate drag using your CFD computation. How does your result compare with the near-field computation? Which is more accurate for projection method approximations?

Predict which boxy truck configuration has the least drag.

Add improvements to the truck model by replacing the vertical windscreen of the boxy truck with an inclined ramp windscreen. How should curved boundaries be handled in the projection method code? Add improvements to the model by fairing the boxy model into a more realistic curved shape. How much does the shape have to be modified to detect a change in your CFD computations?

For which configurations, if any, of the truck shape does the flow field around the truck moving in a steady ambient flow reach a steady state?

How big must the computational domain be relative to the truck size to eliminate (for all practical purposes) boundary effects due to confining the free-stream flow to a bounded computational domain?

Make a movie to show the evolution of the flow field as the truck speed is varied from zero to 70 miles per hour. Refine the grid size until there is no change in the computational results or the computational expense exceeds available computer power.

How does the flow field change when two parameters are altered: the relative length of the truck bed and the height of the tailgate? What tailgate height for a given truck bed-length produces the least drag? Solid results on these problems could produce useful research publications. See [72] for a modern view of the physics of aerodynamic drag and reread the discussion of lift on page 331.

Do an engineering study of fuel cost for a trucking company. Is reducing aerodynamic drag a significant factor in operating costs?

19.6 LOW REYNOLDS NUMBER FLOW

(a) Consider the Stokes flow as in Section 11.3. Learn how to model Stokes flow past a sphere immersed in a hypothetical body of water with infinite extent (see, for example, [60]) and report on the corresponding exact solution of Stokes' equations in this context. Write a proof of Stokes' drag formula for this case: the drag force is $6\pi\mu r u$, where μ is the viscosity and r is the radius of the sphere. Explain how Robert A. Millikan (in his oil drop experiment preformed in 1909) used this result to determine the charge on an electron. Discuss Stokes' paradox.
(b) Consider two-dimensional lid-driven cavity flow for a highly viscous fluid: flow of a fluid filling a box where the lid of the box is moving with nonzero constant velocity relative to the velocity of the box. A Navier–Stokes flow in this setting is depicted in Fig. 16.4. Approximate such a two-dimensional flow numerically using the Stokes model and display your results in a comparable figure. Hint: You may wish to read Chapter 16.

19.7 FLUID MOTION IN A CYLINDER

(a) Address the following numerical (and analytical) challenge: Imagine a cylindrical bottle full of a viscous fluid and capped at both ends. Suppose the bottle is rotating with constant velocity about its central axis. Describe the motion of the fluid inside the bottle. There are numerous variations on this theme. Perhaps one or both of the caps are fixed during the rotation, the rotation speed of the cylinder is zero but one of both of the caps are rotating, the speed of rotation is increasing with time, the bottle is instead rotating end over end, and so on. What can you say? Report on analysis and numerics. Hint: For some of the suggested scenarios, a rotating coordinate system (as discussed in Chapter 14) may be helpful. These problems have received some attention (see, for example, [105]). Why would anyone care about solutions to such problems?

(b) Imagine the fluid-filled cylinder in flight. To model this situation in a realistic manner requires at least coupling Newton's second law of motion for the cylindrical body with the fluid equations. More realistically, lift and

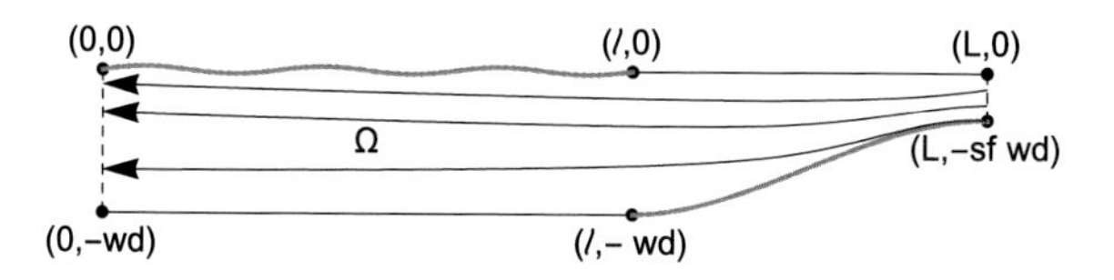

Fig. 19.2 Schematic diagram of flow through a diffuser inlet over a flat bottom, with a free surface, and through a free outlet. Fluid fills the entire region Ω.

drag due to the cylinder moving through air (or water) must be taken into account. How would the motion of the fluid inside the cylinder affect the flight of the cylinder? This is a nontrivial modeling problem. Do some research (literature search) on what is known, write a report discussing your findings, and write a detailed account of at least one mathematical result or numerical experiment that makes a useful prediction from a model.

19.8 FREE-SURFACE FLOW

19.8.1 Free-Surface Flow via Dubreil–Jacotin Transform

The ideal water wave model [Eqs. (15.1)–(15.4)] does not include vorticity and does not lead naturally to a boundary value problem (BVP) where the unknown free surface is incorporated into a fixed domain. An alternative that does allow for these properties, at least in the case of steady flow, is formulated by considering stream functions instead of potentials.

The project is to determine the free-surface profile of flow over a flat bottom past a diffuser (see Fig. 19.2).

Begin with some mathematical considerations.

Recall Exercise 13.6 and show that an incompressible flow has a stream function. Of course, the fluid velocity field (u, v) can be recovered when the stream function ψ is known:

$$u = \psi_y, \qquad v = -\psi_x. \tag{19.14}$$

A useful idea, for some flow problems, is to find a closed set of equations for the stream function and use these instead of the Euler equations for the velocity field. One of the goals of this project is to determine such a set of equations for the diffuser problem.

Show, by simply using its definition, that the vorticity ω is given by

$$\omega = -\Delta\psi.$$

Also show that if the vorticity vanishes everywhere, then $\Delta\psi = 0$ (as in the ideal water wave equations) and the flow is given by a potential. Show that the potential and stream function are related by

$$\phi_x = \psi_y, \qquad \phi_y = -\psi_x. \tag{19.15}$$

Before considering the geometry of the diffuser problem, suppose that the fluid resides over a horizontal bottom $\mathcal{B} := \{(x, y) : y = -\,\mathrm{wd}\}$ and its surface is given by $\mathcal{S} := \{(x, y) : y = \eta(x)\}$ for some unknown function η. Also, let Ω denote the region occupied by the fluid. The main objective is to determine η from the boundary and initial data of the physical problem being modeled.

The simplest case is zero vorticity where the stream function formulation may be derived directly from the ideal water wave equations and the system of PDEs (19.15). There is a constant C (uniform over the entire flow) such that

$$\begin{aligned} \Delta\psi &= 0 && \text{on } \Omega, \\ \psi_y\eta_x + \psi_x &= 0 && \text{on } \mathcal{S}, \\ \psi_x &= 0 && \text{on } \mathcal{B}, \\ \frac{1}{2}(\psi_x^2 + \psi_y^2) + g\eta &= C && \text{on } \mathcal{S}. \end{aligned} \tag{19.16}$$

The last equation is an expression of Bernoulli's law

$$\frac{1}{2}(\psi_x^2 + \psi_y^2) + gh + \frac{p}{\rho} = \text{constant}, \tag{19.17}$$

where g is the magnitude of the acceleration due to gravity ($9.8\,\mathrm{m}\,/\,\mathrm{sec}^2$) and h is the height above some reference surface. In the model, surface pressure is assumed to be held at a constant atmospheric pressure and the motion of the water is assumed not to affect the atmospheric pressure. Surface tension, a topic not covered in this book, is ignored. Constant terms are moved to the right-hand side to obtain the last equation in display (19.16).

Measurements are required to determine the Bernoulli constant C. For instance, the depth of the water at some horizontal position together with

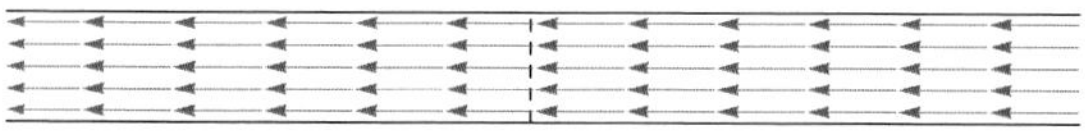

Fig. 19.3 The figure depicts some schematic streamlines, the top and bottom of the fluid domain, and a (dashed) cross section.

the surface flow speed are enough to determine this value. This number is simply treated as a parameter for this project.

Imagine the pattern of streamlines for the flow. They are curves in Ω that are the images of solutions of the autonomous ODE system corresponding to the vector field (19.14). These curves—also called integral curves of the vector field with components u and v—never intersect. Why? Moreover, these curves are level sets of the stream function ψ. Prove this.

An entire region of fluid might be a single level set of the stream function. How can this happen? In the interesting cases where the fluid is flowing over the bottom, the usual streamline pattern would be akin to that shown in Fig. 19.3. The dashed line in the figure is a cross section of the flow; that is, a curve (in this two-dimensional case) situated such that the flow velocity field is never one of its tangents. What is the corresponding notion of cross section for a three-dimensional flow?

In the ideal case depicted in Fig. 19.3, the positions of fluid particles are determined by a coordinate value on the cross section (at the point of intersection of the streamline containing the particle and the cross section) and a second coordinate measuring horizontal position along this streamline. True, but not yet interesting. To tie the new coordinate system more directly to the flow, consider choosing the coordinate at a point on the cross section to be the value of the stream function ψ at this place. Then points elsewhere in the fluid have two coordinates: the value of the stream function and the usual horizontal coordinate. Following the notation in the excellent paper [57], the stream function is adjusted (by adding a constant if necessary) so that its value on $\mathcal{S}$ is zero, and its value on $\mathcal{B}$ is a constant p_0 to be determined.

The new coordinates are

$$q = x, \qquad p = -\psi(x, y);$$

the coordinate transformation is called the Dubreil-Jacotin (DJ) transformation (Marie-Louise Dubriel-Jacotin 1905–1972, see [32]). The minus sign ensures that the vertical coordinate has a negative value at the bottom of the fluid. Because the surface of the fluid is a streamline, its second DJ coordinate is constant (equal to zero). Likewise, the bottom boundary of the fluid corresponds to the constant second coordinate p_0. Thus, a wonderful property of the DJ transformation is revealed: It transforms the fluid domain Ω into a new domain bounded by parallel lines. In particular, the (unknown) free surface in DJ coordinates is a horizontal line.

Why were the DJ coordinates chosen to be (q, p) and not (p, q), which might seem more natural? The answer is not important for this project, but searching for the answer might open new doors. Hint: Learn about Hamiltonian mechanics.

Once the stream function is adjusted (by adding a constant) so that its value on the fluid surface is zero, the value of the stream function is also determined at the bottom of the fluid. To see this, consider the flux of the fluid velocity field through a vertical cross section. The convention in this project is (as in Fig. 19.3) flow from right to left. Thus, it is natural to consider the outer unit normal on a vertical section to be the vector with components $(-1, 0)$. The flux over the cross section is

$$\text{flux} = \int_{-\,\text{wd}}^{\eta(x)} \begin{pmatrix} u \\ v \end{pmatrix} \cdot \begin{pmatrix} -1 \\ 0 \end{pmatrix} dy.$$

Show that this flux is a positive quantity equal to $\psi(x, -\,\text{wd})$ for $0 \le x \le \ell$. Moreover, on the entire level set $\mathcal{B}$ that is taken to be a streamline, this quantity is the same constant p_0. In particular, the flux does not depend on the horizontal coordinate x. The DJ coordinates along the bottom are $(q, -p_0)$ due to the minus sign in the definition of the DJ transformation.

Consider the diffuser geometry in Fig. 19.2. The specified (x, y) coordinates use the length parameters $L > \ell > 0$, the water depth $\text{wd} > 0$, and the scale factor $0 < \text{sf} < 1$. The free surface is on the interval $[0, \ell]$. The entire bottom is solid and the upper boundary between the points $(\ell, 0)$ and $(L, 0)$ is solid. The inlet is the far-right dashed vertical line and the outlet is the left-most dashed vertical line.

In the figure, a cubic polynomial function G provides the profile of the curved solid lower boundary of the diffuser; it is chosen to have zero slope at its end points. What is the precise formula for this diffuser profile function? Of course, you are free to try different diffuser profiles to predict their influence on the flow.

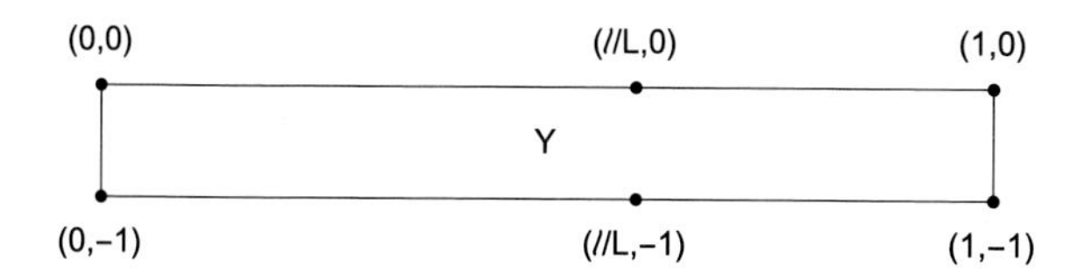

Fig. 19.4 A schematic diagram of fixed domain Υ using DJ transformation of Ω is depicted.

Flow at the inlet must be specified. The obvious simple choice is uniform horizontal flow

$$\psi_y(L, y) = -a, \qquad \psi_x(L, y) = 0 \tag{19.18}$$

with the constant $a > 0$. Recall that atmospheric pressure is specified to have value zero. The pressure at the inlet must also be specified. The simplest, but perhaps unrealistic choice, is uniform pressure given by some positive number p_{in}.

For computation, the nonlinear boundary condition on the free surface S is

$$\frac{1}{2}(\psi_x^2 + \psi_y^2) + g\eta = C, \tag{19.19}$$

where the parameter C may be positive, negative, or zero.

The most problematic boundary condition is at the outlet. There is no way to determine it precisely without taking into account the actual fluid motion beyond the outlet. Correct specification of outflow is a major problem in fluid mechanics that does not have a simple solution. You may wish to try different approaches to appreciate the problem and to see how different choices might affect the predicted flow. One approach is the do-nothing boundary condition: the flow is the same just before and just after the outlet. When the outlet is sufficiently far downstream from the part of the flow under consideration, this boundary condition makes sense as the flow should have no memory of what happened far upstream. At least this is true for viscous flow (which includes dissipation). For the Eulerian flow considered here, the do-nothing condition is less physically realistic but perhaps reasonable for determining a first approximation to the actual flow. Can you do better?

With the physical free-surface problem defined, take advantage of the DJ transformation to determine a fixed BVP for the change of coordinates. A minor change in the choice of DJ coordinates is useful to make the problem dimensionless and to incorporate the inlet flux into the equations as a control parameter:

$$q = \frac{x}{L}, \qquad p = -\frac{\psi(x,y)}{p_0}. \tag{19.20}$$

The inverse transformation is taken to be

$$x = Lq, \qquad y = LH(q,p) \tag{19.21}$$

for an unknown dimensionless function H. The key equation is

$$y = LH(\frac{x}{L}, -\frac{\psi(x,y)}{p_0}). \tag{19.22}$$

Using the chain rule, an easy computation relates the partial derivatives of the stream function and the unknown function H:

$$\psi_x(x,y) = \frac{p_0 H_q(\frac{x}{L}, -\frac{\psi(x,y)}{p_0})}{LH_p(\frac{x}{L}, -\frac{\psi(x,y)}{p_0})}, \quad \psi_y(x,y) = \frac{-p_0}{LH_p(\frac{x}{L}, -\frac{\psi(x,y)}{p_0})}. \tag{19.23}$$

Going further, show that Poisson's equation $\Delta\psi = 0$ in Ω is transformed to

$$H_p^2 H_{qq} - 2H_p H_q H_{qp} + (H_q^2 + 1)H_{pp} = 0 \tag{19.24}$$

on the strip

$$\Upsilon = \{(q,p) : 0 < q < 1, \ -1 < p < 0\}.$$

Thus, the PDE for the inverse coordinate transform H is defined on a *fixed* domain. This is of course the most important feature of the DJ coordinates.

To obtain a PDE BVP for H, the boundary conditions must be set on the strip Υ (see Fig. 19.4).

The solid bottom of the fluid domain corresponds to the level set $\{(x,y) : \psi(x,y) = -p_0\}$. From the key equation and using the diffuser profile function G, $y = LH(q,-1)$ on this set; therefore,

$$H(q,-1) = \begin{cases} -\frac{\text{wd}}{L}, & 0 < q < \frac{\ell}{L}, \\ \frac{G(Lq)}{L}, & \frac{\ell}{L} \le q < 1. \end{cases} \tag{19.25}$$

Likewise, the top boundary corresponds to the level set $\{(x,y) : \psi(x,y) = 0\}$ and the key equation reduces to $y = LH(q,0)$. On the solid part of the boundary $\ell/L < q < 1$, the boundary condition is simply

$$H(q, 0) = 0. \tag{19.26}$$

On the free surface, the boundary condition using the same methodology would be

$$H(q, 0) = \eta(Lq)/L. \tag{19.27}$$

But, this equation alone is not a viable boundary condition because we do not know the free-surface elevation η, which is the most important unknown to be determined. Instead, Eq. (19.27) together with the nonlinear free-surface boundary condition [Eq. (19.19)], where η appears in an equation, is used.

Note that the inlet flux is sf wd a and therefore

$$p_0 = \text{sf wd}\, a. \tag{19.28}$$

Also, define the dimensionless parameters

$$\alpha := \frac{C}{Lg}, \qquad \gamma := \frac{(\text{sf wd}\, a)^2}{2gL^3}. \tag{19.29}$$

Here, α is a measure of the inlet fluid pressure and γ (which is a nonnegative parameter) measures the product of the inlet fluid velocity and the diffuser size. This formulation suggests that a smaller diffuser opening and a larger velocity would produce the same flow as a larger diffuser opening and a smaller velocity.

After some algebraic manipulation using boundary conditions (19.19) and (19.27), Eq. (19.28), partial derivatives (19.23), and dimensionless parameters (19.29), the free-surface nonlinear boundary condition on the set

$$\{(q, p) : 0 < q < \frac{\ell}{L}, p = 0\}$$

is

$$\gamma(H_q^2 + 1) + H_p^2(H - \alpha) = 0. \tag{19.30}$$

In view of partial derivatives (19.23) and Eq. (19.28), the inlet condition (19.18) is recast in the form

$$H_p(1,p) = \frac{\text{sf wd}}{L}. \tag{19.31}$$

Due to the scaling, the control parameter a does not appear. For numerics, the inlet condition is best expressed as

$$H(1,p) = \frac{\text{sf wd}}{L}p, \tag{19.32}$$

where the constant of integration is taken to be zero so that H vanishes on the solid top surface of the diffuser.

A do-nothing boundary condition may be viable at the outlet. Perhaps the simplest possibility is to suppose that the fluid velocity does not change in the horizontal direction near the outlet. More precisely, we may suppose that there is some positive ϵ such that η is constant on the interval $[0, \epsilon]$ and

$$\psi_y(0,y) = \psi_y(x,y)$$

on the set $\{(x,y) : 0 \le x \le \epsilon, \ -\text{wd} < y < \eta(0)\}$. Integration of this expression yields the equality $\psi(0,y) = \psi(x,y) + \omega(x)$ for an arbitrary function ω, which is the constant of integration. Under the further assumption that $(0,y)$ and (x,y) are on the same streamline for $0 \le x \le \epsilon$, the function ω vanishes and

$$\psi(0,y) = \psi(x,y).$$

Such a condition is not mathematically elegant. It implies (the Neumann condition) $\psi_x(0,y) = 0$, which might be tried instead. Using the DJ change of coordinates,

$$H(0, -\frac{\psi(0,y)}{p_0}) = H(\frac{x}{L}, -\frac{\psi(x,y)}{p_0})$$

for every fixed choice of y. In case $(0,y)$ and (x,y) are on the same streamline (as they are under the assumption) and the DJ coordinates $p = -\psi(x,y)/p_0$ and $q = x/L$ are used, the corresponding do-nothing boundary condition at the outlet is

$$H(0,p) = H(q,p) \tag{19.33}$$

for small $0 < q < \epsilon/L$. So, again, this condition might be replaced by $H_q(0, p) = 0$.

A fixed BVP for H has been defined. Although the outlook for making a numerical approximation seems promising, there is a major unresolved mathematical consideration: Do solutions of the BVP exist and are they unique? To approach this problem requires a sophisticated mathematical analysis. As applied mathematicians, we should not try to approximate a solution of a BVP until we know it exists and is unique. Unfortunately, the required mathematical analysis is often difficult or simply beyond what is currently known. Indeed, exactly this issue lies at the heart of many basic problems of fluid dynamics: existence and uniqueness are generally unknown. Of course, even if the analytic problem is known to have a unique solution, the corresponding discrete problem addressed in a computer code may not have a unique solution and vice versa. Nonlinear problems are difficult, but they are important and a source of endless fascination. Numerical experiments may be performed, but caution is advised. One other point is worth mentioning: The full Navier–Stokes equations—at least as long as they are considered to be the correct model for fluid motion—should be used for (numerical) approximations in fluid simulations, and simplifications of this model (like the irrotational Stokes flow considered here) should be used for (mathematical) analysis to gain insight into the true motion of fluids. But sometimes making numerical approximations of approximate models is useful; here, doing so certainly provides good practice in making numerical experiments. Are there other reasons why the full Navier Stokes model might not be viable for numerical experiments when modeling a physical problem?

The project challenge is to explore the behavior of the model flow over the parameter space (especially considering the parameters a and c) by numerical approximations of solutions of the BVP for H as a method to obtain approximate streamlines in the physical domain. Is there any interesting behavior on the free surface? There is unlimited opportunity for experimentation and analysis. New, carefully obtained results that exhibit interesting phenomena might be part of a publishable research article. Of course, the ultimate project would be to compare numerical predictions using the model equations with physical experiments.

Here, the diffuser is viewed as the outlet of some covered channel containing flowing water and the downstream free-surface profile is to be

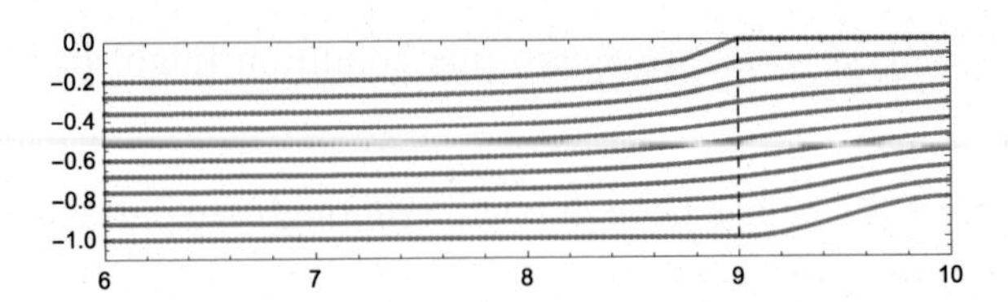

Fig. 19.5 The figure depicts computed streamlines for a cubic diffuser profile for irrotational flow with system parameters $L = 10.0$, $\ell = 9.0$, $\mathrm{wd} = 1.0$, $\mathrm{sf} = 0.8$, $a = 100.0$, $c = a^2/2$, and $g = 9.8$. The DJ fixed domain BVP was solved on a 40×40 grid and the stream function was approximated from the computed solution. In the figure flow is from right to left, and the dashed line marks the downstream outlet of the diffuser.

determined. But, this is not the usual situation. Diffusers are used in gas flow to manipulate pressure in ducts. In this case boundary layers and vorticity play an important role. This subject is ripe for investigation as part of the suggested project. Of course, many other free-surface problems in hydraulics could be approached using the DJ transformation (for instance, flow over a weir as in Fig. 19.6). Also, vorticity may be included in the model (see, for example, [57]).

A few programming notes might prove helpful. Set up a rectangular grid for the fixed BVP on the rectangle $[0, 1] \times [-1, 0]$ and use the PDE for H and the boundary conditions to write one equation for each computational node on the grid. In case your grid has m horizontal and n vertical subdivisions, there will be $(m + 1)(n + 1)$ nodes. The computational nodes are the union of the subset of these, consisting of all interior nodes ($(m - 1) \times (n - 1)$ of them) and the boundary nodes corresponding to the free surface, which may be taken to be the nodes with indices $(i, n + 1)$, $i = 2, 3, \ldots, \mathrm{floor}(\ell m/L)$. In this scheme, the upper left node is not in the computational domain; it is used to implement the do-nothing boundary condition on the downstream end of the free surface: the value of H is the same at the last two downstream nodes. With this convention, there are exactly $N := (m - 1)(n - 1) + (\mathrm{floor}(\ell m/L) - 1)$ nonlinear equations in N variables.

The equations (rearranged so that their right-hand sides are zero) may be viewed as defining a function $F : \mathbb{R}^N \to \mathbb{R}^N$. Your job is to find a zero of this function. There should be exactly one zero, but this fact is not obvious. Newton's method (or one of its variants) can be used to approximate this root. Once it is in hand, you have an approximation of the function H over the rectangular domain. This approximation can be used to construct streamlines in the physical domain. Strive for graphical representations such as in Fig. 19.5.

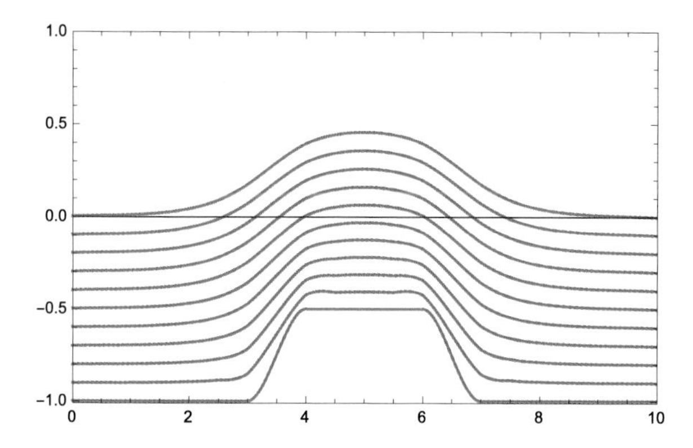

Fig. 19.6 Computed streamlines are shown for flow over a weir with flat top and cubic sides. The flow is from right to left with constant horizontal inlet velocity specified and do-nothing boundary conditions imposed at the outlet.

Implementation of Newton's method to solve $F(x) = 0$ requires the derivative DF evaluated at each iterate and a method to approximate the solution of a system of linear equations. The basic algorithm for approximating a root of a nonlinear function $F : \mathbb{R}^N \to \mathbb{R}^N$ should be familiar from reading this book (if not, see Appendix A.14): Make an initial guess x^0. Suppose that the iterate x^k, for some $k \geq 0$, has been computed. To obtain the next iterate x^{k+1}, solve the linear system

$$DF(x^k)y = F(x^k)$$

and set $x^{k+1} = x^k - y$. Computing and coding the derivative $DF(x^k)$ is feasible for this project. An iteration technique, such as SOR, can be used to solve the linear system at each step; Gaussian elimination is a viable (perhaps better) alternative. With a faithful implementation, convergence is rapid, but the computational overhead per step is high and encoding the components of the derivative is tedious. You might wish to learn about automatic differentiation or computer algebra if you wish the computer to do this task. The number of equations is on the order of mn and thus becomes large as the grid is refined. Taking advantage of the banded structure of the matrix corresponding to the derivative allows finer grids to be employed.

One way to avoid computing the derivatives is to approximate them using difference quotients (divided differences). A good way to make the approximation from a subroutine that approximates F, is to approximate the partial derivatives of F by viewing them as directional derivatives in the directions of the usual basis vectors:

$$\frac{d}{dt}F(x^0 + te_i)\big|_{t=0} = DF(x^0)e_i.$$

The suggested approximation of the first column of the matrix $DF(x^0)$ in the usual basis is, for some small $t > 0$,

$$DF(x^0)e_i \approx \frac{1}{t}(F(x^0 + te_i) - F(x^0)). \tag{19.34}$$

Is there a good way to chose t? Comment on this issue. When t is too large, the truncation error is large; when t is too small, significant digits can be lost (that is, the roundoff error can be large) due to subtraction of nearly equal quantities (which is one of the insidious problems in numerical computation).

Another better alternative is to use a quasi-Newton method, for example Broyden's method (Charles George Broyden, 1965) as explained in Appendix A.14.

All this being said, there is a major difficulty in implementing a viable code: the Jacobian matrix $DF(x^k)$ is highly ill-conditioned (nearly singular). This is the reason for choosing the perhaps nonphysical inlet speed of 100 meters/second. For the parameters used to make Fig. 19.5 the ill-conditioning is manageable. At least, the DJ BVP suggested in this project is an excellent example of ill-conditioning arising in a realistic model.

A remedy for the ill-conditioning (which goes by the name regularization) is beyond the scope of this book, but is well worth your attention. Very briefly, imagine a linear system $Ax = b$ that does not have a solution. The matrix A is singular and b is not in the range of this operator, or perhaps A has more rows than columns as in the least squares problem discussed in Appendix A.12. An approximate solution is sought as the minimizer of $||Ax-b||^2$. In an ill-conditioned problem where A is a nearly singular square matrix, it can be advantageous to seek solutions of the original linear system as the minimizer of the function

$$x \mapsto ||Ax - b||^2 + \lambda^2||x||^2$$

for a *carefully chosen* positive real number λ called the regularization parameter. The idea is to penalize a vector x for which the norm of the residual ($||Ax - b||$) is small whenever the norm of this vector (scaled by λ) is too large. Minimizing this function for fixed λ can easily be transformed to a least squares problem and solved efficiently using the singular value decomposition also discussed in the same appendix. Regularization is an important subject that goes far beyond the basic statements made here (see, for example, [47]).

19.8.2 Free-Surface Flow via the Dirichlet-to-Neumann Operator

Reconsider the ideal water wave model [Eqs (15.1)–(15.4)], and recall that it is a free boundary value problem. A clever reformulation (originally due to V. E. Zakharov [124] and used to advantage by many other authors, for example, [22] and [119]) replaces the free BVP for the unknown free-surface profile function η and fluid velocity potential ϕ by an equivalent fixed BVP for η and the new auxiliary function ζ defined by $\zeta(x,t) = \phi(x,\eta(x,t),t)$.

To formulate the new BVP, ignore for the moment the definition of ζ and simply consider it to be an arbitrary smooth function of (x,t). Using the surface profile η and the potential ϕ define the Dirichlet-to-Neumann operator G by

$$G(\eta)\zeta = \phi_y - \phi_x\eta_x,$$

where ϕ is the solution of the PDE

$$\begin{aligned} \phi_{xx} + \phi_{yy} &= 0 \quad \text{on } \Omega, \\ \phi_y &= 0 \quad \text{on } \mathcal{B}, \\ \phi &= \zeta \quad \text{on } \mathcal{S} \end{aligned} \tag{19.35}$$

together with as yet unspecified boundary conditions on the left and right boundaries of the fluid-filled domain Ω. The operator G takes η and ζ as arguments and produces a new function of (x,t).

The desired reformulation of the ideal water wave model is the system of PDEs

$$\begin{aligned} \eta_t &= G(\eta)\zeta, \\ \zeta_t &= -g\eta - \frac{\zeta_x^2}{2} + \frac{(G(\eta)\zeta + \eta_x\zeta_x)^2}{2(1+\eta_x^2)} \end{aligned} \tag{19.36}$$

together with appropriate initial and boundary data.

The first part of the project is to show the equivalence of the two systems of PDEs. At least, show that if η and ϕ solve the ideal water wave model, then η and ζ solve the first-order system of differential equations (19.36). The first differential equation in this system follows immediately from the definition of the operator G; the second differential equation is obtained by first differentiating both sides of the definition of ζ with respect to time (see [22]).

The second part of the project is to apply the reformulated model to approximate the solution of a free boundary problem. Although the steady state diffuser problem discussed in the previous subsection would be an excellent choice, a simpler problem is suggested for an initial approach to numerical computation with model (19.36).

Imagine two-dimensional ideal flow over a weir as in Fig. 19.6, where the flow is from right to left. To ensure a fixed BVP, remove the boundary by assuming periodic boundary conditions; that is, the state variables are all exactly the same at both ends of the fluid domain. Or, if you like, the flow is on the surface of a cylinder. Specify two functions $p(x)$ and $q(x)$, which respect the periodic boundary conditions, and assign the initial data $\eta(x,0) = p(x)$ and $\zeta(x,0) = q(x)$. Numerically approximate the pair $(\eta(x,t), \zeta(x,t))$ by integrating forward in time. Does the approximation appear to reach a steady state? If so, make a figure similar to Fig. 19.6 to display the approximate steady state. Does the steady state depend on the initial data? Each forward time step of the numerical scheme will require the (approximate) solution of BVP (19.35). The latter task should be relegated to a subroutine designed to approximate solutions of Laplace's equation on a fixed domain with mixed boundary data.

Periodic boundary conditions may not be physically realistic. For example, suppose the inflow at the right end of the two-dimensional fluid domain is specified for all time $t \geq 0$ and the problem is to determine the downsteam flow. Although the inlet boundary conditions are set by the physical problem, formulating physically realistic boundary conditions at the outlet is a genuine challenge and a worthy modeling project. Of course, inlet and outlet boundary conditions are required to solve for the potential potential ϕ used to define the Dirichlet-to-Neumann operator.

For simplicity, assume that the inlet velocity field is horizontal with constant magnitude. Note that the outlet boundary conditions must at least respect conservation of mass; that is, the total outlet flux must be inlet velocity a times water depth wd. Under the assumption that the outlet has horizontal coordinate $x = 0$ and outlet velocity is horizontal, mass conservation requires that

$$-\int_{-\mathrm{wd}}^{\eta(0,t)} \phi_x(x,y,t)\,dy = a\,\mathrm{wd}\,.$$

In case the horizontal velocity component is constant (say $\phi_x(0,y,t) \equiv -b$),

$$b = \frac{a\,\mathrm{wd}}{\eta(0,t) + \mathrm{wd}}.$$

The corresponding boundary condition is

$$\phi_x(0,y,t) \equiv -\frac{a\,\mathrm{wd}}{\eta(0,t) + \mathrm{wd}}.$$

Of course, the value of b changes as the system evolves. For numerical time stepping, the value $\eta(0,t)$ can be used to set the boundary condition during the computation of $(\eta(x,t+\Delta t), \zeta(x,t+\Delta t))$.

To reiterate, the proposed problem is to obtain an approximation of steady state flow by marching to a steady state by integrating the system forward in time. The key advantage of the Dirichlet-to-Neumann operator in this scenario is the avoidance of a free BVP at each time step. Start with an initial specification of the functions η and ζ over the horizontal extent of the computational domain, solve the fixed BVP (19.36) for the potential ϕ in the computational domain with top and bottom horizontal boundary conditions set by the choice of η and ζ and left and right boundary conditions set by assumption, use the computed potential together with η and ζ to advance one step in time to produce updated η and ζ, treat these updates as the new initial data, and repeat until the system settles to a steady state.

The solution of the Laplace equation at each time step can be computationally expensive. With a deeper mathematical analysis, more efficient algorithms to compute the Dirichlet-to-Neumann operator (at least for the case of periodic boundary conditions) can be constructed (see, for example, [119]). A direct approach to the Laplace BVP requires treatment of the curved upper boundary given by the graph of $y = \zeta(x,t)$ for some fixed t. Curved boundaries can be confronted directly by taking appropriate finite differences as parts of the boundary pass through rectangular computational cells, the boundary can be approximated by a curve consisting only of horizontal and vertical line segments that are boundaries of cells, or the computational domain can be transformed to a rectangle at the price of modifying the PDE. The latter method might be implemented by using the DJ transformation discussed in the previous subsection.

Consider applications of the Dirichlet-to-Neumann operator approach to the diffuser problem or some other flow problem that you find interesting?

A theoretical problem, which is not too difficult, is to redo the Dirichlet-to-Neumann operator approach using the stream function instead of the velocity potential. Discuss the advantages or disadvantages of using the stream function.

An understanding of the mathematical context in which a model is set often leads to advantageous reformulations. At least, thinking should always precede computing.

19.9 CHANNEL FLOW TRAVELING WAVES

Consider the dimensionless channel flow model [Eqs. (17.74)]:

$$\begin{aligned} \mathcal{A}_s &= -\mathcal{Q}_\xi + p, \\ \mathcal{Q}_s &= -(1-\frac{\iota}{2})(\frac{\mathcal{Q}^2}{\mathcal{A}})_\xi - (\frac{\mathcal{A}^2}{2})_\xi + \frac{1}{\delta}(\mathcal{Q}_{\xi\xi} - \delta(\frac{\mathcal{Q}\mathcal{A}_\xi}{\mathcal{A}})_\xi) - c\frac{\mathcal{Q}^2}{\mathcal{A}^2} \\ &\quad + \sigma(\epsilon\frac{Q}{A} - (\frac{Q}{A})^2) + \mathcal{A}\sin(\vartheta) - \mathcal{A}\mathcal{B}_\xi. \end{aligned} \tag{19.37}$$

Does this system support traveling waves in the form of fronts or pulses?

In pure form, the traveling wave problem would be for a channel flow with no external flow ($p = 0$), a flat bottom ($\mathcal{B}_\xi = 0$), and no inclination ($\theta = 0$). As usual, the basic idea is to seek solutions for the reduced model of the form

$$A(\xi, s) = F(\xi - Cs), \qquad Q(\xi, s) = G(\xi - Cs),$$

where C is some unknown wave speed.

Using the first equation of the model, show that the wave profiles must be related via

$$G(t) = CF(t) + b,$$

where t is used here as a new independent variable and b is the constant of integration. There are two undetermined constants, but using the relation, only one (second-order) ODE. This is obtained from the second PDE by introducing F and G, differentiating, and then using the relation to eliminate G. At this stage, the translation $\xi - Cs$ may be replace by t and, for beauty, F may be set to x and F' to y. This introduces the equation $x' = y$ and the second-order ODE may be replaced by a system of the form

$$\dot{x} = y, \qquad \dot{y} = f(x, y).$$

Find f explicitly.

A traveling wave front exists when both limits $\lim_{t\to\pm\infty} x(t)$ exist and are distinct real numbers; a traveling pulse exists when both limits exist and are the same. Explain this statement.

For the required limits to exist, solutions must approach rest points of the system of ODEs in the phase plane. In particular, rest points of appropriate types must exist. Moreover, physically realistic traveling waves correspond to orbits $t \mapsto (x(t), y(t))$ where $x(t) > 0$ for $-\infty < t < \infty$.

Start with the case where the wave speed is $C > 0$, $C\delta < 1$, and $\sigma = 0$, and show there are no traveling wave solutions. Something surprising happens when rest points are analyzed. What is this surprise? Recall that the term in the equation of motion with factor σ models the rate of change of the downstream component of velocity with respect to the vertical coordinate. Due to the crude approximation used to model this effect, this term was previously ignored.

Suppose $\sigma > 0$. Show that traveling wave fronts exist for $C > 0$ and $C\delta < 1$.

What would such a front look like with respect to the original state variables of the channel flow model?

CHAPTER 20

Classical Electromagnetism

Two jewels of classical physics are James Clerk Maxwell's field equations (1861–1862) and the Lorentz force law (Henrick Lorentz, 1892) for electrodynamics. Although this field theory is not exact in some extreme situations, it is bedrock physics that has been used successfully to solve many problems in applied science. Maxwell compiled the known laws of electrodynamics introduced by Gauss and Faraday, and he corrected Ampère's law to produce the full set of field equations. The Lorentz force law is used to determine the motion of a charged particle moving in an electromagnetic field.

As fundamental laws, the equations of electrodynamics are simple to understand. The complete theory, in its essential features, will be explained in just a few paragraphs. Of course, this is not the entire story. In applications to the physical world, first-principle applications are usually impossible due to the existence of unknown or approximately known charge and current densities in materials. For practical applications, constitutive laws and approximate forms of the theory are used. Circuit theory is a prime example.

The main theme of this chapter is the application of electromagnetism to transmission lines. This realistic and practical application serves to illustrate the use of differential equations in an important application of electromagnetism.

The chapter is meant to be self-contained, but it is certainly not a complete course of study in electrodynamics. There is more emphasis on the mathematics and practical application than the underlying physics. In particular, the field equations are stated without motivation; they are not derived from an action principle, and their relativistic invariance—which is another milestone in the history of physics—is ignored.

20.1 MAXWELL'S LAWS AND THE LORENTZ FORCE LAW

Four time-dependent vector fields in three-dimensional space are considered in Maxwell's theory: The electric field E, magnetic field H, the displace-

An Invitation to Applied Mathematics: Differential Equations, Modeling, and Computation.
http://dx.doi.org/10.1016/B978-0-12-804153-6.50020-8, Copyright © 2017 Elsevier Inc. All rights reserved.

ment field D, and the magnetic induction field B. In SI units, E is measured in volts per meter, H in ampere's per meter, D in coulombs per square meter, and B in webers per square meter. In addition, the charge density is denoted by ρ and the current flux density by J with respective units of coulombs per cubic meter and ampere's per square meter.

An important note is to beware of different systems of units used in electromagnetism. The SI units, which at present are the most popular, are used here. A consistent choice of units is of course essential when mathematical models are compared with experiments.

Gauss's law states that the electric flux through a closed surface is the total charge enclosed by the surface; in equations,

$$\int_{\partial\Omega} D \cdot dS = \int_{\Omega} \rho \, dV. \tag{20.1}$$

For magnetism, Gauss's law states that the magnetic flux through a closed surface is zero; that is,

$$\int_{\partial\Omega} B \cdot dS = 0. \tag{20.2}$$

Faraday's law of induction states that the line integral of the electric field around a loop equals the negative time derivative of the magnetic flux through the surface bounded by the loop:

$$\int_{\partial\Omega} E \cdot d\ell = -\frac{d}{dt} \int_{\Omega} B \cdot dS. \tag{20.3}$$

And, Maxwell's correction of Ampère's law states that the integral of the magnetic field around a loop equals the current flux through the surface bounded by the loop plus the time derivative of the electric flux through the surface bounded by the loop:

$$\int_{\partial\Omega} H \cdot d\ell = \int_{\Omega} J \cdot dS + \frac{d}{dt} \int_{\Omega} D \cdot dS. \tag{20.4}$$

Maxwell's correction of Ampère's law together with Gauss's law implies that charge is conserved; that is, the current flux through a closed surface is the negative time derivative of the charge enclosed by the surface:

$$\int_{\partial\Omega} J \cdot dS = -\frac{d}{dt} \int_{\Omega} \rho \, dV. \tag{20.5}$$

In case the fields are sufficiently smooth, the integral forms of Maxwell's equations can be recast as equivalent partial differential equations (PDEs)

$$\nabla \cdot D = \rho, \tag{20.6}$$

$$\nabla \cdot B = 0, \tag{20.7}$$

$$\nabla \times E = -\frac{\partial B}{\partial t}, \tag{20.8}$$

$$\nabla \times H = J + \frac{\partial D}{\partial t}, \tag{20.9}$$

and the conservation of charge (also called the continuity equation) is given by

$$\frac{\partial \rho}{\partial t} + \nabla \cdot J = 0. \tag{20.10}$$

The flux densities are related to the field intensities by constitutive relations. In linear, homogeneous, and isotropic materials (where the polarization is parallel to the electric field, the material is the same everywhere and in all directions), the constitutive relations are

$$D = \epsilon E, \qquad B = \mu H, \tag{20.11}$$

where ϵ is the (electric) permittivity and μ is the (magnetic) permeability of the medium in which the fields reside. Also, the current density is related to the electric field by the constitutive relation (called Ohm's law)

$$J = \sigma E, \tag{20.12}$$

where σ is the (specific) conductivity of the medium. The new quantities ϵ, μ, and σ may be functions of space and time as well as the fields E and H. The constitutive laws given here are in the simplest forms that are used in practice for real materials. The reader should be aware that the fields in some composite materials are related (perhaps by the design of the material) by much more complex relations. Indeed, much of the difficulty in applications of electromagnetism is dealing with more complex materials and their electromagnetic properties. Materials where D is proportional to E and B is proportional to H are called linear.

For simplicity, the materials considered here are linear and isotropic (that is, materials whose electromagnetic properties are the same in all directions). Under this assumption, ϵ, μ, and σ may depend on time but *not* on position. For simplicity a linear and isotropic material will simply be called isotropic.

Constitutive equation (20.12) is used to define conductors and dielectrics. A conductor is a material with large conductivity; a dielectric is a material with small conductivity. A perfect conductor is a material with infinite σ; that is, $E = 0$. An insulator is a material with $\sigma = 0$; that is, $J = 0$.

The speed of light in a material is (defined to be)

$$c = \frac{1}{\sqrt{\epsilon\mu}}$$

and the impedance by

$$Z = \sqrt{\frac{\mu}{\epsilon}}.$$

In free space,

$$\epsilon = \epsilon_0 \approx 8.854 \times 10^{-12}\,\text{farad}\,/\,\text{m}, \qquad \mu = \mu_0 \approx 4\pi 10^{-7}\,\text{henry}\,/\,\text{m}\,.$$

Of course, the free space speed of light $c_0 = 1/\sqrt{\epsilon_0\mu_0}$ has the usually quoted value $c_0 \approx 2.9979 \times 10^8\,\text{m}\,/\,\text{sec}$.

Using constitutive relations (20.11) and assuming that ϵ and μ are constants, Maxwell's equations take the (perhaps more familiar) form

$$\nabla \cdot E = \frac{1}{\epsilon}\rho, \tag{20.13}$$

$$\nabla \cdot B = 0, \tag{20.14}$$

$$\nabla \times E = -\frac{\partial B}{\partial t}, \tag{20.15}$$

$$c^2 \nabla \times B = \frac{1}{\epsilon} J + \frac{\partial E}{\partial t}. \tag{20.16}$$

Because $\nabla \cdot B = 0$, there exists a vector potential A such that

$$B = \nabla \times A. \tag{20.17}$$

Clearly, A is not uniquely determined by B (for example, the addition of the gradient of any function to A would not affect B). Combining Eqs. (20.8) and (20.17), we obtain

$$\nabla \times (E + \frac{\partial A}{\partial t}) = 0, \tag{20.18}$$

which implies that there exists a scalar potential φ such that

$$E = -\nabla \varphi - \frac{\partial A}{\partial t}. \tag{20.19}$$

Again, the choice of φ is not unique. In particular, given the potentials A and φ, if f is an appropriately smooth function of space and time, then the redefined potentials

$$A' = A + \nabla f \tag{20.20}$$

and

$$\varphi' = \varphi - \frac{\partial f}{\partial t} \tag{20.21}$$

yield the same fields E and B in Eqs. (20.17) and (20.19). This is called the gauge freedom for the potentials A and φ.

Combining Eqs. (20.16), (20.17), and (20.19), and using the vector identity

$$\nabla \times \nabla \times A = \nabla(\nabla \cdot A) - \Delta A \tag{20.22}$$

(which defines the vector Laplacian ΔA in this context), we obtain the following PDE for the vector potential:

$$\frac{1}{c^2}\frac{\partial^2 A}{\partial t^2} - \Delta A = \frac{J}{\epsilon c^2} - \nabla(\nabla \cdot A + \frac{1}{c^2}\frac{\partial \varphi}{\partial t}). \tag{20.23}$$

Assuming the Lorentz gauge condition

$$\nabla \cdot A + \frac{1}{c^2}\frac{\partial \varphi}{\partial t} = 0, \tag{20.24}$$

Eq. (20.23) reduces to the nonhomogeneous wave equation

$$\frac{1}{c^2}\frac{\partial^2 A}{\partial t^2} - \Delta A = \mu J. \tag{20.25}$$

Combining Eqs. (20.6), (20.19), and (20.24), we obtain the nonhomogeneous wave equation for φ:

$$\frac{1}{c^2}\frac{\partial^2 \varphi}{\partial t^2} - \Delta\varphi = \frac{\rho}{\epsilon}. \tag{20.26}$$

If initial conditions for A and φ (and their first time derivatives) are imposed in the distant past with sufficient spatial decay, then Eqs. (20.25) and (20.26) have the retarded potential solutions

$$A(r_1, t) = \frac{\mu}{4\pi}\int_{\mathbb{R}^3} \frac{J(r_2, t - \frac{r_{12}}{c})}{r_{12}}\, dV(r_2) \tag{20.27}$$

and

$$\varphi(r_1, t) = \frac{1}{4\pi\epsilon}\int_{\mathbb{R}^3} \frac{\rho(r_2, t - \frac{r_{12}}{c})}{r_{12}}\, dV(r_2), \tag{20.28}$$

where $r_1, r_2 \in \mathbb{R}^3$ and $r_{12} = |r_1 - r_2|$, where $|\cdot|$ denotes the usual Euclidean norm.

The observation—made by Maxwell—that the electromagnetic fields all satisfy wave equations—which is an obvious corollary because the vector and scalar potentials satisfy wave equations—is one of the most important moments in the history of science. The upshot is that Maxwell's theory predicts the existence of electromagnetic waves that travel at the speed of light. Guglielmo Marconi (building on earlier work of Heinrich Hertz) verified this prediction with his discovery (*circa* 1896) of the long-distance transmission of radio (frequency) waves. He was awarded a Nobel prize (shared with Karl Ferdinand Braun) in 1909 for the body of his work on telecommunications.

A charged particle of mass m, charge q, and velocity v moves in the electromagnetic field (at least for velocities that are small compared with the speed of light) according to Newton's second law of motion and the Lorentz force law:

$$\frac{d}{dt}(mv) = q(E + v \times B). \tag{20.29}$$

Exercise 20.1. Show that conservation of charge is a consequence of Maxwell's equations. Hint: Compute the divergence of the Maxwell–Ampère law and Gauss's law.

Exercise 20.2. Use the Lorentz force equation to determine the motion of a single charged particle, with unit mass and negative unit charge, in two cases: (a) A constant electric field E and (b) A constant magnetic induction field B.

20.2 BOUNDARY CONDITIONS

Although the electromagnetic fields are assumed to be continuous in each isotropic region of a medium in which the fields reside, the fields are *discontinuous* at material interfaces in the presence of surface charges or currents at these boundaries. Maxwell's equations imply that the discontinuities are not arbitrary. Rather, definite boundary conditions are imposed at an interface.

The simplest case is for the magnetic flux B. Indeed, consider a volume Ω, taken for simplicity to be a solid cylinder, with the interface between the two media cutting the cylinder transverse to its central axis. Using Eq. (20.7) and the divergence theorem, we have that

$$0 = \int_{\Omega} \nabla \cdot B \, dV = \int_{\partial\Omega} B \cdot \eta \, dS,$$

where η is the outward unit normal on the surface of the cylinder. With the configuration just described, the top and bottom of the cylinder are essentially parallel to the interface. These surfaces converge to the area equal to the intersection of the interface with the cylinder as the lateral surface height of the cylinder shrinks to zero. To take advantage of this limit, write

$$0 = \int_{\partial\Omega} B \cdot \eta \, dS = \int_{\text{top}} B \cdot \eta \, dS + \int_{\text{bot}} B \cdot \eta \, dS + \int_{\text{lat}} B \cdot \eta \, dS,$$

where η denotes the outward unit normal restricted to the corresponding part of the cylinder boundary. By passing to the limit as the lateral surface height shrinks to zero, the last equation becomes

$$0 = \int_{\text{top}} B \cdot \eta_{\text{top}} \, dS + \int_{\text{bot}} B \cdot \eta_{\text{bot}} \, dS.$$

Choose a unit normal for the surface interface of the two materials and note that in the limit it must be η_{top} or η_{bot}. Suppose the unit normal is η_{top}, which is of course $-\eta_{\text{bot}}$. We then have

$$0 = \int_{\text{top}} B \cdot \eta_{\text{top}} \, dS - \int_{\text{bot}} B \cdot \eta_{\text{top}} \, dS.$$

The tangential component of B along the interface plays no role in the formula; it might be discontinuous across this boundary. But, by shrinking the surface to some point in this set, the equality of the integrals implies that the normal components of the B field in the two adjacent media (which are labeled a and b with the interface normal pointing from a to b) must be the same at the chosen point. Thus, the normal component of the field is continuous across the shared interface boundary. The same result is encoded in the formula

$$(B_b - B_a) \cdot \eta = 0, \tag{20.30}$$

where B_a denotes the limiting value of the B field in the region marked a at the boundary of this set and B_b is defined similarly. This continuity statement is called the electromagnetic boundary condition for the B field.

Similar (perhaps more difficult) arguments imply boundary conditions for the remaining fields.

By choosing the unit normal η at the materials interface to point from medium a toward medium b, the cross product of the difference of the E fields (taken in the appropriate order) with the unit normal vanishes:

$$(E_b - E_a) \times \eta = 0, \tag{20.31}$$

and the difference of the normal components of the D field taken in the appropriate order is equal to the surface charge density:

$$(D_b - D_a) \cdot \eta = \rho_{\text{interface}}. \tag{20.32}$$

With the same unit normal η as for the D field boundary condition, the H field boundary condition is

$$(H_b - H_a) \times \eta = J_{\text{interface}}. \tag{20.33}$$

Eqs. (20.30)–(20.33) are called the (usual) electromagnetic boundary conditions.

20.3 AN ELECTROMAGNETIC BOUNDARY VALUE PROBLEM

Maxwell's laws [Eqs. (20.6)–(20.9)] are eight scalar equations for the twelve unknown components of the fields. Thus, at the outset, the system of equations is underdetermined. The constitutive relations reduce the number of unknown field components to the six components of the two fields B and E. With this modification, the new system appears to be overdetermined: eight equations for six unknowns. But, for given charge and current densities that satisfy the continuity equation in an isotropic, homogeneous, and linear medium (constant permittivity $\epsilon > 0$ and permeability $\mu > 0$), the system

$$\begin{aligned}\frac{\partial B}{\partial t} &= -\nabla \times E,\\ \frac{\partial D}{\partial t} &= \nabla \times H - J,\\ D &= \epsilon E,\\ B &= \mu H,\end{aligned} \tag{20.34}$$

which includes the constitutive relations, has the same number of equations and unknowns (12 equations in 12 unknowns). By substituting the constitutive relations, this system is reduced to

$$\begin{aligned}\frac{\partial B}{\partial t} &= -\nabla \times E,\\ \epsilon\mu\frac{\partial E}{\partial t} &= \nabla \times B - \mu J.\end{aligned} \tag{20.35}$$

It has six equations and six unknowns.

In physical applications, E and B fields that satisfy system (20.35) can always be modified to produce a solution of Maxwell's equations that satisfies the electromagnetic boundary conditions. An outline of the proof of this result is given in the remainder of this section.

Consider some homogeneous medium that fills a region Ω in space that has a piecewise smooth boundary $\partial\Omega$. Suppose that $\partial\Omega$ is the interface to some medium where the electromagnetic field $(\tilde{B}_b, \tilde{E}_b)$ is given. The medium in Ω is labeled a and the medium outside of Ω is labeled b as in the discussion of the electromagnetic boundary conditions, and the unit normal defined on $\partial\Omega$ points into the b medium.

Suppose that B and E are twice continuously differentiable vector fields that satisfy Eqs. (20.35) and extend continuously to $\partial\Omega$.

By applying the divergence operator to both sides of Eqs. (20.35), using the continuity equation ($\rho_t + \nabla \cdot J = 0$), and using the vector identity $\nabla \cdot (\nabla \times X) = 0$, we have that

$$\frac{\partial(\nabla \cdot B)}{\partial t} = 0,$$
$$\frac{\partial(\nabla \cdot \epsilon E - \rho)}{\partial t} = 0.$$

In other words, $\nabla \cdot B$ and $\nabla \cdot \epsilon E - \rho$ are scalar functions on Ω that are *independent of time*. For notational simplicity, define

$$f := \nabla \cdot B,$$
$$g := \epsilon \nabla \cdot E - \rho.$$

Recall the Helmholtz decomposition (which is also called the fundamental theorem of vector analysis): Let Ω be an open set in three-dimensional space with (piecewise) smooth boundary $\partial\Omega$. If (1) h is a smooth function and Y a smooth divergence-free vector function defined on Ω, (2) γ is a smooth function and Υ a smooth vector function defined on $\partial\Omega$, then there is a unique vector function X defined on the closure of Ω such that

$$\nabla \cdot X = h, \quad \nabla \times X = Y$$

on Ω and either

$$X \cdot \eta = \gamma$$

or

$$X \times \eta = \Upsilon$$

on $\partial\Omega$. Moreover there is a scalar function ϕ and a vector function Z defined on Ω such that

$$X = \nabla \phi + \nabla \times Z.$$

The latter equality is called a Helmholtz decomposition of X (for a proof, see [30] or [14]).

According to the Helmholtz theorem, there are (time-independent) vector fields F and K defined on Ω such that

$$\nabla \cdot F = -f, \qquad \nabla \times F = 0,$$
$$\nabla \cdot K = -g, \qquad \nabla \times K = 0,$$

and on $\partial\Omega$,

$$F \cdot \eta = (\tilde{B}_b - B_a) \cdot \eta, \qquad K \times \eta = (\tilde{E}_b - E_a) \times \eta.$$

In Ω, define the modified fields

$$\tilde{B}_a := B + F, \qquad \tilde{E}_a := E + K.$$

Gauss's law is satisfied for these fields. Indeed,

$$\nabla \cdot \tilde{B}_a = \nabla \cdot (B + F) = f + \nabla \cdot F = 0,$$
$$\epsilon \nabla \cdot \tilde{E}_a = \epsilon \nabla \cdot (E + K) = \epsilon(\nabla \cdot E - g) = \rho.$$

Using that F and K are time independent, Eqs. (20.35) remain valid for the fields $\tilde{B}_a$ and $\tilde{E}_a$. In fact,

$$\frac{\partial \tilde{B}_a}{\partial t} = \frac{\partial B}{\partial t} + \frac{\partial F}{\partial t} = \frac{\partial B}{\partial t} = -\nabla \times E = -\nabla \times (\tilde{E}_a - K) = -\nabla \times \tilde{E}_a,$$
$$\mu\epsilon \frac{\partial \tilde{E}_a}{\partial t} = \mu\epsilon \frac{\partial E}{\partial t} = \nabla \times B - \mu J = \nabla \times (\tilde{B}_a - F) - \mu J = \nabla \times \tilde{B}_a - \mu J.$$

At the boundary,

$$(\tilde{B}_b - \tilde{B}_a) \cdot \eta = (\tilde{B}_b - (B + F)) \cdot \eta = (\tilde{B}_b - B) \cdot \eta - (\tilde{B}_b - B_a) \cdot \eta = 0,$$
$$(\tilde{E}_b - \tilde{E}_a) \times \eta = (\tilde{E}_b - (E + K)) \times \eta = (\tilde{E}_b - E) \times \eta - K \times \eta = 0.$$

Thus, our goal is achieved: The fields $\tilde{B}$ and $\tilde{E}$ satisfy the full Maxwell's equations together with the electromagnetic boundary conditions. Also, with the definitions

$$\tilde{D}_a := \epsilon \tilde{E}_a, \qquad \tilde{H}_a := \frac{1}{\mu} \tilde{B}_a,$$

the interface charge and current densities are modeled by

$$(\tilde{D}_b - \tilde{D}_a) \cdot \eta = \rho_{\text{interface}}, \qquad (\tilde{H}_b - \tilde{H}_a) \times \eta = J_{\text{interface}}.$$

In case the surface charge and current densities are given, the boundary value problem (BVP) should be solved for the D and H fields. The E and B fields are then defined by the constitutive relations.

System (20.35) may be difficult to solve for the unknown fields B and E. In practice, physical intuition is often used to suggest solutions of some special type (sinusoids, for example) that are expected to satisfy the boundary conditions and the field equations (including the Gauss laws) but not necessarily the initial data. The solution of the field equations is sought as a linear combination (superposition) of solutions of the special type that does satisfy the initial data. This strategy is often successful. If not, numerical methods are used to approximate the desired solution.

A procedure has been outlined that can be used to determine a solution of Maxwell's equations that satisfies the electromagnetic boundary conditions. Perhaps a different procedure would produce a different solution. This would not be satisfactory for a physical theory. The BVP should produce a unique solution once the charges and currents are specified and the initial values of the fields are given. In fact, under these conditions the fields are unique.

To show the uniqueness of solutions, suppose there are two electromagnetic fields (B_1, E_1) and (B_2, E_2) that satisfy the full Maxwell system together with the boundary and initial data. In particular, these electromagnetic fields satisfy system (20.35). The boundary data is assumed to include the assumption that there is a specified external field E_b so that both E_1 and E_2 if taken as E_a satisfy the electromagnetic boundary condition [Eq. (20.31)].

Define new fields $B := B_1 - B_2$ and $E := E_1 - E_2$. The strategy is to prove that (B, E) is zero.

By substitution into system (20.35), it follows immediately that

$$\frac{\partial B}{\partial t} = -\nabla \times E, \qquad \epsilon\mu\frac{\partial E}{\partial t} = \nabla \times B.$$

Define a new quantity

$$Q := \int_\Omega \big(\frac{\epsilon}{2} E \cdot E + \frac{1}{2\mu} B \cdot B\big)\, d\mathcal{V}$$

and note that by using the divergence theorem and a vector identity,

$$\begin{aligned}\frac{dQ}{dt} &= \int_\Omega \left(\epsilon E\cdot\frac{\partial E}{\partial t} + \frac{1}{\mu}B\cdot\frac{\partial B}{\partial t}\right) d\mathcal{V}\\ &= \frac{1}{\mu}\int_\Omega \big(E\cdot\nabla\times B - B\cdot\nabla\times E\big)\, d\mathcal{V}\\ &= -\frac{1}{\mu}\int_\Omega \nabla\cdot(E\times B)\, d\mathcal{V}\\ &= -\frac{1}{\mu}\int_{\partial\Omega} E\times B\cdot\eta\, d\mathcal{S}\\ &= \frac{1}{\mu}\int_{\partial\Omega} B\cdot E\times\eta\, d\mathcal{S}.\end{aligned}$$

By subtracting the electromagnetic boundary condition for E_1 and E_2, we have that $E\times\eta = 0$ on the boundary of Ω. Thus, Q must be a time-independent scalar function. At the initial time, both (B_1, E_1) and (B_2, E_2) satisfy the same initial data. This means the electromagnetic field (B, E) vanishes at the initial time. Hence, Q vanishes for all time. By inspection of the integrand in the definition of Q (which is the square of a length of the electromagnetic field with respect to a new inner product), we must have as desired that (E, B) is the zero electromagnetic field at every instant of time that it exists.

The field $E\times B$ is called the Poynting vector; it plays a role in problems involving electromagnetic energy, which is given by the quantity Q. Note that the uniqueness argument includes a proof that electromagnetic energy is conserved.

From a mathematical perspective, an initial boundary value problem (IBVP) is called well posed when it has a unique solution that depends continuously on the initial data and the boundary data. The idea for this definition should be clear. The continuous dependence requirement is a precise way of saying that small changes in the data result in small changes in the solution. In a physical problem, the data might be measured by experiment. For a well posed problem, small errors in measurements of boundary or initial data would be expected to change the solution by a comparable small amount. Of course, to quantify what is meant by a small amount requires further analysis that is beyond the scope of this book.

Exercise 20.3. Suppose that f is defined and smooth on a rectangular solid box Ω in three-dimensional space whose boundary has outer unit normal η. Show that there is a vector function F defined on Ω such that $\nabla \cdot F = -f$ and $F \cdot \eta = 0$. Hint: Look for F as the gradient of a scalar function and use Fourier series.

20.4 COMMENTS ON MAXWELL'S THEORY

Maxwell's theory is a description of the fields produced by charged particles: The fields are produced once the charge and current densities are known. One of the main applications of the theory occurs under the assumption that the charge and current densities are specified (along with initial and boundary conditions and the constitutive laws for the behavior of the fields in some medium); the problem is to determine the electromagnetic fields. In addition, the Lorentz force law specifies how charged particles interact with given electromagnetic fields. Under the assumption that the fields are specified, the motion of charged particles is determined using mechanics; that is, force—including the Lorentz force—equals mass times acceleration for slow particles and, for particles moving with nearly the speed of light, force equals the time rate of change of mass times velocity v divided by the Lorentz factor $\sqrt{1 - v^2/c^2}$, where c is the speed of light. In both cases, the motions of the charged particles is governed by a system of ordinary differential equations. These two aspects of Maxwell–Lorentz theory are the bedrock of applied electromagnetism and they are enormously successful in making useful predictions. Both, however, are not correct at a fundamental level where the fields and the motions of charges must be coupled and solved simultaneously.

Suppose there were exactly two charges in the universe and in some inertial coordinate system one of the charges is accelerating relative to the other. It produces an electromagnetic field that moves at the *finite* speed of light. The other particle interacts with this field (after the delay required for the field to reach it). The second particle produces an electromagnetic field that will eventually interact with the first particle. The particles also interact with the fields that *they* produce. So, to determine the motion of the two-particle system requires a coupling of the fields produced by the moving charges and the interactions (via the Lorentz law) of these fields with the particle motions. There does not seem to be a complete solution of this electrodynamic two-body problem. Indeed, when the self force (radiation reaction) is taken into account, the theory leads to apparent contradictions that have not been resolved (see, for example, [37]).

Fortunately, excellent results that agree with experiments are obtained by making approximations and ignoring small interaction effects. A strong electromagnetic field can be produced and used to determine the motions of a few (relative to the strength of the field) charges by simply ignoring the interaction of the fields produced by the test charges on the strong field. Likewise, the fields around a conductor, for example, can be measured without taking into account the interactions of these fields with the strong current in the conductor. For most applications it is not necessary to use the full coupling of the fields and the charges. But, it is wise to recognize and understand the approximations that are employed.

One tool for avoiding the Maxwell–Lorentz coupling is to assume Ohm's (constitutive) law: current density is the conductance times the electric field: $J = \sigma E$. It replaces the Lorentz force law by specifying the coupling of the current density and the electric field.

20.5 TIME-HARMONIC FIELDS

Under the assumption that all fields can be represented as a superposition of sinusoidal fields, many problems in electrodynamics are simplified by considering sinusoidal fields oscillating at some fixed frequency. From a more sophisticated mathematical viewpoint, which will not be considered here, we simply transfer the fields from the time domain to the frequency domain via the Fourier transform. The naive approach discussed here produces the same results.

Using the generic complex vector field X, suppose all fields under consideration when evaluated at a point in space-time have the from

$$X(x, y, z, t) = \tilde{X}(x, y, z)e^{-i\omega t},$$

where the circular frequency ω does not depend on its position in space-time. As is customary, let us write X for $\tilde{X}$ when the harmonic time-dependence is understood. Also, the physical field is taken to be the real part of X. Using these conventions, system (20.34) for time-harmonic fields (including the current density J) is recast in the form

$$\begin{aligned} i\omega B &= \nabla \times E, \\ i\omega D &= J - \nabla \times H, \\ D &= \epsilon E, \end{aligned}$$

$$B = \mu H. \tag{20.36}$$

Of course, the main simplification is removal of the time derivatives.

For a material where the current density obeys Ohm's law [Eq. (20.12)], system (20.36) may be rewritten for the E and B fields in the form

$$\begin{aligned} i\omega B &= \nabla \times E, \\ i\epsilon\omega\mu E &= \mu\sigma E - \nabla \times B. \end{aligned} \tag{20.37}$$

By computing the curl of both sides of the first equation, using the definition of the vector Laplacian [Eq. (20.22)], and eliminating $\nabla \times B$, the result is a PDE for the E field:

$$\Delta E + \omega\mu(\epsilon\omega + \sigma i)E = \nabla(\nabla \cdot E).$$

Apply the divergence operator ($\nabla\cdot$) to both sides of the second equation in display (20.37) to obtain

$$(\sigma - i\epsilon\omega)\nabla \cdot E = 0,$$

and note that therefore $\nabla \cdot E = 0$. This result implies that E satisfies Helmholtz's equation

$$\Delta E + \omega\mu(\epsilon\omega + \sigma i)E = 0. \tag{20.38}$$

Also, under our assumptions (time-hamonic, Ohm's law, isotropic medium) and Gauss's law, the charge density is zero in the medium. If Ohm's law holds and charge is conserved, then the charge density decreases exponentially fast (see Exercise 20.4). The additional assumptions require the charge density has already decayed to zero.

The B field in the present context also satisfies the same Helmholtz equation:

$$\Delta B + \omega\mu(\epsilon\omega + \sigma i)B = 0. \tag{20.39}$$

Exercise 20.4. Show that in an isotropic material where Ohm's law holds, the charge density decreases exponentially fast. What is the decay constant?

Exercise 20.5. In the context of Eq. (20.38), show that the B field also satisfies Helmholtz's equation.

CHAPTER 21

Transverse Electromagnetic (TEM) Mode

The electromagnetic fields outside two ideal conductors can be defined consistently in a special configuration called a transverse electromagnetic (TEM) mode, which is discussed in this section.

A conductor (usually made of metal) has free charges that are confined to the material. In an ideal conductor, the free charges (which must be all of the same sign) are assumed to repel each other so that they *instantly* move to an equidistant configuration on the surface of the conductor in such a way that the surface density produces a zero electric field inside the conductor. In reality, the charges penetrate the conductor in a very thin layer such that the outer boundary of the layer is the conductor's surface.

For the remainder of this section, all conductors are assumed to be ideal.

When charges are in motion on the surface of a conductor with outward unit normal η, the electric field E vanishes in the conductor. By the first and fourth equations in display (20.36), they produce a magnetic field H that also vanishes inside the conductor. And, by employing the electromagnetic boundary conditions, there is a corresponding surface current

$$\eta \times H = J_{\text{surface}}.$$

Because the electric field vanishes in the conductor, the electric field boundary condition implies that the electric field in the exterior of the conductor has zero tangential component on the interface boundary. Thus, the external electric field E must be perpendicular to the boundary of the conductor.

As $H = 0$ inside the conductor, $B = 0$ inside the conductor. The B field boundary condition implies that this field (at the interface) is tangential to the boundary. Thus, the magnetic field H in this situation is tangential at the interface.

Under the conditions just described, the electric field lines meet the two conductors orthogonal to their surfaces. The magnetic field lines are

An Invitation to Applied Mathematics: Differential Equations, Modeling, and Computation.
http://dx.doi.org/10.1016/B978-0-12-804153-6.50021-X, Copyright © 2017 Elsevier Inc. All rights reserved.

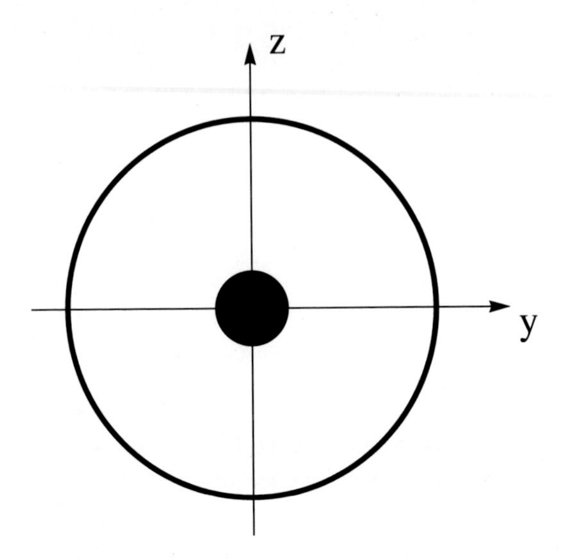

Fig. 21.1 The figure depicts a schematic cross section of a coaxial cable running in the direction of the x-axis in a Cartesian coordinate system. The inner disk represents the inner conductor; the outer circle represents the inner surface of the outer conductor.

tangential to these surfaces; thus, they are transverse to the electric field lines near the conductors (that is, the tangent vectors B and E are not parallel at crossing points of the field lines near the conductor).

For definiteness and in preparation for a discussion of transmission lines (the main application of this chapter), assume the two conductors are coaxial (as suggested in Fig. 21.1) and the inner conductor is a circular cylinder whose radius is smaller than the outer coaxial cylindrical conductor.

A TEM mode is a choice of the electromagnetic fields that is consistent with Maxwell's equations, satisfies the electromagnetic boundary conditions at the inner surface of the outer conductor and the outer surface of the inner conductor, and is such that the field components vanish in the direction of the axis of the conductors. For rectangular coordinates chosen such that the axis of the conductors is the x-axis of the coordinate system, the electromagnetic fields are allowed to have nonzero (time-varying) components in the y and z directions, but not in the x direction. To be clear, the generic field of this type X represented in the usual components relative to the rectangular (x, y, z) coordinate system has the form $(0, X_2(x, y, z, t), X_3(x, y, z, t))$.

Let us seek an axial plane wave time-harmonic TEM mode between two conductors (for example, in the annulus between coaxial conductors whose cross section is depicted in Fig. 21.1) in case the homogeneous and isotropic material between the conductors is such that the current density in the material obeys Ohm's law $J = \sigma E$.

The generic complex field X is an axial plane wave when it has the special component form

$$X(x, y, z, t) = (0, X_2(y, z)e^{i(kx-\omega t)}, X_3(y, z)e^{i(kx-\omega t)}), \tag{21.1}$$

where (the wave number) k might be complex. The physical field is the real part of X.

The wave number cannot be arbitrary; it is determined using system (20.37). Indeed, by substituting the special forms of the B and E fields into this system, the second component of the first equation becomes

$$\omega B_2 = -kE_3, \tag{21.2}$$

and the third component of the second equation is

$$ikB_2 = (\mu\sigma - i\omega\epsilon\mu)E_3.$$

By eliminating B_2, a simple computation can be used to show that

$$k^2 = \omega\mu(\omega\epsilon + i\sigma), \tag{21.3}$$

which is exactly the coefficient in Helmholtz's equations (20.38) and (20.39).

Substitute the special B and E fields into these equations and use Eq. (21.3) to show that

$$\begin{aligned}
\frac{\partial^2 B_2}{\partial y^2} + \frac{\partial^2 B_2}{\partial z^2} &= 0,\\
\frac{\partial^2 B_3}{\partial y^2} + \frac{\partial^2 B_3}{\partial z^2} &= 0,\\
\frac{\partial^2 E_2}{\partial y^2} + \frac{\partial^2 E_2}{\partial z^2} &= 0,\\
\frac{\partial^2 E_3}{\partial y^2} + \frac{\partial^2 E_3}{\partial z^2} &= 0.
\end{aligned} \tag{21.4}$$

It is not necessary to solve all four of the Eqs. (21.4). In fact, it suffices to solve one of them. For example, B_2 and E_3 are related by Eq. (21.2). A full set of relations is

$$\omega B_2 = -kE_3, \quad \omega B_3 = kE_2, \quad \frac{\partial E_2}{\partial z} = \frac{\partial E_3}{\partial y}, \quad \frac{\partial B_3}{\partial y} = \frac{\partial B_2}{\partial z}. \tag{21.5}$$

Thus, it suffices, to determine E_2. One way to do so is to solve the two-dimensional Laplace equation for E_2 with appropriate boundary conditions and recover the remaining fields using these relations.

Inspection of the full set of equations in display (21.5) suggests an alternative method for determining the fields. Converting the fourth equation in this set to an E field equation via the first pair of equations produces the partial differential equations (PDEs)

$$\frac{\partial E_3}{\partial y} = \frac{\partial E_2}{\partial z}, \qquad \frac{\partial E_3}{\partial z} = -\frac{\partial E_2}{\partial y}, \tag{21.6}$$

which are exactly the Cauchy–Riemann equations for the pair of functions (E_3, E_2) viewed as functions of (y, z). Equivalently, the curl of the field $(0, E_2(y, z), E_3(y, z))$ vanishes. This fact implies something we already know: both functions E_2 and E_3 are harmonic with respect to (y, z); that is, the Laplacian with respect to these variables of each function vanishes. Moreover, the vector field

$$E^*(y, z) := (E_2(y, z), E_3(y, z)),$$

which more properly should be defined as a column vector, must be the gradient of a harmonic potential ϕ. In symbols, and in keeping with the usual sign conventions in electrodynamics where E is the negative gradient of the electric potential, these statements are written

$$E^* = -\nabla\phi, \qquad \Delta\phi = 0. \tag{21.7}$$

To be clear, in this context E^* is viewed as a two-dimensional vector field with components (E_2, E_3), and the Laplace equation for the scalar function $\phi = \phi(y, z)$ defined in a cross section of the medium between the conductors is

$$\frac{\partial^2\phi}{\partial y^2} + \frac{\partial^2\phi}{\partial z^2} = 0.$$

Polar coordinates are advantageous for the circular geometry of the coaxial conductors. Recall that the polar coordinate representation of Laplace's equation is

$$\frac{1}{r}\frac{\partial\psi}{\partial r} + \frac{\partial^2\psi}{\partial r^2} + \frac{1}{r^2}\frac{\partial^2\psi}{\partial\theta^2} = 0, \tag{21.8}$$

where $y = r\cos\theta$, $z = r\sin\theta$, and $\psi(r,\theta) := \phi(r\cos\theta, r\sin\theta)$ (see Exercise 21.1).

The boundary conditions for the E^* field are specializations of the electromagnetic boundary conditions to the geometry of the coaxial conductors. The outer normal on the inner conductor in Cartesian components is $\eta = (0, \cos\theta, \sin\theta)$, where θ is the polar angle at a boundary point $(0, y, z)$. The outer normal (which points toward the inner conductor) is $-\eta$. Because the E field vanishes in the conductors and the corresponding electromagnetic boundary condition is $E \times \eta = 0$, it follows that

$$E_2 \sin\theta - E_3 \cos\theta = 0 \tag{21.9}$$

on both boundaries. In particular, E has zero tangential component at the boundaries. In addition, $D = \epsilon E$ in the medium bounded by the conductors. Thus, the boundary condition for D (which is $D \cdot \eta = \pm\rho_{\text{interface}}$) gives a second boundary condition for E. On the inner boundary

$$E_2 \cos\theta + E_3 \sin\theta = \frac{\rho_{\text{inner}}}{\epsilon}$$

and on the outer boundary

$$E_2 \cos\theta + E_3 \sin\theta = -\frac{\rho_{\text{outer}}}{\epsilon}.$$

For the potential function ϕ, we have the Neumann boundary condition

$$\nabla\phi \cdot \eta = -\frac{\rho_{\text{inner}}}{\epsilon}, \qquad \nabla\phi \cdot \eta = \frac{\rho_{\text{outer}}}{\epsilon}, \tag{21.10}$$

where, as previously mentioned, $\eta = (0, \cos\theta, \sin\theta)$ is the unit vector in the radial direction.

The voltage V at point P measured relative to a point G on a reference ground is defined to be

$$V(P) = -\int_\gamma E \cdot d\ell,$$

where γ is a curve connecting G to P through a region where the electric field is defined. Of course, the integral is independent of path when $E = -\nabla\phi$, for some potential ϕ. Indeed, in this case and for γ parameterized (for instance) on the interval $[0, 1]$ with $\gamma(0) = G$ and $\gamma(1) = P$,

$$V(P) = -\int_\gamma E \cdot d\ell = \int_0^1 \nabla\phi(\gamma(s)) \cdot \gamma'(s)\,ds = \phi(P) - \phi(G), \tag{21.11}$$

and the voltage is exactly the electric potential difference at the two points.

Returning to the Neumann boundary data (21.10), there is a compatibility condition for the charge densities on the inner and outer conductors: Their strengths must be equal and opposite to support the static electric field $-\nabla\phi$. More precisely, the condition is

$$-\alpha\frac{\rho_{\text{inner}}}{\epsilon} = \beta\frac{\rho_{\text{outer}}}{\epsilon}, \tag{21.12}$$

where α is the outer radius of the inner conductor and β is the inner radius of the outer conductor. To see why this is true, note that in the static case the total charge sums to zero and the charges on each surface form an equidistant configuration. The total charge on the inner and outer surfaces (with cylinder length ℓ) is

$$2\pi\alpha\ell\rho_{\text{inner}} + 2\pi\beta\ell\rho_{\text{outer}} = 0,$$

which is a restatement of the compatibility condition.

In its purest form, the boundary value problem (BVP) for E^* on a cross section of the coaxial configuration in polar coordinates (where $\psi(r,\theta) = \phi(r\cos\theta, r\sin\theta)$) is

$$\frac{1}{r}\frac{\partial\psi}{\partial r} + \frac{\partial^2\psi}{\partial r^2} + \frac{1}{r^2}\frac{\partial^2\psi}{\partial\theta^2} = 0 \tag{21.13}$$

with the boundary conditions

$$\frac{\partial\psi}{\partial r} = -\frac{\rho_{\text{inner}}}{\epsilon}, \qquad \frac{\partial\psi}{\partial r} = \frac{\rho_{\text{outer}}}{\epsilon}, \tag{21.14}$$

where, of course, the first equation holds for $r = \alpha$, the second for $r = \beta$, and they satisfy the compatibility condition. The normal derivatives in display (21.10) become radial derivatives in polar coordinates.

Due to the symmetry of the problem, the solution is radial (that is, it does not depend on the angle θ). By elementary ordinary differential equation

(ODE) theory, the general solution of the radial differential equation

$$\frac{1}{r}\frac{\partial \psi}{\partial r} + \frac{\partial^2 \psi}{\partial r^2} = 0 \tag{21.15}$$

is

$$\psi(r, \theta) = k \ln r + K, \tag{21.16}$$

where k and K are constants (see Exercise 21.2). Using the inner boundary condition, which suffices due to the compatibility condition [Eq. (21.12)],

$$\psi(r, \theta) = -\frac{\alpha \rho_{\text{inner}}}{\epsilon} \ln r + K,$$

where K is an arbitrary constant.

The BVP does not have a unique solution. But, this result is physically reasonable. A potential should only be defined up to an additive constant. Because the electric field is the negative gradient of the potential, the additive constant does not appear in the electric field. Finally, at least in the static case, voltage is always the *difference* in electric potential at two points; therefore, the additive constant does not appear in measured voltages. For all the stated reasons, we may as well take the potential ψ to be

$$\psi(r, \theta) = -\frac{\alpha \rho_{\text{inner}}}{\epsilon} \ln r. \tag{21.17}$$

This function does not depend on θ.

The (transpose of the desired complex) electric field between the two coaxial conductors is

$$E(x, y, z, t) = (0, \frac{\alpha \rho_{\text{inner}}}{\epsilon} \frac{y}{y^2 + z^2} e^{i(kx - \omega t)}, \frac{\alpha \rho_{\text{inner}}}{\epsilon} \frac{z}{y^2 + z^2} e^{i(kx - \omega t)}). \tag{21.18}$$

Of course, the physical electric field is the real part of E.

Using the definition of voltage [Eq. (21.11)], the (complex) voltage V across the conductors (with the outer conductor as the reference ground) is

$$V = V(x, t) = (\psi(\alpha, \theta) - \psi(\beta, \theta)) e^{i(kx - \omega t)} = \frac{\alpha \rho_{\text{inner}}}{\epsilon} \ln \frac{\beta}{\alpha} e^{i(kx - \omega t)} \tag{21.19}$$

(see Exercise 21.3).

By the relations (21.5),

$$B(x,y,z,t) = (0, -\frac{k\alpha\rho_{\text{inner}}}{\omega\epsilon}\frac{z}{y^2+z^2}e^{i(kx-\omega t)}, \frac{k\alpha\rho_{\text{inner}}}{\omega\epsilon}\frac{y}{y^2+z^2}e^{i(kx-\omega t)}). \tag{21.20}$$

To obtain the current in the conductors, recall that the (signed) current I through a surface Σ is exactly the flux of the current density through Σ with respect to the direction specified by one of its unit normals η:

$$I = \int_\Sigma J \cdot \eta \, dS.$$

Also, Ampère's law for time-harmonic fields is

$$i\omega\epsilon E = J - \frac{1}{\mu}\nabla \times B.$$

Thus, we have (for our isotropic material)

$$I = \int_\Sigma J \cdot \eta \, dS = \frac{1}{\mu}\int_\Sigma (\nabla \times B) \cdot \eta \, dS + i\omega\epsilon \int_\Sigma E \cdot \eta \, dS.$$

Consider a curve Γ in a cross section of the coaxial arrangement of conductors that surrounds the inner conductor and has the outer conductor in the exterior of the bounded surface Σ whose boundary is Γ. For the outer conductor, there is an annular surface $\tilde{\Sigma}$ also in the cross-sectional plane whose inner boundary is Γ and whose outer boundary is some curve Γ^* that lies in the outer conductor.

The TEM mode electric field E is everywhere orthogonal to η (which is pointing in the axial direction of the cylinder) on both surfaces Σ and Σ^*. Thus, the surface integral of E vanishes. By Stokes's theorem, the current in the inner conductor is

$$I = \frac{1}{\mu}\int_\Sigma (\nabla \times B) \cdot \eta \, dS = \frac{1}{\mu}\int_\Gamma B \cdot d\ell.$$

Taking Γ to be a circle with radius R such that $\alpha < R < \beta$ oriented clockwise to be compatible with the normal to the cross section that points in the positive direction of the axial coordinate, an easy computation can be used to show that the (complex) current in this direction is

$$I = I(x,t) = 2\pi\frac{k\alpha\rho_{\text{inner}}}{\omega\mu\epsilon}e^{i(kx-\omega t)}. \tag{21.21}$$

See Exercise 21.5 for the current in the outer conductor.

There are several relations between I and V. The most important of these is a system of first-order PDEs that have these functions as one of its solutions.

For notational simplicity, define the voltage

$$V_0 = \frac{\alpha \rho_{\text{inner}}}{\epsilon} \tag{21.22}$$

and differentiate in Eqs (21.21) and (21.19) to obtain the partial derivatives

$$I_x = ik^2 \frac{2\pi V_0}{\omega\mu} e^{i(kx-\omega t)}, \qquad V_t = -i\omega V_0 \ln\frac{\beta}{\alpha}\, e^{i(kx-\omega t)} \tag{21.23}$$

Also, recall that

$$k^2 = \omega\mu(\omega\epsilon + i\sigma). \tag{21.24}$$

There are two *real* numbers

$$C := \frac{2\pi\epsilon}{\ln(\beta/\alpha)}, \qquad G := \frac{2\pi\sigma}{\ln(\beta/\alpha)} \tag{21.25}$$

such that

$$CV_t = -I_x - GV. \tag{21.26}$$

Because C and G are real, the real parts of I and V (which are the physical fields) satisfy this PDE. The parameter C has the units of farads per meter and G is measured in siemens per meter. Thus, these quantities are immediately identified with the capacitance per meter between the two conductors and the conductance per meter of the media between the conductors.

Differentiate again in Eqs (21.21) and (21.19) to obtain the partial derivatives

$$I_t = -i\frac{2\pi k V_0}{\mu} e^{i(kx-\omega t)}, \qquad V_x = ikV_0 \ln\frac{\beta}{\alpha}\, e^{i(kx-\omega t)} \tag{21.27}$$

and define the real number

$$L = \frac{\mu}{2\pi} \ln\frac{\beta}{\alpha}, \tag{21.28}$$

which has the units of henries per meter. The current and voltage satisfy the PDE

$$LI_t = -V_x, \tag{21.29}$$

where L is the inductance per meter of the conductor.

The system of Eqs. (21.26) and (21.29) for voltage and current is a direct consequence of the TEM mode assumption for time-harmonic fields for an isotropic and ohmic medium between two coaxial conductors. Exactly the same equations, albeit with different expressions for the capacitance, conductance and inductance, can be derived for more general configurations of two conductors separated by an isotropic and ohmic medium; that is, the form of the equations is exactly the same for all transmission lines with two (not necessarily circularly cylindrical or coaxial) conductors. The system

$$CV_t = -I_x - GV, \qquad LI_t = -V_x \tag{21.30}$$

is called the (ideal) transmission line equations.

The (perhaps complex) quantity

$$Z := \frac{V}{I} \tag{21.31}$$

is constant in the TEM mode; Z is called the (line) impedance.

The wave number k is a function of frequency. Thus, in general, the wave speed depends on frequency. For this reason, relation (21.24),

$$k^2 = \omega\mu(\omega\epsilon + i\sigma),$$

is called the dispersion relation.

TEM modes can exist with attenuated electromagnetic waves. This occurs when k is not real; that is, when $\sigma \neq 0$. In this case, the medium between the conductors has nonzero conductance, and from Ohm's law ($J = \sigma E$), there is a nonzero current in the direction of the E field. Attenuation occurs because of the factor $\exp(-\operatorname{Im}(k)x)$, where Im is the projection onto the imaginary part of a complex number. This exponential factor grows or decays with position according to the sign of $\operatorname{Im}(k)$.

In case $\sigma = 0$ and k is real, the wave speed takes a familiar form. By substituting the value of the wave number $k = \omega\sqrt{\mu\epsilon}$ from the dispersion relation into the exponent $i(kx - \omega t)$ and simplifying, it follows that

$$i(kx - \omega t) = i\omega\sqrt{\mu\epsilon}(x - \frac{1}{\sqrt{\mu\epsilon}}t).$$

Therefore, the corresponding wave speed is

$$c := 1/\sqrt{\epsilon\mu},$$

a number that *does not* depend on the frequency. When $\sigma = 0$, there is no dispersion in TEM waves.

Of course, the wave nature of the fields must be reflected in the transmission line equations. In fact, both V and I satisfy (damped) wave equations. For example,

$$V_{tt} + \frac{G}{C}V_t = \frac{1}{LC}V_{xx}. \tag{21.32}$$

In the realm of physical measurements for engineering applications, the culmination of the theory of TEM modes is the (ideal) transmission line model [Eqs. (21.30)]. The derivation of this model, which relies on Maxwell's theory, is on solid physical grounds under assumptions that often closely approximate reality: perfect conductors, isotropic media, Ohm's law, and time-harmonic fields. Voltages and currents predicted by this model agree with experimental measurements. Of course, the approximations built into these assumptions can be improved to construct more accurate physical models. Making predictions from more accurate models that adhere strictly to basic physics (Maxwell's laws) is generally a much more difficult process. There is, however, a simple and reasonable way to relax the perfect conductor assumption with a simple modification of the ideal transmission line equations to account for (line) resistance in real conductors. It is discussed in the next chapter.

Exercise 21.1. (a) Derive the polar coordinate representation for the Laplace equation in two-dimensional space. (b) Derive the cylindrical coordinate representation for the Laplace equation in three-dimensional space.

Exercise 21.2. Derive the general solution of ODE (21.15).

Exercise 21.3. The voltage across the conductors of the transmission line discussed in this Chapter is $(\psi(\alpha, \theta) - \psi(\beta, \theta))e^{i(kx-\omega t)}$. Alternatively, this quantity is the negative of the integral of E field (21.18) along a path connecting the outer and inner conductors.

Show with all details of the computations that both approaches lead to the voltage V in display (21.19).

Exercise 21.4. Check that V in Eq. (21.19) has units of volts and I in Eq. (21.21) has units of amperes.

Exercise 21.5. Using the curves and surfaces defined in preparation for the formula for the induced inner conductor current [Eq. (21.21)], compute the induced current in the outer conductor.

Exercise 21.6. In case the conductance $\sigma \neq 0$, there is a current I leaking through the ohmic material between the inner and outer conductors where $J = \sigma E$. Choose a position with coordinates (x, y, z) along the inner conductor and consider a portion of a circular cylinder Σ whose radius is R (with $\alpha < R < \beta$) and whose ends are in the cross-sectional planes of the coaxial conductors at the points $x - \Delta x$ and $x + \Delta x$. Note that the E field restricted to Σ is everywhere parallel to the outer unit normal η on Σ, which is radial along the cylinder. (1) Show that the total current through Σ in the direction of η is

$$I_{\text{total}} = 2\pi\sigma \frac{\alpha \rho_{\text{inner}}}{\epsilon} \int_{x-\Delta x}^{x+\Delta x} e^{i(k\xi - \omega t)}\, d\xi.$$

and the total current per length at (x, t) is

$$I(x,t) = 2\pi\sigma \frac{\alpha \rho_{\text{inner}}}{\epsilon} e^{i(kx - \omega t)}. \tag{21.33}$$

(2) Compare this result with the current due to conduction corresponding to the voltage term in the transmission line equation (21.26). Hint: Use Ohm's law $V = IR$, where R is the resistance.

Exercise 21.7. (1) Derive the (damped) wave equation (21.32). (2) Derive a similar wave equation for I. (3) Determine the predicted wave speed and compare with the wave speed of TEM waves.

CHAPTER 22

Transmission Lines

22.1 TIME-DOMAIN REFLECTOMETRY MODEL

Electromagnetic waves may move along a transmission line (think coaxial cable such as those used in TV cable systems) according to the material properties of the dielectric between the wire at the center of the cable and the (usually braided) metal cylinder surrounding it (see Fig. 22.1). The nature of such a wave is disrupted (perhaps partially reflected) when the uniformity of the dielectric material is broken. Avoidance of reflected waves is often a desired property that leads to serious design challenges. For example, connectors between transmission lines and electronic devices always disrupt transverse electromagnetic (TEM) waves. On the one hand, well designed (perhaps expensive) connectors are designed to minimize the effects of this disruption. On the other hand, the physical effects, such as wave reflection, due to changes in the dielectric can be used to advantage in other applications. One of these is discussed in some detail in this section.

A common method of testing for faults in a transmission line is called time-domain reflectometry (TDR). A pulse or voltage step is generated and sent into the line, it moves as a wave and is reflected from places along the line where a fault in the dielectric occurs. Perhaps the line is cracked, it has been pierced by a nail, or it is disrupted by a loose connection. By measuring the time of return of the wave and using the known transmission wave speed, the distance to a fault is easily computed.

Using the same idea for a different application, a test chamber in the form of a transmission line can be constructed so that the dielectric is some new substance, perhaps moist soil, whose electric properties (permittivity, for example) are unknown. These might be measured by analyzing the shape

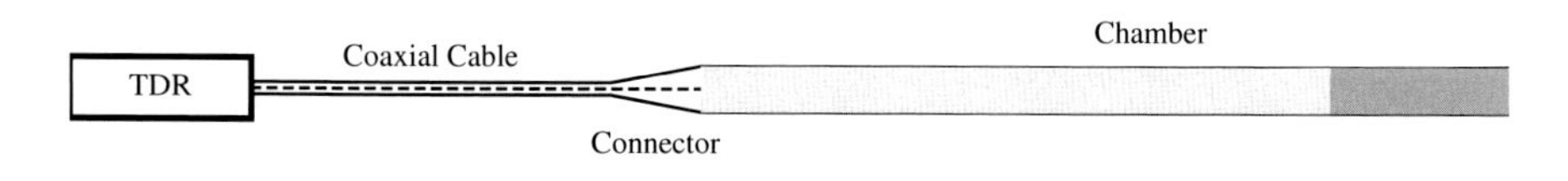

Fig. 22.1 The figure depicts a schematic TDR with test chamber filled with layers of dielectric materials.

An Invitation to Applied Mathematics: Differential Equations, Modeling, and Computation.
http://dx.doi.org/10.1016/B978-0-12-804153-6.50022-1, Copyright © 2017 Elsevier Inc. All rights reserved.

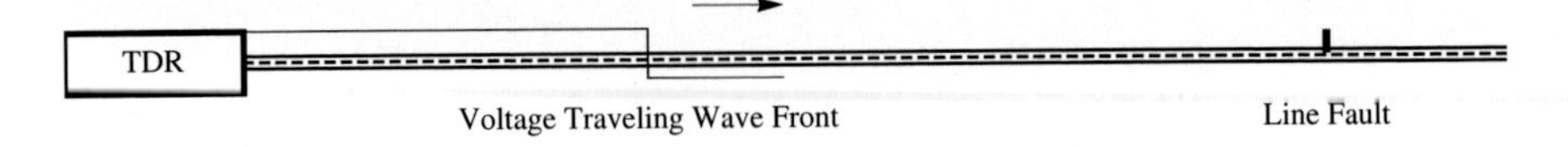

Fig. 22.2 Time of voltage wave generation is recorded. At fault, part of the wave is reflected and part transmitted. Time of return of reflected wave is detected as a jump in recorded voltage. Wave speed in the coaxial cable is known (perhaps given by the manufacturer at $0.8c$). The distance to the fault can be computed simply from $d = r \times t$.

of the returning electromagnetic wave after it passes through the soil and is reflected by a known load at the end of the test line. A typical application is the measurement of soil moisture as a function of soil depth. This is an example of an inverse scattering problem whose full solution is not known.

When the electric properties of the substance in the chamber are known, solving for the electromagnet wave involves the transmission line equations and a proper choice of boundary conditions. This is called the forward problem. Recovering the coefficients of the model partial differential equation (PDE) from reflected waves is called the inverse problem. Some of the (applied) mathematical features of the forward and inverse problems, along with some insight into their solutions, will be discussed. Typical challenge problem: "Drive a transmission line into the ground, run a TDR test, and determine the depths, compositions, and moisture contents of the layers of silt, sand, gravel, and clay at the test site." Of course there are many other applications.

The transmission line (or telegrapher's) equations derived by Oliver Heaviside (1885) for the voltage v (across the outer and central conductor measured as a function of position x and time t) and current i (which is equal and oppositely oriented in these two conductors and also measured as a function of position and time) is the system of PDEs given by

$$
\begin{aligned}
v_t + \frac{1}{C} i_x &= -\frac{G}{C} v, \\
i_t + \frac{1}{L} v_x &= -\frac{R}{L} i.
\end{aligned}
\tag{22.1}
$$

The quantities appearing in the model have units

$$
[i] = \text{ampere}, \qquad [v] = \text{volt}, \qquad [R] = \frac{\text{ohm}}{\text{meter}},
$$
$$
[L] = \frac{\text{henry}}{\text{meter}}, \quad [C] = \frac{\text{farad}}{\text{meter}}, \quad [G] = \frac{\text{siemen}}{\text{meter}}.
$$

These equations are a generalization of the (lossless) transmission line model [Eqs. (21.30)] of the previous section, which are derived from Maxwell's laws. Note: Lower case i is used here to denote current in keeping with convention in electrical engineering. Elsewhere in this book $i = \sqrt{-1}$, which is usual in mathematics; engineers use $j = \sqrt{-1}$.

To see how Heaviside's equations can be derived more directly from circuit theory, imagine a conductor with resistance and inductance distributed along its length connected to another reference conductor, which may be viewed as a ground. The quantities of interest are the voltage along the conductor measured relative to the second conductor and the current along the conductor. Let x be the coordinate that measures length along the line and t the temporal variable. At two nearby points x and $x + \Delta x$ along the line, there are two voltages $v(x, t)$ and $v(x + \Delta x, t)$. The voltage change $v(x + \Delta x, t) - v(x, t)$ along this section of the line is determined by the distributed voltage drops due to the resistance per length R and the inductance per length L. Ohm's law in this context states that the voltage drop across a resistor is the product of current and resistance. For an inductor, the voltage drop is the product of the inductance and the time rate of change of the current through the inductor. In the distributed case,

$$v(x + \Delta x, t) - v(x, t) = -\int_x^{x+\Delta x} Ri(\xi, t)\, d\xi - \int_x^{x+\Delta x} Li_t(\xi, t)\, d\xi. \tag{22.2}$$

The minus sign takes into account the direction of the current. For the positive node $v(x + \Delta x, t)$ (accumulation of positive charge) and the negative node $v(x, t)$, current flows from the position with coordinate $x + \Delta x$ toward the position with coordinate x. The voltage drop in this direction is $v(x, t) - v(x + \Delta x, t)$, which must be equal to the sum of the voltage drops given by the integrals. In case the signs of the nodes are reversed, current flows in the opposite direction and the same analysis leads to Eq. (22.2).

What about losses in the current? Two losses are considered in the segment of the transmission line from x to $x + \Delta x$: leakage of current to the other conductor (the ground) through a resistor with resistance per length $\tilde{R}$ or through a capacitor with capacitance per length C. The loss of current in this scenario is given by

$$i(x + \Delta x, t) - i(x, t) = -\int_x^{x+\Delta x} \frac{1}{\tilde{R}} v(\xi, t)\, d\xi - \int_x^{x+\Delta x} Cv_t(\xi, t)\, d\xi.$$

With $G := 1/\tilde{R}$, called the conductance per length, the current loss is

$$i(x + \Delta x, t) - i(x, t) = -\int_x^{x+\Delta x} Gv(\xi, t)\, d\xi - \int_x^{x+\Delta x} Cv_t(\xi, t)\, d\xi. \tag{22.3}$$

By dividing both sides of Eqs. (22.2) and (22.3) by Δx and passing to the limit as Δx goes to zero, we recover Heaviside's transmission line model [Eqs. (22.1)].

To obtain a physically relevant solution of the transmission line equations in an applied problem, the PDEs must be augmented with initial and boundary data.

Natural initial data

$$v(x, 0) = v_0(x), \qquad i(x, 0) = i_0(x) \tag{22.4}$$

specifies the voltage and current along the line at time $t = 0$. For many applications, the initial state of the line is zero voltage and zero current.

Boundary conditions depend on the application. For the TDR problem, input to the chamber is produced by a wave generator. A viable simple model is obtained by deriving the left-hand boundary condition (which, with an appropriate choice of spatial coordinate, may be taken at $x = 0$) for the transmission line from a circuit consisting of a voltage source, a resistor, and the voltage across the line at $x = 0$. By Ohm's law and Kirchhoff's loop law, the voltage across the resistor plus the voltage across the transmission line ($R_{\text{in}} i(0, t) + v(0, t)$) is equal to the input voltage v_{in} due to the generator, which is defined for $t \geq 0$. Thus, we have the left-hand boundary condition

$$R_{\text{in}} i(0, t) + v(0, t) = v_{\text{in}}(t). \tag{22.5}$$

The right-hand boundary condition is determined by the device at the end of the transmission line, which (in principle) could be an arbitrary circuit. A simple model is provided by specifying the i-v relation at the end of the line:

$$v(\ell, t) = f(i(\ell, t)).$$

For many applications, including the TDR test chamber, a viable model is obtained by supposing the inner and outer conductors are connected by a resistor, that is,

$$v(\ell, t) = R_{\text{end}} i(\ell, t). \tag{22.6}$$

The initial boundary value problem (IBVP) for PDE (22.1), initial data (22.4), and boundary data (22.5)–(22.6) is expected to predict a unique voltage and current on the line for all $t > 0$. Of course, this is a mathematical assertion that requires a proof. At this point, we will simply assume that the system has a unique solution.

Two important right-end boundary conditions are the extremes: right-end shorted by a perfect conductor connecting the inner and outer conductors ($R_{\text{end}} = 0$) and right-end open ($R_{\text{end}} = \infty$). In case the line is shorted, the right-end boundary condition is $v(\ell, t) = 0$; in case it is open, the boundary condition is $i(\ell, t) = 0$.

22.2 TDR MATRIX SYSTEM

A compact form of the transmission line model is obtained by viewing the state of the system (voltage and current) as the vector $u(x, t) := (v(x, t), i(x, t))$ so that the transmission line equations become the matrix system

$$u_t + A(x)u_x = -B(x)u, \tag{22.7}$$

where

$$A = \begin{pmatrix} 0 & \frac{1}{C} \\ \frac{1}{L} & 0 \end{pmatrix}, \quad B = \begin{pmatrix} \frac{G}{C} & 0 \\ 0 & \frac{R}{L} \end{pmatrix} \tag{22.8}$$

with space-dependent capacitance C, inductance L, resistance R, and conductance G.

The matrix A has two real eigenvalues $\pm(LC)^{-1/2}$. As we will see, the function

$$c := \frac{1}{\sqrt{LC}}$$

gives the wave speed (depending on spatial position) for the electromagnetic waves in the transmission line. Here, c is used for the wave speed, which might not be the speed of light in vacuum. The meaning of c should always be clear from the context. Another important quantity is the impedance

$$Z := \sqrt{\frac{L}{C}}.$$

A matrix T of eigenvectors (whose columns are in the order corresponding to the eigenvalues $-c$ and c) and its inverse are

$$T := \begin{pmatrix} -Z & Z \\ 1 & 1 \end{pmatrix}, \qquad T^{-1} = \frac{1}{2Z} \begin{pmatrix} -1 & Z \\ 1 & Z \end{pmatrix}. \tag{22.9}$$

The matrix A is diagonalized by the similarity transformation

$$T^{-1}AT = \begin{pmatrix} -c & 0 \\ 0 & c \end{pmatrix}. \tag{22.10}$$

22.3 INITIAL VALUE PROBLEM FOR THE IDEAL TRANSMISSION LINE

As a first step in the analysis of the transmission line equations, consider the initial value problem (IVP) for the two-dimensional system

$$u_t + Au_x = 0 \tag{22.11}$$

where u is an unknown vector function, A is a given constant matrix with real distinct eigenvalues, and u is defined on the whole real line at some fixed initial time t_0 by $u(x, t_0) = u_0(x)$.

The matrix A is diagonalizable. Indeed, let the distinct eigenvalues of A be $c_1 < c_2$, let R be the matrix whose columns are the corresponding eigenvectors r_1 and r_2, and let the rows of the inverse matrix R^{-1} be ℓ_1 and ℓ_2, the left eigenvectors corresponding to these eigenvalues. The matrix $R^{-1}AR$ is diagonal with c_1 and c_2 its diagonal elements in the given order.

The vector function $p := R^{-1}u$ satisfies the decoupled system of PDEs

$$p_t + R^{-1}ARp_x = 0;$$

that is,

$$p_t^1 + c_1 p_x^1 = 0, \qquad p_t^2 + c_2 p_x^2 = 0.$$

Also, employing the usual inner product in two-dimensional Euclidean space as well as the definitions of p and the vectors ℓ_1 and ℓ_2,

$$p^1 = \langle \ell_1, u \rangle, \qquad p^2 = \langle \ell_2, u \rangle.$$

The general solution of the uncoupled system is a pair of waves whose profiles are given by as yet undetermined functions ϕ_1 and ϕ_2:

$$p^1(x,t) = \phi_1(x - c_1 t), \qquad p^2(x,t) = \phi_2(x - c_2 t).$$

The unknown wave profiles are determined by the initial data.

To determine $p^1(x,t)$ at some point $(x,t) = (\xi, \tau)$, consider the coordinate plane with horizontal coordinate x and vertical coordinate t, and the line $x - c_1 t = \xi - c_1 \tau$ (called a characteristic line) that passes through the point (ξ, τ). The function p^1 is constant along this line; in fact, if (x,t) is a point on this line, then $p^1(x,t) = \phi_1(x - c_1 t) = \phi_1(\xi - c_1 \tau)$. More generally, p^1 is constant on every line of the form $x - c_1 t = k$ where k is some real number. To determine $p^1(\xi, \tau)$, simply note that the point where the characteristic line through (ξ, τ) meets the line $t = t_0$ is $(x,t) = (\xi + c_1(t_0 - \tau), t_0)$ and set

$$p^1(\xi, \tau) = p^1(\xi + c_1(t_0 - \tau), t_0) = \langle \ell_1, u_0((\xi + c_1(t_0 - \tau)) \rangle,$$

where u_0 is the given initial value of u. The same result may be recast as

$$\phi_1(x - c_1 t) = p^1(x,t) = \langle \ell_1, u_0((x + c_1(t_0 - t)) \rangle = \langle \ell_1, u_0((x - c_1 t) + c_1 t_0) \rangle.$$

Hence, the wave profile ϕ_1 is given by

$$\phi_1(s) = \langle \ell_1, u_0(s + c_1 t_0) \rangle.$$

Likewise,

$$\phi_2(s) = \langle \ell_2, u_0(s + c_2 t_0) \rangle.$$

The solution of the original IVP (22.11) is $u = Rp$, or

$$u(x,t) = \langle \ell_1, u_0((x - c_1 t) + c_1 t_0) \rangle r_1 + \langle \ell_2, u_0((x - c_2 t) + c_2 t_0) \rangle r_2. \quad (22.12)$$

22.4 THE INITIALLY DEAD IDEAL TRANSMISSION LINE WITH CONSTANT DIELECTRICS

An important special case is the ideal transmission line with no resistance or conductance and constant capacitance and inductance. It is modeled by the PDE

$$u_t + Au_x = 0$$

for $t \geq 0$ and $0 \leq x \leq \ell$, where

$$A = \begin{pmatrix} 0 & \frac{1}{C} \\ \frac{1}{L} & 0 \end{pmatrix}$$

is constant, and the components of u are the voltage and current. The basic problem is to determine the solution of this PDE with the left-hand boundary condition

$$R_{\text{in}} i(0,t) + v(0,t) = v_{\text{in}}(t) \tag{22.13}$$

(where, for convenience in writing some of the formulas to follow, v_{in} is defined to be zero for $t < 0$), the right-hand boundary condition

$$v(\ell,t) = R_{\text{end}} i(\ell,t), \tag{22.14}$$

and the initial condition $u(x,0) = 0$, for $0 \leq x \leq \ell$.

Using the change of variables $u = Tw$ (where T is defined in display (22.9)) the system decouples. Indeed, we have that

$$Tw_t + ATw_x = 0$$

and

$$w_t + T^{-1}ATw_x = 0.$$

For the vector state $w = (p,q)$, this latter system reduces to two uncoupled PDEs

$$p_t - cp_x = 0, \qquad q_t + cq_x = 0.$$

These one-way wave equations have the general solutions

$$p(x,t) = \phi(x+ct), \qquad q(x,t) = \psi(x-ct),$$

where ϕ and ψ are arbitrary (continuously differentiable) functions. The solution p is a wave traveling to the left with speed c; likewise, q is a wave traveling to the right with the same speed c.

The corresponding voltage and current $u = Tw$ are

$$v(x,t) = -Z\phi(x+ct) + Z\psi(x-ct), \quad i(x,t) = \phi(x+ct) + \psi(x-ct). \tag{22.15}$$

The voltage and current are both zero at time $t = 0$. Thus,

$$-\phi(x) + \psi(x) = 0, \qquad \phi(x) + \psi(x) = 0. \tag{22.16}$$

It follows that $\phi(x) = 0$ and $\psi(x) = 0$ for $0 \leq x \leq \ell$. For a spatial coordinate x in this interval, $\phi(x+ct)$ vanishes for $0 \leq t \leq (\ell - x)/c$ and $\psi(x - ct)$ vanishes for $0 \leq t \leq x/c$.

The left-hand boundary condition is $R_{\text{in}} i(0,t) + v(0,t) = v_{\text{in}}(t)$. At $x = 0$ and $0 \leq t \leq \ell/c$,

$$v(0,t) = Z\psi(-ct), \qquad i(0,t) = \psi(-ct).$$

Over this time interval,

$$R_{\text{in}}\psi(-ct) + Z\psi(-ct) = v_{\text{in}}(t).$$

Thus, with $s := -ct$

$$\psi(s) = \frac{1}{R_{\text{in}} + Z} v_{\text{in}}(-s/c)$$

over the interval $-\ell \leq s \leq 0$, and the state of the system is

$$\begin{aligned} v(x,t) &= \frac{Z}{R_{\text{in}} + Z} v_{\text{in}}(-(x-ct)/c) = \frac{Z}{R_{\text{in}} + Z} v_{\text{in}}(t - x/c), \\ i(x,t) &= \frac{1}{R_{\text{in}} + Z} v_{\text{in}}(-(x-ct)/c) = \frac{1}{R_{\text{in}} + Z} v_{\text{in}}(t - x/c) \end{aligned} \tag{22.17}$$

for $0 \leq t \leq (\ell - x)/c$.

Using the right-hand boundary condition,

$$-Z\phi(\ell + ct) + Z\psi(\ell - ct) = R_{\text{end}}(\phi(\ell + ct) + \psi(\ell - ct)).$$

As $\psi(\ell - ct) = 0$ on the interval $0 \leq t < \ell/c$,

$$-Z\phi(\ell + ct) = R_{\text{end}}\phi(\ell + ct), \tag{22.18}$$

and because $R_{\text{end}} + Z > 0$, we have that $\phi(\ell + ct) = 0$ for $0 \le t \le \ell/c$, or $\phi(s) = 0$ on the interval $\ell \le s < 2\ell$.

So far, ϕ is defined on the interval $[0, 2\ell]$ and ψ on $[-\ell, \ell]$.

For $0 \le t \le \ell/c$ a wave of voltage and current travels to the right along the transmission line. What happens when the wave reaches the right-hand boundary? Answer: The voltage and current (given by the general solution of the transmission line equations) must satisfy the right-hand boundary condition for later times at $x = \ell$. In fact, we must have

$$v(\ell, t) = -Z\phi(\ell + ct) + \frac{Z}{Z + R_{\text{in}}} v_{\text{in}}(t - \ell/c),$$
$$i(\ell, t) = \phi(\ell + ct) + \frac{1}{Z + R_{\text{in}}} v_{\text{in}}(t - \ell/c)$$

on the interval $0 \le t \le 2\ell/c$. The boundary condition requires $v(\ell, t) = R_{\text{end}} i(\ell, t)$. By imposing this condition and rearranging the resulting equation, it follows that

$$\phi(\ell + ct) = \frac{Z - R_{\text{end}}}{Z + R_{\text{end}}} \frac{1}{Z + R_{\text{in}}} v_{\text{in}}(t - \ell/c)$$

on the same time interval. Thus, the function ϕ is defined by

$$\phi(s) = \frac{Z - R_{\text{end}}}{Z + R_{\text{end}}} \frac{1}{Z + R_{\text{in}}} v_{\text{in}}((s - 2\ell)/c)$$

for $2\ell \le s < 3\ell$. At $s = 3\ell$ the left-hand boundary condition must be employed.

The voltage and current are given on the time interval $0 \le t \le (\ell + x)/c$ by

$$v(x, t) = -\frac{Z - R_{\text{end}}}{Z + R_{\text{end}}} \frac{Z}{Z + R_{\text{in}}} v_{\text{in}}(t - (2\ell - x)/c) + \frac{Z}{Z + R_{\text{in}}} v_{\text{in}}(t - x/c),$$
$$i(x, t) = \frac{Z - R_{\text{end}}}{Z + R_{\text{end}}} \frac{1}{Z + R_{\text{in}}} v_{\text{in}}(t - (2\ell - x)/c) + \frac{1}{Z + R_{\text{in}}} v_{\text{in}}(t - x/c).$$

Note that the incoming wave is reflected at the right boundary (which is the end of the transmission line). The amplitude of the reflected wave is the amplitude of the incoming wave times the reflection coefficient $(Z - R_{\text{end}})/(Z + R_{\text{end}})$, and of course, the reflected wave travels in the opposite direction, which in this case is toward the beginning of the transmission line.

The part of the solution obtained so far simply uses the boundary conditions in turn to extend given portions of the solution. This suggests a more systematic approach: Use the general solution [Eqs. (22.15)] of the ideal transmission line equations and the general corresponding boundary conditions. Together, they imply a system of functional differential equations for the unknown functions ϕ and ψ:

$$(R_{\text{in}} - Z)\phi(ct) + (R_{\text{in}} + Z)\psi(-ct) = v_{\text{in}}(t),$$
$$(R_{\text{end}} + Z)\phi(\ell + ct) + (R_{\text{end}} - Z)\psi(\ell - ct) = 0.$$

Set $s = ct$ in the first equation and $\sigma = \ell + ct$ in the second to obtain the system

$$(R_{\text{in}} - Z)\phi(s) + (R_{\text{in}} + Z)\psi(-s) = v_{\text{in}}(\frac{s}{c}),$$
$$(R_{\text{end}} + Z)\phi(\sigma) + (R_{\text{end}} - Z)\psi(2\ell - \sigma) = 0. \tag{22.19}$$

These equations must hold for all $s \geq 0$ and $\sigma \geq \ell$. In particular, the second equation holds for $\sigma > 2\ell$. For such σ, we have $\sigma - 2\ell \geq 0$; thus, this quantity may be substituted for s in the first equation. After solving for $\psi(2\ell - \sigma)$ in both equations, equating the results, and rearranging, we have two cases: For $R_{\text{in}} - Z \neq 0$, the functional equation

$$\phi(\sigma) = -\frac{(R_{\text{in}} - Z)(R_{\text{end}} - Z)}{(R_{\text{in}} + Z)(R_{\text{end}} + Z)}\Big(\frac{v_{\text{in}}((\sigma - 2\ell)/c)}{(R_{\text{in}} - Z)} - \phi(\sigma - 2\ell)\Big) \tag{22.20}$$

and for $R_{\text{in}} - Z = 0$, the explicit equation

$$\phi(\sigma) = -\frac{(R_{\text{end}} - Z)}{(R_{\text{in}} + Z)(R_{\text{end}} + Z)}v_{\text{in}}((\sigma - 2\ell)/c) \tag{22.21}$$

for ϕ.

To obtain ϕ from functional equation (22.20), recall that (using Eqs. (22.16) and (22.18)) this function vanishes on the interval $0 \leq \sigma < 2\ell$. It follows that the function $\sigma \mapsto \phi(\sigma - 2\ell)$ vanishes on the interval $2\ell \leq \sigma < 4\ell$. Using functional equation (22.20), ϕ is defined on this latter interval by

$$\phi(\sigma) = -\frac{(R_{\text{end}} - Z)}{(R_{\text{in}} + Z)(R_{\text{end}} + Z)}v_{\text{in}}((\sigma - 2\ell)/c).$$

By the same process, ϕ may be defined for $4\ell \leq \sigma \leq 6\ell$, and continuing in the same manner, ϕ is defined for all positive real numbers.

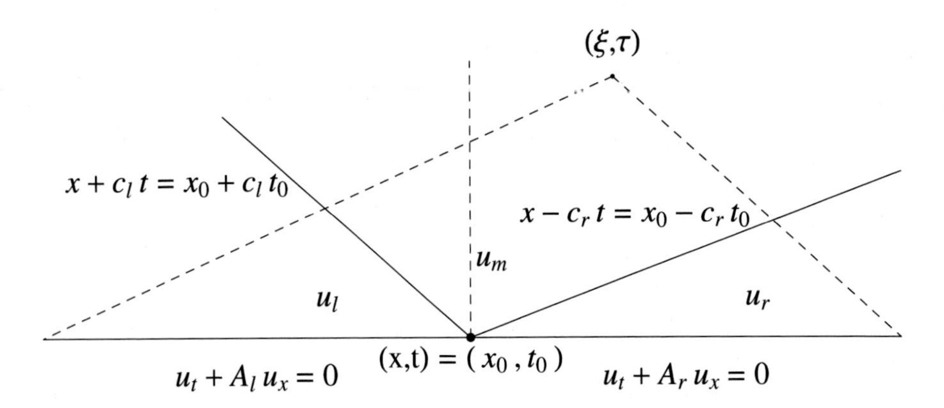

Fig. 22.3 A schematic diagram of the Riemann problem for the ideal transmission line equation $u_t + Au_x = 0$ is depicted. The linear equations are for the solid lines emanating from the point (x_0, t_0).

The function ψ is obtained from ϕ using the second equation in display (22.19). With $\tau := 2\ell - \sigma$ and some rearrangement, ϕ is defined for all *negative* real numbers by the functional equation

$$\psi(\tau) = -\frac{R_{\text{end}} + Z}{R_{\text{end}} - Z}\phi(2\ell - \tau). \tag{22.22}$$

The voltage and current are determined from ϕ and ψ by the formulas in display (22.15).

In case $R_{\text{in}} - Z = 0$, a similar construction can be used to define ϕ and ψ.

Thus, a solution is constructed for the IBVP for the ideal transmission line equations on the spatial interval $0 \le x \le \ell$ for all time $t \ge 0$.

Exercise 22.1. Write out the voltage and current using Eqs. (22.22) and (22.15).

Exercise 22.2. Suppose that the input voltage v_{in} is a *bounded* smooth function and the ideal transmission line is initially dead. (a) Show that the voltage and current remain bounded for every finite time. (b) Do the voltage and current remain bounded for all time? Hint: This question has not been answered by the author.

22.5 THE RIEMANN PROBLEM

A key to further understanding of the general solution of the transmission line equations for layered media and numerical methods for approximating solutions is the analysis of the Riemann problem on the whole real line for the ideal transmission line equation

$$u_t + A(x)u_x = 0,$$

where u is the vector whose components are voltage and current, A is defined in display (22.8), and the following (perhaps discontinuous data) is given on the real line: For some $t_0 \geq 0$, real coordinate $x = x_0$, and given vector-valued functions u_ℓ and u_r of one real variable,

$$u(x, t_0) = u_\ell(x),$$

for $x < x_0$ and

$$u(x, t_0) = u_r(x),$$

for $x > x_0$. The Riemann problem is to determine $u(x, t)$ for (almost) all x (that is, except for a set of measure zero) and all $t > t_0$. Usually, the vector functions u_ℓ and u_r are assumed to be constant and distinct. This restriction is not necessary here.

To begin, suppose that the matrix A, with structure as in displays (22.9) and (22.10), is constant and recall that A is diagonalizable with distinct eigenvalues $\pm c$. Let the columns of the matrix T be the eigenvectors r_1 and r_2 corresponding to $-c$ and c (respectively), and let the rows of the inverse matrix T^{-1} be ℓ_1 and ℓ_2, the left eigenvectors corresponding to these eigenvalues.

With $p := T^{-1}u$, the vector function p satisfies the decoupled system of PDEs

$$p_t + T^{-1}ATp_x = 0;$$

that is,

$$p_t^1 - cp_x^1 = 0, \qquad p_t^2 + cp_x^2 = 0.$$

Also, employing the usual inner product in two-dimensional space,

$$p^1 = \langle \ell_1, u \rangle, \qquad p^2 = \langle \ell_2, u \rangle.$$

As usual, the general solution of the uncoupled system is a pair of waves

$$p^1(x, t) = \phi_1(x + ct), \qquad p^2(x, t) = \phi_2(x - ct)$$

whose profiles ϕ_1 and ϕ_2 are to be determined from the initial data of the Riemann problem (see Fig. 22.3 but read $A_\ell = A$ and $A_r = A$, $c_\ell = c$, and $c_r = c$).

The functions p^1 and p^2 are determined at each point (ξ, τ) with $\tau > t_0$ by tracing back along characteristics, which are lines of the form $x - ct = k$ or $x + ct = k$ for some constant k. As depicted in Fig. 22.3, there are three regions above the horizontal line $t = t_0$ bounded by the characteristics emanating from the point (x_0, t_0). The point (ξ, τ) in this figure is in the middle region between the two aforementioned characteristics. One characteristic through this point meets the line $t = t_0$ to the left of (x_0, t_0), the other characteristic meets this line on the right of this point. The characteristics through a point (in the left region) below and to the left of the characteristic $x + ct = x_0 + ct_0$ both meet the line $t = t_0$ to the left of (x_0, t_0); likewise, the characteristics through a point (in the right region) below and to the right of the characteristic $x - ct = x_0 - ct_0$ both meet the line $t = t_0$ to the right of (x_0, t_0). Using this information and the method in Section 22.3, the complete solution of the Riemann problem is easily constructed: For (x, t) in the left region,

$$u(x,t) = \langle \ell_1, u_l((x+ct) - ct_0)\rangle r_1 + \langle \ell_2, u_\ell((x-ct)+ct_0)\rangle r_2. \tag{22.23}$$

In the middle region,

$$u(x,t) = \langle \ell_1, u_r((x+ct) - ct_0)\rangle r_1 + \langle \ell_2, u_\ell((x-ct)+ct_0)\rangle r_2. \tag{22.24}$$

In the right region,

$$u(x,t) = \langle \ell_1, u_r((x+ct) - ct_0)\rangle r_1 + \langle \ell_2, u_r((x-ct)+ct_0)\rangle r_2. \tag{22.25}$$

In case the initial functions u_ℓ and u_r on the line $t = t_0$ have constant (vector) values denoted by the same function names, the solution u is constant in each region; in fact, for (x, t) in the left region,

$$u(x,t) = \langle \ell_1, u_l\rangle r_1 + \langle \ell_2, u_\ell\rangle r_2 = u_\ell; \tag{22.26}$$

in the middle region,

$$u(x,t) = u_m := \langle \ell_1, u_r\rangle r_1 + \langle \ell_2, u_\ell\rangle r_2, \tag{22.27}$$

and in the right region,

$$u(x,t) = \langle \ell_1, u_r\rangle r_1 + \langle \ell_2, u_r\rangle r_2 = u_r. \tag{22.28}$$

If $u_l \neq u_r$, then the piecewise constant solution u is discontinuous across the characteristics emanating from the point (x_0, t_0). The solution is continuous on the portion of the vertical line $x = x_0$ for $t > t_0$.

The jump across the characteristic $x - ct = x_0 - ct_0$ is (by definition)

$$\begin{aligned} u_r - u_m &= \langle \ell_1, u_r \rangle r_1 + \langle \ell_2, u_r \rangle r_2 - \langle \ell_1, u_r \rangle r_1 - \langle \ell_2, u_\ell \rangle r_2 \\ &= (\langle \ell_2, u_r \rangle - \langle \ell_2, u_\ell \rangle) r_2. \end{aligned} \tag{22.29}$$

Note that the jump, which has just been shown to be a scalar multiple of r_2, is an eigenvector of A corresponding to the eigenvalue c. Likewise, the jump $u_m - u_\ell$ is an eigenvector of A corresponding to $-c$. This is a version of the Rankine–Hugoniot jump condition for systems of conservation laws.

For layered media, the matrix A is piecewise constant. As in Fig. 22.3, the value of A is the constant matrix A_ℓ to the left of (x_0, t_0) and the constant A_r to the right of this point. The solution u will be constructed for the case where the initial functions u_ℓ and u_r are constants. The solution u has constant values u_ℓ and u_r in the left and right regions, respectively, exactly as for the case where A is constant except that the eigenvalues and eigenvectors are computed using A_ℓ in the left region and A_r in the right region.

Using the Rankine–Hugoniot condition directly instead of the analysis just completed, there would be unknown scalars λ_ℓ and λ_r such that

$$u_r - u_m = \lambda_r \begin{pmatrix} Z_r \\ 1 \end{pmatrix}, \qquad u_m - u_\ell = \lambda_\ell \begin{pmatrix} -Z_\ell \\ 1 \end{pmatrix}. \tag{22.30}$$

To determine these unknown constants, simply add the two equations and note that

$$u_r - u_\ell = \begin{pmatrix} -Z_\ell & Z_r \\ 1 & 1 \end{pmatrix} \begin{pmatrix} \lambda_\ell \\ \lambda_r \end{pmatrix} =: T_{\ell r} \lambda. \tag{22.31}$$

Thus, the unknown vector λ is determined by

$$\lambda = T_{\ell r}^{-1} (u_r - u_\ell), \tag{22.32}$$

where all quantities on the right-hand side are known. For (x, t) in the left region,

$$u(x, t) = u_\ell; \tag{22.33}$$

in the middle region,

$$u(x,t) = u_m := u_\ell + \lambda_\ell \begin{pmatrix} -Z_\ell \\ 1 \end{pmatrix}, \tag{22.34}$$

and in the right region,

$$u(x,t) = u_r. \tag{22.35}$$

This is a complete solution of the Riemann problem for the layered media case.

For reference, u_m is given in components by

$$\begin{pmatrix} v_m \\ i_m \end{pmatrix} = \begin{pmatrix} v_\ell \\ i_\ell \end{pmatrix} + \frac{Z_r(i_r - i_\ell) - (v_r - v_\ell)}{Z_r + Z_\ell} \begin{pmatrix} -Z_\ell \\ 1 \end{pmatrix}, \tag{22.36}$$

$$\begin{pmatrix} v_m \\ i_m \end{pmatrix} = \begin{pmatrix} v_r \\ i_r \end{pmatrix} + \frac{Z_\ell(i_r - i_\ell) + v_r - v_\ell}{Z_r + Z_\ell} \begin{pmatrix} Z_r \\ 1 \end{pmatrix}. \tag{22.37}$$

22.6 REFLECTED AND TRANSMITTED WAVES

Suppose that the dielectric media along a transmission line changes its properties at x_0, and a wave front of voltage and current moves to the right and reaches the position x_0 on the line at time t_0. The change in dielectric properties is modeled by assuming the matrix A is piecewise constant as in Section 22.5 with a single jump discontinuity at x_0, where the impedance Z changes from some value Z_ℓ on the left to Z_r on the right. For simplicity, assume that the profile of the traveling wave front is the function ϕ given by $\phi(s) = u_\ell$ for $s < 0$ and $\phi(s) = u_r$ for $s > 0$, and the wave is $u(x,t) = \phi(x - ct - (x_0 - ct_0))$. What happens to the wave for $t > t_0$?

In Section 22.5, the left and right initial values u_ℓ and u_r may be chosen arbitrarily because no assumption is made about the state of the line for $t < t_0$. For the case at hand where the line has a traveling wave front, the jump $u_r - u_\ell$ at the wave front cannot be arbitrary. In fact, this jump in the wave, which is supposed to be moving to the right, must satisfy the Rankine–Hugoniot jump condition along the characteristic $x - c_\ell t = x_0 - c_\ell t_0$ for $t \le t_0$ corresponding to this wave front up to and including the moment that the wave meets the interface between the two dielectrics at x_0. For this reason, there is a constant β such that

$$u_r - u_\ell = \beta \begin{pmatrix} Z_\ell \\ 1 \end{pmatrix}; \tag{22.38}$$

that is, the jump must be an eigenvector of A corresponding the eigenvalue c_ℓ.

From Eq. (22.32) and using the extra condition (22.38), the vector λ (whose components give the strengths of the outgoing waves) is given by

$$\lambda = T_{\ell,r}^{-1}(u_r - u_\ell) = \beta T_{\ell,r}^{-1} \begin{pmatrix} Z_\ell \\ 1 \end{pmatrix} = \frac{\beta}{Z_r + Z_\ell} \begin{pmatrix} Z_r - Z_\ell \\ 2Z_\ell \end{pmatrix},$$

and its components are

$$\lambda_\ell = \frac{Z_r - Z_\ell}{Z_r + Z_\ell}\beta, \qquad \lambda_r = \frac{2Z_\ell}{Z_r + Z_\ell}\beta.$$

The constant value of the middle state (from Eq. (22.34)) is

$$u_m := u_\ell + \lambda_\ell \begin{pmatrix} -Z_\ell \\ 1 \end{pmatrix} = u_\ell + \beta \frac{Z_r - Z_\ell}{Z_r + Z_\ell} \begin{pmatrix} -Z_\ell \\ 1 \end{pmatrix}. \tag{22.39}$$

Using the first equation in display (22.30), this same state is

$$u_m := u_r - \lambda_r \begin{pmatrix} Z_r \\ 1 \end{pmatrix} = u_r - \beta \frac{2Z_\ell}{Z_r + Z_\ell} \begin{pmatrix} Z_r \\ 1 \end{pmatrix}. \tag{22.40}$$

Suppose an observer at $x_* < x_0$ measures the voltage across the transmission line. The observer will record the voltage u_ℓ^1 (the first component of u_ℓ) after the right-going wave passes x_* up to time $t_* = (x_0 - x_*)/c_\ell + t_0$, which is later than t_0, when the wave front corresponding at the jump across the characteristic $x + c_\ell t = x_0 + c_\ell t_0$ due to reflection at x_0 reaches x_*. After this time, the reflected wave front has passed to the left of the observer who measures the new voltage

$$u_\ell^1 + \beta \frac{Z_r - Z_\ell}{Z_r + Z_\ell}(-Z_\ell).$$

In view of relation (22.38)

$$u_r^1 - u_\ell^1 = \beta Z_\ell,$$

this measured voltage is

$$u_\ell^1 - \frac{Z_r - Z_\ell}{Z_r + Z_\ell}(u_r^1 - u_\ell^1). \tag{22.41}$$

In other words, the observed voltage after the reflected wave front passes is the original voltage minus the reflection coefficient times the jump in voltage in the original wave. Likewise, for an observer at a point $x_* > x_0$, the voltage measured after passage of the transmitted wave front, which moves to the right, is

$$u_r^1 - \frac{2Z_r}{Z_r + Z_\ell}(u_r^1 - u_\ell^1). \tag{22.42}$$

The quantity $(Z_r - Z_\ell)/(Z_r + Z_\ell)$ is called the voltage reflection coefficient and $2Z_r/(Z_r+Z_\ell)$ the voltage transmission coefficient. Warning: Formulas and sign conventions for the reflection and transmission coefficients differ for different situations. Check the context before using textbook formulas in computer codes.

For a front arriving from the right at x_0 at time t_0 where u_r is the constant state to the right and u_ℓ is the constant state to the left, the Rankine–Hugoniot jump condition (for some new scalar β) is

$$u_r - u_\ell = \beta \begin{pmatrix} -Z_r \\ 1 \end{pmatrix}$$

and, exactly as before,

$$\lambda = T_{\ell r}^{-1}(u_r - u_\ell).$$

By combining these relations, we have that

$$\lambda_\ell = \frac{2\beta Z_r}{Z_\ell + Z_r}, \qquad \lambda_r = \beta\frac{Z_\ell - Z_r}{Z_\ell + Z_r}.$$

The middle state is

$$u_r - \lambda_r \begin{pmatrix} -Z_r \\ 1 \end{pmatrix};$$

the middle voltage (the voltage reflected to the right) is

$$u_r^1 + \frac{Z_\ell - Z_r}{Z_\ell + Z}(u_r^1 - u_l^1).$$

Using the alternative representation of the middle state,

$$u_\ell + \lambda_\ell \begin{pmatrix} -Z_\ell \\ 1 \end{pmatrix},$$

the transmitted voltage is

$$u_\ell^1 + \frac{2Z_\ell}{Z_\ell + Z_r}(u_r^1 - u_\ell^1).$$

The reflection coefficient $(Z_r - Z_\ell)/(Z_r + Z_\ell)$, for the right-going wave ranges in size from -1 to 1. For $Z_r = 0$ and the value is -1. This corresponds to a shorted line. The reflected voltage is v_r. At the other extreme, $Z_r = \infty$, the value is 1, the line is open, and the reflected voltage is $2v_\ell - v_r$. For $Z_r = Z_\ell$, the (impedance of the) line is matched at the interface and there is no reflection.

Exercise 22.3. Show that transmitted and reflected voltages from a dielectric break are equal.

Exercise 22.4. Find the current reflection and transmission coefficients.

Exercise 22.5. Consider an infinitely long ideal transmission line with three dielectric segments; that is, the line has a finite middle segment with wave speed c and impedance Z, the infinite segment to the left has wave speed c_ℓ and impedance Z_ℓ, and the infinite segment to the right has wave speed c_r and impedance Z_r. There is a voltmeter somewhere in the finite segment that measures the voltage across the line. A wave comes from the left, hits the left boundary of the finite segment of the line, and is transmitted into this segment. The wave in the segment has constant left and right voltages v_ℓ and v_r at its front. This wave front will eventually be reflected from the right boundary. In fact, the wave with be reflected from the right and left boundaries an infinite number of times. What is the voltage measured by the meter after a long time (relative to the wave speed in the finite segment)?

22.7 A NUMERICAL METHOD FOR THE LOSSLESS TRANSMISSION LINE EQUATION

General solution methods for conservation laws remain to be addressed. In this section, the idea is to take full advantage of the special form of the ideal (lossless) transmission line equations. In particular, note that in each layer of the dielectric medium (where the matrix A is constant) *the wave speeds are equal.* This fact allows an (almost) exact numerical solution of the lossless transmission line PDE with initial and boundary conditions. It is wise to consider the principle that every applied problem has some special features that can and should be exploited. Often key special features are not obvious, but they might be uncovered by careful consideration of the mathematical context in which the problem at hand is posed.

Assume that the transmission line has total length ℓ and a coordinate x is chosen so that $x = 0$ is at the left boundary of the line. The right boundary resides at $x = \ell$. In addition, suppose that the interior boundaries of the layers in the medium are at the coordinate values $\{x_1, x_2, x_3, \ldots, x_m\}$, with $x_1 = 0$ and $x_m = \ell$. The interval (x_k, x_{k+1}) is defined to be the kth layer. Also, the flux matrix A (which carries the information on dielectric capacitance and line inductance for the lossless model $u_t + A(x)u_x = 0$) is constant for $x_k < x < x_{k+1}$. In this layer, each matrix $A(x)$ has eigenvalues $\pm c_k$, corresponding to the constant wave speed c_k. Waves may move in either direction, but their speeds are the same.

More notation: let ℓ_k denote the length of the kth interval; that is, $\ell_k := x_{k+1} - x_k$. Define the electrical length of the kth interval to be the elapsed time $T_k := \ell_k/c_k$ a wave takes to traverse this interval. To take full advantage of the constant wave speeds in each layer, the spatial discretization along the transmission line should correspond to intervals of equal electrical lengths (instead of the usual equal lengths). Although exact electrical length discretizations do not exist in general, they always exist up to a predictable error that can be made arbitrarily small by refining the mesh.

Choose an integer $n \geq 1$ (the number of subdivisions in the electrically shortest layer), let $T_{\text{min}} := \min\{T_1, T_2, T_3, \ldots T_{m-1}\}$ be the shortest electrical length, and define p_k to be the nearest integer to the real number nT_k/T_{min}. With this choice, $T_k = p_k T_{\text{min}}/n + \delta_k$ and $|\delta_k| < T_{\text{min}}/n$. Using T_{min}/n as the unit of time measurement, a perfect way to proceed would be to chop the ith original layer into pieces each of length $c_k T_{\text{min}}/n$. Because the original lengths may not be evenly divisible by pieces of these specified lengths, let us accept errors in the computational lengths of the layers by redefining the spatial computational domain and its discretization as follows: Starting at $x = 0$, define $p_1 + 1$ new nodes $kc_1 T_{\text{min}}/n$, for $k = 0, \ldots, p_1$; p_2 new nodes $p_1 c_1 T_{\text{min}}/n + kc_2 T_{\text{min}}/n$, for $k = 1, \ldots, p_2$; p_3 new nodes $p_1 c_1 T_{\text{min}}/n + p_2 c_2 T_{\text{min}}/n + kc_3 T_{\text{min}}/n$, for $k = 1, \ldots, p_3$; and so on. This new spatial mesh determines intervals (between the nodes) that are traversed in equal times and it fills out a computational domain

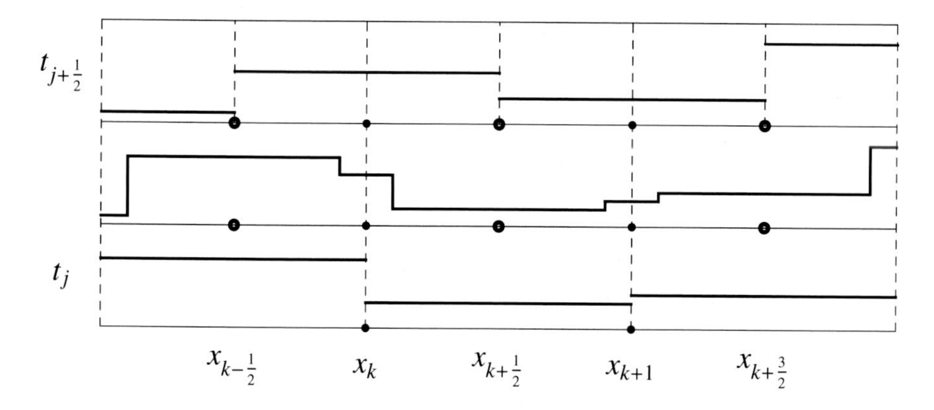

Fig. 22.4 At t_j, graphs of the discontinuous voltage profile is depicted on the bottom third of the figure. The middle third shows the voltage profile shortly after time t_j, and the upper third shows the voltage profile at time t_j plus half the electrical length divided by the wave speed.

that approximates the actual lengths of the layered medium. The differential equation model stays the same except that the spatially varying matrix $A(x)$ is redefined to have corresponding constant values over each of the approximate layers that are equal to the corresponding constant values on the original layers.

The modified problem is an approximation of the original IBVP. The accuracy of the approximation clearly depends on n, the number of subdivisions of the electrically shortest layer. In fact, the error goes to zero as n increases without bound, but there is some work involved to prove this fact.

Suppose that the modified problem has been set so that every interval (cell) in the discretization has the same electrical length and these correspond to contiguous sets of intervals corresponding to the layers of the medium such that A is constant on each layer.

According to the analysis of the Riemann problem in Section 22.5, the exact solution of the transmission line model for layered media is given in Eqs. (22.33), (22.34), and (22.35) at least in the case of two layers. In the multilayer case, exactly the same solution formulas are true as long as the time advance from the starting time to the new time does not exceed half of the smallest electrical length of the cells.

Why is this true?

Consider the cell with end nodes x_{k+1} and x_k and recall from Section 22.5 that the left, middle, and right solutions are determined in the regions bounded by the characteristics. Suppose the starting time is $t = 0$. At x_k the characteristic with positive slope (corresponding to a wave moving to the right) is given by $x - ct = x_i$ and the characteristic at x_{k+1} with negative slope has equation $x + ct = x_{k+1}$. In both cases, the wave speed c is the constant wave speed in the cell with boundaries x_k and x_{k+1}. The Riemann problem solution is valid in forward time $t > 0$ as long as these two characteristics do not meet. As time passes beyond the time at which they cross, new regions are formed bounded by these characteristics. When do they cross? To determine this time, simply solve the two equations simultaneously for t. This value is

$$t = \frac{x_{k+1} - x_k}{2c},$$

exactly half the electrical length of the interval. Because every cell has the same electrical length, the Riemann solution at each cell boundary is valid as long as the forward advance in time is less than half of this constant electrical length.

Fig. 22.4 depicts the time evolution of the graphs of one component of the solution u (for example the voltage) that is determined by solving the Riemann problem at each cell interface, as time increases from bottom to top from t_j at the start to t_{j+1} at the end of the evolution where the final time is t_j plus half of the electrical length of the cells. Due to the discontinuities at the cell interfaces, waves travel left and right from each interface as time increases. The wave profiles are constant step functions as shown in the figure. The constant values of the new voltages and currents at the interfaces are given exactly by Eq. (22.36). At the final time, the waves coming from left and right meet exactly in the middle of each cell as shown in the top row of the figure. The end state u is again piecewise constant (with values given via Eq. (22.36)) in the staggered mesh determined by the midpoints of the cells.

What happens when the solution is advanced in time to an additional half of the electrical length of the cells. Fortunately, the Riemann problem solution again determines the new piecewise constant end state. The discontinuities occur at the mid points of cells where A is constant. At the interfaces at the original cells, there are no discontinuities in the state. There are discontinuities in A at these later interfaces. But the waves leaving them

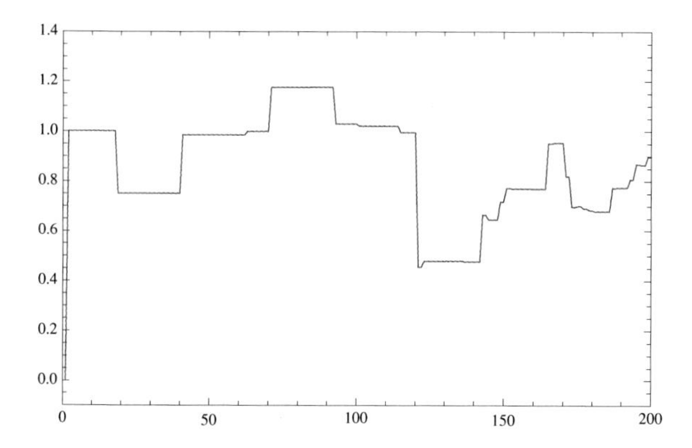

Fig. 22.5 Simulated TDR voltage versus time (in nanoseconds) is plotted for an open line (infinite resistance at the end of the line) of electrical length 100 nanoseconds and impedance profile as follows: 50 ohms for first nine nanoseconds, 30 for 11 nanoseconds, 50 for 15 nanoseconds, 75 for 11 nanoseconds, 50 for 14 nanoseconds, 10 for 11 nanoseconds, and 50 for the remainder of the line. The input voltage is one volt and the TDR trace is the voltage in the first of the 100 cells. The wave speed is assumed to be constant (at the speed of light) along the line.

do not have jumps; they remain at the same constant values as the initial state but travel at different speeds. The discontinuities at the cell midpoints produce left- and right-going waves that move at the same speed within each original cell and meet the original cell boundaries at exactly the end time advance of half of the electrical length of the cells. At the end of this time, the discontinuities in the state are exactly at the original cell boundaries, as in the bottom row of Fig. 22.4. Again the state u is constant in each cell and these constant values are determined by Eq. (22.36) or (22.37), where the left and right impedances (Z_ℓ and Z_r) are both equal to the impedance in the cell where the midpoint resides.

Boundary conditions are respected by specifying the boundary states in ghost cells to the right and left of the end points of the computational domain.

At the end of the line, a useful implementation of boundary condition (22.6), which is simply Ohm's law in the form $v(\ell, t) = R_{\text{end}} i(\ell, t)$, depends on the size of the ohmic resistance R_{end}.

Recall that the last node in the discretization is $x_m = \ell$. The initial data, usually zero voltage and zero current, should be compatible with the end-of-line boundary condition; that is, the values assigned to v_m and i_m at time zero satisfy the relation $v_m = R_{\text{end}} i_m$. Assume such an assignment is made.

For large resistances, which might include the open line where the ideal resistance is infinite, the intermediate voltage V_m at the end of the line is computed using the first component of vector equation (22.36), and the intermediate current I_m is set to

$$I_m = \frac{V_m}{R_{\text{end}}}.$$

For small resistances, which might include the shorted line where the ideal resistance is zero, the intermediate current is set using the second component of Eq. (22.36), and the voltage is assigned, in accordance with the boundary condition, by

$$V_m = R_{\text{end}} I_m.$$

The left-end ghost cell is assumed to take on this assigned voltage and current over its entire length, which might be imagined to have infinite leftward extent. Also, the intermediate voltages are assigned to v_m and i_m in preparation for the intermediate step on the computational domain.

In the ghost cell at the beginning of the transmission line, the initial voltage and current are set together with the voltage generator so that the voltage v_0 and current i_0 in the ghost cell are compatible with the boundary condition

$$v_0 + R_{\text{in}} i_0 = v_{\text{in}}(0).$$

Usually, v_0, i_0 and $v_{\text{in}}(0)$ all vanish at $t = 0$. The intermediate current I is determined from the second component of vector equation (22.37) and assigned as the updated i_0. The new ghost cell voltage is defined to be

$$v_0 = v_{\text{in}}(t) - R_{\text{in}} i_0,$$

where t is set to the end value of t at the current time step.

The complete algorithm for advancing the voltage and current (defined at time t on the computational domain cells (x_k, x_{k+1}) for $k = 1, 2, 3, 4, \ldots, m-1$, the left-end ghost cell, and the right-end ghost cell) for one time step to time $t + \Delta t$, starts with the computation of the new intermediate states in the ghost cells.

Using the second component of the vector equation (22.37) and capital letters for intermediate voltages and currents, the intermediate current in the left-end ghost cell is

$$I_0 = i_1 + \frac{R_{\text{in}}(i_1 - i_0) + (v_1 - v_0)}{R_{\text{in}} + Z_1},$$

where the impedance in this ghost cell is (as it should be by definition) the ohmic resistance of the circuit connection modeled in this cell. Taking into account the boundary condition, the ghost cell voltage is

$$V_0 = v_{\text{in}}(t + \Delta t) - R_{\text{in}} I_0.$$

At the right-end ghost cell, the size of R_{end} is tested; for example, in relation to the impedance Z_{m-1} in the last computational cell. For $R_{\text{end}} \geq Z_{m-1}$, the ghost cell intermediate voltage is set to

$$V_m := v_{m-1} - Z_{m-1} \frac{R_{\text{end}}(i_m - i_{m-1}) - (v_m - v_{m-1})}{R_{\text{end}} + Z_{m-1}}, \tag{22.43}$$

and in accordance with the boundary condition, the intermediate current I_m is

$$I_m = \frac{V_m}{R_{\text{end}}}. \tag{22.44}$$

For $R_{\text{end}} < Z_{m-1}$, the intermediate current is defined to be

$$I_m = i_{m-1} + \frac{R_{\text{end}}(i_m - i_{m-1}) - (v_m - v_{m-1})}{R_{\text{end}} + Z_{m-1}}$$

and the voltage is

$$V_m = R_{\text{end}} I_m.$$

The ideal open line may be approximated by assigning R_{end} to some value much larger than Z_{m-1}. Alternatively, the open line is modeled exactly by first taking limits as $R_{\text{end}} \to \infty$ in the assignment formulas for $R_{\text{end}} \geq Z_{m-1}$. In this case, the current vanishes in the ghost cell. The shorted line is modeled by taking $R_{\text{end}} = 0$. In this case, the voltage vanishes in the ghost cell.

The values of v_0, i_0, v_m, and i_m are reset to V_0, I_0, V_m, and I_m.

New intermediate states in the computational domain are

$$\begin{pmatrix} V_k \\ I_k \end{pmatrix} = \begin{pmatrix} v_{k-1} \\ i_{k-1} \end{pmatrix} + \frac{Z_k(i_k - i_{k-1}) - (v_k - v_{k-1})}{Z_k + Z_{k-1}} \begin{pmatrix} -Z_{k-1} \\ 1 \end{pmatrix} \tag{22.45}$$

for indices $k = 1, 2, 3, \ldots, m-1$.

And, for these same indices, new voltages and currents on the computational domain cells are

$$\begin{pmatrix} v_k \\ i_k \end{pmatrix} = \begin{pmatrix} V_k \\ I_k \end{pmatrix} + \frac{Z_k(I_{k+1} - I_k) - (V_{k+1} - V_k)}{2Z_k} \begin{pmatrix} -Z_k \\ 1 \end{pmatrix}. \tag{22.46}$$

This completes the time step.

After advancing the time to exactly the current time plus the electrical length of the cells, using the two-step process just described, the process can be repeated in increments of the constant electrical length of the cells. As n grows large, the electrical lengths go to zero. Thus, the number of steps required to reach some fixed future time increases as it should to produce more accurate solutions.

A typical simulated TDR voltage signal is depicted in Fig. 22.5.

An important special case is the launch of a front into the transmission line. At time $t = 0$ the line is dead (voltage and current are zero). Sometime between zero and Δt the input voltage (modeled by the function v_{in}) rises to some nonzero value, which (by an abuse of notation) we denote by v_{in}, and it is held at this constant value as time increases. According to the algorithm, at the beginning of the first time step $v_0 = 0$ and $i_0 = 0$. The intermediate values are

$$I_0 = 0, \qquad V_0 = v_{\text{in}}.$$

The intermediate state in the first cell is given by

$$V_1 = v_{\text{in}} - v_{\text{in}} \frac{R_{\text{in}}}{R_{\text{in}} + Z_1}, \qquad I_1 = v_{\text{in}} \frac{1}{R_{\text{in}} + Z_1},$$

and the voltage in this cell at the end of the first time step is

$$v_1 = \frac{v_{\text{in}}}{2}\Big(1 - \frac{R_{\text{in}}}{R_{\text{in}} + Z_1} + \frac{Z_1}{R_{\text{in}} + Z_1}\Big).$$

In the special case where the voltage generator is (impedance) matched to the line, that is, $R_{\text{in}} = Z_1$, the voltage is exactly

$$v_1 = \frac{v_{\text{in}}}{2}.$$

A traveling voltage wave front with this amplitude is propagated down the ideal transmission line. Thus, to launch a voltage front of some desired amplitude in case the voltage generator is matched to the line, the model requires the generator to produce exactly twice the desired voltage with a rise time to this voltage less than the duration of one time step.

Exercise 22.6. Show that a traveling current wave front is propagated down the line when sometime between zero and Δt the input voltage (modeled by the function v_{in}) rises to some nonzero value and is held at this constant value. Determine the amplitude of this front.

Exercise 22.7. Suppose a voltage-current wave front reaches the end of the transmission line terminated as in the model discussed in this section. (a) Determine the amplitudes of the reflected and transmitted waves using the algorithm discussed in this section. Consider separately the open line and the shorted line. (b) Does the algorithm suggested for computing the time evolution of voltages and currents produce reflected and transmitted waves at the end of the line that agree with the exact predictions of the model equations?

22.8 THE LOSSY TRANSMISSION LINE

The full transmission line model [Eq. (22.7)] encodes resistance and conductance along the line into the matrix function B. The inclusion of these effects introduce dissipation of energy (damping) into the system. In the engineering literature, the adjective lossy is often used to describe the model. Of course, all real transmission lines are lossy.

As stated previously, a theme of this chapter is to take advantage of the special features of applied problems. For the lossless transmission line, an important feature is that the wave speeds are the same in both directions in cells where A is constant. This fact led to the exact solution of the IBVP discussed in Section 22.7. For the lossy line, there are two special features: B does not depend on time (at least in the model considered here) and $B(x)$ is diagonal for each spatial coordinate x. How can we take advantage of these facts?

Rewrite the model PDE in the form

$$u_t = -A(x)u_x - B(x)u. \tag{22.47}$$

This formulation suggests that u might be viewed as the solution of an ordinary differential equation (ODE), which would have the form $u_t = F(u)$ for some function F. The theory of ODEs is understood much better than the theory of PDEs. Thus, there is a powerful incentive to view PDEs as ODEs. Perhaps, at first glance, there is no obvious way to do this, but in fact, this idea has been very fruitful for understanding the behaviors of an important class of PDEs. A glimpse into the ramifications of this powerful idea is provided in this section.

To view PDE (22.47) as an ODE $u_t = F(u)$, the function F must be given by

$$F(u) = -A(x)u_x - B(x)u.$$

Something is wrong: The right side involves u_x and x appears explicitly on the right-hand side. If A were zero and B were constant, then there would be no problem as $F(u) = -Bu$ makes perfect sense. In this special case, F is simply a linear transformation and $u_t = -Bu$ is a matrix ODE with solution

$$u(t) = e^{-tB}u(0).$$

This is not quite right. With $A(x) \equiv 0$ the PDE becomes an ODE, but the solution of the original PDE should depend on both x and t. This problem is simply resolved. In fact,

$$u(x,t) = e^{-tB}u(x,0)$$

is a solution of $u_t = -Bu$ viewed as an ODE for u in the space of all (two-dimensional vector-valued) functions defined for $0 \le x \le \ell$. Instead of the usual setting for the ODE $\dot{u} = f(u)$ with f a function from some Euclidean n-dimensional space into itself, which has solutions that are functions $t \to u(t)$ defined on some time interval $0 \le t < T$, the correct way to view $u_t = -Bu$ is in the form $u_t = f(u)$ where f is defined on a space of (vector-valued) functions themselves defined on $[0, \ell]$. A solution $t \to u(t)$ is now a curve in this function space. For each t, $u(t)$ is a vector-valued function defined on $[0, \ell]$ starting at the vector-valued function $x \mapsto u(x, 0)$. Its value at x may be denoted by $u(x, t)$, or if you wish, its value at x may be denoted $u(t)(x)$. In the latter interpretation, $u(t)$ is the name of a function and its value at x is as usual its name followed by x enclosed in parenthesis. This is exactly what we mean when we write $f(x)$. The name of the function is f and its value at x is $f(x)$.

There is no problem when B is a matrix function of x. We may still view $u_t = -B(x)u$ as a PDE whose solution is

$$u(x,t) = e^{-tB(x)}u(x,0).$$

Again, the correct interpretation of the solution of $u_t = B(x)u$ is a curve of functions defined for $0 \leq x \leq \ell$ and parameterized by t starting at $t = 0$ at the initial vector-valued function $x \mapsto u(x,0)$.

In case B is the zero matrix function and $F(u) = -A(x)u_x$, we may again consider F as a function defined on a space of vector-valued functions defined on $[0,\ell]$. The value of F on such a function involves the derivative of this function. But, this does not preclude F itself from being a function. To each u it assigns a unique function $F(u)$ as long as u is differentiable. The solution of $u_t = -A(x)u_x$ is the solution $u(x,t)$ derived previously. It can be interpreted as a curve in a function space: for each t the function $x \mapsto u(x,t)$ is the point in the function space on the curve at time t.

This discussion is purposely vague in the specification of the function space of functions defined on $[0,\ell]$. The specification of this space is essential in the discussion of the well posedness of the PDE. To say that a unique solution exists, for example, requires a precise definition of what we mean by a solution. In the context of transmission line problems, the matrix functions A and B are likely to be discontinuous. Solutions of the PDE will also be discontinuous. As a result, solutions will not be differentiable. This means u_t and u_x do not always exist. How is it possible that we have already found solutions of the ideal transmission line equation? A glib answer is that we simply did not require solutions to be differentiable at all points. Solutions were allowed to be piecewise differentiable. The space of functions in which solution curves of our ODE reside must include piecewise differentiable functions. A precise definition of this function space and the precise meaning of a solution requires the notion of a weak solution, which will be postponed until a later section. We may proceed, as before, but under the warning that the discussion here is informal. Fortunately, all stated results can be made precise. Also, to avoid awkward references to the unspecified function space, let's give it a name and call it $\mathcal{L}(0,\ell)$. Elements of this set are vector-valued functions defined on the interval $[0, \ell]$. Although we do not specify which functions to include, let us assume that $\mathcal{L}(0, \ell)$ is a vector space. This ensures that linear combinations of elements of $\mathcal{L}(0, \ell)$ with real coefficients are elements of $\mathcal{L}(0, \ell)$.

Recall the concept of a splitting method for approximating the solution of an ODE as in Chapter 7. The basic example is a linear ODE $\dot{z} = Pz + Qz$, where P and Q are constant matrices and z is a vector variable. The solution of this ODE is

$$z(t) = e^{t(P+Q)} z(0).$$

In case e^{P+Q} is more difficult or time-consuming to compute than e^{tP} and e^{tQ}, there would be a great saving if $z(t)$ were equal to $e^{tP} e^{tQ} z(0)$. The solution could then be determined by applying e^{tQ} followed by e^{tP}. This strategy is almost never viable because e^{P+Q} is equal to $e^P e^Q$ only if P and Q commute; that is, $PQ = QP$. Nonetheless, there is a definite relation between the two exponentials. For example, the Trotter product formula states that

$$e^{t(P+Q)} = \lim_{n\to\infty} (e^{t/nP} e^{t/nQ})^n. \tag{22.48}$$

This result suggests approximating the solution of the ODE by a finite product $z(t) \approx (e^{t/nP} e^{t/nQ})^n z(0)$ for some positive integer n.

Important note: In case P and Q commute and the solution is given by $z(t) = e^{tP} e^{tQ} z(0)$, where the Q part of the splitting is applied for time t and then the P part is applied again for time t, we might believe that the solution should have been advanced to time $2t$. This is not the case: $e^{ta} e^{tb} = e^{t(a+b)}$. After all, we are advancing only part of the original differential equation in each step.

The transmission line model [Eq. (22.47)] may be written in the form

$$u_t = Pu + Qu \tag{22.49}$$

with the linear transformations P and Q defined on $\mathcal{L}(0, \ell)$ by

$$(Pw)(x) = A(x) w_x(x), \qquad Q(w)(x) = B(x) w(x).$$

Both operators are indeed linear (for instance, $Q(aw + by) = aQ(w) + bQ(y)$ whenever w and y are in $\mathcal{L}(0, \ell)$ and a and b are real numbers) on the subsets of $\mathcal{L}(0, \ell)$ where they are defined. The linear operator P is called a differential operator because its action on functions involves their derivatives.

Under the expectation that the solution of the transmission line PDE behaves similar to $e^{t(P+Q)}u(x,0)$ with an appropriate definition of the exponential, it is reasonable to expect that the solution may be well-approximated by a finite product $(e^{t/nP}e^{t/nQ})^n u(x,0)$. In other words, the solution might be closely approximated by starting at the initial data, solving $w_t = -B(x)w$ for a short time t/n and reaching the evolved state $w(t/n)$, followed by solving $y_t = -A(x)y_x$ for a short time with initial data $w(t/n)$ to reach the new state $y(2t/n)$, solving again for w starting at this new state, then for y, and continuing in this manner until the alternate directions reach time t. The final state in this process is the approximation to $u(x,t)$.

The alternate direction method is viable, but its implementation as a numerical method requires surmounting at least one more problem: boundary conditions must be incorporated.

As an explicit example, consider the PDE

$$U_t + cU_x = -b(x)U \tag{22.50}$$

defined for $-\infty < x < \infty$ and $t \geq 0$ with initial data $U(x,0) = U_0(x)$ defined on the whole real line.

To reveal the operator splitting as in Eq. (22.49), note first that the general solution in case b vanishes is $U(x,t) = \phi(x - ct)$ for some function ϕ. By imposing the initial condition

$$U(x,0) = \phi(x),$$

the function ϕ is determined and the solution of the PDE is

$$U(x,t) = U_0(x - ct).$$

In this example, the operator P (as in Eq. (22.49)) is given by $(Pw)(x) = cw_x(x)$, where the function w is defined on the whole real line. The definition of the solution operator e^{tP} in this case is

$$(e^{tP}w)(x) = w(x - ct).$$

In particular,

$$(e^{tP}U_0)(x) = U_0(x - ct).$$

The exponential is a suggestive notation that is not to be taken literally. On the other hand, it does convey the correct intuition. For example, it is easy to check that

$$(e^{(t+s)P}U_0)(x) = U_0(x - c(t+s)) = (e^{sP}e^{tP}U_0)(x).$$

With c equal to zero, the general solution is

$$U(x,t) = e^{-tb(x)}U_0(x),$$

where the exponential *is* to be taken literally. In other words,

$$(e^{tQ}w)(x) = e^{-tb(x)}w(x).$$

Do the operators P and Q commute? If they do, we can expect the general solution of the full equation to be

$$U(x,t) = (e^{tQ}e^{tP}U_0)(x) = e^{-tb(x)}U_0(x - ct).$$

By substitution of this function U into the PDE, it is easy to show that it is a solution if and only if $b'(x) = 0$; that is, b is a constant function. In this case and with b used to denote the constant value of the function,

$$U(x,t) = e^{-tb}U_0(x - ct)$$

is the unique solution, the operators P and Q commute, and $e^{tQ}e^{tP} = e^{t(P+Q)}$.

Although the solution of PDE (22.50) cannot in general be expressed in such a simple form when b is not a constant, there are some special cases with exact solutions. For instance, the PDE

$$U_t + U_x = -2xU \tag{22.51}$$

with initial data $U_0(x) = e^{x-x^2}$ has the solution

$$U(x,t) = e^{-x^2}e^{x-t}.$$

This is a good example to test operator splitting as a numerical method.

For PDE (22.51),

$$(e^{t/nQ}e^{t/nP}U_0)(x) = e^{x-x^2}e^{-t/n-t^2/n^2}$$

and

$$((e^{t/nQ}e^{t/nP})^n U_0)(x) = e^{x-x^2}e^{-t-t^2/n}. \tag{22.52}$$

Hence, as expected,

$$\lim_{n\to\infty}((e^{t/nQ}e^{t/nP})^n U_0)(x) = e^{-x^2}e^{x-t}. \tag{22.53}$$

The Trotter product formula is valid for the operators P and Q.

As a numerical method, the solution at time $t > 0$ would be approximated by taking n steps, for some positive integer n, as in Eq. (22.52). The error is

$$|e^{x-x^2}e^{-t-t^2/n} - e^{x-x^2}e^{-t}| = e^{x-t-x^2}|e^{-t^2/n} - 1| \le te^{-x^2}e^{x-t}\frac{t}{n}; \tag{22.54}$$

therefore, the method is order one (see Exercise 22.8).

Splitting approximations are improved by replacing $e^{t/nQ}e^{t/nP}$ with

$$e^{t/(2n)P}e^{t/nQ}e^{t/(2n)P}.$$

For PDE (22.51), the improvement is dramatic; in fact, using this splitting produces the exact solution

$$((e^{t/(2n)P}e^{t/nQ}e^{t/(2n)P})^n U_0)(x) = e^{-x^2}e^{x-t}. \tag{22.55}$$

Although the operator splitting method seems viable in the setting of a conservation law with a source, its application to the lossy transmission line requires at least one more important consideration: the boundary conditions.

Consider the toy example $U_t + U_x = -bU$, where b is a constant and the boundary condition is $U(0,t) = f(t)$ for some function f defined on the whole real line. The solution of this BVP is

$$U(x,t) = e^{-bx}f(t-x).$$

Following the splitting prescription, consider solving the PDE

$$W_t + W_x = 0 \tag{22.56}$$

with the boundary condition $W(0,t) = f(t)$ and then the ODE

$$Y_t = -bY \tag{22.57}$$

with initial data $W(x,t)$ forward for time t. The result should be an approximation (or perhaps the exact solution) of the original PDE at time t. Unfortunately, this simple approach is not correct. Note that $W(x,t) = f(t-x)$ and $Y(x,t) = e^{-bt}Y_0$; therefore, the approximation is $U(x,t) = e^{-bt}f(t-x)$. The boundary condition for the original PDE is not satisfied because the ODE does not preserve the boundary condition that W satisfies. As usual, boundary conditions are troublesome.

A cure for operator splitting applied to the toy problem is to set the boundary condition for the PDE $W_t + W_x = 0$ so that the correct boundary condition for the original PDE is met after the ODE is applied. To determine this boundary condition, note that the general solution of PDE (22.56) is

$$W(x,t) = \phi(x-t)$$

for ϕ an arbitrary function defined on the whole real line. ODE (22.57) applied to this initial data yields the approximation $U(x,t) = e^{-bt}\phi(x-t)$. It satisfies the boundary condition at $x = 0$ provided that $e^{-bt}\phi(-t) = f(t)$. Hence, $W(x,t) = \phi(x-t)$ must satisfy the boundary condition

$$W(0,t) = \phi(-t) = e^{bt}f(t)$$

so that

$$W(x,t) = e^{-b(x-t)}f(t-x).$$

By solving PDE (22.56) with boundary condition $W(0,t) = e^{bt}f(t)$ followed by solving ODE (22.57) with initial data $W(x,t)$, the operator splitting in this form gives the exact solution of the toy problem:

$$U(x,t) = e^{-bt}\phi(x-t) = e^{-bt}(e^{-b(x-t)}f(t-x)) = e^{-bx}f(t-x).$$

For the lossy transmission line equation, operator splitting is viable, but the boundary conditions for the PDE $W_t + A(x)W_x = 0$ must be set so that after completing a time step by advancing the solution of this PDE by the ODE $Y_t = -B(x)Y$ the boundary conditions for $u_t + A(x)u_x = -B(x)u$ are satisfied. As a general principle for the case of a splitting into a PDE and an ODE, a three-step procedure should be employed: Move the boundary condition backward one time step via the ODE, advance the current states of the line voltage and current one time step using the lossless-line PDE and the modified boundary conditions, and advance the resulting line current and voltages one time step using the time-advance map of the ODE.

The fundamental matrix solution at $t = 0$ of the linear ODE is $e^{-tB(x)}$. Thus, the time advance for the autonomous ODE is the same matrix exponential $e^{-\Delta t B(x)}$, where Δt is the duration of the time advance from the current time. For the transmission line PDE, the matrix $B(x)$ is diagonal; therefore, the time-advance map is also diagonal. It is given by the matrix

$$\begin{pmatrix} e^{-\Delta t G(x)/C(x)} & 0 \\ 0 & e^{-\Delta t R(x)/L(x)} \end{pmatrix}. \tag{22.58}$$

In other words, at the position with coordinate x on the line, the ODE part of the splitting advances the voltage $v(x)$ forward an additional time Δt to $e^{-\Delta t G(x)/C(x)} v(x)$ and the current $i(x)$ to $e^{-\Delta t R(x)/L(x)} i(x)$.

For some important transmission line applications, the operator splitting method does not require a boundary condition adjustment at the left end of the line. Usually, the voltage generator connection to the transmission line is manufactured to be impedance matched to the connection (which is itself a segment of the entire transmission line), and in addition, the conductance of the insulator at the connection and the ohmic resistance along the connection are manufactured to be so small that they are negligible at the connection. In this case, the matrix B may be taken to vanish at the left end of the line.

The right-end boundary condition for the lossless-line calculation usually requires an adjustment to ensure the system boundary condition is met. Currents and voltages passed to this part of the operator splitting algorithm are assumed to satisfy the full system right-end boundary condition. A viable modification of the ideal-line algorithm at Eqs. (22.43)–(22.44) is to compute the intermediate ghost cell current I_m using Eq. (22.43), store

$$i_m = e^{\Delta t G(\ell)/C(\ell)} I_m,$$

replace I_m by i_m (so that these two quantities are equal), compute

$$V_m = e^{-\Delta t R(\ell)/L(\ell)} R_{\text{end}} I_m,$$

and store $v_m = V_m$. These values of V_m, v_m , I_m, and i_m are used in the second step of the lossless-line algorithm. At the end of the time step, the ghost cell voltage and current satisfy the modified boundary condition

$$e^{\Delta t R(\ell)/L(\ell)} v_m = R_{\text{end}} e^{\Delta t G(\ell)/C(\ell)} i_m.$$

A similar modification is made for the case $R_{\text{end}} < Z_{m-1}$.

The updated voltages and currents in the cells of the computation domain together with v_m and i_m are passed to the lossy part of the operator splitting algorithm. In the kth cell, the resistance divided by the inductance is denoted here by ω_k and the conductance divided by the capacitance by μ_k, and in the ghost cell, ω_m and μ_m are computed by the ratios of the same quantities evaluated at $x = \ell$. The voltage in the kth cell (including $k = m$) is multiplied by the exponential $e^{-\Delta t \omega_k}$ and the current by $e^{-\Delta t \mu_k}$. The new ghost cell voltage and current satisfy the system right-end boundary condition. They are passed to the lossless-line algorithm in the next time step. Thus, the system boundary conditions are satisfied by the approximate solution obtained via operator splitting.

Exercise 22.8. Show that $(1 - e^{-x})/x < 1$ for $0 < x < \infty$ and use this result to establish the inequality in estimate (22.54).

Exercise 22.9. Does the order of operation $e^{tP}e^{tQ}$ or $e^{tQ}e^{tP}$ in the splitting method affect the accuracy of the approximation?

Exercise 22.10. Determine all the solutions of PDE (22.50) that can be expressed as a product $U(x, t) = f(x)g(t)$ with separated variables.

Exercise 22.11. Is Eq. (22.55) true for every solution of PDE (22.50) that is of the form $U(x, t) = f(x)g(t)$?

Exercise 22.12. (a) What are the fundamental units to measure resistance divided by inductance? (b) What about conductance divided by capacitance?

Exercise 22.13. [Historically Important Transmission Line Application] Heaviside's transmission line model played an important role in the history of telegraph communications via the early transatlantic cables. The first transatlantic cable communication was made on April 16, 1858. Due to a controversy over high-voltage versus low-voltage transmissions, the insulation of the cable was compromised and the cable was soon inoperative. A viable new cable was laid in 1866. Although communication was established, the information transmission rate was slow: on the order of a few words per minute. Faster transmission rates were realized after Heaviside's theoretical work—which was controversial at the time for economic reasons related to laying expensive cables—explained the origin of signal distortion, the main obstacle to improving information transmission rates. His work is a gem of applied mathematics. An important result, which you will derive in this exercise, is Heaviside's condition for distortionless transmission.

(a) Use the transmission line model [Eqs. (22.1)] to derive the telegrapher's wave equation for voltage:

$$v_{tt} + \Big(\frac{R}{L} + \frac{G}{C}\Big)v_t - \frac{1}{LC}v_{xx} = -\frac{RG}{LC}v.$$

(b) The objective is to transmit signals as waves that have a definite profile (shape) that is maintained along the line; that is, these waves propagate with no distortion. It is too much to ask for waves that maintain their strength because there will be losses due to the conductance G of the dielectric between the conductors and the resistance R of the conductors. The model equation is linear, so it is reasonable to expect losses to be modeled by exponential decay. Thus, it is reasonable to expect voltage waves of the form

$$v(x,t) = e^{-\gamma x} f(x - at), \tag{22.59}$$

where f is the wave profile, $\gamma > 0$ is the loss rate per distance along the line, and a is the wave speed. Are waves of this type solutions of the telegrapher's wave equation?
(c) The answer to part (b) is yes. Heaviside asked a more important question: What condition would have to be satisfied by the system parameters C, G, L, and R so that every profile f (which we might wish to send down our cable) is not distorted? In other words, is there a condition on these parameters so that *all* profiles f can be transmitted as distortionless waves (of the form given in Eq. (22.59))? Show that the answer is yes provided that Heaviside's condition

$$\frac{C}{L} = \frac{G}{R}$$

is satisfied, the wave speed is $a = (LC)^{-1/2}$ and the loss per length is $\gamma = (RG)^{1/2}$. Hint: Substitute the distortionless wave form [Eq. (22.59)] into the telegrapher's equation and determine conditions to make this equation hold for *all* profiles f.

For a typical cable manufactured in the 19th century, RC was much greater than LG. The natural, or so it seemed, direction was to build cables that would decrease RC. Heaviside realized that this was unnecessary and argued that LG should be increased, a controversial position at the time. This could be done by increasing the inductance L at a relatively low cost. It is also possible to increase G by using higher frequency signals, but to understand this possibility more fully requires more physics and engineering than space allows in this book. Heaviside's theory is based on correct physics (Maxwell's equations). So, of course, his theory eventually became the industry standard. (See [80] for much more on Heaviside's legacy.)

22.9 TDR APPLICATIONS

Fig. 22.6 depicts TDR data collected from an experimental apparatus as in Fig. 22.1 where the chamber is filled with water. Fig. 22.7 shows the result of a simulation that depicts a similar graph of voltage versus time. All the conceptual ingredients for reproducing the simulated data in the figure have been discussed. The required parameters are given in this section.

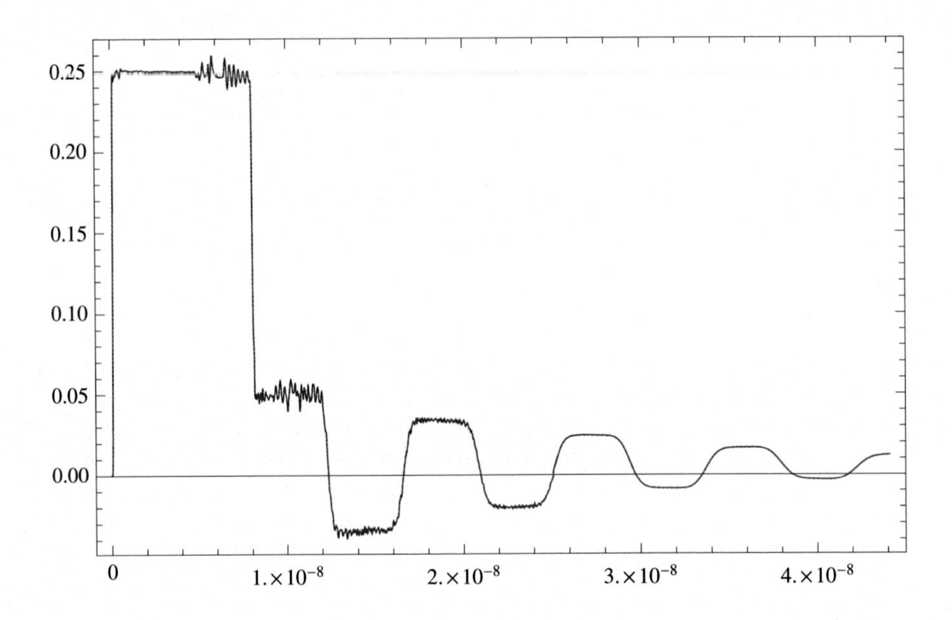

Fig. 22.6 Experimental TDR output of voltage (in volts) versus time (in seconds) for a coaxial cable connected to a dielectric filled chamber is depicted.

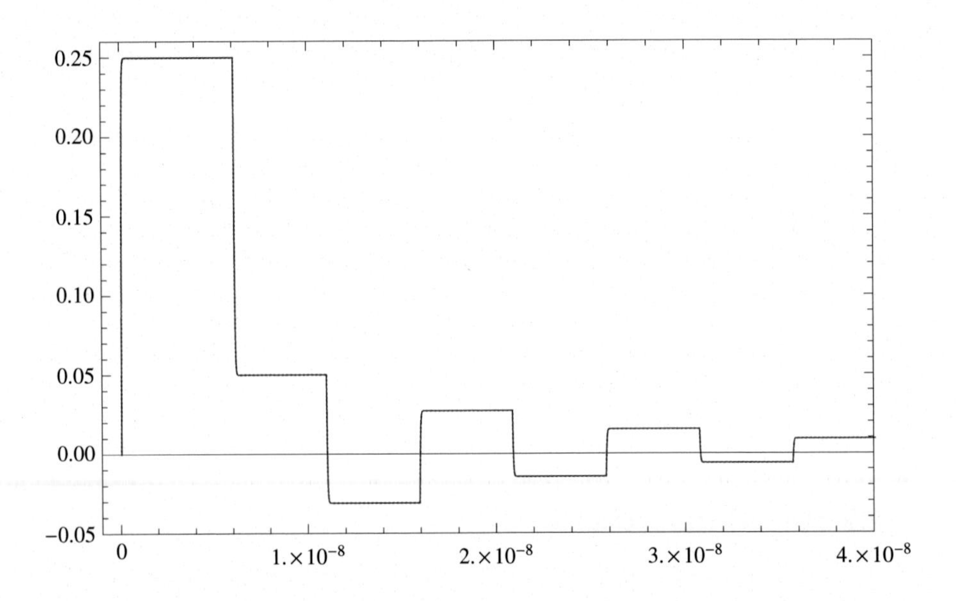

Fig. 22.7 Simulated TDR output of voltage (in volts) versus time (in seconds) for a coaxial cable connected to a dielectric filled chamber is depicted.

From basic physics, recall the free-space permittivity ϵ_0 and permeability μ_0 have numerical values

$$\varepsilon_0 = 8.85 \times 10^{-12} \frac{\text{farad}}{\text{meter}}, \qquad \mu_0 = 4\pi \times 10^{-7} \frac{\text{henry}}{\text{meter}}.$$

The permittivity and permeability of materials are usually specified as $\varepsilon = \varepsilon_r \varepsilon_0$ and $\mu = \mu_r \mu_0$ using dimensionless relative permittivity ϵ_r and permeability μ_r. Of course, for free-space $\epsilon_r = \mu_r = 1$.

For coaxial cable with core conductor diameter a and shield (inner) diameter b the inductance and capacitance are

$$L = \frac{\mu}{2\pi} \ln \frac{b}{a}, \qquad C = \frac{2\pi}{\varepsilon} (\ln \frac{b}{a})^{-1}.$$

The impedance and (lossless line) wave speed are

$$Z = \sqrt{\frac{L}{C}} = \frac{1}{2\pi} \sqrt{\frac{\mu}{\varepsilon}} \ln \frac{b}{a}, \qquad c = \frac{1}{\sqrt{LC}} = \frac{1}{\sqrt{\mu\varepsilon}}.$$

The free-space impedance is (approximately) 376.73 ohm and the wave speed c is the speed of light 2.99792×10^8 meter / second.

Consider a TDR experimental apparatus as in Fig. 22.1. For definiteness, suppose the reflectometer is attached to the test chamber with a manufacturer certified 50 ohm coaxial cable (that is, the cable's impedance is 50 ohm) that has electrical length 3.0 nanosecond and physical length 0.75 meter.

The coaxial test chamber has length 0.3 meter. Its cylindrical central conductor has outer diameter 3.175 cm and its cylindrical outer conductor has inner diameter 7.2898 cm. Also, the chamber is shorted with a metallic cap whose total resistance is measured to be 0.3 ohm. For a first approximation, the electrical properties of the connector between the coaxial cable and the chamber can be safely ignored.

The chamber is filled with a dielectric. Treated as a transmission line, the filled chamber has electrical length 2.5 nanosecond and impedance is 5.57 ohm. Both the ohmic resistance of the chamber and the conductance of the dielectric are known to be small and can be ignored. The inductance and capacitance of the dielectric filled chamber can be computed from its impedance and electromagnetic wave speed.

Exercise 22.14. Determine the inductance in nanohenries and the capacitance in picofarads for a 50 ohm coaxial cable with electromagnetic wave speed $0.6733c$.

Exercise 22.15. (a) Using the transmission line equations and boundary conditions as in Section 22.1 together with the data provided in this section, reproduce Fig. 22.7 under the assumption that a quarter-volt step wave is launched with a rise time of 10 picosecond into the coaxial cable by the TDR. (b) Typical resistance and inductances the chamber and the dielectric filling it are in the microsiemens per meter and and ohm per meter range. Quantify the change in the results of the simulation for resistances and inductances of this size.

22.10 AN INVERSE PROBLEM

One purpose of TDR is to determine the position of a fault on a transmission line. The position of the fault is easily determined from the known dielectric parameters and the time at which a reflection from the fault reaches the TDR monitor. A much more challenging problem is to determine the dielectric parameters of an unknown substance (or substances) that occupy the space between the two conductors of the transmission line from TDR data. This idea is pervasive in many scientific and engineering problems: shine radiation on something and determine what that something is from the return signal.

A TDR voltage trace contains two kinds of information: the voltages and the times of reflections. The voltages can be used to estimate the impedances; the reflection times can be used to determine wave speeds. From these two pieces of information we can deduce the inductance and the capacitances of the dielectrics in case the resistance of the coaxial transmission line is known and the conductance of the dielectric is small enough to be ignored.

More precisely, imagine an experimental apparatus, such as the chamber depicted in Fig. 22.1. A typical experiment is to fill the chamber with some substance and attempt to determine its dielectric constants from TDR measurements. The four dialectic parameters C, L, G and R are typically computed for the connections to the test chamber. Indeed, a typical apparatus would be calibrated to a 50 Ohm cable. Coaxial cables and connectors would be matched to this fixed impedance as closely as possible to avoid spurious reflections due to the apparatus. The resistivity of the chamber is a property of its conducting inner and outer walls. Thus, R should be known before samples are tested.

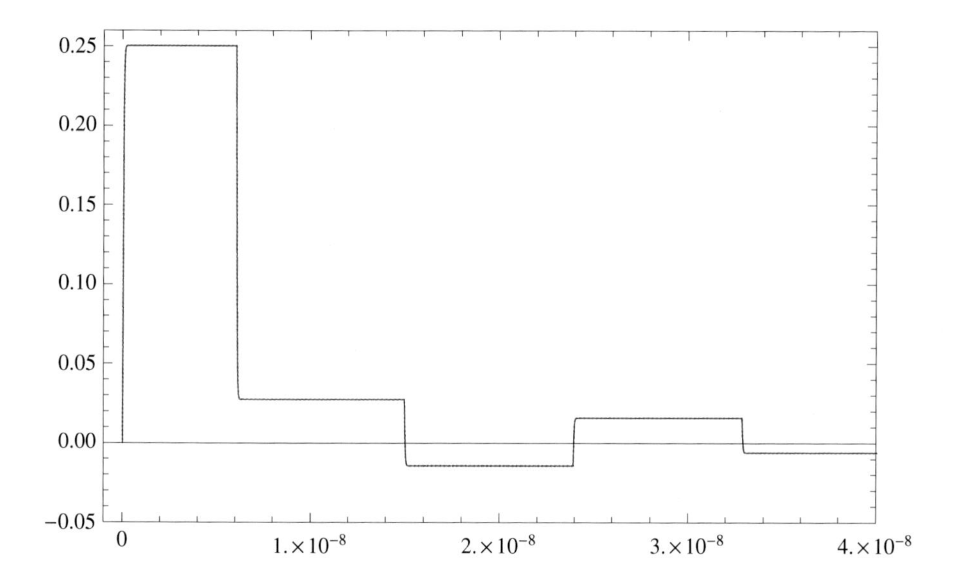

Fig. 22.8 The graph of a simulated TDR time trace (volts) versus time (seconds) is shown.

The basic problem of TDR is to determine C, L, and G for a test sample (or perhaps a layered test sample) that fills a portion of the chamber.

To discuss an important specific application (which allows a further simplification), suppose that test samples are known to be nonmagnetic; that is, the magnetic permeability of the test samples are known to be close to free-space permeability μ_0. Using the circular geometry of the test chamber, the inductance (from Eq. (21.28)) is then known to be

$$L = \frac{\mu_0}{2\pi} \ln \frac{b}{a}, \tag{22.60}$$

where $a < b$ are the radii of the inner and outer walls inside the test chamber. Thus, the inverse problem of determining the dielectric constants is reduced to two parameters: C and G, the capacitance and conductance of the sample. In case the general class of samples is known to have small conductance, the problem is reduced to determining the capacitance. A typical inductance assumed here is 1.5×10^{-7} henry / meter. Given this value, the capacitance can be computed from the impedance $Z := \sqrt{L/C}$.

Exercise 22.16. [TDR Inverse Problem] Fig. 22.8 shows a TDR output of voltage (in volts) versus time (in seconds) for the apparatus discussed in this chapter. The coaxial cable from the TDR to the chamber has electrical length 0.3 nanosecond and the chamber, filled with some unknown dielectric substance, is 0.3 meter long. Approximate the electrical length of the dielectric filled chamber and the capacitance of the unknown substance. Hint: A close approximation to this graph can be made by simply joining the points in the following list with line segments.

$$\begin{array}{ll}
(3.125 \times 10^{-11}, 0.000), & (2.394 \times 10^{-8}, -0.015), \\
(4.062 \times 10^{-10}, 0.250), & (2.416 \times 10^{-8}, 0.016), \\
(6.062 \times 10^{-9}, 0.250), & (3.287 \times 10^{-8}, 0.016), \\
(6.219 \times 10^{-9}, 0.028), & (3.306 \times 10^{-8}, -0.006), \\
(1.500 \times 10^{-8}, 0.027), & (4.178 \times 10^{-8}, -0.006), \\
(1.519 \times 10^{-8}, -0.015). &
\end{array}$$

Write a code for a numerical function that takes as input the electrical length and impedance of the dielectric and returns a (discretized) TDR voltage versus time function. Also, write a code that takes as input the output of the first code and outputs a real number that measures the least squares fit to the function obtained by connecting the points in the list. Recall that after a spatial discretization $t_1, t_2, t_3, \ldots, t_n$ of the time interval over which the TDR trace is taken, the square of the least squares distance between the simulated TDR time trace f and the given TDR time trace g is $\sum_{j=1,n} |f(t_j) - g(t_j)|^2$. A strategy for determining the best fit is to minimize the sum over all feasible electrical lengths and impedances. For one exercise, trial and error might work without undue effort, but an automated approach is of course preferred.

What is an efficient algorithm for minimizing a scalar function on a finite-dimensional space? This question does not have a simple answer. Discussions of available algorithms would likely fill an entire book. When derivatives are easily computed, the methods of calculus (find the first derivative and set it to zero) are preferred. If the zeroes of the derivative are not easy to locate, the method of steepest descent (going in the direction of the negative gradient) can be used to find a zero. Even better, Newton's method can be employed in some cases to speed up the convergence. In case derivatives are not readily available or their zeros are difficult to locate, a direct search algorithm can be effective. Before describing a direct search method that might be useful for minimizing the least squares fit for the TDR inverse problem, there is an essential point that must be addressed: a major problem with all methods of minimization or root finding (when there are multiple roots) is that the chosen algorithm may converge to a root or a local minimum that is not the desired root or the global minimum. There is no known way to avoid this problem in general. Sometimes a local minimum, for example, is good enough. More thorough searches can be made by searching several times with different starting points, altering the parameters in the search algorithm, employing randomization, or mimicking biological evolution. The latter idea, which incorporates random mutations, leads to methods called genetic algorithms.

Simple direct search algorithms are based on an obvious idea: Choose some tolerance $\epsilon > 0$ that will be used to stop the algorithm, a search increment $\delta > 0$, and an integer $M > 0$ that is defined to be the largest number of iterations allowed. Also choose a unit basis $v_1, v_2, v_3, \ldots, v_n$ for the vector space containing the feasible set and define an augmented set of vectors given by $w_j := v_j$ for $j = 1, 2, 3, \ldots, n$ and $w_j := -v_{j-n}$ for $j = n+1, n+2, n+3, \ldots, 2n$. Start the search in the feasible set (in the TDR problem this is the first quadrant in two-dimensional space). Compute the function F to be minimized at the starting point p_0 and compute the function F at the points $p_0 + \delta w_j$, for $j = 1, 2, 3, \ldots, 2n$. If $|F(p_0) - F(p_0 + \delta w_j)| < \epsilon$ for all j, choose the point $p_0 + \delta w_j$ corresponding the smallest computed value of F and return it as the approximation of the point where F reaches its minimum. Also, the function value at this point is an approximation of the desired minimum. Otherwise, replace p_0 by $p_0 + \delta w_j$ with j chosen so that this point gives the smallest value of F over $j = 1, 2, 3, \ldots, 2n$. Repeat this process until a minimum is found or the maximum number of iterations is reached.

The algorithm just described is viable, but it can be made more efficient in several different ways; for example, by using a smaller set of search vectors than the set $\{w_j\}_{j=1}^{2n}$. How should a smaller set be chosen? Can you devise other modifications to carry out a direct search that might be more efficient? Write codes to implement your ideas, test and compare algorithms against minimization problems with known solutions, and apply the best methods to the TDR problem. Remark: Standard test problems abound in the literature on optimization.

CHAPTER 23

Problems and Projects: Waveguides, Lord Kelvin's Model

A waveguide is a hollow (usually metal) tube that is used to transport electromagnetic energy. A waveguide is similar to a transmission line except that there is no central conductor. One important application is the transport of electromagnetic fields from a high-frequency generator to an antenna. The design and plumbing of waveguides remains a thriving industry.

Recall that (under the assumptions that the material in the tube is ohmic and isotropic, and the conductors are perfect) the transverse electromagnetic (TEM) mode in a transmission line is determined by the scalar potential [Eq. (21.17)] defined in a cross section of the waveguide and this potential is proportional to the natural logarithm of the radial coordinate. If there were no central conductor, the electric field would have to be defined and continuous at the radial coordinate $r = 0$. This is not possible because the potential blows up at this point. For this reason, there are no TEM modes in a waveguide satisfying the same assumptions.

The most important modes in a waveguide are the transverse electric (TE) and transverse magnetic (TM) modes. These are defined exactly as expected: TE modes have electric fields with zero components in the axial direction of the waveguide; TM modes have magnetic fields with zero components in the axial direction.

Most practical waveguides have rectangular, circular, or elliptical cross sections. They are manufactured from metal pipes that are usually coated on the inside with copper or gold to ensure the inside skin is nearly a perfect conductor.

It turns out that the design of a waveguide determines the frequency of the electromagnetic waves that it will transport most efficiently. The first project is an outline of the steps required to appreciate this fact in the context of a circular cylindrical waveguide, which is chosen to build on the knowledge base in this chapter on circular transmission lines. The second project is to repeat the entire program for rectangular waveguides, which are in fact the

An Invitation to Applied Mathematics: Differential Equations, Modeling, and Computation.
http://dx.doi.org/10.1016/B978-0-12-804153-6.50023-3, Copyright © 2017 Elsevier Inc. All rights reserved.

most common type in practical use. These projects require some (perhaps) new mathematical analysis that is useful in many other contexts.

23.1 TE MODES IN WAVEGUIDES WITH CIRCULAR CROSS SECTIONS

Consider a circular pipe with inner radius b and choose coordinates so that positions along the central axis of the pipe are measured with respect to the first coordinate x of a rectangular coordinate system so that each cross section of the pipe is a disk centered at the origin in the second and third coordinates y and z. A TE mode is an electromagnetic field inside the pipe such that the first component of the E field with respect to these coordinates vanishes. The first component of the B field is not required to vanish; in fact, the first component of the B field of a nonzero TE mode is *not* zero.

Consider a TE mode in the pipe where the E and B fields are time-harmonic plane waves moving in the axial direction of the pipe. The E field has the form of generic field (21.1). Under the assumption of no electromagnetic sources in the pipe, Faraday's law and the Ampère–Maxwell law are given by

$$\nabla \times B = i\omega\mu\epsilon E, \qquad \nabla \times E = i\omega B.$$

Using these vector identities and time-harmonic plane waveforms for the E and B fields—not assuming yet that $E_1 = 0$, show that

$$\begin{aligned}
E_2 &= -\frac{i\omega}{\omega^2\epsilon\mu + k^2}(B_{1z} + \frac{k}{\omega}E_{1y}),\\
E_3 &= -\frac{i\omega}{\omega^2\epsilon\mu + k^2}(-B_{1y} + \frac{k}{\omega}E_{1z}),\\
B_2 &= \frac{i\omega\epsilon\mu}{\omega^2\epsilon\mu + k^2}(E_{1z} + \frac{k}{\omega\epsilon\mu}B_{1y}),\\
B_3 &= \frac{i\omega\epsilon\mu}{\omega^2\epsilon\mu + k^2}(-E_{1y} + \frac{k}{\omega\epsilon\mu}B_{1z}).
\end{aligned}$$

Thus, the electromagnetic field is determined once E_1 and B_1 are known. In the TE mode, the electromagnetic field is determined from B_1.

From previous results, the first component B_1 of the time-harmonic magnetic flux must satisfy Helmholtz's equation

$$\Delta B_1 + \omega^2\mu\epsilon B_1 = 0 \tag{23.1}$$

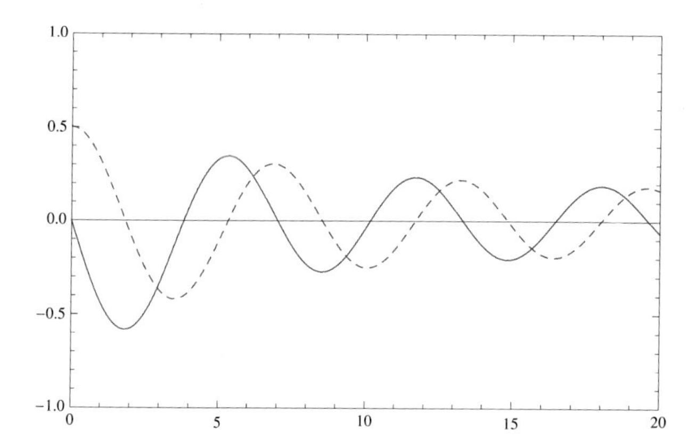

Fig. 23.1 A computer-generated (solid line) plot of the derivative of the Bessel function of the first kind of order zero ($\mathcal{J}_0'$) is depicted together with a (dashed line) plot of the derivative of the Bessel function of the first kind of order one ($\mathcal{J}_1'$).

with zero Neumann boundary condition at $r = b$.

Using the circular geometry, show that the first component in polar coordinates $\tilde{B}_1(r,\theta)$, where the temporal coordinate is suppressed, is given by

$$\tilde{B}_1(r,\theta) = R(r)\Theta(\theta),$$

whenever there is a constant c such that

$$\Theta''(\theta) + c\Theta(\theta) = 0, \qquad r^2R''(r) + rR'(r) + (\omega^2\epsilon\mu r^2 - c)R(r) = 0$$

and the following boundary conditions are satisfied: (1) Θ is a 2π periodic function, (2) $R'(b) = 0$, and (3) $R(0)$ is finite.

Show that c must be the square of an integer ($c = n^2$). Also, by the change of variables $r = s\omega\sqrt{\epsilon\mu}$, it is enough to solve Bessel's differential equation $s^2\tilde{R}''(s) + s\tilde{R}'(s) + (s^2 - n^2)\tilde{R}(s) = 0$ for $\tilde{R}(s) := R(s/(\omega\sqrt{\epsilon\mu}))$ with the boundary conditions $\tilde{R}'(b\omega\sqrt{\epsilon\mu}) = 0$ and $\tilde{R}(0)$ finite. Here $R(r) = \tilde{R}(r/(\omega\sqrt{\epsilon\mu}))$.

Bessel functions have been studied extensively because they arise in many different important applications with circular symmetry (via the Laplacian in polar coordinates). The extensive scope of the theory is apparent in [117].

Bessel's equation is a second-order linear ordinary differential equation (ODE) with nonconstant coefficients. As such, there is a fundamental set of two solutions (on the interval $s > 0$) whose linear combinations give all other solutions. One choice of fundamental solutions consists of the Bessel function of the first kind of order n and the Bessel function of the second kind of order n. These functions can be represented as power series. In fact, the Bessel function of the first kind of integer-order n is given by

$$\mathcal{J}_n(s) = \sum_{j=0}^{\infty} \frac{(-1)^j}{j!(j+n)!}\Big(\frac{s}{2}\Big)^{2j+n}. \tag{23.2}$$

Verify that this power series converges for all $s \geq 0$ and satisfies Bessel's equation. In particular, this function (for each fixed n) extends analytically to give a finite value at $s = 0$, which is required by the second boundary condition for the waveguide problem.

A second solution can be found by reduction of order. Show that every choice of an independent second solution blows up at $s = 0$. Thus, the solution to our boundary value problem (BVP) (if it exists) is a constant multiple of $\mathcal{J}_n$. We may as well take the constant to be unity. Why?

The original BVP for R is solved provided that

$$R'(b) = \omega\sqrt{\epsilon\mu}\mathcal{J}_n'(b\omega\sqrt{\epsilon\mu}) = 0. \tag{23.3}$$

As suggested in the numerical plots in Fig. 23.1, derivatives of integer order Bessel functions of the first kind have infinitely many zeros. Let $j'_{n,m}$ denote the mth nonnegative zero of $\mathcal{J}_n'$. Show (using power series) that $j'(0,1) = 0$ and using a numerical method that $j'(1,1) \approx 1.84$. It is possible to show that $j'(1,1)$ is the smallest such positive zero among integer order Bessel functions. This choice results in the $\text{TE}_{1,1}$ mode (Bessel function of order one, first zero), which is the usual mode of operation in practical waveguides with circular cross sections.

Why not work in the $\text{TE}_{88,100}$ mode?

To support an electromagnetic wave with the lowest possible nonzero frequency ω (called the cutoff frequency) in an air-filled waveguide with radius b meters and the speed of light c in meters per second, requires that

Fig. 23.2 A schematic cross-shaped waveguide is depicted.

the cutoff frequency be

$$\omega = \frac{j'(1,1)}{b\sqrt{\epsilon\mu}} \approx 1.84cb \approx \frac{1.84}{b} \times 2.9979 \times 10^8 \text{ Hz}, \tag{23.4}$$

In view of the frequencies in the electromagnetic spectrum, discuss what this result means for the manufacture of practical waveguides.

Determine the B and E fields for the TE mode.

The number $k^2 := \omega^2\epsilon\mu$ in Helmholtz's equation (23.1) should be viewed as an eigenvalue of the negative Laplacian $-\Delta$ with Helmholtz's equation written in the form

$$-\Delta B_1 = k^2 B_1$$

with zero Neumann boundary condition. In the present case,

$$k^2 = 2\pi\frac{j'(1,1)}{P},$$

where P is the perimeter of the circular boundary of a cross section. In this language, the square of the TE mode cutoff frequency for a waveguide with circular cross section is

$$\omega^2 = 2\pi\frac{j'(1,1)}{\epsilon\mu P}.$$

This result suggests an interesting question. How does the cutoff frequency depend on the shape of the cross section when the perimeter is held constant? Or, put as a more abstract mathematical question, how does the smallest nonzero eigenvalue of the (negative) Neumann Laplacian depend on the shape of a planar domain whose boundary has some fixed perimeter P? Which domain has the smallest such eigenvalue? Perhaps it would be possible to manufacture waveguides with exotic shapes whose cutoff frequencies are small compared to waveguides with circular cross sections.

The design of waveguides provides a wonderful use of mathematics to solve an applied problem.

23.2 RECTANGULAR WAVEGUIDES AND CAVITY RESONATORS

Exercise 23.1. [Transverse Fields in Rectangular Waveguides] (1) Discuss TE modes in rectangular waveguides. (2) Discuss TM modes in rectangular waveguides. (3) How do cutoff frequencies compare with those for circular waveguides?

Exercise 23.2. Suppose someone decides to build a waveguide whose cross section is in the form of a cross: the boundary consists of portions of two rectangles crossing at right angles with the center of each rectangle at the center of the cross (see Fig. 23.2). Before the waveguide is built and tested, you are asked to determine the TE mode theoretically. The key step is to solve Helmholtz's equation $\Delta B_1 + \omega^2\mu\epsilon B_1 = 0$ with zero Neumann boundary condition. Generally, the cutoff frequency is the smallest nonzero ω for which this Neumann BVP has a solution. Approximate the smallest nonzero real number k such that $\Delta B_1 + k^2 B_1 = 0$ has a solution with this boundary condition for cross-shaped waveguides and use this result to determine the cutoff frequency. Hint: Use four parameters to determine the dimensions of the waveguide. Choose values for the dimensions and work numerically. Make graphs and charts to suggest how the cutoff frequency varies with the parameters, perhaps taking them one or two at a time. This project seems to require some new mathematics not covered in this book. The basic difficulty is finding a viable method for approximating the smallest positive eigenvalue of the negative Neumann Laplacian (that is, the smallest $k > 0$ such that $-\Delta u = k^2 u$ has a nonzero solution that satisfies the boundary conditions). One approach is to employ the method of Rayleigh. The desired eigenvalue is obtained as the minimum of the functional

$$u \mapsto \frac{\int_\Omega |\nabla u|^2\, dS}{\int_\Omega |u|^2\, dS}$$

over all functions u that satisfy the following three conditions: (1) $\int_\Omega |u|^2\, dS \neq 0$, (2) the integral $\int_\Omega |\nabla u|^2\, dS$ is defined, and (3) $\int_\Omega u\, dS = 0$. Explain why this method produces the desired result and create a numerical method based on it that produces numbers that would in principle converge to the minimum. A useful starting point for understanding Rayleigh's method is to consider a symmetric matrix A and use Lagrange multipliers to minimize the function

$$x \mapsto \frac{\langle Ax, x \rangle}{|x|^2}$$

over $\{x : |x|^2 = 1\}$. What is the minimum? Are there other local minima?

Exercise 23.3. [Cavity Resonators] Can an electromagnetic field exist inside a closed rectangular box made of a conducting material? If so, how does the frequency of the field depend on the dimensions of the box? Can you imagine a reason why someone would want to build a devise that supports only waves of a fixed frequency? Hint: The key phrase is cavity resonator.

Exercise 23.4. How much of Exercise 22.16 can be done by hand? Using the approximation of a lossless transmission line, the underlying equations are simply wave equations that can be solved exactly. The only issues are the boundary conditions and keeping track of the reflections and transmissions of the waves.

Exercise 23.5. [Lord Kelvin's Transmission Line Model and the Heat Kernel] Before Heaviside's transmission line model, which leads to the telegrapher's equation

$$LCv_{tt} + (RC + GL)v_t - v_{xx} = -RGv$$

for the voltage v as in Exercise 22.13, the accepted model for telegraph transmission was derived by Lord Kelvin in 1854 (see [80]). He ignored the inductance of the line and the conductance of the insulator. Under these assumptions he correctly derived the (too crude) model

$$RCv_l - v_{xx}$$

Fourier's theory of heat had already been published. Kelvin's PDE was the same as for heat flow; thus, for a while, some people believed (wrongly) that electricity flowed in a transmission line by a diffusion process. Heaviside's model eventually convinced the scientific community that transmissions were more accurately viewed as (damped) waves.

Although Lord Kelvin's model is not viable for transmission lines, the theory of the one-dimensional heat equation certainly is important. Some of its properties and exact solutions were derived by Kelvin, George Gabriel Stokes, and Heaviside in contexts other than heat transfer. These results remain important contributions because the heat equation arises, as we have seen, in many different applied problems. This project outlines some basic theory for constructing solutions of the heat equation on the line and on the half-line. The outline should contain enough sign posts for the reader to fill in the details. One goal of the project is to solve the model problem

$$\text{(PDE)} \quad v_t = \frac{1}{RC} v_{xx}, \qquad x > 0, \quad t > 0;$$

$$\text{(BC)} \quad v(0,t) = g(t), \qquad t > 0;$$
$$\text{(IC)} \quad v(x,0) = 0.$$

It may be viewed as the model for a telegraph signal on an initially dead line (IC) with an input (BC), perhaps induced by pressing a telegraph key, that might be in the form of an impulse or a step that is zero for large t. This is exactly the problem solved by Stokes in 1854. To recover Stokes's solution requires the development of some important theory.

(a) [*The heat equation on the whole real line*]

Consider the heat equation on the whole real line $-\infty < x < \infty$, which will from now on be written

$$u_t = k^2 u_{xx}, \tag{23.5}$$

together with the initial condition $u(x,0) = f(x)$. The value $u(x,t)$ is supposed to model the temperature at position x measured at time t. The function f models the initial temperature along the line.

To be a viable physical model, the initial temperature f should be propagated (by solving the heat equation with this initial condition) to a new function $x \mapsto u(x,t)$ that models the temperature along the line for each time $t > 0$. This process is analogous to the action of the flow of an autonomous ODE, which propagates an initial state to a new state for each $t > 0$. For example, the ODE $\dot{y} = ay$ propagates the initial condition $y(0) = x$ forward in time via its flow by the rule $y(x,t) = e^{at}x$. More precisely, the flow should be viewed as a group of transformations $x \mapsto \phi(x,t)$, one for each t. Using this notation, $\phi(x,t) = e^{at}x$. The same equation is sometimes written $\phi_t(x) = e^{at}x$ or $\phi^t(x) = e^{at}x$ to more clearly suggest that ϕ_t or ϕ^t is a transformation (or propagator) acting on x, a point in the space of states of the system.

If there is justice in the mathematical universe, the heat equation should also produce a flow. The only technical difference is that the initial data (representing the initial state of the system) is no longer a number x; it is a function f. The set of such functions represent the possible states of the system, and the flow or propagator for the PDE (also denoted ϕ) assigns to each initial state f and each time $t > 0$ a new function $\phi(f,t) : \mathbb{R} \to \mathbb{R}$. In particular, the propagated state at time t is the function $x \mapsto \phi(f,t)(x)$ on the real line. The problem is to determine the flow ϕ for the heat equation.

Where to start? The first idea is to simply play with the heat equation. Not having a better idea, you might try to make the equation dimensionless. Define a scaling

$$u = av, \qquad x = b\xi, \qquad t = c\tau$$

so that v, ξ, and τ are dimensionless. Check that under this transformation

$$u(x,t) = av(\frac{x}{b}, \frac{t}{c})$$

and

$$v_\mathcal{T} = \frac{ck^2}{b^2} v_{\xi\xi}.$$

By choosing b and c such that $ck^2/b^2 = 1$, the dimensionless equation is reduced to $v_\mathcal{T} = v_{\xi\xi}$. Notice that this calculation reveals another interesting fact: If $c = b^2$, then v is a solution of the original form of the PDE, $v_\mathcal{T} = k^2 v_{\xi\xi}$, where only names have been changed. In other words, if u solves the heat equation, then so does

$$v(x,t) := u(bx, b^2 t)$$

for every choice of b. Check this fact. What does it mean? A solution u of the heat equation must be a function of x and t such that when x is multiplied by b, and t is multiplied by b^2, its value is unchanged.

Is there a function that has this invariance property? Check that the function $(x,t) \mapsto x/\sqrt{t}$ has the required invariance, and as a corollary, so does $(x,t) \mapsto h(x/\sqrt{t})$ whenever $h : \mathbb{R} \to \mathbb{R}$.

There must be a choice of h so that $(x,t) \mapsto h(x/\sqrt{t})$ is a solution of the PDE. This is indeed the case. Show that if h is a solution of the ODE

$$k^2 \frac{d^2 h}{d\sigma^2} + \frac{\sigma}{2} \frac{dh}{d\sigma} = 0,$$

then $(x,t) \mapsto h(x/\sqrt{t})$ is a solution of heat equation (23.5). Make the second-order (nonautonomous) ODE into a first-order system by defining $p = dh/d\sigma$ and show that

$$k^2 \frac{dp}{d\sigma} + \frac{\sigma}{2} p = 0.$$

Solve the latter equation to obtain, for an arbitrary constant d,

$$p(\sigma) = de^{-\sigma^2/(4k^2)}, \qquad h(\sigma) = h(0) + d \int_0^\sigma e^{y^2/4}\, dy.$$

Using the function h, check that a solution of the heat equation is given by

$$u(x,t) := h(0) + d \int_0^{x/\sqrt{t}} e^{y^2/(4k^2)}\, dy.$$

Is there a way to avoid the integration so as to make the solution formula more explicit? Suppose that u is an arbitrary solution of the PDE $u_t = k^2 u_{xx}$, differentiate both sides with respect to x, and check that

$$(u_x)_t = k^2 (u_x)_{xx}.$$

Thus, if u is a solution, so is u_x, at least when u is sufficiently smooth. How smooth? Check that, for a sufficiently smooth solution u, every partial derivative of u is a solution of the heat equation. Apply this result to see that

$$\tilde{U}(x,t) := d\frac{1}{\sqrt{t}}e^{-x^2/(4k^2t)}$$

is a solution of the heat equation for every real number d. This solution—called a similarity solution—is usually normalized so that its total mass is unity for every time $t > 0$; that is, so that

$$\int_{-\infty}^{\infty} \tilde{U}(x,t)\,dx = 1.$$

Show that this normalization holds for the choice $d = 1/(2k\sqrt{\pi})$ and produces the (most important) solution of the heat equation:

$$U(x,t) := \frac{1}{\sqrt{4\pi k^2 t}}e^{-x^2/(4k^2t)}. \tag{23.6}$$

This solution is defined on the whole real line for each $t > 0$; it is called the heat kernel or the fundamental solution.

The heat kernel is not defined for $t = 0$. But, show that

$$\lim_{t\to 0+} U(x,t) = \begin{cases} 0, & \text{if } x \neq 0; \\ \infty, & \text{if } x = 0. \end{cases}$$

Also, recall that U has unit mass. This should suggest the Dirac delta function plays a role. Recall that the delta function is defined by its action on test functions, which are smooth functions with compact support (zero off a closed bounded set). The delta function maps the set of test functions to the real numbers via the formula

$$\delta(\psi) = \psi(0).$$

This is often written as the formal statement

$$\int_{-\infty}^{\infty} \delta(x)\psi(x)\,dx = \psi(0),$$

which is interpreted by saying the delta function is the unit mass measure at the origin. The unit mass measure at some other point, say the point with coordinate y on the real line, is denoted δ_y and often viewed as the translate of the Dirac delta (whose mass is concentrated at the origin) to allow the formal statement

$$\int_{-\infty}^{\infty} \delta(x-y)\psi(x)\,dx = \psi(y).$$

With this definition, we are tempted to write $\lim_{t\to 0+} U(x,t) = \delta$ as convenient shorthand for

$$\lim_{t\to 0+}\int_{-\infty}^{\infty} U(x,t)\psi(x)\,dx = \psi(0),$$

or more generally,

$$\lim_{t\to 0+}\int_{-\infty}^{\infty} U(y-x,t)\psi(y)\,dy = \psi(x).$$

Of course, these statements can be proved (with some careful analysis) by computing the limits. Check that they are correct.

Using results from advanced calculus about differentiation under the integral sign, check that (for a test function f) the function u given by

$$u(x,t) = \int_{-\infty}^{\infty} U(x-y,t)f(y))\,dy = \frac{1}{\sqrt{4\pi k^2 t}}\int_{-\infty}^{\infty} e^{-(x-y)^2/(4k^2t)} f(y)\,dy \tag{23.7}$$

is a solution of the heat equation for all x and $t > 0$ with the initial data $u(x,0) = f(x)$ in the sense that the initial condition is satisfied via the limit

$$\lim_{t\to 0+} u(x,t) = f(x).$$

The singularity due to the function u being undefined at $t = 0$ is removable. At least the function u can be extended to a continuos function on $0 \le t < \infty$ by assigning the value $u(x,0) = f(x)$.

One important note is that the result concerning initial data is true (with a slight modification) for much more general initial functions f. It is enough to have good convergence properties of the integral in Eq. (23.7). The exponent $-x^2$ causes the heat kernel to decrease rapidly to zero. Thus, the integral will converge as long as $|f|$ does not increase too fast toward ∞. Also, the differentiation required to satisfy the heat equation is related to the function u, not f. Thus, the initial function could, for example, be discontinuous. In such a case, the state $x \mapsto u(x,t)$, for $t > 0$, is not only continuous; it is infinitely differentiable. The diffusion process immediately smooths out rough initial data. To make all this precise, the required modification at a jump discontinuity at $x = a$ is to allow a slight change in the limit as t goes to zero by not insisting the limit be $f(a)$ and instead taking the limit $\lim_{t\to 0+} u(a,t)$ to be the average of the right- and left- hand limits at this point. Check this fact.

It might be more pleasing to have a solution of the initial value problem (IVP) without the removable singularity at $t = 0$, but no such solution exists. To see this, suppose that u is a solution of the heat equation that vanishes as x approaches $\pm\infty$, and consider the (energy) function E defined by

$$E(t) := \int_{-\infty}^{\infty} u^2(x,t)\,dx.$$

Make an easy computation using the heat equation and use one integration by parts to show (at least for $t > 0$) the statement

$$E'(t) = -2k^2 \int_{-\infty}^{\infty} u_x^2(x,t)\,dx.$$

Conclude that E does not increase. For two solutions u_1 and u_2 of the heat equation with the same initial data $u(x,0) = f(x)$, show that $u := u_1 - u_2$ is also a solution but with initial data $u(x,0) = 0$. For this u, suppose there is some $\tau > 0$ and some number ξ such that $u(\xi,\tau)$ is not zero. Prove that $E(\tau) > 0$. This inequality contradicts the statement $E'(t) \le 0$ for $0 < t \le \tau$. Conclude that solutions are unique. This method of proving uniqueness is called the energy method.

For completeness, consider the IVP for the heat equation with a source on the whole real line:

$$\begin{aligned} &\text{(PDE)}\quad v_t = k^2 v_{xx} + h(x,t), \qquad -\infty < x < \infty, \quad t > 0; \\ &\text{(IC)}\quad v(x,0) = f(x), \qquad x > 0, \end{aligned} \tag{23.8}$$

where h models a source for v. In the case of heat transfer, what units would be assigned to h?

The corresponding homogenous problem is

$$\begin{aligned} &\text{(PDE)}\quad v_t = k^2 v_{xx}, \qquad -\infty < x < \infty, \quad t > 0; \\ &\text{(IC)}\quad v(x,0) = f(x), \qquad x > 0. \end{aligned} \tag{23.9}$$

There is a beautiful way to use solutions of the homogeneous initial value problem to solve the nonhomogeneous problem that is a special case of a general principle, which is often called Duhamel's principle (Jean-Marie Duhamel, 1845): If you can solve a linear differential equation for all initial data, then you can solve the inhomogeneous problem. You should have already seen an important example of this principle in action in the study of nonhomogeneous linear ODEs, where the IVP is

$$\dot{v} = Av + h(t), \qquad v(0) = \eta. \tag{23.10}$$

Here, A is a constant $n \times n$ matrix, η is an n vector, and h is an n vector-valued function of t. Perhaps $n = 1$, A and η are real numbers, and h is a scalar function. Check that the general solution of the homogeneous problem is $t \mapsto e^{tA}\eta$. In other words, the flow ϕ of the ODE propagates a state $\eta \in \mathbb{R}^n$ after time t to the new state

$$\phi(\eta,t) = e^{tA}\eta. \tag{23.11}$$

The inhomogeneous problem [Eq. (23.9)] is solved by variation of parameters. Indeed, check that

$$v(t) := \phi(\eta, t) + \int_0^t \phi(h(\tau), t - \tau)\, d\tau \tag{23.12}$$

is the unique solution of IVP (23.10). To see where the variation of parameters formula comes from, suppose that the solution of the inhomogeneous problem is given by $v(t) = \Phi(z(t), t)$ for some unknown function z (which is called a variation of the initial parameter η) and find out which differential equation is solved by z (see Appendix A.15). The moral of the story is that the solution of the inhomogeneous problem is constructed using solutions of the homogeneous problem.

The homogeneous heat equation has the (heat kernel) propagator

$$\phi(f, t) := \int_{-\infty}^{\infty} U(x - y, t) f(y))\, dy, \tag{23.13}$$

again a function of two variables where the first variable is taken from a space of appropriate functions (perhaps the space of test functions) and the second variable is a real number. Just like in the finite-dimensional (vector space) case, a point in the function space is propagated to a new point in the function space via this flow. A natural expectation (in view of Eq. (23.12)) is that the solution of the inhomogeneous problem [Eq. (23.8)] is given by

$$v(x, t) := \phi(f, t)(x) + \int_0^t \phi(h(x, \tau), t - \tau)\, d\tau. \tag{23.14}$$

This formula, which is also called Duhamel's formula, is the infinite-dimensional (function space) version of variation of constants. Show that v is indeed a solution of inhomogeneous BVP (23.8).

What does this model (23.8) predict in case $f = 0$ and $h(x, t) = 1$? A more interesting choice is $f = 0$ and

$$h(x, t) := \frac{1}{2}(1 - H(t - t_0))(H(x + x_0) + H(x - x_0))$$

for positive t_0 and x_0 where H is the Heaviside step function defined by $H(t) = 0$ for $t < 0$ and $H(t) = 1$ for $t \geq 0$. Interpret this choice and the corresponding prediction physically.

(b) [*The heat equation on the half-line*]

In part (a) you showed that the IVP for the heat equation on the whole real line has unique solutions. The IVP

$$\begin{aligned} &\text{(PDE)} \quad v_t = k^2 v_{xx}, \qquad x > 0, \quad t > 0; \\ &\text{(IC)} \quad v(x, 0) = f(x), \qquad x > 0 \end{aligned}$$

does not have unique solutions. It is posed on a domain $0 < x < \infty$ that has a finite boundary; namely, $x = 0$. To have unique solutions an appropriate boundary condition must be imposed at $x = 0$. The general inhomogeneous, initial boundary value problem (IBVP)

$$\begin{aligned} &\text{(PDE)}\quad v_t = k^2 v_{xx} + h(x,t), \qquad x > 0, \quad t > 0;\\ &\text{(BC)}\quad v(0,t) = g(t), \qquad t > 0;\\ &\text{(IC)}\quad v(x,0) = f(x), \qquad x > 0 \end{aligned} \tag{23.15}$$

does have unique solutions. It is more general than necessary to solve Lord Kelvin's problem for telegraph signals. But problems of this general type can all be solved via integral representations for known functions f, g, and h. The methods are elementary, but not obvious.

The homogeneous heat equation is linear. Thus, the sum of two solutions is a solution. This superposition is not valid in the presence of a nonzero source or a nonzero boundary condition. For this reason, it is almost always a good idea to work with homogeneous (zero) boundary conditions. In the present context where a Dirichlet boundary condition is imposed, there is a simple way to make an equivalent problem with a homogeneous boundary condition: subtract the boundary condition from the required solution. More precisely, define

$$u = v - g(t)$$

and show that v solves IBVP (23.15) if and only if u solves the IBVP

$$\begin{aligned} &\text{(PDE)}\quad u_t = k^2 u_{xx} + h(x,t) - g'(t), \qquad x > 0, \quad t > 0;\\ &\text{(BC)}\quad u(0,t) = 0, \qquad t > 0;\\ &\text{(IC)}\quad u(x,0) = f(x) - g(0), \qquad x > 0. \end{aligned} \tag{23.16}$$

A natural approach is to somehow convert to a problem on the whole real line and use the known solution for the latter IVP. Making this idea work requires some care: the boundary condition must be taken into account.

As stated, the functions p given by $(x,t) \mapsto h(x,t) - g'(t)$ and q by $x \mapsto f(x) + g(0)$ are defined for $x > 0$ and $t > 0$. To work on the whole line, these functions must be extended to $-\infty < x < \infty$. There are several possible choices: For example, the zero extension of p defined to be p for $x \geq 0$ and zero for $x < 0$; the odd extension given by

$$P(x) = \begin{cases} p(x), & x > 0;\\ 0, & x = 0;\\ -p(-x), & x < 0; \end{cases}$$

or the even extension where $P(x) = p(-x)$ for $x < 0$. One can simply start working by trying the various possibilities. It turns out that the odd extensions of P and Q are the correct choices.

For arbitrary extensions (that are nice enough functions for the integrations and differentiations to be defined) the IVP corresponding to IBVP (23.16) is extended to the whole line and it has the unique solution

$$u(x,t) := \phi(Q,t)(x) + \int_0^t \phi(P(x,\tau), t-\tau)\, d\tau. \tag{23.17}$$

What about the boundary condition? This is the key point. Show that without a boundary condition the initial value on the half-line has many different solutions. The odd extensions lead to solutions that satisfy the boundary condition.

The simplest computation is for the homogeneous case

$$\begin{aligned}
&\text{(PDE)} \quad u_t = k^2 u_{xx}, \qquad x > 0, \quad t > 0;\\
&\text{(BC)} \quad u(0,t) = 0, \qquad t > 0;\\
&\text{(IC)} \quad u(x,0) = f(x) - g(0), \qquad x > 0.
\end{aligned} \tag{23.18}$$

Using the odd extension Q for the initial data, the solution is $v(x,t) = \phi(Q,t)(x)$. Show that the solution on the whole line is

$$\begin{aligned}
u(x,t) &= \frac{1}{\sqrt{4\pi k^2 t}} \int_{-\infty}^{\infty} e^{-(x-y)^2/(4k^2 t)} Q(y)\, dy\\
&= \frac{1}{\sqrt{4\pi k^2 t}} \int_0^{\infty} \left(e^{-(x-y)^2/(4k^2 t)} - e^{-(x+y)^2/(4k^2 t)}\right) Q(y)\, dy.
\end{aligned}$$

Check that the boundary condition is satisfied when this function is restricted to the half-line and that it is the unique solution of the homogeneous IBVP. A similar argument shows that the second summand of the claimed solution [Eq. (23.17)] solves the inhomogeneous IBVP

$$\begin{aligned}
&\text{(PDE)} \quad u_t = k^2 u_{xx} + h(x,t) - g'(t), \qquad x > 0, \quad t > 0;\\
&\text{(BC)} \quad u(0,t) = 0, \qquad t > 0;\\
&\text{(IC)} \quad u(x,0) = 0, \qquad x > 0
\end{aligned} \tag{23.19}$$

when the odd extension P is used to extend it to the whole line.

Returning to Lord Kelvin's problem, the solution is given by Eq. (23.17) but with h and f both zero in the definitions of the extensions P and Q. More explicitly, check that the solution is

$$v(x,t) = g(t)(1 - \tilde{H}(x)) + \frac{x}{2k\sqrt{\pi}} \int_0^t (t-\tau)^{-3/2} e^{-x^2/(4k^2(t-\tau))} g(\tau)\, d\tau, \tag{23.20}$$

where $\tilde{H}(x) = 0$ for $x = 0$ and $\tilde{H}(x) = 1$ for $x > 0$. Here the value of the unit step at $x = 0$ is important to capture the correct boundary condition $v(0,t) = g(t)$. Reducing Eq. (23.17) (with $k^2 = 1/(RC)$) to the solution [Eq. (23.20)] of the model problem

$$\begin{aligned} &\text{(PDE)} \quad v_t = k^2 v_{xx}, && x > 0, \quad t > 0; \\ &\text{(BC)} \quad v(0,t) = g(t), && t > 0; \\ &\text{(IC)} \quad v(x,0) = 0 && \end{aligned} \tag{23.21}$$

takes some careful and lengthy calculations, including an integration by parts. Carrying out the details is an instructive project.

There are more efficient methods that may be used to solve the model problem. Stokes, the first person to find this solution, used the Fourier transform (see [111, p. 391]). Can you fill in the details of his argument? Note that Stokes did not include the first summand of Eq. (23.20). Why not?

Suppose that g is an impulse at $t = t_0 > 0$. It can be modeled by the delta function translated to t_0. For $x > 0$ and $t > t_0$, the voltage in the infinite line is

$$v(x,t) = \frac{x}{2k\sqrt{\pi}} (t - t_0)^{-3/2} e^{-x^2/(4k^2(t-t_0))}. \tag{23.22}$$

Check that an observer at x sees the maximum voltage at time

$$t = t_0 + \frac{1}{6} RCx^2.$$

As discussed in [80], this result—sometimes called the Stokes–Thompson square law—caused some difficulty in the development of undersea telegraph cables: The peak voltage (for an observer at the end of the line who is looking for a telegraph pulse signal) seemed to depend on the *square* of the length of the cable. Thus, to send a signal over a cable of length twice the length of a given cable would take four times as long. This prediction did not bode well for the utility of long undersea cables. Fortunately, the diffusion model is not correct. The Heaviside transmission line theory (that takes into account the inductance and conductance) gives the correct result because it captures the correct physics: the wave nature of impulse signals on a telegraph line. What is the speed of transmission, with respect to maximum voltage, for the diffusion model? Note that t is proportional to x^2 or x is proportional to $\sqrt{t}$. This is a manifestation of the scaling invariance of the heat equation used previously to construct a similarity solution.

Approximate the solution of model (23.21) numerically and verify the square law. Choose g to be a short duration pulse. Work on a finite interval that is long relative to the duration of the pulse, and use zero Dirichlet boundary conditions. The interval should be long enough so that the boundary conditions do not affect (too much) the diffusion of the pulse. Choose some points on the interval and check that the maximum voltage at these observation points occurs according to the square law.

Reinterpret Lord Kelvin's model in a physical context where the underlying physics *is* correct and discuss the Stokes–Thompson square law for the new model.

Mathematical and Computational Notes

A.1 ARZELA–ASCOLI THEOREM

A set $\mathcal{S}$ of continuous functions defined on a compact set Ω is uniformly bounded if there is some number $M > 0$ such that $\sup_{x\in\Omega}|f(x)| \leq M$ for every $f \in \mathcal{S}$. The set is equicontinuous if for every $\epsilon > 0$ there is a $\delta > 0$ such that $|f(x) - g(y)| < \epsilon$ whenever f and g are in $\mathcal{S}$ and $|x - y| < \delta$. For a proof see, for example, [1].

Theorem A.1. The closure of a uniformly bounded and equicontinuous set is compact. In particular, every sequence has a uniformly convergent subsequence.

The limit of the uniformly convergent subsequence may not be in the set $\mathcal{S}$.

A.2 C^1 CONVERGENCE

Suppose that $\{f_n\}_{n=1}^{\infty}$ is a sequence of continuous functions defined on the closure $\bar{\Omega}$ of an open set Ω in a Euclidean space. If $\{f_n\}_{n=1}^{\infty}$ converges uniformly to a function f defined on Ω, then f is continuous and extends to a continuous function on $\bar{\Omega}$. If, in addition, each f_n is differentiable on Ω and the sequence of these derivatives $\{Df_n\}_{n=1}^{\infty}$ converges uniformly to a function g defined on Ω, then f is differentiable and $Df = g$. Moreover g is uniformly continuous and thus extends continuously to $\bar{\Omega}$. This theorem is proved (in various forms) in books on advanced calculus.

A.3 EXISTENCE, UNIQUENESS, AND CONTINUOUS DEPENDENCE

If $f : \mathbb{R}^n \times \mathbb{R} \times \mathbb{R}^k \to \mathbb{R}^n$ is a continuously differentiable function, $x_0 \in: \mathbb{R}^n$, $t_0 \in \mathbb{R}$, and $\lambda_0 \in \mathbb{R}^k$, then the initial value problem

$$\frac{dx}{dt} = f(x, t, \lambda_0), \qquad x(t_0) = x_0$$

has a unique solution defined for t in some open interval containing t_0. More generally, there is a continuously differentiable function $\phi(t, \xi, \lambda)$ defined in some product neighborhood of t_0, x_0, and λ_0 such that $t \mapsto \phi(t, \tau, \xi, \lambda)$ is the solution of the initial value problem

$$\frac{dx}{dt} = f(x, t, \lambda), \qquad x(\tau) = \xi$$

defined in some open interval containing τ.

In short, smooth ordinary differential equations (ODEs) always have solutions that depend continuously on initial data and parameters. The solutions are as smooth as the function f. Solutions may be defined only for a short time. On the other hand, solutions exist until they reach the boundary of the (spatial) domain of definition of f, which might not be defined on all of $\mathbb{R}^n$ or until they blow up to infinity.

A.4 GREEN'S THEOREM AND INTEGRATION BY PARTS

Theorem A.2 (Green's Theorem/Divergence Theorem). Let Ω be a bounded open subset of $\mathbb{R}^n$ whose boundary $\partial\Omega$ is a piecewise smooth hypersurface. If F is a continuously differentiable function $F : \mathbb{R}^n \to \mathbb{R}^n$ defined on the closure of Ω (that is, Ω together with $\partial\Omega$), and η is the outer unit normal on $\partial\Omega$, then

$$\int_\Omega \operatorname{div} F \, d\mathcal{V} = \int_{\partial\Omega} F \cdot \eta \, dS,$$

where $d\mathcal{V}$ is the Euclidean volume and dS is the surface area element induced on the hypersurface via the outer unit normal.

Proof. See a book on advanced calculus. □

Corollary A.3. With the same notation as in Green's theorem, suppose that $f : \mathbb{R}^n \to \mathbb{R}$ is continuously differentiable and i is an integer in the range $1 \le i \le n$. Then,

$$\int_\Omega f_{x_i} \, d\mathcal{V} = \int_{\partial\Omega} f \eta_i \, dS.$$

Proof. Define a vector-valued function F whose component functions are all zero except for the ith component that is defined to be f and apply Green's theorem. □

Corollary A.4 (Integration by Parts). With the same notation as in Green's theorem and the first corollary, suppose that $g : \mathbb{R}^n \to \mathbb{R}$ and $h : \mathbb{R}^n \to \mathbb{R}$ are continuously differentiable. Then,

$$\int_{\Omega} g_{x_i} h \, d\mathcal{V} = -\int_{\Omega} g h_{x_i} \, d\mathcal{V} + \int_{\partial\Omega} g h \eta_i \, d\mathcal{S}. \tag{A.1}$$

Proof. Define f to be the product of g and h and apply the first corollary. □

A.5 GERSCHGORIN'S THEOREM

Let $A = a_{ij}$ be an $n \times n$ matrix and let Γ_i denote the closed disk in the complex plane with center at a_{ii} and radius $\sum_{j=1, j\neq i}^{n} |a_{ij}|$. The eigenvalues of A are contained in the set $\cup_{i=1}^{n} \Gamma_i$. See [12] for a proof of this result and additional information.

A.6 GRAM–SCHMIDT PROCEDURE

Let $\{v_i\}_{i=1}^N$ be a basis for an inner product space H whose inner product is denoted by angle brackets. An orthogonal basis consists of the set of vectors $\{g_i\}_{i=1}^N$ given by

$$g_1 := v_1, \qquad g_i := v_i - \sum_{j=1}^{i-1} \frac{\langle v_i, g_j \rangle}{\langle g_j, g_j \rangle} g_j.$$

An orthonomal basis is given by $\{\frac{1}{|g_i|} g_i\}_{i=1}^N$, where

$$|g_i| = \sqrt{\langle g_j, g_j \rangle}.$$

A.7 GROBMAN–HARTMAN THEOREM

A smooth system of (autonomous) differential equations is locally conjugate to its linearization at a hyperbolic rest point (that is, all eigenvalues have nonzero real parts). See [20] for a proof.

A.8 ORDER NOTATION

Given two functions f and g, we say that $f(x) = g(x) + O(x^n)$ if

$$\frac{|f(x) - g(x)|}{|x^n|}$$

is bounded by a positive constant for $|x|$ sufficiently close to zero. We say that $f(x) = g(x) + o(x^n)$ if

$$\lim_{x \to 0} \frac{|f(x) - g(x)|}{|x^n|} = 0.$$

For example, we have that

$$x - \sin x = O(x^3), \qquad x - \sin x = o(x^2).$$

A.9 TAYLOR'S FORMULA

Taylor's formula is an essential tool in mathematical analysis for approximating the values of functions. Suppose that $f : \mathbb{R}^m \to \mathbb{R}^n$ is smooth (at least C^2 for this appendix). Let x and h be elements of $\mathbb{R}^m$ and $t \in \mathbb{R}$. We have that

$$\frac{d}{dt} f(x + th) = Df(x + th)h.$$

After integration on the interval $0 \leq t \leq 1$, we have the identity

$$f(x + h) = f(x) + \int_0^1 Df(x + th)h \, dt,$$

or equivalently,

$$f(x) = f(a) + \int_0^1 Df(x + t(x - a))(x - a) \, dt.$$

It follows immediately that

$$f(x + h) = f(x) + Df(x)h + \int_0^1 (Df(x + th)h - Df(x)h) \, dt,$$

or equivalently,

$$f(x) = f(a)+Df(a)(x-a)+\int_0^1 (Df(x+t(x-a))(x-a)-Df(a)(x-a))\,dt.$$

A.10 LIOUVILLE'S THEOREM

Suppose that $t \mapsto \Phi(t)$ is a matrix solution of the homogeneous linear system $\dot{x} = A(t)x$ on the open interval J, $\det$ denotes determinant, and tr denotes trace. If $t_0 \in J$, then

$$\frac{d}{dt}\det\Phi(t) = \operatorname{tr} A(t)\det\Phi(t)$$

and

$$\det\Phi(t) = \det\Phi(t_0)e^{\int_{t_0}^t \operatorname{tr} A(s)\,ds}.$$

See [20] for a proof.

A.11 TRANSPORT THEOREM

Let ϕ_t denote the flow of the system $\dot{x} = f(x)$, $x \in \mathbb{R}^n$, and let Ω be a bounded region in $\mathbb{R}^n$. Define

$$V(t) = \int_{\phi_t(\Omega)} dx_1 dx_2 \cdots dx_n$$

and recall that the divergence of a vector field $f = (f_1, f_2, \ldots, f_n)$ on $\mathbb{R}^n$ with the usual Euclidean structure is

$$\operatorname{div} f = \sum_{i=1}^{n} \frac{\partial f_i}{\partial x_i}.$$

Liouville's theorem and the change of variables formula for multiple integrals can be used to prove that

$$\dot{V}(t) = \int_{\phi_t(\Omega)} \operatorname{div} f(x) dx_1 dx_2 \cdots dx_n.$$

In particular, the flow of a vector field whose divergence is everywhere negative contracts volume.

Suppose that $g : \mathbb{R}^n \times \mathbb{R} \to \mathbb{R}$ is smooth and, for notational convenience, let $dx = dx_1 dx_2 \cdots dx_n$. The (Reynolds) transport theorem states that

$$\frac{d}{dt}\int_{\phi_t(\Omega)} g(x,t)\,dx = \int_{\phi_t(\Omega)} g_t(x,t) + \operatorname{div}(gf)(x,t)\,dx. \tag{A.2}$$

A standard proof has three steps: The change of variables formula is used to freeze the integration over the set Ω so that the time-derivative may be taken across the integral sign; differentiation with respect to t is carried out under the integral sign, Liouville's theorem is used to write the time-derivative of the Jacobian as the product of the trace of the derivative of the flow field (which is the divergence of f) and the Jacobian, and the change of variables formula is applied to obtain the form given in the statement of the theorem.

A useful alternate form of the transport theorem is obtained by applying the divergence theorem to the second term in the integrand on the right-hand side of formula (A.2):

$$\frac{d}{dt}\int_{\phi_t(\Omega)} g\,d\mathcal{V} = \int_{\phi_t(\Omega)} g_t\,d\mathcal{V} + \int_{\partial\phi_t(\Omega)} gf\cdot\eta\,d\mathcal{S}, \tag{A.3}$$

where η is the outer unit normal on the boundary $\partial\phi_t(\Omega)$.

A.12 LEAST SQUARES AND SINGULAR VALUE DECOMPOSITION

The basic problem of linear algebra is to solve for the unknown vector x in the system of linear equations $Ax = b$, where A is a matrix and b is a vector. In case A is a square matrix that is nonsingular (its determinant is not zero or its columns are linearly independent), there is a unique solution $x = A^{-1}b$. In general, the worst possible way to compute the solution x is to compute the matrix inverse. Two central principles of numerical linear algebra state: never compute a determinant and never compute the inverse of a matrix. The best solution methods for linear systems are based on Gaussian elimination or iteration. Some iterative methods are discussed in this book.

There are important problems in applied mathematics that require a solution of $Ax = b$ in case A is not a square matrix or A is singular. The prime example is one-dimensional linear regression where the matrix is generally not square: We are given a finite set of points in the plane (x_i, y_i) for $i = 1, 2, 3, \ldots, N$ and asked to find the best fitting line. More precisely, the problem is to find the line with equation $y = mx + \beta$ such that the sum of the squares of the deviations at the x_i from the line to y_i is minimized; that is,

$$\min_{(m,b)\in\mathbb{R}^2} \sum_{i=1}^{N} |mx_i + \beta - y_i|^2.$$

This problem may be recast into the abstract form

$$\min_{x\in\mathbb{R}} |Ax - b|^2,$$

where the vertical bars denote the Euclidean norm; A is the $N \times 2$ matrix with first column the transpose of the row vector $(x_1, x_2, x_3, \ldots, x_N)$ and second column the N vector all of whose entries are one; b is the transpose of the row vector $(y_1, y_2, y_3, \ldots, y_N)$; and x is the column vector of unknowns (m, β). The corresponding matrix equation $Ax = b$ is a prototypical example of an overdetermined system of linear equations (more equations than unknowns). This equation has a solution exactly when every coordinate pair (x_i, y_i) lies on the same line. The purpose of linear regression is to find a line that best fits the data when the data points do not all lie on the same line.

The reason for the squares is to simplify the mathematics. For instance, the problem $\sum_{i=1}^{N} \min_{m,b}|mx_i + b - y_i|$ is more difficult. The key point is that the Euclidean norm is defined by an inner product. In fact, using the usual inner (dot) product $\langle v, w\rangle := \sum v_i w_i$, the length of a vector v is defined to be $|v| = \sqrt{\langle v, v\rangle}$. The square of the length is just the inner product.

Assume that A is $m \times n$ with $m \geq n$ and define $f : \mathbb{R}^n \to \mathbb{R}$ by $f(x) = |Ax - b|^2 = \langle Ax - b, Ax - b\rangle$. We wish to minimize f over all of $\mathbb{R}^n$. From calculus we know that the derivative of f must vanish at a minimum. To compute the derivative $Df(x)$ abstractly, which is best in this case, recall that $Df(x)$ is a linear transformation from $\mathbb{R}^n$ to $\mathbb{R}$. Let v denote an arbitrary vector in $\mathbb{R}^n$. By the chain rule,

$$\frac{d}{dt} f(x + tv)\Big|_{t=0} = Df(x)v.$$

Thus, we have that

$$Df(x)v = \frac{d}{dt}\langle A(x + tv) - b, A(x + tv) - b\rangle\Big|_{t=0}$$

$$= \langle Av, Ax - b\rangle + \langle Ax - b, Av\rangle$$
$$= 2\langle Ax - b, Av\rangle.$$

Let us denote the matrix transpose (interchanging rows and columns of A) by A^T. An important property (which is easy to check by a computation in components) is that $\langle w, Az\rangle = \langle A^T w, z\rangle$ for every pair of vectors w and z. Alternatively, we may *define* the matrix transpose of A to be the unique matrix A^T that satisfies the inner product identity and then prove that in coordinates A^T is obtained from A by interchanging its rows and columns. Using the transpose, it follows that

$$Df(x)v = 2\langle A^T Ax - A^T b, v\rangle.$$

Suppose that $\langle u, v\rangle = 0$ for every v. Then, in particular, $|u|^2 = \langle u, u\rangle = 0$ and $u = 0$. Hence, if the derivative $Df(x)$ is the zero matrix, then

$$A^T Ax - A^T b = 0.$$

This latter equation is called the normal equation for the least squares problem. Because A is $m \times n$, the matrix $A^T A$ is $n \times n$.

For simplicity, let us make an additional assumption (which is the case for linear regression): The columns of A are linearly independent. In this case, $A^T A$ is invertible. To prove this fact, suppose that v is a vector and $A^T Av = 0$. Taking the inner product with respect to v, we have that

$$0 = \langle A^T Av, v\rangle = \langle Av, Av\rangle = |Av|^2.$$

Because the columns of A are linearly independent and Av is a linear combination of these columns, $Av = 0$ only if $v = 0$. This means that $A^T A$ has linearly independent columns and hence is invertible.

Under the assumption that the columns of A are independent, the normal equation has a unique solution

$$x = (A^T A)^{-1} A^T b.$$

Thus, our function f has a unique critical point, which must be a minimum because $f(x) \geq 0$ for every x and $f(x)$ grows to infinity as $|x|$ grows without bound.

The quantity $(A^T A)^{-1} A^T$ is called the pseudoinverse of A. Using the pseudoinverse, the linear regression problem is solved: the transpose of the vector (m, b) is exactly $(A^T A)^{-1} A^T b$ for the given $N \times 2$ matrix A and the N vector b.

In practice, the pseudoinverse is not computed directly. The normal equations are solved by elimination or an iterative method, or better yet, the pseudoinverse is computed using the singular value decomposition (SVD) of A. The reason for not simply solving the normal equations is that these equations may be ill-conditioned; for example, the matrix $A^T A$ may be nearly singular.

The SVD of an $m \times n$ real matrix A, which always exists, has the form

$$A = U \Sigma V^T,$$

where U is an $m \times m$ orthogonal matrix (that is, $U^T U = U U^T = I$), Σ is an $m \times n$ diagonal matrix, and V is an $n \times n$ orthogonal matrix. The square roots of the diagonal elements of Σ are called the singular values of A. This decomposition provides a natural way to look at the action of A as a linear operator. To see why, note that $A : \mathbb{R}^n \to \mathbb{R}^m$. Thus, the columns of V provide an orthonormal basis for the domain and the columns of U are an orthonormal basis of the space containing the range. Choose a column, say v_i of V and check that

$$A v_i = \sigma_i u_i$$

where σ_i is the ith singular value of A. In this sense the bases given by the columns of U and V are the right bases for $\mathbb{R}^m$ and $\mathbb{R}^n$ with respect to the matrix A. It follows immediately that the null space (kernel) of A is spanned by the columns of V corresponding to zero singular values and the range of A is the span of the columns of U corresponding to nonzero singular values.

Using the SVD and the properties of its factors, the normal equation may be written in the form

$$V \Sigma^T U^T U \Sigma V^T x = V \Sigma^T U^T b,$$

and simplified to

$$\Sigma^T \Sigma V^T x = \Sigma^T U^T b.$$

In case the matrix A has linearly independent columns, its singular values are all positive. The matrix $\Sigma^T \Sigma$ is square and its diagonal elements are the squares of the singular values of A. Because $A^T A = V \Sigma^T \Sigma V^T$, the squares of the singular values are also the eigenvalues of A^TA. Note that the inverse of $\Sigma^T\Sigma$ is diagonal and its diagonal elements are simply the reciprocals of its diagonal elements. An easy calculation shows that $\Sigma^+ := (\Sigma^T\Sigma)^{-1}\Sigma^T$ is diagonal (but $n \times m$ so usually not square) with diagonal elements the reciprocals of the singular values of A. It follows that the least squares minimum is achieved at

$$x = V\Sigma^+U^Tb.$$

The matrix $V\Sigma^+U^T$ is the SVD pseudoinverse of A. The ease of the inversion of $\Sigma^T\Sigma$ and the efficiency of the numerical algorithms available to calculate the SVD make the method presented here the most used numerical method for computation of least squares problems. Of course, the SVD has many other applications.

Algorithms for the efficient numerical computation of the SVD are presented in all books on numerical linear algebra.

A.13 THE MORSE LEMMA

If a class C^∞ function $f : \mathbb{R}^n \to \mathbb{R}$ has a nondegenerate critical point a (that is, $Df(a) = 0$ and zero is not an eigenvalue of the symmetric linear transformation representing the quadratic form $D^2f(a)(x,x)$), then there is a C^∞ change of coordinates defined in a neighborhood of a that transforms f to the function $\xi \mapsto f(a) + D^2f(a)(\xi,\xi)$.

A general proof of Morse's lemma is given in [1]. The proof for the one-dimensional case is elementary. Reduce to the case where the function f is given by $f(x) = x^2h(x)$ and $h(0) > 0$. Define $g(x) = x\sqrt{h(x)}$ and prove that there is a function k such that $g(k(x)) = x$. The desired change of coordinates is given by $x = k(z)$.

A.14 NEWTON'S METHOD

Newton's method can be used to approximate solutions of systems of n equations in n unknowns. It is the premier algorithm for this task whenever the first partial derivatives of the functions that define the system of equations are known.

Consider a system of equations of the form

$$\begin{aligned} f_1(x_1, x_2, \ldots, x_n) &= 0, \\ f_2(x_1, x_2, \ldots, x_n) &= 0, \\ &\vdots \\ f_n(x_1, x_2, \ldots, x_n) &= 0, \end{aligned}$$

and let $F : \mathbb{R}^n \to \mathbb{R}^n$ denote the function whose components are $(f_1, f_2, \ldots, f_n)$. Suppose that $r \in \mathbb{R}^n$ is a solution (that is, $F(r) = 0$). The idea underlying Newton's method is to guess an approximation $a \in \mathbb{R}^n$ of r and improve this approximation using linearization. We must assume that F is smooth (at least class C^1 to implement the method and at least C^2 for the method to perform as it should).

For $a \in \mathbb{R}^n$ and using the notation of Appendix A.8,

$$0 = F(r) = F(a + (r - a)) = F(a) + DF(a)(r - a) + O(|r - a|^2).$$

In case a is close to r so that $|r - a|^2$ is small compared to $|r - a|$, the first two terms of the Taylor expansion closely approximate zero; that is, $F(a) + DF(a)(r - a) \approx 0$. By solving for r under the assumption that $DF(a)$ is invertible, we might expect the vector $a - [DF(a)]^{-1}F(a)$ to be a better approximation to r than a. This process can be repeated. More precisely, Newton's method is to guess an initial approximation $x^0 := a$ of the desired root and seek to improve this approximation via the iteration process

$$x^{j+1} = x^j - [DF(x^j)]^{-1}F(x^j). \tag{A.4}$$

The point $x^{j+1} \in \mathbb{R}^n$ is expected to be closer to the desired root r than the approximation x^j.

Newton's method has a special feature that makes it very effective when the initial guess is sufficiently close to a root of F: the sequence of iterates produced by the method converge at a quadratic rate to the root. To see what is meant by quadratic convergence, note first that an alternative way to view Newton's method is to define a new function $G : \mathbb{R}^n \to \mathbb{R}^n$ by

$$G(x) = x - [DF(x)]^{-1}F(x) \tag{A.5}$$

and observe that the Newton iterates are obtained by composition: $x^1 = G(x^0)$, $x^2 = G(x^1) = G(G(x^0))$, and so on. Indeed, we have $x^{j+1} =$

$G(x^j)$ and this point can be obtained by composing G with itself $j+1$ times and evaluating the composition at x^0. If Newton's method converges to the root r, it follows that $r = G(r)$; that is, r is fixed point of G. The converse is also true: a fixed point of G is a root of F.

Suppose that a is an approximation of r and G happens to be a function with a fixed point at r. Maybe this G does not come from Newton's method as in Eq. (A.5). Using Taylor's formula in Appendix A.9, we have the identity

$$|G(a) - r| = |G(a) - G(r)| = |\int_0^1 DG(r + t(a - r))(a - r)\, dt|, \quad \text{(A.6)}$$

and in case the norm of the derivative of G is bounded by M, the estimate

$$|G(a) - r| \le M|a - r|. \qquad \text{(A.7)}$$

If we are lucky and $M < 1$, then $G(a)$ is closer to r than a. Continuing in the same manner

$$|G(G(a)) - r| \le M^2|a - r|,$$

and so on. Hence, the iterates of a obtained by applying G do indeed converge to r as the number of iterates goes to infinity.

In case G is a scalar function of a scalar variable ($n = 1$), G' is a continuous function, and

$$0 < |G'(r)| < 1,$$

Eq. (A.6) implies the equaltiy

$$|G(a^j) - r| = |\int_0^1 G'(r + t(a^j - r))\, dt||a^j - r|. \qquad \text{(A.8)}$$

Using that $G'(r)$ is not zero and passing to the limit as $j \to \infty$,

$$\lim_{j\to\infty} \frac{|G(a^j) - r|}{|a^j - r|} = |G'(r)|.$$

The convergence rate is called linear (or first order) because each iterate decreases the error (which is the absolute value of the difference between the iterate and the root) by a factor of

$$|\int_0^1 G'(r + t(a^j - r))\, dt| \approx |G'(r)|$$

as stated in Eq. (A.8). Analogous results for $G : \mathbb{R}^n \to \mathbb{R}^n$ are more complicated because iterates are not confined to one direction of approach to the root. But when the derivative of G does not vanish, some choices for the starting value will lead to iterations that converge linearly to the root.

For Newton's method, the key observation about G is that $DG(r) = 0$, a fact that is easy to check. Because $DG(r) = 0$, the size of DG can be made as small as we like as long as DG is evaluated at points near r. In particular, the number M in the above analysis can be made less than 1, and as long as the stating point is close enough to the root r, the iterates of G will converge to r; that is, if $DF(r)$ is invertible so that G is defined and the starting value is close enough to the desired root, then Newton's method converges to this root.

How fast does Newton's method converge? Because $DG(r) = 0$ and using Taylor's formula again, note that

$$|G(a)-r| = |G(a)-G(r)| = |\int_0^1 DG(r+t(a-r))(a-r)-DG(r)(a-r)\, dt|. \tag{A.9}$$

Taylor's formula also applies to the function DG. Under the assumption that the second derivative of G is bounded by M, it is not difficult to prove the inequality

$$|DG(r + t(a - r)) - DG(r)| \leq M|r - a|.$$

Using this estimate in Eq. (A.9), note that

$$|G(a) - r| \leq M|a - r|^2.$$

In this case, where $DG(r) = 0$, the convergence is faster than linear; in fact, the ratio

$$\frac{|a^{j+1} - r|}{|a^j - r|} = \frac{|G(a^j) - r|}{|a^j - r|} \leq \frac{M|a^j - r|^2}{|a^j - r|}$$

goes to zero as j increases to infinity (which means the numerator is much smaller than the denominator). As least, the error $|a^{j+1} - r|$ compared to the previous error $|a^j - r|$ is cut by a factor of $M|a^j - r|$. As the iterates approach the root, $|a^j - r| < 1$; thus, in this case, the contraction factor $M|a^j - r|$ is much smaller than M. For this reason, the convergence rate is faster than linear; it is at least second order (or quadratic).

In general and more precisely, we say a convergent sequence $\{a^j\}_{j=1}^{\infty}$ has order of convergence $\alpha > 0$ if

$$\lim_{j\to\infty} \frac{|a^{j+1} - r|}{|a^j - r|^\alpha}$$

is a positive number. Newton's method has order of convergence at least 2. The order of convergence is greater in case the second derivative of G vanishes.

For some examples, check that $\{\frac{1}{2^j}\}_{j=1}^{\infty}$ is linearly convergent and $\{\frac{1}{2^{2^j}}\}_{j=1}^{\infty}$ is quadratically convergent.

An important result that gives sufficient conditions for the convergence of Newton's method is the Newton–Kantorovich theorem. This result has several variants; two of them are stated here. A proof of the first version is given in [52]; the second version is proved in [56].

Theorem A.5. Suppose that $F : \Omega \to \mathbb{R}^n$ is a differentiable function from the open subset Ω in $\mathbb{R}^n$ to $\mathbb{R}^n$, $\omega \in \Omega$, and the derivative $DF(\omega) : \mathbb{R}^n \to \mathbb{R}^n$ is invertible. Define $v = -[Df(\omega)]^{-1}(\omega)F(\omega)$ and U the ball of radius $|v|$ centered at $\omega + v \in \mathbb{R}^n$. If $U \subset \Omega$, there is a positive number M such that

$$\|Df(x) - Df(y)\| \le M|x - y| \text{ for all } x \text{ and } y \text{ in } U,$$

and

$$M|F(\omega)|\,\|[Df(\omega)]^{-1}(\omega)\|^2 \le \frac{1}{2},$$

then F has a unique zero in U and for every starting point in U Newton's method converges to this zero.

Theorem A.6. Suppose that

(1) $F : \Omega_r \to Y$ is a twice continuously differentiable function from the open ball Ω_r of radius $r > 0$ of the Banach space X to the Banach space Y;

(2) A is a bounded linear operator $A : X \to Y$;

(3) $\omega \in \Omega_r$ and there are positive constants η, δ, and C such that

- **(a)** $\|AF(\omega)\| \le \eta$,
- **(b)** $\|ADF(\omega) - I\| \le \delta < 1$, and
- **(c)** $\|AD^2F(x)\| \le C$ for all $x \in \Omega_r$.

Define

$$h = \frac{C\eta}{(1-\delta)^2}, \quad \alpha = \frac{(1-\sqrt{1-2h})\eta}{h(1-\delta)}, \quad \beta = \frac{(1+\sqrt{1-2h})\eta}{h(1-\delta)}.$$

If $\alpha \le r < \beta$ and $h < 1/2$, or $\alpha \le r \le \beta$ and $h = 1/2$, then F has a unique zero $\omega^\infty \in \Omega_r$, the sequence $\{\omega^j\}_{j=1}^\infty$ of Newton approximates converges to ω^∞, and

$$\|\omega^\infty - \omega^j\| \le \frac{(2h)^{2^j}\eta}{2^j(1-\delta)}.$$

In the second version of the theorem, A is usually taken to be $[DF(\omega)]^{-1}$ in case this operator is bounded.

To implement Newton's method as a numerical algorithm, computation of inverses of matrices is avoided by recasting Eq. (A.10) to the form

$$DF(x^j)(x^{j+1} - x^j) = -F(x^j), \tag{A.10}$$

approximating the solution z of the matrix system $DF(x^j)z = -F(x^j)$, and then defining $x^{j+1} = z + x^j$.

For large-scale problems, Newton's method has some weaknesses: coding the derivative DF can be a major problem, solving a large system of linear equations at each step is computationally expensive, and choosing a starting value close enough to the desired root is not trivial. Given the central importance of solving nonlinear systems in applied mathematics, all of these problems have been addressed with some success. These issues are discussed in textbooks on scientific computing (see [82] for an excellent treatment).

As a general rule, computing and coding the derivative is good practice when approximating the solution of a particular problem. Writing a general purpose code is another matter. Quasi-Newton methods are often used in general purpose software to avoid computation of the derivative. One of the most popular quasi-Newton methods is due to Broyden (compare the discussion in [12] or [82]); it employs secant approximations of the derivative DF.

As in Newton's method, start with an initial guess x^0 for the desired root. Suppose, for the moment, that $j \geq 2$ and the iterates x^{j-1} and x^j have been computed. Using Taylor's formula,

$$F(x^j) - F(x^{j-1}) \approx DF(x^j)(x^j - x^{j-1})$$

and this approximation improves as the length of the vector $x^j - x^{j-1}$ approaches zero. Because function values will be part of the iteration process, all the data in the latter formula, except for $DF(x^j)$, is available. To avoid computation of this matrix, Broyden's proposal is to substitute a new matrix A_j, which is close to $DF(x^j)$, so that

$$F(x^j) - F(x^{j-1}) = A_j(x^j - x^{j-1}) \tag{A.11}$$

and use this A_j in place of $DF(x^j)$ for a (quasi) Newton iteration step—solving a linear system of equations—to produce x^{j+1}. The challenge is to choose A_j to be a good approximation of the derivative. Consider the very first iterate x^1. How is it to be computed when only the initial guess x^0 is available? A possibility is to use Newton's method for this step. As mentioned, this is expensive, but Broyden shows (by doing some mathematics) that there is a big payoff: a good approximation of $DF(x^0)$ can be used to avoid approximations of subsequent derivatives.

After taking one Newton iterate, $DF(x^0)$ may be stored along with x^0, x^1, and $F(x^0)$. The function value $F(x^1)$ is computed with one additional function call. Broyden's (exceptionally good) answer to approximating $DF(x^1)$ is to determine A_1 using two assumptions: (1) A_1 satisfies Eq. (A.11) for $j = 1$ and (2) $A_1 v = DF(x^0)v$ whenever v is a vector orthogonal to $x^1 - x^0$. These assumptions uniquely determine A_1. By simple linear algebra, A_1 must be of the form $A_1 = DF(x^0) + B$ where the unknown matrix B is zero on every vector orthogonal to $x^1 - x^0$. This observation suggests that $By = v \cdot (x^1 - x^0)^T y$ for some unknown vector u. Indeed, u is determined as a consequence of assumption (1); in fact,

$$A_1 = DF(x^0) + \frac{F(x^1) - F(x^0) - DF(x^0)(x^1 - x^0)}{|x^1 - x^0|^2}(x^1 - x^0)^T. \tag{A.12}$$

Note that the second term on the right can be expressed as a $(p \times p)$ matrix.

The matrix A_1 is used instead of $DF(x^1)$ to perform a (quasi) Newton step that produces x^2. Subsequent iterations proceed as expected: The matrix

$$A_j = A_{j-1} + \frac{F(x^j) - F(x^{j-1}) - A_{j-1}(x^j - x^{j-1})}{|x^j - x^{j-1}|^2}(x^j - x^{j-1})^T \quad (A.13)$$

is used instead of $DF(x^j)$ to perform a (quasi) Newton step that produces x^{j+1}.

Broyden's method is effective, but as for Newton's method, each iteration is completed by solving a system of linear equations. This computational overhead can be mollified by employing a matrix inversion formula called the Sherman–Morrison formula (Jack Sherman and Winifred J. Morrison, approximately 1950), which is a special case of Max Woodbury's formula (approximately 1950). Tracking down and writing the true history of these formulas would be an interesting project.

To derive the Sherman–Morrison formula, recall from second semester calculus that, whenever $|z| < 1$,

$$(1 - z)^{-1} = 1 - z + z^2 - z^3 + \cdots .$$

Formally (that is, without regard to convergence) the formula is simply obtained by long division. The same formula is true when z is a matrix, the number 1 is replaced by the identity matrix (except in the exponent of course), the left-hand side is viewed as a matrix inverse, and the absolute value is replaced by a matrix norm (which will not be used here).

Let u and v be vectors in $\mathbb{R}^p$. Their combination u^Tv is a scalar and uv^T is a $(p \times p)$ matrix. A bit of *formal* play leads to something interesting:

$$\begin{aligned}
(I + uv^T)^{-1} &= I - uv^T + (uv^T)(uv^T) - (uv^T)(uv^T)(uv^T) + \cdots \\
&= I - uv^T + u(v^Tu)v^T - u(v^Tu)^2v^T + \cdots \\
&= I - u(1 - v^Tu + (v^Tu)^2 - (v^Tu)^3 + \cdots)Iv^T \\
&= I - u(1 + v^Tu)^{-1}Iv^T \\
&= I - \frac{1}{1 + v^Tu}uv^T, \qquad (A.14)
\end{aligned}$$

at least under the hypothesis that $v^T u \neq -1$. Thus, assuming that the identity is correct under this hypothesis, the matrix inverse is given by a simple formula. The formula can be proved simply from the definition of the inverse of a matrix.

The desired Sherman–Morrison formula is a corollary of this result. Suppose that A is an invertible matrix, v and w are vectors, and $v^T A^{-1} w \neq -1$. Set $u = A^{-1} w$ in Eq. (A.14) and check that

$$(I + A^{-1} w v^T)^{-1} = I - \frac{1}{1 + v^T A^{-1} w} A^{-1} w v^T.$$

Factor out A^{-1} on the left-hand side and do the algebra to obtain the desired Sherman-Morrison formula: In case A is invertible and $v^T A^{-1} w \neq -1$,

$$(A + w v^T)^{-1} = A^{-1} - \frac{1}{1 + v^T A^{-1} w} A^{-1} w v^T A^{-1}. \qquad \text{(A.15)}$$

Carefully inspect Eq. (A.13) until you realize that its right-hand side is a matrix in the form $A + w v^T$, exactly what is required to invert the matrix A_k on the left-hand side. Check that the procedure would be

$$v := x^k - x^{k-1}, \qquad w := \frac{1}{|v|^2} F(x^k) - F(x^{k-1}) - A_{k-1} v$$

followed by

$$(A_k)^{-1} = (A_{k-1})^{-1} - \frac{1}{1 + v^T (A_{k-1})^{-1} w} (A_{k-1})^{-1} w v^T (A_{k-1})^{-1}.$$

Each new matrix inverse is obtained by multiplying vectors by previously computed matrices. No costly matrix inversions are necessary after the first matrix inverse is computed. Is the condition $v^T A^{-1} w \neq -1$ satisfied? If not, a check will be needed in a numerical code based on this formula.

There is one remaining problem: $(A_0)^{-1}$ is required to start the iteration; that is, the matrix $DF(x^0)$ must be inverted. This can be done numerically by solving p linear systems: $DF(x^0) y = e_i$, for the usual basis vectors e_i and $i = 1, 2, 3, \ldots, p$.

Broyden's method together with the Sherman–Morrison formula makes a beautiful algorithm that is numerically efficient (fast), at least after the first iterate is completed. It converges to a root most of the time. Although it is not quadratically convergence, its convergence is superlinear; that is, its order of convergence lies in the open interval $(1, 2)$.

A.15 VARIATION OF PARAMETERS FORMULA

Proposition A.7 (Variation of Parameters Formula). Consider the initial value problem

$$\dot{x} = A(t)x + g(x, t), \qquad x(t_0) = x_0 \tag{A.16}$$

and let $t \mapsto \Phi(t)$ be a fundamental matrix solution for the homogeneous linear system $\dot{x} = A(t)x$ that is defined on some interval J_0 containing t_0. If $t \mapsto \phi(t)$ is the solution of the initial value problem defined on some subinterval of J_0, then we have (the variation of parameters formula)

$$\phi(t) = \Phi(t)\Phi^{-1}(t_0)x_0 + \Phi(t)\int_{t_0}^{t} \Phi^{-1}(s)g(\phi(s), s)\, ds. \tag{A.17}$$

Proof. Define a new function z by $z(t) = \Phi^{-1}(t)\phi(t)$. We have

$$\dot{\phi}(t) = A(t)\Phi(t)z(t) + \Phi(t)\dot{z}(t).$$

Thus,

$$A(t)\phi(t) + g(\phi(t), t) = A(t)\phi(t) + \Phi(t)\dot{z}(t)$$

and

$$\dot{z}(t) = \Phi^{-1}(t)g(\phi(t), t).$$

Also note that $z(t_0) = \Phi^{-1}(t_0)x_0$.

By integration,

$$z(t) - z(t_0) = \int_{t_0}^{t} \Phi^{-1}(s)g(\phi(s), s)\, ds,$$

or in other words,

$$\phi(t) = \Phi(t)\Phi^{-1}(t_0)x_0 + \Phi(t)\int_{t_0}^{t} \Phi^{-1}(s)g(\phi(s), s)\, ds. \qquad \square$$

A.16 THE VARIATIONAL EQUATION

Let $t \mapsto \phi(t, \xi, \lambda)$ denote the solution of the differential equation

$$\dot{x} = f(x, t, \lambda) \tag{A.18}$$

such that $\phi(0, \xi, \lambda) = \xi$, where λ is a parameter in $\mathbb{R}^k$, $x \in \mathbb{R}^n$, and $\xi \in \mathbb{R}^n$. The first (matrix) variational equation (also called the linearization) along the solution ϕ is given by

$$\dot{W} = D_x f(\phi(t, \xi, \lambda), t, \lambda)W$$

with initial condition $W(0) = I$, where I is the $n \times n$ identity matrix and D_x denotes the derivative of f with respect to x. The second variational equation is

$$\dot{U} = Df(\phi(t, \xi, \lambda), t, \lambda)U + D_\lambda(f\phi(t, \xi, \lambda))$$

with initial condition $U(0) = 0$, where D_λ denotes the derivative of f with respect to λ.

These equations are important for several reasons. The first variational equation is the differential equation for the derivative of ϕ with respect to the initial condition; in fact,

$$W(t) = D_\xi \phi(t, \xi, \lambda).$$

The second variational equation is the differential equation for ϕ with respect to the parameter; in fact,

$$U(t) = D_\lambda \phi(t, \xi, \lambda).$$

These facts follow immediately by differentiating both sides of differential equation (A.18) with x replaced by ϕ. The key observations are that

$$D_\xi D_t = D_t D_\xi \quad \text{and} \quad D_\lambda D_t = D_t D_\lambda,$$

under the assumption that f is class C^1

The first variational equation also arises as a linearization. Indeed, let ψ denote a solution of differential equation (A.18) and consider its deviation η from the given solution ϕ; that is,

$$\psi = \phi + \eta.$$

We have that

$$\begin{aligned}\dot{\eta} &= \dot{\phi} - \dot{\psi} \\ &= f(\phi, t, \lambda) - f(\psi, t, \lambda) \\ &= f(\phi, t, \lambda) - f(\phi + \eta, t, \lambda) \\ &= D_x f(\phi, t, \lambda)\eta + O(\eta^2),\end{aligned}$$

where the last equation is obtained by expanding the function $x \mapsto f(\phi + x, t, \lambda)$ in its Taylor series at $x = 0$. This suggests that the solution of the differential equation

$$\dot{\eta} = D_x f(\phi(t, \xi, \lambda), t, \lambda)\eta$$

provides a good approximation to the deviation η when ψ is close to ϕ. The initial condition is $\eta(0) = \xi - \psi(0)$.

A.17 LINEARIZATION AND STABILITY

For a differential equations on $\mathbb{R}^n$ of the form

$$\dot{u} = Au + g(u) \qquad \text{(A.19)}$$

where A is an $n \times n$ matrix and $g : \mathbb{R}^n \to \mathbb{R}^n$ is a smooth function such that $g(0) = Dg(0) = 0$, the origin $u = 0$ is a rest point whose linearization is $\dot{w} = Aw$. The system matrix A of the linearized system can be used to determine the stability of the rest point.

Theorem A.8. If all the real parts of the eigenvalues of the system matrix of the linearization at a rest point of an autonomous system have negative real parts, then the rest point is (locally) asympotically stable.

Theorem A.8 is a basic result. One method used to prove it is based on Lyapunov's direct approach, which is the content of another basic result in stability theory for ODEs.

Consider a rest point x_0 for the autonomous differential equation

$$\dot{x} = f(x), \qquad x \in \mathbb{R}^n. \qquad \text{(A.20)}$$

A continuous function $V : U \to \mathbb{R}$, where $U \subseteq \mathbb{R}^n$ is an open set with $x_0 \in U$, is called a *Lyapunov function* for the differential equation at x_0 if

(i) $V(x_0) = 0$,
(ii) $V(x) > 0$ for $x \in U \setminus \{x_0\}$,
(iii) the function V is continuously differentiable on the set $U \setminus \{x_0\}$, and on this set, $\dot{V}(x) := \operatorname{grad} V(x) \cdot f(x) \leq 0$.

The function V is called a *strict Lyapunov function* if, in addition,

(iv) $\dot{V}(x) < 0$ for $x \in U \setminus \{x_0\}$.

Theorem A.9 (Lyapunov's Stability Theorem). If there is a Lyapunov function defined in an open neighborhood of a rest point of differential equation (A.20), then the rest point is stable. If, in addition, the Lyapunov function is a strict Lyapunov function, then the rest point is asymptotically stable.

For Eq. (A.19), assume that every eigenvalue of A has negative real part, and for some $a > 0$, there is a constant $k > 0$ such that (using the usual norm on $\mathbb{R}^n$)

$$|g(x)| \leq k|x|^2$$

whenever $|x| < a$. To show that the origin is an asymptotically stable rest point, it suffices to construct a quadratic Lyapunov function. To do this, let $\langle \cdot, \cdot \rangle$ denote the usual inner product on $\mathbb{R}^n$, and A^* the transpose of the real matrix A. Suppose that there is a real, symmetric, positive-definite $n \times n$ matrix that also satisfies Lyapunov's equation

$$A^*B + BA = -I$$

and define $V : \mathbb{R}^n \to \mathbb{R}$ by

$$V(x) = \langle x, Bx \rangle.$$

The restriction of V to a sufficiently small neighborhood of the origin is a strict Lyapunov function. To see this, make an estimate using the Schwarz inequality. To complete the proof show that

$$B := \int_0^\infty e^{tA^*} e^{tA}\, dt$$

is a symmetric positive definite $n \times n$ matrix that satisfies Lyapunov's equation. This claim is proved with a few observations. First, note that A^* and A have the same eigenvalues, all of which are in the open left-half of the complex plane. There is some number $\lambda > 0$ such that all eigenvalues of both matrices have real parts less than $-\lambda$. In this case,

$$\|e^{tA}x\| \le Ce^{-\lambda t}\|x\|$$

for all $t \ge 0$ and all $x \in \mathbb{R}^n$. This estimate is used to prove that the integral converges.

Alternatively, we can solve Lyapunov's equation using the following outline: Lyapunov's equation in the form $A^*B + BA = S$, where A is diagonal, S is symmetric and positive definite, and all pairs of eigenvalues of A have nonzero sums, has a symmetric positive-definite solution B. In particular, under these hypotheses, the operator $B \mapsto A^*B + BA$ is invertible. The same result is true without the hypothesis that A is diagonal. This fact can be proved using the density of the diagonalizable matrices and the continuity of the eigenvalues of a matrix with respect to its components.

Theorem A.10 (Routh–Hurwitz Criterion). Suppose that the characteristic polynomial of the real matrix A is written in the form

$$\lambda^n + a_1\lambda^{n-1} + \cdots + a_{n-1}\lambda + a_n,$$

let $a_m = 0$ for $m > n$, and define the determinants Δ_k for $k = 1, 2, \ldots, n$ by

$$\Delta_k := \det \begin{pmatrix} a_1 & 1 & 0 & 0 & 0 & 0 & \cdots & 0 \\ a_3 & a_2 & a_1 & 1 & 0 & 0 & \cdots & 0 \\ a_5 & a_4 & a_3 & a_2 & a_1 & 1 & \cdots & 0 \\ \vdots & \vdots & \vdots & \vdots & \vdots & \vdots & \vdots & \vdots \\ a_{2k-1} & a_{2k-2} & a_{2k-3} & a_{2k-4} & a_{2k-5} & a_{2k-6} & \cdots & a_k \end{pmatrix}.$$

If $\Delta_k > 0$ for $k = 1, 2, \ldots n$, then all roots of the characteristic polynomial have negative real parts.

A.18 POINCARÉ–BENDIXSON THEOREM

Two versions of the Poincaré–Bendixson theorem and an important corollary are stated here (see, for example, [20]).

Theorem A.11. If a closed and bounded positively invariant set for a C^1 autonomous system in the plane contains no rest points, then it contains at least one limit cycle.

Theorem A.12. If a closed bounded positively invariant set in the plane for a C^1 autonomous system contains no periodic orbits and exactly one hyperbolic asymptotically stable rest point (the real parts of the eigenvalues of the system matrix of the linearization at the rest point are negative), then every solution starting in the invariant set is asymptotic to the rest point.

Theorem A.13. A periodic orbit for a C^1 autonomous system in the plane surrounds at least one rest point.

A.19 EIGENVALUES OF TRIDIAGONAL TOEPLITZ MATRICES

A tridiagonal matrix is a banded matrix with bandwidth three: only the main diagonal, the first subdiagonal, and the first superdiagonal have nonzero components. A Toeplitz matrix has constant diagonals; that is, each diagonal has all its components equal. A useful fact is that the eigenvalues (and eigenvectors) of a tridiagonal Toeplitz matrix can be computed explicitly.

Let A be an $N \times N$ tridiagonal Toeplitz matrix such that every component on the first subdiagonal is the real number α, every component on the main diagonal is β, and every component on the first superdiagonal is γ. The eigenvalues of A are

$$\beta + \sqrt{\alpha\gamma}\cos\frac{k\pi}{N+1}, \quad k = 1, 2, 3, \ldots, N. \tag{A.21}$$

In case $\alpha\gamma < 0$, the square root is taken to be $i\sqrt{|\alpha\gamma|}$.

If α or γ is zero, the formula states that β is an eigenvalue of algebraic multiplicity N, a fact that is easy to check. Thus, we may assume that $\alpha\gamma \neq 0$. To derive this formula, suppose that $Av = \lambda v$ and note that the components of the vector v must satisfy a boundary value problem (BVP) for a difference equation:

$$\alpha v_{j+1} + \beta v_j + \gamma v_{j-1} = \lambda v_j, \quad j = 1, 2, 3, \ldots, N; \quad v_0 = 0, \quad v_{N+1} = 0.$$

Here, the boundary components v_0 and v_N are fictitious, but using them is a convenient way to state the exact problem to be solved: find λ and $\{v_j\}_{j=1}^{j=N}$ that satisfy the BVP. The vector v, with components $\{v_j\}_{j=1}^{j=N}$, is an eigenvector with eigenvalue λ for the matrix A.

Look for the components of v in the form $v_j = r^j$ for some undetermined number r. By substitution of this guess into the difference equation, it follows that a necessary condition for such a solution to exist is

$$\alpha + (\beta - \lambda)r + \gamma r^2 = 0. \tag{A.22}$$

Thus, the two roots of this quadratic equation, say r and s, provide candidate solutions $v_j = r^j$ and $v_j = s^j$. By the linearity of the difference equation, every linear combination of these two vectors is a candidate solution of the BVP. This suggests the solution of the BVP

$$v_j = ar^j + bs^j.$$

The boundary conditions are satisfied when a and b solve the linear system

$$a + b = 0, \qquad ar^{N+1} + bs^{N+1} = 0.$$

It has nontrivial solutions when

$$r^{N+1} = s^{N+1}.$$

In fact, these solutions are the pairs (a, b), where $b = -a$. But also r and s are roots of the quadratic polynomial (A.22). The product of these roots is α/γ. Using this fact, $s = \alpha/(r\gamma)$, and the first equation implies that

$$\left(\frac{r^2\gamma}{\alpha}\right)^{N+1} = 1.$$

Allowing for complex roots of unity and a yet unspecified eigenvalue, the two possible roots of the quadratic are given by

$$r = \sqrt{\frac{\alpha}{\gamma}} e^{k\pi i/(N+1)}, \quad s = \sqrt{\frac{\alpha}{\gamma}} e^{-k\pi i/(N+1)}, \qquad k = 1, 2, 3, \ldots N.$$

The sum of the roots of the quadratic polynomia is $(\lambda - \beta)/\alpha$. Thus, the desired eigenvalues are given by

$$\lambda = \beta + 2\gamma\sqrt{\frac{\alpha}{\gamma}}\,\frac{e^{k\pi i/(N+1)} + e^{-k\pi i/(N+1)}}{2},$$

a formula that simplifies to the desired quantity [Eq. (A.21)].

An eigenvector corresponding to the eigenvalue with some fixed k has components

$$v_j = r^j - s^j.$$

A.20 CONJUGATE GRADIENT METHOD

The conjugate gradient method is designed to approximate solutions of linear systems $Ax = b$ or minimize the function $E : \mathbb{R}^n \to \mathbb{R}$ given by $E(x) = \frac{1}{2}\langle x, Ax\rangle - \langle b, x\rangle$ in case the $n \times n$ matrix A is symmetric and positive definite, and the angled brackets denote the usual inner product (also called the dot product) on $\mathbb{R}^n$.

Solving the matrix system and minimizing the function are equivalent problems. This fact is proved using calculus.

The critical points of E are determined by computing the first derivative of the (polynomial) function E and setting it equal to zero. The derivative of E at x (the gradient at this point) is $DE(x) = Ax - b$ and the second derivative (the Hessian) is given by the positive definite matrix A. Thus, the critical points of E occur exactly at the solutions of the linear system $Ax = b$ and each critical point is a local minimum. Because A is positive definite, there is exactly one such local minimum. If there were two solutions of $Ax = b$, their difference y would be a nonzero vector in the null space of A (that is, $Ay = 0$) in contradiction to the inequality $\langle y, Ay\rangle > 0$ for nonzero y. Equivalence of the two problems will be proved if the unique local minimum of E is a global minimum of E. This will be true unless the minimum of E were to occur at infinity. In fact, $E(x) \to \infty$ whenever $|x| \to \infty$ and the validity of this claim would complete the proof. One way to prove it is to use some important facts from matrix theory, in particular, the spectral theorem: *A real symmetric matrix can be diagonalized by an orthogonal transformation.* By this result, there is a real orthogonal matrix Q such that $Q^T AQ = \Lambda$, where Q^T denotes the transpose of Q and Λ is a diagonal matrix. The elements on the main diagonal of Λ are the eigenvalues of A and these are all positive real numbers because A is positive definite. Using the new variable $y = Q^T x$, the function Λ is given by $\lambda(y) = \Lambda(Qy)$ and

$$\begin{aligned}\lambda(y) &= \frac{1}{2}\langle Qy, AQy\rangle - \langle b, Qy\rangle \\ &= \frac{1}{2}\langle y, Q^T AQy\rangle - \langle b, Qy\rangle \\ &= \frac{1}{2}\langle y, \Lambda y\rangle - \langle b, Qy\rangle.\end{aligned}$$

It is easy to see that if $c > 0$ is smaller than the smallest eigenvalue of A, then $\frac{1}{2}\langle y, \Lambda y\rangle \geq c|y|^2$. And, by the Cauchy–Schwarz inequality $\langle b, Qy\rangle \leq |Q^T b||y|$. Thus, $\lambda(y) \geq c|y|^2 - |Q^T b||y|$. The right-hand side of this inequality is a quadratic polynomial whose second-degree term has a positive coefficient. Thus, the right-hand side goes to infinity as $|y|$ grows without bound. This completes the proof.

The word "gradient," which appears in the name "conjugate gradient method," refers to the gradient already computed: DE, which when evaluated at x is the vector $Ax - b$. Recall from calculus that the function E increases in the direction of its gradient (in fact, the gradient direction at a point is the direction of maximal increase). To search for the minimum of E, which is the basic problem to be solved, it seems reasonable to start at some vector x_0 in $\mathbb{R}^n$ and go in the direction of the negative gradient; namely, the vector $r_0 = b - Ax^0$. More precisely, this idea would be implemented by considering the function $f_0(\alpha) := E(x_0 + \alpha r_0)$. It decreases for α in some interval with zero as its left-hand end point. Perhaps the ray $x_0 + \alpha r_0$ passes through the vector that minimizes E. In this case, the minimum of f is the desired minimum of E. But, more likely, f decreases, reaches a minimum, and then increases again. It must eventually increase because E increases without bound as the absolute value of its argument goes to infinity. For this reason, a good idea is to find the α_0 that minimizes f and define $x_1 = x_0 + \alpha r_0$. The value of E at the new vector x_1 is smaller than $E(x_0)$. In this sense, the line search (as it is called) improves the estimate x_0 and the line search strategy can be applied again starting at x_1 and proceeding in the direction of the negative gradient at this new point. The next step is to minimize $f_1(\alpha) := E(x_1 + \alpha r_1)$, where now $r_1 = b - Ax_1$. Clearly this iterative procedure produces a sequence of vectors $x_0, x_1, \ldots$ such that $E(x_{j+1}) < E(x_j)$ unless there is exceptional luck: after a finite number of steps the absolute minimum is reached. The latter case would be detected by the appearance of a zero residual $r_j := b - Ax_j$, which is just another way to say that $Ax_j = b$ or E is minimized at x_j. This particular line search method is called steepest descent. It can be used to approximate solutions of the minimization problem. In practice, this method works but the convergence can be (very) slow.

A great idea for improving the convergence speed is to do line searches in a sequence of nonzero directions $d_0, d_1, d_2, \ldots, d_{n-1}$ that form a basis of the vector space $\mathbb{R}^n$ and perhaps have extra properties so that, after completing n line searches with exact arithmetic, the last vector x^n obtained is the solution of $Ax = b$. To implement this algorithm concept in an efficient manner, one of the most important ideas is to take advantage of the natural inner product for the problem: $(x, y) = \langle x, Ay\rangle$. This function from $\mathbb{R}^n \times \mathbb{R}^n \to \mathbb{R}$ has the same essential properties as the usual inner product given by the pointed brackets. Indeed, the function denoted by round brackets is such that (1) $(x, x) \geq 0$ for all $x \in \mathbb{R}^n$ and equal to zero only if $x = 0$, (2) $(x, y) = (y, x)$ for all x and y in $\mathbb{R}^n$, and (3) $\lambda(x, y) = (\lambda x, y)$ for every real number λ and all x and y in $\mathbb{R}^n$. The essential features of the problem that make these properties true are the symmetry and positive definiteness of A. Two vectors x and y are called orthogonal if $\langle x, y\rangle = 0$. Similarly, these vectors are A-conjugate if $(x, y) = 0$; at least this is the origin of the word "conjugate." Perhaps a better terminology for this notion is A-orthogonal, which will be used here.

Suppose that $\{d_0, d_1, d_2, \ldots, d_{n-1}\}$ is an A-orthogonal set of vectors; that is, every pair of vectors in the set are A-orthogonal. Choose some vector x_0 in $\mathbb{R}^n$ and do line searches in these directions. The first search produces a vector $x_1 := x_0 + \alpha_0 d_0$ such that

$$E(x_1) = \min_{\alpha \in \mathbb{R}} E(x_0 + \alpha d_0).$$

The $j + 1$st search produces $x_{j+1} = x_j + \alpha_j d_j$ such that

$$E(x_{j+1}) = \min_{\alpha \in \mathbb{R}} E(x_j + \alpha d_j).$$

After n steps the vector x_n produced in this manner is the unique solution of $Ax = b$ and the point where E has its absolute minimum. The main part of the proof is to show that if the A-orthogonal set is chosen appropriately, then the vector x_{j+1} produced at the $j + 1$st line search is (in addition to being the minimizer for the search along the line) the minimizer over the span of the search directions $\{d_0, d_1, d_2, \ldots, d_j\}$. The last search produces the minimizer over $\text{span}\{d_0, d_1, d_2, \ldots, d_{n-1}\}$, which is the entire space. Thus, after n steps (and with exact arithmetic) the global minimizer is obtained. Assume this result is true for some as yet unspecified choice of the A-orthogonal set.

The scalar α_j that minimizes the $j + 1$st line search is given by a simple and important formula. To derive it, use calculus to find the minimizer of $f_j(\alpha) = E(x_j + \alpha d_j)$. As usual, the desired critical point is a zero of the derivative

$$f_j'(\alpha) = DE(x_j + \alpha d_j)d_j = \langle A(x_j + \alpha d_j) - b, d_j \rangle$$

found by solving the equation $\langle A(x_j + \alpha d_j) - b, d_j \rangle = 0$ for α. Using the properties of the inner product, the minimizer is

$$\alpha_j = \frac{\langle r_j, d_j \rangle}{(d_j, d_j)}, \tag{A.23}$$

where as before the $r_j := b - Ax_j$ is the residual. Note the usual inner product appears in the numerator and the A inner product in the denominator.

Although some facts are not yet proved, a description of the desired algorithm is almost complete: Determine an appropriate A-orthogonal set containing n vectors (so that it spans the entire space) and do successive line searches in the A-orthogonal directions. The only remaining ingredient is producing an appropriate A-orthogonal set.

The correct choices are made iteratively: Given a starting vector x_0, the starting direction d_0 is the residual $r_0 = b - Ax_0$. This is the same choice as for steepest descent. As before, the line search in this direction produces

$$x_1 = x_0 + \alpha_0 d_0,$$

where

$$\alpha_0 = \frac{\langle r_0, d_0 \rangle}{(d_0, d_0)}.$$

Instead of making the next search in the direction of the negative gradient evaluated at x_1 (given by the residual $r_1 = b - Ax_1$), this direction is made A-orthogonal (or, in other words, A-conjugate) to d_0 by subtracting off a scalar multiple of d_0. More precisely, consider making a choice of scalar β

so that the vector $r_1 - \beta d_0$ is A-orthogonal to d_0. This is easy. To make $(r_1 - \beta d_0, d_0) = 0$, simply solve for β to obtain

$$\beta_0 = \frac{(r_1, d_0)}{(d_0, d_0)}.$$

This time the numerator and denominator are round bracket inner products. The line search in the direction $d_1 := r_1 - \beta_0 d_0$ is used to obtain

$$x_2 := x_1 + \alpha_1 d_1,$$

where

$$\alpha_1 = \frac{\langle r_1, d_1 \rangle}{(d_1, d_1)}.$$

This completes the second step.

The third and subsequent steps (up to the nth step) are the same. This is the conjugate gradient method

At the third step the search direction is obtained as prescribed by

$$d_2 := r_2 - \beta_1 d_1$$

with

$$\beta_1 = \frac{(r_2, d_1)}{(d_1, d_1)}.$$

This choice is made so that $(d_2, d_1) = 0$. The miracle of the conjugate gradient method is that $(d_2, d_0) = 0$. In other words, the new search direction is orthogonal to all the previous search directions. This fact is not obvious, but it makes the conjugate gradient method an efficient algorithm. The work involved is simply the computation of a few inner products, which involve multiplications by A. It is not necessary to make a more extensive calculation that would ensure the new search direction is orthogonal to all previous directions (see, for example, [116] for complete proofs).

A.21 NUMERICAL COMPUTATION AND PROGRAMMING GEMS OF WISDOM

This book invites the reader to do numerical experiments as a method to gain insight into the prediction of mathematical models. A few of the gems of wisdom that come with experience are shared here.

A.21.1 There Is Only One Way to Debug a Numerical Code

On its face this statement is false. But, countless hours of valuable time are wasted reading code to find mistakes. The only method that works is to isolate some block of code—perhaps a single line—and *print out* the input and output of that block. The basic principle is to ascertain what the code actually produces rather than deluding yourself by believing the code produces what you intended.

A.21.2 Write Modular Programs

Break up a complicated program into subprograms (subroutines) that each perform a single task. Call the subroutines to build the main program.

A.21.3 Test Code against Known Solutions

This is a general principle. But, for differential equations specifically, use exact solutions of perhaps simplified models to test code. There are some clever ways to obtain exact solutions in case none are available for the problem at hand. A typical example might be a differential equation $\dot{x} = f(x,t)$ with some initial data $x(0) = x_0$. If no suitable exact solution is available, pick your favorite function y, perhaps a function that has some features expected of the solution (growth, oscillation, asymptotes, and so on), and simply substitute y into the differential equation to obtain $\dot{y} - f(y,t) = R(t)$, where R is defined by the left-hand side. You now have an exact solution of the differential equation $\dot{x} = f(x,t) + R(t)$ with initial condition $x(0) = y(0)$. This is not exactly the original problem, but it is close enough to gain some insight into the performance of your code. The same idea works for partial differential equations (PDEs).

A.21.4 Use Dimensionless Models

In realistic applied problems there are often natural parameters that vary in size by many orders of magnitude. This fact can cause serious problems for numerical codes, especially when subtracting nearly equal quantities and dividing by very small numbers. There is no general way to avoid these problems, but the first line of defense is to change to dimensionless variables. Taking this step at least allows meaningful size comparison of pure numbers. Of course, making a model dimensionless is not a unique process. After a change to dimensionless variables, further changes of variables might help to tame the differences in the magnitudes of parameters. A lot of research is related to this subject. It turns out that a small parameter multiplying a term that does not contain the highest-order derivative (as, for example, in the ODE $\dot{x} = f(x,t) + \epsilon g(x,t)$ for some ϵ such that $|\epsilon|$ is small) is usually not a serious issue. Mathematicians call this a regular perturbation problem. The other case, for example, the system $\epsilon\dot{x} = f(x,y,t)$ and $\dot{y} = g(x,y,t)$, is much worse. It is called a singular perturbation problem. Special methods are used for analysis and numerical approximation of such problems. Division by ϵ makes the right-hand side of the first equation much larger than the right-hand side of the second equation. Evolution on widely different timescales can also occur in nonlinear systems of differential equations that are not caused by coefficients that differ widely in magnitude. Instead the differing timescales occur due to nonlinearity. Such systems are called stiff by numerical analysts. Special numerical methods not discussed in this book are used to overcome (sometimes) this difficulty.

A.21.5 Why Worry about Numerical Difficulties When Commercial Codes Are Available

Commercial codes for approximating solutions of ODEs and PDEs have reached a high level of sophistication. They often test for underlying numerical problems such as stiffness and switch methods internally to try to overcome internally diagnosed problems. High-order accuracy is available from carefully written, extensively tested, and efficient codes. The only shortcomings of commercial codes are due to their very nature: they are general purpose codes. The applied problem at hand is usually special in some way. Taking advantage of special features can inform the decision on how to write a successful code to approximate solutions. Also, writing your own code can help to gain insight into the nature of the underlying model. In research problems this can be a key to understanding. Good decisions on the use of a commercial code or a code written by hand are made after an assessment of the purpose of making an approximation. Will a bridge fall down, an airplane crash, or a control device fail if the approximation is not accurate? Does the code have to be run only a few times to obtain a desired approximation, or is the code intended to be used often on a long-term basis? Do you understand the basic features of the underlying model, or are you treating the model as a black box?

A.21.6 Document Code

A code should contain at least as many comments as commands. The necessity for explanatory comments becomes immediately apparent when you return to one of your codes that you have not read recently. It will likely be incomprehensible unless it is well documented. At the very least, each subroutine should be commented with a statement of its purpose and a description of all its arguments on input and output. A code that might be used by people other than its author should also include external documentation containing carefully written instructions on how to implement, compile, and run the code along with a description of all its functionality. Of course, basic and perhaps advanced examples with actual input and output should be included. Manuals can be frustrating documents because they never seem to contain an explanation for the problem *you* wish to address. Perhaps you can write a user-friendly manual for your code.

Answers to Selected Exercises

Page 14

Ex. 2.2 (c) The long-term behavior is sinusoid. After the transient is too small to measure, the response is

$$x(t) = \frac{64}{257}\sin t - \frac{4}{257}\cos t.$$

The amplitude of the steady state response is $4/\sqrt{257}$.

Page 16

Ex. 2.5 The monthly payment M (assuming continuously compounded interest) is

$$M = \frac{rP}{12(1 - e^{-rT})}.$$

Ex. 2.4 The limit as time grows without bound is 10. One solution method is to separate the variables and use partial fractions. A second method is to use the Bernoulli transformation $Q = 1/P$.

Page 18

Ex. 2.11 $f(\phi, \theta) = (A\sin\theta + B\cos\theta)\left(\frac{1-\cos\theta}{\sin\theta}\right)$. After separation of variables we find the ODE $\sin\phi\, p'' + \cos\phi\, p' - p = 0$, which can be written in the form $(\sin\phi\, p')' = p/\sin\phi$. This suggests the substitution $u = \sin\phi\, p'$, and so on.

Ex. 2.12 (g) The answer is probably not known. But, with a high level of confidence, it seems that the fate of the solution with initial data $x(0) = 0$ and $\dot{x}(0) = 1$ converges to the constant solution $x(t) = 1$.

Page 19

Ex. 2.12 (d) Determine the energy of the saddles. The corresponding energy curve in the phase plane meets the line $x = 0$ at $\dot{x} =$

$1/\sqrt{2} > 7/10$. Thus, the solution with the given initial condition is pcriodic.

Page 28

Ex. 2.27 (a) The solutions obtained by the power series method are multiples of the polynomial $y = 1 - 2r + r^2 - r^3/6 + r^4/120$. (b) The reduction of order method is to seek a solution $y = zy_1$, where y_1 is the polynomial solution of part (a). This leads to the system of ODEs

$$w' + (\frac{2-r}{r} + 2\frac{y_1'}{y_1})w = 0, \qquad z' = w.$$

The first equation can be solved using an integrating factor to obtain the nonzero solution

$$w = \frac{e^r}{r^2 y_1^2(r)}.$$

Thus, z is given by an antiderivative of w. The result is $z = -1/r + 5\log r + O(r)$. The second solution is the same at leading order; hence, it blows up (like $1/r$) as $r \to 0$. The sign of the leading term is not important. (d) $y(1) \approx -0.09893$. There is an issue worth mentioning: By ODE theory, power series can be used to represent the solution of the IVP in some interval containing $r = 2$. In other words, there is a power series in powers of $(r-2)$ that convergences to the solution of the IVP in some interval containing $r = 2$. The problem is proving that $r = 1$ is contained in this interval of convergence. There is no easy way to check. In the present case, numerics agree with the power series truncated to a polynomial of degree 15. You should check the approximations by truncated power series at several lower and higher orders to see if the approximation seems to be converging. Whenever a sequence of approximations is available, it is a good idea to check for convergence of the desired value instead of simply choosing a member of the sequence to make the approximation.

Page 81

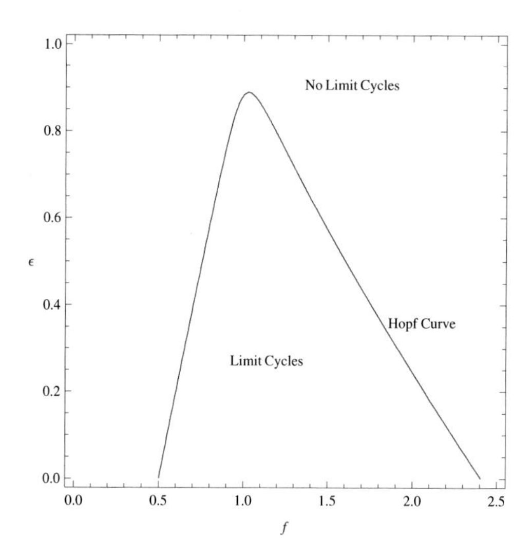

Fig. A.1 Bifurcation Diagram for Ex. 4.19.

Ex. 4.19 The bifurcation diagram is depicted in Fig. A.1.

Page 113

Ex. 5.26 $u(0.4, 60) \approx 0.66$.

Page 160

Ex. 5.42 (b) The steady state has the form $U(\xi) = (a\xi + b)^{1/(\gamma+1)}$ where a and b are determined from the boundary conditions. Note the identity

$$u(u^\gamma)_x = \frac{\gamma}{\gamma + 1}(u^{\gamma+1})_x.$$

Using it and the dimensionless quantity a, the steady state flux through the downstream end is

$$-a\frac{\ell p_0 \rho_0 k\gamma}{\mu(\gamma + 1)},$$

which is approximately 4.835×10^{-3} kg / s.

Page 184

Ex. 5.47 Suppose that it takes N steps to reach L. The global error is proportional to Nh^{n+1}. This quantity is equal to

$$N\left(\frac{L}{N}\right)\left(\frac{N}{L}\right)h^{n+1} = L\frac{1}{h}h^{n+1} = Lh^n.$$

Page 188

Ex. 5.61 The data were generated using $a = 0.001$ and $b = 1.23$ by rounding off the values of f to two decimal places.

Page 189

Ex. 5.61 The data were generated using $a = 0.44$ and $b = 888$ by rounding off the values of f to three decimal places.

Page 239

Ex. 8.7 The dynamical system is given by the mass balance: rate of change of mass equals mass in minus mass out. For $f(t)$ the mass flow rate into the tank and $\rho A\sqrt{2gz}$ the mass flow rate through the drain (where A is the area of the drain cross section), the mass balance differential equation is

$$\frac{d}{dt}(\rho\pi a^2 z) = f(t) - \rho A\sqrt{2gz}.$$

The valve control mechanism actuates the control by changing the area A of the drain cross section. There does not seem to be an obvious way to implement the control; that is, to define the radius of the drain cross section as a function of the proportional error $k(z_{\text{set}} - z)$. This function would be used to run the actuator that moves the valve mechanism. Control design requires experience and ingenuity. One possible actuation function is

$$A = \pi[\frac{r}{2\pi}(\pi - \arctan(k(z_{\text{set}} - z))]^2,$$

where the controller gain is assumed to be positive. When the water level exceeds the set-point depth, the argument of the arctangent is negative and approaches $-\pi$ as the depth of the water in the tank increases; thus, the area of the drain cross section

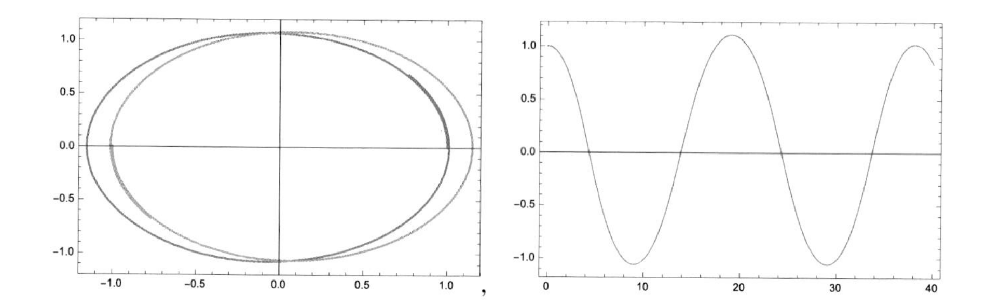

Fig. A.2 The left-hand panel shows the motion of the first two particles in the horizontal coordinate plane; the right-hand panel is a graph of the third coordinate of the third particle versus time. System parameters are those of Exercise 10.9 part (f). The numerical method is Störmer-Verlet with $\Delta t = 0.005$.

increases toward r. The opening rate of the orifice increases as the control gain k increases or the water depth increases. Likewise, the area decreases as the depth or control gain decreases. Note: We have implemented the inherently linear P control as the argument of a nonlinear function! By assumption, the mass inflow rate is less than $\rho\pi r^2\sqrt{2gh}$, which with the given choice of parameters is approximately $491\,\mathrm{kg}\,/\,\mathrm{m}$.

Numerical experiments show that it is not possible to maintain the set-point depth for an inflow of $300\,\mathrm{kg}\,/\,\mathrm{sec}$. Why not? In practice, it is usually better to analyze the system before doing numerical experiments. But, the results of computation might suggest that something in the control design is not working as expected.

The problem is apparent by looking at the steady state(s) of the control system. The steady state of the model is not at the set point. For the given data, the steady state is $z_{\mathrm{SS}} \approx 7.23$. Thus, there is good reason to incorporate an I control. Because I controls are explicitly time-dependent their presence makes the control process nonautonomous; no steady states exist unless the depth is at the set point over the entire time the integral controller is functioning.

Page 285

Ex. 10.9 Fig. A.2 shows the planar two-body motion of the first two masses and the motion of the third particle along the third coordinate axis versus time over 40 time units. The first two particles orbit each other while the third particle oscillates perpendicular to this plane passing through the center of mass of the two-body motion. After 10 units of time, and a bit more numerical analysis, the computed positions of the three particles (to three decimal places) are

$$x^1 = (-0.326, 1.038, 0.000), \qquad x^2 = (0.326, -1.038, 0.000),$$
$$x^3 = (0.000, 0.000, -1.008).$$

Page 326

Ex. 12.4

$$\text{PD} = \text{VFR} \times \frac{8\ell\mu}{\pi a^4}.$$

Page 331

Ex. 13.3 Choose a Cartesian coordinate system whose horizontal axis is in the downstream direction of the river bottom and whose vertical is in the direction of the river's surface. Let x and z be names of the coordinates and u and v the corresponding components of the fluid velocity. According to the assumptions $v = 0$ and $u_x = 0$. Thus, the continuity equation is satisfied automatically. The gravitational force per volume is given in these coordinates by $(\rho g \sin\theta, -\rho g \cos\theta)$, and the steady state two-dimensional Navier–Stokes equations are given by

$$0 = \mu u_{zz} + \rho g \sin\theta, \qquad 0 = -p_z - \rho g \cos\theta. \tag{A.24}$$

It follows immediately that there are constants c and d such that

$$\mu u = -\frac{\rho g \sin\theta}{2} z^2 + cz + d. \tag{A.25}$$

The no-slip boundary condition at the river bottom (where $z = 0$) implies $d = 0$. Under the assumption that the flow is Eulerian near the river's surface, Bernoulli's law holds and there is a constant C such that

$$p + \frac{1}{2}\rho u^2 + g\rho z = C.$$

By differentiation with respect to z, and using the second component of the Navier–Stokes equations (A.24),

$$2\rho u u_z = g\rho(\cos\theta - 1).$$

The angle θ measuring the tilt of the river bottom is small; therefore, $\cos\theta - 1 \approx 0$. To balance the equation, u_z must be small. Taking $u_z = 0$ as an approximation, the unknown constant c in Eq. (A.25) is determined, and the formula for u stated in the exercise is obtained with a simple algebraic manipulation from Eq. (A.25).

Note: How small is small for $\cos\theta - 1$? Surely, to make a comparison, quantities should be made dimensionless. Use a characteristic speed V and a characteristic length L to make the quantities in the equation $2uu_z = g(\cos\theta - 1)$ dimensionless. What would you take for V and L for the Mississippi river? Is the formula applicable in this case?

Page 335

Ex. 13.5 The push forward of the vector field $(c, 0)$ by Q is

$$\Big(\frac{c\sigma}{2(\sigma^2+\tau^2)}, -\frac{c\tau}{2(\sigma^2+\tau^2)}\Big).$$

This vector field is not divergence free; therefore, it is not a steady state solution of the incompressible Euler's equations.

Page 503

Ex. 16.39 Using the predictor-corrector scheme [Eq. (16.138)], compute the local truncation error in three steps: Apply Taylor series expansions to obtain

$$\begin{aligned} y(t+\Delta t) &= y(t) + y'(t)\Delta t + \frac{y''(t)}{2}\Delta t^2 + O(\Delta t^3) \\ &= y(t) + f(y(t))\Delta + \frac{1}{2} f'(y(t)f(y(t))\Delta t^2 + O(\Delta t^3) \end{aligned}$$

and

$$y_{n+1} = y_n + \Delta t f(y_n + \frac{\Delta t}{2} f(y_{n-1/2})$$

$$= y_n + \Delta t(f(y_n) + \frac{1}{2}f'(y_n)f(y_{n-1/2})\Delta t) + O(\Delta t^3),$$

estimate the error by

$$|y(t+\Delta t)-y_{n+1}| = \frac{1}{2}|f'(y_n f(y_n) - f'(y_n)f(y_{n-1/2})|\Delta t^2 + O(\Delta t^3),$$

and use the mean value theorem to estimate the important difference

$$\begin{aligned} |f(y_n) - f(y_{n-1/2}| &= |f'(\xi)||y_n - y_{n-1/2}| \\ &\leq |f'(\xi)|(|y_n - y_{n-1}| + |y_{n-1} - y_{n-1/2}|). \end{aligned}$$

The predictor-corrector formulas for the previous step can be used to show that $|y_n - y_{n-1}|$ and $|y_{n-1} - y_{n-1/2}|$ are $O(\Delta t)$.

Page 541

Ex. 17.10 The upstream and downstream momenta modeled by $Q^2/A + 2gA^2$ are both equal to

$$\frac{2g(B_D - B_U)A_U A_D}{A_D - A_U}$$

and $A_D \neq A_U$. It is not physically rcalistic for the momenta to be zero (because $A > 0$) and $B_D = B_U$.

Page 588

Ex. 18.1 We have that

$$\begin{aligned} \rho\dot{v}_i &= (\lambda\varepsilon_{kk}\delta_{ij} + 2\mu\varepsilon_{ij})_j + \rho b_i \\ &= \lambda\frac{1}{2}(u_{k,kj} + u_{k,kj})\delta_{ij} + \mu(u_{i,jj} + u_{j,ij}) + \rho b_i \\ &= \lambda u_{k,kj}\delta_{ij} + \mu(u_{i,jj} + u_{j,ij}) + \rho b_i \\ &= \lambda u_{k,ki} + \mu(u_{i,jj} + u_{j,ij}) + \rho b_i \\ &= \lambda u_{j,ij} + \mu(u_{i,jj} + u_{j,ij}) + \rho b_i. \end{aligned}$$

The quantity ∇u is a 3×3 matrix. Also recall that $\nabla\dot{u} = \nabla^T u$. These two facts are all that is needed to show that $\nabla \cdot (\nabla u) = u_{j,ij}$.

Ex. 18.15 To prove that the set $\mathcal{L}$ of square integrable functions on the interval $[0, L]$ is not finite-dimensional, let us suppose that $\mathcal{L}$ has a generating set consisting of $n \geq 1$ elements. Recall that $e^k := \sin \frac{k\pi x}{L}$ is in $\mathcal{L}$ for each positive integer k. For the case $n = 1$, suppose there is a function g^1 that generates every element in $\mathcal{L}$. Then there are two nonzero numbers a and b such that $e^1 = ag$ and $e^2 = bg$. The functions e^1 and e^2 are orthogonal. Thus, $(ag, bg) = 0$. Equivalently, $ab\|g\|^2 = 0$. But this is impossible because $\|g\| > 0$.

Suppose there is a generating set with $n > 1$ elements, say $\{g^i\}_{i=1}^n$. There are real numbers $\{b_\ell\}_{\ell=1}^n$ such that $e^{n+1} = \sum_{\ell=1}^n b_\ell g^\ell$. By renumbering the elements in the generating set if necessary and noting that $e^{n+1} \neq 0$, we may assume that $b_1 \neq 0$. Also, there is an $n \times n$ array of numbers a_{ij} such that $e^i = \sum_{j=1}^n a_{ij} g^j$, for $i = 1, 2, 3, \ldots n$. The function e^{n+1} is orthogonal to each such e^i. Hence, we have that

$$0 = (\sum_{\ell=1}^n b_\ell g^\ell, \sum_{j=1}^n a_{ij} g^j) = \sum_\ell^n \sum_{j=1}^n b_\ell a_{i,j} (g^\ell, g^j) = \sum_{j=1}^n a_{i,j} b_j.$$

By rearranging, dividing by $b_1 \neq 0$, and defining $c_j = -b_j/b_1$, for $j = 2, 3, 4, \ldots, n$, we obtain the identities

$$a_{i1} = \sum_{j=2}^n c_j a_{ij}, \qquad i = 1, 2, 3, \ldots, n.$$

Note that, by substitution for a_{i1},

$$e^i = \sum_{j=1}^n a_{ij} g^j = (\sum_{j=2}^n c_j a_{ij}) g^1 + \sum_{j=2}^n a_{ij} g^j = \sum_{j=2}^n a_{ij} (c_j g^1 + g^j).$$

Thus, the set of n independent vectors $\{e^i\}_{i=1}^n$ is generated by the set G_{n-1} of $n - 1$ functions

$$c_2 g^1 + g^2, \quad c_3 g^1 + g^3, \quad c_4 g^1 + g^4, \ldots, c_n g^1 + g^n.$$

The element e^n is orthogonal to each e^i, for $i = 1, 2, 3, ..., n - 1$. There must be at least one nonzero function in G_{n-1}. Thus, by renumbering if necessary, we may assume that the first coefficient of the linear combination of functions in G_{n-1} equal to e^n is not zero. The same argument used to obtain G_{n-1} can be used

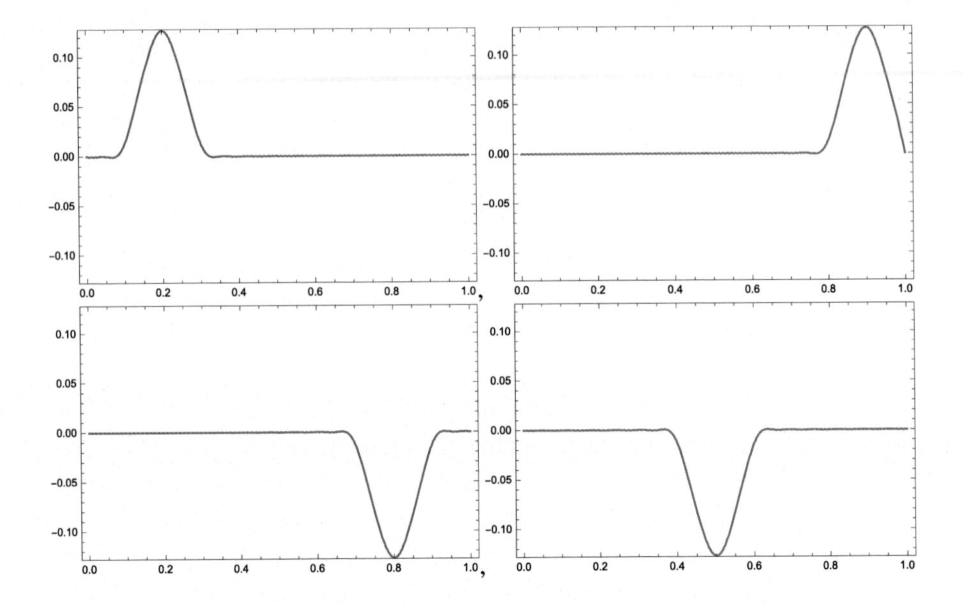

Fig. A.3 The graphs (for Exercise refex:wavemod) are numerical approximations to a wave traveling (from left to right) along a rope tied to a tree. The wave is fully developed in the top left panel, reaches the tree in the top right panel, is reflected from the tree in the bottom left panel, and approaches the position of the hand in the bottom right panel.

to show that the set of $n-1$ independent vectors $\{e^i\}_{i=1}^{n-1}$ is generated by a set G_{n-2} of $n-2$ functions. Continuing in this manner, we may conclude that e^1 and e^2 are generated by one function, which we have shown to be impossible. This contradiction completes the proof.

Page 607

Ex. 18.20 Some graphs for the solution of the rope wave model are depicted in Fig. A.3. These are made using the spectral method described in the problem. Thirty Fourier modes are used to approximate the wave profiles, which are in scaled variables. The physical length of the rope is $8\,\mathrm{m}$ and the wave speed is $2\,\mathrm{m}/\mathrm{sec}$, and the scaling is described in the statement of the problem. The profile function g was constructed with cubic polynomials to be zero outside the interval $[0.125, 0.125]$ (obtained by imagining hand motion of $1\,\mathrm{sec}$ duration), have maximum height of 0.127

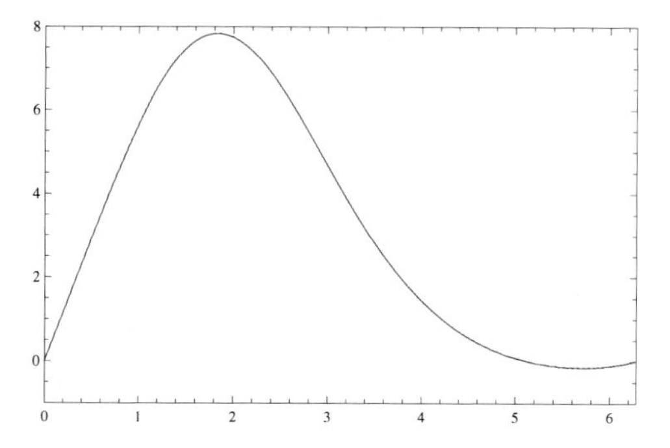

Fig. A.4 Graph of solution Ex. 18.32.

(corresponding to a 4-inch displacement) at the origin, and be symmetric about the origin.

Page 645

Ex. 18.32 For part (g), a graph of the solution of the boundary value problem $u_{xx} + \sin(x)u = 1$ with $u(0) = 0$ and $u(2\pi) = 0$ is depicted in Fig. A.4.

Page 640

Ex. 18.28 For part (a) guess that $\gamma(x) = 2$ for $0 \leq x \leq 1/2$ and $\gamma(x) = -2$ for $1/2 < x \leq 1$ is the desired weak derivative. To check this, compute first

$$\int_0^1 \gamma\psi\,dx = 2\int_0^{1/2} \psi\,dx - 2\int_{1/2}^1 \psi\,dx.$$

Next, using integration by parts, compute

$$-\int_0^1 \gamma\psi'\,dx = -\int_0^{1/2} 2x\psi'(x)\,dx - \int_{1/2}^1 (2-2x)\psi'(x)\,dx$$
$$= -\psi(\frac{1}{2}) + 2\int_0^{1/2} \psi\,dx + \psi(\frac{1}{2}) - 2\int_{1/2}^1 \psi\,dx$$

and note that the cancellation of the term $\psi(\frac{1}{2})$ would not occur unless the function ϕ is continuous (as it is here). This proves that γ is the weak derivative of ϕ. The H^1-norm of ϕ is $\sqrt{13/3}$.

For part(b), the given function is square integrable but it does not have a weak square integrable derivative.

Page 749

Ex. 22.5 Answer:

$$\frac{(1+b)v_\ell - (1+a)bv_r}{1-ab},$$

where

$$a = \frac{Z_\ell - Z}{Z_\ell + Z}, \qquad b = \frac{Z_r - Z}{Z_r + Z}.$$

Derivation: A dynamical systems approach to this problem is to view the wave reflected off the right-hand boundary as a function of the voltage-front vector v whose components are the incoming constant voltage values v_r and v_ℓ; that is, $v = (v_r, v_\ell)$. The reflected voltage-front vector is the product of the matrix

$$\mathcal{R} = \begin{pmatrix} -b & 1+b \\ 0 & 1 \end{pmatrix}$$

with the transpose of the incoming vector v. Likewise, the reflection off the left-hand boundary of a wave traveling left (given again by the vector whose components are the right and left voltages at the front) is determined by multiplication of the incoming vector by the matrix

$$\mathcal{L} = \begin{pmatrix} 1 & 0 \\ 1+a & -a \end{pmatrix}.$$

The voltage-front vector evolves by repeated application of $\mathcal{R}$ followed by $\mathcal{L}$. After n right and left reflections, the voltage-front is $(\mathcal{LR})^n v$. The product matrix $\mathcal{LR}$ has two eigenvalues 1 and ab with corresponding eigenvectors

$$\begin{pmatrix} 1 \\ 1 \end{pmatrix}, \qquad \begin{pmatrix} 1+b \\ (1+a)b \end{pmatrix}.$$

Becuase all impedances are positive, it follows that the second eigenvalue has $|ab| < 1$. The initial vector v is represented in the eigenbasis by

$$\begin{pmatrix} v_r \\ v_\ell \end{pmatrix} = \frac{(1+b)v_\ell - (1+a)bv_r}{1-ab} \begin{pmatrix} 1 \\ 1 \end{pmatrix} + \frac{v_r - v_\ell}{1-ab} \begin{pmatrix} 1+b \\ (1+a)b \end{pmatrix}.$$

After applying $(\mathcal{LR})^n$, we have the state

$$(\mathcal{LR})^n \begin{pmatrix} v_r \\ v_\ell \end{pmatrix} = \frac{(1+b)v_\ell - (1+a)bv_r}{1-ab} \begin{pmatrix} 1 \\ 1 \end{pmatrix} + (ab)^n \frac{v_r - v_\ell}{1-ab} \begin{pmatrix} 1+b \\ (1+a)b \end{pmatrix}. \quad \text{(A.26)}$$

In the long run, the factor $(ab)^n$ approaches zero and the desired result follows.

Page 771

Ex. 22.16 Answer: The electrical length is 4.5 nanosecond, the impedance is 2.88 ohm, the wave speed is 6.667×10^7 meter / second, and the capacitance is 5.21×10^{-9} farad / meter.

REFERENCES

[1] Abraham, R., J. E. Marsden and T. Ratiu (1988). *Manifolds, Tensor Analysis, and Applications.* 2nd Ed. New York: Springer-Verlag.

[2] Acton, F. S. (1990). *Numerical Methods That Work.* Washington D.C: Mathematical Association of America.

[3] Alhumaizi, K. and R. Aris (1995). *Surveying a dynamical system: a study of the Gray-Scott reaction in a two-phase reactor.* Pitman Research Notes in Mathematics Series, 341. New York: Longman, Harlow.

[4] Arnold, V. I. and A. Avez (1968). *Ergodic Problems of Classical Mechanics.* New York: W. A. Benjamin, Inc.

[5] Asmar, N. (2004). *Partial Differential Equations and Boundary Value Problems with Fourier Series (2nd Edition).* New York: Prentice-Hall.

[6] Balmforth, N. J. and S. Mandre (2004). Dynamics of roll waves. *J. Fluid Mech.* **514**, 1–33.

[7] Basmadjian, D. (2003). *Mathematical Modeling of Physical Systems.* New York: Oxford University Press.

[8] Batchelor, G. K. (1973). *An Introduction to Fluid Mechanics.* Cambridge: Cambridge University Press.

[9] Blasius, H. (1908). Grenzschichten in Flüssigkeiten mit kleiner Reibung, *Zeitschrift für Mathematik und Physik*, Band 56, Heft 1. Translated: The boundary layers in fluids with little friction. *National Advisory Committee For Aeronautics*. Tech. Memorandum 1256, 1950.

[10] Bornemann, F., D. Laurie, S. Wagon and J. Waldvogel (2004). *The SIAM 100-digit Challenge.* Philadelphia: Society for Industrial and Applied Mathematics.

[11] Boyce, W. and R. Diprima (2001). *Elementary Differential Equations and Boundary Value Problems (7th Edition).* New York: John WIley & Sons, Inc.

[12] Burden, R. and J. Faires (1997). *Numerical Analysis, 6th Edition.* New York: Brooks/Cole Pub. Co.

[13] Beyn, W-J. and V. Thümmler (2004). Freezing solutions of equivariant evolution equations. *SIAM J. Applied Dynamical Systems.* **3** (2) 85–116.

[14] Bhatia, H., G. Norgard, V. Pascucci, and P-T. Bremer (2013). The Helmholtz–Hodge decomposition—a survey. *IEEE Trans. Vis. Comput. Graph.* **19** (8), 1386–1404.

[15] Bruneau, C-H. and M. Saad (2006). The 2D lid-driven cavity problem revisited. *Computers & Fluids.* **35**, 326–348.

[16] Cartwright, J., P. Oreste, and I. Tuval (2009). Fluid dynamics in developmental biology: moving fluids that shape ontogeny. *HFSP J.* **3** (2), 77–93.

[17] Chang, H-C. (1994). Wave evolution on a falling film. *Annu. Rev. Fluid Mech.* **26**, 103–136.

[18] Chang, H-C. and E. A. Demekhin (2002). *Complex Wave Dynamics on Thin Films.* New York: Elsevier.

[19] Chaudhry, M. H. (2008). *Open-Channel Flow.* 2nd ed. New York: Springer.

[20] Chicone, C. (2006). *Ordinary Differential Equations and Applications.* 2nd ed. New York: Springer-Verlag.

[21] Chorin, A. J. and J. E. Marsden (1990). *A Mathematical Introduction to Fluid Mechanics.* 2nd ed. New York: Springer-Verlag.

[22] Craig, W. and C. Sulem (1993). Simulation of gravity waves. *J. Comp. Physics.* **108**, 73–83.

[23] Clough, R. W. and E. L. Wilson (1999). Early finite element research at Berkeley. *Proc. 5th US National. Conf. Comp. Mech,, Boulder, CO, August 1999.*

[24] Coppel, W. A. (1960). On a differential equation of boundary layer theory. *Philosophical Transactions of the Royal Society of London.* Series A, **253** (1023), 101–136.

[25] Courant, R, K. Friedrichs, and H. Lewy (1967). On the partial difference equations of mathematical physics. *IBM J. Res. Develop.* **11**, 215–234.

[26] Courant, R. and D. Hilbert (1953). *Methods of Mathematical Physics.* New York: Interscience Publishers.

[27] Cowper, G.R. (1966). The shear coefficient in Timoshenko's beam theory. *J. Appl. Mech.* **33** (2), 335–340.

[28] Crank. J. and R. Nicolson (1947). A practical method for numerical integration of solutions of partial differential equations of heat conduction type. *Proc. Camb. Phil. Soc.* **43**, 50–67.

[29] Deaton, M. L. and J. J. Winebrake (2000). *Dynamic Modeling of Environmental Systems (Modeling Dynamic Systems).* New York: Springer Science and Media, Inc.

[30] Denaro, F. M. (2003). On the application of the Helmholtz-Hodge decomposition in projection methods for incompressible flows with general boundary conditions. *Intl. J. Numerical Methods in Fluids.* **43**, 43–69.

[31] Doelman, A., R. A. Gardner, and T. J. Kaper (1998). Stability analysis of singular patterns in the 1D Gray-Scott model: a matched asymptotics approach. *Phys. D.* **122** (1–4), 1–36.

[32] Dubreil-Jacotin, M-L. (1934). Sur la détermination rigoureuse des ondes permanentes périodiques d'ampleur finie. *J. Math. Pures Appl.* **13**, 217–291.

[33] Ellner, S. P. and J. Guckenheimer (2006). *Dynamic Models in Biology.* Princeton: Princeton University Press.

[34] Erickson, L. (1990). *Panel Methods-An Introduction.* NASA Technical Paper 2995, Ames Research Center, California.

[35] Evans, L. (1998). *Partial Differential Equations.* Providence: American Mathematical Society.

[36] Feldman, Y. and A. Yu. Gelfgat (2010). Oscillatory instability of a three-dimensional lid-driven flow in a cube. *Physics of Fluids.* **22** (9), 093602.

[37] Feynman, R. P., R. B. Leighton and M. Sands (1964). *The Feynman Lectures on Physics I–III.* Reading: Addison–Wesley Pub. Co.

[38] Fisher, R. A. (1937). The wave of advance of advantageous genes. *Annals of Eugenics.* **7**, 355–369.

[39] FitzHugh, R. (1968). Motion picture of nerve impulse propagation using computer animation. *Journal of Applied Physiology.* **25** (5), 628–630.

[40] Fowler, A. (2011). *Mathematical Geoscience.* New York: Springer-Verlag.

[41] Gershenfeld, N. (1999). *The Nature of Mathematical Modeling.* Cambridge: Cambridge University Press.

[42] Ghil, M and S. Childress (1987). *Topics in Geophysical Fluid Dynamics: Atmospheric Dynamics, Dynamo Theory, and Climate Dynamics.* New York: Springer-Verlag.

[43] Grier, D. (2001). Dr. Veblen takes a uniform mathematics in the First World War. *Amer. Math. Mon.*, **108**, 992–931.

[44] Gragg, W. B. (1965). On extrapolation algorithms for ordinary initial-value problems. *SIAM J. on Numerical Analysis.* **2**, 384–403.

[45] Gray, P. and S. K. Scott. (1990). *Chemical Oscillations and Instability: Non-linear Chemical Kinetics.* Oxford: Oxford University Press.

[46] Griebel, M., T. Dornseifer and T. Neunhoeffer (1998). *Numerical Simulation in Fluid Dynamics.* Philadelphia, SIAM.

[47] Hansen, Per Christian (1998). *Rank-Deficient and Discrete Ill-Posed Problems.* Philadelphia, SIAM.

[48] Harlow, F. and J. E. Welsh (1965). Numerical simulation of time-dependent viscous flow of fluid with free surface. *Phys. Fluids.* **8**, 2182–2189.

[49] Henrici, P. (1974). *Applied and Computational Complex Analysis, Vol. 1: Power Series-Integration-Conformal Mapping-Location of Zeros.* New York: Wiley.

[50] Hicks, N. J. (1965). *Notes on Differential Geometry.* Princeton: D. Van Nostrand Co. Inc.

[51] Hoover, W. G. (2006). *Smooth Particle Applied Mechanics: The State of the Art.* New Jersey: World Scientific.

[52] Hubbard, J. and B. Hubbard (2002). *Vector calculus, linear algebra, and differential forms: a unified approach.* Upper Saddle River, NJ: Prentice Hall.

[53] Izhikevich, E. M. (2001). Synchronization of elliptic bursters. *SIAM Rev.* **43** (2), 315–344.

[54] Johnson, M, K. Zumbrun, and P. Noble (2011). Nonlinear stability of viscous roll waves. *SIAM J. Math. Anal.* **43** (2), 577–611.

[55] Johnson, R. S. (1997). *A Modern Introduction to the Mathematical Theory of Water Waves.* Cambridge: Cambridge University Press.

[56] Kantorovich, L. V., and G. P. Akilov (1982). *Functional Analysis.* New York: Pergamon Press.

[57] Ko, J., and W. Strauss (2008). Large-amplitude steady rotational water waves. *European J. Mech B/Fluids.*, **27**, 96–109.

[58] Kolmogorov, A. N., I. Petrovskii and N. Piscounov (1937). A study of the diffusion equation with increase in the amount of substance, and its application to a biological problem. In V. M. Tikhomirov, editor, Selected Works of A. N. Kolmogorov I, pages 248–270. Kluwer 1991. Translated by V. M. Volosov from Bull. Moscow Univ., Math. Mech. 1, 1–25, 1937.

[59] Koseff, J. R. and R. L. Street (1984). The lid-driven cavity flow: a synthesis of qualitative and quantitative observations. *Trans. AMSE.* **106**, 390–398.

[60] Kundu, P. K. and I. M. Cohen (2004). *Fluid Mechanics.* 3rd ed. Amsterdam: Elsevier Academic Press.

[61] Kreiss, H-O., N. A. Petersson and J. Yström (2002). Difference approximations for the second order wave equation. *SIAM J. Numer. Anal.*, **5**, 1940–1967.

[62] Landau, L. D. and E. M. Lifshitz (1976). *Mechanics.* 3rd ed. New York: Pergamon Press.

[63] Landau L. D. and E. M. Lifshitz (1987). *Fluid Mechanics.* 2nd ed. J. B. Sykes and W. H. Reid, trans. New York: Pergamon Press.

[64] Lax, P. D. (1973). *Hyperbolic systems of conservation laws and the mathematical theory of shock waves.* Philadelphia : Society for Industrial and Applied Mathematics.

[65] Lin, C. C. and L. A. Segel (1974). *Mathematics Applied to Deterministic Problems in the Natural Sciences.* New York: Macmillan Publishing Co., Inc.

[66] Liu, G. R. and M. B. Liu (2003). *Smoothed Particle Hydrodynamics: A Meshfree Particle Approach.* New Jersey: World Scientific.

[67] Logan, J. D. (1997). *Applied Mathematics.* 2nd ed. New York: John Wiley&Sons, Inc.

[68] Lucy, L. B. (1977). A numerical approach to the testing of the fission hypothesis. *The Astronomical J.* **82**, 1013–1024.

[69] Marsden, J. E. (1983). *Mathematical Foundations of Elasticity*. New York: Dover.

[70] Marsden, J. E. and M. McCracken (1976). *The Hopf Bifurcation and Its Applications*. New York: Springer-Verlag.

[71] Macfarlane, A. (1979). The development of frequency-response methods in automatic control. *IEEE T. Automat. Contr.* **AC-24**, 250–265.

[72] McLean, D. (2012). *Understanding Aerodynamics: Arguing from the Real Physics.* New York: John Wiley & Sons.

[73] Mehrmann, V. and C. Schröder (2016). Eigenvalue analysis and model reduction in the treatment of disc brake squeal. *SIAM News.* **49** (1), 1–3.

[74] Moler, C. and C. Van Loan (2003). Nineteen dubious ways to compute the exponential of a matrix, twenty-five years later. *SIAM Rev.* **45** (1), 3–49.

[75] Monaghan, J. J. (1992). Simulating free surface flows with SPH. *J. Comp. Phys.* **110**, 399–406.

[76] Monaghan, J. J. (1999). Solitary waves on a Cretan beach. *Journal of Waterway, Port, Coastal, and Ocean Engineering.* May/June, 145–154.

[77] Monaghan, J. J. (2005). Smoothed particle hydrodynamics. *Reports on Progress in Physics.* **68** (8), 1703–1759.

[78] Morgan, D. S. and Ta. J. Kaper (2004). Axisymmetric ring solutions of the 2D Gray-Scott model and their destabilization into spots. *Phys. D.* **192** (1–2), 33–62.

[79] Murray, J. D. (1980). *Mathematical Biology.* New York: Springer-Verlag.

[80] Nahin, P. J. (2002). *Oliver Heaviside: The Life, Work, and Times of an Electrical Genius of the Victorian Age.* London: The Johns Hopkins University Press.

[81] Noble, P. (2007). Linear stability of viscous roll waves. *Communications in Partial Differential Equations.* **32**, 1681–1713.

[82] O'Leary, D. P. (2009). *Scientific Computing with Case Studies.* Philadelphia: SIAM.

[83] Pearson, J. E. (1993). Complex patterns in a simple system. *Science.* (5118)**261**, 189–192.

[84] Pao, C. V. (1992). *Nonlinear Parabolic and Elliptic Equations.* New York: Plenum Press.

[85] P. J. E. Peebles (1993). *Principles of Physical Cosmology.* Princeton: Princeton University Press.

[86] Pollard, H. (1972). *Applied Mathematics: An Introduction.* Reading: Addison-Wesley Pub. Co.

[87] Pontryagin, L. (1962). *Ordinary Differential Equations.* New York: Addison Wesley Publishing Company, Inc.

[88] Prandtl, L. (1904). On the motion of fluids with very little friction. *Early Developments of Modern Aerodynamcis.* (2001) J. A. K. Ackroy, B. P. Axell and A.I . Ruban Eds. Oxford (UK): Butterworth-Heinemann.

[89] Robinson, J. A. (2001). *Infinite-Dimensional Dynamical Systems*. Cambridge (UK): Cambridge University Press.

[90] Rothwell, E. J. and M. J. Cloud (2001). *Electromagnetics.* Boca Raton: CRC Press.

[91] Rudin, W. (1966). *Real and Complex Analysis.* New York: McGraw-Hill, Inc.

[92] Saibel, E. A. and N. A. Macken (1973). The fluid mechanics of lubrication. *Annual Review of Fluid Mechanics*, **5**, 183–212.

[93] Schlichting, H. (1960). *Boundary Layer Theory*. New York: McGraw-Hill Book Company, Inc.

[94] Shandarin, S. F. and Ya. B. Zeldovich (1989). The large-scale structure of the universe: Turbulence, intermittency, structures in a self-gravitating medium. *Reviews of Modern Physics.* **61** (2), 185–220.

[95] Shampine, L. F. (2009). Stability of the leapfrog/midpoint method. *Applied Math. and Comp.* **208** (1), 293–298.

[96] Shapiro, A. (1962). Bath-tub vortex. *Nature.* **196**, 1080–1081.

[97] Shkadov, V. Ya. (1967). Wave flow regimes of a thin layer of viscous fluid subject to gravity. *Fluid Dynamics.* **2** (1), 29–34.

[98] Shkadov, V. Ya. (2013). A two-parameter model of wave regimes for viscous liquid film flows. *Moscow Univ. Mech. Bull.* **68** (4), 86–93.

[99] Smoller, J. (1980). *Shock Waves and Reaction-Diffusion Equations.* New York: Springer-Verlag.

[100] Stetter, H. J. (1970), Symmetric two-step algorithms for ordinary differential equations. *Computing.* **5** 267–280.

[101] Stoker, J. J. (1957). *Water Waves: The Mathematical Theory with Applications.* New York: Interscience.

[102] Strang, G. and G. Fix (2008). *An Analysis of the Finite-Element Method.* 2nd ed. Philadelphia: SIAM.

[103] Strauss, W. A. (2008). *Partial Differential Equations.* New York: John Wiley&Sons, Inc.

[104] Strauss, W. A. (2010). Steady water waves. *Bull. Amer. Math. Soc.*, **47** (4), 671–694.

[105] Stewartson, K. (1959). On the stability of a spinning top containing liquid. *J. Fluid Mech.* **5** (4), 577–592.

[106] Strogatz, S. H. (1994). *Nonlinear Dynamics and Chaos.* New York: Perseus Books Pub, LLC.

[107] Taylor, G. I. (1921). Experiments with rotating fluids. *Proc. Roy. Soc. London. Series A.*, **100** (703), 114–121.

[108] Taylor, G. I. (1923). Experiments on the motion of solid bodies in rotating fluids. *Proc. Roy. Soc. London. Series A.* **104** (725), 213–218.

[109] Taylor, G. I. (1953). Dispersion of soluble matter in solvent flowing slowly through a tube. *Proc. Roy. Soc. London. Series A.* **219**, 186–203.

[110] Temnykh, A. (1997). Vibrating wire field measuring technique. *Nuclear Instruments Meth. Phy. Res. Series A.* **399**, 185–194.

[111] Thompson, W. (1854). On the theory of the electric telegraph. *Proc. Roy. Soc. London.* **7**, 382 399.

[112] Trefethen, L. and D. Bau (1997). *Numerical Linear Algebra.* Philadelphia: SIAM.

[113] A. M. Turing (1952). The chemical basis of morphogenesis, *Phil. Trans. Roy. Soc. London.* **B 237**, 37–72.

[114] R. S. Varga (1962). *Matrix Iterative Methods.* New Jersey: Prentice Hall.

[115] H. Wendland (1995), Piecewise polynomial, positive definite and compactly supported radial functions of minimal degree. *Advances in computational Mathematics.* **4** (1), 389–396.

[116] Watkins, D. S. (2010). *Fundamentals of Matrix Computations.* 3rd ed. New York: John Wiley & Sons.

[117] Watson, G. N. (1966). *A Treatise on the Theory of Bessel Functions.* Cambridge University Press.

[118] Whitham, G. B. (1974). *Linear and Nonlinear Waves.* New York: John Wiley & Sons.

[119] Wilkening, G. and V. Vasan (2015). Comparison of five methods of computing the Dirichlet–Neumann operator for the water wave problem. *Contemp. Math.* **635**, 175–210.

[120] Wu, S. (1997). Well-posedness in Sobolev spaces of the full water wave problem in 2-D. *Invent. Math.* **130** (1), 39–72.

[121] Wu, S. (1999). Well-posedness in Sobolev spaces of the full water wave problem in 3-D. *J. Amer. Math. Soc.* **12** (2), 445–495.

[122] Yih, C-S. (1977). *Fluid Mechanics: A Concise Introduction to the Theory.* Ann Arbor: West River Press.

[123] Zagaris, Antonios, Hans G. Kaper and Tasso J. Kaper (2005). Two perspectives on reduction of ordinary differential equations. *Math. Nachr.* **278** (12–13), 1629–1642.

[124] Zakharov, V. E. (1968). Stability of periodic water waves of finite amplitude on the surface of a deep fluid. *Zhurnal Prikladoni Mekhaniki i Tekhnicheskoi Fiziki.* **8**, 86–94.

INDEX

C^1 convergence theorem, 793
H^1
  inner product, 634
  norm, 631
L^2
  inner product, 214
  norm, 214, 596, 600
∇, 294

Courant–Friedrichs–Lewy number, 428
lid-driven cavity flow, 681
Störmer–Verlet method, 608

absolute stability, 186
acceleration
  Aitken's Δ^2, 168
  Richardson's, 142
acid dissociation, 39
action at a distance, 100
Adams-Bashforth method, 501
aerodynamic, 682
agent-based model, 278
Airy's equation, 389
Aitken's Δ^2-method, 168
Ampère, 703
Arzela–Ascoli theorem, 793
asymptotically stable, 187
attenuation, 728

backward Euler method, 190
banded matrix, 174, 183, 421
basis, 602
Belousov–Zhabotinsky reaction, 67
BEM, 450
Bendixson's theorem, 51
Bessel function, 777
bifurcation
  homoclinic loop, 126
  Hopf, 71, 124, 569
  saddle-loop, 575
  saddle-node, 120
Big O, 796
binomial coefficient, 247
Blasius
  equation, 533
  problem, 531
  solution, 530
Blasius, H, 528
body force, 295
  conservative, 312
boundary condition, 116
  Dirichlet, 116
  do-nothing, 689
  electromagnetic, 710
  essential, 630, 649
  natural, 630, 649
  Neumann, 116
  no penetration, 315
  periodic, 116
  zero flux, 89
boundary element method, 450
boundary layer, 315, 398, 504
  equations, 351
  separation, 535
boundary value problem
  exterior Neumann, 343
  ill-posed, 349
  two-point, 26
Boussinesq approximation, 387
Brownian motion, 288
  mathematical, 264
Broyden's method, 696, 808
Brusselator, 82
buckling, 24, 27
buffering, 39
Burgers's equation, 389, 431, 544

calculus of varaitions, 615
Cauchy sequence, 632
Cauchy's equation, 299
Cauchy–Riemann equations, 333, 722
cavity resonator, 781
center, 395
central limit theorem, 265
centrifugal force, 363, 374
CFD, 403, 442
CFL condition, 99, 422
change of variables formula for integrals, 662
channel flow, 511
  control volume, 512
  discharge, 513
  prismatic, 513
  roll wave, 565
characteristic
  curve, 545
  length, 313
  velocity, 314
chemotaxis, 277
circuit theory, 198
circulation, 335
classical solution of PDE, 550

closed loop, 234
closing a model, 156
compartment model, 32
complete function space, 632
complex potential, 334
computational fluid dynamics, 403, 442
concentration gradient, 277
condition error, 134
conditional stability, 133
conductor, 706
conjugate gradient method, 818
conservation law, 431
  with source, 557
constitutive law, 3, 13, 301
continuity equation, 87, 295
control
  cruise control, 278
  example where integral control is required, 831
  fluid level in a tank, 239
  gain scheduling, 239
  heater/cooler, 234
  open loop, 239
  PDE, 229
  PID, 229
  temperature in chamber, 229
  tuning, 235
control volume (channel flow), 512
convection, 87
convergence
  quadratic, 166
Coriolis force, 363, 374
Courant–Friedrichs–Lewy condition, 99, 133, 139, 163, 422
Crank–Nicolson
  numerical stability, 165
cross product, 361
cruise control, 278
cubic spline, 640
cutoff frequency, 779

d'Alembert's
  paradox, 342, 528
  solution, 591
DAE, 74, 216, 217, 562
Darcy's law, 156
data fitting, 567
data structure, 659
deformation, 577
del operator, 294
derivative
  weak, 637
determinism, 11
dielectric, 706
diffeomorphism, 662
differential algebraic system, 74, 216, 217, 562
differential equation, 11
  continuity, 87
  Euler's
    fluid motion, 315
  for fluids, *see* fluid dynamics
  Navier–Stokes, 314
  reaction diffusion, 86
diffuser, 685
diffusion, 85, 87, 241, 277, 671
  equation, 87, 88
  kernel, 192
dilatation scaling, 531
dimensional analysis, 108
Dirac delta function, 263, 784
direct search minimization, 772
Dirichlet
  boundary condition, 101, 107, 116, 207, 319, 406, 591, 604, 630, 788
Dirichlet–Newmann map, 697
discharge (channel flow), 513
discretization error, 118
discretize, 117
dispersion
  relation, 389, 728
  Taylor, 677
displacement gradient, 580
dissociation constant, 42
distortionless transmission, 766
distribution
  Gaussian, 263
  normal, 263
distributional derivative, 637
divergence, 88
  theorem, 794
do-nothing boundary condition, 689
drag, 335, 682
  pressure, 335
  viscous, 335
Duffing's equation, 390
Duhamel's principle, 786
Dulac
  function, 52
  theorem, 54
dynamic viscosity, 309
dynamical system, 187
  stable fixed point, 187

eigenvalues
  and bifurcation, 569
  and CFL, 99
  and CFL condition, 422
  and convergence of iteration, 282
  and convergence of SOR, 181
  and rows of matrix, 795
  and stability, 35, 70, 132, 162, 393
  and wave speed, 735
  of symmetric matrix, 100
  singular value decomposition, 802
  tridiagonal matrix, 816
Einstein's summation convention, 318, 584
elasticity, 577
electrical length, 750
electrodynamics, 703

electromagnetic boundary conditions, 710
energy method, 93, 786
enthalpy, 327
entropy condition, 555
enzyme kinetics, 53
equation
  Burgers's, 431
  Cauchy's, 299
  continuity, 87, 295
  Duffing's, 19, 390
  Euler's, 302
  Fokker–Planck, 257
  full potential, 355
  Helmholtz, 718
  ideal fluid, 302
  KdV, 388
  Korteweg–de Vries, 387
  Langevin, 288
  Lyapunov's, 814
  momentum balance, 297, 516
  Navier-Stokes, 311
  of state, 156, 479
  Poisson, 406
  Prandtl–Glauert, 354
  Saint-Venant, 542
equicontinuous, 793
error
  propagation, 99
  relative, 135
essential boundary condition, 630, 649
Euler's equations, 302, 303
  fluid dynamics, 315
Euler's method, 117, 129, 501
  backward Euler method, 190
  explicit improved, 171
  implicit improved, 161
Eulerian fluid motion, 475
explicit improved Euler, 171
exponential of matrix, 221
exterior Neumann problem, 343
extrapolation
  Richardson, 142

Falkner–Skan solution, 535
Faraday, 703
fast time, 75
FEM, 638
fictitious force, 363
finite volume method, 432
finite-element method, 624
first integral, 48, 62, 393
Fisher model, 105
fitting data, 567
FitzHugh–Nagumo model, 196
flow of an ODE, 224, 786
fluid dynamics
  Chorin projection method, 408
  Bernoulli's
    equation, 327, 329
  Bernoulli's equation, 327
  boundary conditions, 303
  boundary layer, 504
  circulation, 335
  corner flow, 334
  drag, 335
  enthalpy, 327
  equations of motion, 293
  Euler's equations, 315
  flow in a pipe, 321
  geostrophic flow, 379
  incompressible, 311, 383
  inviscid, 383
  irrotational, 383
  irrotational flow, 329
  isentropic flow, 327
  kinematic viscosity, 314
  lift, 335
  Navier–Stokes equations, 293, 311, 403
  plug flow, 324
  potential flow, 329, 383
  pressure equation method, 405
  Proudman–Taylor theorem, 380
  rotation, 357
  stream function, 333
  viscosity, 311
  vorticity, 329, 383
  water waves, 384
flux function, 431
Fokker–Planck equation, 257
formula
  Lie–Trotter product, 221
  variation of parameters, 811
forward difference, 118
Fourier series, 470
  L^2 minimization property, 603
  convergence of, 603
Fourier's law of heat flow, 230
free-surface flow, 685
Frobenius inner product, 650
front
  shock, 563
  traveling wave, 700
  versus pulse, 113
Froude number, 391
full matrix, 466
full potential equation, 355

gain scheduling, 239
Galërkin method, 646
Gauss, 703
Gauss–Seidel method, 179
Gaussian, 263, 675
  distribution, 263
  elimination, 174, 466
  profile, 681
geometric singular perturbation theory, 74
geostrophic flow, 379
Gerschgorin theorem, 164, 428, 795

ghost cell, 413
Gibbs phenomenon, 101, 605
globally asymptotically stable, 51
gradient, 88
  coordinate change, 371
  in cylindrical coordinates, 355
  in polar coordinates, 345
  operator, 294
Gram–Schmidt orthogonalization, 795
gravitational potential, 675
Green's
  first identity, 497
  second identity, 451
  theorem, 51, 794
grid
  ghost cell, 413
  staggered, 413
Grobman–Hartman theorem, 795

Hölder function, 458
Hagen–Poiseuille law, 326
harmonic function, 333
  mean value property, 455
harmonic oscillator, 4
heat equation, 87, 88, 782
  with source, 786
heat kernel, 192, 784
Heaviside
  condition, 766
  step function, 787
Helmholtz
  decomposition, 407, 587, 712
  equation, 718
heteroclinic
  orbit, 210
Hilbert matrix, 647
Hodge decomposition, 407, 587
homoclinic
  loop, 396, 573
  loop bifurcation, 126
  orbit, 210
Hooke's law, 4, 577, 586
Hopf bifurcation, 71, 124, 205, 569
  stability index, 126
Hopf–Cole transformation, 444
hydraulic
  diameter, 570
  jump, 539
  radius, 558, 564

IBVP, 735
ideal fluid, 302, 303
ideal gas law, 157
ill-conditioned, 134, 647
ill-posed
  BVP, 349
  model, 626
impedance, 728
improved Euler method, 161
incompressible, 311, 383
inertial coordinates, 357
infinite dimensional, 602
infinitesimal rotation, 581
inner product
  properties, 601
inverse problem, 732, 771
inviscid, 383
irrotational, 328, 383
isentropic, 327
isotropic material, 706

Jacobi method, 191
Jacobian, 662
Jordan curve theorem, 51

KdV
  equation, 387, 388
Kelvin's
  circulation theorem, 349
  transmission line model, 781
Kirchhoff's rules, 198
Kolmogorov–Petrovskii–Piscounov model, 105
Korteweg–de Vries equation, 387
Kutta–Zhukovsky theorem, 338

Lagrange–Green tensor, 578
Lagrangian
  coordinates, 476
  flow map, 476
  form of Euler's equations, 478
  marker, 476
  SPH formulation, 477
Lamé constants, 590
Langevin equation, 288
Laplacian, 311
  coordinate change, 371
  in cylindrical coordinates, 355
  in polar coordinates, 343
  vector, 707
law
  Ampère's, 704
  Bernoulli's, 330, 384
  constitutive, 3
  constitutive stress, 301
  Coulomb's, 197
  Darcy's, 86
  electrodynamic laws, 703
  Faraday's, 200
  Fick's, 86, 277
  Fourier's, 86, 231
  fundamental, 3
  Gauss's, 704
  Hooke's, 13, 309, 577
  hyperbolic conservation, 558
  ideal gas, 157
  Kepler, 2
  Kirchhoff's laws, 198
  Lorentz force, 198

Maxwell's, 711
Newton, 2
Newton cooling, 229, 231
Newton's second, 293
nonlinear conservation, 431
of large numbers, 249
of mass action, 115
Ohm's, 705
scalar conservation, 434
system of conservation laws, 542
leap frog method, 283, 674
least squares, 772, 798
normal equation, 800
length scale, 377
Lennard–Jones force, 504
lid-driven cavity flow, 418
lift, 335
limit cycle, 205
linear material, 705
linearization, 352, 813
Liouville's theorem, 797
Lipschitz function, 458
little o, 796
local truncation error, 162
localization, 482, 639
functions, 497
of basis, 639
localizing sequence of functions, 483
Lorentz force law, 198
LU decomposition, 175
Lyapunov
equation, 814
function, 814
stability theorem, 814

Mach number, 354
mass action, 40, 115
master equation, 250
material derivative, 477, 582
matrix exponential, 221
Maxwell, 703
Maxwell-Lorentz theory, 716
mean value property, 455
mesh-free method, 624
method of lines, 191
Michaelis–Menton kinetics, 53
midpoint method, 283, 674
minimization
direct search, 772
via pattern search, 472
Minkowski's inequality, 600
mode
fundamental, 609
model
acid dissociation, 39
amplifier, 616
Belousov–Zhabotinsky, 67
boundary layer flow, 527
Boussinesq, 387
buckling, 24
Chézy, 558
chemotaxis, 277
closed loop, 30
compression of block, 620
constitutive, 3
convection, 85
cruise control, 278
dam break, 506
diffusion, 85
drag, 335
draining sink, 375
excitable media, 195
Fisher's, 105
FitzHugh–Nagumo, 195, 196
flow over plate, 448
fluid motion, 293
for time domain reflectometry, 731
free surface flow, 685
fundamental, 3
Gray–Scott, 113
growth, 15
heat flow, 88
heated chamber, 229
Heaviside's, 766
Hookean, 5
hurricane rotation, 375
ill-posed, 626
inverse problem, 771
Kelvin's transmission line, 781
Kirchhoff's circuit, 198
Kolmogorov–Petrovskii–Piscounov, 105
Korteweg–de Vries, 387
Langevin's, 289
lid-driven cavity flow, 418
lift, 335
linear elasticity, 577
logistic growth, 15
longitudinal waves, 611
Manning, 558
mass action, 39
molecular dynamics, 287
mutant gene, 104
Oregonator, 67
oscillating reaction, 67
Pearson's, 273, 283
pollutant, 31
porous medium, 155
Prandtl–Glauert, 354
reaction, 85
resonance horn, 620
river flow, 331, 558, 564
rope tied to tree, 607
Saint-Venant, 526
ship steering, 30
spring, 12
surface waves, 565
taut wire, 589
Taylor dispersion, 677

thermoelastic damping, 673
Tiger fountain, 397
titration, 54
transmission line, 731
truck with tailgate, 411
unicycle, 282
vibrating wire sensor, 609
Volterra-Lotka, 54
water waves, 383
molecular dynamics, 287, 288
momentum balance equation, 297, 516
Monte Carlo integration, 192
Morse lemma, 802
moving coordinates, 357
multiple timescales, 567
multistep method, 284

nabla, 294
natural boundary condition, 630, 649
Navier–Stokes equations, 311, 314, 403
Neumann
boundary condition, 89, 101, 116, 138, 236, 347, 410, 617, 626, 651, 723
Laplacian, 780
Neumann–Richmeyer artificial viscosity, 493
Newton's
equation, 285
law of cooling, 230
method, 165, 184, 186, 696, 772, 802
second law of motion, 297
Newton–Kantorovich theorem, 806
Newtonian fluid, 490
no-penetration boundary condition, 315
no-slip boundary condition, 303
norm, 214, 596
H^1, 631
L^2, 596
supremum, 596
normal
distribution, 263
equation for least squares, 800
normally hyperbolic, 76
numerical
linear algebra, 174
method, 117
numerical analysis
discretization error, 118
truncation error, 118
numerical gems of wisdom, 823
numerical instability, 284, 428
numerical method
Adam's–Bashforth, 503
Aiken's Δ^2, 169
alternate direction, 761
backward Euler method, 190
BEM, 465
BLAS, 186
boundary element, 624
Broyden's, 696, 808
central difference, 429
Chorin's, 409
conjugate gradient, 819
Crank–Nicolson, 161
efficient, 136
Euler's method, 118
far field, 683
finite difference, 94
finite element, 624
finite volume, 432
for Burgers's equation, 431
for computational fluid dynamics, 403
for DAEs, 217
for linear elasticity, 647
for system of linear equations, 171
for transmission lines, 749
for traveling waves, 211
for water waves, 446
forward Euler, 129
Galërkin's, 646
Gauss–Seidel method, 179
Gaussian elimination, 174
iterative for linear equations, 175
Jacobi's, 191
LAPACK, 186
leap frog, 283
least squares, 798
LU decomposition, 175
mesh-free, 624
method of lines, 191
midpoint, 283
Monaghan's, 502
Monte Carlo integration, 192
multigrid, 421
multistep, 284
near field, 683
Newton one variable at a time, 185
Newton's, 165, 802
of lines, 235
order of convergence, 167
pattern search, 472
predictor-corrector, 503
projection method, 408
quadratically convergent, 167
quasi-Newton, 808
Rayleigh's, 781
Richardson extrapolation, 142
singular value decompositon, 798
smoothed particle hydrodynamics, 475
SOR, 179
spectral, 608
splitting, 221, 760
Störmer–Verlet method, 286
stability of ODE solvers, 283
steepest descent, 772
Steffensen's method, 169
successive relaxation, 175
trapezoidal, 161
unconditionally stable, 165

upwind, 431

Ohm's law, 705
open loop, 234, 239
order notation, 796
order of convergence, 166
Oregonator, 66
orientation, 376
orthogonal
  Gram–Schmidt orthogonalization, 795
  in function space, 601
  matrix, 581
  polynomial, 606, 607
  transformation, 359, 363, 819
  unit basis, 309
orthonormal
  set, 599

panel flutter, 672
parameter estimation, 33
partial differential equation, 18
  reaction-diffusion, 85
partial differential
    equations
  fluids, *see* fluid dynamics
path line, 328
pattern search, 472
periodic boundary condition, 116, 698
perturbation theory, 73
  geometric singular, 74
phenomenological model, 3
PID control, 37, 229
plane wave, 721
plug flow, 324, 333
Poincaré map, 128, 188
Poincaré–Bendixson theorem, 816
pointwise convergence, 604
Poiseuille flow, 325, 326
Poisson equation, 406
polarization identity, 581
population model, 54
porosity, 155
porous medium, 155
  equation, 157
positive spanning set, 472
positively oriented, 376
potential
  electric, 722
  flow, 383
  gravitational, 675
  vector, 706
power law, 156
Prandtl, L, 315, 528
Prandtl–Glauert equation, 354
preconditioner, 179
pressure, 302, 582
  drag, 335
  equation method, 405
  Poisson equation, 406
principle of mass action, 40
prismatic channel, 513
projection method, 408
Projects
  1-d Finite Element Coding, 645
  Channel Flow Traveling Wave, 700
  Aerodynamic Drag, 682
  Agent-Based Modeling, 278
  Alternative Galëkin method, 646
  Approximation and Orthogonal Polynomials, 606
  Automatic Control of Steering a Ship, 29
  Barbell-Shaped Rod, 671
  Beam Theory, 673
  Body versus Traction Force, 671
  Boundary Element Code I, 474
  Boundary Element Code II, 475
  Cavity Resonators, 781
  Channel Flow Modeling, 574
  Chemotaxis, 277
  Clamped Plate, 672
  Concentration Gradient, 277
  Convection-Diffusion, 674
  Cruise Control, 278
  Dam Break with SPH, 508
  Discrete Dynamical Systems, 187
  Elastic Plate with Elliptical Hole, 670
  Exasticity vesus Rod Equation, 671
  Falling Fluid Films, 538
  FEM 3-d Coding, 670
  Fireplace Heating and Cooling, 193
  Fluid Motion in Cylinder, 684
  Free-Surface Flows, 685
  Gravitational Potential, 675
  Heat Fluctuations in a Bar, 238
  Heat Kernel, 782
  Heaviside Transmission Line, 766
  Intermittent Fountain, 671
  Iteration and Eigenvalues, 282
  Lid Driven Cavity Flow, 446
  Linearly Tapered Rod, 671
  Lord Kelvin's Transmission Line Model, 781
  Low Reynolds Number Flow, 684
  Method of Lines, 191
  Mississippi River Flow Rate, 564
  Modeling and Control, 239
  Molecular Dynamics, 287
  Newton's Method for an ODE Model, 170
  Newton's Method in Mountain Terrain, 169
  Nonlinear Eigenvalue Problem via DAEs., 218
  Numerical Integration of Newton's Equation, 285
  Numerical Pulse Type Traveling Waves, 219
  ODE Nonlinear BVPs, 188
  Oscillations Carried by Diffusion, 104
  Panel Flutter, 672
  Pattern Formation, 291
  Pearson's Random Walk, 283
  Rayleigh Method, 780

Resonance Horn Amplification, 620
Splines and FEM, 640
Stability of Numerical ODE Solvers, 219
Stability of ODE Solvers, 283
Still Water with SPH, 508
Taylor Dispersion, 677
TDR Inverse Problem, 771
TE Modes in Waveguides, 776
Thermoelastic Damping, 673
Unicycle Control, 282
Upwinding, 445
Vibrating Wire Sensors, 609
Wave Modeling and Numerics, 607
propagator, 786
Proudman–Taylor theorem, 380
pseudo inverse of a matrix, 800
pulse
traveling wave, 209, 575
versus front, 113
punctured plane, 343

quadratic convergence, 166
quasi-Newton method, 696, 808
Broyden's method, 696

random walk, 241
master equation, 250
Rankine–Hugoniot jump condition, 552, 745
Rayleigh method, 780
reflection coefficient, 748
regression
linear, 606
regular perturbation, 74
relative error, 135
rescaling
in Navier–Stokes
equations, 314
resonance
horn, 619
lenght, 671
response, 610
ultrasonic, 619
rest point
center, 395
hyperbolic saddle, 69
Reynolds number, 314, 377
low limit, 316
Richardson extrapolation, 142
Riemann problem, 742, 751
roll wave, 565
root mean square, 160, 600
Rossby number, 378
rotation
in three dimensions, 364
Routh–Hurwitz criterion, 815

saddle
loop, 573
point, 69, 111, 120, 210, 393, 572
saddle-node bifurcation, 120
Saint-Venant equations, 526
scalar field
coordinate change, 368
scale
length, 377
velocity, 377
Schwarz inequality, 600, 631
semiflow, 225
sensor
vibrating wire, 609
separation of variables, 89
shape function, 643
shear
flow, 677
modulus, 588, 590
strain, 306
strain rate, 306
strain rate tensor, 308
stress, 525
Sherman–Morrison formula, 809
shock
front, 563
speed, 555
traveling wave, 700
wave, 355, 493, 549, 552, 555
shooting method, 26, 188, 535
similar matrices, 365
similarity solution, 531, 784
simply connected, 329
singular perturbation, 73, 74, 380, 823
singular value decomposition, 798, 801
skew-symmetric matrix, 360, 581
slow
manifold, 77
relaxation oscillation, 572
time, 75, 151
timescale, 205
smoothed particle hydrodynamics, 475
SOR, 179
sound speed, 494
sparse matrix, 174, 646, 669
spectral
method, 608
radius, 283
theorem, 818
SPH, 475
splines, 640
and FEM, 640
and interpolation, 640
spring equation, 4
square brackets [ ], 47
square integrable, 214
Störmer–Verlet method, 286
stability, 813
absolute, 186
index, 125, 126
numerical, 428
stable

fixed point, 187
manifold, 69
staggered grid, 413
standard map, 187
steepest descent, 820
Steffensen's method, 169
stiff ODE, 186
stochastic differential equation, 289
Stokes
flow, 684
theorem, 338, 341
strain, 305, 577
rate tensor, 308
volumetric strain rate, 310
stream
function, 333
line, 328, 333
stress, 582
field, 612
shear, 564, 573
strain relation, 610
tensor, 296, 490, 515
vector, 296
subcritical flow, 540
successive
overrelaxation, 179
relaxation, 175
supercritical flow, 540
superlinear convergence, 188
support of a function, 550
supremum norm, 596

Taylor
and Proudman theorem, 379
dispersion, 677
expansion and linearization, 353
formula, 796
multivariate series, 271
series, 118
series in h^2, 142
theorem, 95
TDR, 731
TE mode, 776
telegrapher's wave equation, 766
temperature control, 229
tensor field, 580
test function, 550, 637
theorem
C^1 convergence, 793
Arzela–Ascoli, 793
Bendixson's, 51
Boundary integral, 458
central limit, 265
divergence, 794
Dulac's, 54
Gerschgorin, 164, 428, 795
Gerschgorin's, 133
Green's, 51, 794
Grobman–Hartman, 394, 795
Helmholtz-Hodge, 407
Hopf bifurcation, 211
Implicit Function, 64, 73
Kelvin's circulation, 349
Kutta–Zhukovsky, 338
Law of large numbers, 249
Lie–Trotter product, 221
linearized stability, 813
Liouville's, 797
Lyapunov
stability, 814
Newton–Kantorovich, 806
Poincaré–Bendixson, 51, 127, 816
Proudman–Taylor, 379, 380
Routh–Hurwitz, 815
spectral, 818
Stokes, 338, 341
Taylor's, 95
Transport, 295, 797
topology
nontrivial, 347
traction, 582
transmission coefficient, 748
transmission line, 731
equations
ideal, 728
Kelvin's model, 781
Transport theorem, 797
trapezoidal method, 161
traveling wave, 109, 208, 429, 568, 700
front, 113, 700
periodic, 573
pulse, 209, 573, 575, 700
tridiagonal matrix, 816
Trotter product formula, 221, 760
truncation error, 118
Turing's principle, 114

ultrasonic resonance, 619
unconditionally stable, 163
uniform convergence, 604
uniformly bounded, 793
uniqueness
and lift, 346
and Neumann boundary data, 462
heat equation, 786
Helmholtz decompotion, 587
via energy method, 93, 786
unstable method, 284
upwind numerical scheme, 431

variation of parameters, 786, 811
variational equation, 812
vector field
coordinate change, 368
vector Laplacian, 707
velocity profile, 334
vibrating wire sensor, 609
viscosity, 311

dynamic, 308
viscous
drag, 335
Volterra-Lotka model, 54
volumetric strain rate, 310
vorticity, 328, 383, 504

water wave equations, 384
water waves, 685
wave equation, 588, 589
one-way, 428
wave number, 728
waveguide
cutoff frequency, 779
weak solution, 551, 634
well posed, 715
Wendland localization, 498

zero flux boundary condition, 89
Zhukovsky, 338

Printed in the United States
By Bookmasters